Semiconductor-Device Electronics

The Holt, Rinehart and Winston Series in Electrical Engineering
published by Saunders College Publishing

M. E. Van Valkenburg, Senior Consulting Editor
Adel S. Sedra, Series Editor/Electrical Engineering
Michael R. Lightner, Series Editor/Computer Engineering

ALLEN AND HOLBERG
CMOS Analog Circuit Design
BELANGER, ADLER AND RUMIN
Introduction to Circuits with Electronics: An Integrated Approach
BOBROW
Elementary Linear Circuit Analysis, 2/e
BOBROW
Fundamentals of Electrical Engineering
CHEN
Linear System Theory and Design
CHEN
System and Signal Analysis
COMER
Digital Logic and State Machine Design, 2/e
COMER
Microprocessor-Based System Design
COOPER AND McGILLEM
Probabilistic Methods of Signal and System Analysis, 2/e
GHAUSI
Electronic Devices and Circuits: Discrete and Integrated
HOSTETTER, SAVANT AND STEFANI
Design of Feedback Control Systems, 2/e
HOUTS
Signal Analysis in Linear Systems
JONES
Introduction to Optical Fiber Communication Systems
KARNI AND BYATT
Mathematical Methods in Continuous and Discrete Systems
KENNEDY
Operational Amplifier Circuits: Theory and Application
KUO
Digital Control Systems
LASTMAN AND SINHA
Microcomputer-Based Numerical Methods for Science and Engineering
LATHI
Modern Digital and Analog Communication Systems, 2/e

LEVENTHAL
Microcomputer Experimentation with the IBM PC
LEVENTHAL
Microcomputer Experimentation with the Intel SDK-86
LEVENTHAL
Microcomputer Experimentation with the Motorola MC6800 ECB
McGILLEM AND COOPER
Continuous and Discrete Signal and System Analysis, 3/e
NAVON
Semiconductor Microdevices and Materials
PAPOULIS
Circuits and Systems: A Modern Approach
RAMSHAW AND VAN HEESWIJK
Energy Conversion: Electric Motors and Generators
SADIKU
Elements of Electromagnetics
SCHWARZ
Electromagnetics for Engineers
SCHWARZ AND OLDHAM
Electrical Engineering: An Introduction
SEDRA AND SMITH
Microelectronic Circuits, 3/e
SINHA
Control Systems
VAN VALKENBURG
Analog Filter Design
VRANESIC AND ZAKY
Microcomputer Structures
WARNER AND GRUNG
Semiconductor-Device Electronics
WASSER AND FLYNN
Introduction to Arithmetic for Digital Systems Designers
WOLOVICH
Robotics: Basic Analysis and Design
YARIV
Optical Electronics, 4/e

Semiconductor-Device Electronics

R. M. Warner, Jr.
Electrical Engineering Department
University of Minnesota

B. L. Grung
Systems and Research Center
Honeywell, Inc.

Holt, Rinehart and Winston
Philadelphia   Fort Worth   Chicago   San Francisco
Montreal   Toronto   London   Sydney   Tokyo

Copyright © 1991 by Saunders College Publishing, a division of Holt, Rinehart and Winston, Inc.

All rights reserved. No part of this publication may be reproduced or transmitted in any form or by any means, electronic or mechanical, including photocopy, recording, or any information storage and retrieval system, without permission in writing from the publisher.

Requests for permission to make copies of any work should be mailed to Copyrights and Permissions Department, Holt, Rinehart and Winston, Inc., Orlando, Florida 32887.

Text Typeface: Times Roman
Compositor: York Graphic Services
Acquisitions Editor: Barbara Gingery
Managing Editor: Carol Field
Project Editor: Mary Patton
Copy Editor: Andy Potter
Manager of Art and Design: Carol Bleistine
Art Director: Christine Schueler
Art and Design Coordinator: Doris Bruey
Text Designer: Rita Naughton
Cover Designer: Lawrence R. Didona
Text Artwork: Rolin Graphics
Director of EDP: Tim Frelick
Production Manager: Charlene Squibb
Marketing Manager: Denise Watrobsky

Cover Credit: The cover diagram is based on Figure 5-48(b).

Printed in the United States of America

SEMICONDUCTOR-DEVICE ELECTRONICS

ISBN 0-03-009559-X

Library of Congress Catalog Card Number: 90-053333

1234 016 987654321

THIS BOOK IS PRINTED ON **ACID-FREE, RECYCLED** PAPER

To Tom, Lee, Mark, Gary, and Alison,
and to their marvelous spouses and offspring,
and to Helen E. Ruff,
our one-and-only Aunt Helen in her 91st year,
the kind of Aunt *everyone* needs.

                                        RMW

To my wife, Sandi.

                                        BLG

Preface

Semiconductor-Device Electronics is a textbook designed for a core course on devices, a course that is usually offered in electrical-engineering and computer-science curricula during the junior year. The book was class-tested in nine quarters at the University of Minnesota before publication and has had a gratifying reception. This was a ten-week course, first in an electronics sequence. In ten weeks one can cover only the most basic material, and so the extension to a one-semester course would be easy. However, there is plenty of material in the textbook for a full year, so it is well adapted for use in a senior elective course or in a graduate program.

ORIGIN AND CONCEPT

We have seen an urgent need for a device-oriented electronics text of a particular kind. A student addressing this subject needs foundation knowledge that is extensive. Most electrical engineering (EE) students, at least, have been exposed to most of the necessary topics, but our experience indicates that for all but the most capable students, these topics remain at the fringes of understanding.

For this reason we have included foundation subject matter. This review material has been composed in a way that is consistent with a deeply held pedagogical conviction. *It proceeds from the specific to the more general.* The simplest, clearest example is given first as a stepping stone toward more complex cases. While a person of long experience in a subject can appreciate the elegance of a development that proceeds from the general to the particular, most newcomers cannot. Instead, they need specific and manageable components for constructing their own conceptual frameworks.

To illustrate our application of this principle, we can point to field theory.

Rather than starting with Maxwell's equations, we address a series of one-dimensional problems of a kind especially relevant to device theory. We stress the importance of gaining an intuitive, visceral understanding of Poisson's equation in these simple contexts and of some of the other, more elusive classical concepts such as electric displacement and dielectric-relaxation time. The device theory that is the subject of this book does not require us to go beyond one-dimensional descriptions, and space limitations do not permit it. But the theory provides a firm foundation for the student who subsequently faces two- and three-dimensional problems and situations requiring more general field theory. In a similar way, the Bohr atom is treated in terms that endeavor to relate electron behavior in an isolated atom to electron behavior in a crystalline solid.

Crystallography, on the other hand, is a topic that most EE students have not encountered before, but nonetheless it qualifies as a foundation topic in the sense that device engineers use its concepts almost daily. Additional fundamental subjects addressed are basic to all branches of engineering and science, but are often neglected. These include general problem-solving procedures and the skillful, consistent use of units.

Our primary aim in writing this textbook was to achieve clarity. We have made liberal use of analogies and heuristic descriptions and have avoided unnecessary jargon. In explanations, we have tried to choose a level that can be understood on first exposure. Intermediate equations have been included in most derivations so that the student can follow them easily, rather than deferring equation reading until "later."

One of the authors has taught device electronics to somewhat over 20 classes of EE juniors at the University of Minnesota and, in the same time interval, to 10 in-house classes for industrial firms at various locations throughout the country. Many of the university classes were large, numbering over 200 students, a factor that poses its own special problems. The least of these is speaking a bit more loudly or writing on the blackboard with larger letters.

As everyone who has had the experience knows, it is the administrative burden that is most troublesome with a very large class, even with adequate and competent assistance. Our book offers an important measure of relief in such a situation. It provides 229 analytic problems and an accompanying Solutions Manual. Given these analytic problems and solutions, one can choose the approach that fits the immediate situation best. The method frequently used, assigning problems and providing solutions later, is of course an option. But there is at least one other valuable option—an approach that we have used successfully for several years: Make the problems *and solutions* available to the students on the first day of class. Inform them that quiz and exam problems will be closely related to these, and that the students must therefore understand these problems and solutions thoroughly. This approach eliminates the burden of grading homework papers, which after the first offering of a particular course often becomes largely copied homework. (There are few academic activities that are more wasteful than the grading of copied homework.) This further frees the instructor to assign computer and design problems as homework.

In our experience, the approach of supplying both problems and solutions leads to a highly desirable focus on *understanding* on the part of the students. Discussion of problems and associated concepts dominates the recitation hours and student visits during office hours. To reinforce the instructor's serious intent in this regard, we often place a problem in an early quiz that is taken directly from the book, without even the alteration of numbers.

To use such a system, of course, the closed-book approach is necessary in quizzes and exams. This further reinforces the importance of understanding (rather than the skill of a file clerk). And this system does not make unfair or unreasonable demands upon a student's memory. The equations the students must know are extremely simple, typically involving only three or four symbols. If students truly understand, for example, that diffusive flux is proportional to density gradient, they can write down the relevant equation effortlessly. For more complex expressions, such as the continuity equations, we state the needed equation in the problem. Typically, the most complicated equations that we insist the student must know are the transport equations.

A further example along these lines is pertinent. The equations for depletion-layer thickness in various kinds of step junctions are important but can be looked up readily by engineers on the job. We consider memorizing these to be a relatively unprofitable investment of energy. But we *do* insist that students clearly understand that the depletion-layer thickness in *any* step junction varies as the square root of the potential difference between its two sides, and why this relationship exists. On this basis it is easy to devise problems that test understanding rather than memory. We have selected topics to emphasize fundamentals rather than state-of-the-art concepts. Only two major devices—the bipolar junction transistor and the MOS field-effect transistor—are addressed here because a thorough knowledge of these devices enables one to understand all other important IC devices. State-of-the-art devices are by their nature constantly evolving, and these changing facts are best learned on the job.

CONTENT AND ORGANIZATION

Chapter 1 presents the fundamentals of electricity in fresh fashion, as well as unit manipulation and problem solving, the Bohr model, and crystallography.

Chapter 2 treats equilibrium and nonequilibrium bulk properties of semiconductors, with silicon receiving heavy emphasis. It introduces band theory and explains how semiconductors, conductors, and insulators differ. The Fermi-level concept and its application is next, along with the most basic approximations frequently used in semiconductor work. The nature and consequences of ''doping'' lead into further fundamental equations and laws—mass action, the neutrality equation, and the Boltzmann relation. Carrier transport, recombination and generation, and the continuity equations for carrier-behavior analysis complete the chapter.

Chapter 3 deals comprehensively with the *PN* junction, at equilibrium and under bias. After the basic junction concepts are introduced, the depletion approximation is applied to a carefully selected sequence of examples and is subsequently expanded beyond step-junction cases. Static theory is augmented by a set of meticulously recorded experimental data taken from the literature for the case of a particular silicon diode. The treatment of breakdown phenomena goes beyond the usual textbook treatment.

Unique to our book, however, is a new and integrated treatment of the dynamic properties of the *PN*-junction diode. It is used subsequently as the foundation for treating the dynamic properties of the BJT and the MOSFET. Also, a general treatment of step-junction and semiconductor-surface problems is included, which is found only in one other book. The high-low junction as ohmic contact is discussed—a feature found in vast numbers of semiconductor devices but ignored in most electronics textbooks. The principles of SPICE numerical analysis are given; in some cases, the detail presented exceeds that found even in SPICE manuals.

Chapter 4 offers BJT rudiments, basic device theory, biasing practice, and circuit-configuration options, stressing the properties of each. Structures and properties of real devices come next, followed by a detailed survey of high-level effects in the BJT. These effects are omitted from most texts, but because the BJT is routinely used under high-level conditions, this omission cannot be justified. Our section on the Ebers-Moll model stresses clarity and an appreciation of the physical significance of each step and is liberal with application examples. The small-signal-dynamic modeling of the BJT starts with the hybrid model, relating it to device physics, and proceeds through the hybrid-pi and other models, the charge-control model, and figures of merit. SPICE modeling is presented in unusual detail, large-signal and small-signal, with parasitic properties and thermal effects included.

Chapter 5 deals with the MOS capacitor and the MOSFET. After presentation of the elementary theory and inverter options, we describe the numerous phenomena that must be treated in an MOS capacitor. These are modeled carefully, using equivalent circuits and the general semiconductor-surface analysis introduced in Chapter 3. We include physical and analytic treatments of the capacitance-component interplay in the MOS capacitor and in the junction diode and then compare the two devices with respect to capacitive properties. Advanced modeling of the MOSFET follows, and the SPICE treatments of small-signal and large-signal problems follow that. The concluding section of the chapter, and of the book, is a new and detailed look at MOSFET-BJT performance comparisons that is not found in any other textbook.

LEARNING AIDS

Immediately following the text of each chapter are two features designed to provide a firm qualitative grasp of the subject matter. The first is a Summary that endeavors to encapsulate the essential elements of the chapter. The second is a set of review questions called "Topics for Review" (averaging 115 per chapter) that lets the student know very specifically whether key points have been mastered or, possibly, have been missed altogether. The most basic of the questions posed in this section are also useful for quiz and examination purposes.

An additional feature of our book is the in-text exercise and solution. On average about 60 of these appear per chapter. We have endeavored here to anticipate questions that could reasonably pop into the head of an alert reader. (We are eager to obtain feedback on our success in reaching this goal.) Then the more aggressive or ambitious reader can ponder the issue and try to supply his or her own answer, before simply reading ahead into the solution.

There are other respects in which our book differs from most engineering texts. Topics in which spatial relationships are important are numerous in this subject area, with the space lattices of Chapter 1 providing a good example. In such cases, perspective drawings have been employed, replacing the primitive orthogonal-isometric drawings usually encountered.

Our accrediting agency, ABET, has in recent years placed a valid and growing emphasis on design skills, since engineers on the job do more synthesis than analysis. For this reason we have provided design problems at the end of each chapter in addition to the analytic problems discussed earlier. In a similar way and for similar reasons, computer problems also accompany each chapter, averaging more than two per chapter. Here we have selected problems *requiring* numerical treatment. In the last three chapters, these problems emphasize SPICE modeling. Normally we assign design and computer problems as homework. We have chosen to program the solutions using PASCAL, but other options are obviously available to the users of this book.

Finally, the Solutions Manual that accompanies our book has been prepared with unusual attention to legibility and accuracy. Of particular importance is the careful treatment of dimensions, a point stressed to reinforce the lesson on this subject in Chapter 1.

REFERENCE FEATURES

As much care and thoroughness have gone into assembly of the Subject Index as into the text material. It is unusually detailed, and important topics are liberally cross-referenced. While we can appreciate the fatigue factor that sometimes causes delegation of index preparation to nontechnical clerical people, we feel that extra effort invested here will enhance considerably the value of the book as a study and reference resource. In a similar vein, our policy on references has been to supply a substantial number, but far less than encyclopedic listings. (If a measure of chauvinism has crept into the selection process, we hope that our readers will be understanding.) We believe that the book will be used as a reference as well as a text. In summary, we have tried to select and emphasize the most basic and unchanging topics; then we have sought the clearest possible presentation. We hope that the result will provide entry to the eclectic discipline of solid-state-device electronics for many future students.

ACKNOWLEDGMENTS

We are deeply indebted to past students, approximately 1000 of them, who have used our manuscript as a text and who have given us their comments and numerous corrections. Also we acknowledge with appreciation the long-term encouragement provided by Prof. Alfons Tuszynski, and the fact that Prof. Ronald D. Schrimpf supplied the insight embodied in Exercise 3-54 and Figure 5-48(b). Mae Warner deserves special mention for endless checking at every stage of the process and for entering most of the manuscript into the computer. The professionalism and patience of Mary Patton, our Project Editor during much of the production phase, is enormously appreciated. We feel the same appreciation for Becca Gruliow, who completed the process. Also, the skill and knowledge of our copy editor, Andy Potter, produced a smoother, clearer result than we could have achieved otherwise.

We are grateful to the reviewers of our manuscript, not only for their predominately positive feedback, but also for sparing us the embarrassment of certain lapses that would otherwise be present in the book. They are Alan Marshak (Louisiana State), Donald Wilson (San Diego State), Larry Burton (Virginia Tech), Robert Engleken (Arkansas State), Hisham Massoud (Duke), D. K. Reinhard (Michigan State), and H. P. D. Lanyon (Worcestor Polytech).

Finally, we thank the three acquisitions editors who served in sequence on our project during its four years of preparation. They are Deborah Moore, Bob Argentieri, and Barbara Gingery.

R. M. Warner, Jr.
B. L. Grung
January 1991

Contents

Chapter 1 Foundations of Modern Electronics *1*

  1-1 Electric Charge, Field, and Energy *1*
    1-1.1 The Electric-Field Concept *1*
    1-1.2 Work and Energy in an Electrical Context *4*
    1-1.3 Electrostatic Potential *5*
    1-1.4 Lines of Force *7*
    1-1.5 Potential Energy and Kinetic Energy *9*

  1-2 Unit Manipulation and Problem Solving *13*
    1-2.1 The Unity Factor *13*
    1-2.2 Problem-Solving Procedure *14*
    1-2.3 Unit and Variable Symbols *17*
    1-2.4 One-Dimensional Problems *18*
    1-2.5 Normalization *18*

  1-3 Equations Dealing with Moving and Motionless Charges *20*
    1-3.1 Conductivity and Resistivity *20*
    1-3.2 Ohm's Law in Terms of Electric Field *22*
    1-3.3 Dielectric Materials, Permittivity, and Polarization *22*
    1-3.4 Electric Displacement *26*
    1-3.5 Displacement Current *28*
    1-3.6 Dielectric Relaxation *30*
    1-3.7 The Meaning of Poisson's Equation *32*

  1-4 The Bohr Model of the Hydrogen Atom *35*
    1-4.1 The Planetary Analogy *35*

1-4.2 Electromagnetic Radiation and Quanta *37*
1-4.3 Classical Components of the Bohr Model *40*
1-4.4 The Bohr Postulates *44*
1-4.5 Model Predictions *45*
1-4.6 Refinements to the Bohr Model *50*

1-5 Crystallography   *57*

1-5.1 The Lattice *58*
1-5.2 The Unit Cell and Primitive Cell *61*
1-5.3 The Space Lattice *63*
1-5.4 Relating Lattices and Crystals *67*
1-5.5 The Silicon Crystal *70*
1-5.6 Atomic Planes and Crystal Directions *73*

Summary *77*

Tables *79*

References *81*

Topics for Review *83*

Analytic Problems *86*

Computer Problems *96*

Design Problems *98*

Chapter 2 Bulk Properties of Semiconductors   *100*

2-1 Energy Bands   *100*

2-1.1 Oscillator Analogies *101*
2-1.2 Band Structure versus Atom Spacing *103*
2-1.3 Relating Bands and Bonds *105*
2-1.4 Electrons and Holes *106*
2-1.5 Energy Gap *109*
2-1.6 Conductors *110*

2-2 Electron Distributions in Conductors and Intrinsic Silicon   *111*

2-2.1 Fermi Level *112*
2-2.2 Density of States in a Conduction Band *114*
2-2.3 Band-Symmetry Approximation *116*
2-2.4 Equivalent-Density-of-States Approximation *117*
2-2.5 Intrinsic Carrier Density *121*

2-3 Impurity-Doped Silicon   *123*

2-3.1 Donor Doping and Hydrogen Model of a Donor State *123*
2-3.2 Uniform Doping *127*
2-3.3 Acceptor Doping *129*
2-3.4 Impurity Compensation *132*
2-3.5 A Fermi-Level "Computer" *133*

2-4 Analyzing Bulk-Semiconductor Problems 137
- 2-4.1 The Neutrality Equation *138*
- 2-4.2 The Boltzmann Approximation *138*
- 2-4.3 The Law of Mass Action *141*
- 2-4.4 Band Diagrams in Terms of Electrostatic Potential *143*
- 2-4.5 Carrier Densities in Terms of Electrostatic Potential *144*
- 2-4.6 The Boltzmann Relation *146*

2-5 Carrier Transport 147
- 2-5.1 Carrier Scattering by Phonons and Ions *147*
- 2-5.2 Drift Velocity *152*
- 2-5.3 Conductivity Mobility *154*
- 2-5.4 Velocity Saturation *158*
- 2-5.5 The Conductivity Equation *160*
- 2-5.6 Carrier Diffusion *163*
- 2-5.7 The Transport Equations *167*
- 2-5.8 The Einstein Relation *167*

2-6 Carrier Recombination and Generation 169
- 2-6.1 Excess Carriers *169*
- 2-6.2 Low-Level Recombination Rate *175*
- 2-6.3 Time-Dependent Recombination *176*
- 2-6.4 Carrier Lifetime *178*
- 2-6.5 Recombination Mechanisms *181*
- 2-6.6 Relative and Absolute Carrier Densities *186*

2-7 Continuity Equations 188
- 2-7.1 Constant-E Continuity-Transport Equations *188*
- 2-7.2 Continuity-Equation Applications *191*
- 2-7.3 Haynes-Shockley Experiment *195*
- 2-7.4 Surface Recombination Velocity *198*
- 2-7.5 Recombination-Based Ohmic Contacts *201*
- 2-7.6 Comparing Equilibrium and Steady-State Conditions *202*

Summary *203*

Tables *208*

References *210*

Topics for Review *212*

Analytic Problems *215*

Computer Problems *230*

Design Problems *231*

Chapter 3 *PN* Junctions 233
3-1 Junction Concepts 233
- 3-1.1 Space Charge at a Junction *233*

3-1.2 Dipole Layer *236*
3-1.3 Field and Potential Profiles *236*
3-1.4 Band Diagram for a Junction *237*
3-1.5 Carrier Profiles Through a Junction *239*
3-1.6 Symmetric Step Junction *242*
3-1.7 Current-Density Profiles in the Junction *242*

3-2 Depletion Approximation *244*

3-2.1 Assuming Total Depletion *244*
3-2.2 Charge-Density Profile *246*
3-2.3 Electric-Field Profile *246*
3-2.4 Electrostatic-Potential Profile *249*
3-2.5 Contact Potential *250*
3-2.6 Asymmetric Step Junction *253*
3-2.7 One-Sided Step Junction *256*
3-2.8 Comparing the Step Junctions *256*

3-3 Junction Under Bias *259*

3-3.1 Algebraic-Sign Convention *259*
3-3.2 Reverse Bias *262*
3-3.3 Forward Bias and Boltzmann Quasiequilibrium *267*
3-3.4 Law of the Junction *271*

3-4 Static Analysis *272*

3-4.1 Forward Current–Voltage Characteristic *273*
3-4.2 Reverse and Overall Characteristics *278*
3-4.3 Defining *Models* and Related Terms *281*
3-4.4 Piecewise-Linear Model *284*
3-4.5 Charge-Control Model *284*
3-4.6 Characteristic of a Real Silicon Junction *287*
3-4.7 High-Level Forward Bias *290*

3-5 Junctions other than *PN* Step Junctions *292*

3-5.1 PIN Diode *292*
3-5.2 Linearly Graded Junction *294*
3-5.3 Diffused Junctions *298*
3-5.4 High-Low Junctions and Ohmic Contacts *305*

3-6 Breakdown Phenomena *309*

3-6.1 Avalanche Breakdown *309*
3-6.2 Tunneling *314*
3-6.3 Punchthrough *319*

3-7 Approximate-Analytic Model for the Step Junction *327*

3-7.1 Poisson-Boltzmann Equation *328*
3-7.2 Debye Length *330*
3-7.3 First Integration of Poisson-Boltzmann Equation *333*
3-7.4 Second Integration of Poisson-Boltzmann Equation *337*
3-7.5 Depletion-Approximation Replacement *340*
3-7.6 Inversion Layer and Accumulation Layer *342*

3-8 Small-Signal Dynamic Analysis *347*
- 3-8.1 Small-Signal Conductance *348*
- 3-8.2 Diffusion Capacitance *354*
- 3-8.3 Depletion-Layer Capacitance *358*
- 3-8.4 Junction-Capacitance Crossover *364*
- 3-8.5 Coexisting Phenomena and Multiple Time Constants *365*
- 3-8.6 Small-Signal Equivalent-Circuit Model *372*
- 3-8.7 Effective Lifetime and Diffusion Capacitance *382*
- 3-8.8 Small-Signal Charge-Control Analysis *383*
- 3-8.9 Linear Differential Equations *385*

3-9 Advanced Dynamic Analysis *387*
- 3-9.1 Survey of Analytic Techniques *387*
- 3-9.2 Device-Physics Charge-Control Analysis *390*
- 3-9.3 Circuit-Behavior Charge-Control Analysis *394*
- 3-9.4 Device-Physics Exact Analysis *406*
- 3-9.5 Circuit-Behavior Exact Analysis *418*
- 3-9.6 SPICE Analysis *424*
- 3-9.7 A Numerical Example *427*

Summary *438*

Tables *444*

References *449*

Topics for Review *454*

Analytic Problems *459*

Computer Problems *479*

Design Problems *482*

Chapter 4 The Bipolar Junction Transistor *484*

4-1 BJT Rudiments *485*
- 4-1.1 Structure and Terminology *485*
- 4-1.2 Biases and Terminal Currents *488*
- 4-1.3 Carrier Profiles *489*
- 4-1.4 Typical Dimensional and Doping Values *492*
- 4-1.5 One-Dimensional Electron Current *495*

4-2 Elementary Device Theory *498*
- 4-2.1 Internal Current Patterns *499*
- 4-2.2 Parasitic Internal Currents *502*
- 4-2.3 Common-Emitter Current Gain *505*
- 4-2.4 The Gain Mechanism *507*

4-3 Biasing and Using the BJT *510*
- 4-3.1 Basic Bias Circuit *510*

4-3.2 Static Equivalent-Circuit Model 513
4-3.3 Rudimentary BJT Amplifier 515
4-3.4 Saturation 518
4-3.5 Other Operating Regimes 524
4-3.6 Other Circuit Configurations 529

4-4 Structures and Properties of Real BJTs 534

4-4.1 Electrochemical Potential 535
4-4.2 Nonuniform Base-Region Doping 539
4-4.3 The Gummel Number 544
4-4.4 Breakdown Voltage 546
4-4.5 Output Conductance 554
4-4.6 Structural Variations 556
4-4.7 Forward and Reverse Current Gain 563

4-5 High-Level Effects 566

4-5.1 Rittner Effect 566
4-5.2 Webster Effect 568
4-5.3 Ambipolar Effect 571
4-5.4 Kirk Effect and Quasisaturation 577
4-5.5 Lateral Voltage Drops in the Base Region 582
4-5.6 High-Level Effects in Combination 582
4-5.7 General Base-Region High-Level Analysis 585

4-6 Ebers-Moll Static Model 591

4-6.1 Gummel-Poon Reformulation 591
4-6.2 Assumptions and Problem Definition 592
4-6.3 Equations in Transport Form 594
4-6.4 Equations in Original Form 600
4-6.5 Applications 604
4-6.6 Equivalent-Circuit Model 606

4-7 Small-Signal Dynamic Models 610

4-7.1 Low-Frequency Hybrid Model 610
4-7.2 Hybrid Model and Device Physics 615
4-7.3 BJT Transconductance 619
4-7.4 Hybrid-Pi and Other Models 620
4-7.5 Improving Model Accuracy 629
4-7.6 Charge-Control Model 629
4-7.7 Base-Charging Time 639
4-7.8 Figures of Merit 644

4-8 SPICE Models 645

4-8.1 Model Equations 645
4-8.2 Series-Resistance Effects 647
4-8.3 Early Effect 648
4-8.4 High-Current Effects 648
4-8.5 Nonideal-Diode Effects 651
4-8.6 Capacitance Effects 652

4-8.7 Example of Small-Signal Analysis *653*
4-8.8 Example of Large-Signal Analysis *657*
4-8.9 Thermal Resistance *665*

Summary *667*

Tables *674*

References *682*

Topics for Review *686*

Analytic Problems *691*

Computer Problems *708*

Design Problems *709*

Chapter 5 The MOSFET *712*

5-1 Basic MOSFET Theory *713*
5-1.1 Field-Effect Transistors *713*
5-1.2 MOSFET Definitions *716*
5-1.3 Rudimentary Analysis *718*
5-1.4 Current–Voltage Equations *723*
5-1.5 Universal Transfer Characteristics *727*
5-1.6 Transconductance *734*
5-1.7 Inverter Options *736*

5-2 MOS-Capacitor Phenomena *741*
5-2.1 Oxide–Silicon Boundary Conditions *742*
5-2.2 Approximate Field and Potential Profiles *744*
5-2.3 Accurate Band Diagram *748*
5-2.4 Barrier-Height Difference *751*
5-2.5 Interfacial Charge *760*
5-2.6 Oxide Charge *766*
5-2.7 Calculating Threshold Voltage *774*

5-3 MOS-Capacitor Modeling *776*
5-3.1 Exact-Analytic Surface Modeling *777*
5-3.2 Comparing MOS and Junction Capacitances *780*
5-3.3 Small-Signal Equivalent Circuits *783*
5-3.4 Ideal Voltage-Dependent Capacitance *789*
5-3.5 Real Voltage-Dependent Capacitance *794*
5-3.6 Physics of MOS-Capacitance Crossover *799*
5-3.7 Analysis of MOS-Capacitance Crossover *804*

5-4 Improved MOSFET Theory *806*
5-4.1 Channel–Junction Interactions *807*
5-4.2 Ionic-Charge Model *813*
5-4.3 Body Effect *819*

5-4.4 Advanced Long-Channel Models *823*

5-5 SPICE Models   *824*
5-5.1 Level-2 Parameters *824*
5-5.2 Level-2 Model *829*
5-5.3 Small-Signal Applications of Model *833*
5-5.4 Large-Signal Applications of Model *840*

5-6 MOSFET–BJT Performance Comparisons   *850*
5-6.1 Simple-Theory Transconductance Comparison *851*
5-6.2 Subthreshold Transconductance Theory *854*
5-6.3 Calculating Maximum MOSFET g_m/I_{out} *857*
5-6.4 Transconductance versus Input Voltage *859*
5-6.5 Physics of Subthreshold Transconductance *860*

Summary *868*

Tables *874*

References *879*

Topics for Review *886*

Analytic Problems *889*

Computer Problems *900*

Design Problems *903*

Symbol Index   *I-1*
Subject Index   *I-6*

Foundations of Modern Electronics

Here we review topics that underlie the electronics of solid-state devices. Most students who use this book have been exposed previously to the concepts, relationships, procedures, and subjects that we now address, possibly excepting crystallography. But experience shows that reinforcement is in order. A working understanding of these topics is essential to competent handling of solid-state electronics. Therefore, this chapter presents the necessary review and emphasizes the problems most relevant to solid-state devices.

The subject of units, or dimensions, will receive substantial attention in this chapter. We shall, however, introduce them individually on an as-needed basis, along with discussion of the associated physical entities. The skillful and consistent use of units is a powerful aid for avoiding errors in calculations, and emphasis will be placed on so doing. In addition, systematic problem-solving procedures will be formulated.

1-1 ELECTRIC CHARGE, FIELD, AND ENERGY

The sources of an electric field are electrical charges. Once a field has been established within a given region, another charge placed within that region typically experiences a force. If the charge is free to move in response to the force, it can acquire kinetic energy from the field. Let us now examine these relationships in some detail.

1-1.1 The Electric-Field Concept

Electric field is an entity having magnitude and direction, and hence is, in short, a vector quantity. We now introduce the concept of a *uniform electric field*. Visualize a volume wherein all of the field vectors are parallel to one another—pointing, for example, in the positive-x direction. Let the electric field have the same magnitude everywhere within the volume. This is not simply an abstraction; rather, the uniform field can easily be approximated in the laboratory, using a familiar device—the parallel-plate capacitor. When voltage is applied to the capacitor, charges of opposite sign are distributed on its two plates. As a result, the volume between the plates constitutes a region of uniform electric field. A caveat is necessary, however, because conditions become more complicated as one approaches the outer edges of the capacitor plates, where the field is no longer uniform. To avoid this complication we can define the region of interest as that lying "inside" the edges by a distance amounting to several times the plate spacing. This uniform-field approximation is fair when "several" has the value three, and excellent when its value is ten.

Let the symbol **E** represent the electric field so established. Now introduce an electrical charge Q. We shall use this symbol for a general charge, reserving the symbol q to stand exclusively for the magnitude of the charge on the electron, as is customary. Specifically, $q = 1.602 \times 10^{-19}$ C, where C is the symbol for the coulomb, the unit of electrical charge in the RMKSA (rationalized meter-kilogram-second-ampere) system that was widely adopted in 1948. Returning to the general charge Q, we position it within the region of uniform field just described; it will then experience a force in the amount of

$$\mathbf{F} = Q\mathbf{E}. \tag{1-1}$$

Force, like electric field, is a vector quantity and Q is a scalar quantity; Equation 1-1 informs us that the force vector is aligned with the field vector and points in the same direction. That is, a *positive* charge placed in an electric field experiences a force *in the direction of the field*.

This very fundamental relationship is useful as a defining equation for electric field. That is, we shall assert that electric field is defined as *force per unit charge,* or the quotient of force by charge when the charge in question is placed within the field region. But now a problem arises. We noted earlier that electrical charges are the sources of electric field. The field in the parallel-plate capacitor stems from the charges located on the plates. Therefore, the act of placing the charge Q within the region of uniform field produced by the capacitor will cause a distortion, or perturbation, of the field. In brief, the field will no longer be uniform. The solution needed is a familiar one: We observe the quotient of force by charge as the magnitude of the charge is decreased. In the limit of infinitesimally small charge, electric field is defined by

$$\mathbf{E} \equiv \lim_{Q \to 0} \frac{\mathbf{F}}{Q}. \tag{1-2}$$

The ideas contained in Equation 1-2 are represented heuristically in Figure 1-1.

Actually, Equation 1-2 involves two concepts (as does Figure 1-1). The first idea is that of a *test charge,* which can be regarded as a measurement technique in the present thought experiment. To measure an electric field at a particular point, introduce a small charge at that point and observe the force upon the charge. The

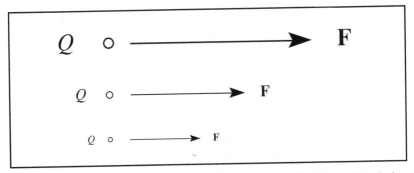

Figure 1-1 The quotient of force by charge approaches a limit that is identical to the electric field.

resulting quotient of force by charge is an approximate value of the electric field at that point before the charge was introduced. The second concept is that of *limit*, which most readers have encountered before in the study of mathematics. In a physical context, such as the present one, the challenge is to measure a certain quantity without altering its value in the process of measurement. Thus one employs successively smaller test charges, plots the F/Q quotient, and extrapolates the curve to zero charge to obtain an accurate value of electric-field magnitude.

At this point an illuminating subtlety is worth noting. There is one case in which one can determine electric field accurately and directly from knowledge of the force on a finite charge—even a charge of arbitrary magnitude! This is the case wherein the charges responsible for the initial field are fixed firmly in position. This is not the usual case. In a typical physical situation, the responsible charges are free to move. Consider once more the region within the parallel-plate capacitor, the uniform-field situation that is easiest to achieve. When a charge is introduced within that volume, charges on both plates redistribute themselves; it is this charge readjustment that accounts for the field "perturbation" alluded to earlier.

In semiconductor work, the use of RMKSA units is well established. To introduce the units relevant at this point, let us drop the vector notation in Equation 1-1 and then rearrange it:

$$E = \frac{F}{Q}. \tag{1-3}$$

The corresponding MKS units are

$$\frac{\text{volts}}{\text{meter}} = \frac{\text{newtons}}{\text{coulomb}}, \tag{1-4}$$

and the correct symbols for these units are

$$\frac{[V]}{[m]} = \frac{[N]}{[C]}. \tag{1-5}$$

We shall use square brackets in this manner to distinguish equations involving unit symbols from those involving the symbols for ordinary variables. The minor respect in which semiconductor practice deviates from the use of pure RMKSA units is in the use of *centimeter* in place of *meter* for distance measurement, a step that will be taken in a later section and will be used thereafter.

A point on distance-measurement terminology is worth making. The smallest dimensions, or "feature sizes," in the structures of devices in solid-state electronics have decreased so dramatically in recent decades that they are now of the order of one millionth of a meter, or a *micrometer,* abbreviated μm. It is important to remember that $1\ \mu\text{m} = 10^{-4}$ cm. This state of affairs literally validates the term *microelectronics*. The older term for micrometer was "micron."

1-1.2 Work and Energy in an Electrical Context

Once again visualize a region of uniform electric field. Furthermore, let the field be created by charges that are fixed in position, so that the problem of field perturbation is eliminated. And let the field be aligned with a spatial axis—let us say, the x axis. The vast majority of field problems addressed in this book involve a unidirectional field and an axis aligned therewith.

Now we come to the second of a long series of arbitrary but essential conventions with respect to algebraic sign. (The first was the choice that electric field points in the direction of the force on a *positive* charge.) Let us refer field direction to the spatial axis to assign its algebraic sign. That is, we will say that an electric field in the positive-x direction is positive.

To introduce the work concept, let us place a charge $+Q$ in a region arranged as shown in Figure 1-2. The quantity known as work can be defined as the integral of force through distance. What remains, of course, is to define limits and, once more, algebraic sign. Taking the latter first, we employ the convention that positive work is work done **on** an object. In Figure 1-2 it is evident that to do positive work on the charge $+Q$ we must move it leftward, *against* the force F. Thus, a particular increment of positive work done on the body in that situation can be written

$$\text{work} = \int_{x_1}^{x_0} F\ dx. \tag{1-6}$$

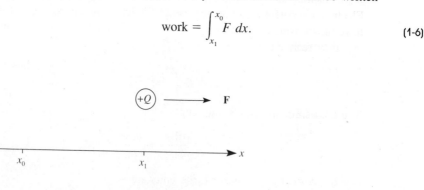

Figure 1-2 To move the charge Q from the position x_1 to the position x_0, we must do work **on** it, which by convention is taken to be positive work.

We usually prefer, however, to integrate in the positive-x direction, so let us write instead

$$\text{work} = -\int_{x_0}^{x_1} F\, dx. \tag{1-7}$$

The RMKSA unit for work is the *joule* (symbol, J)—the same as for energy, because energy is the ability to do work. To assess the right-hand side of Equation 1-7 with respect to units, one ignores both the integral sign and the d in the differential. The differential quantity is an infinitesimal increment of distance, which (for the moment) we measure in meters, so it becomes evident that a joule is equal to a newton-meter, or

$$[J] = [N][m]. \tag{1-8}$$

1-1.3 Electrostatic Potential

The symbol for electrostatic potential is ψ (the lowercase Greek letter psi). For reasons that will become evident in a moment, we shall first define an *increment* in electrostatic potential $\Delta\psi$, which can be taken as the work done per unit charge in moving a charge from one position to another. If an electric field is present, the charge will experience a force, so moving the charge will involve positive or negative work. (The latter term means simply that the charge is permitted to move in response to the force upon it.) In a region that is field-free, no work is involved in moving the charge, because we are not at present considering any other kind of force. In particular, gravitational force is usually neglected in such a problem. In the presence of an electric field, then, the increment in electrostatic potential accompanying an increment of displacement, or motion, is

$$\Delta\psi = \frac{\text{work}}{Q}. \tag{1-9}$$

Electrostatic potential, as well as an increment in electrostatic potential, is measured in volts, for which the symbol is V. Thus the consistent RMKSA units corresponding to Equation 1-9 are

$$[V] = \frac{[J]}{[C]}. \tag{1-10}$$

In view of Equations 1-3 and 1-7, Equation 1-9 can be rewritten as

$$\Delta\psi = \frac{-\int_{x_0}^{x_1} F\, dx}{Q} = -\int_{x_0}^{x_1} \frac{F}{Q}\, dx = -\int_{x_0}^{x_1} E\, dx. \tag{1-11}$$

The magnitudes F and E of the vector quantities **F** and **E** are employed in Equation 1-11 because directions are evident in the problem posed in Figure 1-2. And although E was taken as constant throughout the region represented there, so that F

was also constant, such a restriction is not necessary. That is, Equation 1-11 is equally valid when E and F are functions of x.

To define an absolute electrostatic potential ψ, all that is necessary is to choose a point (that is, a position) to serve as reference (or origin, or zero). It is of utmost importance to realize that this point or position is a matter of *arbitrary* choice. There are situations wherein one point may be mathematically more convenient than another, but the selection remains arbitrary. At present, let us make this selection:

$$\psi(x_0) \equiv 0. \tag{1-12}$$

And let us make this equally arbitrary choice of spatial origin (or reference, or zero):

$$x_0 \equiv 0. \tag{1-13}$$

Subject to these two definitions, then, absolute electrostatic potential can be written

$$\psi = -\int_0^x E\, dx', \tag{1-14}$$

where x' is a dummy variable for integration.

The matter of the dummy variable is one that some students and practitioners tend to regard as a "frill." In fact, the dummy variable is essential to meaning, as can be illustrated easily.

Exercise 1-1. Suppose that one omits the dummy-variable notation, as is often incorrectly done, writing

$$f(x) = \int_0^x x\, dx.$$

Then suppose someone asks the legitimate question, "What is the value of the integral when $x = 2$?" Respond to the question in a literal way to illustrate the importance of the dummy variable.

Using the given expression and instruction literally yields

$$f(2) = \int_0^2 2\, d2,$$

which is a meaningless collection of symbols. If, on the other hand, a dummy variable has properly been used, we would have

$$f(x) = \int_0^x x'\, dx'.$$

Then the answer to the same question becomes

$$f(2) = \int_0^2 x'\, dx' = \left.\frac{(x)^2}{2}\right|_0^2 = 2.$$

An important result is obtained by differentiating Equation 1-14. Recall that to differentiate an integral with respect to its upper limit, when the lower limit is constant, one simply writes the integrand with the upper limit replacing the variable of integration, dropping the integral sign and the differential. In this case the result is $(d\psi/dx) = -E$, or

$$E = -\frac{d\psi}{dx}, \qquad (1\text{-}15)$$

which indicates that the vector quantities, electric field and potential gradient, have equal magnitudes and opposite directions. Because a gradient points "uphill," it is evident (see Figure 1-3) that the electric-field vector points from a region of higher potential to a region of lower potential. And since the force on a positive charge in an electric field has the direction of the field vector, it follows that electric-field vectors originate in regions of positive charge and terminate in regions of negative charge; the force, after all, arises from the mutual repulsion of like charges and the mutual attraction of unlike charges.

1-1.4 Lines of Force

Problems involving electric field can be made graphic and palpable by employing a fictional entity—the line of force. This is a directed line that is "launched" by a positive charge and that terminates on a negative charge of equal magnitude; the charge magnitude involved is a matter of arbitrary definition. That is, the line is seen to originate on a positive charge of a magnitude one may specify at will, and to end on an equal and opposite charge. In the general case, these lines are curved. The electric-field vector at any point along a line of force is tangent to the line of force at that point and has a direction (sense) that is the same as that of the line of force.

The magnitude of the electric field at a particular position is proportional to the *density of lines of force* at that position (that is, to their degree of "packing"). Since the amount of charge required to "create" a line of force is arbitrary, it follows that the relationship between line density and field magnitude is also a matter of arbitrary definition.

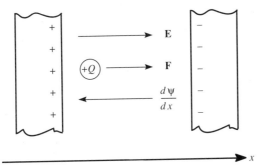

Figure 1-3 Here it is emphasized that force on a positive charge is aligned with electric field, pointing from a region of higher potential to a region of lower potential.

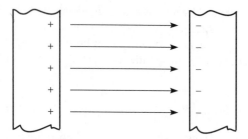

Figure 1-4 The uniform electric field deep inside a parallel-plate capacitor involves parallel, linear *lines of force* of fixed spacing, represented here in two dimensions.

Returning to the uniform-field concept introduced at the beginning of this chapter, we can claim that it is obvious that the simultaneous requirements of field-vector parallelism everywhere and field-magnitude equality (line-density equality) everywhere can be met only by parallel and equally spaced lines of force. These conditions in turn require uniform (areal) charge densities on the plates of a parallel-plate capacitor when this means is used for creating the region of uniform field, a situation illustrated in two-dimensional fashion in Figure 1-4.

Exercise 1-2. For an air-dielectric parallel-plate capacitor, let us make the arbitrary definition (3 lines of force/cm^2) $\equiv$ (1 V/cm) = E, as illustrated in Figure 1-5. Calculate the amount of charge necessary to launch one line of force.

The elementary expression for parallel-plate capacitance is

$$C = \frac{A\epsilon_0}{d},$$

where A is the area under consideration on one plate, d is the plate spacing, and ϵ_0 is the *permittivity of free space* = 8.85×10^{-14} F/cm, a topic addressed in consid-

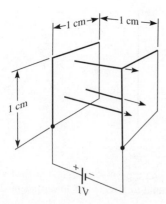

Figure 1-5 Line-of-force *density* (number per cm^2) is proportional to electric-field magnitude, with the proportionality factor being a matter of arbitrary definition.

erably more detail in Section 1-3.3. Thus capacitance per unit area is given by

$$\frac{C}{A} = \frac{\epsilon_0}{d}.$$

From the charge-voltage capacitor law, $Q = CV$, we can obtain an expression in C/A by dividing both sides by A:

$$\frac{Q}{A} = \frac{C}{A}V.$$

Hence the desired quantity, Q/A, is given by

$$\frac{Q}{A} = \frac{\epsilon_0 V}{d}.$$

Let us take $V = 1$ V and $d = 1$ cm, which gives us a field of 1 V/cm. Then

$$\frac{Q}{A} = \frac{(8.85 \times 10^{-14} \text{ F/cm})(1 \text{ V})}{1 \text{ cm}} = 8.85 \times 10^{-14} \text{ C/cm}^2.$$

(If V and d are altered by the same factor, Q/A is unaffected, as the last expression shows.) Dividing the last result by 3 lines/cm² yields 2.95×10^{-14} C/line, which is the answer sought.

As an extension, determine the number of electrons necessary to terminate a line of force so defined:

$$\frac{2.95 \times 10^{-14} \text{ C/line}}{1.602 \times 10^{-19} \text{ C/electron}} = 1.84 \times 10^5 \text{ electrons/line}.$$

As already noted, uniform-field conditions do not exist near the edges of a capacitor, where the lines of force curve outward, an effect often described as "fringing." But this further illustrates the utility of the line-of-force picture. In the representation of Figure 1-6, several qualitative features of the field distribution become immediately apparent. It is evident that "crowding" of the lines occurs at the inner corners of the plate edges, corresponding to a local increase in field intensity, or magnitude. But in the median plane of the capacitor, field magnitude diminishes as the plate edges are approached and passed. Furthermore, a question as to field direction at a point such as P is answered by the vector **E**. Thus the line-of-force representation conveys electric-field information with respect to both magnitude and direction.

1-1.5 Potential Energy and Kinetic Energy

Once a reference has been chosen for potential ψ, then Equation 1-9 becomes simply $\psi = (\text{work}/Q)$, which we can rewrite as

$$\text{work} = Q\psi. \tag{1-16}$$

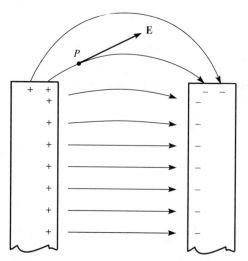

Figure 1-6 The "fringe" effects at the edges of capacitor plates, illustrating line-of-force nonlinearity, electric-field concentration (closely spaced lines of force) at a corner, and gradual electric-field decrease as one departs the capacitor along the median plane.

Equation 1-16 tells us the amount of work we must do *on* the charge Q to move it from the selected origin (where electrostatic potential is by definition zero) to a point where electrostatic potential has the value ψ. Under these conditions, the *potential energy* (P.E.) of Q is defined as the ability of Q to do work in the process of "relaxing" from the point where electrostatic potential has the value ψ, back to the origin. That is,

$$\text{P.E.} \equiv Q\psi. \tag{1-17}$$

An obvious feature of the quantity P.E. is the arbitrariness in its value—a value that changes with changes in the position of the arbitrary electrostatic-potential origin. This feature, though, is completely analogous to the corresponding feature of the mechanical (rather than electrical) P.E. of an object with mass M in the gravitational field, which is possibly more familiar. In this case, P.E. = Mgh, where h is height above an equally arbitrary horizontal plane. If you hold an object at eye level, you may calculate its mechanical P.E. with respect to the floor, your waistline, the ceiling (in which case its P.E. becomes negative), or any other horizontal surface that strikes your fancy.

If we do indeed have to do work on the charge Q to move it away from the origin, then in the process of relaxing to the origin once more, the charged object can acquire energy from the electric field. Let us assume that no forces exist other than that attributable to the electric field. Under these conditions, the "final" energy of the charged object is *kinetic energy, or energy of motion*, which is written as

$$\text{K.E.} = \tfrac{1}{2}Mv^2, \tag{1-18}$$

where v is the velocity of the object. Let us use M for the mass of the charged object, or a general mass, and reserve m to stand for the mass of the electron.

Under the conditions assumed here, furthermore, the total energy of an object can be written as the sum of its kinetic and potential energies, or

$$\xi = \text{P.E.} + \text{K.E.} = Q\psi + \tfrac{1}{2}Mv^2 \tag{1-19}$$

in the electrical context. The kinetic-energy term is free of arbitrariness, but the potential-energy term is not. The idea that we have described in terms of reference or origin selection is sometimes equivalently expressed by saying that an arbitrary constant is associated with the potential-energy term. And from this it is evident that total energy ξ contains the same arbitrary constant.

A *conservative system* is one in which total energy ξ is constant. Here again it may be helpful to look briefly at a more familiar mechanical example. Figure 1-7 shows a ball of mass M constrained to roll on a curved track. Suppose that it is placed at position 1 and is released. Just before release, its velocity is zero and so, too, is its kinetic energy. Its potential energy Mgh depends, once more, on the reference selection. A logical, but not mandatory, choice is the lowest position the center of the ball can reach, as has been done in Figure 1-7, so that

$$\text{P.E.} = Mgh_1 \tag{1-20}$$

initially. When the ball has rolled down to position 2, its P.E. has vanished. It now has a velocity v, however, and

$$\text{K.E.} = \tfrac{1}{2}Mv_2^2. \tag{1-21}$$

If the force of gravity is the only force present (with frictional forces specifically being negligible), then the system is conservative, and then it becomes evident from Equation 1-19 that for the invariant energy ξ we have

$$\xi = Mgh_1 = \tfrac{1}{2}Mv_2^2. \tag{1-22}$$

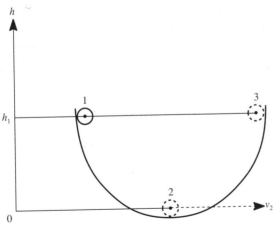

Figure 1-7 Ball rolling on a frictionless curved track, illustrating a conservative system in the mechanical context.

In principle, the ball continues to roll back and forth forever, with its energy cycling from 100% P.E. to 100% K.E.

To illustrate a conservative system electrically, let us turn to the vacuum diode represented schematically in Figure 1-8. On the left is the conventional symbol for the cathode, or normally negative electrode. It is capable of emitting an electron, represented there as a small sphere. The anode, "plate," or positive electrode is at the right, and will attract the electron after it has been emitted. The electrodes are inside an evacuated vessel, so that gas molecules do not interfere with electron trajectory. Let us arbitrarily and unconventionally apply a "bias" of one volt to the diode, as illustrated in Figure 1-8. Further, let us choose the plate (or anode) potential as our reference. In an electrical circuit, it is common (but not mandatory!) to choose the circuit-ground potential as reference, so we attach to the plate a ground symbol.

The electron in this system exhibits conservative properties. For the choices made here, potential energy will have a maximum value when the electron is at the cathode. As it is drawn to the plate after being released, it will be accelerated, gaining velocity and kinetic energy, while its potential energy declines. Just before striking the plate, its kinetic energy is at a maximum, and its potential energy has dropped to zero. After the electron strikes the plate, it loses energy rapidly under conditions that are no longer conservative, so we rule this more complex portion of the sequence out of consideration. When the electron has just left the cathode we have these conditions:

$$\xi = \text{P.E.} + \text{K.E.} = Q\psi + 0 \qquad (1\text{-}23)$$
$$= (-q)(-1 \text{ V}) = (-1.602 \times 10^{-19} \text{ C})(-1 \text{ V})$$
$$= 1.602 \times 10^{-19} \text{ C}\cdot\text{V} = 1.602 \times 10^{-19} \text{ J}.$$

And just before impact we have, in view of the fact that the system is conservative,

$$\xi = \text{P.E.} + \text{K.E.} = 0 + \tfrac{1}{2}mv_2^2 \qquad (1\text{-}24)$$
$$= 1.602 \times 10^{-19} \text{ J} \equiv 1 \text{ eV}.$$

The electron in this problem has "fallen through" a potential difference of one volt, and in so doing has acquired a tiny amount of kinetic energy, 1.602×10^{-19} J. So

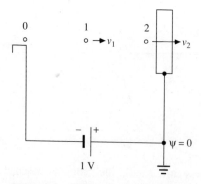

Figure 1-8 Schematic representation of a vacuum diode. The electron in transit from cathode (0) to plate (2) constitutes a "conservative system" in the electrical context.

important is this miniscule energy in electronics, however, that it has been separately identified as a unit. It is named the *electron volt,* with the symbol eV, as can be seen in Equation 1-24.

Notice in Equation 1-23 that we follow a quirk of convention that is now too well established to be altered. We use q for the *magnitude* of the electronic charge, and when the actual electronic charge is of interest, write $-q$, showing the negative sign explicitly. The symbol ψ, on the other hand, is one receiving the more usual treatment; included within the symbol is the algebraic sign, which in the present problem is determined by the arbitrary choices outlined earlier.

1-2 UNIT MANIPULATION AND PROBLEM SOLVING

Most of the analytical problems encountered in solid-state electronics can be addressed systematically by using a procedure that we outline here. Near the end of the process, and just before the calculator is brought into play, is a dimension-verification step that involves the cancellation of units. This step is facilitated by using the simple idea we address first.

1-2.1 The Unity Factor

Virtually everyone has had the experience of performing an informal unit conversion, often mentally, and discovering later that multiplication was in order, instead of division, or vice versa. The same thing occurs in more formal calculations. There is, however, an unfailing technique for avoiding this trap—namely, use of the *unity factor*. This entity can be defined as a fraction having the same physical quantity in numerator and denominator. An example is (100 cm/1 m). To use it, we simply apply it or its inverse as a factor in a calculation, choosing the orientation of the fraction so that undesired units will cancel.

Exercise 1-3. In Section 1-1.5 we learned that 1 eV $\equiv 1.60 \times 10^{-19}$ J. This, then, becomes the basis of a unity factor. Let us pose this problem: Convert 1 billion electron volts to joules.

The first step is to write the given quantity; the second is to multiply it by a unity factor that causes the unwanted units (eV) to cancel, thus:

$$(10^9 \, \cancel{\text{eV}}) \left(\frac{1.60 \times 10^{-19} \text{ J}}{1 \, \cancel{\text{eV}}} \right) = 1.60 \times 10^{-10} \text{ J}.$$

It is evident that to make this system work, one must develop the habit of substituting numbers *and units* in place of variable symbols when performing calculations.

In a typical calculation, more than one unity factor will be necessary. This point is worth a further illustration.

Exercise 1-4. Calculate the equivalent of one mile in terms of the angstrom ($1 \text{ Å} = 10^{-10}$ m).

The first step is to write 1 mile, and the second is to follow it by an appropriate series of unity factors, thus:

$$(1 \text{ mile})\left(\frac{5280 \text{ ft}}{1 \text{ mile}}\right)\left(\frac{12 \text{ in}}{1 \text{ ft}}\right)\left(\frac{2.54 \text{ cm}}{1 \text{ in}}\right)\left(\frac{1 \text{ Å}}{10^{-8} \text{ cm}}\right) = 1.6 \times 10^{13} \text{ Å}.$$

This example employs four unity factors in "cascade."

To make this system work, we reemphasize, one must *always* associate units with numerical values in the process of problem solving. To do otherwise is to omit important information and, all too often, to get the wrong answer. One more example is appropriate, involving the unit conversion that is peculiar to semiconductor electronics, the conversion of m to cm.

Exercise 1-5. Given an electric field of 10^5 V/m, find the equivalent in terms of kV/cm.

First, write the given quantity, and follow it by appropriate unity factors:

$$\left(\frac{10^5 \text{ V}}{\text{m}}\right)\left(\frac{1 \text{ kV}}{10^3 \text{ V}}\right)\left(\frac{1 \text{ m}}{10^2 \text{ cm}}\right) = 1 \text{ kV/cm}.$$

Next, let us look at the problem-solving steps that normally precede unit conversion, and then assemble all of the steps.

1-2.2 Problem-Solving Procedure

There are five basic steps in solving an analytical engineering problem in solid-state electronics, and in other disciplines as well. They are these:

1. Draw a diagram and write one or more equations, as appropriate. If a diagram is not appropriate, simply write the equation(s).
2. Solve the equation(s) for the desired quantity using variable symbols, and *not* using numerical values. The motive here is to save writing and to avoid the

errors that accompany unnecessary writing. Compare, for example, the labor in writing q with that required to write 1.602×10^{-19} C.

3. Into the final expression, solved for the variable of interest, substitute numbers *and* units.
4. If a unit conversion is needed, use a unity factor to eliminate the unwanted unit. For example, suppose the final equation involves an energy value in J and another in eV; use the unity factor employed in Exercise 1-3 to eliminate the undesired energy unit. Then verify that the final result has appropriate units. This verification procedure provides a potent weapon against errors by signaling a missed unit conversion, an erroneous starting equation, or an error in the preceding algebraic manipulations.
5. Here, and only at this point, carry out the necessary numerical calculation.

Let us illustrate these five steps by means of a near-trivial but illuminating example:

Exercise 1-6. Find the voltage applied to a parallel-plate capacitor having a plate spacing of $d = 1.0$ cm and a uniform electric-field magnitude in its central region of $|E| = 10{,}000$ V/m.

---

1. This is a case where a diagram is an especially appropriate starting point because it enforces and records relevant arbitrary definitions. Let us orient the plates normal to a spatial axis, the x axis as before. Next, let us assign spatial and potential references.

   It is often convenient (but not obligatory) to let the two positions coincide. In the absence of special considerations, this constitutes a reasonable starting point—and it can readily be altered subsequently if necessary. Figure 1-9(a) shows the capacitor and the two "zero" assignments. Yet another arbitrary decision enters with respect to the polarity of applied voltage V. Let us choose the positive sign. It follows, then, that **E** is directed leftward, and hence is negative according to a convention cited earlier. The relevant equation here is a variant of Equation 1-11:

$$V = \Delta\psi = -\int_0^d E\, dx.$$

2. Algebra comes next. We see that our equation is already solved for the unknown variable, but some manipulation (namely, an integration) is to be done with symbols, putting the expression into simplest possible form and into terms of given quantities. We employ the knowledge that E is uniform in this problem, and hence can be factored out of the integral.

$$V = -E\int_0^d dx = -Ed.$$

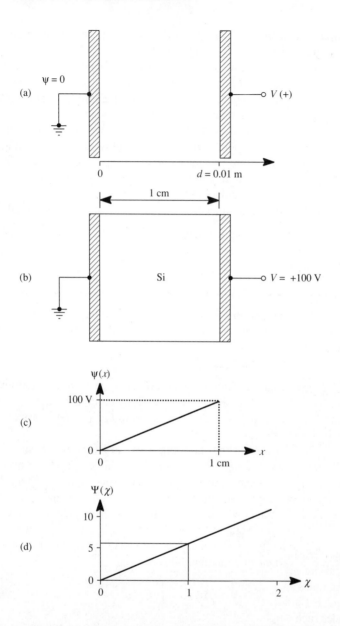

Figure 1-9 A sample diagram for problem solving. (a) A parallel-plate capacitor with arbitrary reference assignments. (b) Physical representation of a sample in a problem involving normalized variables. (c) Graphical solution with unnormalized variables. (d) Graphical solution using normalized variables.

3. Substituting numbers *and* units gives

$$V = -\left(-\frac{10^4 \text{ V}}{\text{m}}\right)(1.0 \text{ cm}).$$

4. It is evident that unit conversion is necessary. Applying an appropriate unity factor gives

$$V = \left(\frac{10^4 \text{ V}}{\cancel{\text{m}}}\right)(1.0 \cancel{\text{ cm}})\left(\frac{1 \cancel{\text{ m}}}{100 \cancel{\text{ cm}}}\right).$$

5. The numerical result, then, obtainable by inspection in this case, is $V = 100$ V.

It will be worthwhile for the reader to practice these five steps until they become automatic, initially making up examples as simplistic as this one. When problems at the end of the chapter are addressed, they should always be done in this way.

1-2.3 Unit and Variable Symbols

At this point we warn against an error that is very commonly made by beginning students, one that engenders major confusion. The most common kind of equation is one that relates variable symbols, examples being Equations 1-1, 1-3, 1-9, 1-17, and 1-18. Another legitimate kind of equation relates unit symbols, with examples for this kind being Equations 1-5, 1-8, and 1-10. The caution we now emphasize is that one should *never* mix the two. In particular, one should *never* set a collection of variable symbols equal to a collection of unit symbols. Note in the previous examples that the different kinds of equations are written on different lines. If they must be written on the same line, it is essential that they be separated by an appropriate punctuation mark, such as a semicolon.

There are major and minor reasons for which this rule must be followed without fail. The minor reason is that the stockpile of symbols in common use is limited, so many of them are pressed into double use—use as variable symbols *and* as unit symbols. Examples are A (area) and A (ampere), C (capacitance) and C (coulomb), F (force) and F (farad), J (current density) and J (joule), and m (mass) and m (meter). Under such conditions, mixing the two kinds of symbols in one expression would lead to inevitable confusion.

A stronger reason, however, is that the variable symbol subsumes the unit (unless the quantity is dimensionless), and therefore mixing of the two sets of units is simply incorrect. In fact, a variable symbol (except for a few examples that we shall cite particularly as they arise) stands for *three* entities: It contains within it (1) an algebraic sign, (2) a numerical value, and (3) dimensions. For example, we have

$$E = -10 \text{ V/cm}. \tag{1-25}$$

The one exception to this rule noted so far is the symbol q (which indeed is a constant and not really a variable), standing for only *two* entities—a numerical value and a unit. As noted earlier, it is the magnitude of the electronic charge, $q = 1.602 \times 10^{-19}$ C, and it is customary to show the algebraic sign explicitly. That is, when the charge Q of interest is that of the electron, we write $Q = -q$.

Occasionally one encounters a writer who, in a spirit of helpfulness, writes a symbolic expression for a variable and follows it immediately by units. This practice is not helpful; it leads to inevitable confusion. What is worse, it is incorrect.

1-2.4 One-Dimensional Problems

Most of the problems we shall address are *one-dimensional* in the sense we shall now explain. First, this term is usually applied to the spatial dependence of some variable. Second, it means that the variable in question is changing in one direction only. In most cases we will orient the three mutually orthogonal rectangular coordinates so that the variation occurs along the x axis, and not along the y and z axes.

The dependence of electrostatic potential ψ in the parallel-plate capacitor affords a good example. Suppose that we let the x axis be normal to the plates, as was done earlier. Then ψ varies in the x direction only, and does not vary along the y and z axes, insofar as "fringing effects" can be ignored. Putting the matter another way, we can assert that the interior of a parallel-plate capacitor, the region away from the edges of the plates, constitutes a *one-dimensional problem* with respect to electrostatic potential ψ.

The matter of spatial-origin selection, illustrated for a one-dimensional problem in Figure 1-9, deserves a further word. In Sections 1-1.3 and 1-2.2 it was noted that although the choice is arbitrary, some options are more convenient than others. Now we make a stronger assertion, to the effect that there are cases wherein the choice is *far* from trivial. For example, origin relocation in the transition from the Ptolemaic to the Copernican model of the solar system converted a murky aggregation of epicycles to a set of nearly circular and nearly coplanar concentric orbits.

1-2.5 Normalization

It is often convenient analytically and graphically to employ *normalized variables*. This means that each variable has been divided by a fixed quantity having the *same dimensions* as the variable. As a result, one can now work with *dimensionless variables*.

As a simple example, suppose that the function of interest is $\psi(x)$, or electrostatic potential as a function of position in a one-dimensional problem. Let the physical situation, equally simple, be a silicon sample 1 cm thick to which 100 V is applied, as shown in Figure 1-9(b). The function $\psi(x)$ is plotted using customary (unnormalized) variables in Figure 1-9(c). It is evident that the applicable equation is $\psi = (100 \text{ V/cm})x$, where ψ is in V and x is in cm.

Next, let us deal with the same problem through normalization. The variables to be normalized are ψ and x. Take the independent variable first. Usually one employs as the "normalizing constant" a quantity that has physical meaning and significance. Such a constant arises for the case of distance in Section 1-5, on the subject of crystallography. There we learn that a silicon crystal can be resolved into a large number of identical elementary cubes known as *unit cells*. The edge dimension of each of these cubes is known as the *lattice constant of silicon*, or $a_0 = 5.43$ Å $= 5.43 \times 10^{-8}$ cm. Let us use this quantity for defining a new normalized (and hence dimensionless) position variable. Let us call it χ (lowercase Greek chi). Then evidently $\chi = x/a_0$. The total thickness of the sample in terms of the new normalized distance variable is $\chi = [(1 \text{ cm})/(5.43 \times 10^{-8} \text{ cm})] = 1.84 \times 10^7$.

Now turn to the dependent variable ψ. There exists at least one "natural" unit of voltage that is widely used for normalization. But since it cannot be conveniently introduced until later, let us illustrate the further point that arbitrary constants are sometimes employed for normalization. Suppose we are interested in a microscopic presentation of conditions in the silicon and therefore will find a small value of voltage a convenient choice, let us say 1 μV. Thus the total applied voltage in the new normalized variable Ψ (uppercase Greek psi) will be $\Psi = [(100 \text{ V})/(10^{-6} \text{ V})] = 10^8$.

From the results thus far it is evident that the number of normalized voltage units per normalizing distance unit in the present problem is given by $[(10^8)/(1.84 \times 10^7)] = 5.43$. Hence, the problem can be represented graphically in normalized fashion, as has been done in Figure 1-9(d), where we have chosen to represent a microscopically thin portion of the sample near its left-hand face.

One advantage of normalization is that it lends a degree of universality to a solution, whether analytic or graphic. The solution becomes *independent* of the *system of units chosen,* provided of course that the same physical quantities are retained as normalizing constants. This is evident in Figure 1-9(d), where no units are cited. It is equally evident in the analytic statement of the function of interest, which now becomes $\Psi = 5.43 \chi$. Normalization, in other words, has changed the physical problem into one that is purely mathematical.

A related advantage of normalization is a reduction in the number of mathematical operations required to deal with a given equation. This is especially important in computer solutions. For example, if we adhere to the unnormalized variable x in an equation involving x/a_0, then for each new value assigned to x, a division by a_0 must be performed. But by using the normalized variable $\chi \equiv x/a_0$, we avoid the division step. In iterative procedures involving many repetitions of a given calculation, the saving of time and money achieved through normalization can be appreciable.

In fairness, however, we must admit that normalization sometimes has an obscuring effect. The submerging of units is desirable from the mathematician's point of view, but can be a loss to the engineer who derives physical insight from dimensions. Further, there are cases where a given quantity chosen as a normalizing divisor has differing relevance in various solution "regimes." Under such conditions, a simple relationship between unnormalized variables can become blurred by normalization in a regime where the normalizing constant has lesser relevance.

1-3 EQUATIONS DEALING WITH MOVING AND MOTIONLESS CHARGES

It is of course the net motion or "transport" of electrical charge that constitutes an electrical current, and control of this current is the function of an electronic device. Behavior of the moving charge in such a device is often conditioned by the presence of other charges that are fixed in position, or *static*. Thus the laws governing both motionless and moving charges are relevant, and are summarized here in the forms that will apply most frequently in the discussions of device electronics that follow.

1-3.1 Conductivity and Resistivity

Once again the parallel-plate capacitor provides a convenient starting point, primarily because it guarantees (in its central region again) one-dimensional conditions. We have seen that the electric field present when a voltage V is applied to plates of spacing d, is simply the quotient, $|E| = |V|/|d|$. No matter what kind of material is introduced into the volume between the plates, E is still fixed by V and d alone, *provided the material is homogeneous throughout that entire volume*. The proviso is important. In Section 1-3.4 we shall examine one of the simplest nonhomogeneous cases, namely a layered structure lying between a pair of plates.

If the material so introduced contains "free" charges, or *carriers* of electricity that move continuously in response to an electric field, then a direct current will exist in the medium between the plates. *Conductors* of electricity, such as metals, contain large populations of free carriers. *Semiconductors*, on the other hand, contain fewer. It is of course semiconductor properties that will largely occupy us through the remainder of the book. *Insulators* contain no free carriers, or so few that only a negligible current accompanies an electric field.

Once more, arbitrary convention enters. Current in the direction of the electric field is taken to be positive. If the carriers are positive charges, they move in the direction of the field in response to the force they experience. In the more common case of negative carriers, electrons, the "particle" current is opposite to the direction of the *conventional current*.

Dividing the current I that flows in the presence of a given steady-state electric field E by the cross-sectional area A of the sample gives us *current density*, for which the symbol is J. Thus the RMKSA units of J are A/m^2. (In this case, A is of course the unit symbol for *ampere*, so identified by the foregoing words; this underscores the important point made in Section 1-2.3 concerning clear separation of unit and variable symbols, since A for *area* entered the preceding sentence.) Let us now, however, shift permanently to the modified RMKSA system that is universal in semiconductor work, the one in which the centimeter is used in place of the meter. Thus the units of J become A/cm^2.

The magnitude of the current density J that accompanies a given electric field E depends upon carrier availability (density), as we have seen, and also upon the responsiveness of the carriers to the field. These two factors determine the resistance of a given material to the passage of a current, or more precisely, its *resistivity*. To define this quantity, let us shift attention to a bar of uniform material, with perfect (nonresistive) electrical contacts made to its ends, as shown in Figure 1-10. The fact that the length L of the sample is large compared to the sample's cross-sectional dimensions does not upset the (much-desired) one-dimensional character of this problem, because current is confined to the interior of the sample. That is, there are no "fringing effects" in this physical situation. It is self-evident that when this sample is used as a resistor by passing a current through it, we will find its resistance to be directly proportional to its length L, and inversely proportional to its cross-sectional area A. That is, doubling L places two resistors in series; doubling A places two in parallel. Therefore, introducing a constant of proportionality ρ, on a purely intuitive basis we can write

$$R = \rho(L/A), \tag{1-26}$$

where ρ is known as *resistivity*. Solving this expression for ρ yields RA/L, so that the dimensions of resistivity evidently are

$$\frac{[\text{ohm}][\text{cm}^2]}{[\text{cm}]} = [\text{ohm}\cdot\text{cm}]. \tag{1-27}$$

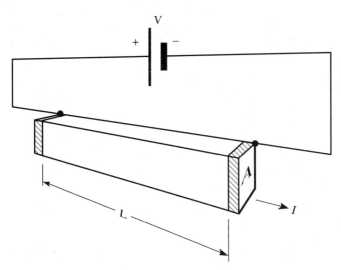

Figure 1-10 A uniform macroscopic sample of length L and cross-sectional area A for defining resistivity ρ. The current I flows one-dimensionally, and the end-to-end voltage drop in the sample is V.

Notice that ρ is measured in ohms *times* centimeters (and not "ohms per centimeter").

The inverse of resistivity is *conductivity*, σ:

$$\sigma \equiv (1/\rho). \tag{1-28}$$

Obviously the dimensions of conductivity are 1/ohm·cm, or S/cm, where S is the symbol for siemens, the unit for conductance.

1-3.2 Ohm's Law in Terms of Electric Field

Armed with the units of resistivity ρ, next examine the dimensions of the product of ρ and current density J:

$$[\text{ohm·cm}]\frac{[\text{A}]}{[\text{cm}^2]} = \frac{[\text{V}]}{[\text{cm}]}. \tag{1-29}$$

From this we infer that the passage of a current having a density J through a uniform material of resistivity ρ is accompanied by the presence of an electric field E. Or,

$$E = J\rho. \tag{1-30}$$

This, in fact, is a form of Ohm's law, as can be seen by yet another dimensional argument. Writing the unit equation corresponding to Equation 1-30 and multiplying both sides by cm yields

$$[\text{V}] = [\text{A}][\text{ohm}], \tag{1-31}$$

the unit equation corresponding to $V = IR$, Ohm's law. An important feature of Ohm's law written as in Equation 1-30 is that all three of the quantities involved can be defined at a mathematical point.

1-3.3 Dielectric Materials, Permittivity, and Polarization

The parallel-plate capacitor that was so useful for discussing electric field (and to a lesser degree, resistivity) has further basic contributions to make, and is intrinsically important as well. In general, a capacitor can be defined as a two-terminal device that is capable of charge storage when a voltage difference is applied from terminal to terminal. Specifically, the applied voltage V causes a charge $+Q$ to be stored on one plate, and $-Q$ on the other. Insofar as the storage is associated with internal field, these charges will always be equal in magnitude. (In fact one must go to considerable trouble to store unequal charges on the two plates of such a capacitor, introducing significant electric field outside the device.) The next point to realize is that "stored charge" refers to the charge on *one* plate. Let us arbitrarily take the negative plate as voltage reference, and then focus on the positive voltage applied to the other plate, and its corresponding positive charge. As noted in Exercise 1-2, the charge stored is proportional to the voltage applied. Thus the capacitor is a linear device, and its governing law is

$$Q = CV. \tag{1-32}$$

The constant of proportionality between these two variables is of course *capacitance*, for which the unit is the *farad*, or equivalently, a coulomb per volt. Now we can be more specific on the plate whose charge is to be considered here. For algebraic-sign consistency in Equation 1-32, we must employ the charge Q on the *same* plate to which the voltage V is being applied, with the other plate taken as voltage reference.

In the capacitors we considered in Section 1-1, the space between the plates was devoid of atoms, molecules, and charges. In short a *vacuum* existed between the plates. Somewhat more loosely we often use the term *air-dielectric capacitor* to denote the same device, because the presence of air molecules (mainly nitrogen and oxygen) between the plates has little effect on the device's properties through wide ranges of temperature, pressure, and applied voltage. The classical term denoting vacuum—or an approximation to it in the electrostatic context—is *free space*. (If the workers of a century ago had shared the modern picture of interplanetary and even interstellar space as a low-density "soup" composed of many species, some neutral and some charged, they probably would have preferred the term *vacuum*.)

In Section 1-1.4, Exercise 1-2, we found it convenient to divide both sides of Equation 1-32 by plate area A, yielding charge per unit area (C/cm^2) on the left-hand side of the resulting equation. Let us write a unit equation corresponding to Equation 1-32 as thus modified:

$$\frac{[C]}{[cm^2]} = \frac{[F]}{[cm]} \frac{[V]}{[cm]}. \tag{1-33}$$

Note that we have chosen to divide each factor on the right by cm, with the result that the second factor on the right corresponds dimensionally to electric field E. And the first factor on the right, farads per centimeter, corresponds dimensionally to *permittivity*, previously encountered in Section 1-1.4, Exercise 1-2. Specifically, for the air-dielectric case it is the permittivity of free space ϵ_0 that is appropriate, so that charge per unit area on one of the plates is given by

$$\frac{Q}{A} = \epsilon_0 E. \tag{1-34}$$

Therefore, permittivity can be regarded as a factor that, by multiplication, converts electric field at a particular place (since field can vary from place to place) into charge per unit area on the plates of a capacitor introduced locally to produce the same electric field.

Permittivity has the same meaning when the electric field under scrutiny is located in the practically important class of *dielectric* media rather than in a vacuum. This adjective usually implies *nonconducting* or *insulating*. That is, a dielectric material of ideal properties contains no free carriers that will move continuously in the presence of an electric field. With a dielectric material filling the space between capacitor plates, such continuous charge motion would constitute a "leakage current," normally considered to be parasitic in a capacitor.

Even with continuous charge motion ruled out, however, there is a second kind of charge motion possible in a dielectric medium that makes it useful in a capacitor, because the resulting effect is an enhancement of the capacitor's ability to store charge. This second kind of motion is termed *polarization*. In the presence of an electric field in a *polarizable* medium, constituent positive and negative charges on an atomic or molecular scale undergo a slight separation. Figure 1-11(a) shows the

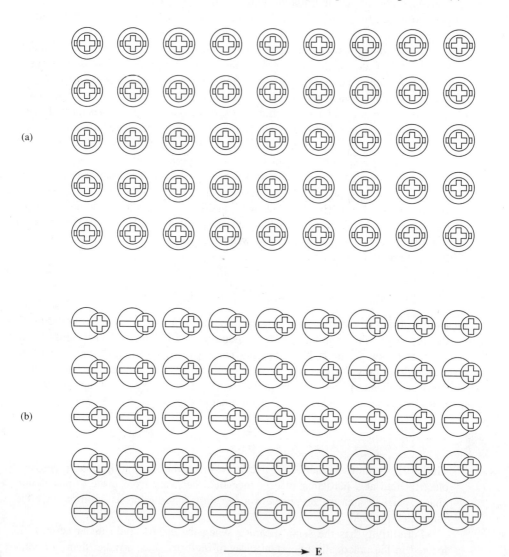

Figure 1-11 Heuristic representation of an ideal polarizable material. (a) With no electric field present, positive and negative charge populations neutralize each other throughout the sample on an atomic or molecular scale. (b) In the presence of an electric field, charge separation, or *polarization,* occurs. Net charge density in an arbitrarily chosen interior volume is still zero, but sheets of opposite net charge appear on the faces normal to field direction.

two kinds of charges schematically under the condition of no electric field. In Figure 1-11(b) is shown equally schematically the same material after the local charge separation, or *polarization,* has taken place, this in response to the presence of the electric field E indicated in the diagram. Notice that if one chooses an arbitrary volume inside the polarized material that has large dimensions (compared to atomic dimensions), then no net charge is found inside the arbitrary volume. But at the outer faces normal to the polarizing field, there is a sheet of net charge, negative on the left and positive on the right in Figure 1-11(b). It is this thin sheet of net charge on either side that accounts for the storage-enhancing effect of placing a polarizable medium in a capacitor.

To see how this works, consider the air-dielectric capacitor depicted in a charged state in Figure 1-12(a). Let the applied voltage be V. In Figure 1-12(b), then, we show the effect of placing a particular polarizable (dielectric) material between the plates. The applied voltage is still V, and the electric field E is unaltered, as the unaltered density of lines of force indicates. However, charge storage on the plates of the capacitor has doubled in the example arbitrarily chosen. This is a consequence of the polarization and the resulting monolayers of additional charge existing just outside the conducting plates and inside the contiguous dielectric me-

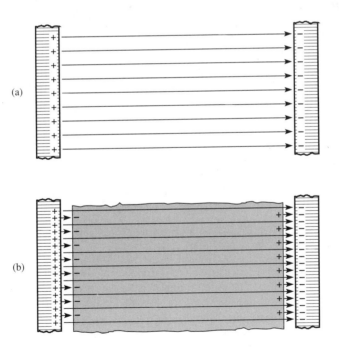

Figure 1-12 The ability of a polarizable (dielectric) material to enhance charge storage in a capacitor. (a) Charge storage in the plates of an air-dielectric parallel-plate capacitor. (b) Doubled charge storage that results when a slab of material having a dielectric constant $\kappa = 2$ is introduced into the capacitor of part (a) with the same applied voltage.

dium. In Figure 1-12(b) we have omitted the complex representation of Figure 1-11(b) to improve clarity, but we intend that the same kind of situation is being presented there. Further, caution is needed because Figure 1-12 offers a two-dimensional representation of a three-dimensional problem.

In this example, charge storage has been doubled by the presence of the polarizable medium, so according to the description of permittivity already given, we can say that the permittivity of this material is twice that of free space. In other words, the *relative dielectric permittivity* of this material is

$$\kappa \equiv (\epsilon/\epsilon_0) = 2 \tag{1-35}$$

in this case. (The value $\kappa = 2$ was chosen in a completely arbitrary way for convenience in drawing Figure 1-12.) The everyday term for κ is *dielectric constant*. Charges in the metal, free to come and go from the adjacent circuitry, are sometimes termed *free* charges, and those that make up the charge sheet in the dielectric material, *bound* charges.

1-3.4 Electric Displacement

Refer again to Figure 1-12. The act of introducing a dielectric material with a dielectric constant of two caused the charge storage to double, but did not alter the electric field E, a point emphasized at the beginning of Section 1-3.1. The doubling of charge storage can also be expressed by saying that the product of permittivity and electric field has doubled, remembering that the dimensions of this product are C/cm^2, an areal charge density. To identify this important quantity, the term *displacement vector* was chosen by Maxwell [1], for which the symbol is **D**. In the simple situation being treated here, the displacement vector can be written as

$$\mathbf{D} \equiv \epsilon \mathbf{E}. \tag{1-36}$$

The term *displacement* (or the term *electric displacement* that is sometimes used) calls forth a picture of the local charge separation that constitutes polarization, the phenomenon represented in Figure 1-11(b). But it is important to realize that because of the way it is defined, Equation 1-36, the areal charge density represented by displacement magnitude is the *sum* of the charge stored with the dielectric material absent *and* the additional charge stored with the dielectric material present. That is, in Figure 1-12(a), we have charge storage in an air-dielectric capacitor, and in Figure 1-12(b), we have that charge plus the added charge contributed by the dielectric material's presence. The definition of D is based upon the *total* charge per unit area. (Here we have dropped the vector symbol for displacement, just as we did previously for electric field.) Therefore, *displacement* is defined and has meaning even in the air-dielectric case.

A number of complications can alter the simple picture just outlined, complications that are treated at length in standard texts on electricity and magnetism [2–4]. One complication is that it is possible to have nonuniform polarization in some materials, with the result that net charge can exist within the volume of the material, and not just at the external faces [as in Figure 1-11(b)]. When this situation exists, the problem is obviously more complicated, because net charge within the volume would affect capacitor-plate charge density, and hence the magnitude of *D*. We shall avoid this situation by confining ourselves to *homogeneous* materials. A second complication is that for reasons involving the detailed structure of the dielectric material on an atomic or molecular scale, the charge separation may occur along an

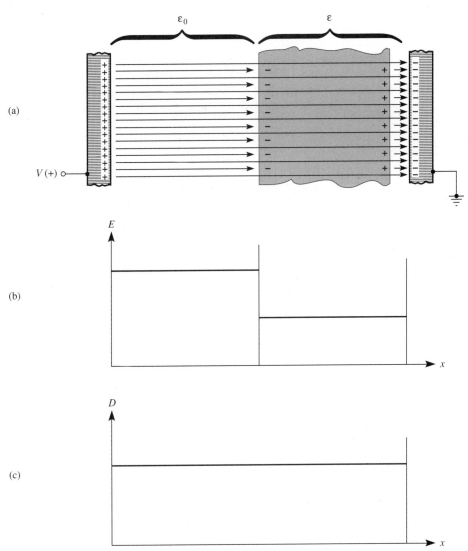

Figure 1-13 A capacitor having a layered dielectric region. (a) Representation of a charged parallel-plate capacitor with $\kappa = 1$ in the left layer and $\kappa = 2$ in the right. (b) Electric-field profile in the charged capacitor, as inferred from the densities of lines of force. (c) Electric-displacement profile as inferred from charge per unit area on the capacitor plates.

axis that is not aligned with the electric field. Such a material is *nonisotropic* in this respect; we rule out this complication by confining ourselves to *isotropic* materials, a term based upon the Greek roots *iso* (the same), and *tropo* (angle). An isotropic material is one that is "the same through all angles."

It was noted at the beginning of Section 1-3.1 that when a capacitor's dielectric region is not homogeneous, we no longer have the field strength given simply by V/d. A nonhomogeneous case of substantial relevance to solid-state electronics involves a layered dielectric region, for which a simple example is depicted in Figure 1-13(a). In this case, the left half of the region between the plates is free space, while the right half is a dielectric material with $\kappa = 2$ once more. By inspection of the lines-of-force density, it becomes evident that the electric-field "profile" has the form depicted in Figure 1-13(b). Displacement, however, is constant throughout the region between the plates; it is evident that both plates have the same areal charge density, which is an interpretation we have placed upon displacement magnitude.

A more complicated case is one in which the **E** and **D** vectors are not normal to the interface between regions of differing permittivity. In such a case, the electric-field lines of force are refracted. The laws governing this situation are simple; the tangential component of the vector **E** is continuous through the interface, and the normal component of the vector **D** is continuous through the interface. Such problems, of course, are at least two-dimensional and fortunately these are rare for us. Note, however, that the laws just cited are valid in the special case of Figure 1-13; the tangential component of **E** is zero on both sides of the interface, and the normal component of **D** is continuous through the interface, as one infers from areal charge density on the plates.

1-3.5 Displacement Current

Suppose that we are given a bar of conducting material with a thin gap normal to its axis, as represented in Figure 1-14(a). Then suppose that a time-dependent voltage $V(t)$ is applied at the left-hand terminal. Furthermore, let $V(t)$ be a slowly and linearly increasing function of time, as represented in Figure 1-14(b), a waveform often described as a voltage "ramp." The gap in the conductor clearly constitutes a parallel-plate capacitor. Let us once again neglect fringing effects and assume that a one-dimensional treatment of the capacitor is adequate. The charge it stores will also increase linearly with time. Negative charge builds up on the right-hand plate as electrons are extracted from "ground." Positive charge effectively builds up on the left-hand plate, in a process that actually involves the extraction of electrons from the left-hand plate and from the left-hand terminal. At an arbitrary position in the conductor, such as the position x_1, the passage of charge that causes linearly increasing charge storage on the capacitor can be written $d(Q/A)/dt$, where A is the cross-sectional area of the conducting bar and Q/A is the areal charge density on the capacitor plate at any given instant. But in a trivial manipulation we have

$$\frac{d(Q/A)}{dt} = \frac{dQ/dt}{A} = J, \qquad (1\text{-}37)$$

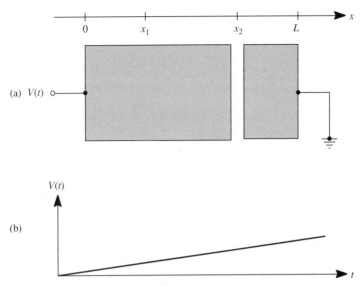

Figure 1-14 Descriptively relating conduction current to displacement current. (a) A one-dimensional sample that is a uniform conductor, except for an air gap forming a parallel-plate capacitor. (b) Ramp voltage applied to the sample in part (a).

which is the *conduction-current density* at any position in the bar, in response to the applied voltage ramp.

Next, let us remind ourselves that an alternate way to write areal charge density (Q/A) on a plate of the capacitor is $D = \epsilon_0 E$, for the present air-dielectric case. It follows that the rate of change of areal charge density can hence be written as dD/dt at the position x_2. It was Maxwell [1] who first noted a relationship between time rate of change of electric displacement and current density. And it is convenient to introduce a symbol J_D for the *displacement-current density*. Combining the definition of J_D and the interpretation we have given for electric displacement then yields

$$J_D = \frac{dD}{dt} = \frac{d(Q/A)}{dt}. \tag{1-38}$$

Now, comparing the last expression in Equation 1-38 with the first in Equation 1-37 shows that in the experiment of Figure 1-14, the *conduction-current density* J in the bar is equal to the *displacement-current density* J_D in the gap. Note that it is perfectly possible for a displacement current to exist in a vacuum; the presence of a polarizable material is not necessary. This is true because D is defined in terms of *total* areal charge density, as emphasized in Section 1-3.4.

The situation depicted in Figure 1-14 is a highly special case wherein current is steady-state, or non-time-varying, whereas voltage is not. Further analysis of this problem is easy because the problem is so special. For example, we know that $V(t)$ has two components, one being the "ohmic" voltage drop in the bar associated with the passage of a conduction current having the density J. From Equation 1-29, it is given approximately by

$$V_{\text{ohmic}} = J\rho L. \tag{1-39}$$

The second component is the voltage drop on the capacitor given by

$$V_{\text{cap}}(t) = V(t) - V_{\text{ohmic}}(t), \tag{1-40}$$

and also by

$$V_{\text{cap}}(t) = \frac{Q(t)}{C}, \tag{1-41}$$

where C in turn is $\epsilon_0 A/d$.

Note that (constant) conduction-current density J in the bar is accompanied by a constant electric field E, given by $E = J\rho$. Thus, an electric displacement $D = \epsilon E$ exists in the conducting bar as well. But because D is constant there, displacement current vanishes in the bar for the assumed conditions.

Setting aside this special case, and turning to the general situation, we find conduction and displacement currents coexisting in time-dependent problems, so that there exists a total current density J_T that is the sum of the conduction and displacement currents,

$$J_T = J + J_D, \tag{1-42}$$

a concept employed by Maxwell in his famous equations [1].

We should also point out that many materials are both polarizable and conducting. A notable example is silicon, the dominant material in today's solid-state electronics. It has a dielectric constant of 11.7, and conducts sufficiently to be classed as a *semiconductor*.

1-3.6 Dielectric Relaxation

Visualize a charge Q placed on a ground-connected block of solid material. The charge will eventually be neutralized because opposite-polarity charges arrive at the site as a result of the mutual attraction that exists, and because the elementary charges making up Q spread out as a result of their mutual repulsion. The time scale for "eventually" may be years or a tiny fraction of a picosecond (a trillionth of a second), depending upon the permittivity and resistivity of the material medium involved. Resistivity plays a dominant role because the resistivities of materials in common engineering use exhibit an enormous range—from about 10^{-6} ohm·cm for conductors, to 10^{22} ohm·cm for insulators. The physical situation just cited evokes in general a complicated three-dimensional picture. Fortunately, the problem can be posed in a one-dimensional form that has the further virtue of practical significance as well. Once again we are rescued by the parallel-plate capacitor.

When a resistor R is placed from terminal to terminal of a charged capacitor C, voltage falls exponentially with a characteristic time given by the product RC. If C amounts to one farad (an enormous capacitance), and R has a value of one ohm, then the RC product is one second, or

$$[\text{ohm}][\text{F}] = [\text{s}]. \tag{1-43}$$

In this amount of time, capacitor voltage has fallen to a value that is $1/e$ of its initial value.

There is, however, an instructive yet different way [5] to create an analogous situation. This time, instead of using an external resistor R, let the parallel-plate capacitor incorporate a dielectric layer that has finite resistivity (rather than the infinite resistivity of an ideal dielectric material), and assume an ohmic contact (see Section 3-5.4) between each plate and the intervening material. Silicon, with finite resistivity and appreciable permittivity, is a perfectly satisfactory choice as a material. The silicon layer itself then becomes the discharging resistor, with a value given by Equation 1-26. Thus the device can be described either as a silicon resistor or a "leaky" capacitor. Substituting d for L, since we are dealing with the geometry of a capacitor, yields

$$R = \rho(d/A). \tag{1-44}$$

And the capacitance involved, given the presence of the dielectric layer exhibiting a permittivity ϵ, is given by

$$C = \epsilon(A/d). \tag{1-45}$$

Now let the device terminals be removed from the power supply at $t = 0$. It is evident that the characteristic time for discharge, the RC product, is

$$t_D = \rho\epsilon. \tag{1-46}$$

This is the *dielectric relaxation time,* an important material property.

We noted at the end of Section 1-3.5 that silicon exhibits appreciable polarizability as well as a finite resistivity. The latter value depends upon the densities and identities of impurities in the silicon. These impurities usually are intentionally introduced for the express purpose of adjusting resistivity and other electrical properties of the material, and normally their densities can vary through huge ranges, with the result that resistivity does also. A fairly typical resistivity for silicon, however, is $\rho = 1$ ohm·cm. The permittivity of silicon is approximately $\epsilon = 1$ pF/cm (1 picofarad per centimeter, or 10^{-12} F/cm), and is independent of resistivity. Thus for this "typical" silicon case, $t_D \approx 1$ ps. This is such a short time interval that the relaxation phenomenon is considered instantaneous throughout this book, and in most prior literature. But in recent years, other important characteristic times in devices, such as switching times, have become so short that this assumption is not always warranted, greatly complicating device theory.

Exercise 1-7. Determine the current waveforms $J(t)$, $J_D(t)$, and $J_T(t)$ in the capacitor-discharge experiment just described.

It is evident that for $t > 0$,

$$J(t) = J(0) \exp(-t/t_D),$$

where

$$J(0) = \frac{V(0)}{AR} = \frac{V(0)}{\rho d}.$$

But $J_D(t)$ has *opposite* sign; the current, field, and displacement vectors all have the same direction, but declining displacement means a negative displacement current. Thus,

$$J_D(t) = -J(0)\exp(-t/t_D),$$

and

$$J_T(t) = J(t) + J_D(t) = 0.$$

Current in the external terminals for $t > 0$ is not possible in any case.

1-3.7 The Meaning of Poisson's Equation

In Section 1-1.1 we emphasized that charge is the "source" of electric field in the sense that a mythical line of force emanates from a positive charge of arbitrarily designated magnitude, and terminates on a negative charge of the same magnitude. All of the discussion thus far has dealt with thin layers of charge, such as those in Figure 1-13(a), for which the appropriate density description is *areal* charge density, measured in C/cm^2. In the general case, however, we encounter charges distributed arbitrarily in three dimensions; there the applicable description involves *volumetric* charge density, measured in C/cm^3. The traditional symbol for this density is the Greek letter rho, but we shall employ ρ_v for volumetric charge density to avoid confusion with the equally well-established symbol for resistivity. This distinction is particularly important because the two quantities frequently enter the same calculation or analysis.

Volumetric charge density can be defined at a point by the device of computing a limiting charge-to-volume quotient, as the volume is permitted to vanish at the point of interest. It follows, then, that ρ_v can vary from point to point. The general expression that relates electric field (or electric displacement) to charge in this potentially complex situation is Poisson's equation, a contribution by Simeon Denis Poisson (pwa-son'). This equation is usually stated in terms of electrostatic potential (from which we can of course obtain electric field by differentiation):

$$\nabla^2 \psi = -\rho_v/\epsilon. \tag{1-47}$$

In spite of, or perhaps because of, the elegant simplicity of this equation, some explanation of its meaning is needed. As usual, let us first consider the one-dimensional case, so that the left-hand side becomes $d^2\psi/dx^2$. But recalling Equation 1-15, $E = -d\psi/dx$, we can convert Equation 1-47 to

$$\frac{dE}{dx} = \frac{\rho_v}{\epsilon}, \tag{1-48}$$

the form of Poisson's equation we shall most frequently employ. Notice also an even simpler version of the same equation:

$$\frac{dD}{dx} = \rho_v. \qquad (1\text{-}49)$$

Poisson's equation is equally valid for moving and motionless charges, but there are particularly important and frequently encountered problems in solid-state electronics that involve motionless or *static* charges, so we shall focus on these cases. Visualize a solid medium with charges distributed through its volume, and let us choose a particular practical example of considerable significance. Consider the medium to be silicon dioxide, SiO_2. In its *amorphous,* or "glassy," form it is a superb insulator ($\rho \approx 10^{15}$ ohm·cm) and is extensively used in transistors and integrated circuits. (The same material is the primary constituent of ordinary window glass. See the end of Section 1-5.6.) It is possible, however, for carriers to pass through thin layers of SiO_2, provided their energy is sufficient. In the process, some of them often become "stuck," or *trapped,* where they can remain for many years if the sample is protected from light and high temperatures. Because the SiO_2 layers are typically thin, a one-dimensional treatment of the problem is totally justified. As a final simplification, let us assume that the charge density ρ_v is uniform throughout the region of interest.

The situation to be examined can be described in terms of the "one-plated capacitor" represented in Figure 1-15(a). Here we have assumed the trapped charges to be positive. And for graphical convenience, we have shown them with positional regularity; it will be explained in Chapter 2 that this situation is sufficient but not necessary for compliance with the uniform-density condition. Assume that the device is isolated in the sense that it is a long way from other objects, but let its single plate, initially uncharged, be connected to earth ground. At the instant of grounding, electrons will be drawn into the plate in a number exactly equal to the number of positive charges trapped in the dielectric material. The result, then, is a one-dimensional but nonconstant field arrangement, or as it is often called, "field distribution." The latter point is verified by noting that the density of lines of force is a function of x; the former point, by the parallelism of all lines of force. Plotting the density of lines of force (which is to say, the electric-field strength) in Figure 1-15(a) as a function of position x yields the electric-field *profile* (or *distribution*) shown as the solid line in Figure 1-15(b). Thus it is evident that the gist of Poisson's equation, Equation 1-48, is that *the slope of the field profile (dE/dx) is proportional to charge density (ρ_v).*

Imagine, now, that we have modified the structure of Figure 1-15(a) by placing a second dielectric layer between the first dielectric layer and the metallic plate, and assume that the second layer has half the charge density of the first. It is evident, then, that at the interface position, the field profile would exhibit a sharp break, with its slope falling by a factor of two. If the second layer were charge-free, the profile would be horizontal there. If it had twice the charge density, that would cause the field profile to double in slope, and so on. It is straightforward to generalize this picture to an arbitrary (but still one-dimensional!) charge distribution. Heuristically, one can shrink oneself to microscopic dimensions, "walk" through

34 FOUNDATIONS OF MODERN ELECTRONICS

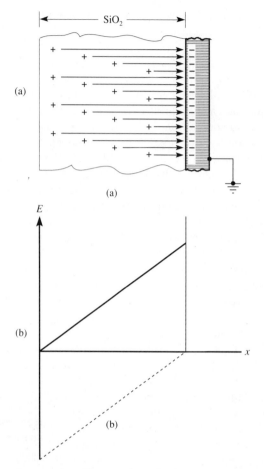

Figure 1-15 The meaning of Poisson's equation in relation to a region of uniform space-charge density. (a) A "one-plated capacitor" with a dielectric region containing a uniform, positive space-charge density. (b) The electric-field profile in the device of part (a), exhibiting a uniform slope as a consequence of the uniform space-charge density.

the sample in the positive-x direction, and observe the number (or density) of lines of force at every x position.

Exercise 1-8. Suppose we place the field plate of the one-plated capacitor in Figure 1-15(a) on the left-hand side of the SiO$_2$ region, so that all of the lines of force will be directed leftward. How can Poisson's equation, Equation 1-48, still be valid under these conditions?

According to our convention, a field direction opposite to that of the spatial x axis is considered negative. Under the revised conditions, its magnitude would then have its maximum value at the left face of the SiO$_2$ slab, and would vanish at the right face. Thus it would appear like the dashed line in Figure 15(b), and Poisson's equation remains fully valid. The kind of change described in this exercise merely involves a change in the boundary conditions that are applied to the solution of the differential equation.

Exercise 1-9. Repeat Exercise 1-2, using a shorter method. How is the result altered if a dielectric material having $\kappa = 10$ fills the space between the capacitor plates?

The areal density of charge on a capacitor plate can be obtained from $D = \epsilon_0 E$. Since this corresponds to 3 lines/cm^2, the desired answer is the quotient of the two expressions, or

$$\frac{\epsilon_0 E}{3 \text{ lines/cm}^2} = \frac{(8.85 \times 10^{-14} \text{ F/cm})(1 \text{ V/cm})}{3 \text{ lines/cm}^2} = \frac{2.95 \times 10^{-14} \text{ C}}{\text{line}}.$$

For the given case of a dielectric constant other than unity, we substitute $\epsilon = \kappa \epsilon_0$ for ϵ_0, which gives us 2.95×10^{-13} C/line.

1-4 THE BOHR MODEL OF THE HYDROGEN ATOM

Writing in the London, Edinburgh, and Dublin *Philosophical Magazine and Journal of Science* in 1913, Niels Bohr put forward a theory "On the Constitution of Atoms and Molecules" [6]. It brilliantly combined the quantum theory contributed by Planck and the nuclear theory of the atom contributed by Rutherford. Bohr's mastery of English composition as exemplified in this landmark paper could well be emulated by today's engineers and scientists, and is the more remarkable since English was not his native tongue.

There are several reasons for addressing the Bohr model at this point. First, assembling and describing the ideas embodied in his model provides an excellent opportunity to apply a number of the principles developed in the preceding sections. Second, understanding the behavior of electrons in the isolated atom (the major concern of the Bohr model) helps one to understand their behavior in atoms that interact. Atoms forming the single crystals that are the basis of modern electronics are "interacting" to an extreme degree because of their proximity. And third, the synthesis embodied in Bohr's model has a very prominent place in the development of modern physical thought.

1-4.1 The Planetary Analogy

Bohr's starting place was the plateau of understanding achieved by Ernest Rutherford [7], in whose laboratory Bohr worked as a young man. Significantly, Bohr chose this working location after a brief sojourn at Cambridge with J. J. Thomson, discoverer of the electron, because he felt that the nuclear atom was "right," while Thomson's effort to press nineteenth-century physics into the new realm was "wrong" [8]. Through incisive observations and analysis, Rutherford had concluded that, in Bohr's words, "[an atom consists] . . . of a positively charged nucleus surrounded by a system of electrons kept together by attractive forces from the nucleus; the total negative charge of the electrons is equal to the positive charge of the nucleus. Further, the nucleus is assumed to be the seat of the essential part of the mass of the atom, and to have linear dimensions exceedingly small compared with the linear dimensions of the whole atom. The number of electrons in an atom is deduced to be approximately equal to half the atomic weight" [6]. Thus the atomic nucleus could be compared to the sun, about which planet-like electrons circulate, with electrical forces in the atom taking the place of the gravitational forces in the solar system.

Rutherford sought to explain the puzzling observations of Geiger and Marsden on the deflection or "scattering" of alpha particles, which are helium-atom nuclei. Thus, alpha particles exhibit a mass roughly four times that of the hydrogen atom (or of a proton) and a charge of $+2q$. In experiments that directed a beam of these particles at a thin foil of a heavy metal, such as gold, the surprising observation was that some were almost reversed in direction, and a larger-than-expected number experienced large-angle changes in direction. Treating the heavy-metal nucleus as a point charge, Rutherford reasoned that an intense electric field would exist in its vicinity. If we once again invoke the lines-of-force concept as an aid to visualizing the electric-field pattern, we obtain the picture shown in Figure 1-16(a). Recalling that field strength is proportional to the density of lines of force gives us the field-versus-radius result shown qualitatively in Figure 1-16(b). Thus a near miss or a direct hit on a gold-atom nucleus by an alpha particle would cause the particle to experience such an extreme value of electrostatic repulsion as to account qualitatively for large-angle deflection. (The phrases usually employed in these connections, "alpha-particle scattering" and "large-angle scattering," make sense if one visualizes a large number of particles incident, so that they are deflected off in many directions.)

Rutherford's further point was that if the constitutents of an atom were more or less uniformly distributed in space—the older view of the atom—then one would expect an alpha particle passing through the metal foil to experience a large number of relatively small deflections, with a substantial amount of statistical cancellation involved. On the other hand, if the atom consisted primarily of empty space, then one would expect a particle in transit to experience few interactions, but on average to undergo a large deflection in each. A gold-atom nucleus in particular, being about 49 times more massive than the alpha particle, and having a charge of $+79q$, would have a major effect on the trajectory of an alpha particle coming near it. These qualitative pictures he then augmented with careful observations of the angular distribution of scattered alpha particles, and a physical and statistical treatment of the angular distributions to be expected as a result of the *coulomb scattering* of point-charge targets. The result was the Rutherford scattering law, which was well matched to the experimental findings.

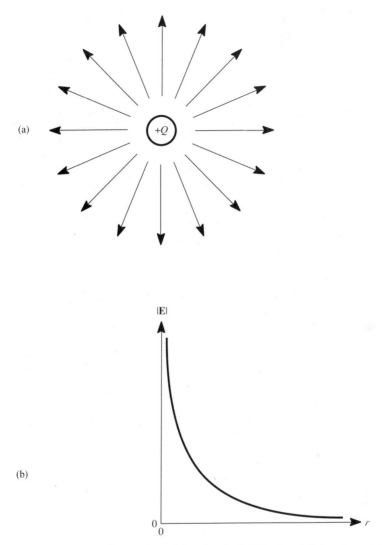

Figure 1-16 A representation of the electric field in the vicinity of a point charge $+Q$. (a) Fictitious lines of force emanate from the positive charge shown, to terminate on negative charges at very remote locations. (b) Profile of the electric-field magnitude as a function of radius r. Because the representation in (a) has spherical symmetry, and inasmuch as field strength is proportional to lines-of-force density, the result is an inverse-square field function.

1-4.2 Electromagnetic Radiation and Quanta

The existence of electromagnetic radiation was postulated by Maxwell [1], following his key assumption that a magnetic field accompanies a displacement current in the same way that it accompanies a conduction current. Generalizing to three spatial dimensions with time as a fourth independent variable, let us write $\partial \mathbf{D}/\partial t$ for displacement current, so that in a situation where displacement current and conduction current coexist, we have

$$\nabla \times \mathbf{H} = \mathbf{J} + \frac{\partial \mathbf{D}}{\partial t}. \tag{1-50}$$

The *magnetic-field* vector $\mathbf{H}$ has dimensions of current per unit length, so we can see the dimensional consistency of this equation by recalling the meaning of the operator ∇:

$$\nabla \rightarrow \mathbf{i}\frac{\partial}{\partial x} + \mathbf{j}\frac{\partial}{\partial y} + \mathbf{k}\frac{\partial}{\partial z}. \tag{1-51}$$

Recall too that $\nabla \times \mathbf{H}$ can also be written as curl $\mathbf{H}$. Equation (1-50) is one of Maxwell's equations. In a vacuum (or "free space") then, where no charge or matter exists, $\mathbf{J}$ must vanish so that Equation 1-50 simplifies to

$$\nabla \times \mathbf{H} = \frac{\partial \mathbf{D}}{\partial t}. \tag{1-52}$$

The companion equation given by Maxwell takes on a remarkably symmetric form when it too is simplified for the case of free space:

$$\nabla \times \mathbf{E} = -\frac{\partial \mathbf{B}}{\partial t}, \tag{1-53}$$

where

$$\mathbf{B} = \mu_0 \mathbf{H} \tag{1-54}$$

is termed the *magnetic induction* or *flux density* and has dimensions of voltage times time per unit area. The quantity $\mu_0 = 4\pi \times 10^{-7}$ V·s/A·m is the *permeability of free space*. To Equations 1-52 and 1-53 we then add

$$\text{div } \mathbf{B} = 0, \tag{1-55}$$

which holds in view of Equation 1-54 because magnetic field is by nature "divergenceless," and

$$\text{div } \mathbf{D} = 0, \tag{1-56}$$

which must hold in charge-free space. (Recall that charge is the source of electric field, and that electric field in turn is proportional to displacement.) Employing these four expressions, Equations 1-52, 1-53, 1-55, and 1-56, in combination with a vector identity yields a pair of differential equations that are characteristic of wave motion [9]; in terms of the magnitudes of the electric and magnetic fields,

$$\nabla^2 E = \epsilon_0 \mu_0 \frac{\partial^2 E}{\partial t^2}; \tag{1-57}$$

$$\nabla^2 H = \epsilon_0 \mu_0 \frac{\partial^2 H}{\partial t^2}. \tag{1-58}$$

Maxwell's brilliant prediction that electromagnetic waves could propagate in free space was verified in 1887 by H. R. Hertz, and such waves were known for years thereafter as *Hertzian waves*.

Thus we see that the electric-field and electric-displacement concepts that we introduced in Sections 1-1.1 and 1-3.4 in the context of a parallel-plate capacitor have extremely significant existence totally apart from that context. An important feature of any useful theory is its ability to predict new and unexpected results. Maxwell's theory predicted that when electric field and electric displacement are time-varying, then magnetic-field and magnetic-induction vectors coexist with these electric-vector quantities in electromagnetic waves that move at the speed of light through matter-free space. The launching and reception of these electromagnetic waves is of course the basis for "wireless" communications, such as radio, television, and radar. Let us look a bit more closely at this important application.

We know that a wire carrying direct current is encircled by "lines" of magnetic field partly analogous to the lines of force we have employed to visualize an electric field, with the sense of the magnetic-field lines being given by a right-hand rule. As long as the current is static, or dc, the magnetic-field pattern is equally static, and no radiation occurs. But when the current is time-varying, the magnetic field is also time-varying, and inevitably an electric field, too, comes into existence in the region about the wire, the latter fact being required by Equation 1-53. The simultaneous creation of electric and magnetic fields in a volume of space means that the conditions for electromagnetic radiation have been created. In terms of the charge involved in the current responsible for the radiation, the feature that had to be present was the *acceleration* of this charge; there is no other way to create a time-varying current. Finally, then, *charge acceleration leads to electromagnetic radiation*.

The next important point is that electromagnetic radiation transmits energy. In the example just outlined, this energy was derived from the energy source responsible for the charge acceleration. In the case of radio or television, a *transmitting antenna* puts forth energy into electromagnetic radiation, a portion of which is then absorbed by a *receiving antenna*.

The pattern of charge acceleration involved in this application is particularly important. The simplest form of transmitting antenna is a long, thin conductor. Charge is caused to oscillate from end to end of the conductor, thus undergoing a cyclic acceleration, and the resulting electromagnetic wave that carries energy away from the antenna exhibits the frequency of the oscillating charge. Thus, the classical physics embodied in Maxwell's elucidation of electricity and magnetism requires that energy loss by radiation inevitably accompany the acceleration of a charge or charges.

The wavelengths encountered in electromagnetic radiation can range through many of orders of magnitude, from radio waves at the long end of the wavelength scale to gamma radiation at the short end, with visible radiation—light—falling roughly at the middle of a logarithmic wavelength scale, in the neighborhood of 0.6 μm. A consequence of the crucial developments that led to this understanding was the integration of optics with electricity and magnetism, which had previously been treated as two separate realms.

One of the modes of energy loss by a heated object is thermal radiation, which is also a form of electromagnetic radiation. The spectrum of this *black-body radiation* is wide. A challenge that occupied leading investigators in the closing years of the last century was attempting to explain the energy distribution in the black-body spectrum on the basis of first principles, involving the properties of elementary harmonic oscillators. Classical theory, however, yielded an absurd result. It was in grappling with this dilemma that Max Planck got the idea of assuming that electromagnetic radiation was not continuous. (In contemplating a source of light, radio waves, or radiant energy, one's assumption of continuousness had always seemed eminently reasonable.) By assuming that the energy was emitted in "packets" or *quanta,* Planck found that he could achieve a satisfactory fit to experimental results. He reported this surprising finding without fully believing it himself [10]. The energy of a quantum, he said, is proportional to its frequency f, and is given by

$$\xi = hf, \qquad (1\text{-}59)$$

where $h = 6.626 \times 10^{-34}$ J·s is *Planck's constant*. It is important to realize that Planck's revolutionary theory represented an empirical modification of classical theory made in order to bring theoretical predictions into harmony with experiment.

Confirmation of Planck's hypothesis was not long in coming, however, because Albert Einstein in 1905 successfully employed quantum ideas to describe the photoelectric effect [11], in the process eliminating another puzzle. Combining the still-new quantum theory with even newer atomic theory, then, awaited Bohr.

1-4.3 Classical Components of the Bohr Model

As we have seen, Rutherford's nuclear atom suggested a planetary analogy. The planets of our solar system move about the sun in orbits that are approximately circular and approximately coplanar, with gravity supplying the necessary forces. In earlier centuries, giants such as Kepler, Copernicus, and Newton worked out the underlying patterns and laws. When we move from an astronomical scale, through the everyday scale, and then down to the atomic scale, electrical forces become dominant and gravitational forces become negligible. The process does not stop there, and the nature and extent of other forces has been a major preoccupation of physical scientists since Bohr's time. But for our purposes, a picture that includes electrical forces and excludes all others is sufficient.

The fundamental relation governing the force exerted on one charged object by another is Coulomb's law, which states that the resulting force is proportional to the product of the charges, and inversely proportional to the square of their separation. Taking the case of point charges for simplicity, we have

$$F \propto \frac{Q_1 Q_2}{r^2}, \qquad (1\text{-}60)$$

where Q_1 and Q_2 are the respective charges and r is their spacing. Acknowledging that force is a vector quantity and using the RMKSA system, we may write

$$\mathbf{F} = \frac{Q_1 Q_2}{4\pi\epsilon_0 r^3} \mathbf{r}, \tag{1-61}$$

where the vector **r** extends from one of the point charges, chosen as reference, to the other, as illustrated in Figure 1-17. The selection of the charge to serve as positional reference is obviously an arbitrary matter. When the two charges have the same sign, the force is repulsive, the case, which we shall define as a positive force, depicted in Figure 1-17.

Next, turn to the simplest atom, the hydrogen atom, and apply the principles outlined in the foregoing sections. The hydrogen atom consists, of course, of a proton, carrying a charge $+q$, and a much less massive electron, carrying a charge $-q$. We visualize the electron in "orbit" about the proton, as shown in Figure 1-18, with the unlike charges causing a force this time that is attractive. Taking the more massive proton as the center of the system, we then have a *central* force acting on the electron, precisely the centripetal force necessary to maintain its circular motion. This force (reverting to scalar notation), from Equation 1-61, is evidently

$$F = -\frac{q^2}{4\pi\epsilon_0 r^2}. \tag{1-62}$$

Applying Equation 1-3, then, yields for the electric field at the position of the electron

$$E = \frac{F}{-q} = \frac{q}{4\pi\epsilon_0 r^2}. \tag{1-63}$$

We have taken the outward direction in this central system as positive for the two vectors position **r**, and force **F**, choices that we now see are consistent with the physics of the problem; Equation 1-63 indicates that the field at the position of the electron, *in the absence of the electron,* is positive, as it must be to be directed away from the positive charge. Also, Equation 1-63 is a quantitative expression for the

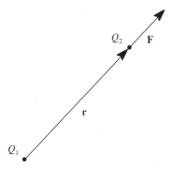

Figure 1-17 The force **F** exerted by a charge Q_1 on a like-sign charge Q_2 has the same direction as the radius vector **r** and is taken to be positive.

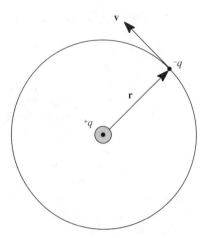

Figure 1-18 Bohr's picture of an electron in orbit about the more massive proton.

field distribution plotted qualitatively in Figure 1-16(b) on the basis of the physical picture offered in Figure 1-16(a) once $+q$ is substituted for $+Q$ in Figure 1-16(a). Notice that both this physical picture and Equation 1-63 make it clear that the electric field approaches zero as r approaches infinity.

Our object is to develop an expression for the energy of the electron in its circular orbit about its proton nucleus. Thus the next step, having calculated electric field E, is to determine electrostatic potential ψ with the aid of Equation 1-14. It is appropriate first to substitute r for x and r' for x' because we deal now with a central system—one having spherical symmetry. Also, it is necessary to assign limits. It is convenient to choose $\psi \equiv 0$ at $r = \infty$ (where E also vanishes), because this will yield a sharply defined energy in the present problem. Thus we have

$$\psi = -\int_\infty^r \frac{q\,dr'}{4\pi\epsilon_0(r')^2} = +\frac{q}{4\pi\epsilon_0 r'}\bigg|_\infty^r = \frac{q}{4\pi\epsilon_0 r}. \tag{1-64}$$

Notice that ψ, like E, increases without limit as $r \to 0$, but that its dependence on r is less sensitive; it is an inverse-r function (hyperbola) rather than being an inverse-square function like electric field.

Having an expression for electrostatic potential, we can immediately write the potential energy of the electron using Equation 1-17:

$$\text{P.E.} = Q\psi = (-q)\left(\frac{q}{4\pi\epsilon_0 r}\right) = -\frac{q^2}{4\pi\epsilon_0 r}. \tag{1-65}$$

The resulting negative sign is a consequence of origin choice; to move the electron from infinite radius to r involves "negative work."

To calculate kinetic energy, Equation 1-18, it is convenient to start with Newton's second law:

$$F = ma. \tag{1-66}$$

The force on the electron was given previously, Equation 1-62; the acceleration is centripetal, so the acceleration vector is directed toward the proton, or is negative with the present sense option, $a = -v^2/r$. Hence Equation 1-66 can be rewritten as

$$-\frac{q^2}{4\pi\epsilon_0 r^2} = m\left(-\frac{v^2}{r}\right). \tag{1-67}$$

Multiplying through by $-r/2$ and interchanging the two sides yields

$$\text{K.E.} = \frac{1}{2}mv^2 = \frac{q^2}{8\pi\epsilon_0 r}. \tag{1-68}$$

The total energy, Equation 1-19, is simply the sum of the two energies just computed, Equations 1-65 and 1-68, or

$$\xi = -\frac{q^2}{4\pi\epsilon_0 r} + \frac{q^2}{8\pi\epsilon_0 r} = -\frac{q^2}{8\pi\epsilon_0 r}. \tag{1-69}$$

Plotting total energy ξ of the electron as a function of r yields the hyperbola shown in Figure 1-19. Visualizing the surface generated by rotating this function about the ξ axis, we perceive a funnel shape that is often labeled an *energy well*. A free electron arriving from a long distance away could "fall into" such a well, giving up energy by some mechanism, and as a result become *trapped* by the coulomb force exerted by the positive nucleus. The electron's final energy, then, is negative, consistent with Equation 1-69, because we arbitrarily chose the energy of an electron at rest at infinity as zero. Conversely, when the the electron in this hydrogen atom receives an amount of energy from some source that is equal to or greater than the energy difference between where it resides (in energy) in the well, and $\xi = 0$,

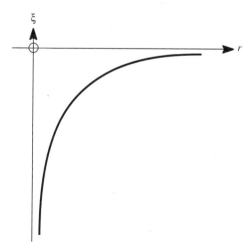

Figure 1-19 The total energy of an electron in a circular orbit about a proton, as a function of orbital radius. Equation 1-69 shows that this curve is a hyperbola.

then it can escape the atom. When this occurs, the electron leaves behind just the proton, or *hydrogen ion*. If the electron received an amount of energy greater than the amount needed to *ionize* the hydrogen atom, the energy difference would be kinetic energy possessed by the escaping electron.

At this point, however, the classical elements combined so far in the hydrogen-atom model lead to difficulties. We have already seen that an electron in orbit about a proton is experiencing unabated centripetal acceleration. In Section 1-4.2 we stressed (1) that when a charge is accelerated, it radiates, and (2) that in radiating the charge experiences energy loss. Hence, in the picture just sketched, an electron placed in a circular path near the top of the energy well, or "funnel," would experience continuous energy loss and would describe a distorted helical path in the funnel, moving ever toward lower energy. (We have elaborated the energy-well picture, plotting the electron's azimuthal position at any instant as well as its radius r.) It is this classical dilemma that leads us to the nonclassical aspects of Bohr's model.

1-4.4 The Bohr Postulates

Bohr encapsulated the nonclassical features of his model in a pair of postulates, a *postulate* being a proposition offered without proof. Although his ideas were based upon brilliant intuition rather than "proof," they could not be dismissed by the best-informed workers at the time as simply musings; his synthesis of the quantum and nuclear-atom theories yielded predictions that were consistent with an appreciable body of information that existed at that time. The information consisted of data on atomic radiation assembled by a series of spectroscopists and empirical formulas contributed by spectroscopy theorists. Subsequent years brought a steady series of refinements and extensions of his original model, with each step being able to explain another feature—of increasing subtlety—of the experimental data. Thus, the Bohr model is less a foundation of great generality, along the lines of Maxwell's equations or relativity theory, and more a bold departure that led to a fruitful new path of inquiry.

Bohr's first postulate addressed head-on the dilemma posed by classical theory, namely, that an electron in an orbit of the kind he had described would radiate continuously because of the continuous centripetal acceleration it experienced, continuously losing energy in the process. He *postulated* that the electron could remain in the circular orbit without radiating, and hence at constant energy. This electronic state he described as a *stationary state,* and sometimes as a *stationary orbit.* He considered various elliptical and circular orbits and various criteria for fixing the size and shape of the orbit, but finally settled upon a circular orbit for the electron about the nucleus, with the orbital radius having a value such that the angular momentum of the electron about the nucleus was an integral multiple of $h/2\pi \equiv \hbar$. That is,

$$mvr = n\hbar, \qquad (1\text{-}70)$$

where $n = 1, 2, 3, \ldots, \infty$. Only later was it recognized that this formulation of Bohr's postulate was consistent with the final interpretation Planck had given for his harmonic-oscillator theory, a theory that Planck had carried through several stages of interpretation [12].

Bohr's first postulate relates directly to the first of two essential features of Planck's original quantum theory. Planck concluded that the harmonic oscillator had available to it only discrete quantum states, with a definite allowed value of energy associated with each. The nonobvious and bold step taken by Bohr was to substitute a circulating electron in an atom for Planck's oscillator, and then to assume the resulting classical dilemma out of existence!

The nonobviousness of relating the classical harmonic oscillator to an electron revolving about a proton can be illustrated thus: The classical oscillator possessed an energy that increased with its frequency of oscillation. For example, consider two mechanical oscillators having equal masses and equal amplitudes, but differing spring constants. The oscillator incorporating the stiffer spring would exhibit the higher frequency and possess the higher energy. But the electron in Bohr's model behaved differently; as it dropped to *lower* values of allowed energy in the potential well, its angular frequency of revolution *increased!* (To be sure, the magnitude of the total energy ξ given by Equation 1-69 increases as radius r is diminished, but this is a negative energy as a result of origin choice.)

Bohr's second postulate dealt with the frequency of radiation emitted by an atom, and again he pondered various options. But he chose once again the option consistent with quantum theory—its second essential feature, which stated that a pulse or *quantum* of radiation was emitted or absorbed when an oscillator jumped from one allowed state to another. That is, the packet had energy

$$hf = \Delta\xi = \xi_i - \xi_f, \tag{1-71}$$

where $\Delta\xi$ is the difference in energy between the two allowed states involved in a transition from an initial state of energy ξ_i to a final state of energy ξ_f. This choice by Bohr had the further virtue of consistency with Einstein's photoelectron theory [11].

1-4.5 Model Predictions

In the first decade of this century, abundant data existed on the characteristic line spectra of a wide range of elements. Further, in a number of cases a set of lines could be identified as a *series* because of a kinship the lines exhibited as observed by one or more spectrographers. A line was usually identified in terms of its wavelength λ. (As we have noted, $\lambda = 0.6\ \mu\text{m}$ falls roughly at the center of the visible spectrum.) Wavelength in turn is converted to frequency f by the most basic equation of wave motion,

$$f = \frac{c}{\lambda}, \tag{1-72}$$

where $c\ (= 3.00 \times 10^{10}$ cm/s), the *velocity of light,* is the velocity applicable in this case. Equation 1-59, then, contributed by Planck, permits a conversion from frequency to the energy of the electromagnetic quantum, known as a *photon.* Combining Equations 1-59 and 1-72 gives us

$$\xi = \frac{hc}{\lambda}. \tag{1-73}$$

Thus the spectrographer's wavelength observation could be converted into photon energy.

Bohr's task, in order to test his theory, was clear. He had stated that a photon was created and emitted when an electron dropped from one stationary state in an atom to another of lower energy. Thus he needed knowledge of the possible energy spacings, or intervals $\Delta\xi$, between allowed states. This became possible through simple subtraction as soon as he could write the energy of each stationary state relative to the energy zero already arbitrarily established. His first postulate provided the necessary means. It gave an expression, Equation 1-69, for electron energy in its circular stationary orbit, this in terms of physical constants and orbital radius r. In addition, the statement of quantized angular momentum in this postulate yielded an expression in the quantizing integer n, the *principal quantum number,* and significantly in r as well, Equation 1-70. Thus he was able to eliminate r from the energy equation: Solving Equation 1-70 for r yields

$$r = \frac{n\hbar}{mv}. \tag{1-74}$$

Placing this in Equation 1-69 yields

$$\xi = -\frac{q^2(mv)}{8\pi\epsilon_0 n\hbar}. \tag{1-75}$$

But an independent relationship between ξ and mv can be obtained by comparing Equations 1-68 and 1-69:

$$\xi = -\frac{mv^2}{2}. \tag{1-76}$$

Evidently,

$$mv = \sqrt{-2\xi m}. \tag{1-77}$$

Placing this expression in Equation 1-75 gives

$$\xi = -\frac{q^2\sqrt{-2\xi m}}{8\pi\epsilon_0 n\hbar}. \tag{1-78}$$

Finally, squaring this equation and dividing both sides by ξ yields

$$\xi = -\frac{q^4 m}{32\epsilon_0^2 n^2 \pi^2 \hbar^2}. \tag{1-79}$$

Assigning to the principal quantum number n its allowed integer values (1, 2, 3, ..., ∞) thus yields an infinite series of allowed energy values for the stationary states, culminating in a state at $\xi = 0$ for which $r = \infty$. In Figure 1-20 we plot the first five energy levels so calculated. Subtracting to obtain a few of the energy spacings, or intervals, we can then use Equation 1-73 to determine the wavelength corresponding to each energy-interval value; these are stated in micrometers for the six examples shown. These six examples include the two highest-energy transitions possible in each of three infinite sets of transitions. Each set, or *series,* of transitions and the corresponding set of spectral lines is identified by the name of a worker who identified and catalogued the spectral lines. From left to right, the three sets shown (and two not shown) are known as the Lyman series, the Balmer series, the Paschen series, the Brackett series, and the Pfund series.

At this point it is convenient to introduce a number of definitions and concepts

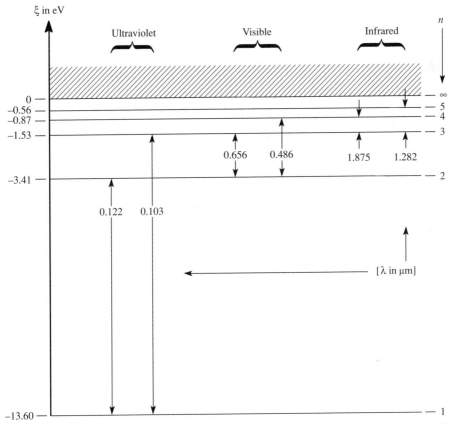

Figure 1-20 A diagram of a few of the energy levels for a hydrogen atom. Six of the infinite number of possible transitions are indicated, and the wavelength of the corresponding radiation is given.

that will prove useful in subsequent discussion. When the single electron of the hydrogen atom resides in its lowest state, or *ground state,* for which $n = 1$, the atom is neutral and *unexcited*. The atom becomes *excited* when its electron acquires energy from an outside source and is consequently elevated to one of the higher discrete states available to it. There are various ways in which the electron can acquire its excitation energy. An important case is that in which a photon impinges on the atom and imparts its energy hf to the atom, disappearing in the process. When the energy hf matches the energy $\Delta \xi$ between the *filled state* (the state containing an electron) and a higher state, the electron can be elevated to the higher state, thus accomplishing excitation. When we say that the energies must "match," what we mean is that the probability of having such an event occur is a sharply peaked function; even a slight mismatch in the two energies greatly diminishes the probability that excitation will occur.

In an analogous way, when an atom is in an excited state, its electron can be induced to make a *downward* transition in energy upon the arrival of a photon whose energy is matched in the same sense as before. In this case, the original photon continues on its way, accompanied by a matching photon emitted by the atom. This phenomenon is sometimes termed *light amplification* or *stimulated emission* and is the basis of the laser. These terms, in fact, account for the first four letters of the acronym *laser,* with the fifth letter standing for *radiation*.

In the usual case, however, the *relaxation* of an atom from an excited state to a lower state occurs spontaneously. An atom can return to its ground state in a single transition, or by means of two or more transitions. In the first case, the photon emitted is essentially a duplicate of the photon that produced excitation in the first place. In the second case, two or more photons will be emitted, one corresponding to each of the constituent transitions. Obviously the emitted photons have lower energy than did the photon that initially caused excitation (adhering for the moment to this excitation mechanism). This pattern of electronic transitions accounts for the phenomenon of *fluorescence*. In a familiar application, the fluorescent light, energetic (high-frequency) ultraviolet photons cause the excitation of atoms that emit lower energy (and lower frequency) visible photons in the relaxation process. When the relaxation transition of interest has a low probability, and therefore requires a relatively long waiting time after excitation, the same phenomenon is termed *phosphorescence*. Objects that glow in the woods at night contain atoms that were excited by daytime ultraviolet light, indicating characteristic times of hours at least.

When the energy delivered to an electron is sufficient to drive it out of the atom, what is left behind is a charged species, or *ion*. The hydrogen ion, therefore, is identically a proton. An important method for producing *ionization* is to place the atoms or molecules of interest (in a gaseous state) in a closed vessel, usually at a pressure lower than one atmosphere, and then to pass a current of electrons through the vessel between a pair of electrodes. Let us consider the gas to be hydrogen. In such a "discharge tube" it is possible to measure the energy of the electrons (that make up the current) necessary to elevate an electron in a hydrogen atom from its ground state to $\xi = 0$, at which it escapes from the nucleus. This is known as the *ionization energy,* and this amount of energy is imparted to the electrons making up the current (or the "discharge") by applying between the electrodes a voltage equal to the *ionization potential.* In Figure 1-20, one can see that for hydrogen, the ionization energy is $\xi_i = 13.6$ eV. If we take an opposite point of view and think of an electron that arrives with negligible kinetic energy in the vicinity of a hydrogen ion, then it can be "trapped" by the proton, falling to the ground state in the potential well (described in connection with Figure 1-19), and radiating in the process. From this point of view we can say that the electron is now "bound" to the nucleus, and that its *binding energy* is 13.6 eV.

When the photon or electron, or other cause of ionization, imparts to an electron in an atom an energy in excess of the ionization energy, the escaping electron carries away kinetic energy equal to the difference of the energy imparted and the ionization energy. Since this final energy, $\xi > 0$, is not discrete as are the energies available to a *bound electron* for which $\xi < 0$, it follows that a *continuum* of possible electron energies must exist in the range $\xi > 0$. This continuum of states is suggested by the shaded band at the top of Figure 1-20.

In 1885, Balmer noted that the wavelengths of the nine (then-known) lines in the hydrogen spectrum could be predicted from an expression of very simple form. Subsequently Rydberg generalized Balmer's formula in a way that fitted it to many other sets of observed lines. Bohr then tested his own theory by comparing its predictions against those of the Rydberg formula. According to Bohr's second postulate, the frequency f of a photon radiated when an electron goes from an initial orbit of quantum number n_i to a final orbit of quantum number n_f can be calculated from Equation 1-71 to be

$$f = \frac{\xi_i - \xi_f}{h}. \tag{1-80}$$

Combining this expression with Equation 1-79, then, yields

$$f = \frac{q^4 m}{64\epsilon_0^2 \pi^3 \hbar^3} \left(\frac{1}{n_f^2} - \frac{1}{n_i^2} \right). \tag{1-81}$$

Rydberg's formula possessed a form similar to that of Equation 1-81. It is, however, usually stated in terms of *wave number,* or reciprocal wavelength, which can be obtained from Equation 1-72 simply as

$$\frac{1}{\lambda} = \frac{f}{c} = R \left(\frac{1}{n_f^2} - \frac{1}{n_i^2} \right), \tag{1-82}$$

where R is the *Rydberg constant.* Comparing Equations 1-81 and 1-82 shows that Bohr's theory yields an expression for Rydberg's empirical constant in terms of *fundamental natural constants* as

$$R = \frac{q^4 m}{64\epsilon_0^2 \pi^3 \hbar^3 c}. \tag{1-83}$$

Evaluating this collection of constants yielded a result that agreed within a small fraction of a percent with Rydberg's constant!

Bohr's original paper [6] also addressed the cases of helium, lithium, and even simple molecules. His new theory was less successful in these cases, however, and found fairly numerous critics as a result. Nonetheless, the agreement between Rydberg's well-established empirical constant and Bohr's novel approach to modeling the hydrogen atom was so remarkable that it could not be ignored. And as we know, Bohr's model prevailed.

1-4.6 Refinements to the Bohr Model

Bohr's theory triggered an extraordinary series of refinements and insights in the ensuing fifteen years. A qualitative understanding of these developments is helpful in dealing with the subject matter of Chapter 2, where we consider the behavior of electrons in solid materials; appreciating certain further aspects of their behavior in isolated atoms helps to bridge the gap. A clear and quantitative description of these developments has been given by Semat [13]. For the reader seeking still more depth, we recommend the book by Joos [14]. This unusually terse and comprehensive book, published originally only nineteen years after Bohr's paper, addresses a full sweep of subject matter, from mathematical foundations through the nuclear physics of that day, and relates the topics we now review to those that came before and after.

First, let us look at an assumption implicit in the original model. By attributing all energy and angular momentum in the atom to the orbiting electron, one assumed an infinite ratio of proton mass to electron mass. But in fact, the ratio is approximately 1836:1. As a result, the system must rotate about a point on a line connecting the centers of the two particles, and very close to the proton. Further, the two particles would share the atom's energy and angular momentum, but with the electron having most of both. It is electron behavior that is of primary interest, and so its energy and angular momentum were calculated by analyzing the mechanics of the two-body system. A convenient way of handling the equations was to attribute a *reduced mass* to the electron, a quantity dependent on the mass ratio, but relatively insensitive to it, and having a value slightly smaller than that of an electron at rest.

This step opened the door to experimental testing. Bohr's formulation of the Rydberg constant, Equation 1-83, contains electron mass in the numerator. What was needed, then, was a series of hydrogen-like atoms of differing mass ratios. These were obtained, in effect, by studying helium in a singly ionized state (having lost one of the two electrons present in the neutral atom), doubly ionized lithium, triply ionized beryllium, and so forth. In more recent times, the process has been extended up to the case of oxygen with seven electrons missing from its normal complement of eight.

Spectrographers were then in a position to scrutinize the spectra of these hydrogen-like atoms, looking for the subtle effect of mass ratio, but only after entering the gross effect of Coulomb's law into Equation 1-83. The other factor in the numerator for the hydrogen case, q^4, is contributed equally by the charges of the proton and electron. For the hydrogen-like cases, this factor required generalization to $(Zq)^2(-q)^2$, where Z is the *atomic number*—the number of positive charges on the nucleus or the number of electrons present in the neutral atom. This done, changes in the Rydberg constant could be observed, amounting to a mere five parts in ten thousand when hydrogen-like oxygen is compared to hydrogen. Studies of this kind produced early on an estimate of the mass ratio in the hydrogen atom very close to that obtained by other means, thus building additional strong support for the modified Bohr theory. Further, such experiments led to the discovery of deuterium, or hydrogen of mass two rather than mass one. Deuterium and its ion, the *deuteron*, played a major role in the ensuing study of atomic nuclei.

Bohr's original theory, which employed only circular orbits, was extended to include elliptical orbits by Sommerfeld. He showed that a, the semimajor axis of any orbit, was determined solely by n, the principal quantum number, as illustrated in Figure 1-21. The symbol k is the *angular* (or *azimuthal*) quantum number, which, when added to a *radial* quantum number must equal n. It was decided that k could never be zero because that would require the electron to pass through the nucleus in a linear oscillation. Therefore the only possible orbits for three values of n are those pictured in Figure 1-21. Generalizing, then, the allowable values for k are $k = 1, 2, \ldots, n$.

Thus the elliptical-orbit concept introduced two quantizing conditions in place of one. But no new allowed energy states (and hence no new spectral lines) were predicted as a consequence. A mathematical analysis of systems having more than one periodic feature entered at this point, showing that when the ratio of two of the periods is a rational number, the associated quantum conditions reduce to a single quantum condition; such a system is described as *degenerate*. Generalizing, once more, the number of independent quantum conditions incorporated in a system is equal to the number of incommensurable periods it embodies.

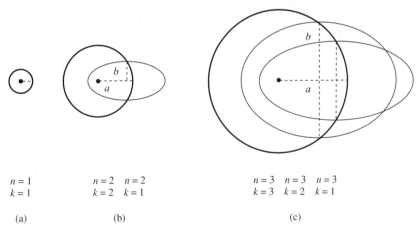

$n = 1$ $\qquad$ $n = 2$ $n = 2$ $\qquad\qquad$ $n = 3$ $n = 3$ $n = 3$
$k = 1$ $\qquad$ $k = 2$ $k = 1$ $\qquad\qquad$ $k = 3$ $k = 2$ $k = 1$

(a) $\qquad\qquad\qquad$ (b) $\qquad\qquad\qquad\qquad$ (c)

Figure 1-21 The elliptical electronic orbits that are possible for three different values of the principal quantum number n. (a) $n = 1$. (b) $n = 2$. (c) $n = 3$. (After Semat [13], with permission.)

Sommerfeld went on to show that an elliptical orbit did introduce a feature causing the atomic system to be *nondegenerate*. The electron in its elliptical path undergoes a cyclic change in linear velocity. But Einstein had shown that mass is velocity-dependent. Taking into account this relativistic mass change leads to a precession of the elliptical orbit in the manner illustrated in Figure 1-22, an observation that also has a planetary analog.

The relativity correction added a small term to Bohr's energy expression, Equation 1-79. The correction term involved the ratio n/k and a factor α^2, where

$$\alpha \equiv \frac{1}{4\pi\epsilon_0} \frac{2\pi q^2}{hc} \approx \frac{1}{137} \tag{1-84}$$

has become known as the *fine-structure constant*.

The experimentally observed fine structure in the line spectra of certain elements, however, was still not well explained by atomic theory at this stage of development. Then Uhlenbeck and Goudsmit proposed that the electron spins like a top and hence possesses *spin angular momentum*. Angular momentum is a vector quantity and so a vector treatment is necessary to handle the combination of orbital angular momentum and spin angular momentum, even in the simplest case, that of hydrogen. In multielectron atoms one deals of course with many vectors, two for each electron in the general case. In the *vector model* of the atom, one employs ℓ as the orbital quantum number, where $\boldsymbol{\ell}$ is the corresponding vector quantity. In a similar way, the spin-angular-momentum quantum number is s, and the corresponding vector quantity is $\mathbf{s}$. Only one magnitude is possible for s, namely $s = 1/2$. When the angular momenta (both orbital and spin) of several electrons are added, the addition must be done vectorially. It is found that with an odd number of electrons, the result is always an odd multiple of 1/2; with an even number of electrons, the magnitude of the vector sum is found to be an integer. From this, it

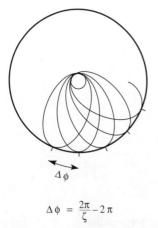

$$\Delta\phi = \frac{2\pi}{\zeta} - 2\pi$$

Figure 1-22 Precession of an elliptic orbit as a result of relativistic mass change. The quantity ζ is a number less than unity. (After Semat [13], with permission.)

is inferred that a given spin vector can only take a position parallel to or antiparallel to the resultant of the other angular-momentum vectors. In other words, we must have $s = \pm 1/2$.

Because orbital angular momentum is that of a charged particle, the circulating electron constitutes a "current" and exhibits a magnetic moment as a result [15]. And because angular momentum is quantized in multiples of $\hbar$, there is a corresponding quantization of the orbital magnetic moment in multiples of the quantity

$$\text{magnetic moment} = \mu_0 \frac{q\hbar}{2m}, \tag{1-85}$$

known as the *Bohr magneton*. The smallest finite (nonzero) values of orbital angular momentum and magnetic moment are therefore $\hbar$ and one Bohr magneton, respectively. In the case of electron spin, as we have just seen, the only possible value of angular-momentum magnitude is $\hbar/2$. A treatment of spin magnetic moment analogous to that involved in the calculation of orbital magnetic moment would require knowledge of electron "shape," an inquiry that has always been rebuffed and never rewarded. Hence it became necessary to resort to experiment in order to determine spin magnetic moment, which proved to be one Bohr magneton also (and not half this value, as one might have guessed). There is thus a departure from parallelism worth noting in the orbital and spin properties of the electron with respect to angular momentum and magnetic moment.

At this point we first mention the new discipline of *wave mechanics* that evolved in the brief but eventful period under discussion, augmenting the older quantum theory. Wave mechanics showed that ℓ, the *orbital quantum number*, could be identified with $(k - 1)$ in the older elliptic-orbit picture. Thus its possible values are $\ell = 0, 1, 2, \ldots, (n - 1)$. The product of ℓ and $\hbar$ yields the orbital magnetic moment. If one places an atom in a magnetic field **H**, a torque develops that tends to orient the magnetic-moment vector. But another factor is also present, one that introduces a fourth quantum number. According to wave mechanics, once more, the orbital magnetic moment may be oriented only in certain specific directions with respect to the external magnetic field, and the directions that the vector ℓ may assume are such that its projection in the direction of the magnetic field must have an integer value, less than or equal to the value of ℓ. Since ℓ can be fully parallel to **H** or fully antiparallel as two possible orientations, it follows that m, the *magnetic quantum number* fixed by the projection of ℓ on the **H** direction, can have the values $m = 0, \pm 1, \ldots, \pm \ell$, as illustrated in Figure 1-23. The effect of m is to impose a fine structure on an electronic state otherwise defined by n and ℓ. And s, in turn, imposes its "doublet" fine structure on a state otherwise defined by the foregoing three quantum numbers. In the presence of the magnetic field **H**, all of the states so defined exhibit slightly different energies, or are *nondegenerate*. The presence of the field has eliminated the *degeneracy* (or energy equality) that would otherwise exist among states differing only in m. The small energy differences among the states having the same value of n, we should emphasize, are vastly smaller than the energy interval between states of differing n. Therefore, the value of n (unassisted) defines a *group* or *shell* of states clustered in energy.

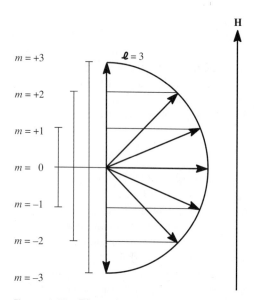

Figure 1-23 The projection of the orbital angular-momentum vector ℓ on the direction of an imposed magnetic field **H** fixes the magnetic quantum number m.

As a separate matter, wave mechanics finds that the quantum number ℓ must be replaced by $\sqrt{\ell(\ell+1)}$ and the quantum number s, by $\sqrt{s(s+1)}$ for accurate calculations. We shall not have need to carry out such calculations in what follows, however, and so will continue to use the simpler quantum-number values.

Before introducing the primary wave-mechanical concept, namely, the substitution of a *wave function* for the rather more concrete orbital picture, let us describe another basic contribution to atomic theory that lends itself to an orbital interpretation, and that in fact was a key step on the way to wave mechanics.

The notion of a wave–particle duality for fundamental entities had existed for centuries, with both properties being attributed to light, specifically. But de Broglie brought the topic sharply to the fore in the mid-1920s by hypothesizing that an electron possesses wave properties. The wavelengths predicted by his hypothesis suggested the use of a single-crystal material as a diffraction grating for a beam of electrons; atomic planes in crystals are spaced at appropriate distances. This experiment was carried out a year or two later by Davisson and Germer, and the wave nature of the electron was confirmed.

Specifically, de Broglie proposed that an electron exhibits a wavelength given by

$$\lambda = \frac{h}{mv}. \tag{1-86}$$

For an electron in an orbit, the wave will be a *standing wave* if the orbital circumference is an integral number of wavelengths, or if

$$2\pi r = n\lambda. \qquad (1\text{-}87)$$

Substituting Equation 1-86 into Equation 1-87 yields

$$2\pi r = \frac{nh}{mv}, \qquad (1\text{-}88)$$

and solving this for angular momentum yields

$$mvr = n\hbar, \qquad (1\text{-}89)$$

which is identical to Bohr's quantization hypothesis.

With this substantial impetus, Schroedinger developed his wave mechanics and applied it to atomic systems more complex than that of hydrogen. In spite of the triumph of the older quantum mechanics in hydrogen-model theory, its application to more involved cases was at best qualitative or semiempirical. Wave mechanics altered the situation. In the new formulation, the electron no longer had a precise location and velocity at a particular instant. Rather, a given electron had a particular probability of residing within the volume defined by a specific closed surface, also calculable from knowledge of the wave function. The size and shape of the surface was fixed by the four quantum numbers specifying an electronic state. These *orbital surfaces* as they were termed, now usually called simply "orbitals," have obvious reference to the earlier picture. When $\ell = 0$, denoting the lowest possible value of orbital angular momentum, the surface exhibits spherical symmetry. When $\ell > 0$, it exhibits axial symmetry—is a surface of revolution—with a toroidal (or doughnut-shaped) surface constituting a good example. An interesting consequence of this situation is that an axial direction is thus defined, with the atom now possessing a "built-in z axis."

It may appear that the concrete planetary model of an atom had been traded for a nebulous "probability-cloud" picture without compensating benefits. But the greater success of the new approach in modeling complex atoms was significant. Moreover, it provided a measure of plausibility to Bohr's intuitive and seemingly arbitrary assumption of the *stationary state*. First, it showed that radiation from various parts of an electronic "cloud" corresponding to an allowed state will cancel by interference, which made Bohr's radical postulate more nearly congruent with classical theory. Further, it was shown that the frequencies existing in the spectrum of an atom are equal to the differences of the frequencies of allowed states, thus supplying background missing when Bohr made his bold transition from the classical harmonic oscillator to an electron in orbit.

Now note once more the four quantum numbers that define an allowed state for an electron in an isolated atom: They are (1) the principal quantum number, $n = 1, 2, \ldots, \infty$; (2) the orbital quantum number, $\ell = 0, 1, \ldots, (n-1)$; (3) the magnetic quantum number, $m = 0, \pm 1, \ldots, \pm \ell$; and (4) the spin quantum number, $s = \pm 1/2$. Wave mechanics, however, does not provide guidance in assigning these numbers. But the *Pauli exclusion principle,* also advanced in the mid-1920s, does provide the necessary guidance. It asserts that *no two electrons in a given atom can exist in the same state*. This, then, permits us to set up the necessary hierarchy of available states in an atom, and to supply one electron to each, starting with the lowest state, until that atom's electrons are fully accommodated. The result is the electronic structure of the chosen atom in its unexcited state. These principles are summarized in Figure 1-24, where the possible values of all four quantum numbers are listed for $n = 1$, 2, and 3. As suggested by the success of Bohr's hydrogen model, and as noted above, it is n that single-handedly nearly fixes the energy of a state, with the variations in ℓ, m, and s providing mere perturbations. Within the shell or group defined by a given value of n are *subgroups* fixed by ℓ. This picture is

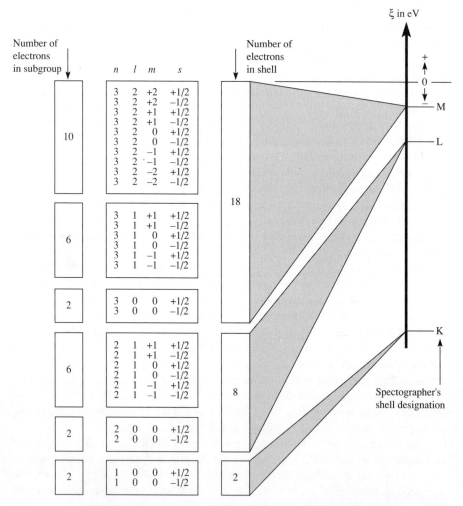

Figure 1-24 An exhaustive listing of the possible electronic states corresponding to $n = 1$, 2, and 3. The principal quantum number n plays the largest part in fixing state energy, with variations in the other three quantum numbers producing small variations.

one that we shall invoke again when we address the element that is dominant in solid-state electronics today—namely, silicon.

Figure 1-24 illustrates an important point. Starting with the L shell and proceeding upward, we see that at the bottom of each shell is a pair of states for which $\ell = 0$, and just above that the six states for which $\ell = 1$. It is this repeating eight-state cluster that accounts for the eight columns of the periodic table presented in its entirety in Table 1-1. Elements falling in a given column have similar chemical properties. For example, hydrogen has one electron in the K shell; lithium has one electron in the L shell; sodium has one electron in the M shell, and so forth. All of these elements are very reactive chemically, and all fall in column I. For these three elements, the K, L, and M shells, respectively, are described as *the outer shell*.

Finally, note that tables have been placed at the end of the chapter for convenient reference. Table 1-1 is the periodic table, as just noted; Table 1-2 gives the values of important physical constants, including the Bohr radius, the electronic charge, the Bohr magneton, the free-space permittivity and permeability, and the masses of the electron and proton.

1-5 CRYSTALLOGRAPHY

A *single crystal* is characterized by geometric regularity in the placement of its constituent atoms, and it is the focus of crystallography. Gem stones are significant examples of single crystals, but far from the only examples. Larger naturally occurring single crystals, such as quartz and calcite, attracted the attention of the curious beginning in early times because of unusual external features—smooth, flat surfaces meeting in sharp edges and corners. In more recent times it was appreciated that such external regularities are a reflection of internal regularity, and scientists set about cataloguing the possibilities. For the reader interested in crystallography for its own sake, there are good standard references, such as those by Barrett [16] and Wyckoff [17].

Until this century, only naturally occurring single crystals of reasonable size were available for study and application. But early in this century, methods for the growth of man-made single crystals were developed, with scientific ends primarily in view. Only a few years later technical uses for such crystals evolved, with electronics providing significant applications. For example, quartz crystals proved useful for precise frequency control. And it was in the 1940s, with the beginning of solid-state electronics as we now know it, that single crystals moved into a position of great importance. The element silicon in single-crystal form, which is the foremost material, probably will retain its position for a long time to come, although other single-crystal materials that have similar properties are acquiring growing significance. So dominant is silicon, however, that its production in single-crystal form is literally described in terms of tonnage. Our aim here is to explain certain rudiments of crystallography that are used daily by many engineers in solid-state electronics.

1-5.1 The Lattice

Dealing in a rigorous way with the atomic regularity of a single crystal has spawned various mathematical abstractions, some rather obscure. For our purposes, however, an intuitive geometrical understanding of crystallographic principles will be adequate, and the relevant abstraction, the lattice, is fortunately easy to grasp. A *lattice* can be defined as *an infinite array of mathematical points arranged so that each point has an identical configuration of neighboring points*. A lattice can be one-, two-, or three-dimensional. It is the last of these that applies to a single crystal, of course, but the other two cases have important illustrative value. The question of how many basically different "wallpaper patterns" are possible is answered by considering two-dimensional lattices. And the one-dimensional lattice provides an excellent starting point. It is intuitively obvious that the only arrangement of points on a line that can meet the lattice definition is a set of equally spaced points. There exists only *one* possible one-dimensional lattice. This statement is not trivial, as it might at first seem, because it illustrates the important principle that altering a lattice by applying a scale factor does not create a new lattice; whether the point spacing is a micrometer or a kilometer in the one-dimensional case, we are still dealing with the same lattice.

There are five possible two-dimensional lattices, pictured and named in Figure 1-25. In each case we depict a small portion of the lattice that can generate the necessary infinite array of points by *parallel repetition*. This term in turn implies that *translation* alone is permitted in the generation process, and that *rotation* is disallowed. Note the areas defined by light lines in Figure 1-25; it is evident that in the lattice-generation process, the ultimately infinite plane will be completely covered by the repeated small areas, leaving no "voids" whatsoever.

The square lattice, Figure 1-25(a), has a high degree of *symmetry*. Consider the

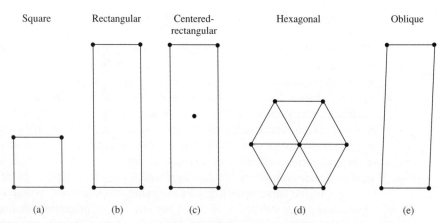

Figure 1-25 Only five two-dimensional lattices, illustrated by the examples given here, are possible. In each case, consider that the points shown are permitted to generate an infinite array by parallel repetition.

lattice to be rotated in its plane about an axis normal to the plane. If the axis passes through a lattice point, we see that each successive 90° rotation will produce a lattice arrangement identical to the starting arrangement. If the points are individually identified, then four successive 90° rotations in the same direction will restore the lattice to its starting position. Therefore the square lattice exhibits *fourfold* symmetry about such an axis. The same holds for an axis through the center of the square. With an axis through the midpoint of a side, two 180° rotations would have the analogous effect of restoring the original situation, thereby indicating *twofold* symmetry about that axis position.

In addition, the square lattice exhibits many examples of *line symmetry,* the two-dimensional analog of mirror symmetry. Let a line be congruent with any side of the square. The "reflection" of any lattice point in that line produces a point that coincides with an already existing lattice point. The same is true of a diagonal line through the square, and of a line parallel to a pair of sides and midway between them. Similarly, the square lattice illustrates abundant examples of symmetry about a point, or *point symmetry* for brevity. As one example, let the point chosen be the center of the square. Then draw a line from any lattice point to the point of symmetry; extend the line an equal distance in the same direction beyond the point, and observe that this too brings us to an existing lattice point. Other points of symmetry are easy to find in the square case.

For our purposes, this kind of intuitive examination of the symmetry properties of lattices is sufficient. However, the study of symmetries and "symmetry groups" provides the most rigorous approach to the subject of crystallography. Although by this means one can answer such questions as "how many distinct lattices exist?" with more logical validity than by geometrical arguments, we shall continue to emphasize geometry because of its intuitive accessibility, and will confine our further comments on symmetry to fairly obvious points. For a good discussion of group theory and the mathematics of symmetry, the interested reader should refer to the treatise by Nussbaum [18].

The rectangular two-dimensional lattice, Figure 1-25(b), possesses obvious symmetries by virtue of its right-angle corners, but fewer than the square lattice. The centered-rectangular case, Figure 1-25(c), shares certain symmetries with the rectangular case, but introduces a new element. Note, as illustrated by dashed lines in Figure 1-26(a), that the centered-rectangular case could with equal logic be termed *rhombic*. The term *centered-rectangular* is preferred, however, because it emphasizes the presence of the 90° angles, which have important symmetry consequences. In addition, notice the angle θ designated in Figure 1-26(a). The angle θ may possess an infinite number of values without altering the fact that the resulting lattice is centered-rectangular. This specification is completely analogous to the earlier statement that altering a lattice by a scale factor does not create a new lattice. The permissible values of θ are in the range $0 < \theta < 90°$, *but* with two restrictions. The first is that $\theta \neq 45°$. It is evident that for this condition, the dashed rhombus in Figure 1-26(a) becomes a square, so the centered-rectangular case has become a simple square lattice oriented at 45° to the original centered-rectangular lattice. When this occurs, the centered-rectangular lattice is said to *degenerate* into a square

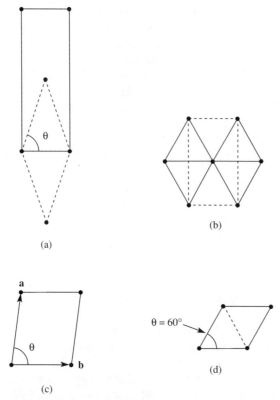

Figure 1-26 Features of particular two-dimensional lattices. (a) A rhombus is associated with the centered-rectangular lattice. (b) The hexagonal lattice is a special case of the centered-rectangular lattice. (c) Vectors defining an oblique lattice. (d) One choice of unit cell for a hexagonal lattice.

lattice. The broad meaning of this term is *to reduce to a simpler or more special case*. (In Section 1-4.6, where the same term entered, it described the coinciding of energy levels that in a *nondegenerate* situation possessed distinct and different values.)

The second restriction on the value of θ in Figure 1-26(a) is that it must not lead to the hexagonal lattice. For example, when $\theta = 60°$, the centered-rectangular case degenerates into the hexagonal case, depicted in Figure 1-25(d). The fact that the latter is a special case of the former is displayed by the dashed lines in Figure 1-26(b). The reason for granting the hexagonal lattice the status of a separate case is once again the higher symmetry it exhibits. It shares twofold symmetry with the centered-rectangular lattice. More importantly, it possesses in addition *sixfold* rotational symmetry about a lattice point. As is evident in both Figure 1-25(d) and Figure 1-26(b), the hexagonal lattice can be regarded as a net of equilateral triangles.

Exercise 1-10. Why does Figure 1-25 not include a "centered-square" case as a sixth possible lattice?

Using pad and pencil, construct a square array of points, and then place a point in the center of each square. It will become evident that the result is a smaller simple-square array oriented at 45° to the original array. Changing a lattice by a scale factor does not create a new lattice, nor does merely changing the orientation of a lattice. Hence the centered-square case is neither new nor distinct. It is, as noted previously, the case encountered by degeneration of the centered-rectangular lattice when $\theta = 45°$.

This brings us then to the oblique lattice, Figure 1-25(e), which in a sense is the "most general" of the five, and exhibits the least symmetry. The vectors **a** and **b**, shown in Figure 1-26(c), serve to define the lattice. A condition necessary in the oblique lattice is that $a \neq b$, since having these vectors equal in magnitude would define a rhombus. In such a case, the oblique lattice would have degenerated into a centered-rectangular lattice, involving a relationship noted above.

Less obvious restrictions exist also in order to avoid degeneration. For example, having $|\mathbf{a}| = |\mathbf{a} - \mathbf{b}|$ is not permitted because this would require the lattice points to define an isosceles triangle, which leads again to a rhombus, and hence to the centered-rectangular lattice. Such restrictions can be described in terms of disallowed combinations of the angle θ and the ratio a/b.

Exercise 1-11. Cite an additional combination of θ and the ratio a/b in the oblique lattice, Figure 1-26(c), that is disallowed.

As one example, if $(a/b) = 2$, and $\theta = \text{arc sec } 2$, then the oblique lattice degenerates into a rectangular lattice.

1-5.2 The Unit Cell and Primitive Cell

The representations in Figures 1-25(a), (b), (c), and (e) are the unit cells of those four lattices. We can define the two-dimensional *unit cell* as *a parallelogram having lattice points at its center and corners, or at its corners only, that can generate the entire lattice by parallel repetition, and that displays the symmetry of the lattice.* The last proviso in this definition, to repeat, explains the choice of a rectangular unit cell for the centered-rectangular lattice. In the case of the hexagonal lattice, the rhombus mentioned earlier is taken as the unit cell, with one possible choice displayed in Figure 1-26(d). The hexagon in Figure 1-25(d) is not elected (even though

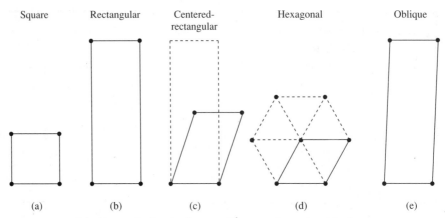

Figure 1-27 Primitive cells for the five possible two-dimensional lattices.

it can generate the lattice by parallel repetition without leaving voids) because it is not a parallelogram—it is more complex than necessary. Also, the centered rectangle illustrated in Figure 1-26(b) is not used because it displays the symmetry of a right angle, rather than the higher symmetry connected with a 60° angle.

There is another important building-block concept—that of the *primitive cell*. It is defined as *a parallelogram with lattice points at its corners only that can generate the entire lattice by parallel repetition*. The requirement for displaying symmetry has been dropped in this case. Figure 1-27 shows a set of five primitive cells corresponding to the five two-dimensional lattices. Compare these to the unit cells depicted in Figure 1-25; note that four of the five unit cells are also primitive.

It is evident that two vectors, the *primitive vectors,* are able to define a primitive cell. The pair of vectors chosen must correspond to adjacent sides of the primi-

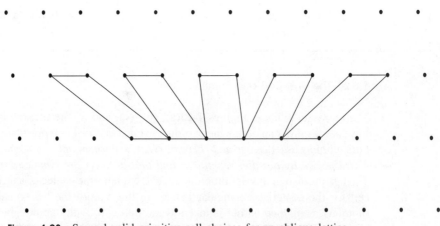

Figure 1-28 Several valid primitive-cell choices for an oblique lattice.

tive-cell parallelogram. Figure 1-26(c) illustrates the primitive cell of the oblique lattice and its accompanying primitive vectors. However, the choice of primitive vectors, and hence the choice of primitive cell as well, is not unique. In Figure 1-28 are shown several primitive-cell options for the oblique case. Customary preference, however, is for the cell in which one vector is the shortest possible and the second is the next shortest, as in the cell in the middle of the series shown. It is also true that this choice yields the most nearly orthogonal primitive cell. As a final point, note also that once the cell has been chosen, there exist four different but equally valid choices of primitive vectors, and that all four choices have the same respective lengths and the same departure from orthogonality. The unit-cell and primitive-cell concepts just illustrated using two-dimensional lattices are equally applicable to three-dimensional lattices, addressed next.

Exercise 1-12. In Figure 1-28 are shown five different but valid primitive-cell options for a particular oblique lattice. How do their areas compare?

The area of a parallelogram is equal to base times height, and since both base and height are unchanged from one case to the next, all five areas are equal.

1-5.3 The Space Lattice

The term *space lattice* has the same meaning as the term "three-dimensional lattice," and is covered by the definition of lattice given in Section 1-5.1. The unit cell of a space lattice is *a parallelepiped that displays the symmetry of the lattice and that has lattice points at its corners only, or at its corners and the center of its body, or at its corners and the centers of two or more faces*. The process of parallel repetition now becomes much like the process of stacking blocks. It is important that no space voids are left unfilled in the process; what is perhaps new to some readers is that the blocks can assume a number of shapes quite different from the familiar cube and still meet the requirement of filling all space. (Let us assume that the more oddly shaped blocks have sticky surfaces to avoid the obvious practical problem in such a stacking exercise.)

The number of possible space lattices is relatively small—fourteen. These configurations were accurately described on an intuitive geometrical basis one and one-half centuries ago by Frankenheim [19]. Sixteen years later they were rigorously defined by the crystallographer Bravais (bra-vay′) [20], and as a result, *Bravais lattice* has become an additional synonym for "three-dimensional lattice." Figure 1-29 depicts the unit cells of nine of the space lattices—those that involve right-angle corners only, or totally orthogonal edges. They fall in the orthorhombic, cubic, and tetragonal systems. Their individual names are indicated in the caption and are self-explanatory. The sequence of the cells and the construction lines in this perspective rendition have been arranged to indicate clearly which faces are square and which are not [21].

Exercise 1-13. Why does a "base-centered tetragonal" unit cell not appear in Figure 1-29?

The reason is closely related to the explanation given in Exercise 1-10. A centered square degenerates into a simple square. Therefore a "base-centered tetragonal" unit cell degenerates into a simple-tetragonal unit cell of the same height, but with a smaller base oriented at 45° to the original.

In Figure 1-30 are shown the two *monoclinic* unit cells, combining two right angles and one oblique angle at each corner. In Figure 1-31(a) is shown the hexagonal prism occurring in the *hexagonal* space lattice. The accented prism having a rhombic base is properly the unit cell. The unit cell of the *trigonal* space lattice is shown in Figure 1-31(b). It has equal edges, thus constituting a rhombohedron. Finally, the *triclinic* case is shown in Figure 1-31(c). It has three unequal angles and possesses the lowest symmetry of all of the space lattices. It is also the most general

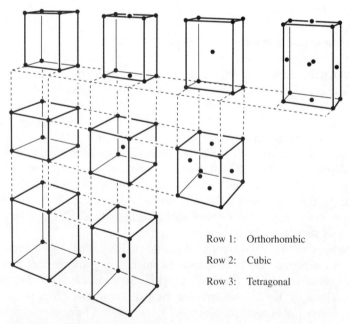

Row 1: Orthorhombic

Row 2: Cubic

Row 3: Tetragonal

Figure 1-29 Of the fourteen space lattices, nine, shown here, have unit cells with orthogonal edges. The names of the individual cells, taking the orthorhombic set from left to right for illustration, are simple, base-centered, body-centered, and face-centered orthorhombic. (After Warner and Grung [21].)

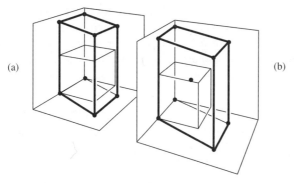

Figure 1-30 Unit cells of the two monoclinic lattices, referred to orthogonal planes for clarity. These cases are termed (a) simple and (b) centered monoclinic. The reason for the abbreviated description of the latter is that the unit cell can be either body-centered or base-centered (where the "bases" in this case are vertical planes), by appropriate choice of the primitive vectors. (After Warner and Grung [21].)

of the fourteen in the sense that every possible space lattice is a "special case" of the triclinic lattice.

A *space-lattice primitive cell,* which can be specified by means of three primitive vectors, is *a parallelepiped having lattice points at its corners only that is capable of generating the entire lattice by parallel repetition.* Although important, the primitive cell is in some cases difficult to visualize or draw. The difficulty, in general, is mitigated to some extent by the fact that half of the unit cells are primitive, as inspection of Figures 1-29, 1-30, and 1-31 confirms. As an example of a

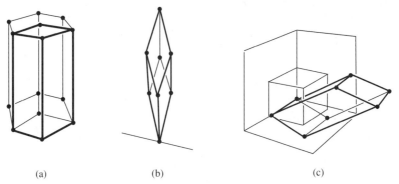

Figure 1-31 Unit cells for the remaining three space lattices. (a) The unit cell (and primitive cell) of the hexagonal lattice, shown as it relates to the corresponding hexagonal prism. (b) The trigonal space lattice has a unit cell that is primitive and a set of primitive vectors of equal lengths, forming equal angles. (c) The triclinic space lattice has a unit cell that is primitive and possesses no right angles. It is referred here to a cube and to orthogonal planes for clarity. (After Warner and Grung [21].)

difficult case, however, examine the primitive cell for the face-centered-cubic (fcc) lattice depicted in Figure 1-32. It is evidently a rhombohedron, but still is appreciably more difficult to "grasp" than is the face-centered cube that constitutes the unit cell. The fcc lattice is the most important by far to an engineer engaged in solid-state electronics. It is the case that applies very specifically to silicon.

Exercise 1-14. Examine the fcc primitive cell depicted in Figure 1-32. Choose as primitive vectors the three vectors corresponding to the cell edges that meet in the lower-left corner of the cube. What angles are formed by these vectors?

Focus on two of the vectors. They extend from the cube corner to the centers of adjacent faces. They define a plane that "lops off" a corner of the cube. The *trace* of this plane on the cube consists of three face diagonals; the plane and the chosen primitive vectors define a tetrahedron with an edge length that is equal to the face diagonal. Therefore, the three primitive vectors mutually form 60° angles.

Exercise 1-15. Examine Figure 1-32 once more and focus on the three edges of the primitive cell that meet at the center of the front face. What are the angles among the three primitive vectors corresponding to these edges?

Each face of the primitive cell is a rhombohedron. Reference to the previous exercise shows that the rhombohedron consists of two equilateral triangles, and therefore possesses two 120° angles as well as two 60° angles. Inspection of Figure 1-32 shows that the primitive vectors of interest form two 120° angles and one 60° angle.

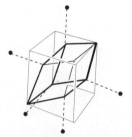

Figure 1-32 The important face-centered-cubic lattice has a primitive cell that is rhombohedral and somewhat difficult to visualize. In this drawing, the face-centered points are emphasized by dotted lines orthogonal to the corresponding faces. (After Warner and Grung [21].)

Exercise 1-16. All of the cubic unit cells exhibit twofold and fourfold symmetries about a number of obvious axes. Is there an axis about which any (or all) of them exhibit threefold symmetry?

Yes. All of the cubic lattices exhibit threefold symmetry about any cube diagonal. An experiment with a child's block confirms this for the simple cubic and fcc cases. For the fcc case, refer again to Figure 1-32, and observe that the primitive cell exhibits axial threefold symmetry.

1-5.4 Relating Lattices and Crystals

Let us move now from the mathematical abstraction of a lattice to the concrete reality of a crystal. The simplest crystal contains atoms of a single element, and places individual atoms at positions in space that can be defined by means of a lattice. For example, single-crystal tungsten has the body-centered-cubic (bcc) form. Single crystals of the precious metals—platinum, gold, silver—adopt the face-centered-cubic (fcc) structure, as do their relatives such as nickel and copper.

The next step up the ladder of complexity still deals with an elemental (that is, a single-element) crystal, but associates two atoms with each lattice site. One that is easy to visualize is the particular hexagonal structure adopted by zinc, cadmium, and magnesium. Refer to the hexagonal prism depicted in Figure 1-31(a), and imagine an atom of zinc at each lattice site; then place an additional zinc atom at the center of one of the six triangular prisms that are shown in Figure 1-31(a). Do the same for two additional alternate triangular prisms, for a total of three added atoms. The resulting structure (after some adjustment of prism height to make the spherical atoms fit nicely) is termed close-packed hexagonal (cph), and represents one of the options for stacking layers of hard spheres, such as marbles. (The other option— and the only other—is a face-centered-cubic arrangement, somewhat unobviously.) Thus we now have *two* atoms associated with a given lattice site. In the case of the atom placed in the center of a triangular prism in Figure 1-31(a), we can select as the "partner atom" one of those associated with the base of that triangular prism, choosing any one of the three. Having made the choice, we then consider that *pair* of atoms as being associated with *one* site of the hexagonal lattice. The entity thus associated with a lattice site, in this case a pair of identical zinc atoms, is termed the *basis* of the crystal. In an earlier example, such as the bcc tungsten crystal, a *single* atom of tungsten can be taken as the basis. (Some writers require a basis to be more complex than a single atom, but we prefer an inclusive definition that permits a single-atom basis.)

Additional insight can be gained by extending the addition of atoms in Figure 1-31(a) to the centers of adjacent prisms that are not shown. When this is done, one can see that the added atoms themselves conform to a simple-hexagonal lattice. As a result, we find that the zinc *structure* can be described as a *pair of interpenetrating simple-hexagonal lattices*. The term *structure* has been emphasized here because the atomic configuration just described is *not* a "zinc lattice," but rather a zinc crystal or a crystal having the zinc structure. Precise use of these terms should be practiced to avoid unnecessary confusion.

As a further step toward complexity, consider the case of sodium chloride, NaCl, or common table salt. Sodium has an atomic number of 11. Reference to Figure 1-24 shows that as a result, sodium normally possesses one electron in the M shell. This places sodium, an alkali metal, in column I of the periodic table, which contains the other elements, including hydrogen, that possess one *outer-shell* electron. These elements, as noted above, are very reactive chemically.

The element chlorine, with atomic number 17, has seven electrons in its M shell. Figure 1-24 shows that this is one too few electrons to complete a subgroup. Argon, with $Z = 18$, has that subgroup completed, and as a result, is chemically inactive. But chlorine, a *halogen,* is extremely reactive. When sodium and chlorine are combined in a crystal, each exhibits an *ionic* property. The sodium atom, having partially lost its lone outer-shell electron, acts like a positive ion; the chlorine atom, having partially acquired its "eagerly sought" eighth outer-shell electron, acts like a negative ion. The consequent coulomb forces give the resulting solid its rigidity, and give us an example of an *ionic crystal*.

As a starting point in sketching the NaCl crystal, construct a cube and place a pair of Na atoms diagonally opposite on one face; on the opposite face, place another pair of Na atoms at the ends of the face diagonal that is not parallel to the first. Then place Cl atoms at the remaining four corners of the cube. Extend this structure until a unit cell is represented, involving eight cubes of the original size. Next, focus on one particular atom, say a Cl atom. Equidistant from it along the orthogonal axes are positioned six Na atoms that are closer to it than any other atoms of the crystal. These six atoms are termed *nearest neighbors*.

The NaCl crystal is only one of several very similar *alkali-halide* crystals, with other examples being potassium chloride, KCl, and potassium bromide, KBr.

Exercise 1-17. What lattice is involved in the NaCl crystal? Give a verbal description of the basis.

The lattice corresponding to the NaCl crystal is fcc. The Na atoms conform to an fcc lattice and so do the Cl atoms. Choose one of the two arrays (Na or Cl), and select a particular atom in that array. Then the basis can be taken as the selected atom of the first type and an atom of the second type displaced away from it by half the unit-cell edge dimension, and along any of the orthogonal directions. Since this can be done by going in either direction along any of the three orthogonal axes, there are six ways to choose the second atom of the basis.

Exercise 1-18. In view of the small cube just described in introducing the NaCl crystal, why is NaCl not "simple-cubic"?

On four sites of the small cube are Na atoms, each with six Cl atoms as nearest neighbors. On the other four sites are Cl atoms, each with six Na atoms as nearest neighbors. Thus the small cube incorporates two different "environments," or configurations of neighboring atoms, and this violates the lattice definition, which permits only one.

Because of the two environments of the NaCl crystal, each involving a different species, this crystal also invites description as a pair of interpenetrating lattices, this time fcc. The two lattices are displaced from one another along the orthogonal axes by an amount equal to half the unit-cell edge dimension.

There are numerous other examples of binary, or two-component, crystal structures. One interesting set involves pairs of metals, and this time a pair of interpenetrating simple-cubic lattices. Each atom is at the center of a cube of atoms of the opposite kind, and thus has eight nearest neighbors. Examples are crystals of aluminum and nickel, of silver and magnesium, and of copper and zinc; the last is known as β *brass.*

As an example that is still more complex, consider calcium fluoride, CaF_2. This time the basis consists of one Ca atom and two F atoms, and once again the relevant lattice is fcc. The placement of the atoms, inferred mainly from x-ray examination of the crystal, forms a rather simple three-dimensional pattern, involving spacings that are multiples of a quarter of the unit-cell edge dimension.

Crystal complexity can grow enormously, but not without limit. First, the number of lattices is finite, and so is the number of elements. When we start combining elements to form complex molecules, the number of possibilities obviously increases hugely. But not every possible molecule is a candidate to serve as basis in an arbitrary lattice, because it turns out that a basis must have certain symmetry properties in common with the lattice involved to make a crystal possible—this restriction being imposed in addition to any chemical restrictions on crystal formation by given molecules or atoms.

From the examples given it is evident that crystals involving cubic lattices—especially the fcc case—are fairly numerous. This is fortunate from the student's point of view because of the simplicities inherent in the cube. Also, the several references to *cube-edge length* imply that for real crystals we are indeed concerned about absolute dimensions. The cube-edge dimension for any cubic crystal is termed the *lattice constant,* and is typically a few angstroms. Thus, crystal studies differ from those concerning an abstract lattice, because in the lattice case, applying a scale factor produced nothing new nor interesting; in short, absolute dimensions were of minor concern there. In treating less regular lattices, we applied more than one scale factor without producing a new result. For example, compare Figures 1-26(c) and 1-27(e), where scale factors were applied to two length dimensions, with no significance so long as degeneration was avoided. Further, with the same proviso, a factor could be applied to the angle θ without producing a new result. But in the physical world, each dimension and angle will be important. Table 1-3 at the end of the chapter summarizes the differences between *lattice* and *crystal*. The two entities are quite distinct, so their labels should be used carefully and precisely.

Another distinction between the lattice and crystal realms is that a lattice is by definition perfect. Any realizable crystal, by contrast, contains imperfections, or *defects*. These come in enormous variety. Foreign atoms are a common example. In silicon crystals, other elements are intentionally introduced to fix electrical properties. If this step is omitted, there still exist other *impurities* that even the most advanced purification methods are unable to eliminate, although impurity concentrations have been reduced to levels that are extremely low in the thrust toward the refinement of solid-state electronics.

If we visualize an elemental crystal containing *no* foreign atoms, it can still have defects in the form of missing atoms, or *vacancies*. Also it can contain extra or *interstitial* atoms. Returning to the case of foreign atoms for a moment, we note that they can also be interstitial, or else *substitutional,* which means holding a site in place of a primary atom.

All of these imperfections constitute *point defects*. But one can also observe a *line defect*. A relatively clear-cut example is an *edge dislocation*, which can be visualized as the edge of an extra plane of atoms inserted part way into an otherwise perfect crystal. Dislocation lines need not be straight, however. Often they wander in three dimensions. There are x-ray methods that can image such imperfections, and they sometimes present the appearance of a tangled mass of yarn.

Even if all the real-life possibilities are eliminated in a thought experiment, however, the fact that a crystal is finite introduces massive defects, namely, surfaces. In the progression above, a surface can be classified as a two-dimensional defect in the event that it is plane, as it often is. A surface constitutes a sharp discontinuity in the nearly perfect periodicity of a single crystal. Surfaces, especially silicon surfaces, have been objects of intense scrutiny for decades and are now reasonably well understood. It was, in fact, the study of crystal surfaces that led in almost direct fashion to the invention, or perhaps discovery is a more appropriate term [22], of the point-contact transistor. The phenomena that cause the point-contact transistor to function are still not understood, but the device is so archaic and so inferior to the transistors treated in later chapters that workers today have little incentive to unravel the remaining mysteries.

1-5.5 The Silicon Crystal

Once again it is the fcc lattice that is relevant; a two-atom basis is applied and the result is the arrangement of atoms depicted in Figure 1-33 [23]. This structure is remarkable in a number of ways. In the fcc unit cell represented, it is evident that atoms A, B, C, and D, for example, reside either at a unit-cell corner or face center.

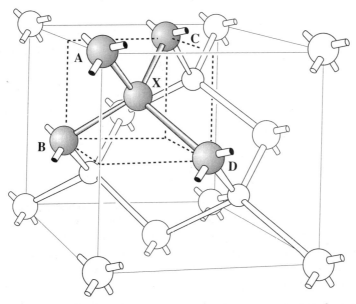

Figure 1-33 The unit cell of silicon. Its lattice constant is 5.43 Å. (After Shockley [23], with permission.)

But there are also atoms such as X that do not. Further scrutiny of Figure 1-33 shows that upon extending the structure, one will encounter equal numbers of atoms in the two categories (on and off fcc lattice sites). But still closer examination of the structure reveals that the off-site atoms themselves are located at the sites of *another* fcc lattice. In short, the silicon structure can be described as a pair of interpenetrating fcc lattices (constituting another elemental example this time). The opening statement above about an fcc structure and a two-atom basis really conveys the same information, but approaching the matter through the concrete geometrical representation of Figure 1-33 imparts more explicit understanding. In particular, it shows *how* the two lattices relate to one another. Let the atom X be considered the corner of a unit cell of the second lattice. This is a totally arbitrary designation, of course; because the points of a lattice are equivalent, *any* point may be designated as the corner of a unit cell. But the choice of atom X makes the relationship of interest between the two lattices clear.

Refer again to the silicon-crystal unit cell depicted in Figure 1-33. For a cubic unit cell, as noted previously, the edge dimension of the unit cell is an important characterizing quantity, and is known as its *lattice constant*. For silicon, the value is $a_0 = 5.43$ Å. The unit cell depicted in Figure 1-33 can be divided into eight smaller cubes. The atom X is located precisely at the body center of one of these smaller cubes. In other language, it is at the midpoint of a diagonal of the smaller cube. Thus the displacement of one fcc lattice from the other is half a small-cube diagonal.

Next focus on the atoms A, B, C, and D. They are situated at four of the eight corners of the same small cube, and they define the four corners of a *tetrahedron,* a solid with four faces that are equilateral triangles. Let us digress briefly to point out that the tetrahedron is one of the five *Platonic solids,* also sometimes called *Archimedean solids.* These are polyhedrons for which the faces are congruent regular polygons. The other four members of the series are the *regular hexahedron,* with six squares as faces (the cube); the regular octrahedron, with eight equilateral triangles as faces (two "Egyptian pyramids" base to base); the dodecahedron, with twelve regular pentagons as faces; and the icosahedron, with twenty equilateral triangles as faces. (The regular dodecahedron is found on many desk tops because its twelve faces invite its use as a calendar.)

The tetrahedral relationship of the atoms in the small cube, with the atom X at the center of the tetrahedron, has great significance, and consequently the structure depicted in Figure 1-33 is often characterized as a *tetrahedral structure.* The significance of *four* in this connection is that the tetrahedral structure is favored by certain atoms from column IV of the periodic table (Table 1-1). In particular, carbon, with $Z = 6$, adopts the tetrahedral structure when it is in *diamond* form. For this reason, yet another term frequently applied to the atomic configuration depicted in Figure 1-33 is *diamond structure.* Silicon, with $Z = 14$, adopts the same structure, as we have seen. So too does germanium, with $Z = 32$; next comes tin with $Z = 50$ in a crystalline form known as *gray tin.*

The explanation for the regularity exhibited in Figure 1-33 lies in the bonding mechanism. Each atom in the tetrahedral or diamond structure has four nearest neighbors. Also, each atom from column IV has four outer-shell electrons. By *sharing* each of these electrons with one of its nearest neighbors, and by sharing equally one electron "belonging to" each of its nearest neighbors, a given atom (and every other atom in the crystal) can achieve the "magic-eight" complement of electrons in its vicinity.

It is valid to think of each of the "sticks" in Figure 1-33 as representing two electrons, shared by the atoms they connect. Each represents a *shared-electron* bond, also known as a *covalent* bond. Such a bond is extremely stable, accounting for certain physical properties of the resulting solid. A germanium crystal is a hard, brittle material. Single-crystal silicon is even harder. And carbon in the form of diamond is the hardest material known.

The fact that covalent bonds can impart hardness to a crystal can be appreciated in part by recalling an electron property mentioned in Section 1-4.6. There it was noted that an electron exhibits spin, and an associated magnetic moment—one Bohr magneton. In short, the electron is a miniature magnet. When two electrons are paired with their spins in opposite directions, as are the electrons in a covalent bond, they are strongly bound together. As an analogy, consider a pair of identical bar magnets placed side by side, north pole to south pole. They adhere so securely that appreciable force is required to pull one away from the other; the integral of that force from zero separation to large separation yields the work that was expended on one magnet to remove it from the other, or the *binding energy* of the pair of magnets in their initial proximity. In an analogous way, it is possible to "break" a covalent bond, or to separate one of the electrons from the other, but significant energy is required to do it. In Chapter 2 we examine ways this can be accomplished, and the consequences.

Now return to the crystal representation in Figure 1-33. This representation sometimes is carelessly called the "silicon lattice" instead of the *silicon crystal*. But *lattice* and *crystal* have totally distinct meanings (Table 1-3, located at the end of the chapter), so this incorrect usage should be avoided.

Exercise 1-19. The regularity of point positioning in the tetrahedral structure, Figure 1-33, leads one to ask why this point configuration does *not* constitute a lattice. Give a clear argument based upon Figure 1-33 that rules out lattice status for this point array.

The existence of two distinct environments in this point array can be shown as follows: When the directed line segment from the center of atom A to the center of atom X is extended an equal distance in the same direction, it terminates on the opposite end of a cube diagonal of the small dashed cube shown. But no atom is located at the far end of that diagonal. Therefore, points A and X have differing arrangements of neighboring points, so the point array is not a lattice.

1-5.6 Atomic Planes and Crystal Directions

Even a brief examination of the lattice representations in Figures 1-29 through 1-32 shows that the corresponding crystals will exhibit repeated parallel planes of atoms, and that the sets of planes exist with various orientations. A system for specifying such sets of planes was devised by W. H. Miller [24], an English crystallographer. His designations are termed *Miller indices*. They are simplest for the cubic cases, which happily are more important to us than any others. Therefore let us use a cubic example, specifically the fcc silicon case, to illustrate his approach.

Choose any silicon atom as origin and construct three mutually orthogonal axes upon it, orienting the axes along the unit-cell edges. Using the lattice constant of silicon, 5.43 Å, as a basic unit, calibrate all three axes. Then specify any plane of interest in terms of its intercepts on the axes. As an example, consider the plane depicted in Figure 1-34(a), having intercepts on the x, y, and z axes, of 1, 3, and 2 units, respectively. The Miller indices for this plane, then, can be found by determining the smallest integers standing in the same ratio as the intercept reciprocals. Since the respective reciprocals in the present case are 1, 1/3, and 1/2, the Miller indices for the plane of interest are 6, 2, and 3, written (623). Summarizing, let us say that the Miller indices of a plane having intercept values of m_x, m_y, and m_z, are the smallest integers h, k, and l that satisfy this proportionality statement:

$$h:k:l = \frac{1}{m_x} : \frac{1}{m_y} : \frac{1}{m_z}. \tag{1-90}$$

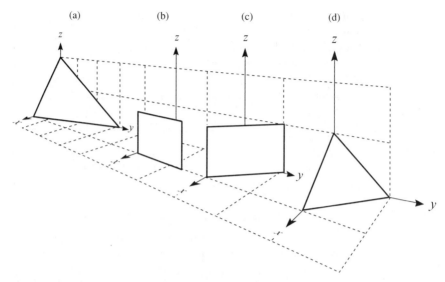

Figure 1-34 Representations of several planes in a cubic system. (a) The Miller indices for this plane are (623). (b–d) Representations of other planes, with Miller indices given in Exercise 1-20. (After Warner and Grung [21].)

It is evident that in applying the Miller-index definition, one must choose origin and plane so that the two do not coincide, because in such a case the intercept reciprocals are all undefined and the indices are as well.

Exercise 1-20. Find the Miller indices for the planes represented in Figures 1-34(b) through 1-34(d).

In Figure 1-34(b), the intercept reciprocals are evidently 1, ∞, ∞. Therefore we have in this case a (100) plane. Proceeding similarly, we obtain for Figures 1-34(c) and 1-34(d), respectively, the indices (110) and (111).

Writing Miller indices in parentheses, as has been done so far, is taken to designate either a particular plane or a full set of parallel planes. In the cubic system, however, a plane or set of planes parallel to *any* of the six faces of the unit cell must for obvious reasons be regarded as *equivalent planes*. When distinctions among the six sets are needed, they can be written (100), (010), (001), ($\bar{1}$00), (0$\bar{1}$0), and (00$\bar{1}$); a bar over an index indicates that it came from a negative intercept. If such distinctions are not necessary, the entire group of sets of equivalent planes can be designated by Miller indices written as {100}.

The perspective diagram in Figure 1-35 illustrates the three most important differing kinds of planes in a cubic crystal, using a somewhat more graphic kind of representation. It is possible to do experiments that produce results much like those depicted in Figure 1-35. Various reagents are capable of etching silicon. Some of these, usually the faster-acting etchants, tend to ignore the existence of atomic planes. Suppose a piece of silicon is exposed to one of these through a small circular opening in a protective layer of some sort that is not attacked by the etchant. In this case the resulting growing depression will tend toward a spherical-surface shape. Such an etchant is termed *isotropic,* with the term having the same meaning as in Section 1-3.4. In the etching context, the antonym for isotropic is (for some reason) *anisotropic*. Such an etchant, usually slow-acting, removes atoms that are more weakly bound in preference to those that are more strongly bound. As a result, the advancing surface develops facets that correspond to crystallographic planes.

In the silicon (or diamond, or tetrahedral) structure, the (111) planes possess unusually high atomic densities. This can be seen by referring to Figure 1-33 once more. Visualize the (111) plane containing the atoms A, B, and C. Note that the atom X and its relatives, equal in number to the atoms of the plane containing A, B, and C, form a second (111) plane that is very close to the first. The covalent bond that connects atoms X and D is normal to those (111) planes. The atom D, then, and the two other face-centered atoms shown form another (111) plane that is spaced away from the plane of atom X by a full bond length. There is yet another (111) plane (of which only one atom is shown) that lies very close to that of atom D. Thus the pattern of (111) planes is two closely spaced and tightly bound planes (observe the density of covalent bonds between them) followed by a relatively large space with a low bond density. As a result, it is possible to *cleave* (or break) a silicon crystal along a surface lying between two pairs of (111) planes. This procedure has

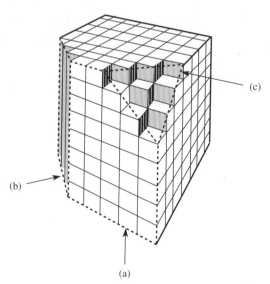

Figure 1-35 Examples of three important crystallographic planes in a cubic crystal: (a) (100), (b) (110), and (c) (111). (After Warner and Grung [21].)

been used extensively for converting a slice of silicon containing hundreds of integrated circuits into individual integrated circuits, or in an earlier time, into individual transistors. Note in Figure 1-35 that if the surface of the starting slice constitutes a (110) plane, then the broken (111)-plane edges will be orthogonal to the surface of the "chip" or *die,* as the individual smaller pieces are called; this orientation option is sometimes elected because right angles are desirable for automatic handling of these small bits of silicon,

For some purposes—for example, in characterizing silicon surfaces that constitute various crystallographic planes—it makes sense to treat the paired (111) planes as a single plane. On this basis, the areal density of atoms appreciably exceeds that of any other plane, and the paired planes do indeed appear to function as one in some situations.

Thus far we have mentioned only elemental crystals having the diamond structure. But it is equally possible for an element from column III of the periodic table, and one from column V, to adopt the closely related *zincblende* structure; the "magic-eight" total of outer-shell electrons can be achieved in this way as well as with all atoms from column IV. There are many such *compound* crystals of growing importance. Among them are gallium arsenide (GaAs), gallium phosphide (GaP), and indium antimonide (InSb). The paired (111) planes represent the two different environments that exist in the diamond structure; therefore, in a binary crystal—let us take GaAs as an example—in order to meet valence requirements, one of these planes consists solely of Ga atoms, and the other, of As atoms. (Inevitable defects involving out-of-place atoms will alter this regularity, but the density of such defects can be extremely low.) A curious result of this state of affairs is that when a slice of such a material (let us consider GaAs again) has surfaces that are (111) planes, one face will consist of Ga atoms, and the other, of As atoms!

Elemental crystals of the diamond structure and binary crystals of the zincblende structure do not exhaust the possibilities, however. In addition to III-V compounds, there are also II-VI compounds, such as cadmium sulfide (CdS), and zinc sulfide (ZnS), also known as zincblende. The last, in fact, has given the name *zincblende structure* as a generic indicator of all of the two-component structures having covalent bonds and a point configuration like that of Figure 1-33.

The index system developed by Miller is also useful for indicating directions in a crystal. Square brackets are used to label a directional designation, so that [111] means "the direction normal to the (111) planes," a relationship that holds, however, *only* in a cubic system.

Exercise 1-21. In Figure 1-34, determine the Miller indices for the y axis and the z axis.

The plane normal to y axis is designated (010), so the y-axis direction becomes [010]. Similarly, the z-axis direction becomes [001].

As noted above, the orthogonality of a plane having a given index and the direction having the corresponding index is lost as soon as we depart from the cubic cases. Nonetheless, the intercept-index principles continue to hold for all but the hexagonal case, which requires more complicated definitions.

To conclude this discussion of crystallography, we introduce a term not used above, but one that is self-explanatory. Perfect crystals exhibit total *long-range order*. The materials that are devoid of long-range order are termed *amorphous* materials. Window glass is an everyday example. Thus crystalline and amorphous materials constitute opposite extremes. Between these extremes are *polycrystalline* materials. These are made up of aggregations of tiny "grains" that are individually monocrystalline, but that have random crystal orientations with respect to one another. Starting in about 1970, these materials began to assume major importance in solid-state electronics. They can be "doped" with impurities that enhance electrical conductivity (as is explained in Section 2-3), and these impurities move very rapidly along the grain boundaries at elevated temperatures. Consequently, polycrystalline silicon (in particular) is often substituted for a metal to realize an electrical interconnection. The result is a more nearly homogeneous device that carries certain advantages over the heterogeneous metal-and-semiconductor device.

Substantial interest focuses currently on yet another class of materials, termed *quasicrystals* [25], that appear to fall somewhere between the crystalline and amorphous classes. They do not possess a unit cell, thus failing a primary crystal test, but they do exhibit long-range order. Also they exhibit fivefold symmetry, a "forbidden" symmetry in classical crystallography. Interestingly, the enigmatic icosahedron, described in Section 1-5.5, also exhibits fivefold symmetry. The icosahedron is an "outcast" among the Platonic solids because it is unrelated to any of the space lattices, but it appears that quasicrystals possess icosahedral configurations. Thus even the discipline of crystallography, in spite of its relative antiquity, has not been so thoroughly canvassed that new possibilities are foreclosed. We offer this illustration because there have been so many occasions in the history of science and technology on which a person has declared an area "closed," only to learn a few months or years later that a new and significant path for investigation was developing.

SUMMARY

Electrical charges are the sources of an electric field. A mythical line of force extends from a positive charge to a negative charge of equal magnitude, with the common magnitude being a matter of arbitrary definition. A charge placed in an electric field experiences a force, and the field, with suitable precaution, can be defined as force per unit charge. Work is the integral of force through distance, with positive work being done *on* an object. When a single point is taken as the origin for calculating both spatial position and electrostatic potential ψ, the value of the latter function at a point of interest can be defined as the negative integral of electric field from the origin to the point of interest. In the electrical context the potential energy of a charge Q is given by the product $Q\psi$. Its kinetic energy is, as usual, $\frac{1}{2}Mv^2$, where M is the mass of the object carrying the charge Q. A conservative system is one in which the sum of the potential and kinetic energies has a constant value.

The central region of a charged parallel-plate capacitor is a region of uniform electric field, in the sense that the lines of force involved are straight, parallel, and equally spaced. The number of lines of force emerging from the positive plate is proportional to the magnitude of the uniform internal field. Taking the spatial x axis to be normal to the plates, one has an example of a one-dimensional problem. Electrostatic potential in this example is a one-dimensional function, depending as it does upon x but not upon y or z.

A problem is converted to normalized form by dividing each variable by a fixed quantity having identical units. The normalizing constants can be arbitrary or can be physical quantities of special significance. A physical problem is converted into a purely mathematical problem by normalization, a step that often improves efficiency in computer solutions.

There are five steps in solving analytic problems of the kind that are of interest here. (1) Draw a diagram and write an equation (or equations). If arbitrary choices or definitions are needed, indicate them on the diagram. (2) Solve for the unknown using *symbols*. (3) Substitute into the resulting expression the appropriate numbers *and* units. (4) Carry out any necessary unit conversion using a unity factor, which is a fraction having the same physical quantity in numerator and denominator. (5) Perform the numerical calculation. In these procedures, variable-symbol and unit-symbol equations must be clearly separated. The predominant system of units in semiconductor work is a pure-RMKSA system, except that the centimeter is used in lieu of the meter.

When any homogeneous material fills the space within a parallel-plate capacitor, electric field can be calculated from applied voltage divided by plate spacing. Positive charge moving in the direction of the field (or negative charge moving in the opposite direction) constitutes positive conventional current. Electric field is related to the density of this conventional current (A/cm^2) by a constant of proportionality called resistivity, ρ, for which the units are ohm·cm. (The reciprocal of resistivity is known as conductivity, σ.) This equation is a form of Ohm's law, one which is valid at a point, or through a volume with uniform, one-dimensional features. A material with no flowing charge in the presence of an electric field (an insulator) can still exhibit charge separation, or *polarization,* which causes enhanced charge storage on capacitor plates when such a material resides between them. The ratio of such charge storage to that existing with a vacuum between the plates is termed the *dielectric constant,* or relative dielectric permittivity, of that material. The quantity *permittivity,* in turn, is the constant of proportionality that relates the charge per unit area on the capacitor plates, known as *electric displacement,* to the electric field between them. The rate of change of displacement is known as *displacement current.* Some materials, silicon being a notable example, possess *both* conductivity and polarizability. The product of resistivity and permittivity for a particular medium is *dielectric relaxation* time, a measure of the time required for a charge deposited in the medium to disappear. Poisson's equation in a one-dimensional situation informs us that the slope of the electric-field profile is proportional to volumetric charge density, with reciprocal permittivity being the constant of proportionality.

The Bohr model for the hydrogen atom postulated that an electron in orbit about a proton, or hydrogen nucleus, can exist in a nonradiative stationary state, and that transitions between allowed states are accompanied by the emission or absorption of a photon of energy equal to the difference in the energies of the initial and final states. Further, the electron in its allowed circular orbit possesses an angular momentum that is an integral multiple ($n = 1, 2, 3, \ldots, \infty$) of Planck's constant divided by 2π. The integral multiple is the principal quantum number. Further refinements of the model defined three additional quantum numbers. Each set of four (allowed) quantum numbers determines a state that can accommodate only *one* electron in an atom. Each value of n defines a set or *shell* of closely spaced allowed electron energies; the average energy in each shell increases with n.

The *lattice* abstraction is an infinite array of mathematical points, each having the same "environment" as neighboring points. The possibilities are limited to a single one-dimensional case, five two-dimensional cases, and fourteen three-dimensional cases. The last, or space-lattice, case serves to describe a *crystal*, in which one or more atoms (the basis) is placed in the same relationship to each point of a lattice within a finite volume of space. The silicon crystal can be described as involving a pair of interpenetrating face-centered cubic (fcc) lattices. The same crystal structure can be assumed by germanium and by carbon in the form of diamond; for this reason such a crystal is often said to have the *diamond structure*. In it, each atom is effectively at the center of a tetrahedron defined by its four nearest neighbors. In spite of the high degree of regularity that thus exists in the structure, its points do not constitute a lattice because half of the tetrahedrons "point" in one direction, and the other half, in the opposite direction. These two collections of tetrahedrons correspond to the two fcc lattices just cited.

Table 1-1 Periodic Table†

Period	Group IA	Group IIA	Group IIIB	Group IVB	Group VB	Group VIB	Group VIIB	Group VIII		
1	H 1 1.01									
2	Li 3 6.94	Be 4 9.01								
3	Na 11 22.99	Mg 12 24.31								
4	K 19 39.10	Ca 20 40.08	Sc 21 44.96	Ti 22 47.90	V 23 50.94	Cr 24 52.00	Mn 25 54.94	Fe 26 55.85	Co 27 58.93	Ni 28 58.71
5	Rb 37 85.47	Sr 38 87.62	Y 39 88.90	Zr 40 91.22	Nb 41 92.91	Mo 42 95.94	Tc 43 (99)	Ru 44 101.07	Rh 45 102.90	Pd 46 106.4
6	Cs 55 132.90	Ba 56 137.34	La 57 138.91	Hf 72 178.49	Ta 73 180.95	W 74 183.85	Re 75 186.2	Os 76 190.2	Ir 77 192.2	Pt 78 195.09
7	Fr 87 (223)	Ra 88 (226)	Ac 89 (227)	Th 90 232.04	Pa 91 (231)	U 92 238.04	Np 93 (237)	Pu 94 (242)	Am 95 (243)	Cm 96 (247)

Table 1-1 Periodic Table† (continued)

Period	Group IB	Group IIB	Group IIIA	Group IVA	Group VA	Group VIA	Group VIIA	Inert Gases
1								He 2 4.00
2			B 5 10.81	C 6 12.01	N 7 14.01	O 8 16.00	F 9 19.00	Ne 10 20.18
3			Al 13 26.98	Si 14 28.09	P 15 30.97	S 16 32.06	Cl 17 35.45	Ar 18 39.95
4	Cu 29 63.54	Zn 30 65.37	Ga 31 69.72	Ge 32 72.59	As 33 74.92	Se 34 78.96	Br 35 79.91	Kr 36 83.80
5	Ag 47 107.87	Cd 48 112.40	In 49 114.82	Sn 50 118.69	Sb 51 121.75	Te 52 127.60	I 53 126.90	Xe 54 131.30
6	Au 79 196.97	Hg 80 200.59	Tl 81 204.37	Pb 82 207.19	Bi 83 208.98	Po 84 (210)	At 85 (210)	Rn 86 (222)
7	Bk 97 (247)	Cf 98 (251)	Es 99 (254)	Fm 100 (253)	Nd 101 (256)	No 102 (254)	Lw 103 (257)	

The Rare Earths

Ce 58	Pr 59	Nd 60	Pm 61	Sm 62	Eu 63	Gd 64	Tb 65	Dy 66	Ho 67	Er 68	Tm 69	Yb 70	Lu 71
140.12	140.91	144.24	(147)	150.35	151.96	157.25	158.92	162.50	164.93	167.26	168.93	173.04	174.97

†Two numbers are associated with each element symbol; the first number (an integer) is atomic number Z, and the second number is atomic weight.

Table 1-2 Physical Constants

Quantity	Symbol	Value
Angstrom	Å	10^{-8} cm = 10^{-4} μm
Atmospheric pressure	—	1.013×10^5 N/m²
Avogadro's number	N	6.022×10^{23}/mol
Bohr magneton	—	1.165×10^{-29} V·s·m
Bohr radius	r_0	0.5292 Å
Boltzmann's constant	k	1.381×10^{-23} J/K = 8.618×10^{-5} eV/K
Charge of electron	q	1.602×10^{-19} C
Electron volt	eV	1.602×10^{-19} J
Gas constant	R	1.987 cal/mol·K
Ionization energy, hydrogen	ξ_{ion}	13.6 eV

Table 1-2 continued on page 81

Table 1-2 (*continued*)

Mass of electron	m	0.9110×10^{-30} kg
Mass of proton	—	1.673×10^{-27} kg
Micrometer	μm	10^{-6} m $= 10^{-4}$ cm $= 10^4$ Å
Permeability, free space	μ_0	$4\pi \times 10^{-7}$ V·s/A·m
Permittivity, free space	ϵ_0	8.854×10^{-14} F/cm $= 8.854 \times 10^{-12}$ F/m
Planck's constant	h	6.626×10^{-34} J·s
Planck's constant/2π	$\hbar$	1.054×10^{-34} J·s
Speed of light in vacuum	c	2.998×10^{10} cm/s $= 2.998 \times 10^8$ m/s
Thermal voltage	kT/q	0.02586 V at 300 K
		$= 0.02566$ V at 297.8 K ($= 24.8$ C)
Thousandth of an inch	mil	25.4 μm

Table 1-3 Important Distinctions Between the Terms *Lattice* and *Crystal*

A lattice is:	A crystal is:
• Abstract	• Real
• Infinite	• Finite
• Perfect	• Imperfect
• An entity in which absolute dimensions are unimportant	• An entity in which absolute dimensions are important

REFERENCES

1. J. C. Maxwell, *Treatise on Electricity and Magnetism*, Cambridge University Press, 1873.

2. E. M. Pugh and E. W. Pugh, *Principles of Electricity and Magnetism*, Addison-Wesley, Reading, Mass. 1960.

3. G. P. Harnwell, *Principles of Electricity and Electromagnetism*, McGraw-Hill, New York, 1983.

4. J. D. Jackson, *Classical Electrodynamics*, Wiley, New York, 1962.

5. P. E. Gray and C. L. Searle, *Electronic Principles: Physics, Models, and Circuits*, Wiley, New York, 1969, p. 46.

6. N. H. D. Bohr, "On The Constitution of Atoms and Molecules," *Philos. Mag.*, **26**, 1 (1913).

7. E. Rutherford, "[The Scattering of Alpha Particles By Metal Foil]," *Philos. Mag.*, **21**, 1 (1911).

8. G. Holton, "Niels Bohr and the Integrity of Science," *American Scientist*, May–June 1986, p. 237.

9. Harnwell, p. 533.

10. M. K. E. Planck, 1901. See M. J. Klein, "Max Planck and the Beginning of Quantum Theory," *Archive For the History of the Exact Sciences*, **1**, Springer, New York, 1962, p. 459.

11. A. Einstein, "[Photoelectric Effect]" *Ann. d. Phys.*, **27**, 132 (1905).

12. F. K. Richtmyer and E. H. Kennard, *Introduction To Modern Physics*, McGraw-Hill, New York, 1947, p. 205.

13. H. Semat, *Introduction to Atomic and Nuclear Physics*, Rinehart & Co., New York, 1958.

14. G. Joos, *Theoretical Physics* (translated from the first German edition by I. M. Freeman), Hafner Publishing Co., New York, 1934.

15. Harnwell, p. 292.

16. C. S. Barrett, *Structure of Metals; Crystallographic Methods, Principles, and Data*, McGraw-Hill, New York, 1952.

17. R. W. G. Wyckoff, *Crystal Structures*, **1**, 2nd ed., Wiley-Interscience, New York, 1963.

18. A. Nussbaum, *Applied Group Theory For Chemists, Physicists and Engineers*, Prentice-Hall, Englewood Cliffs, N.J., 1971.

19. M. L. Frankenheim, *Die Lehre von der Kohäsion*, Breslau, 1835.

20. A. Bravais, *J. de l'École Polytech.* (Paris), **XIX**, 1 (1850); **XX**, 101 (1851).

21. R. M. Warner, Jr. and B. L. Grung, *Transistors: Fundamentals for the Integrated-Circuit Engineer*, Wiley, New York, 1983; reprint edition, Krieger Pub. (P. O. Box 9542, Melbourne, Fl. 32902-9542), 1990, p. 97.

22. Warner and Grung, p. 23.

23. W. Shockley, *Electrons and Holes In Semiconductors*, Van Nostrand, Princeton, N.J., 1950, p. 6.

24. W. H. Miller, *A Treatise on Crystallography*, Cambridge University Press, 1839.

25. D. R. Nelson, "Quasicrystals," *Scientific American*, **255**, 42 (1986).

TOPICS FOR REVIEW

R1-1. What are the *sources* of an electric field?
R1-2. What is a *vector*? Give an example of a physical vector quantity.
R1-3. What is a *uniform* electric field?
R1-4. Name a practical method for achieving a uniform electric field.
R1-5. What is the value of q (two significant figures)?
R1-6. Give a basic definition of *electric field*.
R1-7. What direction is the field-caused force on a positive charge?
R1-8. What caution is necessary when a *test charge* is employed?
R1-9. How is *positive work* defined?
R1-10. What is the RMKSA unit for *work*? For *energy*?
R1-11. Give a fundamental definition for an increment in *electrostatic potential*.
R1-12. State a relationship between electric field and electrostatic potential.
R1-13. Describe *lines of force*.
R1-14. What features of a lines-of-force picture are related to electric field? How?
R1-15. State the expression for (electrical) *potential energy*.
R1-16. What is meant by *conservative system*? Give an example.
R1-17. State the equation for *total energy* ξ.
R1-18. What is an electron volt? State its equivalent value in RMKSA units.
R1-19. Define *unity factor*.
R1-20. What is the first step in problem solving?
R1-21. What is the *last* step in problem solving?
R1-22. A variable symbol normally subsumes what entities?
R1-23. What is meant by *one-dimensional problem*?
R1-24. What is a convenient way to define electric-field algebraic sign in a one-dimensional problem?
R1-25. In a one-dimensional problem, how is electric field determined, irrespective of the homogeneous material involved?
R1-26. What is the definition of *positive conventional current*?
R1-27. What are the units for *current density, J*?
R1-28. What is *resistivity*?
R1-29. What are the units of resistivity?
R1-30. How is *conductivity* defined?
R1-31. State Ohm's law in terms of resistivity.
R1-32. What are the units of *permittivity*?
R1-33. Describe the space-charge conditions in a sample of *polarized homogeneous isotropic* material.
R1-34. What is the ordinary term for *relative dielectric permittivity*?
R1-35. What are the units of *electric displacement*?
R1-36. How is electric displacement related to electric field in the simplest situation?
R1-37. Which component of electric field (normal or tangential) is constant

through an interface between materials of differing dielectric constant? Which component of electric displacement?

R1-38. In view of the fact that the electric-displacement concept arose in the context of a dielectric material, how can a *displacement current* exist in a vacuum?

R1-39. What is the essence of the *dielectric-relaxation* phenomenon?

R1-40. What is the essential meaning of Poisson's equation? (Use a one-dimensional situation to explain.)

R1-41. What are the units of charge density ρ_v?

R1-42. What two landmark theories were combined by Bohr in his hydrogen-atom model?

R1-43. What accounted for the large-angle α-particle ''scattering'' explained by Rutherford?

R1-44. What key assumption involving displacement current was made by Maxwell and to what did it lead?

R1-45. What is the name of **B** and what are its units?

R1-46. What is the name of μ_0 and what are its units?

R1-47. Why is div **B** = 0?

R1-48. Under what condition is div **D** = 0?

R1-49. What results from charge acceleration? Explain.

R1-50. Name a device that exploits this principle.

R1-51. What wavelength falls in the approximate middle of the visible spectrum?

R1-52. *Maxwell's theory* combined what two previously separate disciplines?

R1-53. What is *black-body radiation?*

R1-54. What is the energy of a quantum according to *Planck's theory?*

R1-55. State *Coulomb's law* in words or symbols.

R1-56. What is a *central system?*

R1-57. What is the algebraic-sign convention for a Coulomb force?

R1-58. Where is the energy reference (or zero) placed in Bohr's model? Why?

R1-59. What is meant by *energy well?*

R1-60. Define *ion.*

R1-61. What is a *postulate?*

R1-62. State *Bohr's postulates.*

R1-63. State the basic equation of wave motion.

R1-64. What is meant by *light amplification* or *stimulated emission?*

R1-65. What is meant by *fluorescence?* by *phosphorescence?*

R1-66. What is meant by *ionization energy?* by *binding energy?*

R1-67. In what energy range does a *continuum* of energy states exist in the context of the Bohr model? Why?

R1-68. What was the key factor in gaining acceptance of Bohr's hydrogen model?

R1-69. What is meant by *reduced mass?*

R1-70. What is *atomic number?*

R1-71. What is a *deuteron?*

R1-72. What is the general meaning of the verb *to degenerate?*

R1-73. What are the MKS units of *angular momentum?*

TOPICS FOR REVIEW

R1-74. What is the *Bohr magneton?*
R1-75. What is the smallest finite orbital magnetic moment for an electron?
R1-76. What is the magnitude of the spin magnetic moment of an electron?
R1-77. Name the four quantum numbers and state the significance of each.
R1-78. What was the *de Broglie hypothesis?*
R1-79. How does the de Broglie hypothesis relate to an orbiting electron?
R1-80. State the *Pauli exclusion principle.*
R1-81. What is the subject matter of *crystallography?*
R1-82. Define *lattice.*
R1-83. State the number of possible lattices that are *one-dimensional, two-dimensional,* and *three-dimensional,* respectively.
R1-84. Explain *twofold, threefold, fourfold,* and *sixfold symmetry,* and give a geometric example of each.
R1-85. Name the two-dimensional lattices.
R1-86. What is meant by *symmetry in a line?*
R1-87. What is meant by *symmetry in a point?*
R1-88. Define *unit cell.*
R1-89. Define *primitive cell.*
R1-90. What are *primitive vectors?*
R1-91. State two synonyms for *three-dimensional lattice.*
R1-92. Which lattice has most relevance to solid-state electronics? What is its abbreviation?
R1-93. How many kinds of rotation, or axial, symmetry are exhibited by a cube? Identify one or more axes for each.
R1-94. Name the possible three-dimensional lattices.
R1-95. What is meant by *cph?*
R1-96. Is there a lattice of the same point configuration?
R1-97. What structure is favored by the precious metals?
R1-98. Is there a lattice of the same point configuration?
R1-99. What is meant by *basis?*
R1-100. Give an example of an *ionic crystal.*
R1-101. What is meant by *nearest neighbor?*
R1-102. Give an example of an *alkali-halide* crystal.
R1-103. What is meant by the *lattice constant* for a cubic crystal?
R1-104. What is a typical value?
R1-105. Give several examples of *point defects.*
R1-106. What is meant by the terms *vacancy, substitutional,* and *interstitial?*
R1-107. What is an *edge dislocation?*
R1-108. How do a *lattice* and a *crystal* differ?
R1-109. How can one determine the *Miller indices* of a given plane in a crystal?
R1-110. What is meant by equivalent planes in a crystal?
R1-111. What symbols denote a full set of equivalent planes?
R1-112. What is an *anisotropic* etchant?
R1-113. In the diamond structure, which planes possess the highest atomic density (atoms/cm^2)?
R1-114. What is meant by *zincblende* structure? Give an example.

R1-115. What is a *III-V* material? Give an example.

R1-116. What is a *II-VI* material? Give an example.

R1-117. Give an example of a *IV-IV* binary material.

R1-118. With respect to rotation about an axis normal to the lattice plane, what degree of rotational symmetry is exhibited by an oblique lattice when the axis passes through a lattice point? Through the midpoint of one side of the parallelogram? Through the center of the parallelogram?

R1-119. Is it possible to pass a plane through a cube in such a way that it cuts all six faces of the cube?

R1-120. If such a plane is the perpendicular bisector of a cube diagonal, what is the trace of the cube on the plane?

ANALYTIC PROBLEMS

A1-1. An electron is positioned at the negative plate of a parallel-plate capacitor and is released at $t = 0$. Use the negative plate as the origin (reference) for potential and position. Assume classical nonrelativistic dynamics.

a. Given plate spacing d, electron mass m, and applied voltage V, derive an expression for the acceleration of the electron.

b. What is the P.E. of the electron when it is at $d/2$?

c. Derive an expression for the time T it takes to reach the positive plate.

d. Given that $m = 0.91 \times 10^{-30}$ kg, $d = 1$ mm, and $V = 100$ V, find T.

A1-2. Given the situation shown below wherein an electron is projected at an angle between plane-parallel plates, find the entrance angle θ for which the electron will undergo maximum horizontal displacement before striking the lower plate.

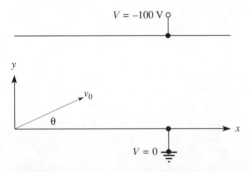

A1-3. An electron is injected with a kinetic energy of 10^{-17} J through a hole in the positive plate of a parallel-plate capacitor heading in a normal direction toward the negative plate. Take the positive plate as spatial and potential reference and assume one-dimensional conditions within the capacitor. The capacitor voltage is 100 V and the plate spacing is $d = 1$ cm.

a. Will the electron strike the negative plate?

b. Sketch the potential-energy profile within the capacitor.

c. When the electron is at $x = 0$, what are its values of K.E. and P.E.?

d. At what point are the K.E. and P.E. values interchanged?
e. What happens subsequent to the time corresponding to the condition in part (d)?
f. The sequence of events just outlined is sometimes described as a "collision by an electron with a potential barrier." Explain.
g. How does the situation change if the plate spacing is changed to $d = 1$ mm, and all other values are held constant?

A1-4. A parallel-plate capacitor has a plate spacing of 1 mm.
a. An electron starts at rest at the negative plate. If a voltage of 100 V is applied, how long will it take the electron to reach the positive plate?
b. What is the magnitude of the force that is exerted on the electron at the beginning and at the end of its path?
c. What is its final velocity?

A1-5. Using unity factors, convert 60 Hz to billions of cycles per year.

A1-6. The capacitor shown is in a vacuum. An electron is projected through a hole in the left-hand plate in the manner indicated below with a velocity of $v = 10^8$ cm/s, and then strikes the right-hand plate. Take the circuit-ground potential as the potential reference.

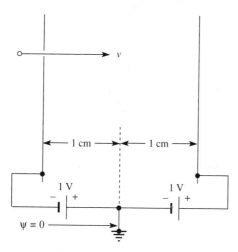

a. As the electron passes through the hole, what is its potential energy, P.E.$_i$, in electron volts?
b. Its kinetic energy, K.E.$_i$, in electron volts?
c. Its total energy, ξ_i, in electron volts?
d. Just before the electron strikes the right-hand plate, what is its total energy, ξ_f, in electron volts?
e. Its potential energy, P.E.$_f$, in electron volts?
f. Its kinetic energy, K.E.$_f$, in electron volts?

A1-7. An airborne dust particle of mass M and charge Q experiences an initial acceleration when an electric field E is switched on.

a. Derive an expression for its acceleration a.

b. Calculate its acceleration in pure-RMKSA units, given that its mass $M = 1~\mu g$, charge $Q = 10~q$, and that the field E is produced by applying 1000 V to a pair of parallel plates with a separation of $d = 1$ mm.

A1-8. An object of mass $M = 1$ g is at rest on a frictionless horizontal surface. A constant horizontal force F acts on its producing the acceleration $a = 10$ cm/s². The force acts for the time T necessary to move the object a distance $x = 10$ cm. Find its energy in joules at the time T. [This is an exercise in unit manipulation and Newtonian basics.]

A1-9. Using unity factors, convert 60 mi/hr to furlongs per fortnight. (A furlong is one-eighth mile.)

A1-10. Give the correct symbol for the pure-RMKSA unit corresponding to each of the physical entities listed in the table below:

Physical Entity	Symbol for Entity	RMKSA-Unit Symbol
Charge	Q	
Potential	ψ	
Electric field	E	
Capacitance	C	
Time	t	
Distance	x	
Force	F	
Energy	ξ	
Mass	M	
Acceleration	a	

A1-11. An electron is projected with a kinetic energy of 10^{-18} J through a hole in the positive plate of a vacuum-dielectric parallel-plate capacitor heading in a normal direction toward the negative plate. Take the negative plate as potential and spatial reference with the x axis directed toward the other plate; assume one-dimensional conditions. The capacitor voltage is 5 V and the plate spacing is 1 cm. Calculate the electron's P.E. and K.E. in eV when

a. $x = 1$ mm.
b. $x = 5$ mm.
c. $x = 1$ cm.

A1-12. A silicon resistor has a cross section of 1 mm² and a length of 1 cm. Find its resistance, given that the conductivity of the silicon is 2/ohm·cm.

A1-13. The resistor of Problem A1-12 is carrying 10 mA.
a. Find the electric field in the silicon.
b. Where are the charges that are the sources of the electric field in part a?
c. Find the voltage applied to the resistor.

A1-14. Find the charge per unit area at the contacts to the resistor of Problem A1-12 when it is in the condition of Problem A1-13.

A1-15. You are given the silicon resistor below with resistivity $\rho = 1$ ohm·cm. With a certain voltage applied, there is an electric field in the silicon of -5 V/cm. For silicon, $\epsilon \approx 1$ pF/cm.

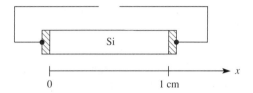

a. Calculate the (conduction) current density J with due regard for algebraic sign.
b. Calculate the applied voltage. Sketch a battery symbol with correct polarity in the circuit diagram above.
c. The resistance $R = 100$ ohm for this resistor. Calculate the cross-sectional area of the resistor.
d. Calculate the magnitude of the charge per unit area on the ohmic contacts.
e. Calculate the displacement-current density J_D, with due regard for algebraic sign.

A1-16. Treating the contacts as a pair of parallel plates, calculate the charge per unit area on the plates for the conditions of Problem A1-13, but with the silicon removed and replaced by air. (Let the plates be expanded in area until the problem becomes one-dimensional.)

A1-17. The four-region volume shown below has $\epsilon = 4\epsilon_0$ throughout. In its four respective regions it has these conditions.

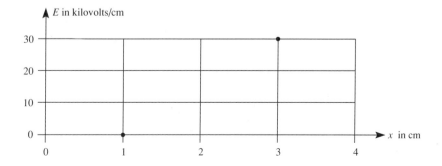

1 and 4:   volumetric charge density = 0
2:   volumetric charge density = ρ_v
3:   volumetric charge density = $2\rho_v$

Electric field has the value $\mathbf{E}(1 \text{ cm}) = 0$, and $\mathbf{E}(3 \text{ cm}) = 30{,}000$ V/cm. Consider the problem to be accurately one-dimensional. All lines of force terminate on a metal plate located at $x = 4$ cm.

a. In the space provided, plot electric field for all four regions carefully and quantitatively.
b. Calculate ρ_v.
c. Calculate the areal charge density on the metal plate.
d. In the top part of the diagram, sketch lines of force consistent with the given conditions. State any arbitrary quantitative assumption you make.

A1-18. The resistivity of ideally pure silicon at room temperature is about 2.8×10^5 ohm·cm. Calculate the dielectric relaxation time for such a sample.

A1-19. A ramp voltage of 100 V/min is applied to an ideal capacitor starting at $t = 0$. The capacitor has $C = 1$ μF, $d = 1$ μm, and a dielectric constant of 8.
a. Calculate the magnitude of the charge per unit area on either capacitor plate at $t = 10$ s.
b. Calculate the displacement-current density at $t = 10$ s.

A1-20. Examine the device and situation of Figure 1-14. With an electric field in the conducting regions, an electric displacement must exist there as well. Comment on displacement current in those regions.

A1-21. Consider a parallelepiped of material having resistivity ρ and permittivity ϵ. Ohmic contacts of area A are applied to opposite faces, and have a spacing of L. The ramp voltage applied as indicated has $V = 0$ at $t = 0$.

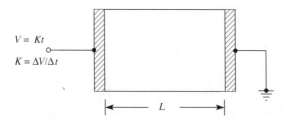

a. Derive an expression for $J(t)$, the conduction-current density as a function of time.
b. Derive an expression for $(dD/dt) \equiv J_D(t)$, the displacement-current density as a function of time.
c. Plot these two current-density functions in the space provided on the next page; calculate and state the significance of the time at which $J = J_D$, if indeed there exists such a time.

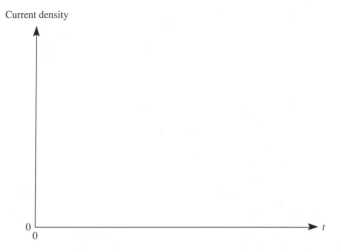

A1-22. Reexamine Figure 1-14 and the description of the response of that structure to a ramp voltage.
a. Will the charge behavior in that physical situation lead to radiation? Support your answer with an explanation.
b. If your answer to part (a) was positive, describe the resulting radiation.
c. If your answer to part (a) was negative, consider this observation: Carriers drifting in the conductor are brought to the static state when they reach a capacitor plate. Doesn't this charge deceleration inevitably lead to radiation?

A1-23. Consider the Bohr model of the hydrogen atom.
a. Determine the Bohr radius r as a function of the principal quantum number n. That is, find $r = r(n)$.
b. Calculate the radius r_1 of the electron in its ground state in the hydrogen atom, for which $n = 1$.
c. Calculate the ionization energy of an electron in the ground state, the energy required to drive the electron to $r = \infty$ and $\xi = 0$.

A1-24. Evaluate the Rydberg constant, applying any unity factors necessary to obtain the answer in m^{-1}.

A1-25. Do the following for the Bohr model of the hydrogen atom.
a. Sketch accurately on the *same* set of axes, P.E. versus r (radius), and ξ (total energy) versus r. Choosing a particular value of r, use an arrow to indicate the energy interval corresponding to K.E.
b. Offer an explanation for the algebraic signs of P.E., K.E., and ξ.

A1-26. As long as the physical dimensions of the system are much larger than the de Broglie wavelength, classical physics is applicable. Examine each of the following cases and determine whether the particle is classical or not. [*Hint:* Calculate velocity first.]

a. An electron accelerated through a potential of 10 V in a device whose dimensions are of the order of 1 cm.
b. An electron in the electron beam of a cathode-ray tube (anode–cathode voltage = 25 kV).
c. An electron in a hydrogen atom.
d. The de Broglie condition imposes a lower limit on electron energy for the valid application of classical physics. What imposes an upper limit?

A1-27. A photon of wavelength 1216 Å excites a hydrogen atom that is at rest.
a. Calculate the momentum imparted by the photon to the atom.
b. Calculate the energy corresponding to this momentum and imparted to the hydrogen atom.
c. How does the energy calculated in b manifest itself?
d. What is the photon energy?
e. Account for the difference between the values calculated in parts b and d—as one must in view of the energy-conservation principle.
f. Comment on the roles of the three entities—photon, electron, and proton—with respect to conservation of momentum and energy.

A1-28. A hydrogen atom, stripped of its electron, has a velocity magnitude in a vacuum of 10^6 cm/s.
a. Calculate its kinetic energy in eV.
b. The velocity vector of the particle in part a lies in the median plane of a parallel-plate capacitor. Write an expression for the potential energy of the particle in terms of its velocity magnitude v and its total energy ξ.
c. The potential difference between the plates is $\Delta\psi = 2$ V, and the more positive plate is taken as the potential reference. Calculate the potential energy of the particle.
d. Assume the more positive plate is then connected to the positive terminal of a 100-V battery, with the other battery terminal connected to earth ground and with $\Delta\psi$ remaining the same. Repeat part c, continuing to use the positive capacitor plate as reference. Explain your answer.
e. Assuming the conditions of part c, calculate the total energy ξ of the particle.
f. Repeat the calculation of part e for the case where the more negative plate is taken as potential reference.
g. Repeat the calculation of part e for the case where the velocity vector is normal to the median plane of the capacitor (but the particle still lies in the median plane). Explain your answer.
h. Is it necessary to consider the relativistic mass change of the particle in part a? Why?
i. If the particle experiences an electric field of $E = 40$ V/cm, calculate the plate spacing d in mm.
j. The plate spacing is doubled, but all of the conditions of parts a, b, and c are held constant. Repeat the calculation of part c. Explain your answer.
k. Imagine that a silicon atom is stripped of all its electrons, and that one electron is then returned to it. What would be the energy of the ground state that this lone electron would "see"?

A1-29. The antenna of an AM radio station radiates 10 kW at 1 MHz.
(a) What is the energy of each radiated quantum (photon)?
(b) How many quanta are radiated per second?

A1-30. Shown here are unit cells for the simple-cubic and simple-tetragonal space lattices. The only other tetragonal lattice is *body-centered*. Why is there no "base-centered-tetragonal" lattice?

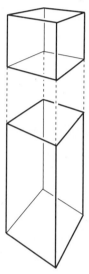

A1-31. Consider a face-centered-cubic unit cell formed by stacking layers of spheres of diameter D. Determine the *lattice constant*. (That is, determine the center-to-center spacing of a pair of corner spheres in the cubic unit cell.)

A1-32. Place identical hard spheres with their centers at the points of the following lattices, and calculate the *packing fraction* when they are packed as closely as possible. This is defined as the volume of the unit cell minus the volume of the void inside the unit cell, all divided by the volume of the unit cell.
a. Simple-cubic lattice
b. Face-centered-cubic lattice
c. Body-centered-cubic lattice

A1-33. There are four space lattices in the orthorhombic family. Why are there only three in the cubic family?

A1-34. How many vertical planes of mirror symmetry can pass through or touch a single unit cell in these cases?
a. The cubic lattices
b. The orthorhombic lattices

A1-35. Consider the close-packed-hexagonal (cph) structure. Suppose that this configuration is produced by stacking layers of identical spheres.
a. Sketch a plan view (top view) of three layers of the cph configuration. (Use distinguishing lines to represent spheres in the three layers or use the numerals 1, 2, and 3 to represent the centers of the spheres in the three layers.)

b. Explain with this or another diagram and with words why the cph structure does not qualify as a lattice.

c. If the top layer of points (or spheres) is rotated in its plane by 180 degrees about the center of one of the layer-2 spheres, the result is a configuration with a volumetric packing density (thinking in terms of spheres) that is equivalent to that of the cph structure, a value that is about 74%. Is the new configuration a lattice?

d. Explain your answer to part (c), using a sketch.

e. If your answer to part (c) was affirmative, identify the lattice. [See N. J. A. Sloan, *Scientific American*, **250** (January 1984), p. 116.]

A1-36. The fcc lattice can be regarded as a special case of another lattice.
a. Name the other lattice.
b. Explain, using a sketch.

A1-37. Visualize an axis normal to the plane of a two-dimensional lattice and passing through a lattice point.
a. Explain the meaning of *n-fold symmetry* about such an axis.
b. State the degree of symmetry (that is, give the value of n) for each of the five two-dimensional lattices with respect to rotation about a lattice point.

A1-38. Find the nearest-neighbor spacing (center-to-center distance) in the silicon crystal.

A1-39. Sketch a cube with one of its diagonals in a vertical position. Indicate on your sketch the positions of the four layers of hard-sphere centers that exist when the centers conform to a face-centered-cubic array.

A1-40. You are given a centered-rectangular two-dimensional lattice for which the unit cell is shown below.

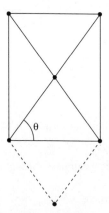

a. What values of the angle θ are disallowed if this is to remain a centered-rectangular lattice?

b. What results when θ takes on the disallowed values? [Be very specific in stating the result for *each* disallowed value of θ.]

c. State the term that labels the phenomenon described in part b.

A1-41. The most fundamental equation in electrostatics is Coulomb's law. In an unrationalized system of units, such as the cgs–esu system, it is written in the simplest possible form as $F = Q_1 Q_2/r^2$. In the RMKSA system, which is rationalized, Coulomb's law (Equation 1-61) is written $F = Q_1 Q_2/4\pi\epsilon_0 r^2$. What benefit is to be derived from rationalization (since it complicates the most basic equations)?

A1-42. An electron is projected on a normal path through a small hole in one plate of a vacuum-dielectric parallel plate capacitor, as shown below. The kinetic energy of the electron is 10^{-17} J as it passes through the hole at $t = 0$. At $t = t_1$, the electron passes through the median plane of the capacitor (the plane positioned at $x = d/2$). You may assume that a uniform field exists within the capacitor.

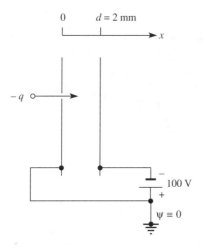

a. Calculate the total energy of the electron at t_1, or $\xi(t_1)$, in joules and electron volts.
b. Calculate the potential energy of the electron at t_1, or P.E.(t_1), in joules and electron volts.
c. Calculate the kinetic energy of the electron at t_1, or K.E.(t_1), in joules and electron volts.
d. With due regard for algebraic sign, calculate the force on the electron at t_1, or $F(t_1)$, in newtons.
e. With due regard for algebraic sign, calculate the force on the electron at $t_1/2$, or $F(t_1/2)$, in newtons.
f. Calculate the time t_2 at which the electron will strike the right-hand plate.

A1-43. Two point charges are positioned on the x axis with a spacing of d. The two-charge system is a very long way from any other charges. The charge Q_1 is positive, and $Q_2 = -2Q_1$. Demonstrate through a quantitative argument presented in *complete sentences* that a line of force departing from Q_1 in the negative-x

direction and tangent to the x axis will also be tangent to the y–z plane passing through Q_2, as indicated qualitatively in the diagram below.

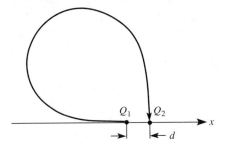

COMPUTER PROBLEMS

C1-1. A one-dimensional distribution of positive-charge density $\rho_v(x)$ in a material of permittivity ϵ has the half-Gaussian profile shown. The lines of force that originate on these positive charges terminate on an equal number of negative charges far to the right, outside the region of interest. We wish to plot the accompanying electric-field profile. The dimensions of ρ_v and ρ_0 are C/cm^3.

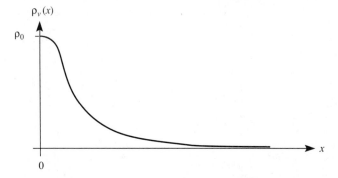

The starting point is the continuity equation known as Poisson's equation. For this one-dimensional problem.

$$(dE/dx) = \rho_v(x)/\epsilon. \tag{1}$$

The full-Gaussian function can be written

$$\rho_v(x) = \rho_0 \exp(-x^2/L^2). \tag{2}$$

Therefore,

$$(dE/dx) = (\rho_0/\epsilon) \exp(-x^2/L^2). \tag{3}$$

By separating variables and integrating, we can get the electric-field profile. By adjusting the limits of integration, we can introduce the desired half-Gaussian charge-density profile. Notice that the argument of the exponential factor is dimensionless, because both x and the constant L are lengths. The variable x is the spatial variable. The variable (*x/L*) is a *dimensionless spatial variable* that is *proportional to x*. It is desirable to convert to the dimensionless spatial variable in the balance of the equation as well. This can be done by multiplying both sides of the equation by *L*:

$$[dE/d(x/L)] = (L\rho_0/\epsilon) \exp(-x^2/L^2). \tag{4}$$

Note the dimensions of $L\rho_0/\epsilon$:

$$\frac{[\text{cm}][\text{C/cm}^3]}{[\text{C/V}\cdot\text{cm}]} = \frac{[\text{V}]}{[\text{cm}]}. \tag{5}$$

Because these are the dimensions of electric field, we can divide both sides of Equation 4 by $(L\rho_0/\epsilon)$, in the process converting electric field to a dimensionless variable proportional to electric field. Call it $\zeta \equiv E/(L\rho_0/\epsilon)$. Then,

$$[d\zeta/d(x/L)] = \exp(-x^2/L^2). \tag{6}$$

Next, as is often done, let λ be a dummy variable corresponding to (x/L), so that

$$\zeta = \int_0^{x/L} \exp(-\lambda^2) d\lambda. \tag{7}$$

Converting thus to dimensionless units is termed *normalization* as pointed out in Section 1-2.5, causing the problem and its solution to be independent of unit systems. Normalizing the charge-density profile given above as well yields this result:

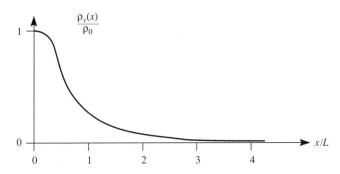

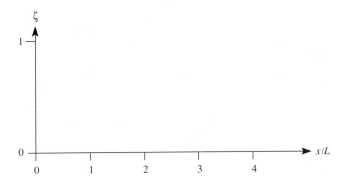

a. Write a program for integrating Equation 7 numerically, since it cannot be integrated analytically. As part of the solution of this problem, submit a well-documented and individualized printout of your program, and also of your numerical data in tabulated form.

b. The second part of the figure shows axes with normalized calibration for electric field versus distance. Find the electric-field profile in these terms. If you have access to a graphics capability for plotting the resulting curve, you may use it. Otherwise, plot the data manually.

c. The integral in Equation 7 is so important that it has been given a name and symbol. What are they?

C1-2. In a cubic crystal, a plane may be described by

$$hx + ky + lz = 1, \qquad (1)$$

where h, k, and l are the Miller indices.

a. Prove that the separation between two adjacent parallel planes (hkl) is given by

$$d = a/\sqrt{h^2 + k^2 + l^2}. \qquad (2)$$

b. Write a computer program to calculate d, given an arbitrary set of Miller indices. Assume that the lattice constant is $a = 5.43$ Å.

DESIGN PROBLEMS

D1-1. Design a 2-kilohm resistor that is to have a width of 1 μm, a thickness of 1 μm, and is to be fabricated from a material having a conductivity of 10^2/ohm·cm.

D1-2. In an ordinary parallel-plate capacitor with plate spacing d and $\rho_v = 0$ in the dielectric material, we have the condition $E =$ constant between the plates.

a. Design a parallel-plate capacitor that has

$$E(x) = Kx^2,$$

where K is chosen so that $E(0) = 0$ and $E(d) = 10$ V/cm.

b. Derive an expression for $E(x)$ versus x and plot it.

c. Derive an expression for $\rho_v(x)$ versus x and plot it.

d. Normalize the diagrams of parts (b) and (c).

e. Physically, how could one achieve the situation described in this problem?

D1-3. A parallel-plate capacitor is to have a two-layered dielectric region designed so that the *net* areal charge density at the interface between these two layers is in magnitude one-sixth that on either plate. One layer is to be silicon dioxide, SiO_2 (glass), for which $\kappa \approx 4$.

a. Determine the dielectric constant of the material needed for the second layer.

b. Choose an appropriate material.

c. Design a formula that gives the net areal density of interfacial charge as a fraction of total areal charge density (magnitude) on either plate.

D1-4. Design a cube of single-crystal material having the diamond structure and including one trillion unit cells.
a. How many unit-cell edges are arrayed along an edge of the cube?
b. Assuming the material is silicon, determine the edge dimension in cm and μm.

D1-5. Pretending that electron mass is a variable, design a hydrogen atom for which the first Bohr orbit (that for the electron in its ground state) is $r_1 = 1$ cm.

D1-6. Design a problem *and* solution on the subject matter of Chapter 1 illustrating the five steps in solving analytic problems.

2 Bulk Properties of Semiconductors

The term *bulk properties* refers to the attributes or qualities of the interior of a sample, well away from the complications and perturbations caused by sample surfaces. The earliest solid-state devices depended primarily on bulk phenomena, with surface effects playing only a secondary role. This was not entirely a result of chance. Mastery of silicon surfaces, in particular, both practically and theoretically, required almost the first two decades of the solid-state era. Even in the current decade, new insights on the surface problem are still being developed. It was the MOSFET that forced the issue—because its operation is *primarily* dependent upon surface phenomena, so a firm grasp on those effects was needed to convert the device from feasibility to practicality. Our discussion of surface properties will likewise be deferred until the MOSFET is addressed.

The term *semiconductors* arose because these materials in pure form possess conductivities roughly halfway (logarithmically) between those of good conductors and good insulators. This fact is not especially important, aside from supplying a convenient term, but semiconductors do have properties that are both distinctive and important, properties that we examine next.

2-1 ENERGY BANDS

In Section 1-4 we described the electronic structure of a *single* atom, unaffected by the proximity of any others. Now we turn to the case of a crystal containing N atoms, all of which interact in a real sense. The contrast between the two situations can be dramatized by noting that in the practical case, the minimum value of N is of the order of 10^{19} atoms, or ten million trillion atoms! It is the combination of this enormous number of atoms and a fixed set of electronic states associated with each atom that gives rise to *energy bands*.

2-1.1 Oscillator Analogies

Let us continue to focus on silicon. Because a silicon atom ($Z = 14$) has four outer-shell electrons, it is the lower states of the M shell that concern us—states that are clearly specified and identified in Figure 1-24. The fact that N atoms interact in a crystal means that *each* of the isolated-atom states of interest will be transformed into N states, and we have seen that N is a huge number. This phenomenon is known as energy-level *splitting*.

An intuitive appreciation of the splitting phenomenon can be derived from a familiar device. Consider the mechanical oscillators shown in Figure 2-1 [1]. Imagine the oscillating masses to be constrained by frictionless guides. Each device will exhibit a specific characteristic frequency that is a function of its mass and spring

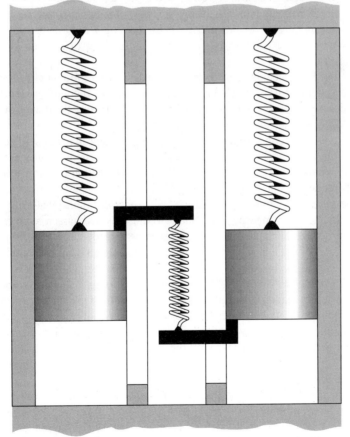

Figure 2-1 A pair of identical mechanical oscillators coupled by a weak, massless spring and illustrating energy-level "splitting." (After Warner and Fordemwalt [1].)

constant. If the two oscillators are identical and completely unconnected, they will exhibit identical characteristic frequencies. In this situation, oscillator frequency is the analog of the energy of an electronic state in an *isolated* atom.

Now let the two oscillators be coupled by means of a relatively weak spring, like that shown in Figure 2-1. Assume that this spring is massless and contributes only by lending additional "stiffness" to the system. The result is that the two-oscillator *system* now exhibits two characteristic frequencies, or modes of motion, or *states*. (An appealing feature of this analogy is having oscillator frequency correlate with state energy.) One of these states corresponds to oscillation in unison. Since the coupling spring was assumed to be massless, it plays no part in this case, and the characteristic frequency is identically the intrinsic oscillator frequency. In the second characteristic state, or mode of motion, the two devices oscillate in opposition. As a result, the added stiffness contributed by the coupling spring gives rise to a new frequency, higher than before. Here we have a splitting into two states because two coupled oscillators are involved. If we now imagine an increase in the stiffness of the coupling spring, we can "observe" a resulting increase in the degree of splitting. In an analogous way, bringing a pair of identical atoms from "infinite" separation (which may in practice be a fraction of a millimeter) into proximity causes a monotonic increase in the degree of splitting of a given isolated-atom electronic state.

The oscillator analogy can be pressed further. Let a third identical oscillator be introduced, and let it be coupled by means of an identical coupling spring to one of the others. The system will now exhibit *three* characteristic states, or modes of motion. In one state, there is unison oscillation and the intrinsic frequency is again displayed; in the second, next-higher state, the two outer masses oscillate in opposition and the center mass is stationary; and in the third, the two outer devices oscillate in unison and the center device oscillates in opposition to them, leading to the highest-frequency state.

The process continues. One can visualize a two-dimensional array of N such identical oscillators, in which case there would exist N characteristic frequencies. Furthermore, these frequencies or states must be regarded as a joint property of the assemblage, for none of the states can be assigned exclusively to individual oscillators.

Other analogies that illustrate energy-level splitting also exist, some of which lend themselves readily to experimental demonstration. One employs identical resonant LC circuits that can interact magnetically, one of which is "driven." An observation of amplitude versus frequency reveals a series of resonance peaks at close but different frequencies, with the number of peaks equal to the number of circuits. The differing resonance conditions are related to differing phase conditions of oscillation in the various circuits.

In the silicon crystal, the overall range of energies, highest to lowest, of the relevant electronic states is of the order of 1 eV. Into this range are packed the small number of lower outer-shell states, each multiplied by the huge number N, which typically falls in the range 10^{19} to 10^{21} atoms. So closely spaced are these manifold states resulting from splitting that they form a virtual *continuum,* and it is this energy-state configuration that has been given the apt label *band*.

2-1.2 Band Structure versus Atom Spacing

The degree of splitting increases with the degree of interaction, as the qualitative arguments just made have shown. On a purely theoretical basis, it is possible to calculate the width of an energy band and the band's position in energy as a function of the degree of atomic interaction. The measure of "interaction," of course, is atom spacing. Figure 2-2 [2] shows the result of such a *thought experiment*—a procedure that cannot be carried out in the real world, but can in the realm of imagination—and is illuminating as well. The abscissa is atom spacing, with large spacing represented at the right, which corresponds to the isolated-atom case. Each *subgroup*, however, is represented by a single line for the case of the isolated atom, because the states in a subgroup are so close in energy. Accordingly, the possible quantum-number combinations are listed at the right; they are identical to the listing

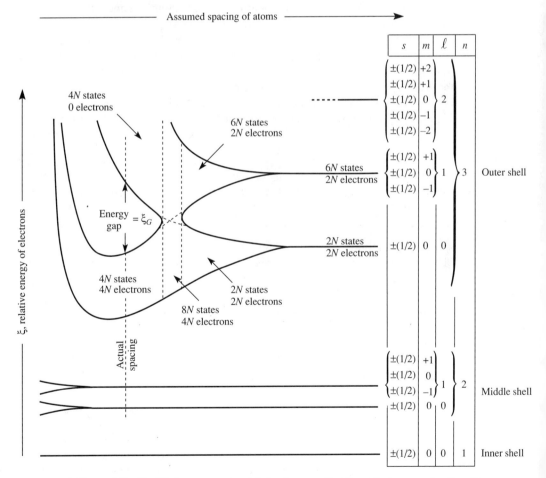

Figure 2-2 Qualitative energy-band behavior as a function of atom spacing for silicon and other diamond-structure materials. (After Warner and Grung [2].)

in Figure 1-24 but in opposite sequence. Unlike Figure 1-24, however, Figure 2-2 shows energy intervals and values that are only qualitatively valid. It has been constructed with silicon in mind; but with a relabeling of the shell structure, it could apply equally to other Column-IV elements that can assume the diamond structure.

The thought experiment is aided by visualizing a silicon crystal with a "knob" on it that controls atom spacing. Starting from large spacing, with the discrete energy levels displayed at the right, we can "turn the knob" and observe the onset of splitting. There are two bands initially. The lower one is formed by the $2N$ states contributed by the lowest subgroup of the M shell, and the upper one by the $6N$ states of the next higher subgroup, where N once more is the number of atoms in the crystal. At a certain value of diminished spacing, then, the two bands merge. Following this is a range of still smaller spacings for which there exists only a single band, containing $8N$ states. This band, it is important to realize, is half filled, containing the $4N$ outer-shell electrons contributed by all of the silicon atoms in the crystal. Before the bands merged (that is, at larger atom spacing), these $4N$ electrons were equally divided between the two initial bands.

Continuing to turn the knob toward smaller spacing, we find that the single band again splits into two. A significant change has occurred, however. There are now $4N$ states in *each* band. Thus, if all electrons occupy the lowest available states (as they do at low temperature), the $4N$ electrons in the crystal will *just fill* the lower band, leaving the upper band empty. At an atom spacing labeled *actual spacing* on Figure 2-2 are represented the conditions that exist in a real silicon crystal.

Exercise 2-1. Observe that the actual spacing does not occur at the spacing value that would minimize the energy of the electrons in the filled, or nearly filled, lower band. What might account for this fact?

The curves of Figure 2-2 give information only on *electronic* energies. At the actual spacing, however, there is significant nuclear repulsion in the picture as well. The actual spacing occurs at the condition that minimizes the energy of the *overall* crystal system.

Note also in Figure 2-2 that the middle-shell (L-shell) states show no splitting at the actual spacing. Those electrons are so tightly bound that they are unperturbed by the proximity of other atoms. Further, for the same reason, the K-shell states show no splitting at any indicated spacing.

It is important to appreciate the theoretical character of Figure 2-2. Experiments involving enormous hydrostatic pressures have been able to produce only small reductions in the atom spacings within real materials. The main message of the theoretical data is that states in the crystal can be related to states in the isolated atom, and that in a real crystal of pure silicon, two energy bands exist. The lower band is nearly filled by the available outer-shell electrons, and the upper band is nearly empty.

2-1.3 Relating Bands and Bonds

The energy-band picture for silicon is usually presented somewhat in the manner of Figure 2-3(a). One can think of this as the result of slicing through Figure 2-2 at the position labeled "actual spacing." Only two band edges are usually shown—the lower edge of the upper band and the upper edge of the lower band—because these edge regions contain the variable electron populations that are most significant. In Figure 2-3(b), a two-dimensional representation of the diamond structure shown three-dimensionally in Figure 1-33 is given. It is a conversion that is easily made, since the main requirement is that each silicon atom must have four nearest neighbors. The lines between atoms represent covalent bonds, of course, and in an even more literal way, let us consider each line to represent one electron.

The connection between the two representations is this: There would be $4N$ electron "lines" in a diagram like Figure 2-3(b) that was made large enough to represent the entire silicon crystal. This assumes the sample is perfect and at a low temperature, which is necessary in order to have all of the covalent bonds intact. These same $4N$ electrons are those that under the same conditions precisely fill the lower band in Figure 2-3(a). The band thus filled with electrons that are incorporated into covalent bonds is also termed the *valence band*.

Now let the crystal, still perfect, be taken to room temperature. The result is the presence in the sample of *thermal energy*, which partly takes the form of silicon-atom vibrations. There is a small but finite probability that enough thermal energy

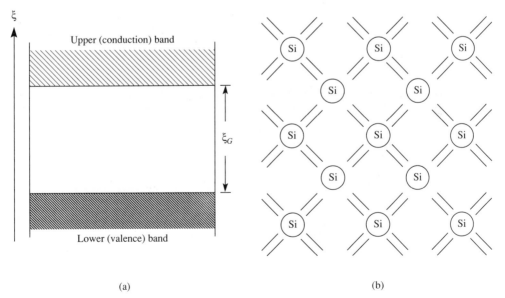

Figure 2-3 Relating energy bands and covalent bonds. (a) A band picture derived from Figure 2-2 by slicing the diagram in a direction normal to the paper at the actual spacing. (b) A two-dimensional representation of the silicon crystal that was shown three-dimensionally in Figure 1-33.

can converge on a covalent bond to eject one of the electrons making up the bond. In pure silicon at room temperature, far fewer than one bond in a trillion will be "broken" in this manner. Even though improbable, such an event is important. The electron thus ejected appears in the upper band and as a result is able to move about the entire crystal. It can respond to an electric field, thus contributing to conduction. For this reason, the upper band also carries the label *conduction band*. The conductivity resulting from bond breaking in pure silicon at room temperature falls in the intermediate range cited earlier that qualifies silicon as a *semiconductor*. The binding energy for an electron in a covalent bond (recall the pair of identical bar magnets) is approximately equal to the band-edge spacing in Figure 2-3(a).

An alternative description of the situation just outlined provides additional insight. Recall Figure 1-19, which was considered to define a funnel-shaped "energy well" that constrains the single electron of the hydrogen atom. In a similar way, the lowest states of the M shell of an isolated silicon atom can be regarded as situated within an energy well. The curve in Figure 2-4(a) is a two-dimensional representation of such an energy well; for silicon it has depth in excess of 8 eV. Each atom of a silicon crystal would be characterized by an identical well if that atom were isolated from all others. But because of atom proximity and mutual interactions, the energy barriers *between* atoms are appreciably lower than those of the isolated atom, as is illustrated in Figure 2-4(b). As a result, any electron that escapes over this lowered barrier and departs from its atom of origin can move throughout the entire crystal and become a conduction electron. In this picture, the bound electrons are those in the lower band that constitute the covalent bonds. The freed or conduction electrons are those that have enough energy to escape over the diminished interatomic barriers within the silicon crystal.

2-1.4 Electrons and Holes

As we have seen, the breaking of a covalent bond delivers an electron to the upper band and contributes to conductivity. But this is only half the story. The same event enables the valence band to contribute to conductivity. The opening in the valence structure left behind by the departing electron permits a net motion or "transport" of valence electrons.

A primitive analogy can contribute to understanding at this point. Visualize a line of automobiles parked bumper to bumper, with immovable barriers at either end of the line. No traffic is possible. But if one vehicle is removed from the middle of the line, then half the vehicles can move one car length. The automobiles that do the moving could be selected by having arranged to park the line of cars on a hill, which is to say, by having "applied a gravitational field." Now suppose the right-hand half of the line of cars moved leftward. By further assuming that instead of moving *en bloc*, they moved sequentially, we find a new way to describe the events that occurred. We can say that the opening created by removing a vehicle from the middle of the line has moved a step at a time from that position to the right-hand end of the line.

ENERGY BANDS

(a)
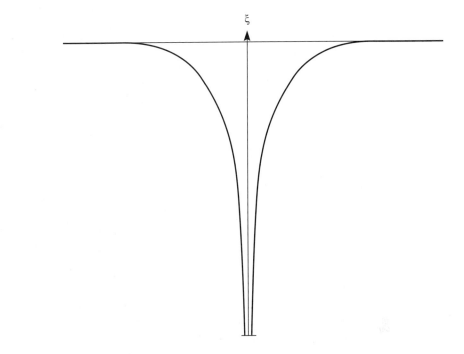

(b)
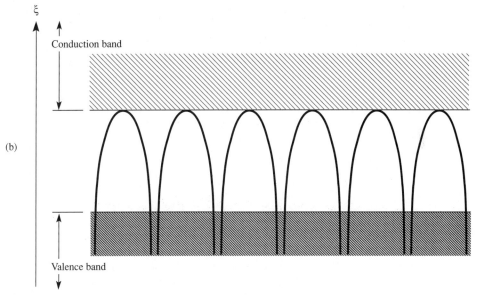

Figure 2-4 An alternative view of silicon's energy bands. (a) The energy well for the outer-shell electrons in an isolated silicon atom has a depth exceeding 8 eV. (b) Well depth is reduced to 1.1 eV (not drawn to scale) for atoms in a crystal because proximity diminishes the height of the interatom barrier. The resulting depth fixes the depth of the valence-band edge below the conduction-band edge.

Now return to the crystal of pure silicon and generate an analogous description. Let the field applied this time be a rightward-directed electric field. As a result, the electrons elevated to the conduction band from thermally broken covalent bonds will be caused to move leftward. So, too, will the valence electrons, which now have "moving space" because a few of them are missing, just as the absent automobile permitted some of the others to change position. But now we focus on the "opening" or missing electron in the valence structure and specifically designate it as a *hole in the valence band*. Under the assumed conditions, the hole will move to the right, in the direction of the electric field. The *hole* behaves, therefore, like a particle carrying a *positive* charge. Furthermore, its effective charge magnitude is q. This is true because a region of silicon with complete covalent bonds is neutral, while a region missing one valence electron lacks the charge $-q$ that would create neutrality, which is the same as saying that the region in question *exhibits* the charge $+q$. The seat of this positive charge, to be sure, is the nucleus of a nearby silicon atom; but as the hole moves about the crystal, the location of the net positive charge moves with it as well.

Reference back to Figure 2-4 may lead one to wonder how a valence electron can move from one atom location to an adjacent location in view of the significant potential barrier between the two sites. The answer is that it does not have to *surmount* the barrier, but can instead penetrate *through* the barrier by the phenomenon of *tunneling*. Recall that a wave function that occupies a certain volume of space yields the best description of an electron. The silicon-atom spacing is of the order of one angstrom—so small that the wave function for an electron associated with one pair of atoms has a nonnegligible value at the location of an adjacent covalent bond. In tunneling, the electron can instantaneously move from one side of a thin potential barrier to the other. Consequently, a given valence electron can very readily move—say, toward the left—to fill an adjacent hole; in the process, the hole moves toward the right. Having introduced the concept and description of the hole, from this point forward we shall use the terms *holes* and *electrons* to mean specifically holes in the valence band and electrons in the conduction band, respectively, unless accompanying modifiers or descriptions indicate otherwise. Both holes and electrons are said to be *carriers* of electricity.

Because the events described as "hole motion" really involve valence-electron motion, one can legitimately inquire why an electron description is not used. The hole in a sense is artificial. A good response to this question is provided by yet another analogy. Think about a bubble in a glass of liquid. It has a shape, and size, and if it is "free," it also has a velocity and acceleration. In short, it has the attributes of a "real" entity, even though it is the absence of liquid. In principle one could describe bubble motion by detailing the motion of adjacent liquid, but to do so would be far more difficult, and less concise as well. Similarly, descriptions of hole behavior are more convenient than descriptions of the equivalent valence-electron behavior.

The hole moves in an electric field in a direction opposite to that of the corresponding electron motion, and hence exhibits an opposite, or positive, *charge*. A bubble moves in the gravitational field in a direction opposite to the corresponding liquid motion, and hence exhibits an opposite, or negative, *mass*. These differences between the two entities of the analogy illustrate the hazards of pressing analogies too far. All analogies must break down at some point. If an analogy does not break down, it is not an analogy, but is the same phenomenon.

Before we leave this topic, however, it is worthwhile to mention a classic analogy offered by Shockley [3] that is related to the earlier automobile description. He likened the band structure of silicon to a two-level garage. When the lower level is parked full and the upper level is empty, no "traffic" is possible on either level. But let one or more vehicles be moved from the lower to the upper level, and then traffic becomes possible on *both* levels.

It is important to understand that a single event, bond breaking, leads to *two* kinds of conductivity, that of the electron and that of the hole. The electron and hole are created as a *pair,* and the phenomenon is termed pair *generation*. There also exists a reverse process known as *recombination,* in which the two particles disappear as a pair. We shall examine this process later, in Section 2-6. The thermal generation described earlier is only one possible mechanism for pair production. Another important mechanism involves the breaking of a bond by a *photon,* or quantum of electromagnetic energy. Any photon from the visible portion of the spectrum is capable of pair generation. In silicon the "cutoff" wavelength, beyond which photons have less than the required 1.1 eV of energy, falls in the near infrared, at approximately 1 μm.

Exercise 2-2. Devise ways of showing electrons and holes in both parts of Figure 2-3.

A small electron symbol, such as a minus sign, can be used in the upper band of Figure 2-3(a), or anywhere in Figure 2-3(b), to represent an electron. A small symbol, such as a "bubble," can be placed in the valence band of Figure 2-3(a) to represent a hole. In Figure 2-3(b), removing an "electron" from one of the covalent bonds can represent a hole.

2-1.5 Energy Gap

The minimum energy required to break a covalent bond is an important characteristic feature of materials in the category being considered. This amount of energy is equal to the interval between the band edges, labeled ξ_G on Figure 2-3(a), and is called the *energy gap*. Other terms with the same meaning are *bandgap* and *forbidden band,* the latter term arising because a pure, defect-free, and infinite sample has no electronic states in that energy range, and so the energy gap is "forbidden" to electrons. The silicon energy gap, approximately 1.12 eV at room temperature, exhibits a weak temperature dependence. For many problems one can ignore this effect without serious error.

Table 2-1 (at the end of the chapter) gives values for a series of other materials of both the diamond (elemental) structure and the zincblende (binary) structure. The populations of holes and electrons in pure samples of these materials decline very steeply with increasing energy gap; consequently, conductivity declines as well. There exist, therefore, fuzzy boundaries on the conductivity and energy-gap scales that separate the "semiconductor" materials from the *insulator* materials. An important accompaniment to this observation is that semiconductors and insulators differ *in degree*. Both exhibit an energy gap between a pair of bands. In both materials, at absolute zero the upper band is completely empty of electrons and the lower band is completely empty of holes. At room temperature, a semiconductor contains moderate numbers of electrons and holes in its respective bands, while the bands of an insulator remain essentially empty.

Carbon in the form of diamond, with an energy gap of approximately 5.5 eV, is normally an insulator. But at the very beginning of the solid-state era, it was noted that a diamond crystal can display a pulse of conductivity (a transient population of holes and electrons) created by an incident energetic particle, giving it utility as a particle detector [4]. This early observation was not merely a curiosity. Devices employing semiconductor diamond are now being seriously contemplated, especially for power applications.

2-1.6 Conductors

While the band structures of insulators and semiconductors differ in degree, that of a conductor differs from both *in kind*. A *conductor* possesses a conduction band that is *partly filled at any temperature*. Recall the hypothetical silicon crystal introduced in Section 2-1.2 that was equipped with a knob for adjusting atom spacing. In the range of spacings where the two bands had merged, this mythical crystal would have functioned as a conductor, possessing $8N$ electronic states and only $4N$ electrons.

Exercise 2-3. Where else in the theoretical diagram of Figure 2-2 are conditions satisfied that define behavior as a conductor?

In the region just to the right of the one just cited we have a band containing $6N$ states and $2N$ electrons. Thus this band is partly filled and would lead to metallic behavior if it had real existence.

A useful analogy for summarizing band-structure differences has been offered by Muller and Kamins [5]. Visualize a pair of closed vessels, as shown in Figure 2-5(a), representing the two bands of an insulator at room temperature. The lower band is completely filled with water, and the upper band is empty. Again, employ a

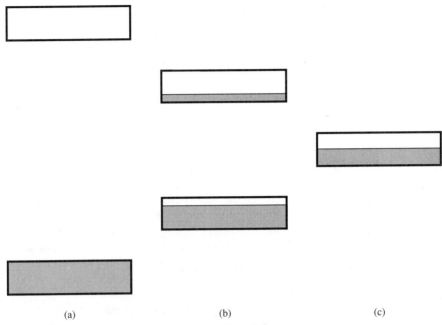

Figure 2-5 A band-structure analogy employing closed vessels that are filled with water, partly filled with water, or empty. The structures represented are those of (a) an insulator, (b) a semiconductor, and (c) a conductor.

gravitational field as the analog of an electric field; that is, tip the two vessels slightly. No water motion or "transport" occurs because of the conditions of filling of the two vessels. Next, consider the room-temperature semiconductor case represented in Figure 2-5(b). Let us lift up the right-hand ends of both vessels slightly. Water ("electrons") will move left in the upper tube, and the air space ("holes") will move right in the lower tube. But in the limit of falling temperature, these tubes ("bands") approach the conditions of the insulator tubes. Finally, then, the single vessel in Figure 2-5(c) represents the conduction band of a conductor. Tipping it will always cause water motion because it is *partly* filled with water. In an actual conductor, the population of outer-shell electrons is sometimes likened to a "gas" that fills the space within the sample surfaces.

2-2 ELECTRON DISTRIBUTIONS IN CONDUCTORS AND INTRINSIC SILICON

The outer-shell electrons in a solid arrange themselves in energy according to a well-established law. In particular, the distribution is temperature-dependent. Qualitative reference to this fact was made in previous sections; now our purpose is to examine the matter quantitatively. The expression that governs the energy distribution of electrons as a function of temperature was worked out in the mid-1920s independently and nearly simultaneously by Fermi [6] and Dirac [7], and is known as the *Fermi-Dirac probability function,* often shortened to simply *Fermi function.* In the sections immediately following, we assume that all samples considered are *isothermal,* or uniform in temperature, and are at *equilibrium,* or protected from all outside disturbances.

2-2.1 Fermi Level

There is yet another water analogy, given by Shive [8], that imparts a good grasp of the electron-distribution problem. For reasons that will be evident in a moment, let us first consider the conduction band of a metal or other good conductor. The analogy is drawn between the way electrons distribute themselves in energy within this conduction band at a particular temperature, and the way water will "distribute" itself with respect to height in a container that is being shaken or agitated. (Visualize the familiar device found in paint stores.) The severity of agitation is the analog of temperature, with no agitation corresponding to absolute zero. Since height in the analogy is proportional to the potential energy, *mgh,* of a drop of water, there is little effort involved in letting height be seen as the analog of electron energy in spite of the dimensional difference in the two quantities.

Figure 2-6(a) depicts the container of water and shows symmetric "waves" that correspond to one particular degree of agitation. Figure 2-6(b) shows a *probability function* that tells the likelihood of finding water at a particular height in the container. Near the top of the container, probability is near zero; near the bottom, there is a certainty, or a probability of unity, that water exists. Electrons in the conduction band of a metal behave similarly, as is heuristically indicated in Figure 2-6(c). Thus the probability function applies to both situations. The abscissa is probability P in both cases, but the ordinate does double duty—height h for water and energy ξ for electrons.

The function given by Fermi and Dirac has very simple form:

$$P(\xi) = \frac{1}{1 + e^{(\xi - \xi_F)/kT}}. \tag{2-1}$$

Note carefully that the energy origin is arbitrary, because this function depends upon an energy *difference*. Focus on the curve for $T = 300$ K in Figure 2-6(d). At this temperature, $kT \approx 0.026$ eV, so that with $\xi - \xi_F = 0.1$ eV, we find $P(\xi) \approx 0.02$, or 2%. For higher energies, the probability rapidly approaches zero. In a similar way, when $\xi - \xi_F \approx -0.1$ eV, $P(\xi) \approx 0.98$, and rapidly approaches unity for more negative energies. These values can be roughly confirmed by inspecting Figure 2-6(d), and thus the character of the probability function is approximately verified.

Of great importance is the energy value ξ_F. At this energy, the numerator of the exponential term in the denominator vanishes, so that $P(\xi_F) = 0.5$. Evidently the probability function exhibits symmetry in the point at which it intersects the energy ξ_F. The energy ξ_F is termed the *Fermi level,* which is defined simply as *the energy*

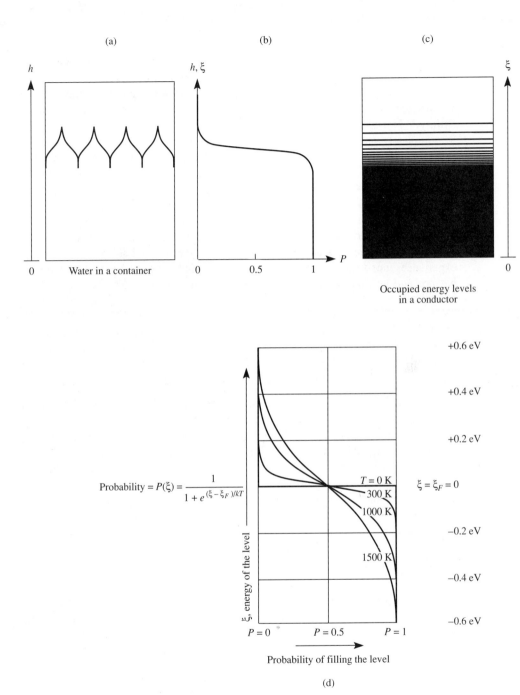

Figure 2-6 A water analogy illustrating the Fermi-level concept. (a) A container of water being agitated to form waves on the surface. (b) Probability of having water (electrons) present as a function of height (energy). (c) A representation of energy-level occupancy as a function of energy in a conductor. (d) Fermi-Dirac probability function plotted for four different temperatures. (After Shive [8], with permission.)

at which the probability of electron occupancy is one half. The water analogy of Figure 2-6 imparts unusually tangible meaning to the Fermi level. At $T = 0$ K, the analogous water container is unagitated, or still, and the top surface of the water is smooth and sharply defined. As agitation is increased, the water surface becomes "fuzzy." Nonetheless there still exists a sharply defined height at which $P = 0.5$. In completely analogous fashion, the Fermi level can be regarded as the "top surface" of the electron distribution in the conduction band of a metal, associated with a probability transition that is perfectly sharp at $T = 0$ K, and progressively less sharp as temperature increases.

We have made the point that the present discussion deals with equilibrium conditions. Beyond this, there is perhaps an implication in Figure 2-6 that we are describing static conditions for the electrons under discussion. But this is not so. The electrons being considered are in violent random motion, and a given electron experiences incessant energy changes. The curves in Figure 2-6 represent a dynamic balance and average of conditions in this seething context.

Exercise 2-4. Devise an analogy to illustrate dynamic equilibrium, and give at least two interpretations of *occupancy probability,* the information provided by the Fermi-Dirac probability function.

A recent student, Mr. Tuong Van Nguyen, after struggling with these ideas for several days, offered this illuminating analogy: Visualize a classroom with 100 chairs and with 10 students in violent random motion from one chair to another. If we focus on one chair, we will find that it is occupied 10% of the time (neglecting travel time); its occupancy probability is 0.1. A different, but fully equivalent description of occupancy probability is to say that a snapshot of the room would show 10% of the chairs occupied at that instant. Either way, we have here an occupancy probability of 0.1.

2-2.2 Density of States in a Conduction Band

The term *density of states* is intended to mean the number of available electronic states per unit energy interval and unit volume of solid as a function of energy. Reverting once more to the water analogy of Figure 2-6, one can say that a density-of-states inquiry is an inquiry into "the shape of the bucket." For the container represented in Figure 2-6(a), the density-of-states function would be rectangular. Near the bottom of the container, the volume per unit height is the same as at the top. In the case of a conduction band, however, the density-of-states function is approximately parabolic, vanishing at the bottom edge of the band. The analogous water container would be a round-bottomed kettle.

The parabolic approximation to the density-of-states function can be verified using a method developed by the pioneering investigator A. H. Wilson. He considered the possible wave functions assignable to an electron confined in an idealized solid sample. Among his simplifying assumptions were infinite energy barriers at the surfaces of the sample, so that the electron was contained in a rectangular potential well, and further, that the bottom of the well was flat and smooth (rather than being "bumpy" like the potential profile displayed in Figure 2-4).

Once the density-of-states function is established, one can multiply it by the function that determines the probability that each state is occupied, the Fermi-Dirac function. The result of this multiplication is a profile of the actual *electron distribution in energy*. These two functions and the result of multiplication are illustrated for a conductor in Figure 2-7. The fact that the Fermi level is far above the bottom of the band marks this conduction band as belonging to a metal or other good

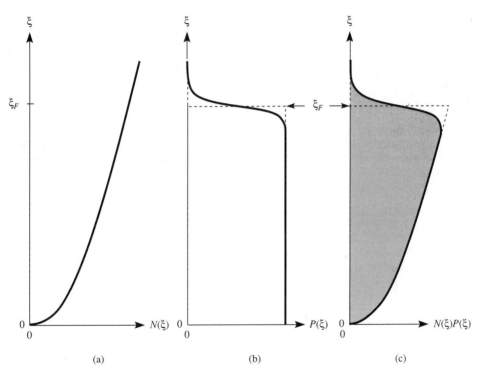

Figure 2-7 Functions determining the distribution of electrons in the conduction band of a metal. (a) The density-of-states function $N(\xi)$, approximately parabolic, giving the number of states for electrons per unit energy interval and unit volume of the solid. (b) The Fermi-Dirac probability function $P(\xi)$ for $T = 0$ K (dashed line) and for a higher temperature. (c) A plot of the product of $N(\xi)$ and $P(\xi)$ for the same two temperatures, yielding the distribution of conduction electrons as a function of energy. (The fact that this diagram applies to a conductor is shown by the position of the Fermi level, well above the bottom of the band.) Since the number of electrons present must be the same at both temperatures (that is, the shaded area under the solid curve must equal the corresponding area involving the dashed curve), it is evident that Fermi level is not very sensitive to temperature in a conductor.

conductor. Elementary considerations, then (greatly aided by the water analogy), indicate that Fermi level in a metal is only slightly temperature-dependent.

Exercise 2-5. Comment qualitatively on the sensitivity to temperature of the mean energy of electrons in the conduction band of a metal.

The mean energy of the electrons in the nonzero-temperature condition depicted in Figure 2-7 can be regarded as the energy position of the centroid of the shaded area. If the density-of-states function were rectangular, centroid position would be temperature-independent because of the symmetry of the Fermi-Dirac function. But because the density-of-states function is approximately parabolic, rather than rectangular, the mean energy of the conduction-band electrons will increase slightly with temperature.

2-2.3 Band-Symmetry Approximation

A silicon crystal that is defect-free and perfectly pure is, as we have seen, a fiction. It is nonetheless a useful fiction, because such a sample would exhibit the inherent or intrinsic properties of silicon and is therefore termed *intrinsic silicon*. This condition is an ideal that cannot be reached but can be approached, and it involves important features and concepts. Virtually all of the discussion of silicon crystals up to this point has implicitly assumed intrinsic samples.

The Fermi-Dirac probability function introduced in Section 2-2.1 is just as applicable to a semiconductor or an insulator as it is to a conductor. The useful water analogy, however, must either be set aside, or must be elaborated to a degree that is unprofitable. The difficulty is that the region of the bandgap must somehow be made "water-free" because it is electron-free; therefore, let us abandon the water analogy and deal directly with electrons in the solid, now intrinsic silicon.

Figure 2-8 shows the probability function plotted alongside a band diagram for intrinsic silicon. We have already noted the symmetry of the former. But the densities of available states in the neighborhood of the conduction-band edge are approximately equal to those near the valence-band edge, and in fact, are often assumed to be exactly the same. This assumed equality is known as the *band-symmetry* approximation. In view of probability-function symmetry and assumed band symmetry, it follows that the Fermi level must fall at the middle of the energy gap, or bandgap, as shown in Figure 2-8. To have the Fermi level fall elsewhere would mean that the number of electrons in the conduction band (recall the calculation involved in Figure 2-7) would be different from the number of holes in the valence band. Since the charges on these particles are equal and opposite, it follows that the sample in question does not exhibit overall neutrality, and that would be an abnormal situation.

ELECTRON DISTRIBUTIONS IN CONDUCTORS AND INTRINSIC SILICON   117

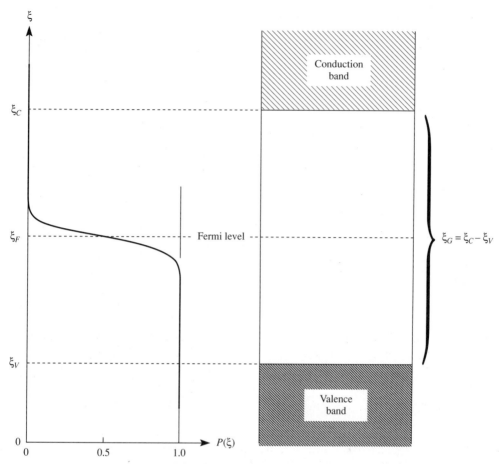

Figure 2-8 The Fermi-Dirac probability function plotted in its correct relationship to a band diagram for intrinsic silicon.

2-2.4 Equivalent-Density-of-States Approximation

The procedure for determining the distribution of electrons in the conduction band of silicon is identical to that discussed in Section 2-2.2 for the conduction band of a metal. A clear distinction in the two cases, however, is that in the case of a metal, the Fermi level is well above the bottom edge of the band, while in the case of intrinsic silicon (which we shall specifically assume), the Fermi level is well below the bottom edge of the band. Figure 2-9 shows a series of diagrams completely parallel to those of Figure 2-7, and a direct comparison of the two sets underscores the important differences of conductor and semiconductor. In Figure 2-9(a) is the density-of-states function $N(\xi)$ for the silicon energy bands. The conduction band is once again roughly parabolic, for the same reasons as before. The Fermi-Dirac probability function $P(\xi)$ is plotted in Figure 2-9(b) with the energy zero arbitrarily

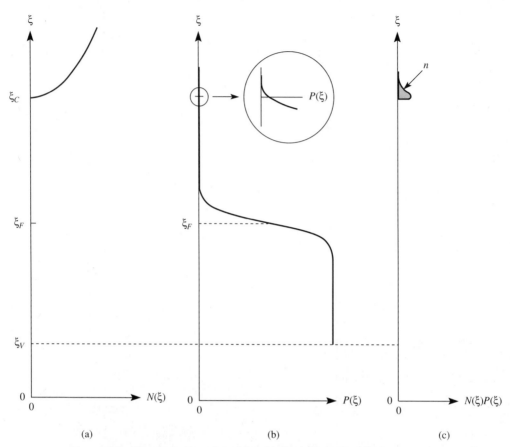

Figure 2-9 Functions determining the distribution of electrons in the conduction band of silicon (assumed intrinsic). (a) The density-of-states function $N(\xi)$, again approximately parabolic. (b) The Fermi-Dirac probability function $P(\xi)$ for room temperature. (A magnified picture of the function is given in the circle.) (c) A plot of the product $N(\xi)P(\xi)$ giving the distribution of conduction-band electrons as a function of energy. The shaded area is n, the density (number per unit volume) of electrons in the conduction band, found by integrating the product function with respect to energy.

placed at a position within the valence band. Note that for the intrinsic case being considered, the Fermi level falls at the middle of the bandgap. The edges of the conduction and valence bands are designated ξ_C and ξ_V, respectively. In the energy range of interest, the range at the lower portion of the conduction band, occupancy probability is extremely small, appearing to be zero in the diagram. In fact, the function $P(\xi)$ can be approximated with excellent accuracy as a simple declining exponential function in the region of interest. When the "strong" exponential function $P(\xi)$ is applied as a factor to the "weak" power law $N(\xi)$, the result is that the product function $N(\xi)P(\xi)$ rapidly approaches zero with increasing energy, as can be seen in Figure 2-9(c).

Our aim is to calculate the volumetric electron density in the conduction band, for which the usual symbol is n because of the electron's negative charge. As the symbol n indicating the small shaded area in Figure 2-9(c) suggests, the desired density is formed by integrating the function $N(\xi)\,P(\xi)$. Thus,

$$n = \int_{\xi_C}^{\infty} N(\xi)\,P(\xi)\,d\xi. \tag{2-2}$$

The electron density n is sometimes termed *electron concentration*.

Exercise 2-6. Write a unit equation corresponding to Equation 2-2.

The density-of-states function $N(\xi)$ describes the number of states per unit energy interval and unit volume, and $P(\xi)$ is dimensionless, so the unit equation becomes

$$\frac{1}{[\text{cm}^3]} = \frac{1}{[\text{eV}][\text{cm}^3]}\,[\text{eV}].$$

Because the function $P(\xi)$ plunges to zero so decisively as ξ increases, the conduction-band electrons are clustered very close to the band edge. This makes possible a simple approximate method for determining the density n. In this approach, the electrons are assumed to reside *right at* the band edge; that is, the function $N(\xi)\,P(\xi)$ in Figure 2-9(c) is replaced by a Dirac delta function positioned at the energy labeled ξ_C in Figure 2-9(a). One then employs a quantity N_C known as the *equivalent density of states*, which can be defined as

$$N_C \equiv \frac{\int_{\xi_C}^{\infty} N(\xi)\,P(\xi)\,d\xi}{P(\xi_C)}. \tag{2-3}$$

An important simplifying fact is that N_C is very nearly constant for wide-ranging conditions. Qualitative consideration of Figure 2-9 makes it evident that the equivalent-density-of-states approximation will break down if the Fermi level comes too close to the band edge. As a practical matter, the constancy of N_C is a good approximation as long as the Fermi level is spaced away from the band edge by an energy interval of more than $4kT$ (or about 0.1 eV), which thus permits the Fermi level to range widely through the central region of the bandgap.

Using N_C, the computation of n becomes simply a matter of evaluating

$$n \approx N_C P(\xi_C), \tag{2-4}$$

as can be seen by comparing Equations 2-2 and 2-3. In short, the evaluation of an integral has been replaced by multiplying a constant N_C by the Fermi-Dirac probability function evaluated at the conduction band edge, or $P(\xi_C)$.

Exercise 2-7. What are the units of the equivalent density of states N_C?

From Equation 2-4 it is evident that N_C has units of reciprocal centimeters cubed because $P(\xi)$ is dimensionless.

The function $N(\xi)P(\xi)$, pictured in Figure 2-9(c) and integrated in Equation 2-2 to find n, shows how electrons are distributed in energy in the conduction band. The energy position of the centroid of the shaded area—Figure 2-9(c) again—represents the mean energy of the conduction-band electrons. Letting $\xi_C \equiv 0$, this mean energy has the value $(3/2)kT$. At room temperature this is a small energy, about 38 milli-electron volts (meV), amounting to merely $(1/29)\xi_G$.

Understanding the significance of this mean energy and the status of electrons in the conduction band is aided by yet another analogy. It is a mechanical analogy—but very different from those presented earlier, and especially different from Figure 1-7—that illustrates a conservative system. This time let the electrons be represented by billiard balls on a billiard table. The balls are not permitted to leave the surface of the table, but are permitted to have wide-ranging velocities. Their potential energy is a constant, with a value depending upon the arbitrary choice of reference. If the table top is selected as reference, then the potential energy of each ball is zero and the total energy of each is its kinetic energy. Conduction-band electrons are situated similarly. An electron at the band edge is at rest. With increasing velocity, it is positioned higher and higher above the conduction-band edge. Using the *mean thermal energy* of $(3/2)kT$, one can calculate a (mean) *thermal velocity* v_t, since the electron's energy is now all kinetic.

Exercise 2-8. Calculate the thermal-velocity magnitude v_t of an electron in the conduction band of a silicon sample that is at equilibrium and at room temperature. Use $T = 297.8$ K as room temperature (for a reason that will be given in Section 2-2.5). Make any necessary assumptions and test their validities at the end of the calculation.

Let us assume the thermal-velocity magnitude is well below the speed of light, $c = 3.0 \times 10^{10}$ cm/s, so that relativistic effects may be neglected. Then

$$\tfrac{3}{2}kT = \tfrac{1}{2}mv_t^2,$$

or

$$v_t = \sqrt{3kT/m} = 1.1 \times 10^7 \text{ cm/s}.$$

Because this result is nearly three orders of magnitude less than c, the initial assumption was justified.

The Fermi level in intrinsic silicon falls, as we have just seen, in a region of the energy-band diagram where there are no states and no electrons. In spite of this fact, the meaning of Fermi level is unaltered. It is, first of all, an *energy position*. It tells us that if a state is located at the Fermi-level energy position, then that state will have an occupancy probability of 0.5. Similarly, it is worthwhile to emphasize that the function $P(\xi)$ does not itself inform us on *electron* or *state* distributions in energy, but instead tells us the occupancy probability of a state at a particular energy, *if* a state exists at that energy. Section 2-3 treats ways of introducing selected states into the bandgap of a semiconductor for the purpose of altering its electrical properties.

2-2.5 Intrinsic Carrier Density

Hole density in the valence band can also be given an equivalent-density-of-states treatment, yielding a density designated N_V. The conduction-band density N_C exceeds N_V by a small factor (approximately 1.76) because of differing effective masses for electrons and holes. The band-symmetry approximation introduced in Section 2-2.3 assumes these densities to be equal, an assumption that simplifies certain analyses appreciably and introduces errors that are usually not serious.

The only other step necessary is to realize that the probability of finding a hole in a state at a particular energy is given by the probability of *not* finding an electron there. Therefore, let us define a hole-occupancy probability:

$$P'(\xi) \equiv 1 - P(\xi). \tag{2-5}$$

The usual symbol for hole density in the valence band is p, chosen because the hole has a positive charge. Hence the desired result is found by evaluating

$$p \approx N_V P'(\xi_V). \tag{2-6}$$

Exercise 2-9. Give a geometrical interpretation of the probability function $P'(\xi)$.

The function $P(\xi)$ gives the probability of electron occupancy at any energy; it falls to zero above the Fermi level and rises to unity below the Fermi level. The function $P'(\xi)$ can be sketched as the reflection of $P(\xi)$ in a vertical line located at $P = P' = 0.5$. Alternatively, one can continue to use the curve $P(\xi)$ and reverse zero and unity on the probability axis to arrive at $P'(\xi)$. The function $P'(\xi)$ gives the probability of hole occupancy at any energy; it rises to unity above the Fermi level, and falls to zero below the Fermi level.

With these assumptions, the hole and electron densities in intrinsic silicon are equal, and the resulting density is important enough to be assigned its own symbol; the symbol n_i stands for *intrinsic carrier density*. Our best data indicate that n_i takes on the nice round value

$$n_i = 1.00 \times 10^{10}/\text{cm}^3 \qquad (2\text{-}7)$$

at a temperature of 24.8 C. Intrinsic density enters calculations so frequently that we have chosen to adopt this temperature as our standard *room temperature*. A value that is often used, 300 K, has nothing to recommend it except that it, too, is "a nice round number." It amounts to 27 C = 80.6 F, not an especially comfortable room temperature. The steep temperature dependence of n_i is illustrated in Table 2-2 (placed at the end of the chapter for convenient reference) by the fact that it has increased to $1.2 \times 10^{10}/\text{cm}^3$ at 27 C.

Exercise 2-10. Given the facts that for silicon $n_i = 1.00 \times 10^{10}/\text{cm}^3$ at 24.8 C, and that $\xi_G = 1.12$ eV, and using the approximations introduced earlier, calculate $N_C = N_V$ to three significant figures.

In intrinsic silicon, $n = p = n_i$. Using Equation 2-4, we can write

$$n_i = n = N_C P(\xi_C),$$

where

$$P(\xi_C) = \frac{1}{1 + e^{(\xi_C - \xi_F)/kT}}.$$

Assuming band symmetry, ξ_F is at the gap center, and $(\xi_C - \xi_F) = \xi_G/2$. Thus

$$N_C = n_i (1 + e^{\xi_G/2kT}).$$

From Table 1-2, $kT = 0.02566$ eV at 24.8 C, so that

$$N_C = N_V = 3.01 \times 10^{19}/\text{cm}^3.$$

Calculations such as this are extremely sensitive to small changes in values employed in the exponent. Nonetheless, this result agrees within a small factor with the best available values for N_C and N_V that were independently determined and that are based upon both experimental data and theoretical analysis.

The importance of the intrinsic carrier density is explained in Section 2-4.3, an explanation that requires additional basic concepts provided in the intervening sections. Also, the analytical tools available at that point provide a means for explaining the temperature dependence of n_i.

Exercise 2-11. The density of atoms in single-crystal silicon is 5×10^{22}/cm^3. Since each atom has four valence electrons, this yields 2×10^{23} valence states/cm^3. Why is the value $N_V = 3.01 \times 10^{19}$/cm^3 calculated in Exercise 2-11 so much smaller?

The quantities $N_C = N_V$ are the equivalent densities of *available* states. The states lying well above the conduction-band edge are almost certain to be empty, and those well below the valence-band edge are almost certain to be filled. In either case, such states are not "available."

2-3 IMPURITY-DOPED SILICON

The judicious introduction of impurity atoms on substitutional sites in an otherwise perfect silicon crystal produces useful modifications of its electrical properties. In particular, impurity elements from Column V favor electron (or *N*-type) conductivity, while those from Column III favor hole (or *P*-type) conductivity. The impurity densities involved are small, typically one part per million. At this low density, the impurity atoms can be regarded as distantly spaced from one another (by 100 atom spacings on the average). The fact that these impurity atoms are largely isolated from one another simplifies their analysis. The intentional introduction of impurities into silicon is termed *doping*.

2-3.1 Donor Doping and Hydrogen Model of a Donor State

The Column-V impurities that are most important in silicon technology are phosphorus, arsenic, and antimony. These can be seen in Table 1-1 in Column V (Group VA), just below nitrogen, and on the same horizontal lines as silicon, germanium, and tin, respectively. The most common Column-V dopant is phosphorus. It has been placed in Column V because it has five outer-shell electrons. When a phosphorus atom is introduced substitutionally into a silicon crystal, it has one more outer-shell electron than is needed to complete the covalent bonds with its nearest neighbors (which are silicon atoms). Although there is not room for the "fifth electron" in the ruling covalent structure that quickly absorbs the first four electrons, still the fifth electron may be bound to the phosphorus atom by the coulomb force associated with its negative charge and the "extra" positive charge (from the silicon viewpoint) that is located on the phosphorus nucleus, with the result depicted heuristically in Figure 2-10.

The idea of having an electron "bound to" or "in orbit about" a positive charge center of charge $+q$ (the phosphorus atom is locked tightly into its crystal position) calls forth the central idea in Bohr's model of the hydrogen atom. The similarity is so great, in fact, that applying Bohr's analysis to the case of a phospho-

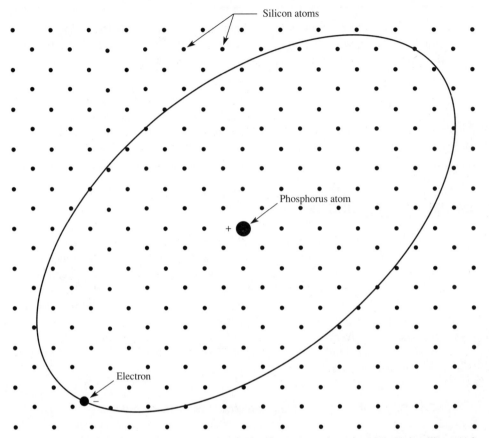

Figure 2-10 The hydrogen model of a donor impurity (phosphorus) in silicon (After Warner and Fordemwalt [1].)

rus atom in a silicon crystal yields a result that is surprisingly accurate. The most important difference in the two cases is that in the hydrogen-atom case, the two equal and opposite charges exist in free space, while in the present case, they are immersed in silicon. Because silicon is a dielectric material ($\kappa \approx 12$), the coulomb force between the pair of charges at a given spacing is smaller than it would be in free space. In Equation 1-62, Coulomb's law, we see that permittivity occurs in the denominator. As a result, the force between the charges is diminished by a factor of κ for the case of a phosphorus atom in silicon, and we can expect the present hydrogen-atom analog to have relatively large dimensions. It is for this reason that there exists an upper limit on the doping density for which the individual atoms are far enough apart to be considered "isolated."

More significantly, we can expect the "fifth electron" to have a binding energy to the phosphorus atom, or an ionization energy, that is appreciably smaller than that of an electron in the ground state of hydrogen. Equation 1-79 and the subsequent discussion showed that the ionization-energy expression has permittivity squared in the denominator. We can therefore use Equation 1-79, with $n = 1$ and with ϵ substituted for ϵ_0, to estimate the ionization energy for the electron in the phosphorus-atom case. Noting that the ionization energy for hydrogen is 13.6 eV, we have

$$\xi_{\text{ionization}} = \frac{(13.6 \text{ eV})}{(12)^2} = 0.09 \text{ eV}. \tag{2-8}$$

Table 2-3 at the end of the chapter lists the experimentally determined ionization energies for a number of impurities in silicon [9]. Note that those for phosphorus, arsenic, and antimony are all approximately half the value just calculated. The fact that such a simple calculation yields a result that is correct within a small factor reveals a basic strength and advantage of solid-state electronics. Its materials are so predictable and *simple* in a real sense that meaningful calculations can be based on elementary models. Further, the fact that these three Column-V impurities have experimental values of ionization energies that are so nearly the same lends strength to the hydrogen-atom picture just described. This calculation illustrates a reason that silicon is the best-understood material on earth, a statement that has for some time been cliché.

The ionization energy calculated in Equation 2-8 is so small that thermal energy in the silicon crystal at room temperature is sufficient to dislodge almost all of the "fifth electrons." When thus dislodged, each becomes a *conduction electron*, free to move about the entire crystal. Equally important, each leaves behind (in our example) a *phosphorus ion* that is fixed (immobile) in the silicon crystal. So probable is the *ionization* of the phosphorus impurity atom that it is termed a *donor*, because of the high probability that the electron it requires for neutrality has been "donated" to the conduction band.

In the unlikely event that the donor state is occupied by an electron (that is, that there is an electron "in orbit" about it), then it exhibits overall neutrality. Viewed from a short distance away it is electrically "invisible." In this condition it is said to be *nonionized*. The word "unionized" is avoided on grounds of ambiguous pronunciation and meaning.

Exercise 2-12. Name the most common donor impurity in silicon technology and *spell it correctly*.

The most common donor impurity, drawn from Column V of the periodic table, is phosphorus.

The hydrogen-atom analogy described in detail above is known as *the hydrogen model of a donor* (or donor state). It invites appropriate elaboration of the silicon band diagram to include this different kind of electronic state. As a first step, note that the ionization energy of a typical donor is about one twenty-fifth of the silicon energy gap. Second, recall that an electron escaping from a donor atom becomes a conduction electron. This suggests placing the energy well for the donor atom just below the conduction-band edge, as shown in Figure 2-11. This simple picture is important for these reasons: The probability of the thermal elevation of an electron from one energy state to a higher state is steeply dependent on the size of the energy interval between the two, with probability decreasing as interval size increases. The probability for thermal elevation of an electron from the valence band to the conduction band, as we have seen, is finite but extremely small—close to zero. On the other hand, the probability for elevation from a donor state to the conduction band is near unity. (It may be helpful here to recall from Figure 2-4(b) that a valence electron resides in an energy well just as truly as an electron bound to a donor state, and that the former is roughly 25 times deeper than the latter.) It is thus clear that donor impurities have the ability to raise electron density, and it is this fact that has made them so important technologically.

Over the past forty years, some half a dozen basically different methods have been devised for introducing impurities into a silicon crystal in controlled fashion. The oldest method for doping (still very important) involves adding precisely determined quantities of impurity to the "melt" from which the silicon crystal is grown. In this approach it is possible to produce a large-volume silicon crystal having an essentially constant impurity density—the condition of *uniform doping*. This topic is treated in more detail in the following section. In the balance of this chapter we shall consider uniformly doped samples almost exclusively.

The symbol N_D stands for the volumetric density of donor impurities, or the number per cubic centimeter. Since most of these donate an electron to the conduc-

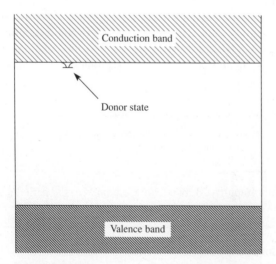

Figure 2-11 The energy well of a hydrogenlike donor atom in relation to a band diagram for silicon.

tion band, to an excellent approximation one can assert that the electron density in a uniformly donor-doped specimen is given by

$$n \approx N_D. \tag{2-9}$$

It is hard to imagine a simpler or more useful engineering equation, and this one is typically in error by less than two percent. In a smaller, but still important range of N_D values, the error is a fraction of one percent.

The typical density range for intentionally introduced impurities is 10^{14} to 10^{21}/cm^3. At $N_D = 10^{16}$/cm^3, which can be taken as a typical value of doping density, at room temperature only one donor atom in about 500 has an electron "in orbit" about it. In other words, only one donor state in 500 is occupied, which corresponds to 99.8% ionization of the donor states. The meaning of Equation 2-9, of course, is that *all* donor atoms are ionized, and so this equation is an algebraic statement of another important approximation—the *100%-ionization approximation*.

Exercise 2-13 In intrinsic silicon, the equal values of hole and electron density are fixed by the thermal generation of pairs. Equation 2-9 makes no mention of thermally generated electrons. How can it be as accurate as claimed?

The answer here is provided by considering magnitudes. Thermally generated carriers are present in donor-doped silicon, but at negligible density. Even for the "light-doping" value of 10^{14}/cm^3, the density of doping-caused electrons is ten thousand times larger than n_i.

A sample with a carrier density much greater than the intrinsic density n_i is said to be *extrinsic*. Even a sample with only one part per billion of impurity can qualify for this description, because this corresponds to a density of approximately 0.5×10^{14}/cm^3. Numerical examples of this kind underscore the technical challenge that was involved in reducing the densities of unwanted impurities sufficiently to make controlled and reproducible light doping possible, an effort that has extended over many years.

2-3.2 Uniform Doping

An approximation greatly simplifies the following discussion and is altogether accurate enough for our purposes. Assume that the silicon crystal has a simple-cubic structure, rather than the diamond structure. Table 2-2 at the end of the chapter informs us that the atomic density of silicon is $N_a = 5 \times 10^{22}$/cm^3. The cube root of this number tells us the number of atoms along one edge of a 1-cm cube of "simple-cubic" silicon, and the reciprocal of that result gives us the "new" lattice constant, or

$$a_0 = \frac{1}{(N_a)^{1/3}} = \frac{1}{(5 \times 10^{22}/\text{cm}^3)^{1/3}} = 2.71 \times 10^{-8} \text{ cm} = 2.71 \text{ Å}. \quad (2\text{-}10)$$

Now let us add the impurity atoms, continuing to consider donors and phosphorus in particular. The first doping requirement is that they occupy substitutional sites. This point is important because the desired electrical behavior of any dopant atom depends upon four surrounding covalent bonds, as we have already seen for the case of donors.

The second present requirement is uniform impurity-atom placement. This does *not* mean perfectly regular atomic placement in the crystal, and what it does mean will be explained in a moment. But for now, let us consider the case of regular placement as an extreme case of "uniformity." Assume that the phosphorus doping density is one part per million. In view of the value of N_a just given, this amounts to $5 \times 10^{16}/\text{cm}^3$. It follows that the number of simple-cubic atom spacings or lattice constants a_0, that lie between phosphorus atoms along a cube edge, can be found from

$$\left(\frac{N_a}{N_D}\right)^{1/3} = \left(\frac{5 \times 10^{22}/\text{cm}^3}{5 \times 10^{16}/\text{cm}^3}\right)^{1/3} = (10^6)^{1/3} = 100. \quad (2\text{-}11)$$

Thus we have a cubic array of phosphorus atoms in the crystal, all separated from nearest neighbors by a distance of $100a_0$. This relatively large spacing is significant. Recall that the hydrogen model of a donor is based upon the properties of an *isolated* hydrogen atom. It follows that the phosphorus atoms must be spaced sufficiently in the silicon crystal to function independently, and must not interact. Since the wave function of an electron bound to a donor atom (or occupying the donor state) spreads out through a dozen or so atomic spacings, the $100\text{-}a_0$ spacing in this example is adequate to guarantee that the phosphorus atoms do not interact. In fact, the requirement of independence holds quite well up to doping densities in the neighborhood of $5 \times 10^{18}/\text{cm}^3$ or $10^{19}/\text{cm}^3$. At these and higher densities, the electron is able to "hop" from one impurity site to another, introducing a new conduction mechanism known as *impurity-band conduction*.

With a transition from the extreme and idealized case of regular impurity placement with constant spacing to the actual case that involves statistical fluctuations in spacing, the true mean spacing is altered a bit, but not drastically. What is now meant by "uniform" is *statistical uniformity*, a concept that can be described in this way. Select a certain volume of the silicon crystal that contains N impurity atoms. Then pick another portion of the crystal having the same volume. Statistically, the probable fluctuation in the number of impurity atoms from one volume to the next is of order $\sqrt{N}$. Thus from volume to equal volume we have an

$$\text{Expected percent fluctuation} = 100\frac{\sqrt{N}}{N} = \frac{100}{\sqrt{N}} \quad (2\text{-}12)$$

in the number of impurity atoms that will be found in these equal volumes. Hence, if $N = 100$, we find a 10% fluctuation, if $N = 10,000$, a 1% fluctuation, and so forth. Samples that display impurity-density fluctuations in agreement with this prescription can be regarded as *uniformly doped*.

Exercise 2-14. How many atoms per unit cell are there in "simple-cubic" silicon?

Visualize an array of cubic volumes matching the crystal structure. Then displace this array of volumes so that one atom is at the center of each volume. Since the unit cell is of the same size as one of these volumes, it is evident that there is one atom per unit cell.

Exercise 2-15. How many atoms per unit cell are there in an actual or diamond-structure silicon crystal? Find the answer in two ways.

One method is to calculate the volume of the silicon unit cell. From Table 2-2, its lattice constant is $a_0 = 5.43$ Å. Thus the volume of the unit cell is

$$a_0^3 = \left[(5.43 \text{ Å})\left(\frac{10^{-8} \text{ cm}}{1 \text{ Å}}\right)\right]^3 = 1.60 \times 10^{-22} \text{ cm}^3/\text{unit cell.}$$

Multiplying this volume by the atomic density of silicon yields

$$(5.0 \times 10^{22} \text{ atoms/cm}^3)(1.6 \times 10^{-22} \text{ cm}^3/\text{unit cell}) = 8.0 \text{ atoms/unit cell.}$$

[In this kind of calculation, entities such as "atom" and "unit cell" are not units in the sense of dimensional analysis, but are tags to assist memory. They can be subjected to the same cancellation and carry-forward rules that apply to true units.]

As a second method, count the atoms shown in the unit cell depicted in Figure 1-33, taking due account of the fact that some atoms are shared with neighboring unit cells. Consider categories of atoms, with N being the number in that category, and f being the fraction of each atom in that category that "belongs to" the cell represented.

Category	N	f	Nf
Corner atom	8	1/8	1
Face-centered atom	6	1/2	3
Octant-centered atom	4	1	4
Total number of atoms per unit cell			8

2-3.3 Acceptor Doping

To enhance hole conductivity, we need to add to the crystal an element from column III of the periodic table, Table 1-1. A number of elements are used, but boron is by far the most common in silicon applications. When a boron atom is placed substitutionally in the silicon crystal, its three outer-shell electrons enter covalent bonds.

But so strong is the affinity of one electron for another in a covalent bond that there is a probability close to unity that an electron from elsewhere will be present to complete a fourth covalent bond, pairing with an electron contributed by the fourth nearest neighbor (a silicon atom) of the boron atom. In the most probable case, this electron has "tunneled over" from an adjacent covalent bond. Once this has occurred, the resulting "hole" in the adjacent bond can continue to move through the crystal. (Recall the automobile analogy.) The effective result, then, is that the boron atom has contributed a hole to the valence band, an event in every way parallel to the contribution of an electron to the conduction band by a donor atom. Since the boron atom has accomplished this by "accepting" a nearby electron, boron and other Column-III impurities are termed *acceptor* impurities.

Figure 2-12 represents the substitutional boron atom in its most probable condition, as a *negative boron ion* [2]. It has in its vicinity one too many negative electronic charges for neutrality—namely, the charge of the electron completing the fourth covalent bond. The next step in the description requires a bit of imagination. We have pointed out that a hole is a carrier in the same sense as an electron, differing mainly in the algebraic sign of its charge. When a hole comes into the vicinity of the boron ion, there is a small but finite probability (typically less than 2%) that the hole will "go into orbit" about the negative charge center that is securely locked in place in the silicon crystal.

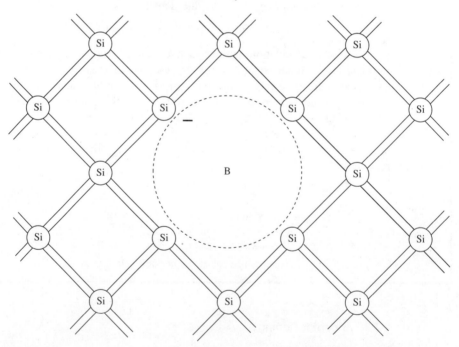

Figure 2-12 The hydrogen model of a boron atom in silicon. When a hole occupies this state, it "goes into orbit" about the extra negative charge associated with a boron atom, yielding a neutral or nonionized state. (After Warner and Grung [2].)

The word picture just sketched is the *hydrogen model of an acceptor state,* parallel to the donor model but with signs reversed. When the hole occupies the acceptor state, the acceptor atom and hole combined exhibit overall neutrality, and the acceptor state is *nonionized*. An energy well can be sketched upside down at the valence band edge, as shown in Figure 2-13. Because a hole "falls up" (in analogy to a bubble) on an energy diagram, there is a small probability that a hole from the valence band will occupy the acceptor state. But it is far more probable that the hole will be "excited down" to the valence band, leaving the acceptor atom ionized.

The ionization energy of boron is the same as that for phosphorus within about 2%. (This agreement further strengthens the hydrogen model.) As a result, an assumption of 100% ionization is just as appropriate here as for phosphorus, giving a parallel approximate equation:

$$p \approx N_A. \tag{2-13}$$

The 100%-ionization approximation is valid for boron to the same degree and through the same doping-density ranges as those cited for phosphorus.

In addition to giving ionization-energy data for column-III and column-V impurities, Table 2-3 lists values for a sampling of elements from all other columns of the periodic table, with silicon's column IV being of course the only exception.

A slightly different way of viewing an acceptor state can be secured by referring back to Figure 2-4. (This view involves a shift back to an "all-electron" picture.) There we show energy wells that constrain valence electrons, with the highest state being at the valence-band edge. The boron state, however, is situated a small distance (0.045 eV) above this edge. Hence, the electron that occupies that state (the one completing a fourth covalent bond) is bound in place almost as securely as a valence electron. The two ionization energies, respectively, are 1.08 eV and 1.12 eV.

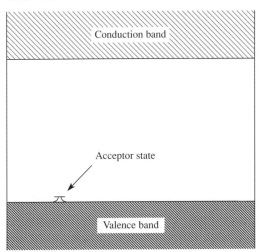

Figure 2-13 The inverted energy well of the hydrogenlike acceptor atom in relation to a band diagram for silicon.

2-3.4 Impurity Compensation

Any real silicon crystal contains both donors and acceptors. Often this situation is created intentionally to achieve a device purpose. Also, some doping methods make this situation unavoidable, although there are others that permit having essentially only one impurity in a given region. Ultimately, though, it is our inability to reduce unwanted impurity densities to zero that enforces impurity combinations.

Visualize a sample containing three times as many donor as acceptor atoms. To be specific, let $N_D = 3 \times 10^{16}/\text{cm}^3$ and $N_A = 1 \times 10^{16}/\text{cm}^3$. Under these conditions the acceptor states, located well below the Fermi level, have a near-unity probability of being filled. The only source of the necessary electrons will be the supply of "fifth electrons" contributed by the donors. Thus a total of $1 \times 10^{16}/\text{cm}^3$ electrons will drop down to fill those states. As a result, only $2 \times 10^{16}/\text{cm}^3$ remain to be elevated thermally to the conduction band. Examining the numbers in this example makes it evident that *each minority impurity has effectively cancelled out one majority impurity*. This behavior of opposite-type impurities is termed *impurity compensation*. As a result, the electron density is evidently

$$n = N_D - N_A. \tag{2-14}$$

The quantity $N_D - N_A$ is termed the *net doping density;* in this case the resulting sample is *N*-type.

The acceptor states are nearly filled and hence are fully ionized, just as they were in the sample doped with acceptors only. Also, the donor states are virtually all ionized, having given up electrons to the conduction band and to acceptor states. Thus approximately *all* impurities are ionized in a compensated sample, as well as in one with a single impurity type.

Next consider the opposite situation, with $N_D = 1 \times 10^{16}/\text{cm}^3$ and $N_A = 3 \times 10^{16}/\text{cm}^3$. In this case, the electrons contributed by the donors drop down into an equal number of acceptor states, which are then unable to perform the "accepting" function. Once more, each minority impurity has cancelled a majority impurity, and the hole density resulting is

$$p = N_A - N_D. \tag{2-15}$$

In this case the net doping is *P*-type.

The terms *majority* and *minority* are most commonly applied to carrier densities. In a sample with net *P*-type doping, *holes* are the *majority carrier*. In a sample with net *N*-type doping, *electrons* are the *majority carrier*. In both cases, minority-carrier density will be far below the intrinsic density, $n_i = 10^{10}/\text{cm}^3$, following a law that we shall develop in Section 2-4.3.

When a semiconductor crystal is doped with equal densities of donors and acceptors, it is evident from the foregoing discussion that a "standoff" results. Just enough electrons are contributed by the donors to drop down and fill all of the acceptor states, deactivating both donor and acceptor impurity atoms in the process. In this case, the Fermi level ends up at "gap center" (the middle of the energy gap), and the resulting sample is termed *compensated-intrinsic*. If the doping densities are low, such a sample truly simulates an actual intrinsic (or perfectly pure) sample. But with higher doping densities, secondary effects enter, clearly separating the two cases.

Carrier-density calculations involve pitfalls that can be avoided only through practice. The first is that calculating a majority-carrier density requires algebraic addition of powers of ten, as in Equations 2-14 and 2-15. It is important to realize that $10^{10} - 10^5 \approx 10^{10}$ (and *not* 10^5). In Section 2-4.3 we shall see that calculating a minority-carrier density involves division, which sometimes causes further confusion because the subtraction of exponents is involved.

2-3.5 A Fermi-Level "Computer"

The unbalanced addition of donors and acceptors to a semiconductor causes the Fermi level to deviate from its midgap position—upward for net *N*-type doping, and downward for net *P*-type doping. In a sense the Fermi level is responding to conditions of electron "supply and demand" in the sample, and adjusting its position so that all are accommodated in the neutral case. In spite of the limitations of the original Fermi-level water analogy, it makes sense that the addition of electrons to the sample by adding donors should *raise* the Fermi level ("surface"), and that the increased demand for electrons of acceptor atoms should *lower* the Fermi level.

A model has been devised to illustrate the interaction of the Fermi-Dirac probability function and impurity states of various densities at various positions within the energy gap [10]. It employs ball-bearing "electrons" and apertures in a layer of plastic as "states," with a template cut in the shape of the Fermi function that can be slid up and down between band edges to accommodate the electrons fully, thus functioning as a primitive sort of analog computer that is able to determine Fermi-level position.

The principles involved can also be illustrated by means of a series of modified band diagrams. In Figure 2-14(a) is shown a band diagram in which small circles represent states available to electrons. The densities of states in the two bands are shown to be equal, and for drawing convenience, both are given rectangular density-of-states functions $N(\xi)$. In a band diagram, the abscissa often represents distance, but here a different assignment is made, namely, sample volume. Let us assume that the sample in question has the typical donor-doping value of $10^{16}/\text{cm}^3$. Then the portion of it represented in Figure 2-14(a) has a volume 10^{15} times smaller than a cubic centimeter, or a volume of 10^{-15} cm^3, because it contains only ten donor states. Assuming that this microscopic volume is cubical in shape, we can find its edge dimension by extracting the cube root of 10^{-15} cm^3, which is 10^{-5} cm, or 0.1 μm.

The next step is to assure sample neutrality by introducing the correct number of electrons. Figure 2-14(b) shows this situation, with black dots representing electrons. The rules are these: Each silicon atom supplies four electrons to four valence states, shown by the black dots in the lower band; at this rate, there are enough electrons for all valence states in the crystal. A conduction state is "a place where

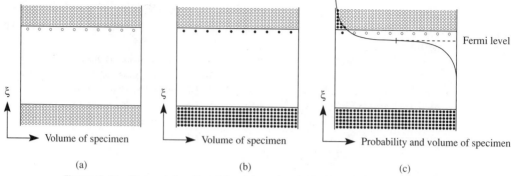

Figure 2-14 Determining Fermi level in a donor-doped sample. (a) Band diagram with circles representing states and the abscissa representing sample volume. (b) The addition of the right number of electrons (black dots) to insure neutrality; the electrons occupy states consistent with $T = 0$ K. (c) Result of analog multiplication of state density and occupancy probability $P(\xi)$, determining Fermi-level position.

an electron can go,'' but it does not supply an electron. Each donor state supplies a "fifth" electron. These electrons are associated with the donor states in Figure 2-14(b) to simplify "accounting." As a result, Figure 2-14(b) fairly represents the states that would be occupied at $T = 0$ K, the case where electrons seek the lowest available states.

The next step is to introduce the room-temperature Fermi function. This has been done in Figure 2-14(c), with zero probability coinciding with zero sample volume and unity probability coinciding with the total volume, 10^{-15} cm³. This permits us to perform an "analog multiplication." That is, for example, with an occupancy probability of 0.1 (or 10%) at the energy where there are ten donor states, as is approximately the case, it follows that one state will be occupied. The other nine electrons contributed by donors are accommodated in the conduction band.

The "computation" of Fermi-level position is accomplished by sliding the Fermi function up or down so that the electrons required for neutrality are just accommodated in states below and to the left of the Fermi function. The shape of the function itself is fixed *only* by temperature and is unaffected by doping densities, and so forth. Had we estimated too low a Fermi level position in Figure 2-14(c), there would not have been "room" in the conduction band for all of the electrons in the neutral crystal. Had we assumed too high a position, there would have been too much room.

The Fermi function in Figure 2-14(c) has been distorted a bit to accommodate the coarseness of the model; for the assumed donor density, the actual degree of ionization would be 99.8%, rather than the 90% shown. However, the indicated Fermi level is approximately correct. It is important to realize also that this diagram represents electron position in *energy only*.

Exercise 2-16. In Figure 2-14(c), there are nine electrons in the conduction band in spite of the fact that there are nine "lower available states" created by donor atoms. How can this situation exist?

The number of electrons present at a particular energy depends on the product of *two* factors—the density of available states, such as N_D in this example, and the corresponding occupancy probability, $P(\xi_D)$.

Figure 2-15 is parallel to Figure 2-14, but for a sample that is acceptor-doped, again at $10^{16}/cm^3$. All of the conventions, assignments, and assumptions are the same as before. Figure 2-15(a) presents the available states, and Figure 2-15(b) illustrates the neutral condition at $T = 0$ K. Acceptor states, like conduction states, can take on electrons, but do not supply them to the crystal. In Figure 2-15(c) then, the room-temperature Fermi function has been adjusted to accommodate all electrons. Again, the impurity states are 90% ionized, meaning that nine of the ten states are filled and negatively charged, while one is empty (or if you prefer, occupied by a hole); there are nine holes in the valence band that have been "excited down" from the filled acceptor states. In Figures 2-14 and 2-15, as well as in Figure 2-16, which follows, the situation for $T = 0$ K is shown for the sake of fuller explanation only. It is not in any sense necessary to cool a sample to $T = 0$ K to make dopants "work."

Now consider a case of compensation. Let $N_D = 1.1 \times 10^{16}/cm^3$, and $N_A = 0.1 \times 10^{16}/cm^3$, values that translate into eleven donor atoms and one acceptor atom in our tiny cube of silicon measuring 0.1 μm on an edge. These are shown in

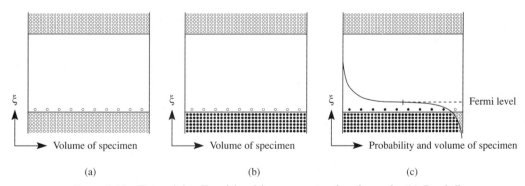

Figure 2-15 Determining Fermi level in an acceptor-doped sample. (a) Band diagram showing available states. (b) The neutrality condition, assured by adding the correct number of electrons. They occupy states consistent with $T = 0$ K. (c) Result of analog multiplication of state density and occupancy probability $P(\xi)$, determining Fermi-level position.

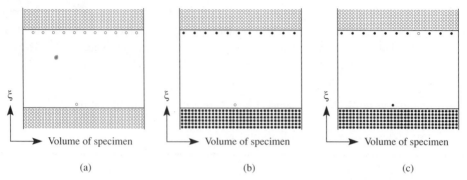

Figure 2-16 Determining Fermi level in a compensated sample. (a) Band diagram showing available states. (b) The neutrality condition, assured by associating electrons with states that contribute them. This "accounting" diagram does *not* correspond to $T = 0$ K. (c) The situation at $T = 0$ K, clearly displaying the effective cancellation of a donor state by the acceptor state.

their proper positions in Figure 2-16(a). In Figure 2-16(b), then, is the "accounting" exercise that guarantees sample neutrality. Donor and valence states contribute electrons, while acceptor and conduction states do not; this diagram does *not* correspond to $T = 0$ K. Finally, Figure 2-16(c) shows the situation at $T = 0$ K. We emphasize it here because it illustrates so clearly the cancellation of a majority impurity by a minority impurity. With a return to room temperature, the electron in the acceptor state has only a trivially greater probability of reaching the conduction band than does a valence electron. As a result, the effective number of donors in the sample is ten rather than eleven, and the majority-carrier density is given by *net* doping.

Exercise 2-17. What Fermi-level position is to be expected in the sample of Figure 2-16?

For the assumptions and conditions just given, the Fermi level in the sample of Figure 2-16 will have a position identical to that in Figure 2-14.

The procedures outlined in this section should impart a good qualitative feeling for the behavior of Fermi level with changes in doping density or temperature. Seeing these trends requires only intuition and geometrical insight in the interpretation of Figures 2-14 through 2-16. For example, in Figure 2-14, a doubling, tripling, or more, of donor density would cause progressive upward shifts in Fermi level so that the additional electrons could be accommodated in the conduction band. Conversely, reducing donor density causes a downward shift and, in the limit, brings the Fermi level to the center of the energy gap. Increasing temperature, on the other hand, causes the Fermi function to become extended along the energy axis, as is illustrated in Figure 2-6(d). Thus, the two branches extend more deeply into the bands; the lower band is a rich source of electrons and the upper band provides matching space for them at high temperatures. As a result, the effect of impurities becomes "washed out," and the Fermi level tends toward gap center. With temperature reduction, on the other hand, the Fermi function takes on a progressively more rectangular shape, causing the Fermi level to settle down in energy in the vicinity of the states created by the dominant impurity type.

A specific example helps to illustrate the important effects of temperature change. Consider a sample having $N_D = 10^{16}/\text{cm}^3$, and one-percent compensation, or $N_A = 10^{14}/\text{cm}^3$. At room temperature the Fermi level resides about 0.2 eV below the conduction-band edge. With rising temperature the Fermi level moves lower, preserving a fixed value of $P(\xi_C)$, with the result that n is temperature-independent through an appreciable temperature range. Only when significant numbers of electrons are elevated into the conduction band will this condition change. For the assumed conditions, most of these electrons will come from the valence band. As n starts to rise and doping effects become less relevant, the sample is said to "go intrinsic" with increasing temperature. A threshold for this condition can be arbitrarily defined as the temperature at which $n \approx 2N_D$. Now return to room temperature and start to reduce temperature. The Fermi level rises, again to preserve the value of $P(\xi_C)$ through a significant range, but then n starts to decline as electrons return to the donor states, a phenomenon often described as carrier "freezeout." Again, a threshold can be defined, this time as the temperature at which $n = N_D/2$, or $\xi_F = \xi_D$. With further temperature decrease, the Fermi level then rises above ξ_D, but declines again, so that at $T = 0$ K, it is only infinitesimally above ξ_D. Increasing net doping has the effect of raising both of the threshold temperatures.

In conclusion, we can say that Fermi-level position is determined by a series of factors that may be summarized as follows:

- Impurity type; whether the impurity atom is a donor or acceptor determines whether it increases or diminishes the supply of electrons available in the crystal.
- Volumetric density of each kind of impurity.
- Position in energy of state introduced by each kind of impurity.
- Temperature; this factor single-handedly determines Fermi-function shape.

2-4 ANALYZING BULK-SEMICONDUCTOR PROBLEMS

The principles developed thus far in Chapter 2 are combined and extended here, yielding tools to apply to an important set of problems. Although these problems are restricted in range, their solutions underlie all of the subsequent subject matter. The restrictions are five in number: (1) Only bulk properties are treated—properties of regions well away from surfaces. (2) Only equilibrium conditions are considered—the samples must be protected from external disturbances such as incident radiation or applied electric fields. (3) Only uniformly doped samples are examined. (4) Only isothermal conditions are permitted. (5) Only regions exhibiting charge neutrality

138 BULK PROPERTIES OF SEMICONDUCTORS

are considered. These restrictions are all relaxed to varying degrees in the balance of the book. Now we examine the last restriction in a bit more detail.

2-4.1 The Neutrality Equation

There are four charged species that must be considered—holes, electrons, donor ions, and acceptor ions. This is a good place to reemphasize an extremely basic point that sometimes evades beginning students: The first two species named are *mobile carriers,* free to move about the entire sample. The last two are *fixed in position,* locked into the crystal structure. The 100% ionization-assumption yields a donor-ion density equal to donor-atom density N_D, and an acceptor-ion density equal to acceptor-atom density N_A. Confining ourselves to *extrinsic* samples means by definition that the density of thermally generated carriers (excited from band to band) is negligible compared to the density of doping-caused carriers. Thus the only electrons that must be considered, those in the conduction band with density n, are of doping origin. In Problem A2-22 we relax this restriction and note the presence of added thermally generated carriers, important at room temperature only in an extremely narrow near-intrinsic range.

The next step, then, is to equate the total density of negative charge centers to that of positive charge centers, yielding the *neutrality equation:*

$$n + N_A = p + N_D. \tag{2-16}$$

Let us assume we are dealing with an extrinsic N-type sample. N_A may or may not be negligible compared to N_D, depending upon the degree of compensation. But the minority-carrier density p *must* be negligible compared to the majority-carrier density n because the sample is extrinsic, a matter treated quantitatively in Section 2-4.3. Dropping p from the equation, then, yields

$$n = N_D - N_A, \tag{2-17}$$

identical to Equation 2-14. Thus, the carrier-density expression derived by considering the physical behavior of compensating impurities is an approximate form of the neutrality equation. In a similar way, we find that for an extrinsic P-type sample, we arrive at Equation 2-15 by dropping n from Equation 2-16.

2-4.2 The Boltzmann Approximation

Four basic approximations are used extensively in semiconductor problems for convenience and clarity. In each case, the correct solution is known, so the approximation is employed not as a matter of necessity but as a simplifying assumption. For these reasons, the terms *approximation* and *assumption* are used somewhat interchangeably in identifying the four items, a custom that is evidently justifiable even though the two terms have quite different meanings.

The first three approximations have already been cited. The first is the band-symmetry approximation described in Section 2-2.3. The second is the equivalent-

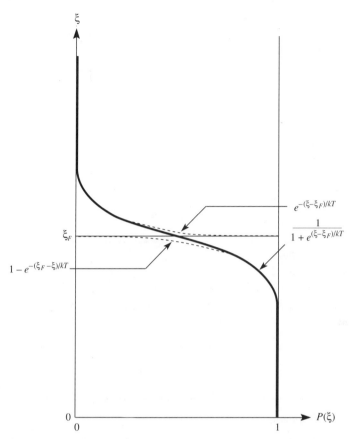

Figure 2-17 The two Boltzmann exponential functions (dashed) needed to approximate the Fermi-Dirac probability function (solid). At the Fermi level the error in either case is a factor of two, but it rapidly diminishes as one moves away from the Fermi level.

density-of-states approximation developed in Section 2-2.4. The third is the 100% ionization approximation explained in Section 2-3.1 and employed in Section 2-4.1. Now we put forward the fourth, an extremely useful simplified version of the Fermi-Dirac probability function known as the *Boltzmann approximation*.

It might be thought that the Fermi function, Equation 2-1, is simple enough that an approximation is unnecessary. However, having a denominator that is the sum of constant and transcendental terms yields some awkwardness. The approximation consists simply of dropping the unity term in the denominator. The result is the simple declining-exponential function plotted as the upper dashed curve in Figure 2-17, thus verifying the description of the Fermi function given in Section 2-2.4, namely, that it has a simple-exponential "tail." The approximate probability function, or Boltzmann approximation, is

$$P(\xi) \approx e^{-(\xi - \xi_F)/kT}. \tag{2-18}$$

BULK PROPERTIES OF SEMICONDUCTORS

Exercise 2-18. Construct a table showing the percentage error introduced by Equation 2-18 at $(\xi - \xi_F)$ equal to 0, kT, $2kT$, $4kT$, and $8kT$.

$(\xi - \xi_F)$	$\dfrac{1}{1 + e^{(\xi - \xi_F)/kT}}$	$e^{-(\xi - \xi_F)/kT}$	Error
0	0.500	1.00	100%
kT	0.269	0.368	37
$2kT$	0.119	0.135	14
$4kT$	0.0180	0.0183	1.8
$8kT$	0.00034	0.00034	0.034

Obviously, Equation 2-18 may only be used above the Fermi level since a probability in excess of unity is without meaning. Therefore a second exponential function is needed to treat energies below the Fermi level; the second function is the lower dashed curve also plotted in Figure 2-17. In terms of the probability of electron occupancy, it can be written

$$P(\xi) \approx 1 - e^{-(\xi_F - \xi)/kT}. \tag{2-19}$$

Exercise 2-19. Derive Equation 2-19.

First derive an approximate expression for hole occupancy $P(\xi)$, using Equation 2-5.

$$P'(\xi) = 1 - P(\xi) = 1 - \frac{1}{1 + e^{(\xi - \xi_F)/kT}} \approx e^{-(\xi_F - \xi)/kT}.$$

The exponential term in the denominator becomes small rapidly with increasingly negative $(\xi - \xi_F)$ and was dropped. Solving the expression above for $P(\xi)$, then, yields Equation 2-19.

Following Equation 2-6 and Exercises 2-10 and 2-19, we usually prefer to use symmetric versions of the two Boltzmann exponential functions. That is, the expression parallel to Equation 2-18 is

$$P'(\xi) \approx e^{-(\xi_F - \xi)/kT}, \tag{2-20}$$

the probability of hole occupancy. The lower dashed curve in Figure 2-17 can represent this function by the device of interchanging zero and unity on the probability scale, as was done in Exercise 2-9. The "symmetric" approach, therefore, discusses electron occupancy for states above the Fermi level, and hole occupancy for states below it. However, there are two other options. One is to use the "electron point of view" implicit in writing Equation 2-19. The other is to use a "hole point of view." In this case, one would describe the valence band as normally almost empty of holes, and the conduction band as "full of holes."

The Boltzmann approximation is used so extensively that often no mention is made of the fact. When a comment is made, it often takes the form of "Boltzmann statistics are assumed."

2-4.3 The Law of Mass Action

Exercise 2-18 shows that the Boltzmann approximation is so accurate for $|\xi - \xi_F| \geq 4kT$ that we are justified in using an equals sign if we restrict its use to those ranges. Now recall that the very same range of validity was specified in Section 2-2.4 for the equivalent-density-of-states approximation. Thus, these two key approximations (Boltzmann and equivalent-density) permit the Fermi level to range widely in the central region of the bandgap, from $\sim 4kT$ above the valence band to $\sim 4kT$ below the conduction band.

Using these two approximations we may rewrite Equation 2-4 as

$$n = N_C \, e^{-(\xi_C - \xi_F)/kT}. \tag{2-21}$$

In a similar way, Equation 2-6 becomes

$$p = N_V \, e^{-(\xi_F - \xi_V)/kT}. \tag{2-22}$$

Forming the product of p and n expressed in these terms yields a significant result:

$$pn = N_C \, N_V \, e^{-(\xi_C - \xi_V)/kT} = N_C \, N_V \, e^{-\xi_G/kT}. \tag{2-23}$$

The Fermi-level energy ξ_F has cancelled, leaving an expression that is *independent of Fermi-level position*. The basic reason for this situation is the symmetry of the Fermi function and of the two Boltzmann functions employed to approximate it. Because these two approximations are valid for energy positions removed some $4kT$ from the Fermi level, we conclude that the pn product is a *constant,* independent of Fermi-level position *anywhere* in the bandgap, except near the band edges. This range obviously includes the center of the gap where the Fermi level resides in intrinsic silicon, the case for which $n = p = n_i$. From this observation follows the expression

$$pn = n_i^2, \tag{2-24}$$

known as the *law of mass action,* a label borrowed from the physical chemist.

The law of mass action has major significance because it enables us to calculate minority-carrier density. At this point, let us elaborate our notation a bit with a subscript indicating the conductivity type (*N*-type or *P*-type) of the sample in question. Thus Equation 2-14 informs us that $n_N = N_D - N_A$, so that the minority-hole density becomes, using Equation 2-24,

$$p_N = \frac{n_i^2}{N_D - N_A}. \tag{2-25}$$

Similarly, in a P-type sample, making use of Equation 2-15, the minority-electron density is

$$n_P = \frac{n_i^2}{N_A - N_D}. \tag{2-26}$$

Equations 2-23 and 2-24 incorporate the Boltzmann, the 100%-ionization, and the equivalent-density-of-states approximations. In addition, the band-symmetry approximation discussed in Sections 2-2.3 and 2-2.5, $N_C = N_V$, is sometimes convenient but *not* necessary. Suppose we let the subscript A denote the *asymmetric* or *actual* values of the equivalent densities. It was noted in Section 2-2.5 that $N_{CA} \approx 1.76\, N_{VA}$. Reformulating Equation 2-23 with N_{CA} and N_{VA} yields

$$pn = N_{CA} N_{VA} e^{-\xi_G/kT}. \tag{2-27}$$

Comparing Equations 2-23 and 2-27 makes it evident that the two can be reconciled with the definition

$$N_C = N_V \equiv \sqrt{N_{CA}\, N_{VA}}. \tag{2-28}$$

(This, in fact, was the basis for calculating $N_C = 3.01 \times 10^{19}/\text{cm}^3$ in Exercise 2-10.) The law of mass action, Equation 2-24, has the same degree of utility as before. We continue to define the intrinsic condition as that for which $p = n$. It follows, therefore, that the true location of the Fermi level in intrinsic silicon must be slightly below the gap center, in order to diminish slightly the electron occupancy probability at the conduction-band edge, and to increase slightly the hole occupancy probability at the valence-band edge. The position of the Fermi level in the intrinsic case is often called the *intrinsic level* ξ_I. The law of mass action highlights the importance of the intrinsic density n_i.

Exercise 2-20. Given $N_{CA} \approx 1.76\, N_{VA}$, the definition $p = n$ for the intrinsic condition, find ξ_I, the true Fermi-level location in intrinsic silicon. Let ξ_H stand for the true center ("halfway") position in the bandgap.

By the methods described in this section we can write $n_i = n = p$ as

$$n_i = N_{CA}\, e^{-(\xi_C - \xi_I)/kT} = N_{VA}\, e^{-(\xi_I - \xi_V)/kT}.$$

This permits us to write the equivalent-density ratio as

$$\frac{N_{CA}}{N_{VA}} = e^{-(\xi_I - \xi_V - \xi_C + \xi_I)/kT},$$

or

$$\ln \frac{N_{CA}}{N_{VA}} = \frac{(\xi_C + \xi_V) - 2\xi_I}{kT}.$$

Hence

$$\xi_I = \frac{1}{2}(\xi_C + \xi_V) - \frac{1}{2}kT \ln \frac{N_{CA}}{N_{VA}}.$$

The first term is the "average" value of ξ_C and ξ_V, which of course is ξ_H, the true gap center. The second term is the correction sought. Evaluating it at $T = 24.8$ C yields

$$-\frac{1}{2}kT \ln \frac{N_{CA}}{N_{CV}} = -\frac{1}{2}(0.02566 \text{ eV}) \ln (1.76) \approx -7 \text{ meV}.$$

The exponential occupancy-probability functions are so strong that the condition $p = n$ is realized by displacing the Fermi level downward through a very small energy interval—7 milli-electron volts.

Finally, let us reiterate the advice given in Section 2-3.4 on the importance of practice to develop facility in manipulating powers of ten in order to handle carrier-density calculations. Equations 2-25 and 2-26 show that to find minority-carrier density it is necessary to divide, which involves the subtraction of exponents. But to find majority-carrier density in a compensated sample, we must subtract *densities*, and *not* exponents.

2-4.4 Band Diagrams in Terms of Electrostatic Potential

It is often convenient to convert from electron-energy (ξ) band representations to electrostatic-potential (ψ) representations. This is because the semiconductor structures that these diagrams describe will ultimately be used in circuits and will have to be related to circuit voltages.

Suppose we move an electron at the valence-band edge to an isolated state somewhere in the energy gap. The electron of course experiences an increment in total energy of $\Delta\xi$ as a result. The agency that imparts this energy increment to the electron must do work on it, and so Equation 1-9 informs us that the corresponding increment in electrostatic potential is given by

$$\Delta\psi = \frac{\text{work}}{Q} = \frac{\Delta\xi}{-q}. \tag{2-29}$$

Recall now that the origin or reference is an arbitrary matter for both ψ and ξ. That is, we can replace $\Delta\xi$ by $\xi - \xi_{\text{REF}}$, and $\Delta\psi$ by $\psi - \psi_{\text{REF}}$; but in order for $\Delta\xi$ and $\Delta\psi$ to differ by a constant factor, as in Equation 2-29, it is necessary for ξ_{REF} and

ψ_{REF} to be chosen in a consistent manner. In other words, it is necessary to have $\psi_{REF} = 0$ when $\xi_{REF} = 0$. Having made this arrangement, we can then convert from familiar and important energy-value designations such as ξ_C, ξ_V, ξ_F, and ξ_I to the corresponding potential-value designations, ψ_C, ψ_V, ψ_F, and ψ_I. Thus the potential ψ corresponding to an arbitrary energy ξ can be written, with the aid of Equation 2-29, as

$$\psi = -\frac{\xi}{q}. \tag{2-30}$$

The band diagrams in terms of ξ and ψ will look the same except for labeling and one additional important feature. The fact that the electronic charge is negative, $-q$ in Equation 2-29, introduces an awkward necessity. We must either let electrostatic potential increase *downward* in the second case, or else flip the band diagram over, placing the valence band at the top. We usually choose the first option, so that the two band diagrams will look the same. Figure 2-18 presents the results for an arbitrarily chosen N-type sample. Notice that Fermi level has been chosen as reference in both cases. This is frequently a convenient choice because it simplifies important algebraic expressions, as will be shown in Section 2-4.5. Notice also that the ξ and ψ axes are oppositely directed.

Consider now the effects of net-doping changes on these band pictures; regard the Fermi level as fixed in position because it has been chosen as reference. For the intrinsic case, the intrinsic level coincides with the Fermi level because the band-symmetry approximation will be in use from this point forward. The conduction-band edge lies half the gap above it; the valence-band edge lies half the gap below it. With increases in net N-type doping, all three of these features move downward together; with changes in the P-type direction, all three move upward together. Heavy-doping effects cause a bandgap reduction ("narrowing") that distorts the simple picture just sketched, but through wide doping ranges, this effect is small enough to ignore. Any one of the three features could be taken as the indicator of electron energy in the first band-diagram case, and as the indicator of electrostatic potential in the second, because both of these variables already involve an arbitrary constant. But by choosing the intrinsic level to represent energy or potential, we can take advantage of symmetry by being "even-handed" with respect to holes and electrons. With these choices, it is evident in Figure 2-18 that ξ in an N-type sample is negative, while ψ in an N-type sample is positive. Thus the intrinsic level in Figure 2-18(b) is *the potential* ψ of the sample. Note also in Figure 2-18(b) that the symbol ϕ (phi) is shown as equivalent to ψ_F. It is a longstanding custom to let ϕ represent Fermi level in terms of electrostatic potential because of the phonic aid to memory.

2-4.5 Carrier Densities in Terms of Electrostatic Potential

Combining Equations 2-21 and 2-30 yields

$$n = N_C \, e^{q(\psi_C - \phi)/kT}, \tag{2-31}$$

ANALYZING BULK-SEMICONDUCTOR PROBLEMS 145

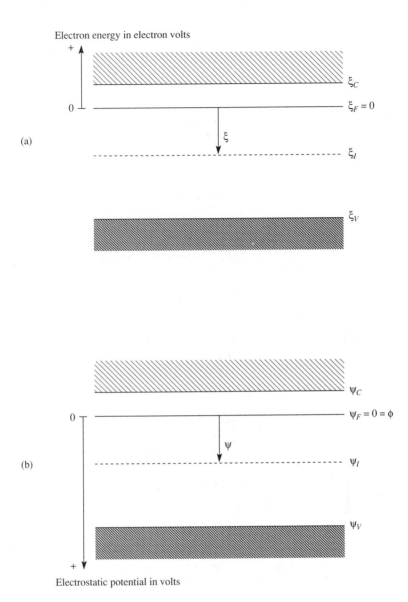

Figure 2-18 Equivalent band diagrams for an N-type sample in terms of (a) electron energy and (b) electrostatic potential. In both cases Fermi level has been taken as reference. Because both diagrams have the same orientation (conduction band above), their ordinate axes are oppositely directed, a consequence of the electron's negative charge.

where the conversion from ψ_F to ϕ has also been made. Now add and subtract ψ ($= \psi_I$) inside the parentheses in the numerator of the exponent of Equation 2-31:

$$n = N_C \, e^{q(\psi_C - \psi + \psi - \phi)/kT} = N_C \, e^{-q(\psi - \psi_C)/kT} e^{q\psi/kT}, \qquad (2\text{-}32)$$

where we have once again let $\phi = 0$, as was done in Section 2-4.4. But since $\psi - \psi_C$, a positive electrostatic-potential increment, is just half the energy gap, it is evident that the coefficient of $\exp(q\psi/kT)$ on the right-hand side of Equation 2-32 is equal to n_i. Hence,

$$n = n_i e^{q\psi/kT}. \qquad (2\text{-}33)$$

Proceeding in a completely parallel fashion, we find that

$$p = n_i e^{-q\psi/kT}. \qquad (2\text{-}34)$$

These results again emphasize the utility of the constant n_i.

2-4.6 The Boltzmann Relation

An immediate consequence of the compact expressions for carrier densities in terms of electrostatic potential developed in Section 2-4.5 is an important statement known as the *Boltzmann relation*. It informs us that the ratio of equilibrium densities at any two locations for a particular carrier type is exponentially related to the potential difference existing between the two positions. Suppose, for example, that in a particular sample of silicon with a one-dimensional doping variation we have $x_2 > x_1$, and $\psi_2 > \psi_1$, where $\psi_2 \equiv \psi(x_2)$ and $\psi_1 \equiv \psi(x_1)$; from Equation 2-33 we immediately determine that

$$[n_0(x_2)/n_0(x_1)] = \exp[q(\psi_2 - \psi_1)/kT]. \qquad (2\text{-}35)$$

The corresponding expression for holes is

$$[p_0(x_2)/p_0(x_1)] = \exp[q(\psi_1 - \psi_2)/kT]. \qquad (2\text{-}36)$$

Exercise 2-21. Find the potential difference between two locations in a silicon sample at which equilibrium electron densities differ by a factor of ten.

From Equation 2-35

$$|\psi_2 - \psi_1| = \frac{kT}{q} \ln \frac{n_0(x_2)}{n_0(x_1)} = (0.02566 \text{ V}) \ln(10) = 0.059 \text{ V}.$$

In view of the law of mass action, the equilibrium hole densities at the two positions also differ by a factor of ten, but inversely.

2-5 CARRIER TRANSPORT

By *transport* is meant a net motion, and the transport of electrons (or holes) gives rise to an electrical current because they are charged particles. There are three functions whose gradients can cause carrier transport in a semiconductor material. A *gradient*, of course, is a rate of change of the function with respect to distance. The first function is electrostatic potential. Its gradient, identically the negative of electric field, causes *drift transport*, to be examined in Sections 2-5.1 through 2-5.5. The second function is carrier density. A density gradient gives rise to *diffusion transport*, addressed in Section 2-5.6. The third function is temperature, but we define temperature gradients out of existence by insisting upon isothermal conditions. Although this choice rules thermoelectric effects out of our discussions, they are treated at length elsewhere [11].

The presence of an electrical current, a necessary consequence of a net carrier transport, is a *nonequilibrium* condition. The balance of this chapter will deal with nonequilibrium phenomena, with one brief exception. The essence of equilibrium (and nonequilibrium) in the present context is best addressed after several features of carrier behavior have been described, so we offer some detail on this basic topic in Section 2-5.2. Drift transport is the first nonequilibrium process to be treated. Thus, of the five restrictions named at the beginning of Section 2.4, one (equilibrium) is being dropped. The four that are retained will confine us to bulk phenomena, neutral conditions, isothermal conditions, and (again with one brief exception) uniformly doped samples.

The practical use of semiconductor devices inevitably involves nonequilibrium conditions. One may legitimately inquire, therefore, why equilibrium conditions have been discussed at such length. The reason is at least twofold: Understanding of equilibrium phenomena is essential to understanding nonequilibrium phenomena. Also, there are important and common practical cases wherein *quasiequilibrium* (or "almost-equilibrium") conditions exist. The significance of this fact is that numerical values that apply to equilibrium may be used for calculations involving nonequilibrium conditions, with accurate results. A good example is the case of *ohmic conduction*, or electrical conduction that obeys Ohm's law, Equation 1-30, which is an important subcategory of drift transport.

2-5.1 Carrier Scattering by Phonons and Ions

In Sections 1-3.1 and 1-3.2 we treated the behavior of a conducting material in the presence of an electric field. The field was created by applying a voltage to plane-parallel contacts arranged to create a one-dimensional problem. The result was net motion (transport) of carriers in the sample in accordance with Ohm's law. The discussion here will depart from the previous discussion in two ways. First, it will deal specifically with silicon, which obeys Ohm's law under some conditions and does not under others. Second, it will take a microscopic view of the conduction process, in contrast to the macroscopic view involved in the definition of conductivity and resistivity in Section 1-3.1. The carriers, holes and electrons, exhibit the wave-particle duality underscored by de Broglie's hypothesis [12], introduced in Section 1-4.6. Both descriptions are relevant to a detailed examination of carrier motion.

First, consider waves. One can accurately attribute wave-guide properties to a single crystal of silicon. Visualize the "channels" formed by perfectly aligned atoms in a single crystal of silicon. In a sample 0.5 mm thick and having the (100) orientation, about one million unit cells will be "stacked up" in the thickness direction. But even if the crystal is defect-free (aside from having surfaces), the wave guide cannot be considered perfect at room temperature because the silicon atoms are vibrating—their manifestation of thermal energy.

A model used in lecture demonstrations is very helpful at this point. It consists of a three-dimensional array, usually cubic, of spherical masses connected by coil springs that are relatively weak, but stiff enough to maintain the shape of the array. With this model it can be shown that displacing and releasing a single mass causes the entire structure to take part in coordinated vibrations. The structure is an array of coupled oscillators, closely related to the one- and two-dimensional oscillator arrays introduced in Section 2-1.1. Extrapolating to millions or trillions of masses, or atoms in a crystal, from the tens of masses in the model, makes it possible to describe the collective motions of the atoms in terms of vibrational waves.

One way of describing the coordinated atomic motions in a crystal is through a superposition of *normal modes* of vibration. A normal mode can be visualized by focusing on one of the three independent directions in a crystal and considering a vibrational standing wave in that direction. The number of half wavelengths in the standing wave, the crystal's length in the chosen direction, and the velocity of sound in the crystal, combine to fix the frequency that characterizes one particular normal mode. Thus there exist certain allowed frequencies, each corresponding to a particular standing-wave pattern. Further, it follows that the number of normal modes existing in a particular crystal specimen is fixed and finite. But the vibrational energy associated with the normal mode of frequency f is *quantized*. In fact, $\xi_n(f) = (n + 1/2)hf$, where $n = 1, 2, \ldots, \infty$. That is, n denotes the nth energy level for a normal mode of frequency f.

Quanta of vibrational energy result from the transition of a normal mode from one energy level to another. These exhibit a statistical behavior identical to that of photons, which in turn result from the kinds of electronic transitions treated in Section 1.4. For this reason, the quanta of vibrational energy have been termed *phonons*. These packets of energy exhibit wave–particle duality and move through the crystal in an erratic fashion that is analogous to carrier motion. A phonon can be described by means of its energy and its *wave vector*. The wave vector has a direction corresponding to the direction of phonon motion, and a magnitude given by a property known as crystal momentum.

Now return to the normal modes that give rise to phonons. The atomic motion in a normal mode can be viewed as a sinusoidal, plane-wave displacement. In the general case, the displacement vector can have any direction with respect to the standing-wave direction, arbitrarily chosen earlier. But in cubic crystals, and hence in the important semiconductor materials, the situation is simpler. Here a standing wave of arbitrary direction can be resolved into waves in three orthogonal crystallographic directions. In each of the three, only two basic displacement directions are possible. The displacement can be *longitudinal,* or aligned with the wave direction (normal to the wave front), in which case compressions and rarefactions occur in the crystal. Or the displacement can be *transverse,* as in a vibrating string.

Recall next the character of the two most important semiconductor-crystal structures, namely the diamond structure exemplified by silicon, and the zinc-blende structure exemplified by gallium arsenide. Each can be described in terms of a pair of interpenetrating fcc lattices, often identified as *sublattices.* Any pair of nearest-neighbor atoms must have one member drawn from each sublattice. The two members of a nearest-neighbor pair may vibrate in phase (in unison), with the associated phonon being known as an *acoustic* phonon, or they may vibrate out of phase, with the associated phonon being known as an *optical* phonon. Furthermore, both possibilities exist for both the longitudinal and transverse kinds of waves. Thus, in a given material we have four categories of phonons for the two crystal-structure types, as is shown in Table 2-4. In one of the eight cases illustrated there (namely, the transverse vibration of nearest neighbors in opposition to one another in a zinc-blende-structure crystal), a fluctuating transverse dipole moment exists. In this case, photon–phonon interaction is very strong, leading to light absorption at the corresponding frequency.

Of great importance is the interaction of phonons with carriers. The phonons with a high probability of doing so are primarily phonons of two types: acoustic phonons, with the relatively low frequencies characteristic of sound waves in the crystal, and optical phonons with small wave vectors, whose frequencies are in the infrared range. The energies of these particular optical phonons are in the neighborhood of 60 meV in silicon, and 33 to 35 meV in gallium arsenide.

The presence of phonons in a crystal interferes with carrier motion. The carrier undergoes a change in direction, energy, and momentum when it collides (let us use the particle picture) with a phonon. As a result, the process of acceleration in an applied electric field must be restarted after each particle interaction. Not surprisingly, the degree of phonon interference with carrier motion increases with temperature.

A second important factor affects the ability of a hole or electron to move freely through a crystal. This factor is the presence of *impurity ions* positioned randomly but securely in the crystal as described in Section 2-3. The coulomb forces acting on a passing hole or electron can also change its direction, energy, and momentum. This picture calls forth another coup of the early twentieth century—Rutherford's analysis of alpha-particle deflection or "scattering" by atomic nuclei. In fact, Conwell and Weisskopf [13] applied the principles of Rutherford's analysis to the problem of impurity-ion effects on carriers and were able to achieve a good explanation of the experimental data. Here, as in the case of Rutherford's experiment, the term *scattering* is descriptive when one considers the behavior of a large number of carriers affected by an ion and going off in all directions as a result.

There is, however, a condition that determines the effectiveness of an impurity ion as a scattering center; that condition is the density of mobile carriers in the sample. Take the case of a donor ion surrounded by mobile electrons in an N-type sample. Because of coulomb attraction, an elevated average density of electrons will exist near the positive ion, a density that declines with distance. That is, if the ion may be regarded as isolated, a negative cloud with spherical symmetry surrounds it. The lines of force emanating from the positive charge go a shorter distance on average when electron density in the overall sample is high than when this density is low. The phenomenon being described has been given the apt label *screening*. In Section 3-7.2 we shall deal with it quantitatively, but in a one-dimensional context rather than in the three-dimensional situation just described. Because of screening, an ionic charge can be "felt" at a greater distance when carrier density is low than when carrier density is high; consequently the ion's effectiveness as a scattering center is larger in the former case.

One graphic treatment of scattering effectiveness places a disk at the position of the scattering center and asserts that an impinging carrier must strike within the area of the disk in order to be deflected. This area has been termed the *scattering cross section* of the center, and obviously it increases as screening decreases, or as carrier density falls.

Donor and acceptor ions are equally effective in the scattering process, as suggested in Figure 2-19(a) for the case of holes passing near both kinds of ions, provided the degree of screening is the same in the two cases. For a closer look at the somewhat subtle interplay of screening and impurity densities, let us turn to a series of specific examples. In Figure 2-19(b) are represented a set of three silicon samples, all extrinsic N-type samples, with doping densities as indicated thereby. Samples 1 and 2 are uncompensated. The total impurity density $(N_D + N_A)$ in sample 2 is five times that in sample 1 (as is the net impurity density). But the amount of carrier scattering in sample 2 is not five times greater than in sample 1 because of the higher degree of screening in sample 2.

Quantitative statistical treatments of problems like this often use the concept of scattering cross section, which we now elaborate somewhat. A disk of a particular area is positioned at a scattering site. A normally incident carrier striking the disk is then considered to undergo a deflection, through an angle that can be statistically elected. If the carrier misses the disk, it passes by unaffected. Generalizing to three dimensions, we can visualize a sphere at the scattering site that presents a cross section of a certain area to an arriving particle, namely the area of the great circle of the sphere that is normal to the trajectory in question. The greater the carrier density, and hence the greater the degree of screening, the smaller is the scattering cross section. In Figure 2-19(b) it is shown that the aggregate cross section of the impurity ions in sample 2 exceeds that in sample 1, but is not five times greater.

Now turn to the compensated case, sample 3. It has a total ion density that matches that of sample 2. Yet it has a majority-carrier density, the factor determining the amount of screening, matching that of sample 1. Hence we see that insofar as impurity scattering is concerned, compensation constitutes (to employ the vernacular) a "double whammy." It increases the number of scattering centers and

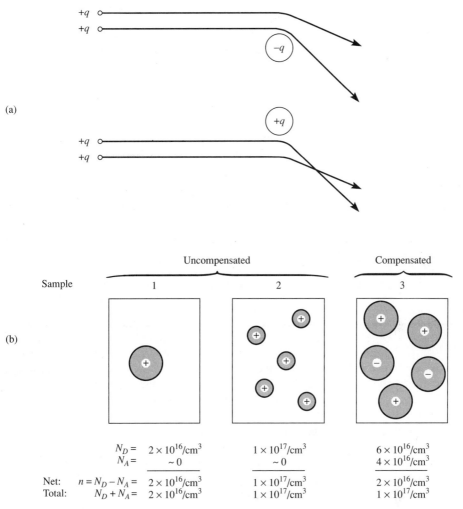

Figure 2-19 Rutherford scattering of carriers. (a) Donors and acceptors are equally effective. (b) Three silicon samples. Net doping (and hence screening) is the same in 1 and 3; total doping is the same in 2 and 3.

makes each more effective. A new ingredient in sample 3 is the *negative* ion, still bathed in negative mobile electrons. Screening in this instance is achieved by the *depression* of electron density in the vicinity of the acceptor ion.

To summarize the screening issue, we can say that in uncompensated silicon, scattering increases with impurity density, but *less* rapidly than impurity density because of screening. In compensated silicon, scattering cross section increases with total net doping, but *more* rapidly than total net doping because of the diminished degree of screening in such a sample.

Finally, in addition to these impurity-scattering considerations, there exists phonon scattering. The last idea, regrettably, is sometimes labeled with the corrupt term "lattice scattering." The connotation intended would be better conveyed by the phrase "crystal scattering," as Table 1-3 emphasizes.

2-5.2 Drift Velocity

Let us continue with the particle picture of holes and electrons. Room-temperature thermal energy, for which the mean value is $\frac{3}{2}kT$, leads to an average *thermal velocity* magnitude v_t, of the order of 10^7 cm/s as calculated in Exercise 2-8.

Exercise 2-22. To emphasize the huge magnitude of v_t, express its value in "automobile units," km/hour.

Using unity factors we have
$$\left(\frac{10^7 \text{ cm}}{\text{s}}\right) \left(\frac{1 \text{ km}}{10^5 \text{ cm}}\right) \left(\frac{3600 \text{ s}}{1 \text{ hour}}\right) = 3.6 \times 10^5 \text{ km/hour}.$$

The incessant deflections experienced by a given carrier and its enormous average speed cause it to execute a kind of "random walk" in the crystal. The velocity, energy, and momentum of the carrier all change from one path segment to the next. The *mean free path* for the carrier is the average length of these linear path segments.

At this point it is appropriate to define *equilibrium*, before giving more detail on the nonequilibrium phenomenon of carrier drift. A concise description of a sample at equilibrium is that *at equilibrium the Fermi level is constant throughout the sample*. Still another time, a water analogy lends concreteness to the discussion. Turning back to the "top surface" interpretation of Fermi level, we can offer a calm pond with its horizontal surface as the analog of an equilibrium case, and the sloping and uneven surface of a cascade, for nonequilibrium.

In addition to the constant-Fermi-level description of equilibrium, there is an equivalent statement that is a simple but very important proposition, the *principle of detailed balance*. It states that *at equilibrium, a given molecular process and its reverse proceed at equal rates*. This may seem like a statement of the obvious, but its importance stems in part from the fact that sometimes it is easier to calculate the forward process than the reverse, or vice versa. The erratically moving carriers introduced above provide good examples of "molecular processes." Visualize a 1-cm^2 area buried deep inside a silicon crystal, and focus on one kind of carrier—say, electrons. Suppose that (in another thought experiment) for one second we count all of the electrons passing rightward through the designated area. In effect we are observing an important quantity termed *flux*, which has dimensions of reciprocal centimeters squared and seconds (1/cm^2·s). We have observed the number of electrons passing in one direction through a unit area in unit time. According to the principle of detailed balance, *if* the sample is at equilibrium, the leftward flux will equal the rightward flux, and the net flux will vanish. But when the net flux is finite, we have a *nonequilibrium* situation.

Exercise 2-23. Consider the converse of the principle of detailed balance. Is it a necessary and sufficient guarantee of equilibrium to have equal forward and reverse molecular processes?

Such forward and reverse equality is necessary for equilibrium but not sufficient to guarantee it. Instead, it guarantees a specialized kind of *steady-state* condition. See Section 2-7.6 for a fuller discussion of this point.

As we consider additional phenomena, we will note several examples of nonequilibrium but steady-state conditions wherein molecular processes are balanced. Meanwhile, we can readily devise primitive analogies to illustrate the point. Consider two water hoses directed in opposition through a hoop. It is straightforward to arrange zero net flux, but the droplets that make up the "system" are most certainly not in equilibrium.

The reason for the adjective "specialized" in Exercise 2-23 is that *steady-state* defines a much broader class of phenomena that are *non-time-varying*. A single hose stream can be steady-state but result in finite flux. Similarly, returning to carriers in a solid, a direct current (dc) is a perfect example of a steady-state condition, and is in fact the one we examine now in detail.

Once again consider the semiconductor sample at equilibrium with its chaotically moving carriers. The effect of superimposing an electric field upon this picture is to cause particle acceleration during each *free path*. After each deflection or "scattering" event, the acceleration process must start all over again because velocity has been randomized. As a result, the path segments that were straight in the absence of a field are now curved, except for the rare paths that are precisely parallel or antiparallel to the field. The situation can be described by means of a picture that is illuminating in spite of being unrealistic. The accompanying thought experiment, represented in Figure 2-20, assumes that we could follow a given carrier—let us say a hole—through a portion of its meandering excursion, then rerun the same excursion segments, this time with a field present, and have each deflection cause a new velocity identical to the corresponding initial velocity in the absence of the field, and with identical elapsed times for corresponding path segments. Such an experiment is clearly impossible; yet it poses no difficulty at all in the realm of thought. The dashed curves in Figure 2-20 represent the "rerun" paths, each perturbed by the presence of the electric field. By arranging to have the field-induced displacement on each path segment small compared to the path-segment length, we guarantee that the carrier remains approximately in *thermal equilibrium*

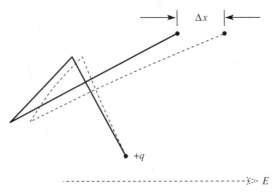

Figure 2-20 Thought experiment for introducing the concept of drift velocity, $v_D \equiv \Delta x/\Delta t$.

with the host crystal. As long as the field-induced increments in average carrier velocity are small compared to v_t, the huge room-temperature average thermal velocity (a magnitude of $\sim 10^7$cm/s), we have the *quasiequilibrium* condition described in the introduction to Section 2-5. As we shall see in Section 2-5.3, this situation also coincides with *ohmic conduction*. Such conditions are commonly encountered in practice because they hold for electric-field values of up to several kilovolts per centimeter.

Under such conditions, then, we note the total displacement Δx in the path portion shown in Figure 2-20, and the corresponding total elapsed time, Δt. The symbol v_D is assigned to the ratio, and is known as *drift velocity:*

$$v_D \equiv \frac{\Delta x}{\Delta t}. \tag{2-37}$$

Thus the drift picture that is drawn is one of a cloud of carriers in violent and erratic motion. The application of an electric field urges them gently in the direction indicated by charge sign and field direction, with negligible effect on the average energy of the carrier aggregation. Thus the coordinated motion described as *drift* is merely a perturbation of the chaotic motion associated with thermal energy under quasiequilibrium conditions.

2-5.3 Conductivity Mobility

Continuing to think about holes because of the convenient positive charge, and continuing with a particle representation, let us imagine a volume of one cubic centimeter inside a large *P*-type silicon crystal. For typical doping, $p = 10^{16}/\text{cm}^3$, so this is the (huge) number of holes under consideration. Applying an electric field now will cause all of them to drift with an average velocity v_D. It is physically and dimensionally evident that the product pv_D yields the *flux* of holes in the field direction. Multiplying the flux by the charge carried by each particle will yield the resulting current density, or

$$J_p = qpv_D. \tag{2-38}$$

Solving for v_D and using Ohm's law (Equation 1-30), yields

$$v_D = \frac{J_p}{qp} = \frac{E}{qp\rho}. \tag{2-39}$$

Evidently drift velocity is proportional to the applied field, a relationship made physically plausible in Figure 2-20. The constant of proportionality, given the name *hole conductivity mobility*, is

$$\mu_p = \frac{1}{qp\rho}, \tag{2-40}$$

so Equation 2-39 becomes

$$v_D = \mu_p E. \tag{2-41}$$

The corresponding equation for electrons is

$$v_D = -\mu_n E; \tag{2-42}$$

the minus sign enters because a rightward field causes a leftward drift. Custom dictates that the algebraic sign of mobility be shown explicitly, the same treatment accorded to the symbol q for the magnitude of the electronic charge. Because mobility is the ratio of drift velocity to the value of the responsible electric field, its dimensions are (cm/s)/(V/cm), or cm^2/V·s.

Exercise 2-24. Examine the dimensions of the expression $1/qp\rho$ from Equation 2-40.

Starting with the dimensions of the three factors in the denominator gives us

$$1/[C][1/cm^3][ohm \cdot cm] = [cm^2]/[C \cdot ohm]$$

$$= [cm^2]/[C][V/(C/s)] = [cm^2/V \cdot s],$$

which is consistent with the units for mobility.

Substituting Equation 2-41 into Equation 2-38 yields an important result:

$$J_p = q\mu_p p E. \tag{2-43}$$

This is the *drift-transport equation for holes*. The companion equation, the *drift-transport equation for electrons*, is

$$J_n = q\mu_n n E. \tag{2-44}$$

Notice that the explicit negative signs on q and μ_n have cancelled. These transport equations are of such basic significance and frequent use that the absence of a sign asymmetry (between the hole and electron transport equations) may account for the peculiar sign treatment accorded to q and μ.

The proportionality of J and E is the *ohmic* condition because it is dictated by Ohm's law. This is the same as requiring that v_D be proportional to E (Equation 2-41); and since mobility is the proportionality factor, it is the same as requiring *constant* mobility. These are alternate ways of stating the requirements for quasi-equilibrium or ohmic conduction.

In Section 2-5.1 we pointed out that phonon scattering declines in importance with falling temperature. Not surprisingly, it disappears altogether at absolute zero. The temperature dependence of mobility in the range where phonon scattering is dominant is proportional to (or "goes as") $T^{-3/2}$, where T is absolute temperature. For less obvious reasons, impurity scattering becomes more pronounced as temperature goes down, leading to a mobility dependence that goes as $T^{3/2}$ at very low temperatures. The result is a maximum in the "profile" of mobility versus temperature, typically occurring well below room temperature. In Figure 2-21 is a curve bearing out this behavior based on data reported by solid-state pioneers Pearson and Bardeen in the 1940s [14].

The reason that ionic or impurity scattering becomes more effective as temperature declines is that the Coulomb force responsible for the effect has a longer time to

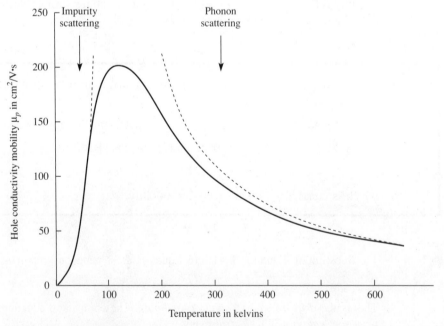

Figure 2-21 Hole mobility in heavily doped silicon. (After Pearson and Bardeen [14], with permission.)

act on a slower-moving particle. Basic physics informs us that the change in momentum imparted to a body by a transient force is equal to the time integral of the force, known as the *impulse*. For a carrier passing near an ion, the impulse increases as carrier velocity decreases, and this accounts for the increasing efficacy of ionic scattering (and hence, declining mobility) with decreasing temperature.

At an impurity-doping density (total density) in the neighborhood of $10^{15}/cm^3$, impurity scattering begins to affect mobility significantly in room-temperature silicon. With further total doping, then, mobility steadily declines. Data on hole and electron mobility in uncompensated silicon are presented in Figure 2-22. These curves were generated recently [15], and are based upon data, primarily experimental, from eight sources. The constant low-density, room-temperature mobility values for holes and electrons are 450 cm²/V·s and 1400 cm²/V·s, respectively, also entered in Table 2-4.

Finally, we note that there are two kinds of mobility other than conductivity mobility, namely, *drift* mobility and *Hall* mobility. These will be described in Section 2-7.3, after the necessary additional concepts are developed.

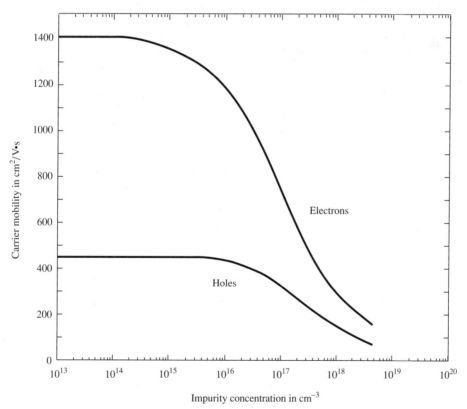

Figure 2-22 Room-temperature (300 K) carrier mobilities in silicon as a function of doping for uncompensated silicon.

2-5.4 Velocity Saturation

Returning to Figure 2-20, one can easily imagine the progressive changes in carrier path as electric field is arbitrarily increased. Under such conditions, mean carrier energy and speed *do* increase significantly. Suppose we focus on electrons. In accord with the billiard-ball description given in Section 2-2.4 for electrons in the conduction band, it follows that these increasingly energetic carriers are positioned farther and farther above the conduction-band edge. The descriptive term "hot electron" is often given to such a carrier—one that has derived appreciable energy from an applied electric field.

A fact that does not change, however, is that even hot carriers experience incessant path interruptions through collisions with phonons and ions. The randomizing of after-collision velocity continues to occur, and this places a cciling on velocity that a carrier can achieve in the direction dictated by the field. In other words, such collisions place a cap on drift velocity. Not surprisingly it is close to the thermal velocity v_t. For electrons in silicon, this *saturation velocity* v_s is very nearly 10^7 cm/s; the saturation is quite sharp, occurring at a field value of 50 kV/cm. For holes, the saturation is appreciably less pronounced, although they, too, exhibit large departures from constant mobility. The word "saturation" is assigned two significant meanings in solid-state electronics. Here it is used in the sense of *approaching a constant value*.

Electron drift velocity as a function of applied field is shown in Figure 2-23.

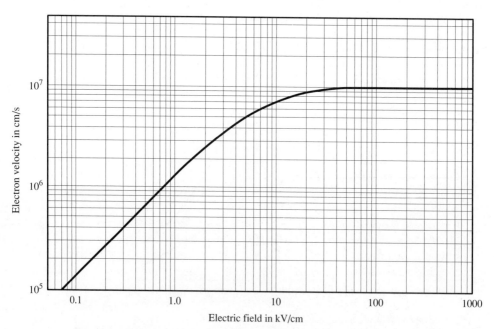

Figure 2-23 Room-temperature electron drift velocity in silicon as a function of electric field.

Such data can readily be converted into a curve of current versus voltage, because velocity and current are proportional on the one hand, while electric field and applied voltage are also proportional on the other. The next step is to observe that current and voltage must in turn be proportional in an *ohmic* material—one that obeys Ohm's law—with the constant of proportionality being resistance or conductance, depending upon formulation. Thus it is evident in Figure 2-23 that ohmic conditions prevail up to field values of a few kilovolts per centimeter in the linear initial portion of the curve. Above that, the drift component of a carrier's velocity can no longer be regarded as a mere perturbation. The carrier's average velocity and energy increase because of the field.

The mobility corresponding to any point on the curve is the slope of a line drawn from the origin to the point in question. Thus mobility is a constant in the ohmic regime, but is *not* constant when the saturated-velocity regime is entered. Dimensions in modern silicon devices have become so small that fields at a given applied voltage are large, and velocity saturation is assuming increasing importance.

Exercise 2-25. Explain qualitatively why the nonohmic effects that are so prominent in silicon are not readily observed in metals. [*Hint:* Compare Figures 2-7 and 2-9.]

Recall that the height of an electron above the bottom edge of the conduction band is a measure of the kinetic energy of that electron. Now compare Figures 2-7 and 2-9. Evidently the mean energy of a conduction electron in a metal is one to two orders of magnitude larger than that of a conduction electron in silicon, and its thermal velocity v_t is correspondingly higher as well. At the same time, the density n of conduction electrons in a metal approximates its atomic density, accounting for its large conductivity. As a result, even at very high current-density values, drift velocity v_D remains small compared to thermal velocity v_t.

Figure 2-23 conveys the important message that there exists a unique relationship between carrier velocity and electric field. There are conditions, however, wherein this relationship does not hold, and these conditions can be described as follows [16]: First, recall that the velocity presented in Figure 2-23 is drift velocity, whose computation involves averaging over several collision or "scattering" events, as is suggested in Figure 2-20, and in a device of extremely small dimensions, a carrier will experience so few scattering events that the drift-velocity concept loses validity. (When there are *no* scattering events, we have ballistic-mode transport.) An alternative approach to the problem, known as the *energy transport method* [17,18] uses the statistical Monte Carlo technique to track the energies of carriers through a limited number of collisions.

Second, even in a sample of large dimensions, problems are encountered in the extremes of high-frequency operation. When the period of the cyclic electric field in

2-5.5 The Conductivity Equation

We have seen that carrier-transport equations come in pairs. Let us arbitrarily choose the case of holes. Combining the hole drift-transport equation (Equation 2-43) and Ohm's law (Equation 1-30) yields

$$J_p = q\mu_p p E = \frac{E}{\rho_P}, \qquad (2\text{-}45)$$

where the subscript on resistivity denotes a resistivity fixed by hole conduction. Dividing Equation 2-45 through by E, noting the definition of conductivity (Equation 1-28), gives

$$\sigma_P = q\mu_p p = \frac{1}{\rho_P}. \qquad (2\text{-}46)$$

The companion equation for electrons is

$$\sigma_N = q\mu_n n = \frac{1}{\rho_N}. \qquad (2\text{-}47)$$

Figure 2-24 gives experimentally determined curves of low-field (ohmic-regime) ρ_P and ρ_N as a function of p and n, respectively, which for the uncompensated samples employed to gather the data can be regarded as equivalent to N_A and N_D, respectively. These curves are based on the experimental data of Irvin [19], with more recent adjustments [20].

It is important to realize that the hole and electron densities in the conductivity equations, Equations 2-46 and 2-47, are *equilibrium* densities, a consequence of the quasiequilibrium nature of ohmic conduction. This is an example of using equilibrium values in a nonequilibrium problem, a possibility noted in the introduction to Section 2.5. To guarantee equilibrium carrier densities, a particular kind of electrical contact to the silicon is needed. Fortunately its structure is straightforward, as will be described in Chapter 3. This property and very low resistance are the two key features of what is known as an *ohmic* contact.

Note that the reciprocity of resistivity and density that is evident in Equations 2-46 and 2-47 is equally evident in Figure 2-24; in the low-doping range where room-temperature mobility is constant, the two curves exhibit slopes of minus one in the log–log plot. Then, when mobility begins to decline (refer back to Figure 2-22), the resistivity curves rise progressively above linear extrapolations of the low-density curves. The crossing of the two curves at high density, if real, has not been explained.

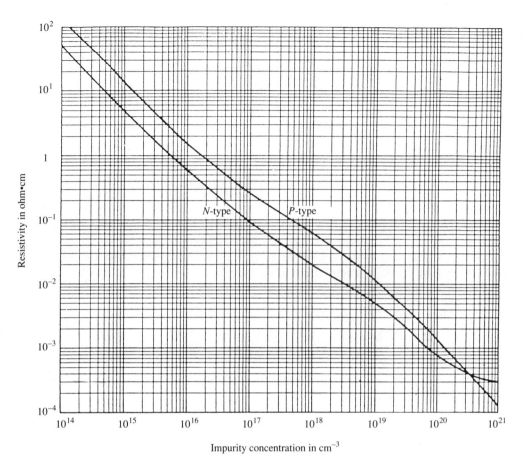

Figure 2-24 Resistivity for uncompensated *P*-type and *N*-type silicon at 300 K. (After Irvin [19], with permission.)

We can summarize the preceding discussions of room-temperature mobility and resistivity (or conductivity) as follows: In the ohmic regime, mobility is constant, while resistivity is a function of doping only. In the nonohmic regime, mobility is a function of electric field, while resistivity is a function of doping and electric field.

For extrinsic *P*-type and *N*-type samples, respectively, Ohm's law can be written

$$J_p = \sigma_P E \approx J, \tag{2-48}$$

and

$$J_n = \sigma_N E \approx J. \tag{2-49}$$

The current density J without subscript is the *total* current density, and the approximate equalities arise because of the vast differences in majority and minority carrier densities in extrinsic samples. In near-intrinsic samples, however, where both carriers contribute to conductivity, it is necessary to write a two-term conductivity equation:

$$\sigma = q(\mu_n n + \mu_p p). \qquad (2\text{-}50)$$

For the strictly intrinsic case this simplifies to

$$\sigma_i = q n_i (\mu_n + \mu_p), \qquad (2\text{-}51)$$

which amounts roughly to 3×10^{-6}/ohm·cm at room temperature, since the low-density values of mobility are obviously appropriate.

Exercise 2-26. Given 1 ohm·cm as a "typical" value of silicon resistivity, and 100 A/cm^2 as a typical current density encountered in an operating silicon device, calculate typical values of electric field and drift velocity.

From Ohm's law we have

$$E = J\rho = \left(\frac{10^2 \text{ A}}{\text{cm}^2}\right)(1 \text{ ohm·cm}) = 100 \text{ V/cm}$$

for a typical electric field. Taking 10^3 cm^2/V·s as an order-of-magnitude value of mobility, we have from Equations 2-41 and 2-42

$$v_D = \mu E = \left(\frac{10^3 \text{ cm}^2}{\text{V·s}}\right)\left(\frac{10^2 \text{ V}}{\text{cm}}\right) = \left(\frac{10^5 \text{ cm}}{\text{s}}\right).$$

Since this amounts to about 1% of the thermal velocity v_t, these "typical" conditions place us well within the ohmic regime.

Exercise 2-27. Devise an analogy that combines the essential features of drift transport.

A modified pinball machine can serve this purpose. Let the large ball bearings or other dense spheres represent carriers, and arrange the inclined plane to have a continuously variable angle. The resulting variation of effective gravitational field is the analog of electric-field variation. Pinball-style "bumpers" provide the necessary impedance to the transport of spheres down the plane, simulating the collisions of drift transport. This apparatus, sufficiently enlarged and refined, will yield a drift velocity proportional to the applied field.

Exercise 2-28. A resistor fabricated from intrinsic silicon is subjected to an electric field directed rightward. (a) State the directions of the hole and electron drift velocities. (b) State the directions of the hole and electron current densities. (c) State the hole current I_p and the electron current I_n as approximate fractions of the total current I, with due regard for sign.

(a) The vector $\mathbf{v_D}$ is rightward for holes and leftward for electrons.
(b) The vectors $\mathbf{J}_p$ and $\mathbf{J}_n$ are both rightward.
(c) The hole current is $I_p \approx \frac{1}{4} I$, and the electron current is $I_n \approx \frac{3}{4} I$, this as a result of the hole and electron mobility difference.

2-5.6 Carrier Diffusion

We have seen that the conditions necessary for drift transport are three in number—a species in random motion, charges on the species, and a potential gradient. The force exerted by an electric field on a charged particle is palpable enough to impart a physical feeling for drift transport. Diffusion transport is more subtle. It has only two requirements: a species in random motion and a density gradient. The more energetic the motion, the faster the diffusion. Electrical charge on the particles involved is not a factor in diffusion *per se*. Analogies do not come readily to mind, so instead of going immediately to the diffusion of carriers, a charged species, let us start by considering a more accessible example involving uncharged species.

Here is another thought experiment. Visualize a large vessel divided into two equal volumes by a thin, impervious plastic sheet. Let one side be filled with nitrogen molecules and the other side with oxygen molecules, both at atmospheric pressure and room temperature. Avogadro's hypothesis informs us that equal numbers of the two kinds of molecules are involved. Also, we know that the gas atoms are in violent random motion, satisfying the first condition of diffusion. Next, imagine that the plastic separator is instantaneously "dissolved." Now the second condition is met because there is a density gradient—in fact, an infinite gradient for both species—at the position occupied an instant before by the separator. Experience and intuition inform us that a spontaneous mixing process will commence, and that after a due waiting period, the vessel will contain uniform concentrations of both gases, with a one-to-one density ratio everywhere. The mixing that occurred in this example was driven by *diffusion*. The *transport* aspect resides in the fact that nitrogen moved into a region where initially there was oxygen only, and vice versa.

Now turn to the specific case of carriers in a silicon sample, arbitrarily choosing holes. The situation to be described is a highly relevant real-life case, but a discussion of how it can be arranged is best deferred until additional phenomena have been introduced and described. We now consider a *steady-state* situation. The desired condition is represented in Figure 2-25(a). The solid line is a linear density profile, or a region of *constant* density gradient (*not* a "linear gradient") in a region where the density function is *one-dimensional* and a region that is also *neutral*. The latter condition involves one of the subtleties for which discussion must be deferred to a later point. In Figure 2-25(b), then, is shown the region of interest, divided into compartments or boxes of equal volume. The number printed in each box represents

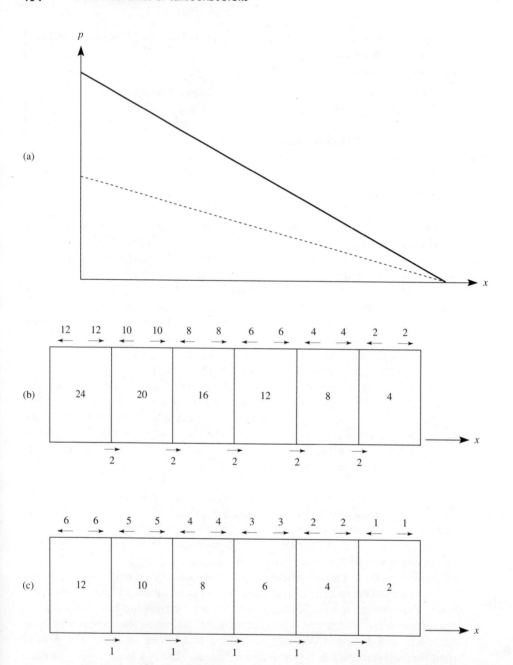

Figure 2-25 The diffusion phenomenon. (a) The solid line shows a region of constant hole gradient. The dashed line involves a gradient half as large. (b) A region inside a sample of one-dimensional properties in which a linear density profile exists. (c) The same region as in (b), but with the hole-density gradient reduced by a factor of two.

the number of holes whose erratic motion will take them *out* of that box in the time interval Δt through an *x*-normal face. The flux of holes through the *y*-normal and *z*-normal faces of a given box can be ignored; the one-dimensional nature of the density function assumed ensures that the inward and outward fluxes through those faces will be balanced, and will therefore not enter into the next step of the process.

Now we assume *equipartition* in the random hole motion through the *x*-normal faces of a given box, or the same number moving in the positive-*x* and negative-*x* directions. It can be argued that for the small numbers of such holes postulated in Figure 2-25(b), this is an unrealistic assumption; statistical fluctuations could produce 100% errors with such small numbers. The answer is that, conversely, the assumption could be made valid to meet any arbitrary criterion by altering geometry so as to increase the numbers—to 24 million, 20 million, and so on. But we will adhere to the small numbers for simplicity, since the principle involved is valid. As the result of equipartition, the numbers of holes moving through an *x*-normal face in each direction from a given box are those indicated with numbers and arrows above that box. Finally, we observe that there was a *net* transport of two holes through each such face, indicated by the number and arrow below each *x*-normal face. Because the gradient was *constant* in the region of interest, the net flux of holes was also constant in the same region. The flux is "down" the gradient.

Now modify the situation by cutting the density gradient in two, yielding the density profile shown with a dashed line in Figure 2-25(a). Repeating the procedure just described, with plausible assumptions along the way, yields the result shown in Figure 2-25(c). Looking at the numbers below the diagram makes it clear that cutting the gradient in half has cut the flux in half. This proportionality between gradient and flux is known as *Fick's first law,* named for the physicist-physiologist who worked it out in the last century. For the present problem it can be written

$$f = -D_p \frac{dp}{dx}. \tag{2-52}$$

The total derivative is appropriate because we have postulated a time-independent (or steady-state) problem that is also one-dimensional. The negative sign enters because rightward transport, or a *positive* flux in our convention, has resulted from the negative gradients represented in Figure 2-25(a). The symbol D_p is the constant of proportionality, and the presence of the negative sign in Equation 2-52 informs us that it is given the same unusual treatment accorded to q and μ; the constant is stated as a magnitude and the minus sign must be shown explicitly. This important constant is named the *diffusivity,* or *diffusion constant,* or *diffusion coefficient,* all equivalent terms. Fick's first law for electrons, the companion equation, is

$$f = -D_n \frac{dn}{dx}. \tag{2-53}$$

At this point it becomes evident that while D has something in common with q and μ, it is also different. That is, the explicit minus sign is needed for *both* holes and electrons because flux and gradient are opposite in sign for either particle. Putting the matter another way, we could say that if D, q, and μ were given the treatment conventionally accorded to algebraic symbols, q and μ would subsume opposite signs for holes and electrons, while in the case of D it would be the same sign, a minus sign.

Exercise 2-29. Find the dimensions of *density gradient*.

In testing the units of a derivative, the rule is to ignore the symbol d. Therefore dn/dx has the dimensions

$$\frac{[1/\text{cm}^3]}{[\text{cm}]} = [\text{cm}^{-4}].$$

(Note that one must avoid the meaningless formulation "$1/\text{cm}^3/\text{cm}$" because it is important to know which is the main fraction line.)

Exercise 2-30. What are the units of diffusivity?

From Fick's first law, Equation 2-52, we have

$$D_n = \frac{-f}{dn/dx}.$$

Therefore,

$$\frac{[1/\text{cm}^2 \cdot \text{s}]}{[1/\text{cm}^4]} = [\text{cm}^2/\text{s}].$$

The graphic device employed in Figure 2-25 can be pressed further, illustrating a point that many find difficult to accept at first. Suppose we add ten holes to each box in Figure 2-25(b), holes that will move out of the box in the time Δt. In other words, we increase hole density overall, while holding density gradient constant. The result will be *no change* in the flux indicated below the diagram. Hence diffusion transport of carriers is independent of *carrier density* and is dependent only on *density gradient*. We will come back to this elusive point a number of times.

Following this discussion of diffusion transport, we can move directly from flux equations to current-density equations. It is only necessary to multiply each flux by the charge carried by each particle, or q and $-q$, respectively, for the two cases. The resulting equations are the *diffusion-transport equations*:

$$J_p = -qD_p \frac{dp}{dx}; \tag{2-54}$$

$$J_n = qD_n \frac{dn}{dx}. \tag{2-55}$$

2-5.7 The Transport Equations

Drift and diffusion phenomena coexist in the general case. Dealing with this situation analytically involves simply adding the two transport equations already developed. The present section has been set apart in spite of its brevity because the resulting equations are so important. They exhibit the sign asymmetry discussed just above. For holes and electrons, respectively, the *transport equations* in one-dimensional form are

$$J_p = q\mu_p p E - qD_p \frac{dp}{dx}; \tag{2-56}$$

$$J_n = q\mu_n n E + qD_n \frac{dn}{dx}. \tag{2-57}$$

Of great importance is the fact that drift transport depends upon *two* variables, carrier density and electric field, while diffusion transport depends upon only *one* variable—density gradient. This fact is responsible for numerous subtleties that are encountered later.

2-5.8 The Einstein Relation

Both mobility and diffusivity are measures of how readily carriers move in response to the relevant gradients. It should not be surprising, therefore, that the two constants of proportionality are themselves proportional. This fact was worked out by Einstein in a study of Brownian motion [21]. In our context it can be illustrated by postulating a sample having a nonuniform net doping, and at equilibrium. Thus, this section embodies both of the brief exceptions just noted—a departure from nonequilibrium conditions and a departure from uniform doping. What follows is not a "proof" or a "derivation" of the Einstein relation; rather, it is a plausibility demonstration. The reason for this caveat is that we employed intuitive rather than rigorous developments of the equations governing drift and diffusion transport.

Visualize a one-dimensional *P*-type sample having a monotonic net-doping density declining from left to right. A constant doping gradient is not obligatory but is quite acceptable. At equilibrium, the hole-density profile will qualitatively resemble the net-doping profile, provided the gradient of net-doping density is small everywhere. Therefore, holes will diffuse from left to right, a phenomenon that is inevitable in the presence of a hole-density gradient. As a consequence, holes will accumulate at the right of the sample, over and above the density required for neutrality, and an accompanying positive space charge will exist there. (Refer back to Section 2-4.1.) By the same token there will be a deficiency of holes at the left of the sample, and a negative space charge there. The result of these space charges, in turn, will be an electric field directed right to left. This field will produce a right-to-

168 BULK PROPERTIES OF SEMICONDUCTORS

left drift transport of holes that counters the left-to-right diffusive transport. At equilibrium, these two current-density components will be precisely balanced at every x position. (Determining the resulting hole-density profile for an arbitrarily assigned net-doping profile is a nice problem, requiring numerical methods in the general case.)

Now choose an arbitrary position in the equilibrium sample and equate the countervailing drift and diffusion components of current. (This is the same as setting $J_p = 0$ in Equation 2-56.) Since

$$J_{p,\text{drft}} = q\mu_p p E, \tag{2-58}$$

and

$$J_{p,\text{diff}} = -qD_p \frac{dp}{dx}, \tag{2-59}$$

setting $J_p = 0$ yields

$$\frac{D_p}{\mu_p} = \frac{pE}{dp/dx}. \tag{2-60}$$

Now, with this constituting the first demonstration of the convenience of being able to express carrier density in terms of potential ψ, we make use of Equation 2-34. Differentiating it yields

$$\frac{dp}{dx} = -\frac{qn_i}{kT} e^{-q\psi/kT} \left(\frac{d\psi}{dx}\right). \tag{2-61}$$

Invoking Equation 2-36 again gives

$$\frac{dp}{dx} = -\frac{q}{kT} p \frac{d\psi}{dx}. \tag{2-62}$$

Substituting this result, along with $E \equiv -d\psi/dx$, into Equation 2-60 gives an expression that simplifies to

$$\frac{D_p}{\mu_p} = \frac{kT}{q}. \tag{2-63}$$

This is the *Einstein relation for holes,* and as usual there is a corresponding expression, the *Einstein relation for electrons:*

$$\frac{D_n}{\mu_n} = \frac{kT}{q}. \tag{2-64}$$

This result underscores the basic significance of the *thermal voltage, kT/q*. For the choice of room temperature used to take the data of Figure 2-22, $T = 300$ K, the thermal voltage amounts to 0.02586 V. As a practical matter, one's most frequent use of the Einstein relation is in calculating diffusivity, given mobility, or vice versa. For the asymptotic low-doping values of mobility in silicon given in Table 2-2, also for $T = 300$ K, the corresponding diffusivity values are $D_p = 11.6$ cm^2/s, and $D_n = 36.2$ cm^2/s. As a second practical matter, note that the Einstein relation is easy to remember because it rhymes, when stated as "*D* over μ equals *kT* over *q*."

The Einstein relation is a principle of considerable generality. A basic requirement is that the energy distribution for the species involved must be Maxwellian, a topic treated by McKelvey [22]. There is another fundamental point to be stressed. The equilibrium condition assumed earlier requires that there be *no* net transport of holes. Invoking precisely balanced drift and diffusion components of transport is a useful but artificial mathematical device. The reality is not countervailing *streams* of holes, but rather, holes undergoing their individual random walks under the influence of both gradients. The test of whether a given current component is *real* or not is whether there is power dissipation associated with it. In the equilibrium situation assumed, net current vanished, as did power dissipation.

Yet another point can be made concerning the physical picture just offered. In terms of the principle of detailed balance introduced in Section 2-5.2, one can regard the rightward diffusion transport of holes at a particular point as a "molecular process," and the leftward drift transport as the "reverse process." This throws a bit more light on the somewhat obscure phrase *molecular process*. The process and its reverse do not have to involve the same *mechanism*.

2-6 CARRIER RECOMBINATION AND GENERATION

The utility of various water analogies suggests that there may be yet another such analogy to illuminate carrier flow, but gross differences in the two kinds of media make this infeasible. Nonetheless, noting the differences is itself enlightening. First, water is essentially incompressible, while carrier densities can be altered from equilibrium values by enormous factors. A trillionfold increase or decrease in density is by no means uncommon, and is not in any sense a limit. Second, hole-electron pairs can be generated by thermal energy in a crystal or by external factors such as incident light, and the carriers can also disappear in pairs through the phenomenon of *recombination*. This process, which has no analog in the more familiar (water) medium, was mentioned in Section 2-1.4, and is examined quantitatively here.

2-6.1 Excess Carriers

Let us deal quite arbitrarily with an *N*-type sample. All of the expressions developed can be written for the *P*-type case by simply reversing the symbols *n* and *p*, *N* and *P*, and *D* and *A*. With carrier densities now permitted to depart from equilibrium values, it is helpful to introduce the subscript zero that is commonly used to identify equilibrium values. Thus majority electrons in the *N*-type sample chosen have the equilibrium density

$$n_{0N} = N_D - N_A, \qquad (2\text{-}65)$$

and minority holes, the density

$$p_{0N} = \frac{n_i^2}{n_{0N}}, \qquad (2\text{-}66)$$

accepting the approximations introduced in Sections 2-2 and 2-4. In addition, we continue the practice of using a subscript P or N to designate the conductivity type of the sample in question. This gives immediate information on which are majority carriers and which are minority carriers. Let us use symbols without a zero subscript to designate actual densities, now different from the equilibrium densities in the general case. We define the *difference* in the actual and equilibrium densities as the density of *excess carriers* (designated by the prime superscript),

$$p'_N = p_N - p_{0N}, \qquad (2\text{-}67)$$

in this case, for minority carriers. In similar fashion, the excess majority-carrier density is

$$n'_N = n_N - n_{0N}. \qquad (2\text{-}68)$$

In the most common situation the sample remains neutral in the presence of excess carriers; recall that carriers can be generated in pairs and can disappear in pairs. Under these conditions, the two excess-carrier densities must match, or

$$p'_N = n'_N \qquad (2\text{-}69)$$

under neutral conditions. It is worth noting that excess-carrier densities, unlike equilibrium and total carrier densities, can be positive *or* negative. A negative excess density simply indicates a total density depressed below the equilibrium value.

The simplest problem involving excess carriers is one in which their densities are constant throughout the sample and are time-independent as well. This situation can be approximated closely by specifying several conditions: Choose a thin sample of N-type silicon, once more being arbitrarily specific. Let it be irradiated on a major face by *penetrating* radiation. This term designates radiation of an energy just barely able to produce pairs; as a result, most of the radiation passes right through the thin sample. Since the critical energy ξ_G is about 1 eV, corresponding to infrared (IR) radiation, it follows that infrared radiation qualifies as "penetrating." (In fact, a very thin sample of silicon held before a source of white light has the appearance of a piece of red glass.) The point of specifying penetrating radiation is to make pair generation improbable, and therefore essentially uniform through the volume of the thin sample. (If more energetic radiation were used, such as ultraviolet (UV) light, the absorption would mainly occur close to the irradiated face.)

The situation just described is depicted in Figure 2-26. By virtue of having (1) a thin sample, (2) penetrating radiation, and (3) steady-state radiation, we have a

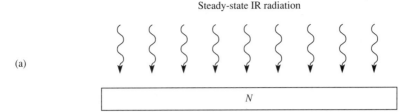

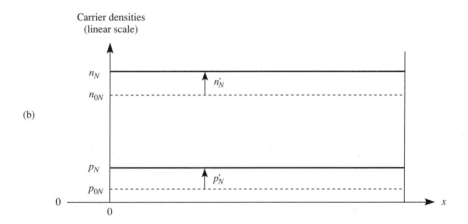

Figure 2-26 The simplest excess-carrier situation. (a) Thin, N-type sample being irradiated by steady-state infrared (penetrating) radiation. (b) The perturbed carrier densities that result, illustrating the equal excess-carrier densities n'_N and p'_N.

situation independent of time and independent of all spatial variables as well. This is another situation in that special steady-state category wherein a forward molecular process (pair generation) and its reverse (pair recombination) are precisely balanced, providing a clear-cut illustration of the conditions invoked in Exercise 2-23.

Exercise 2-31. Estimate the net doping of the sample depicted in Figure 2-26.

Since the density axis is calibrated linearly, we see that $n_{0N} \approx 9 p_{0N}$. The law of mass action thus informs us that here

$$(9 p_{0N}) p_{0N} = n_i^2,$$

or

$$p_{ON} = \frac{n_i}{3}.$$

Thus

$$p_{ON} = (10^{10}/3)/\text{cm}^3,$$

and

$$n_{ON} = 3 \times 10^{10}/\text{cm}^3.$$

The sample is very near intrinsic.

Now let us set the near-intrinsic case aside and examine a more realistic extrinsic case (again N-type). Assume

$$n_{ON} = N_D - N_A = 10^{14}/\text{cm}^3, \qquad (2\text{-}70)$$

so that

$$p_{ON} = 10^6/\text{cm}^3. \qquad (2\text{-}71)$$

Note that the equilibrium majority-to-minority density ratio is 10^8, or 100 million. Suppose we now use a band-diagram representation of this sample, similar to those used in Figures 2-14 through 2-16. Choose the volume to be represented of such a size that it will contain one hole, and therefore 10 electrons, as shown in Figure 2-27(a). The necessary volume is 10^{-6} cm^3.

Next, choose a radiation intensity such that the number of holes in the chosen volume is increased to ten, a tenfold increase because of the additional nine holes now present on the average as a result of the steady-state irradiation. That is, we now have

$$p'_N = p_N - p_{ON} = 9/(10^{-6} \text{ cm}^3). \qquad (2\text{-}72)$$

This situation is shown in Figure 2-27(b). Note particularly that in spite of the tenfold increase in minority-hole density the *relative* increase in majority-electron density is completely negligible. Nonetheless the *absolute* increase in the number of electrons in the chosen volume is the same as for holes—nine, or

$$n'_N = n_N - n_{ON} = 9/(10^{-6} \text{ cm}^3). \qquad (2\text{-}73)$$

Notice especially that even if the hole density had been increased by a factor of 100,000 (by more intense radiation), the *relative* increase in majority-carrier density would still be negligible. But the *absolute* increases in the two carrier populations would still be matched. This interplay of relative and absolute increments is a key concept one must master in order to understand semiconductor materials. As long as the total minority-carrier density remains negligible compared to the equilibrium

CARRIER RECOMBINATION AND GENERATION 173

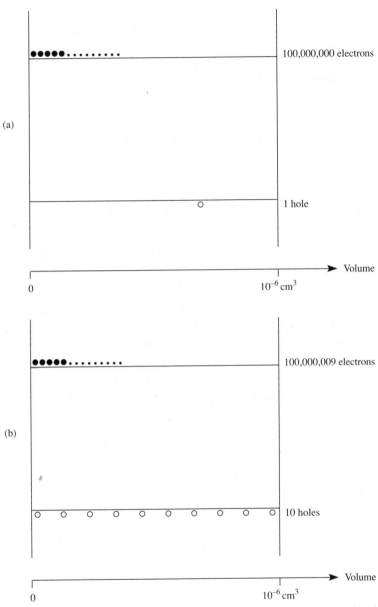

Figure 2-27 A case of low-level excess-carrier generation. (a) A sample with $n_{0N} = 10^{14}/\text{cm}^3$; the volume represented by this equilibrium band diagram is 10^{-6} cm^3, which therefore contains one hole. (b) Weak IR radiation causes a tenfold increase in minority-hole density, but an equal absolute increase in majority-electron density, which nonetheless is utterly negligible on a relative basis.

majority-carrier density, we have what is known as a *low-level* disturbance of the equilibrium condition. When the total minority-carrier density approaches and exceeds the equilibrium majority-carrier density, we have a *high-level* disturbance; high-level disturbances can produce carrier densities exceeding the equilibrium majority-carrier density by enormous factors.

Exercise 2-32. In Figure 2-27, suppose that IR radiation yielded $p'_N = 10^{12}/\text{cm}^3$. Calculate the total minority- and majority-carrier densities. Characterize the conditions existing in the sample.

With an excess-carrier density of 1 trillion, we have

$$p_N \approx p'_N = 10^{12}/\text{cm}^3,$$

and

$$n_N = (1.01)\, n_{0N} = 1.01 \times 10^{14}/\text{cm}^3.$$

With the 1% increase in total majority-carrier density, we have a case on the borderline of high-level conditions.

When the presence of excess carriers significantly changes the *majority-carrier density*, one has the condition known as *conductivity modulation*. Both terms of the conductivity equation, Equation 2-50, are significantly altered by the presence of the excess-carrier populations.

Exercise 2-33. Assuming that $\mu_n \approx 3\mu_p$, calculate the factor of conductivity increase in the sample of Figure 2-27 if radiation causes $p'_N = n'_N = n_{0N}$.

In this case, the electron term of the conductivity equation becomes

$$q\mu_n n_N = 2q\mu_n n_{0N},$$

and the hole term becomes

$$q\mu_p p_N \approx q(\mu_n/3) n_{0N}.$$

Hence the rate of the conductivity values after and before irradiation is

$$\frac{\mu_n n_{0N}[2 + (1/3)]}{\mu_n n_{0N}} = 2\frac{1}{3}.$$

2-6.2 Low-Level Recombination Rate

In the equilibrium situation depicted in Figure 2-27(a), there exists a certain pair-generation rate attributable to thermal energy in the silicon crystal. Let us call this equilibrium value G_0, measured in units of reciprocal volume and time (1/cm^3·s). It is precisely balanced by the rate R_0 at which carrier pairs disappear at equilibrium, also having the units 1/cm^3·s. Now go to the nonequilibrium situation of Figure 2-27(b), with a minority-carrier density that has been increased tenfold. In this case as well as the equilibrium case, the minority density is literally millions of times smaller than majority density; consequently, the recombination rate that is observed is *proportional to the minority-carrier density*. In other words, it is the scarcity of minority carriers that fixes recombination rate. The total recombination rate R is the sum of the equilibrium rate and an incremental rate, or

$$R = R_0 + R'; \tag{2-74}$$

the incremental recombination rate is nine times larger than R_0 and the total rate is ten times larger. In summary, we can say that

$$R = Cp_N, \tag{2-75}$$

where C is a constant, and that, using the *same* constant, we can write

$$R_0 = Cp_{0N}, \tag{2-76}$$

and

$$R' = Cp'_N. \tag{2-77}$$

Exercise 2-34. Find the units of C.

From Equation 2-75

$$R/p_N = C,$$

and so

$$\frac{[1/\text{cm}^3 \cdot \text{s}]}{[1/\text{cm}^3]} = [1/\text{s}].$$

It is also valid to write

$$G = G_0 + G', \tag{2-78}$$

where G is the total generation rate and G' is the incremental generation rate. For the net rate of change of density p_N, we have, under the conditions of Figure 2-27(b).

$$\frac{dp_N}{dt} = G - R = 0, \tag{2-79}$$

because a steady state was explicitly assumed. Next, let us treat a departure from steady-state conditions.

2-6.3 Time-Dependent Recombination

In Equation 2-79 we can write G and R each in terms of an equilibrium rate and an incremental rate, yielding

$$\frac{dp_N}{dt} = (G_0 + G') - (R_0 + R') = 0. \tag{2-80}$$

In the previous section we pointed out that the equilibrium rates G_0 and R_0 must cancel, so Equation 2-80 becomes

$$\frac{dp_N}{dt} = G' - R' = 0, \tag{2-81}$$

or, using Equation 2-77,

$$\frac{dp_N}{dt} = G' - Cp'_N. \tag{2-82}$$

Next suppose that at $t = 0$, the source of infrared radiation is switched off, so that G' is changed abruptly to zero. The result is a transient situation for which the applicable differential equation is obtained from Equation 2-82 by dropping G':

$$\frac{dp_N}{dt} = -Cp'_N. \tag{2-83}$$

From Equation 2-67, it is equivalent to write

$$\frac{dp'_N}{dt} = -Cp'_N; \tag{2-84}$$

the total and excess densities differ by a constant, p_{0N}. Recalling that C has dimensions of reciprocal time, let us define a new symbol

$$\tau \equiv \frac{1}{C}, \tag{2-85}$$

which of course has time dimensions, yielding

$$\frac{dp'_N}{dt} = -\frac{p'_N}{\tau}. \tag{2-86}$$

Next, separate variables in this differential equation and integrate from $t = 0$ to an arbitrary time t, and from $p'_N = p'_N(0)$ to the corresponding excess-minority-carrier density $p'_N(t)$:

$$\int_{p'_N(0)}^{p'_N(t)} \frac{dp'_N}{p'_N} = -\int_0^t \frac{dt_1}{\tau}. \tag{2-87}$$

The symbol t_1 is a dummy variable for integration. Carrying out the integration and substituting limits yields

$$\ln \frac{p'_N(t)}{p'_N(0)} = -\frac{t}{\tau}, \tag{2-88}$$

or

$$p'_N(t) = p'_N(0)e^{-t/\tau}. \tag{2-89}$$

This solution is sketched in Figure 2-28.

From Equations 2-77, 2-85, and 2-86, we find that

$$\frac{dp'_N}{dt} = -\frac{p'_N(t)}{\tau} = -R'. \tag{2-90}$$

The quantity R' is the *net recombination rate,* or the difference between R and R_0, described earlier as an incremental recombination rate. It is often described as a "recombination rate for excess carriers." The problem with that description is the implication that one can distinguish between an "excess" carrier and a "normal" carrier. But carriers are quite indistinguishable, and so the concept of *net rate* is what counts.

The geometrical meaning of Equation 2-90 is made clear in Figure 2-28. The *net recombination rate* at any time t is the *slope* of the density function at that time. A tangent to the function at a particular point displays the slope in question, and it is a property of a declining exponential function that such a tangent intersects the

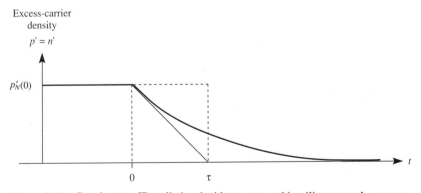

Figure 2-28 Steady-state IR radiation incident upon a thin silicon sample causes constant and equal excess-carrier populations to exist in the sample. Switching off the radiation source at $t = 0$ leads to an exponential return to equilibrium conditions with the carrier lifetime τ characterizing the exponential function.

abscissa at a distance on the axis from the tangency position that is *independent* of the independent variable. In the present case this constant value (known as the *decrement* of a logarithmic function) is the characteristic time τ. This relationship is displayed in Figure 2-28 for $t = 0$, but it holds for any $t > 0$ as well. Figure 2-28 also emphasizes that $p'_N = n'_N$.

It was pointed out in Section 2-6.1 that an excess-carrier density can be positive or negative. A negative value simply means that the corresponding total density is less than its equilibrium value, a situation that frequently occurs. When the disturbance responsible for this condition is abruptly terminated, the total density rises, exponentially approaching its equilibrium value, and once more exhibits the time constant τ in the process.

2-6.4 Carrier Lifetime

An experiment of the kind sketched in Section 2-6.3 above can be used to determine the average life of a carrier. This is the mean time that a carrier identified at an arbitrary instant and location will exist until recombination. It is a concept and characteristic time that applies to all carriers in the sample. But it is most easily measured by observing the incremental recombination rate as an excess-carrier population declines to zero, after an external disturbance (such as the IR lamp) is switched off. The chance population of excess carriers that exists at $t = 0$ will exhibit the average life we wish to determine. That involves summing their individual lives beyond $t = 0$ and dividing by the number at $t = 0$. Continuing to focus on minority holes in an extrinsic N-type sample, we wish to analyze the situation shown in Figure 2-29. Suppose that a microscopic volume identified as (vol) contains ten holes at $t = 0$. The object, then, is to track the product $(\text{vol})p'_N$ as a function of time for $t > 0$. Summing the ten individual life-spans that are indicated heuristically in Figure 2-29 and dividing by the number of incremental holes present at $t = 0$ amounts to evaluating the expression

$$\text{Average carrier life} = \frac{\int_0^\infty (\text{vol})p'_N(0)e^{-t/\tau}\,dt}{(\text{vol})p'_N(0)}, \tag{2-91}$$

which yields

$$(-\tau)e^{-t/\tau}\Big|_0^\infty = \tau. \tag{2-92}$$

For this reason the constant τ is termed *carrier lifetime*. Sometimes the phrase *excess-carrier lifetime* is used because practical experiments usually track the disappearance of incremental carriers. But it is important, once more, to realize that for the simplified conditions treated here, the lifetime τ characterizes *all* the carriers in the sample. Notice that combining Equations 2-91 and 2-92 informs us that the integral of $p'_N(t)$ from $t = 0$ to $t = \infty$ is equal to $\tau p'_N(0)$, an elementary property of

CARRIER RECOMBINATION AND GENERATION   179

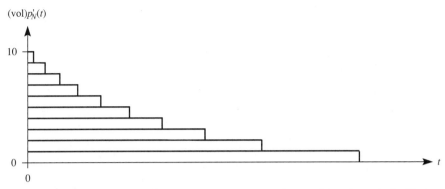

Figure 2-29 At $t = 0$ there are ten minority excess carriers in the volume (vol). Their average life, or carrier lifetime τ, can be determined by adding the individual lives and dividing by the number present at $t = 0$.

the declining exponential function. The rectangle with dimensions $p'_N(0)$ and τ is indicated on Figure 2-28.

Exercise 2-35. Suppose that the ten holes present at $t = 0$ in Figure 2-29 had been created in the volume (vol) at $t = 0$ by ten photons. How would the function (vol)$p'_N(t)$ differ for $t > 0$ in such a case? (After all, the holes present at $t = 0$ have various "ages" in the experiment of Figure 2-29, which involves steady-state irradiation prior to $t = 0$.)

The function (vol)$p'_N(t)$ versus t for $t > 0$ is completely independent of events prior to $t = 0$. All that matters is the *number* of holes at $t = 0$. They may have been produced by ten photons at $t = 0$ in an "impulse" situation, or by the steady-state irradiation of Figure 2-29, which inevitably involved a larger number of photons, or by decay from a larger density at an earlier time. Statistics can predict accurately the number of holes that will exist at a given later time, but cannot predict the longevity of a particular hole. (An actuary faces a highly analogous situation with respect to the survival of people.) Furthermore, the density function for $t > 0$ does not depend in any way on the history of a particular hole, such as its "age" at $t = 0$.

Now let us return to a point made earlier that deserves additional emphasis. This is the point that the all-important *net* rate of change of carrier density can be found by consideration of either the *total* generation and recombination rates, or the *incremental* generation and recombination rates. Once again take the situation depicted in Figure 2-28, wherein lifetime was being measured. In Figure 2-30(a), we represent the situation in terms of (i) total minority-carrier density, (ii) total genera-

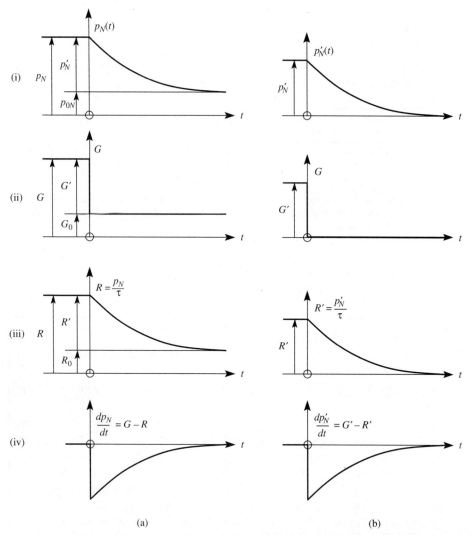

Figure 2-30 Net rate of change of carrier density is the same, whether calculated from (a) total values, or (b) excess and incremental values. (i) Minority-carrier density. (ii) Generation rate. (iii) Recombination rate. (iv) Net rate of change of density.

tion rate, and (iii) total recombination rate. Subtracting the last two yields the desired (iv) net rate of carrier-density change (negative because recombination predominates). In Figure 2-30(b), then, the same situation is described in terms of (i) excess minority-carrier density, (ii) incremental generation rate, and (iii) incremental recombination rate. Subtracting the last two once more yields a (iv) net rate of carrier-density change that is identical to that obtained using total values throughout.

Exercise 2-36. What changes are needed in Figure 2-30 to make the various diagrams apply to majority carriers n_N?

Several changes are needed: (1) Replace the symbol p everywhere by the symbol n. (2) In Figure 2-30(a-i), place a gap in the lower part of the ordinate scale, or else obtain a piece of graph paper that extends approximately to the moon. (3) In Figure 2-30(a-iii), delete the expression n_N/τ, which is without physical relevance and meaning. In the remaining diagrams, no further changes are needed.

Here it is worthwhile to return to a point made in Section 2-5.2, where the principle of detailed balance was cited. Sometimes, it was noted, calculation of the forward molecular process is straightforward, while calculation of its reverse is obscure, or vice versa. But the *principle* gives us assurance that we can calculate *both*, a nontrivial consequence of a seemingly obvious statement. We have here an example. Calculation of the equilibrium generation rate from first principles is not straightforward; one needs detailed knowledge of phonon spectra, and more. But calculation of the equilibrium recombination rate *is* straightforward, from $R_0 = p_0/\tau$ (for an N-type sample). Expanding beyond equilibrium cases to steady-state cases, we note that $G' = R' = p'/\tau$, and $G = R = p/\tau$, for the same sample.

2-6.5 Recombination Mechanisms

Let us postulate two possible ways for recombination to occur, and then examine the consequences in the light of principles already developed. The first and simplest event would involve having a conduction electron lose sufficient energy to "drop" directly into a hole in the valence band, thereby eliminating a pair of carriers. Such an event, which could occur anywhere in the crystal, is described as *band-to-band* recombination.

A second mechanism, by contrast, is one that could occur only at a specific location within the crystal, a site that is labeled a *recombination center*. Such a center is seen as a defect of some kind. It could be an out-of-place atom, a foreign atom, or a dislocation, for example. All such defects are capable of introducing localized (in energy and position) states within the bandgap. The function of such a state when filling the role of recombination center is to "seize" a carrier of one kind and hold it until a carrier of the opposite kind arrives at the site, with recombination being the consequence. The kinds of defects that carry out this function most efficiently are those that introduce states near the middle of the gap. The term *deep states* is applied generically to those located far from the band edges, while the adjective "shallow" is sometimes applied to states contributed by the usual dopants.

At the top of Figure 2-31 is a schematic representation of the two recombination mechanisms, making use of band diagrams. The Fermi-level indications there inform us that the two samples are N-type and have equivalent doping. (We will continue to use the extrinsic N-type case for explanation.) Take first the case of low-level excess-carrier populations. For reasons described in Section 2-6.3, the recombination rate is fixed by the scarcity of minority carriers, holes, and is proportional to their density. As before we can talk in terms of total hole density p_N and total recombination rate R, or incremental density p'_N and rate R'. This time let us choose the former. By increasing the intensity of irradiation enough to double p_N, we produce a doubling of R.

Turning to Figure 2-31(b), we see that the recombination center assumed is a state located at the center of the gap. Since it is well below the Fermi-level "surface," this isolated state will be occupied by an electron most of the time. It is the rare arrival of a hole that triggers recombination. And it follows, therefore, that the low-level recombination R will once more be proportional to p_N.

Now let the light source, or infrared lamp, be increased in intensity so that the

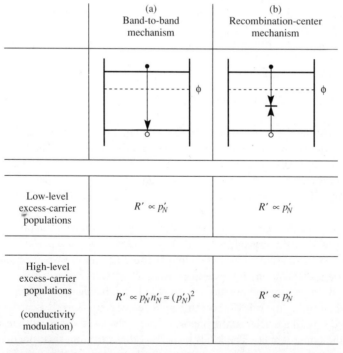

Figure 2-31 Recombination-rate predictions of two possible recombination mechanisms. (a) The band-to band mechanism predicts for low-level and high-level cases respectively, a linear and square-law dependence upon excess-carrier density. (b) The recombination-center mechanism predicts linear dependence for *both* cases, which is observed experimentally.

total minority-carrier density is very large even when compared to the equilibrium *majority*-carrier density n_{0N}. Figure 2-32 gives this situation a graphical representation, by letting bar graphs stand for the relevant density magnitudes. It is evident that all four densities (two total, two excess) are roughly equal under these extreme conditions. Since $p_N \approx n_N$, it is clear that recombination is no longer proportional to a vastly outnumbered species, but instead should be proportional to a density product $p_N n_N$, or approximately to the square of the minority density p_N^2, as shown in Figure 2-31(a).

The recombination-center mechanism predicts a different result, however. With carrier densities that are roughly equal, the recombination center at the middle of the gap will be occupied approximately half time by each species. With the occupancy probability of the recombination center thus equal to about 50%, recombination depends upon the arrival of a hole when the center is occupied by an electron, and vice versa—so the rate R clearly depends *linearly* on the density p_N (or n_N). In fact, it is a high-level recombination rate that is *linear in the densities* that was observed experimentally at an early time, establishing the dominance of the *recombination-center* mechanism in silicon samples [23].

Band-to-band recombination can occur in silicon, but is quite improbable. The reason is that a means must exist to accomplish momentum transfer as well as energy transfer in the hole–electron interaction, making such recombination a kind

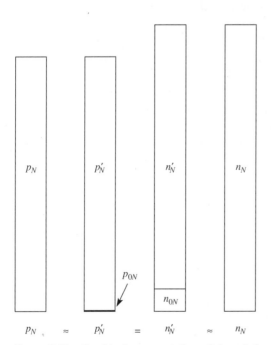

Figure 2-32 Graphical representation of the relationships of total densities and excess densities under high-level conditions that are extreme. All four densities are approximately equal.

of three-body problem. The band structure of silicon is described as *indirect* on account of this requirement. There are other semiconductor materials, however, GaAs being notable among them, in which high densities of electrons at the band edge have momentum values matching those of many holes at the other band edge, so that recombination can occur in a process involving energy loss only, usually by radiation. This makes such recombination very probable. A material of this kind is said to have a *direct* band structure.

At this point a note on terminology is in order. The term *trap* is used in relatively recent literature to designate a state that effectively is in communication with one band only. More precisely, a state constitutes an electron trap if an electron in the state has a higher probability of a transition to the conduction band than to the valence band. A parallel description applies to a hole trap. Just which behavior a particular state exhibits depends upon Fermi-level position, and this is subject to variation. Thus when this criterion is used, a given state can change its description as conditions vary. In the older literature, however, the term trap designates any state located in the bandgap, including a recombination center. In fact, *trap* and *recombination center* were often used interchangeably, a practice we also employ.

Exercise 2-37. According to quantum theory, the probability of an electron or a hole transition from one state to another declines as the energy spacing of the two states increases. What does this tell you about the energy position within the bandgap for a recombination center that will maximize its effectiveness in raising recombination rate R_0 and generation rate G_0?

A valid point of view is that for a recombination center to function, a hole and an electron must meet there. As a matter of further arbitrary definition, let us take the case of an electron transition from ξ_C to ξ_T (the "trap" position) and of a hole transition from ξ_V to ξ_T. Let these respective probabilities be labeled $P_C(\xi_T)$ and $P_V(\xi_T)$. Figure 2-33 shows qualitatively the behavior of these two probabilities as a function of ξ_T. Since a transition by *each* particle is required for recombination, the probability of a recombination event is proportional to the *product* of the two probabilities, or $P_C(\xi_T)P_V(\xi_T)$, which evidently has its maximum value at $\xi_T = \xi_I$.

As is evident in Table 2-3, many impurity elements create deep states in silicon, with heavy metals predominating. In particular, a gold atom introduces two states, one of which is an acceptor state very near the center of the gap. As we have just seen, this makes a substitutional gold atom a particularly effective recombination center. In the early years of silicon technology, gold was often introduced intentionally into devices whose performance was enhanced by low carrier lifetime. (It may be strange to think of gold as an "impurity.") The rapid disappearance of excess carriers often made for faster operation or "switching" of the device. But undesirable side effects also came with *gold doping*, as it was known, and so structure and

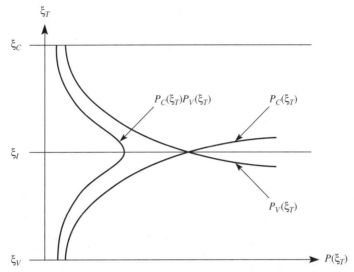

Figure 2-33 The probability of a recombination event or a generation event has its maximum value for a recombination-center location at the middle of the bandgap.

circuit techniques were devised to reduce or avoid excess-carrier introduction in the first place, thus eliminating the need for lifetime-reducing methods.

Exercise 2-38. Consider a sample having a fixed density of traps located precisely at the bandgap center. Now imagine that net doping is varied monotonically from extrinsic N-type to extrinsic P-type conditions. What variation, if any, in lifetime τ would you expect?

Through the extrinsic N-type range, lifetime τ is constant, which is to say, recombination rate is proportional to minority-carrier density. But as the near-intrinsic range is entered, the recombination center or trap must wait a significant time for the arrival of *either* carrier type, and so lifetime τ increases. Then it declines again and becomes constant in the extrinsic P-type range. This behavior was predicted explicitly in the lifetime equation derived by Hall [23], and later in more general form by Shockley and Read [24].

A note is also in order concerning the shallow states contributed by conventional impurities. While these are not efficient recombination centers, the presence of large densities of donors or acceptors can cause other kinds of defects that lead to a reduction in carrier lifetime. One factor is atomic size. A boron atom is smaller than a silicon atom, a phosphorus atom is larger. A high concentration of either impurity can cause strain in the silicon crystal, which in turn can cause crystalline

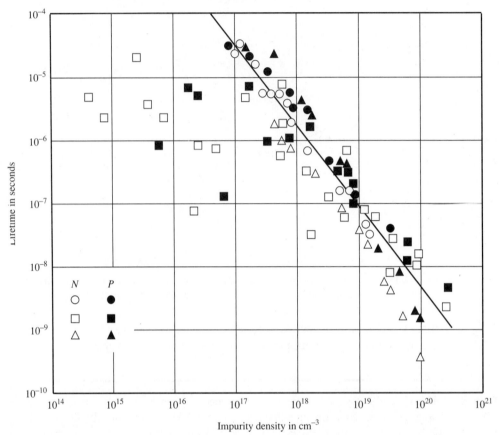

Figure 2-34 Carrier-lifetime data for 300 K obtained from measurements by three groups of investigators on silicon doped with boron and phosphorus. (After a compilation by Slotboom [25], with permission.)

imperfections that serve as recombination centers. As a result, there is a correlation between doping-impurity density and lifetime τ. Typical values based on extensive data collection [25] are shown in Figure 2-34 and they can be seen to range from about 10^{-9} s to 10^{-5} s (1 ns to 10 μs).

2-6.6 Relative and Absolute Carrier Densities

Problems involving excess-carrier densities have both relative and absolute aspects of great importance. Let us take the relative case first. To display densities in relative fashion, a logarithmic scale is useful. For illustration, take an extrinsic N-type sample with $n_{0N} = 10^{15}/\text{cm}^3$, so that $p_{0N} = 10^5/\text{cm}^3$. Again let a thin sample be irradiated with IR photons, producing excess-carrier densities $p'_N = n'_N = 10^6/\text{cm}^3$. In Figure 2-35(a) the sample is represented, and in Figure 2-35(b) the

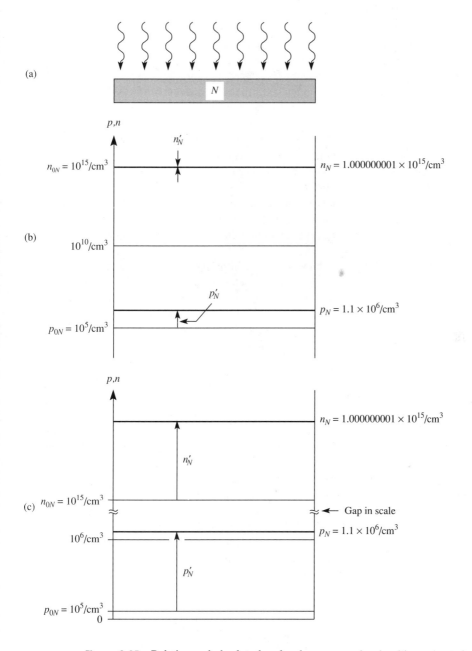

Figure 2-35 Relative and absolute low-level excess-carrier densities presented graphically. (a) Thin sample under penetrating radiation. (b) Logarithmic ordinate showing that excess majority carriers are relatively negligible. (c) Linear ordinate, showing the equality of majority and minority excess carriers needed to assure sample neutrality.

carrier densities are plotted. Notice that the elevenfold increase in hole density is easily seen, whereas n'_N is utterly negligible compared to n_{0N} and cannot be seen in the diagram. Numerically, hole density has changed from $10^5/\text{cm}^3$ to $1.1 \times 10^6/\text{cm}^3$, while electron density has changed from $10^{15}/\text{cm}^3$ to $1.000000001 \times 10^{15}/\text{cm}^3$. When the relative change in majority density is not visible in a diagram such as Figure 2-35(b), low-level conditions prevail.

In Figure 2-35(c) the same density data are presented, but this time with a *linear* ordinate. The presence of an origin or *zero* on the scale and the calibration of the scale both mark it immediately as linear. (It is of great importance to note that the ordinate in Figure 2-35(b) has no origin.) The diagram immediately communicates the equality of p'_N and n'_N, but the price of this is the gap in the ordinate. If the entire diagram were drawn without the gap, but at the indicated scale, it would extend approximately around the earth! The excess majority carriers, so negligible on a relative basis, are nonetheless important in that their presence preserves *space-charge neutrality*.

It is frequently important to convert from logarithmic to linear presentation of the data in order to secure a complete picture of a given situation. The former has the advantage of presenting wide-ranging values on a single diagram, but the disadvantage of distortion. The second has the advantage of undistorted presentation, but requires a gap in the scale to present widely different values.

2-7 CONTINUITY EQUATIONS

The idea expressed in a continuity equation has to do with the unbroken character of a line of some kind, such as a flow line in a hydrodynamic problem or the idea that "what goes into a volume must come out of the volume or disappear within it." Poisson's equation (Section 1-3.7), for example, conveys the idea that the mythical line of force extends from a positive charge to an equal negative charge. The importance of continuity equations in general is that they constitute starting points for analysis. We have already used Poisson's equation in this way (Problem C1-1) and will do so again in Chapter 3. Our present interest is in the continuity of carrier density. Therefore, it is necessary to examine carrier transport in and out of a given volume, and the creation and disappearance of carriers through generation and recombination, respectively, within the volume.

2-7.1 Constant-E Continuity-Transport Equations

A general attack on the problem of carrier-density continuity would yield the densities n and p, each as a function of four independent variables. That is, $n(x, y, z, t)$ and $p(x, y, z, t)$ would constitute the problem solution. In addition, a general analysis would permit the specifying of several functions:

- Temperature $T(x, y, z, t)$
- Electric field $E(x, y, z, t)$
- Carrier-generation rate $G(x, y, z, t)$
- Net doping $N(x, y, z)$
- Carrier lifetime $\tau(x, y, z)$
- Carrier mobility $\mu(x, y, z, t)$
- Carrier diffusivity $D(x, y, z, t)$

The time dependences of μ and D are a consequence of their high-level dependences on n and p, and the fact that n and p are time-dependent.

We will derive a pair of equations, one for electrons and one for holes. Our treatment of the problem, however, will retreat a long way from the degree of generality outlined above. First, for the sake of clarity and simplicity, we will once again reduce the problem to one spatial dimension, thus seeking the solutions $n(x, t)$ and $p(x, t)$. But in a more substantive way, let us add these restrictions by assumption:

- T = constant. The usual isothermal assumption is maintained, with room temperature chosen in most cases.
- E = constant or is negligible. Analytic complexity mounts steeply when electric-field variations are permitted.
- $G = G_0$. Carrier generation caused by external agencies is ruled out in the space and time ranges wherein the continuity equations will be applied, or briefly, $G' = 0$.
- N = constant. No net-doping variations are permitted.
- τ = constant. This specification effectively extends the previous assumption to include N_D = constant and N_A = constant.
- μ = constant. This further restricts electric-field values to the ohmic regime.
- D = constant.

The last two restrictions require that

- Low-level conditions exist.

When certain of these restrictions are tightened further, others can be relaxed. This is a point best illustrated after the continuity equations have been derived. But to apply the equations to every problem that can be devised, all of the restrictions must be observed. Because more general continuity equations exist, we have elected the label *constant-E continuity-transport equations* to identify the restricted expressions [26]. These restricted forms are by far the most commonly used.

To start the derivation, focus on an elementary volume deep inside a silicon sample wherein the density n varies only in the x direction and with time. Figure 2-36 depicts the elementary volume of area A and length dx, located at x_1. It is evident that for the number of electrons inside this volume, the rate of change is given by

$$\text{Rate of change} = A\,dx \left.\frac{\partial n}{\partial t}\right|_{x_1, t}. \tag{2-93}$$

The dimensions in this equation are reciprocal seconds, and the partial derivative is

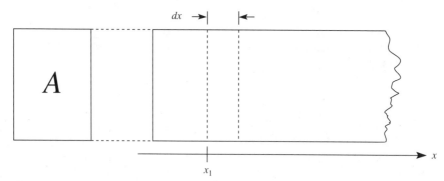

Figure 2-36 A region within a silicon sample where one-dimensional conditions prevail, useful for deriving a continuity equation.

appropriate because there are two independent variables, x and t. There are two broad categories of phenomena that can give Equation 2-93 a finite value: One is having flow into the volume different from flow out; the second (for the restrictions assumed) is having net recombination within the volume. Hence, it is possible to write

$$A dx \left.\frac{\partial n}{\partial t}\right|_{x_1,t} = A dx \left[(\text{Flow in}) - (\text{Flow out}) \right] + A dx \left[-\frac{n'(x_1, t)}{\tau} \right]. \quad (2\text{-}94)$$

To deal with the flow term, let us reintroduce the concept of flux f, which is now a function of x and t. Thus, for the volume face located at $x = x_1$, the flux of electrons into the volume is $f(x_1, t)$. If the electrons happen to have a net flow out, it simply means that $f(x_1, t) < 0$. At the right-hand face of the elementary volume, a rudimentary principle of calculus lets us write the electron flux as

$$f(x_1 + dx, t) = f(x_1, t) + \left.\frac{\partial f}{\partial x}\right|_{x_1,t} dx. \quad (2\text{-}95)$$

(Once more, a negative value simply means that the flux of electrons is leftward rather than rightward.) Multiplying each of these fluxes by area A gives flow into the volume at $x = x_1$ and flow out at $x = x_1 + dx$, and the difference of the two resulting expressions can be substituted for the first term of Equation 2-94, yielding

$$A dx \left.\frac{\partial n}{\partial t}\right|_{x_1,t} = \left\{ A f(x_1, t) - A \left[f(x_1, t) + \left.\frac{\partial f}{\partial x}\right|_{x_1,t} dx \right] \right\} \quad (2\text{-}96)$$

$$+ A dx \left[-\frac{n'(x_1, t)}{\tau} \right].$$

Evidently the two terms in $f(x_1, t)$ cancel, and then the factor $A dx$ cancels throughout, leaving an expression valid anywhere in the one-dimensional sample, or

$$\frac{\partial n}{\partial t} = -\frac{\partial f}{\partial x} - \frac{n - n_0}{\tau}, \quad (2\text{-}97)$$

where the excess-carrier definition has been substituted for n'.

Equation (2-97) is a skeleton form of the desired equation expressed in terms of flux. Introducing the two flux-causing mechanisms comprehended here—drift and diffusion—can be accomplished by using the electron-transport equation:

$$J_n = q\mu_n nE + qD_n \frac{\partial n}{\partial x}. \tag{2-98}$$

It has been repeated here because the partial derivative is needed in the present case. The desired flux expression is found by dividing Equation 2-98 by $-q$, yielding

$$f = -\mu_n nE - D_n \frac{\partial n}{\partial x}. \tag{2-99}$$

Thus, the first term on the right-hand side of Equation 2-97 can be written

$$-\frac{\partial f}{\partial x} = \mu_n E \frac{\partial n}{\partial x} + D_n \frac{\partial^2 n}{\partial x^2}. \tag{2-100}$$

At this point, the important simplifying effect of constant electric field has become evident. Substituting Equation 2-100 into Equation 2-97 yields the constant-E continuity-transport equation for electrons, and an analogous analysis yields that for holes as well:

$$\frac{\partial n}{\partial t} = D_n \frac{\partial^2 n}{\partial x^2} + \mu_n E \frac{\partial n}{\partial x} - \frac{n - n_0}{\tau}; \tag{2-101}$$

$$\frac{\partial p}{\partial t} = D_p \frac{\partial^2 p}{\partial x^2} - \mu_p E \frac{\partial p}{\partial x} - \frac{p - p_0}{\tau}. \tag{2-102}$$

The algebraic-sign asymmetry is attributable to the explicit-sign convention that applies to mobility. Next we turn to important problems that yield to analyses that start from one of these continuity equations.

2-7.2 Continuity-Equation Applications

Each of the four terms of the constant-E continuity-transport equation has an evident origin and identification. Continuing to focus on minority holes in N-type silicon for consistency with the earlier parts of Section 2-6, we can use these labels to identify the four respective terms:

$$\underset{\substack{\text{Accumulation} \\ \text{term}}}{\frac{\partial p_N}{\partial t}} = \underset{\substack{\text{Diffusion} \\ \text{term}}}{D_p \frac{\partial^2 p_N}{\partial x^2}} - \underset{\substack{\text{Drift} \\ \text{term}}}{\mu_p E \frac{\partial p_N}{\partial x}} - \underset{\substack{\text{Recombination} \\ \text{term}}}{\frac{p_N - p_{0N}}{\tau}}. \tag{2-103}$$

Some general observations can be made. In any steady-state case, the accumulation term must obviously vanish. And in the event no excess carriers are present, the recombination term (referring, we repeat, to net recombination) must vanish. In the simple special cases, one or two terms vanish, leaving a differential equation with a well-known solution. One of these cases has already been described, having been used to introduce the concepts of recombination. It involves the thin sample shown in Figure 2-26(a), wherein penetrating radiation causes a uniform presence of excess carriers, independent of position. For such a condition, the spatial derivatives in Equation 2-103 must vanish, so that the diffusion and drift terms drop out. When the radiation is switched off, the accumulation term becomes finite (because of "negative accumulation") and the recombination term becomes finite as well because net recombination exists.

This experiment is one used for measuring lifetime τ, and as usually performed, employs repeated light flashes at intervals that are long (as compared to τ). The decay of excess-carrier population can be observed by sensing the conductivity of the thin sample as a function of time, and the repeating decay curves are superimposed on an oscilloscope screen, providing a steady display. A two-term differential equation applies during the decay phase, consisting of the first and fourth terms of

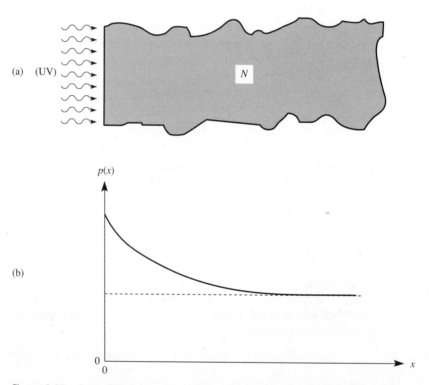

Figure 2-37 A steady-state problem that requires the use of two terms of the constant-E continuity-transport equation. (a) A physical representation of a semi-infinite sample receiving energetic (UV) radiation on its free surface. (b) The minority-carrier profile that results as carriers diffuse to the right, recombining as they go. The characteristic length for this declining exponential function is the minority-carrier diffusion length, in this case L_p.

Equation 2-103. This equation is equivalent to Equation 2-86. Its solution is Equation 2-89, which is depicted in Figure 2-28.

Another very simple yet important case involves only two terms as well. The physical situation is shown in Figure 2-37(a). Arbitrarily, an N-type sample has been chosen, one that is "semi-infinite" in the sense that it has one accessible surface and is very extensive in the positive-x direction, as well as in both y directions and both z directions. The free surface is being irradiated by very energetic photons—ultraviolet (UV) photons in this example—that are absorbed in an exceedingly thin layer of silicon at the surface. The excess minority carriers (holes) then diffuse to the right, recombining as they go, to yield the steady-state profile shown in Figure 2-37(b). If the radiation is not too intense, the sample will be practically field-free.

The desired solution is an analytic expression for p_N as a function of x. As before, Equation 1-103 provides the necessary starting point. Because we have a steady-state problem, the accumulation term is zero. And because the electric field is vanishingly small (a given), we can neglect the drift term. Thus, there is only one independent variable and the governing differential equation is

$$D_p(d^2p_N/dx^2) = (p_N - p_{0N})/\tau. \tag{2-104}$$

The solution for this equation is plotted in Figure 2-37(b), or is

$$p'_N(x) = p'_N(0)\exp(-x/\sqrt{D_p\tau}) \tag{2-105}$$

when it has been recast in terms of excess carriers.

It is evident that the expression $\sqrt{D_p\tau}$ must have length dimensions. It is a very important characteristic length for the declining exponential function that is the solution for this problem, and this length is known as the *minority-carrier diffusion length L_p*.

Exercise 2-39. Verify the dimensions of L_p and L_n.

In general,

$$L = \sqrt{D\tau}.$$

The corresponding unit equation is

$$\sqrt{[cm^2/s][s]} = [cm].$$

Had we been using a P-type sample, the minority electrons would be in charge, and the applicable minority-carrier diffusion length would be $L_n = \sqrt{D_n\tau}$. For doping densities equal to those in the preceding example, but opposite in type (and for the same value of τ), we would find L_n larger than L_p by the factor $\sqrt{D_n/D_p}$, or roughly 1.7. It is important to realize that it is the behavior of *minority* carriers that determines the solution in a situation like this.

Exercise 2-40. In deriving the constant-E continuity-transport equation we simplified the problem by requiring $G' = 0$, which rules out external disturbances, such as incident radiation. Yet it is asserted above that this continuity equation is a valid starting point for the problem of Figure 2-37, which involves radiation on the surface of a sample. How can this paradox be resolved?

Because the energetic radiation in this problem is absorbed in a very thin layer, that layer can be eliminated from consideration without substantially altering the problem. The function of the UV light and the thin layer where the light is absorbed is to fix the boundary condition on $p(x)$ at $x \approx 0$. In the balance of the sample, $G' = 0$ as is required, and Equation 2-103 is applicable.

Exercise 2-41. Why are minority carriers "in charge" in a situation like that of Figure 2-37?

The minority carriers exist in a vast sea of majority carriers for which density values are a billion or a trillion times larger. On a relative basis, the positional readjustment of the majority carriers required to preserve neutrality is utterly trivial, and so the majority carriers "adapt," following "instructions" from the minority carriers.

Exercise 2-42. If the majority carriers adjust their distribution to preserve neutrality, then the excess-majority-carrier profile must match the excess-minority-carrier profile. It follows from examination of Figure 2-37(b) that majority *electrons* must also diffuse to the right, and three times more readily than the holes at that! Yet we completely ignored majority-carrier motion in dealing with this problem. How can *this* paradox be resolved?

The excess-carrier profiles do indeed match so precisely that $(dp/dx) \approx (dn/dx)$ is a very valid approximation at any chosen position. But because of their larger diffusivity, the electrons "outrun" the holes slightly, causing a tiny negative space charge at the right and positive space charge at the left, and hence a tiny positive field. This results in leftward electron flow that precisely balances the rightward majority-electron flow by diffusion.

Exercise 2-43. Why doesn't the electric field just described affect minority carriers?

Drift transport, unlike diffusion transport, is proportional to carrier density. Since the minority-carrier density involved is a billion or a trillion times smaller than the majority-carrier density, the drift of minority carriers is totally negligible.

Exercise 2-44. Take the case of a P-type sample in the situation depicted in Figure 2-37(a), and show that a solution of the form of Equation 2-105 satisfies a differential equation of the form of Equation 2-104.

From Equation 2-101,

$$D_n \frac{d^2 n_P}{dx^2} = \frac{n_P - n_{0P}}{\tau},$$

where the total derivative is appropriate because there is only one independent variable, namely position x. Because n'_P and n_P differ by a constant, this differential equation can be rewritten as

$$D_n \frac{d^2 n'_P}{dx^2} = \frac{n'_P}{\tau},$$

or as

$$(D_n \tau) \frac{d^2 n'_P(x)}{dx^2} = n'_P(x).$$

Letting the function $n'_P(x)$ be parallel to that in Equation 2-105 yields

$$\frac{dn'_P(x)}{dx} = -\frac{n'_P(0)}{\sqrt{D_n \tau}} e^{-x/\sqrt{D_n \tau}}$$

Also,

$$\frac{d^2 n'_P(x)}{dx^2} = \frac{n'_P(0)}{D_n \tau} e^{-x/\sqrt{D_n \tau}}$$

Thus,

$$(D_n \tau) \frac{d^2 n'_P(x)}{dx^2} = n'_P(0) e^{-x/\sqrt{D_n \tau}} = n'_P(x).$$

Now let us illustrate the point that further tightening some restrictions permits relaxation of others. As an example, notice that the "lifetime experiment" cited as the first application causes the diffusion and drift terms to vanish. Hence the high-level carrier-density dependences of μ and D are not of concern, so the the resulting expression, Equation 2-86, could be applied to a high-level problem.

2-7.3 Haynes-Shockley Experiment

Interesting and important applications of the constant-E continuity-transport equation employ a long, thin sample like that depicted in Figure 2-38(a) [27]. One experiment is performed in a field-free sample, so that only the drift term vanishes. Again, repeated light flashes are used, but this time in the manner also shown in Figure 2-38(a). The result is a delta-function distribution of excess carriers at the time t_0 (shorthand for $t = 0$), as depicted in Figure 2-38(b). These carriers then

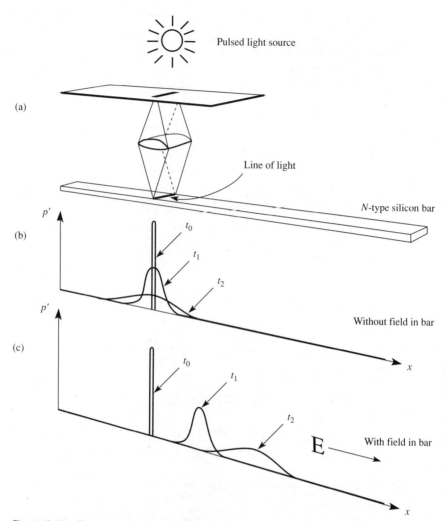

Figure 2-38 Experiment illustrating continuity-equation applications. (a) Physical setup involving a flashing line of light incident transversely on a thin silicon bar.
(b) Experiment involving accumulation, diffusion, and recombination terms of the continuity equation. (c) Experiment involving all four terms of the continuity equation—the *Haynes-Shockley experiment* [28–30]. (After Warner and Grung [27].)

diffuse away in both directions because of the large gradients on both flanks of the delta function, recombining as they go. The carrier profiles at two later times, t_1 and t_2, are also shown in Figure 2-38(b), assuming the form of a time-dependent Gaussian function.

If this experiment is modified by introducing a constant electric field into the silicon sample, the result is an important case known as the *Haynes-Shockley experiment* [28–30]. The electric field is introduced by simply applying a voltage from end to end of the bar depicted in Figure 2-38(a). Once again the result is a decaying (spreading) Gaussian function, but this time the pulse of minority carriers will *drift* in a direction dictated by the electric field. Minority holes have been assumed, so the drift is in the field direction, as shown in Figure 2-38(c). This experiment has appreciable academic interest in that an analytic solution is possible, and also it "exercises" all four terms of the constant-E continuity-transport equation. More important, its historical significance was that it illustrated in tangible terms the behavior of minority-carrier populations.

Exercise 2-45. The carrier-density profiles in Figure 2-38(c) are those of (excess) minority carriers. Describe the majority-carrier profiles.

The ordinate in Figure 2-38(c) could be labeled $n'_N = p'_N$. The minority and majority excess-carrier populations move together in the electric field. A condition of essential neutrality is preserved.

Exercise 2-46. Why does the disturbance in the carrier populations move in a direction dictated by the *minority* carriers?

Remember once more that the minority carriers exist in a "sea" of majority carriers whose density is a billion, a trillion, or more times larger. A virtually negligible adjustment of the majority-carrier population is necessary to neutralize space charge. Hence the minority carriers are in command, and the majority carriers adapt as they respond to the electric field and the perturbed minority-carrier profile.

If low-level conditions are assured by turning down the intensity of the light flash, then the pulse of holes (minority carriers) will drift with a velocity essentially the same as the conductivity drift velocity discussed in Section 2-5.2. A corresponding *drift mobility* is defined for the case of a moving pulse of minority carriers. For the low-level condition just described, this is essentially equal to the *conductivity mobility* of the same carrier, described in Section 2-5.3. But with the onset of high-level conditions, the electric-field profile in the silicon sample (Figure 2-38(a)) is perturbed and drift mobility begins to decline, a topic addressed in Section 4-5.3. For this reason, it is necessary to distinguish between the two kinds of mobility for precise work.

Exercise 2-47. Why do high-level conditions in the Haynes-Shockley experiment perturb the electric-field profile? What results?

In the sample of Figure 2-38(a) with a field present, there is a current density J that is independent of x. But under high-level conditions, there is an upward modification of conductivity σ in the vicinity of the carrier-population disturbance. Ohm's law, $J = \sigma E$, therefore informs us that there must be a corresponding dip in $E(x)$ at that location in order to preserve the continuity (that is, the x-independence or constancy) of current density J. The locally diminished electric field causes a reduction in drift velocity of minority carriers there, in turn causing drift mobility (based upon disturbance velocity) to fall below conductivity mobility.

Yet a third kind of mobility is *Hall mobility*. A mutually orthogonal magnetic field and drift velocity cause a potential difference in the third orthogonal direction because of the magnetic deflection of the moving carriers. The sign of this *Hall voltage* depends upon the carrier type; its magnitude defines a mobility different from (but of the same order of magnitude as) the conductivity mobility for that carrier.

Because of the restrictions embodied in the constant-E continuity-transport equation, the number of possible combinations of finite and zero terms is limited insofar as physical meaning is concerned. The number, in fact, is eight, with four cases involving only two finite terms. Hence the possibilities can be handled in an exhaustive manner [31].

Exercise 2-48. The mercury-cadmium-telluride ternary compound in crystalline form (known informally as *mercadtel*), is a semiconductor with a low bandgap and is useful as a detector of infrared radiation. Why?

The low-energy infrared photons are capable of changing the conductivity of mercadtel through the phenomenon of photoconductivity.

Exercise 2-49. Professor A. van der Ziel and his students have been working with mercadtel at the University of Minnesota and have recently reported that they observed a much smaller diffusion length in their samples at high radiation intensities than at low intensities. Speculate on the reason for this observation.

Mercadtel is a direct-gap material. Accordingly its recombination rate is quadratic in the carrier densities rather than linear in the minority-carrier density, as is the case in silicon. In other words, lifetime τ declines with the intensity of irradiation. As a result, diffusion length $L = \sqrt{D\tau}$ also declines with intensity.

2-7.4 Surface Recombination Velocity

Because the surface of a silicon sample represents a gross interruption of the crystalline near-perfection of its interior, the surface can exhibit recombination and (or) generation properties that are very different from those of the sample's "bulk," or interior region. Visualize a one-dimensional, steady-state situation wherein excess carriers are being created by energetic (nonpenetrating) radiation at the left-hand surface of a sample, and are being transported in the positive-x direction toward the sample's right-hand surface, where they recombine. Since both populations of excess carriers must move to the right, we cannot simply apply an electric field to accomplish the desired transport; the field by itself would move the two carrier populations in opposite directions. But as we have seen in connection with Figure 2-37, with its nearly matched excess-carrier profiles, diffusion transport can cause the two populations to move in the same direction, so let us take this case to examine.

Arbitrarily taking the case of an N-type sample, let us make simplifying assumptions that improve clarity. First, let the radiation be intense enough to have $p'_N \gg p_{0N}$, so that $p'_N \approx p_N$, but let low-level conditions prevail nonetheless. (We have seen that there exists typically an enormous range in which this pair of conditions will hold simultaneously.) Second, let sample thickness X in the x direction be small compared to minority-carrier diffusion length L_p. The result of this assumption is that $p_N(0) \approx p_N(X)$. Thus we have a situation in which a cloud of holes of nearly constant density $p_N(0) \approx p_N(X)$ moves by diffusion toward the right surface, and there the holes recombine with excess electrons that are similarly diffusing from left to right. The net velocity of the minority-carrier population toward the right has been given the symbol s and the name *surface recombination velocity*. The rationale for this label is that the holes are in effect "disappearing into" the right-hand surface; we have a steady-state rightward transport at a velocity s of the carrier cloud of density $p_N(0) \approx p_N(X)$.

We thus have two ways of writing the hole current density: Since transport is by diffusion, one can write

$$J_p = qD_p p_N(0)/L_p. \qquad (2\text{-}106)$$

But elementary considerations also permit writing in terms of s,

$$J_p = qp_N(X)s \approx qp_N(0)s; \qquad (2\text{-}107)$$

equating the expressions in Equations 2-106 and 2-107 yields

$$s = \frac{D_p}{L_p} = \frac{D_p}{\sqrt{D_p \tau}} = \sqrt{\frac{D_p}{\tau}}. \qquad (2\text{-}108)$$

The significance of this equivalent expression for s can be appreciated by referring back to Figure 2-37. There we have a sample characterized by particular bulk values of D_p and τ. Suppose that in a thought experiment, we cut through the sample in a direction normal to the x axis at any value of $x > 0$; if we then adjust the surface recombination velocity at the surface so created to have the value $\sqrt{D_p/\tau}$, the carrier profile to the left of the new surface will coincide precisely with that in the semi-infinite sample, assuming that radiation conditions are unaltered. That is, this choice of s causes the recombination properties of the new surface to be "matched" to those of the sample's interior.

On the other hand, if s has a value smaller than $\sqrt{D_p/\tau}$, then the carrier density profile rises to higher steady-state position. Conversely, a large value of s pulls the profile to a lower position. This latter observation leads to the description of one kind of "ohmic contact," the topic that will be addressed next. For further insight, we should point out that the expression $\sqrt{D_p/\tau}$ is sometimes termed *diffusion velocity*.

Exercise 2-50. In the physical situation just described, hole and electron transport toward the right are equalized by the creation of an electric field that retards the electrons but has a negligible effect on minority holes. Derive an expression for E, with due regard for algebraic sign.

The diffusive flux of holes, obtainable immediately from Equation (2-107) by dropping the factor q, is
$$f_{p,\text{diff}} = D_p p_N(0)/L_p.$$
That for electrons is greater by the mobility ratio, or is
$$f_{n,\text{diff}} = D_n p_N(0)/L_p.$$
Because a low-level case has been specified, the electron profile amounts to $n_N(X) = n_{0N}$. Hence the drift of electrons is
$$f_{n,\text{drft}} = \mu_n n_{0N} E.$$
But this last component must provide the "difference flux," so that
$$(D_n - D_p) p_N(0)/L_p = \mu_n n_{0N} E.$$
Hence,
$$E = \frac{(kT/q)(\mu_n - \mu_p) p_N(0)/L_p}{\mu_n n_{0N}} = \frac{kT}{q} \frac{p_N(0)}{n_{0N} L_p}\left[1 - \frac{\mu_p}{\mu_n}\right].$$
The positive sign for E is correct because a rightward field retards rightward-moving electrons.

Exercise 2-51. Calculate the "matched" value of s for the sample above, assuming it is lightly doped and has $\tau = 1~\mu\text{s}$.

The appropriate value of D_p is
$$D_p = (kT/q)\mu_p = (0.02566~\text{V})(450~\text{cm}^2/\text{V}\cdot\text{s}) = 11.55~\text{cm}^2/\text{s}.$$
Thus,
$$s = \sqrt{D_p/\tau} = \sqrt{(11.55~\text{cm}^2/\text{s})/(10^{-6}~\text{s})} = 3.4 \times 10^3~\text{cm/s}.$$

2-7.5 Recombination-Based Ohmic Contacts

In connection with the thought experiment in Section 2-7.4, it was noted that $p_N(X)$ will be a function of s, rising (for a given sample and a fixed radiation condition) as s is diminished, and falling as s is increased. In the limit of large s, we will have $p_N(X) = p_{0N}$. This point is relevant to the matter of an *ohmic contact*. In ideal form, such a contact to a semiconductor crystal exhibits two basic properties. First, it permits the passage of current into or out of the crystal without placing any resistance of its own in the path. That is, the ideal ohmic contact is resistanceless, as pointed out in Section 1-3.1. Second, such a contact does not perturb carrier densities within the sample. (We shall see in Chapter 3 that passing current through a *PN* junction causes profound changes in nearby carrier densities, so that such a "surface" is in a sense an extreme opposite of an ohmic contact.) It is evident, now, that one can go a long way toward meeting the second ohmic-contact criterion by introducing a surface having a high value of s immediately under the metallic layer that is usually present in an ohmic contact; such a surface will maintain near-equilibrium values of carrier density in its immediate vicinity when current is passed into or out of the crystal through it.

This technique was, in fact, widely practiced in the early years of semiconductor technology. In an ohmic contact region, the semiconductor surface was abraded by one means or another to create a high density of crystalline defects that could serve as recombination centers, and then a metal layer was applied over this disturbed surface. One of the most popular abrasion methods, indeed, was sandblasting! It should not be surprising that these methods have largely been replaced. With the sharply increasing effort in recent decades directed toward maintaining crystalline perfection and avoiding particulate contamination, procedures such as sandblasting became very unattractive.

Fortunately a simple and effective alternative exists. It employs the *high-low junction,* a surface defined by a heavily doped region and an adjacent lightly doped region of the *same* conductivity type, a subject treated further in Section 3-5.4. Such an ohmic contact exhibits very low resistance and causes very slight carrier-density modulation when conducting. But unlike the high-s contact, it "tolerates" excess carriers in its vicinity, a feature that is extremely valuable in some applications. Thus the recombination-based (or high-s) ohmic contact can be regarded as a brute-force solution, in terms of operating principles as well as fabrication methods.

In device modeling, it is sometimes convenient to specify boundary conditions with the assumption that carrier densities have equilibrium values at ohmic contacts. That is, one assumes the high-s principle is operating. For brevity it is often stated that the ohmic-contact surfaces exhibit "infinite recombination velocity." As pointed out by McKelvey in a nice discussion [32], this requirement is excessive because there is a maximum rate at which carriers can disappear at a surface, and it is fixed by the maximum possible diffusion velocity; he goes on to point out that this maximum velocity is about $v_t/2$, where v_t is a carrier's thermal velocity, which is of order 10^7 cm/s. Comparing this to the "typical" diffusion velocity calculated in Exercise 2-51 ($\sim 3 \times 10^3$ cm/s), we see a ratio of about three and one-half decades. This large ratio is significant because he further points out that for $\sqrt{D_p/\tau} \ll v_t/2$, negligible error results from assuming the physically unachievable infinite surface recombination velocity.

2-7.6 Comparing Equilibrium and Steady-State Conditions

Having considered a number of different conditions that can exist in a semiconductor sample, we are now in a position to illustrate and compare equilibrium and steady-state conditions more crisply than before. The comparison is important because these two differing conditions have been subject to some confusion.

In Figure 2-39 is a diagram that illustrates relevant relationships using a device sometimes identified as a set of Euler's circles. The left-hand circle encompasses steady-state conditions, and its area is subdivided into three regions. Thus equilibrium, the first of these (at the left) is clearly a special case of steady-state condi-

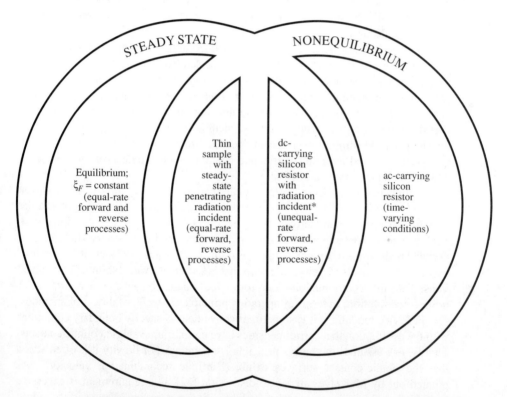

Figure 2-39 Illustrating by means of Euler's circles the relations among four possible conditions in a semiconductor sample. One circle encloses steady-state conditions, and the other, nonequilibrium conditions. One example is given for each of four conditions. Equilibrium is a special case of steady-state conditions. *Radiation is added to this example because a case of ohmic conduction is sometimes termed *quasiequilibrium* on grounds that the carrier densities involved have equilibrium values.

tions. The next two conditions in the steady-state circle are also encircled by the nonequilibrium circle, so they are two nonequilibrium but steady-state cases. One is a special case wherein forward and reverse molecular processes are balanced, the converse of the equilibrium principle of detailed balance that was mentioned in Exercise 2-23. The other is a modified case of direct current. Steady-state incident radiation has been added to set the example apart from pure dc ohmic-flow conditions, which are sometimes regarded as a *quasiequilibrium* condition because equilibrium carrier densities are preserved in ohmic transport.

Finally, then, a simple case of time-varying current is used to round out the nonequilibrium circle. This nonequilibrium, non-steady-state example balances the equilibrium, steady-state region at the far left.

SUMMARY

The resistivity of a good conductor is of the order of 10^{-6} ohm·cm, that of a semiconductor, in the range from 10^{-3} ohm·cm to 10^6 ohm·cm, and that of an insulator, from perhaps 10^{10} ohm·cm upward. Bulk properties of a material are properties that are free of perturbations by surfaces.

An oscillator possesses a characteristic frequency, or mode of motion, and a coupled assemblage of N identical such oscillators exhibits N characteristic modes of motion, a phenomenon known as *splitting*. An electron in an isolated atom resides in a discrete allowed energy state defined by four quantum numbers; the state so defined can accommodate only one electron. When two atoms interact by virtue of proximity, each such state splits into two states. In a silicon crystal containing N atoms, each such state splits into N states, where N is typically an enormous number. The splitting phenomenon thus gives rise to so many states "packed into" an energy range of the order of eV in height, that effectively there exists a continuum of allowed states, termed a *band*. Since bands originating from a number of the discrete states that exist in the isolated atom can overlap in the case of a solid, the result is an even denser continuum of allowed states. In semiconductors and insulators, there are bands separated by a gap, the bandgap, in which no (or very few) states exist. Further, the number of states in the lower of these two bands is precisely equal to the total number of outer-shell electrons in all of the atoms of the solid. In an insulator, the gap amounts to several eV (e.g., 5 eV or more), so that the upper band is empty and there are no carriers of electricity in the solid. In a semiconductor, the gap is smaller (e.g., ~1 eV), so a few electrons at room temperature can reach the upper band from the lower band. These upper-band electrons and the "holes" they have left behind in the lower band lend modest conductivity to the solid. The upper band is the conduction band; its electrons are conduction electrons, or simply *electrons*. The lower band is the valence band; its missing valence electrons are *holes*. At low temperature, the electrons revert to the valence band, eliminating the holes and "disappearing" themselves as a result, so that the semiconductor is converted to an insulator. A conductor, by contrast, has a band that is partly filled at any temperature, and is thus capable of conducting at any temperature.

The probability that an electronic state in a solid is occupied by an electron is a function of the energy position of the state; the relevant expression is known as the *Fermi-Dirac probability function*. This function is symmetric about a point at which probability equals 0.5, a point that defines an energy known as the Fermi level. Above the Fermi level, the occupancy probability (of a state by an electron) approaches zero, below it, unity; for this reason, the Fermi level is a kind of "top surface" of the electron distribution, analogous to the top surface of liquid in a container. A simple-exponential function approximates the Fermi-Dirac function for any energy greater than $4kT$ above the Fermi level. A different exponential function approximates it for any energy more than $4kT$ below the Fermi level. These two exponential expressions are known as the Boltzmann approximations to the Fermi-Dirac function. Because of the "strength" of these exponential functions, most of the occupied states in the conduction band (i.e., most of the conduction electrons) are located close to the conduction-band edge, and most of the holes in the valence band are close to the valence-band edge, provided in both cases that the Fermi level lies between the band edges and more than $4kT$ from either. Because the carriers (electrons and holes) reside close to the band edges, one can define equivalent densities of states N_C and N_V for the conduction and valence bands, respectively, by assuming all available states in the bands to be located right *at* the band edges. This is known as the equivalent-density-of-states approximation. The densities N_C and N_V are constants at a particular temperature (and exhibit only weak temperature dependence), provided the Fermi level is restricted to the energy range cited just above. Although N_C is larger than N_V by a factor lying between one and two, for many purposes one can safely assume that $N_C = N_V$. This is known as the *band-symmetry approximation*. With band symmetry assumed, and because of the actual symmetry of the Fermi-Dirac function, the electron density n and hole density p are equal when the Fermi level is at the center of the gap, and the common density value is known as the intrinsic density, n_i. Because of the symmetry of the Fermi-Dirac function, the product of the hole and electron densities at a particular temperature is a constant, independent of Fermi-level position, a statement known as the law of mass action, $pn = n_i^2$. In perfectly pure, or *intrinsic,* silicon, and with band symmetry assumed, the Fermi level resides at the center of the gap, an energy position often termed the intrinsic level.

Impurities are intentionally introduced into a semiconductor crystal to adjust its electrical properties, in a process termed "doping." The impurities are "dopants." Such atoms are introduced substitutionally, which is to say, in place of silicon atoms. Atoms from Column V of the periodic table are used for doping silicon, with phosphorus being the most commonly used such impurity. Having five outer-shell electrons, these atoms ensure that a neutral silicon sample that includes them will contain more electrons than an intrinsic sample. As a result, the Fermi-level position rises above the center of the gap, just as the top surface of liquid in a container rises when more liquid is added to it. The chosen impurity atoms, fixed in the crystal at randomly positioned sites, introduce localized states within the bandgap. Such impurity atoms and their states can be modeled quantitatively as a hydrogen atom immersed in a dielectric medium. An electron residing in such a state is likened to an electron "in orbit" about a proton. Its binding energy, however, is smaller than that of the electron in hydrogen by a factor of the dielectric constant squared, and so the phosphorus state is located close to the conduction-band edge; an electron driven out of the state becomes a conduction electron (the analog of a free electron). At room temperature, thermal energy in the crystal typically dislodges electrons from over 99% of these impurity states, so that each atom "left behind" carries a charge $+q$, or is ionized. The customary assumption of 100% ionization means that the Column-V impurity atoms have all "donated" their fifth electrons to the conduction band; hence such impurities are termed *donor impurities*. Near-100% ionization also causes electron density in a donor-doped sample to equal donor density; these densities typically exceed the density of thermally generated electrons (the only ones present in an intrinsic sample) by a large factor.

Substitutional Column-III impurities, with boron being by far the most common, also introduce localized states in the silicon bandgap. This occurs because of boron's need for one additional electron to complete the kind of bonding (covalent) that exists in a tetrahedral structure. The needed electron will with high probability be stolen or "accepted" from a neighboring covalent bond. An electron so appropriated is bound in place almost as tightly as a normal valence electron in a covalent bond, this because of the great affinity of electron pairs of opposite spin in such bonds. This fact accounts for the positioning of the *acceptor* state just above the valence band. Such action causes, on average, one hole to be introduced into the valence band for each boron atom, and also causes over 99% of these atoms to possess an ionic charge of $-q$, attributable to the fourth electron that was captured. In the unlikely event that a hole is "in orbit" about such a negative charge center (which is to say, that the boron state is nonionized), then an "upside-down" hydrogen model can be applied because charge signs are now reversed and because holes "fall up," somewhat like a bubble in a liquid.

When both donor and acceptor states are present and the Fermi level lies between them (the usual case), the acceptor states are likely to be occupied and the donor states, empty, which is to say, that *all* are likely to be ionized. In a neutral sample (again, the usual case) with more donors than acceptors, one can consider that each acceptor has received an electron from a donor atom, in the process eliminating the latter as an effective donor atom because it now has no electron to donate to the conduction band. In the reverse situation, each donor atom has contributed an electron to an acceptor state, rendering it unable to accept an electron from an adjacent bond and thus to create a mobile hole. This effective cancellation of a minority impurity by a majority impurity is termed *compensation*. In the case of equal donor and acceptor densities we have compensated-intrinsic silicon.

The most common state of affairs involves a sample of silicon that exhibits overall neutrality. In such a sample the total of positive species, holes and donor atoms (the latter assumed fully ionized), equals the total of negative species, electrons and (fully ionized) acceptor atoms. The resulting expression is a more exact version of the net doping laws that apply to compensated samples, $n \approx N_D - N_A$ and $p \approx N_A - N_D$. Because electron population in a particular small energy range is fixed by a product of available-state density and occupancy probability in that vicinity, the effects of changing doping and temperature can be illustrated by means of a "computer" that performs an analog multiplication.

A conversion from electron energy ξ to electrostatic potential ψ can be accomplished (in a band diagram, for example) by choosing *consistent* references for the two variables and replacing each value of ξ by $\psi = \xi/(-q)$. The direction of increasing ψ is then opposite to that for ξ because of the electron's negative charge. It is particularly advantageous to choose the Fermi level as reference and the intrinsic (midgap) potential as *the potential* ψ, because compact and symmetric expressions for n and p as functions of n_i and ψ result.

The electron and hole populations in a silicon sample at room temperature are in violent random motion, manifesting their shares of available thermal energy. The average value of the "thermal velocity" associated with this violent motion is a bit over 10^7 cm/s. When a net motion of a carrier population is superimposed on its random motion, the phenomenon is called *carrier transport*. Such transport can be caused by the existence of a gradient of electrostatic potential, carrier density, temperature, or any combination of these. We remove temperature gradients from consideration by confining attention to isothermal, or nearly isothermal, problems. The existence of an electric field (a potential gradient) causes drift transport. For field values up to about 3 kV/cm in silicon, drift velocity is proportional to field, with the constant of proportionality being known as (hole or electron) mobility. Electron mobility in silicon is roughly three times greater than that for holes. At higher field values, mobility is no longer constant but becomes a declining function of field. Drift velocity of electrons in silicon levels off ("saturates") at 10^7 cm/s, a value a bit less than average room-temperature thermal velocity, this occurring at 50 kV/cm. Hole mobility shows similar high-field nonconstancy, but the saturation of hole drift velocity is less pronounced than that of electrons.

One factor fixing mobility at a particular field value is temperature, which determines available thermal energy, and hence the density and average energy of vibrational wave packets, or phonons, in the crystal. Two categories of phonons are capable of causing carrier "scattering," or deflection. Acoustic phonons arise from vibrations at sound-wave frequencies in the crystal. Optical phonons arise from vibrations at much higher infrared frequencies. Both kinds of vibrations can be either longitudinal or transverse. Some of the transverse optical phonons in a crystal of zincblende structure also interact strongly with photons.

Other factors affecting mobility are impurity-ion density, with donors and acceptors being equally effective as scattering centers, and screening. The last factor is the tendency of majority carriers to "cluster" slightly near a charge center of opposite sign, reducing the effective range at which its charge can be "felt." In a compensated sample, diminished screening and increased scattering-center density cause especially low mobility. For both carriers, mobility is constant in an uncompensated sample up to a doping density of about $10^{14}/cm^3$, which is to say that phonon scattering dominates at room temperature in purer samples.

The three requirements for drift transport are (1) a charged species (2) that is in random motion, and (3) that is in the presence of an electric field. For diffusion transport, by contrast, there are only two requirements: (1) a species in random motion, (2) that exhibits a density gradient. Diffusive flux is proportional to density gradient, a relationship known as *Fick's first law*. The constant of proportionality is the diffusivity. Of great significance is the fact that drift transport is a function of *two* variables, carrier density and electric field, while diffusion transport is a function of but *one* variable, density gradient. This state of affairs is responsible for subtle relationships in device analysis. For a particular population, the ratio of the two proportionality constants entering the basic equations for the two transport mechanisms of interest (that is, the ratio of diffusivity to mobility) is equal to the thermal voltage kT/q, an equation known as the *Einstein relation*.

The density of excess carriers is the difference between actual and equilibrium densities for the species in question. It can be positive or negative. In the most common circumstances and for a neutral sample, excess-electron and excess-hole densities are equal. For this reason, a many-fold change in minority-carrier density can result in a negligible *relative* change in majority-carrier density. When the majority-density change is significant, the high-level regime has been entered. In the graphical presentation of excess-carrier populations under low-level conditions, a logarithmic scale emphasizes the important *relative* information that $n_N \approx n_{0N}$, or $p_P \approx p_{0P}$, while a linear scale conveys the *absolute* information that $p' = n'$. The latter choice, however, requires introducing an enormous gap in the scale. Both the band-to-band and recombination-center mechanisms predict low-level recombination rate to be proportional to minority-carrier density. The latter mechanism predicts the same result for high-level conditions, a result confirmed experimentally for the case of silicon, thus indicating the dominance of this mechanism in silicon. Electronic states that are localized spatially and far from the band edges in energy are termed *deep states*, and can be produced in silicon by using heavy-metal atoms for doping. The term *trap* is also used to identify deep states. Such states are effective recombination centers, their effectiveness increasing with nearness of the state to the gap center. Gold, which has an acceptor state 0.54 eV from the conduction-band edge, is particularly effective. Crystalline imperfections that inevitably accompany conventional doping also serve as recombination centers, and the density of such centers increases with total doping, causing carrier lifetime to decline. Typical lifetime values in silicon range from 1 ns to 10 μs.

The rate of change of the density of a carrier at a point (or the accumulation of carriers in a region) can be equated to the combined effects of transport, generation, and recombination at that point (or near that region), with the result being a continuity equation. Its importance is as a starting point for analysis. A four-term form known as the constant-E continuity-transport equation is particularly useful, but is based upon a significant number of simplifying assumptions. Typical problems require the retention of only two terms. The Haynes-Shockley experiment, however, satisfies all the simplifying assumptions, and yet requires retention of all four terms. It observes the drift of a pulse of minority carriers in a constant electric field, and in principle yields a measure of *drift mobility*. The kind of mobility cited earlier is *conductivity mobility*. The two are equal under low-level conditions. A subtle and important facet of low-level conditions is that the drift and diffusion of majority carriers can coexist in a certain region, while minority carriers in the same region exhibit comparable diffusion, yet (practically) zero drift.

In a steady-state situation wherein minority carriers diffuse toward a surface to recombine, their diffusion velocity $\sqrt{D_p/\tau}$ for holes) is termed the *surface recombination velocity s* for that surface. An electrical contact (to a semiconductor crystal) that embodies a high-s surface can meet the two criteria that define an ohmic contact—low resistance and avoidance of carrier-density disturbances. A better kind of ohmic contact, however, is a high-low junction, which is to say a P^+P junction for the case of a P-type region and an N^+N junction for the case of an N-type region.

Steady-state conditions and nonequilibrium conditions can coexist. But equilibrium conditions are a special case of steady state, falling outside the region of commonality. By the opposite token, time-varying conditions are a special case of nonequilibrium, also falling outside the region of commonality.

Table 2-1 Energy Gaps of Various Materials at $T = 300$ K

BN	~7.5 eV	CdTe	1.56
C (diamond)	5.47	GaAs	1.42
ZnS	3.68	InP	1.35
GaP	2.26	Si	1.12
BP	2.0	GaSb	0.72
CdSe	1.70	Ge	0.66
AlSb	1.58	InAs	0.36

Table 2-2 Properties of Silicon

Property	Value (at 300 K)	Value (at 297.8 K)	Unit
Atom density (N_a)	5.0×10^{22}		cm^{-3}
Atomic number (Z)	14		
Atomic weight	28.09		
Density	2.33		g/cm^3
Relative dielectric permittivity (κ)	11.7 (SiO$_2$:3.9)		
Absolute dielectric permittivity (ϵ)	1.035×10^{-12}		F/cm
Energy gap (ξ_G)	1.119	1.120	eV
Equivalent density of states			
Conduction band (N_{CA})	3.22×10^{19}	3.18×10^{19}	cm^{-3}
Valence band (N_{VA})	1.83×10^{19}	1.81×10^{19}	cm^{-3}
$N_C = N_V$ (for calculation)	3.04×10^{19}	3.01×10^{19}	cm^{-3}
Intrinsic carrier density (n_i)	1.20×10^{10}	1.00×10^{10}	cm^{-3}
Intrinsic Debye length (L_{Di})	26.4	28.8	μm

Table 2-2 continued on page 209

Table 2-2 (*continued*)

Intrinsic mobilities		
Electron (μ_n)	1,400	cm²/V·s
Hole (μ_p)	450	cm²/V·s
Lattice constant (a_0)	5.43	Å
Melting point	1,415	°C
Specific heat	0.7	J/g·C
Thermal conductivity	1.5	W/cm·C°
Thermal-expansion coefficient (linear)	2.6×10^{-6}	(C°)$^{-1}$

Table 2-3 Experimentally Determined Ionization Energies for Various Impurities in Silicon[†]

			Periodic Column				
I	II	III	V	VI	VII	VIII	Rare Earths
			Ionization Energy Measured from Conduction-Band Edge				
Li(D)0.033	Mg(D)0.11		Sb(D)0.039	S(D)0.11	Mn(D)0.53	Ni(A)0.35	Na(D)0.25
Ag(A)0.22	Mg(D)0.25		N(D)0.04	Te(D)0.14		Co(A)0.53	Tm(D)0.29
Au(A)0.54	Hg(A)0.31		P(D)0.044	O(D)0.16		Fe(D)0.55	
	Hg(A)0.36		As(D)0.049	S(D)0.18			
	Zn(A)0.55		Bi(D)0.069	W(A)0.22			
				W(A)0.30			
				Mo(D)0.33			
				S(D)0.37			
				W(A)0.37			
				O(A)0.38			
				S(D)0.61			
			Ionization Energy Measured from Valence-Band Edge				
		Tl(A)0.26	O(A)0.35		Fe(D)0.40		
Cu(A)0.49		In(A)0.16	W(D)0.35		Pt(D)0.37		
Au(D)0.35	Hg(D)0.33	Ga(A)0.065	Mo(D)0.34		Co(A)0.35		
Ag(D)0.32	Zn(A)0.31	Al(A)0.057	W(D)0.31		Pt(D)0.31		
Cu(D)0.24	Hg(D)0.25	B(A)0.045	Mo(D)0.30		Ni(A)0.23		

[†]The symbols (D) and (A) designate donors and acceptors, respectively. (After Neuberger and Welles [9], with permission.)

Table 2-4 Phonon Categories

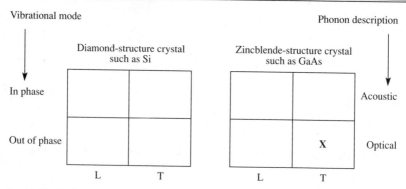

L ≡ longitudinal
T ≡ transverse
X-type phonon is able to interact with photon, and vice versa

REFERENCES

1. R. M. Warner, Jr. and J. N. Fordemwalt (Eds.), *Integrated Circuits; Design Principles and Fabrication,* McGraw-Hill, New York, 1965, pp. 4, 17.

2. R. M. Warner, Jr. and B. L. Grung, *Transistors: Fundamentals for the Integrated-Circuit Engineer,* Wiley, New York, 1983; reprint edition, Krieger Publishing Company (P. O. Box 9542, Melbourne, Florida 32902-9542), 1990, pp. 110, 136.

3. W. Shockley, *Electrons and Holes in Semiconductors,* Van Nostrand, Princeton, N.J., 1950, p. 8.

4. D. E. Wooldridge, A. J. Ahearn, and J. A. Burton, "Conductivity Pulses Induced in Diamond by Alpha-Particles," *Phys. Rev.* **71,** 913 (1947).

5. R. S. Muller and T. I. Kamins, *Device Electronics for Integrated Circuits,* Wiley, New York, 1977, pp. 7–10.

6. E. Fermi, "Zur Quantelung des Idealen Einatomigen Gases," *Z. Physik* **36,** 902 (1926).

7. P. A. M. Dirac, "On the Theory of Quantum Mechanics," *Proc. Roy. Soc. (London)* **112,** 661 (1926).

8. J. N. Shive, *The Properties, Physics and Design of Semiconductor Devices*, Van Nostrand, Princeton, N.J., 1959, pp. 305–307.

9. M. Neuberger and S. J. Welles, *Silicon*, Electronic Properties Information Center, Hughes Aircraft Company, Culver City, Calif., October 1969.

10. R. M. Warner, Jr., "Fermi Level Demonstration," *Am. J. Phys.* **29,** 529 (1961).

11. A. Nussbaum, *Semiconductor Device Physics*, Prentice-Hall, Englewood Cliffs, N.J., 1962, p. 212; see also the paper by Domenicali cited by Nussbaum on p. 222.

12. L. de Broglie, "A Tentative Theory of Light Quanta," *Philos. Mag.* **47,** 446 (1924); L. de Broglie, "Recherches sur la Théorie des Quanta," *Ann. Phys. (Paris)* **3,** 22 (1925).

13. E. Conwell and V. F. Weisskopf, "Theory of Impurity Scattering in Semiconductors," *Phys. Rev.* **77,** 388 (1950).

14. G. L. Pearson and J. Bardeen, "Electrical Properties of Pure Silicon and Silicon Alloys Containing Boron and Phosphorus," *Phys. Rev.* **75,** 865 (1949).

15. Warner and Grung, p. 169.

16. M. Shur, private communication.

17. R. K. Cook and J. Frey, "An Efficient Technique for Two-Dimensional Simulation of Overshoot Effects in Si and GaAs Devices," *COMPEL* **1,** 66 (1982).

18. N. Goldsman and J. Frey, "Efficient and Accurate Use of the Energy Transport Method in Device Simulation," *IEEE Trans. Electron. Devices* **35,** 1524 (1988).

19. J. C. Irvin, "Resistivity of Bulk Silicon and of Diffused Layers in Silicon," *Bell Syst. Tech. J.* **41,** 387 (1962).

20. Warner and Grung, p. 162.

21. A. Einstein, "On the Movement of Small Particles Suspended in a Stationary Liquid Demanded by the Molecular-Kinetic Theory of Heat," *Ann. Phys. (Leipzig)* **17,** 549 (1905).

22. J. P. McKelvey, *Solid State and Semiconductor Physics*, Harper and Row, New York, 1966, p. 324.

23. R. N. Hall, "Germanium Rectifier Characteristics," *Phys. Rev.* **83,** 228 (1951).

24. W. Shockley and W. T. Read, Jr., "Statistics of the Recombination of Holes and Electrons," *Phys. Rev.* **87,** 835 (1952).

25. J. W. Slotboom, "Analysis of Bipolar Transistors," Ph.D. Thesis, Eindhoven Technical University, October 1977.

26. Warner and Grung, p. 198.

27. Warner and Grung, p. 206.

28. J. R. Haynes and W. Shockley, "Investigation of Hole Injection in Transistor Action," *Phys. Rev.* **75,** 691 (1949).

29. J. R. Haynes and W. Shockley, "The Mobility and Life of Injected Holes and Electrons in Germanium," *Phys. Rev.* **81,** 835 (1951).

30. J. R. Haynes and W. C. Westphal, "The Drift Mobility of Electrons in Silicon," *Phys. Rev.* **85,** 680 (1952).

31. Warner and Grung, p. 199.

32. McKelvey, p. 346.

TOPICS FOR REVIEW

R2-1. What is meant by *bulk properties?*
R2-2. Describe energy-level *splitting.*
R2-3. How does the splitting phenomenon lead to *energy bands?*
R2-4. The characteristic frequency of a mechanical oscillator is analogous to what in an atom?
R2-5. Describe the band structure of the column-IV elements discussed in Section 2-1.2.
R2-6. Draw a two-dimensional representation of a silicon crystal.
R2-7. What feature of this diagram is related to the lower band of silicon?
R2-8. In a silicon crystal at room temperature, how does thermal energy manifest itself?
R2-9. Explain how the conduction and valence bands can both contribute to conductivity in silicon.
R2-10. What is the algebraic sign of hole charge? Of hole mass?
R2-11. How can a valence electron fill an adjacent hole without surmounting the intervening energy barrier?
R2-12. Describe an analog of a hole in the valence band. Indicate analogous quantities and important limitations of the analogy.
R2-13. What is meant by *energy gap?*
R2-14. Distinguish clearly among *conductor, semiconductor,* and *insulator* in terms of band structure. Describe a related water analogy.
R2-15. State the *Fermi-Dirac probability function.*
R2-16. What is meant by *Fermi level?*
R2-17. Describe an analog of *Fermi level in a conductor.*
R2-18. Describe the Fermi function at $T = 0$ K.
R2-19. What is a *density-of-states function?*
R2-20. Explain qualitatively the differing sensitivities of Fermi-level positions in a conductor and a semiconductor.
R2-21. What is *intrinsic silicon?*

R2-22. What is the *band-symmetry approximation*?
R2-23. What is the *equivalent-density-of-states approximation*?
R2-24. Of what order of magnitude is the room-temperature thermal velocity v_t of an electron in silicon.
R2-25. How does v_t compare with c, the velocity of light? What is the significance of the relative magnitudes?
R2-26. What is meant by *intrinsic carrier density*?
R2-27. What is the value of n_i for silicon at 24.8 C?
R2-28. How much do N_{CA} and N_{VA} differ, if at all?
R2-29. Give an approximate value for $N_C \approx N_A$.
R2-30. What is meant by *doping*?
R2-31. What is a *donor impurity*?
R2-32. Describe the hydrogen model of a donor state.
R2-33. Name the common donor impurities used in silicon.
R2-34. What is a *donor ion*?
R2-35. What is an *acceptor* impurity?
R2-36. Name the most common acceptor impurity used in silicon.
R2-37. Describe an *acceptor ion*.
R2-38. Describe a *nonionized* acceptor.
R2-39. What is meant by *uniform doping*?
R2-40. What is the *100%-ionization approximation*? In what temperature range, roughly, is it valid and why?
R2-41. How many atoms are there per unit cell of silicon?
R2-42. What is *impurity compensation*?
R2-43. What is *compensated-intrinsic silicon*?
R2-44. Name the four factors that determine Fermi-level position. How does Fermi-level position respond to changes in each?
R2-45. Name all of the positive and negative species in a sample of doped silicon.
R2-46. How is the *Boltzmann approximation* derived from the Fermi-Dirac probability function?
R2-47. Sketch the two Boltzmann functions needed to approximate the Fermi function.
R2-48. State the *law of mass action*.
R2-49. What is meant by the *intrinsic level*?
R2-50. Where is the intrinsic level in silicon?
R2-51. What approximation places the intrinsic level at the center of the energy gap?
R2-52. Sketch equivalent band diagrams in terms of electron energy ξ, and electrostatic potential ψ.
R2-53. State electron and hole densities as a function of electrostatic potential.
R2-54. What is meant by *carrier transport*?
R2-55. What is *quasiequilibrium*?
R2-56. Give an example of a quasiequilibrium condition.
R2-57. How can electron waves interact with a perfect silicon crystal at room temperature?
R2-58. What is a *phonon*? Identify four kinds of phonons.

R2-59. What is meant by carrier *scattering* and what can cause it?
R2-60. What is meant by charge *screening?*
R2-61. How do donor and acceptor ions compare in scattering effectiveness?
R2-62. What is meant by *mean free path* and *mean free time?*
R2-63. State the *principle of detailed balance.*
R2-64. Distinguish between *steady-state* and *equilibrium* conditions.
R2-65. What is *conductivity mobility?* What are its dimensions?
R2-66. State the *drift-transport equations* for holes and electrons.
R2-67. What is *ohmic conduction?*
R2-68. State whether mobility is nonconstant or constant in ohmic and nonohmic conduction.
R2-69. What kind of scattering is dominant at high temperatures? Why?
R2-70. What kind of scattering is dominant at low temperatures? Why?
R2-71. State the hole and electron conductivity mobilities in intrinsic silicon.
R2-72. What is meant by "hot electron?"
R2-73. What kind of electron energy (kinetic or potential) is proportional to its distance from the conduction-band edge?
R2-74. What is meant by the *saturation velocity* of an electron? What is its approximate value?
R2-75. Why are nonohmic effects much more pronounced in silicon than in a conductor?
R2-76. What are the two primary properties of an *ohmic contact?*
R2-77. State the conductivity equation.
R2-78. What is the approximate conductivity of intrinsic silicon at room temperature?
R2-79. Consider a sample of silicon subjected to an electric field. State the directions of the hole and electron current-density vectors. Do the same for the hole and electron drift-velocity vectors.
R2-80. State the conditions necessary for diffusion transport.
R2-81. State Fick's first law.
R2-82. What is *diffusivity?* What is another name for diffusivity?
R2-83. What are the dimensions of diffusivity?
R2-84. What are the dimensions of density gradient?
R2-85. State the transport equations.
R2-86. State the Einstein relation.
R2-87. What is the value of the thermal voltage at 24.8 C?
R2-88. Give a concise definition of equilibrium.
R2-89. What is meant by the term *excess carriers?*
R2-90. What is *penetrating radiation?*
R2-91. With respect to carrier densities, what is a *low-level* disturbance? A *high-level* disturbance?
R2-92. What are the dimensions of recombination rate R?
R2-93. How can *carrier lifetime* be observed?
R2-94. What is meant by *deep states?*
R2-95. What is meant by *trap?*
R2-96. What is a *recombination center?*

R2-97. What recombination-rate predictions are made by the band-to-band and recombination-center mechanisms, respectively, for low-level and high-level cases? What is observed?

R2-98. Explain *direct* and *indirect* band structures.

R2-99. State the strengths and weaknesses of linear and logarithmic graphical representations.

R2-100. Name at least two continuity equations dealing with topics other than carrier transport. Indicate their meanings.

R2-101. Describe the *Haynes-Shockley experiment*. What kind of mobility can be observed in such an experiment?

R2-102. Describe two other kinds of mobility measurements.

R2-103. What is meant by the term *surface recombination velocity*?

R2-104. What are the properties of an *ohmic contact*?

R2-105. How can an ohmic contact be realized?

ANALYTIC PROBLEMS

A2-1. Shown in part b is a pair of axes with dimensionless coordinates.

a. Calculate the coordinates of three points on the Fermi-Dirac function and enter the coordinates in this table:

$\dfrac{\xi - \xi_F}{kT}$	$P(\xi)$

b. Plot the same three points on the following axes and sketch the Fermi-Dirac function with reasonable accuracy.

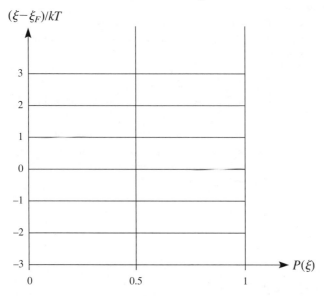

A2-2. In a certain silicon sample at equilibrium, the Fermi level resides 0.400 eV above the center of the bandgap.

a. Calculate the occupancy probability for a lone isolated state located right at the center of the bandgap.

b. This sample contains donor impurities and no acceptor impurities. The donor states are situated 0.044 eV below the conduction-band edge. Find the occupancy probability of the donor states.

c. Comment on the validity of the assumption of 100% ionization of the donor states in the present situation.

d. Derive an approximate form of the Fermi-Dirac probability function that could be applied in (b) with reasonable validity.

e. Use your approximate expression to recalculate the probability in (b) and find the percentage difference in the exact and approximate results.

f. Sketch the exact probability function (ξ versus P) accurately by plotting several points. Superimpose on the same diagram an accurate sketch of the approximate expression of (d).

g. Comment on conditions wherein use of the approximate expression is justified.

h. How much error results when the approximate expression is used to calculate occupancy probability for a state located right at the Fermi level?

A2-3. The conduction band can be characterized by a state density (number of states per cm^3) of $N_C = 3.01 \times 10^{19}/\text{cm}^3$, with these states assumed to be situated right at the conduction-band edge.

a. Using this assumption, calculate the conduction-electron density n (number of electrons per cm^3) for the conditions of Problem A2-2.

b. The valence band can be characterized by a state density of $N_V \approx N_C$, with these states assumed to be situated right at the valence-band edge. Using this assumption, calculate the hole density p.

c. Calculate the pn product using the results from (a) and (b).

A2-4. Determine the approximate density of donor states N_D (number of donor states per cm³) for the silicon sample of Problems A2-2 and A2-3. Explain your reasoning.

A2-5. Boron as a substitutional impurity in silicon introduces a state at 0.045 eV above the valence-band edge. In a certain sample at room temperature, the Fermi level is 0.10 eV above the valence-band edge.

a. Calculate the percentage ionization of its boron states.

b. Comment on the assumption of 100% ionization in this case.

A2-6. A certain uniformly doped silicon sample at room temperature has $n = 10^6/\text{cm}^3$ and $N_A = 10^{15}/\text{cm}^3$.

a. Find p.

b. Find N_D.

c. Using unity factors, calculate the volumetric density of covalent bonds in this sample.

A2-7. Using the Boltzmann approximation to the Fermi-Dirac probability function (obtained by dropping the unity term in its denominator), find the Fermi-level position relative to the conduction-band edge for a sample having $n = 5 \times 10^{16}/\text{cm}^3$ at room temperature and at equilibrium.

A2-8. Given that the majority impurity in the foregoing problem is phosphorus, find the occupancy probability at the donor level. Comment on the assumption of 100% ionization in this case.

A2-9. An intrinsic silicon sample contains a trace (meaning so little that the Fermi level is unaffected) of the rare-earth element thulium. (See Table 2-3.) What is the occupancy probability of the states so introduced? Take room temperature as 297.8 K. Use the Boltzmann function if doing so is justified, and comment on your choice in the matter.

A2-10. A certain silicon sample has $N_A = 10^{14}/\text{cm}^3$ and $N_D \approx 0$. Find its

a. majority-carrier density.

b. minority-carrier density.

c. conductivity σ.

A2-11. A certain silicon sample has $p = 2 \times 10^{10}/\text{cm}^3$. Find its

a. electron density n.

b. resistivity ρ.

A2-12. A certain silicon sample has $N_D = 5.020 \times 10^{16}/\text{cm}^3$ and $N_A = 5.010 \times 10^{16}/\text{cm}^3$. Find its

a. majority-carrier density.

b. minority-carrier density.

c. conductivity σ, using $\mu_n = 390$ cm^2/V·s, and $\mu_p = 200$ cm^2/V·s.
d. Another silicon sample has $N_D = 1.1 \times 10^{17}$/cm^3, and $N_A = 10^{17}$/cm^3. Find p.
e. Find n.
f. Find σ using $\mu_n = 400$ cm^2/V·s and $\mu_p = 200$ cm^2/V·s.
g. Find ρ.
h. Another silicon sample has $N_D = 0.97 \times 10^{12}$/cm^3, $N_A = 0.99 \times 10^{12}$/cm^3. Find ρ.

A2-13. A silicon sample is doped with 1.0×10^{15} phosphorus atoms per cubic centimeter and 0.6×10^{15} boron atoms per cubic centimeter, and is at equilibrium.
a. Find the majority-carrier density.
b. Find the minority-carrier density.
c. Calculate the sample's conductivity, using $\mu_n = 1220$ cm^2/V·s.
d. Calculate its resistivity.

A2-14. A silicon sample has $N_D = 10^{16}$/cm^3 and $N_A = 0$.
a. Find p.
b. What is the conductivity type of the sample in a?
c. Another sample has $N_D = 0.1 \times 10^{14}$/cm^3 and $N_A = 10^{15}$/cm^3. Find p.
d. Find n in the sample of c.
e. Another sample has $N_D = 1.1 \times 10^{17}$/cm^3 and $N_A = 10^{17}$/cm^3. Find p.
f. Find n in the sample of e.
g. Another sample has $N_D = 0.9 \times 10^{14}$cm^3 and $N_A = 10^{14}$/cm^3. Find p.
h. Find n in the sample of g.

A2-15. You are given this uncompensated silicon sample:

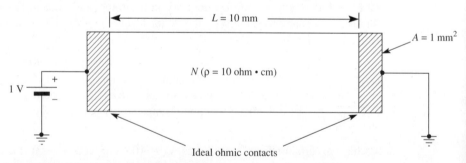

a. Calculate electron current density J_n.
b. Calculate hole current density J_p.
c. How is J_n affected by doubling sample length L and keeping A, ρ, and applied voltage V the same as in part a?
d. How is J_n affected by doubling cross-sectional area A and keeping ρ, applied voltage V, and sample length L the same as in part a?

A2-16. Ohmic contacts are made to the ends of a silicon resistor having a length $L = 0.5$ cm. The resistor has a cross-sectional area A that is given by the product of its width W and thickness X, where $W = 1$ mm and $X = 1$ μm. The silicon is uniformly doped with $N_A = 1 \times 10^{14}$ cm^3 and $N_D = 3.0 \times 10^{17}$/cm^3. For this doping density, $\mu_p = 230$ cm²/V·s and $\mu_n = 510$ cm²/V·s. Calculate the resistance R.

A2-17. A silicon sample contains 3.0×10^{16}/cm³ of one impurity type and a negligible amount of the opposite type. It exhibits a resistivity of 0.21 ohm·cm at room temperature.
a. Determine majority-carrier mobility.
b. Is the sample N-type or P-type? Explain your reasoning.

A2-18. A major advantage available to the semiconductor engineer is the opportunity to infer accurate chemical information from simple electrical measurements.
a. Given an extrinsic but lightly doped N-type silicon resistor R of length L and cross-sectional area A, derive an expression for its net impurity density (a chemical property).
b. For the resistor in a, $R = 1$ kilohm, $L = 5$ mm, and $A = 1$ mm². Evaluate the expression derived in a.
c. What is the probable majority impurity in the resistor? (Spell correctly.)

A2-19. Minority carriers in a particular silicon sample drift 1 cm in 100 μs when $E = 10$ V/cm.
a. Determine drift velocity v_D.
b. Determine minority-carrier diffusivity D. Include units!
c. Determine the conductivity type (N or P) of the sample and explain your reasoning.

A2-20. A sample of heavily doped P-type silicon has a drift-current density of 100 A/cm². Hole drift velocity is 60 cm/s. Find hole density p.

A2-21. The concept of carrier mobility is as valid for a conductor as for a semiconductor. The resistance of No. 18 copper wire (having a diameter $d = 1.03$ mm) is 6.51 ohm/1000 ft. The density of conduction electrons in copper is $n = 8.4 \times 10^{22}$/cm³.
a. Given that the current in the wire is 1 A, calculate the current density J.
b. Find the magnitude of the drift velocity v_D of the electrons in cm/s and cm/hr. (As a test of intuition, guess at the answer before performing the calculation and then compare your guess to the calculated result.)
c. Calculate the resistivity ρ of the wire.
d. Calculate the electric field E in the wire.
e. Calculate the mobility magnitude $|\mu_n|$ of the electrons in copper.

A2-22. Assume complete ionization.
a. Combine the neutrality equation and the law of mass action to obtain an accurate expression for n in near-intrinsic N-type silicon.
b. Use the expression obtained in a to calculate n in a sample having $N_D = 0.99 \times 10^{12}$/cm³ and $N_A = 0.97 \times 10^{12}$/cm³.
c. Comment on the accuracy of the approximate equation $n \approx N_D - N_A$ for ordinary doping values.

A2-23. Given a hole concentration gradient of -10^{20}/cm⁴ in a lightly doped sample, calculate the corresponding hole diffusion-current density.

A2-24. A lightly doped field-free sample of silicon that is 1 μm thick in the x direction exhibits a majority-electron gradient of $(dn/dx) = -10^{15}/\text{cm}^4$.
a. Calculate $J_{n,\text{diff}}$.
b. If the electron density at the higher-density (left) face is $10^{14}/\text{cm}^3$, what is it at the right face?
c. In a thought experiment, we supply an additional population of electrons throughout the sample in the amount of $10^{14}/\text{cm}^3$. Repeat calculation a.
d. Return to a and b. Suppose that without any other changes an electric field of $E = 0.05$ V/cm is superimposed on the sample in the positive-x direction. Calculate $J_{n,\text{drft}}$ approximately.
e. Calculate J_n (total for electrons) for the conditions of d.

A2-25. Given that $p(x)$ versus x is linear, $p(0) = 0.9 \times 10^{16}/\text{cm}^3$, and $(dp/dx) = -10^{19}/\text{cm}^4$ in a silicon sample that is lightly doped,
a. Sketch $p(x)$ versus x in the space provided.
b. Calculate $J_{p,\text{diff}}$.

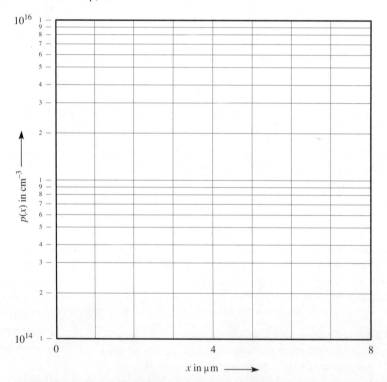

A2-26. In a certain lightly doped silicon sample there exists a hole current-density value $J_{p,\text{diff}} = -1.85$ A/cm². Calculate the corresponding density gradient.

A2-27. A certain silicon sample has negligible acceptor doping and a donor doping that is linearly graded from $1.0 \times 10^{15}/\text{cm}^3$ at the left surface to $2 \times 10^{14}/\text{cm}^3$ at the right surface. The two surfaces are 100 μm apart.

a. Assuming that $n(x) \approx N_D(x)$ throughout, calculate $J_{n,\text{diff}}$ at the middle of the sample.

b. In a thought experiment, we add $1.0 \times 10^{15}/\text{cm}^3$ of donors uniformly to the sample of a; recalculate $J_{n,\text{diff}}$.

c. In a second thought experiment we add $1.0 \times 10^{14}/\text{cm}^3$ of acceptors to the sample of a; recalculate $J_{n,\text{diff}}$.

A2-28. A thin silicon sample receives steady-state radiation that produces excess carriers uniformly throughout the sample in the amount $p' = n' = 10^{10}/\text{cm}^3$. Excess-carrier lifetime in the sample is 1 μs. At $t = 0$, the radiation source is turned off. Calculate the excess-carrier density and recombination rate at

a. $t = 0.5$ μs;

b. $t = 2.5$ μs;

c. $t = 5$ μs.

A2-29. A thin N-type sample of silicon having an equilibrium minority-carrier density p_0 is subjected to penetrating radiation with the radiation source turned on at $t = 0$. At $t = \infty$, $p = p(\infty)$.

a. Write the differential equation appropriate to this situation.

b. Find the solution for the differential equation of a under the given boundary conditions.

A2-30. Uniform, steady-state ultraviolet radiation impinges on the surface of a semi-infinite silicon sample in which $n_0 = 10^{14}/\text{cm}^3$, producing an excess-carrier density at the surface of $p'(0) = n'(0) = 10^{11}/\text{cm}^3$. Given further that $\tau = 1$ μs, and that the spatial origin is at the irradiated surface,

a. calculate $J_{p,\text{diff}}$ at $x = 0$;

b. calculate $J_{n,\text{diff}}$ at $x = 0$;

c. calculate $J_{p,\text{diff}}$ and $J_{n,\text{diff}}$ at $x = L_p$. Sketch vectors to scale representing these current-density components.

d. Since the sample is open-circuited, the total current density at $x = L_p$ must be zero, $J = 0$. In fact, a tiny electric field accounts for the "missing" current-density component. Speculate on the cause of the electric field.

e. Calculate the magnitude and direction of the field cited in (d) at the position $x = L_p$.

f. Does the presence of the field destroy the validity of the J_p calculations that were based upon a pure-diffusion picture? Explain.

g. Derive an approximate symbolic (not numeric!) expression for E.

A2-31. A thin silicon sample receives steady-state penetrating radiation that produces a uniform excess-carrier density of one million per cm^3. Minority-carrier lifetime is 1 μs. If the light is switched off at $t = 0$, at what time has p' fallen to one hundred per cm^3? The radiation source was turned on a long time before $t = 0$.

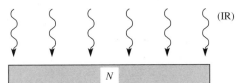

A2-32. This trace is observed on an oscilloscope in a lifetime experiment and from it we learn that $\tau = 10^{-6}$ s. Compute $(dn'/dt)|_{t=1\ \mu s}$.

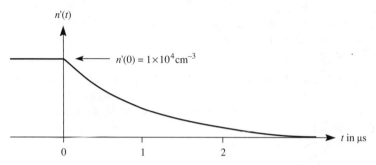

A2-33. A thin N-type silicon sample with $\tau = 5\ \mu s$ is subjected to infrared radiation for a long period. At $t = 0$, the radiation source is turned off. At $t = 15\ \mu s$, excess-carrier recombination rate is observed to be $5 \times 10^{13}/cm^3/s$. Calculate G', the steady-state generation rate caused by the radiation.

A2-34. A sample of N-type silicon is being irradiated on one face with energetic (nonpenetrating) light. The scale of the ordinate is 1 cm per (corresponding to) a density of $10^4/cm^3$, and the minority-carrier profile is plotted against it.

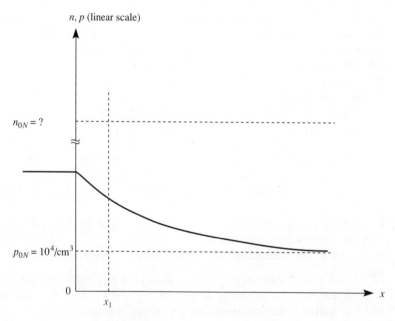

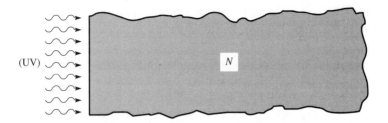

a. Compute and enter n_{0N}.
b. Sketch $n_N(x)$.
c. A gap is shown in the ordinate. Compute its magnitude in km for the given scale.
d. Using a scale of your choice, draw vectors representing for the position $x = x_1$ the vector quantities $J_{p,\text{diff}}$, $J_{p,\text{drft}}$, $J_{n,\text{diff}}$, and $J_{n,\text{drft}}$.
e. On your profile sketch, indicate $n'(x_1)$ and $p'(x_1)$.

A2-35. In a certain thin silicon sample for which $n_0 = 10^{15}/\text{cm}^3$, the total equilibrium recombination rate is $R_0 = 10^{11}/\text{cm}^3\cdot\text{s}$. Penetrating radiation produces an excess-carrier density of $n' = p' = 10^{16}/\text{cm}^3$.
a. By what factor has total recombination increased?
b. What is the carrier lifetime in this sample?

A2-36. A thin silicon sample has $N_D - N_A = 10^{16}/\text{cm}^3$ and $\tau = 1\ \mu\text{s}$. With steady-state penetrating radiation incident, the excess minority-carrier density is $10^6/\text{cm}^3$. The radiation is switched off at $t = 0$.
a. What is the recombination rate immediately after the radiation is switched off for the excess minority carriers?
b. What is the rate for the excess majority carriers?
c. What is the recombination rate at $t = \tau$ for the excess minority carriers?
d. What is the recombination rate at $t = 10\tau$ for the excess minority carriers?
e. Just after the radiation is turned off, what is the total minority-carrier density?
f. What is the total majority-carrier density?

A2-37. Using a sheet of linear graph paper and a sheet of three-cycle semilog graph paper (ordinate logarithmic and abscissa linear), replot the respective electron-density profiles shown in the sketches numbered I and II.

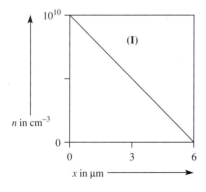
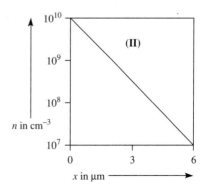

a. Plot profile I on chart II, and plot profile II on chart I.
b. Evaluate $(dn/dx)|_{x=0}$ for profiles I and II.

A2-38.

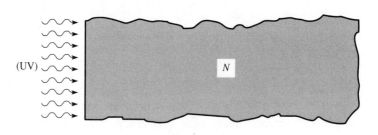

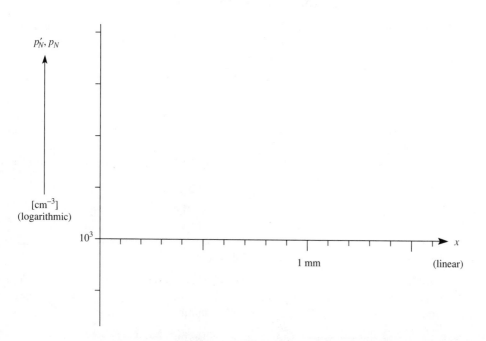

As indicated above, the free surface of a semi-infinite sample of silicon is irradiated using a uniform, steady-state source of ultraviolet light. The sample has $p_{0N} = 10^3/\text{cm}^3$ and $L_p = 100\ \mu\text{m}$. Excess-carrier density declines exponentially toward the right according to

$$p'_N(x) = p'_N(0)\exp(-x/L_p).$$

a. Given the information that $p'_N(1\text{ mm}) = p_{0N}$, find $p'_N(0)$. Calibrate the logarithmic ordinate above appropriately and plot $p'_N(x)$ using a dashed line.

b. Calculate $p_N(0)$. Calculate at least two other points of the function $p_N(x)$ and plot them on the given diagram. Draw a smooth solid line through the points, superimposing this curve on the previous curve.

c. Calibrate the logarithmic ordinate below appropriately for plotting $n_N(x)$. Use the same calibration as in part (a). That is, use the same interval per decade. Plot $n_N(x)$ using a solid line.

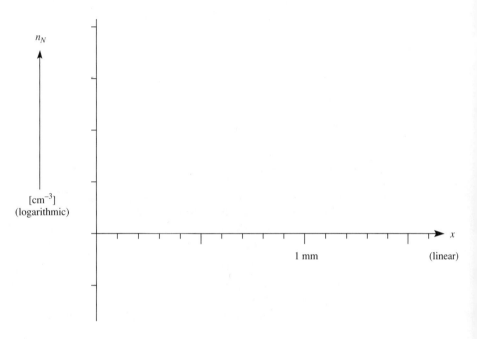

d. With nothing else changed, the intensity of the ultraviolet source is altered so that $p'_N(0) = 10^6/\text{cm}^3$. Calculate $J_{p,\text{diff}}$ at $x = L_p$, given that $\mu_p = 390 \text{ cm}^2/\text{V·s}$ in this sample.

e. Assuming that $\mu_n \approx 3\mu_p$, draw vectors in the following spaces that represent the current-density components indicated, correctly representing the directions and relative magnitudes of the various components.

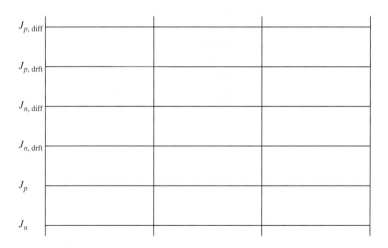

A2-39. A certain thin silicon sample has $N_A = 1.0 \times 10^{17}/\text{cm}^3$, and $N_D \approx 0$. When subjected to steady-state penetrating radiation, its excess-carrier density is $n' = 2 \times 10^{17}/\text{cm}^3$.
a. Calculate equilibrium conductivity σ_0.
b. Calculate its modulated (or nonequilibrium) conductivity σ, and as a measure of conductivity modulation, calculate the ratio σ/σ_0.

A2-40. Assuming literal validity for the hydrogen model of a donor atom in silicon, determine the ratio of each following quantity in the donor case to the corresponding quantity in the original hydrogen case as presented by Bohr. Express all ratios in terms of the silicon dielectric constant, κ.
a. The radius of the first Bohr orbit r_0.
b. The mutually attractive force F between the two particles.

A2-41. Four basic approximations are widely used in semiconductor analysis.
a. Identify each approximation, using words and equation(s).
b. Which of these approximations will worsen rapidly as the Fermi level moves into the valence band or conduction band? Explain.
c. Why is it that using these approximations collectively predicts a value for n or p that departs rather slowly from the true value as the Fermi level moves past relevant band edge?

A2-42. In a certain silicon sample it is valid to assume 100% ionization of the donor and acceptor impurity states.
a. Write a neutrality equation for this sample in the form density of positive charge centers = density of negative charge centers.
b. Given the information that $n = 10^5/\text{cm}^3$ and $N_D = 10^{15}/\text{cm}^3$ in this sample, calculate the values of the other charge-center densities that enter into the neutrality equation.

A2-43. A sample of silicon contains phosphorus as its only impurity, and the phosphorus states are 90% ionized at room temperature.
a. Calculate the energy interval $(\xi_C - \xi_F)$.
b. Draw a portion of a band diagram with reasonable quantitative accuracy, displaying the relative positions of the Fermi level, band edge, and impurity states, and labeling the energy intervals relevant to part a.
c. State in percentage terms the approximate magnitudes of the errors involved in using, for the sample of part a,
  (1) the Boltzmann approximation for determining n;
  (2) the 100%-ionization assumption for determining n.
d. Another silicon sample has $N_D \gg N_A$, where the donors are phosphorus, and has $(\xi_C - \xi_F) = 110$ meV. [1 meV = 0.001 eV.] Calculate the electron density to three significant figures.
e. Assuming band symmetry, calculate donor density to three significant figures for the sample of part d.

A2-44. You are given a thin sample of silicon having one of its major faces subjected to uniform, steady-state, penetrating radiation. The radiation source has been

on for a long time. Doping density in the sample is $N_D = 9 \times 10^{16}/\text{cm}^3$; $N_A = 1 \times 10^{17}/\text{cm}^3$; and carrier lifetime is $\tau = 1~\mu\text{s}$. The total generation rate in the sample is $G = 4 \times 10^{10}/\text{cm}^3\cdot\text{s}$.
a. Calculate the equilibrium recombination rate R_0.
b. Calculate the total recombination rate R.
c. Calculate the incremental recombination rate R'.
d. Calculate the incremental generation rate G' resulting from the incident radiation.
e. The net generation rate can be written as

$$\frac{dp}{dt} = \frac{dp'}{dt} = \frac{dn'}{dt} = \frac{dn}{dt}.$$

What is the net generation rate for the conditions of this problem?

A2-45. In the uncompensated silicon sample shown, $L_n = 115~\mu\text{m}$ and $p_{0P} = 10^{14}/\text{cm}^3$. Also given is the current-density component $J_{n,\text{diff}}(x_1) = -10^{-9}$ A/cm^2. The radiation causes a low-level disturbance of equilibrium.

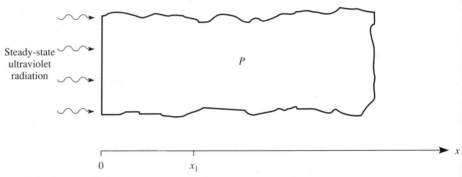

a. Calculate $n_P(x_1)$.
b. Calculate $J_{p,\text{diff}}(x_1)$ with due regard for algebraic sign.
c. Calculate $J_{p,\text{drft}}(x_1)$ with due regard for algebraic sign.
d. Calculate $E(x_1)$.

A2-46. A thin N-type silicon sample has been subjected to steady-state penetrating radiation for a long time, causing its total minority-carrier density to be inflated to $p = 10^{12}/\text{cm}^3$. Net doping is $10^{16}/\text{cm}^3$ and carrier lifetime is $1~\mu\text{s}$. The radiation source is turned off at $t = 0$. Calculate the incremental recombination rate at $t = 3~\mu\text{s}$.

A2-47. In a certain sample of silicon are donor states located 0.044 eV below the conduction-band edge. The Fermi level is 0.10 eV below the conduction-band edge.
a. Using the Fermi-Dirac probability function, calculate the fractional occupancy (decimal fraction) of the conduction-band states, assumed to be concentrated right at the conduction-band edge.
b. In a similar way, using the Fermi-Dirac probability function, calculate the fractional occupancy of the donor states.
c. Calculate the percentage ionization of the donor states.

d. In the space provided immediately below, indicate the value of ξ_D clearly, and plot $P(\xi)$ for $\xi = \xi_C, \xi_D,$ and ξ_F. Sketch a smooth curve through the three points using a solid line.

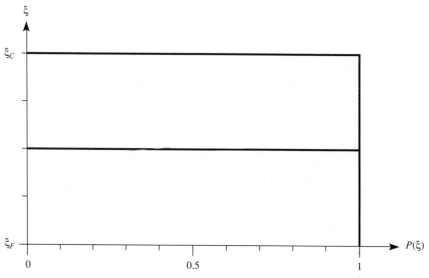

e. Calculate the fractional occupancy of the donor states using the Boltzmann approximation.

f. Calculate the percentage error in using the Boltzmann approximation for calculating donor-state occupancy. [That is, compare the results in parts b and e.]

g. Using a dashed line, sketch the Boltzmann probability function on the diagram of part d.

A2-48. The band diagram that follows represents a microscopic sample of silicon having a uniform net doping of $10^{13}/cm^3$. The abscissa represents sample volume. Small circles represent electronic states.

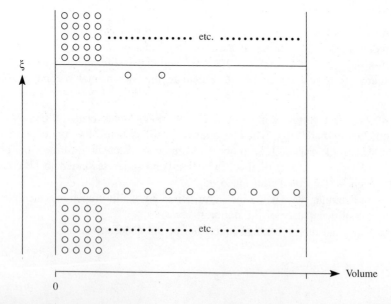

a. What is the volume of the sample that is represented?
b. Assuming the silicon sample to be cubic in shape, calculate its edge dimension in centimeters.
c. Convert the edge dimension into micrometers.
d. What is the conductivity type of this sample (N-type or P-type)?
e. What is the most probable donor impurity? (Spell it correctly.)
f. What is the most probable acceptor impurity? (Spell it correctly.)
g. The sample is cooled to absolute zero. Indicate by blackening circles which states will be occupied by electrons.

A2-49. A certain silicon sample is uniformly doped with boron at $4.320 \times 10^{16}/cm^3$, and phosphorus at $4.310 \times 10^{16}/cm^3$. For this doping density, $\mu_p = 300\ cm^2/V \cdot s$ and $\mu_n = 750\ cm^2/V \cdot s$.
a. Calculate its conductivity.
b. State the adjective that describes the doping condition of this sample.

A2-50.
a. In the following chart, indicate by means of an X the state of charge for each of the four possibilities listed.

		+q	-q	0
1	Donor state occupied by an electron			
2	Donor state empty of an electron			
3	Acceptor state occupied by an electron			
4	Acceptor state empty of an electron			

b. In a similar way, indicate for each of the four states of occupancy whether the impurity atom is ionized or nonionized.

	Ionized	Nonionized
1		
2		
3		
4		

A2-51. The semi-infinite silicon sample shown here is uniformly doped with boron only, with a density of $1 \times 10^{14}/cm^3$; the sample exhibits a carrier lifetime of $1\ \mu s$. Its free surface receives weak, steady-state, ultraviolet radiation, causing total minority-carrier density at $x = 0$ to be $3 \times 10^6/\text{cm}^3$. Calculate the minority-carrier diffusion-current density at a position one diffusion length from the surface, with due regard for algebraic sign.

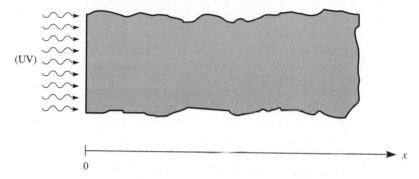

A2-52. A certain silicon sample, shown here, is doped with acceptor impurities only and is at equilibrium. It has a constant impurity gradient (a linearly varying doping profile) and you may assume that $p(x) \approx N_A(x)$. Also, $(dp/dx) = (dN_A/dx) = 10^{17}/\text{cm}^4$ and $N_A(0) = 10^{13}/\text{cm}^3$.

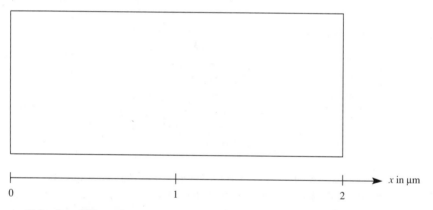

a. Calculate $E(1 \ \mu\text{m})$.
b. Using complete sentences, state what relevance, if any, the principle of detailed balance has to this problem.

COMPUTER PROBLEMS

C2-1. It is shown in Section 2-2.4 that the electron density in the conduction band can be found by integrating the product of the density-of-states function and the Fermi-Dirac probability function written in Equation 2-2 as

$$n = \int_{\xi_C}^{\infty} N(\xi) \, P(\xi) \, d\xi. \tag{1}$$

This integral equation is well approximated by

$$n = N_C e^{-(\xi_C - \xi_F)/kT} \quad (2)$$

for $(\xi - \xi_F) \geq 4kT$, where N_C is the equivalent density of conduction-band states. However the approximation worsens as $(\xi_C - \xi_F)$ approaches and passes zero. For a more accurate determination of n, it is necessary to evaluate Equation 1, which can be written as

$$n = N_C(2/\sqrt{\pi}) F_{1/2}, \quad (3)$$

where

$$F_{1/2} \equiv \int_0^\infty \frac{\sqrt{\eta}\, d\eta}{1 + \exp(\eta - \eta_F)}, \quad (4)$$

and

$$\eta \equiv (\xi - \xi_C)/kT, \quad (5)$$

and

$$\eta_F \equiv (\xi_F - \xi_C)/kT. \quad (6)$$

The integral in Equation 4 is known as the *Fermi-Dirac integral*. [See J. S. Blakemore, *Electronic Communications* **29,** 131 (1952).]

a. Write a computer program to evaluate $F_{1/2}$ versus η_F. Evaluate the integral for at least eleven values of η_F in the range $-5 < \eta_F < +5$. Choose a finite upper-limit value of η for the integration, and test to verify that the value you have chosen is large enough. To verify, choose a different value for the upper limit and observe whether the limit change causes a change in the value of the integral. You may be surprised to find out how small a number qualifies as "infinity." (Such verification is a routine part of numerical integration that involves an infinite limit.)

b. Tabulate the results in the form η_F versus $F_{1/2}$.

c. Tabulate and plot η_F versus n using Equation 3 and the value $N_C = 3.01 \times 10^{19}/\text{cm}^3$. Use a semilog diagram, this time with linear ordinate and logarithmic abscissa. For comparison, plot η_F versus n on the *same* diagram using Equation 2 and the same value of N_C. Comment on the different, if any.

C2-2. Experimental data on carrier mobilities as a function of doping was given in Figure 2-22. This data can be approximately represented by

$$\text{MU\_N} = 65.4 + 1265/(1 + (N/8.5E16)^{0.72}); \quad (1)$$

$$\text{MU\_P} = 47.7 + 447.3/(1 + (N/6.3E16)^{0.76}). \quad (2)$$

[See Caughy and Thomas, *Proc. IEEE* **55,** 2192 (1967).] Write a computer program to evaluate these expressions for arbitrary doping N.

DESIGN PROBLEMS

D2-1. Design a doping formula that will yield a uniformly doped sample in which holes contribute 100 times more to sample conductivity than do electrons. Consider that densities below $10^{14}/cm^3$ cannot be controlled with sufficient precision to accomplish this result.

D2-2. A one-dimensional N-type sample of length L is desired that will yield a *linear* hole profile that goes from $1 \times 10^4/cm^3$ at $x = 0$ to $2 \times 10^4/cm^3$ at $x = L$. You may assume that L is large enough and hence dn/dx and dp/dx are small enough so that quasineutrality in the sample is preserved. Design the necessary one-dimensional doping profile.

D2-3. Design a demonstration of the principles of diffusion.

D2-4. Design an experimental method for measuring high-level carrier lifetime.

D2-5. Design a silicon sample for use in a Haynes-Shockley experiment. Choose reasonable values of current, voltage, and sample resistivity and dimensions.

D2-6. A resistor can be formed at the surface of a silicon sample and isolated from the balance of the sample by several methods. In a 1-kilohm resistor that is to be fabricated, the resistive layer will have a resistivity of $\rho = 10^{-2}$ ohm·cm and a thickness of $X = 1$ μm.

a. Calculate the resistor's length L (which is in the direction of current) in terms of its width W.

b. Explain qualitatively why the resistance (for fixed ρ and X) depends only on the *ratio* of L to W.

c. The length and width dimensions of the resistor can be controlled only within ± 10 μm because of process-related factors. Also, the nature of the fabrication process is such that $\Delta L \approx \Delta W$. (That is, an error of $+6$ μm in L will be accompanied by an error of $+6$ μm in W, and so forth.) Find the minimum length and width of a 1-kilohm resistor that has no more than a $\pm 10\%$ variation (tolerance) in its resistance value. Assume that control of resistivity and layer thickness is very good. Comment on any engineering trade-off between size (which is proportional to cost) and resistance tolerance in the design of such a resistor. [*Hint:* The way to address a tolerance problem is to write the total differential of R as a function of its several variables. Then make the engineering assumption that $dR \approx \Delta R$, $dL \approx \Delta L$, etc.]

d. Given the information that temperature coefficient of resistivity decreases as doping increases, justify qualitatively the choice of resistivity that was proposed for fabricating this resistor, and the additional engineering trade-off that is involved.

3 PN Junctions

An interface lying between a *P*-type region and an *N*-type region in a semiconductor single crystal is termed a *PN junction*. The problem posed by such a junction has elements in common with the problems posed by a semiconductor surface and by a metal–semiconductor junction, the latter often called a *Schottky junction*. The understanding of all three of these entities has evolved over about half a century [1–15], and operation of all of the most important semiconductor devices is based upon one or more of them.

3-1 JUNCTION CONCEPTS

No matter how gradual the transition from *P*-type to *N*-type material, the junction surface is always sharply defined—the surface where $N_D - N_A$ vanishes. A plane surface is the simplest form that a junction can take, and happily, it is by far the most important case. It is the only case we shall consider in this chapter.

3-1.1 Space Charge at a Junction

A thought experiment is helpful in appreciating several key features of the *PN* junction. Visualize two uniformly doped blocks of silicon with plane surfaces, one block of each conductivity type, as shown in Figure 3-1(a). To be specific we have arbitrarily assumed that each block has a doping of $10^{16}/\text{cm}^3$, a "typical" doping value. For even further simplicity, let us assume only donors are present on the *N*-type side, and only acceptors on the *P*-type side. It is then evident that the carrier densities are as stated in the diagram. Now imagine that these two blocks of silicon can be brought together and joined instantaneously at room temperature to form a

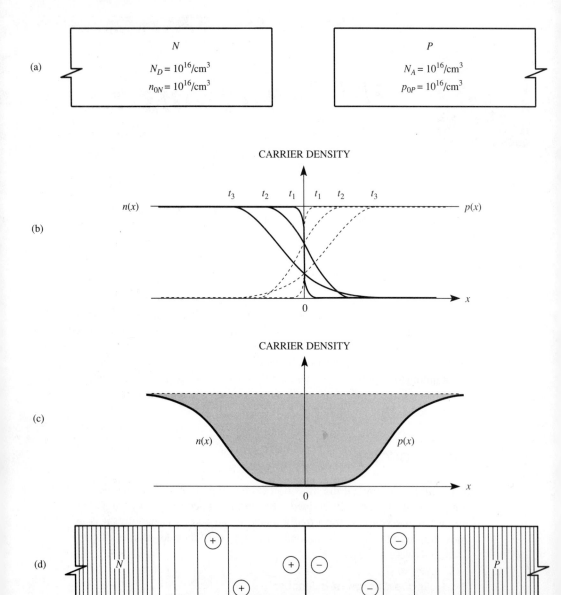

Figure 3-1 A thought experiment on junction formation. (a) A pair of matching blocks of silicon with equal but opposite uniform doping. (b) The carrier profiles at three successive instants after joining the two blocks to form a *PN* junction. (c) The equilibrium carrier profiles in the resulting junction. (d) Physical representation of the junction sample showing ionic charges "exposed" by the depressed carrier densities near the junction.

perfect single crystal. (Such an experiment is unrealistic, but is completely straightforward in the realm of thought. It is equally straightforward to rule out any real-life surface complications.) The result will be infinite carrier gradients located at the junction at the instant of joining, $t = 0$. A brief instant later, at $t = t_1$, the diffusion that inevitably accompanies a huge gradient will "soften" the rectangular carrier profiles to shapes somewhat like those labeled t_2 in Figure 3-1(b). This process will continue, in the manner suggested by the curves labeled t_3, as holes rush to the N-type side by diffusion and electrons similarly rush to the P-type side. Finally, the carriers "settle down" to assume the profiles depicted in Figure 3-1(c) for reasons examined now.

The inevitable consequence of the carrier behavior just described is the "uncovering" of impurity ions near the junction on both sides. The charge of these ions has been neutralized by majority carriers in the individual blocks before joining. Thus an electric field develops. Lines of force start from donor ions on the N-type side of the junction and terminate on an equal number of acceptor ions on the P-type side. The shaded areas in Figure 3-1(c) are proportional to the total amounts of fixed ionic charge on each side of the junction, positive donor ions on the left and negative acceptor ions on the right. Those areas constitute the differences between the constant impurity densities—the horizontal straight dashed lines at the top in Figure 3-1(c)—and the variable carrier densities (the curves below).

Figure 3-1(d), then, is a physical representation of a small portion of the junction-containing sample. The circled symbols represent ionic charges, securely fixed in place in the silicon crystal. These exist in an "exposed" condition only near the junction. Away from the junction, neutral conditions exist because majority carriers are present in just the right numbers to neutralize the ionic charges. In the intermediate region, there is a fuzzy *space-charge-region boundary*, where a transition is made from complete neutrality to space charge, with the values $\rho_v = qN_D$ on the N-type side of the junction plane, and from complete neutrality to the condition $\rho_v = -qN_A$ on the P-type side. The variably spaced lines in Figure 3-1(d) represent this fuzzy boundary. These equations assume complete ionization. In fact, carriers are of course present in the space-charge region. But right at the junction surface, the densities of both carrier types are a million times smaller (in the present example) than N_D and N_A on the two sides of the junction plane and very close to it. Therefore we are justified in ignoring them in writing the charge densities as $+qN_D$ and $-qN_A$, respectively.

Exercise 3-1. Compare the number of majority electrons in the N-type region (or the number of majority holes in the P-type region) before and after the thought experiment on junction formation. How is a difference possible?

The N-type region has fewer electrons after the junction has been formed and has come to equilibrium than it did before the junction was formed. The electrons that are "lost" disappear through recombination; the final equilibrium condition requires a smaller number of electrons than does the initial equilibrium condition. Because a symmetric sample is being considered, completely parallel relationships and values hold for majority holes on the *P*-type side.

3-1.2 Dipole Layer

The term *metallurgical junction* is often used to designate very specifically the mathematical surface at which $N_D - N_A = 0$. The reason for introducing such a term is that *junction* is sometimes used to label a region of finite thickness—the space-charge or transition regions that flank the mathematical surface. The overall thickness of this finite region, it should be pointed out, is typically of the order of a few tenths of a micrometer, so it, too, is often thin compared to other dimensions in a device.

Realize, now, that equal quantities of charge per unit of junction area are present on the two sides of the metallurgical junction. Therefore, the region of electric field is completely *internal*. This is a familiar state of affairs because it suggests conditions existing in the parallel-plate capacitor, a comparison that will be extended in Section 3-8. Terms often applied to thin parallel layers of equal and opposite charge are *double layer* and *dipole layer*.

3-1.3 Field and Potential Profiles

A dipole layer has curious properties. Suppose we focus specifically on the kind of dipole layer shown in Figures 3-1(c) and (d). The electric field is confined to a thin region and vanishes at the space-charge-layer boundaries. It is present spontaneously and under equilibrium conditions (the only conditions considered in Section 3-1). For this reason it is often described as a *built-in field*. Furthermore, the prescription $\psi = -\int E dx$ informs us that the dipole layer will also exhibit a *built-in potential difference*.

These features of the *PN* junction can be appreciated more fully by extending the sequence of ideas presented in Figure 3-1. This has been done in Figure 3-2, which repeats Figure 3-1(d) as Figure 3-2(a) for this purpose, except that lines of force have been added to the diagram. By "counting" lines of force as an indication of their density, we go from first principles to the field profile presented in Figure 3-2(b). The field profile has "tails" on either side that are a consequence of the fuzziness of the space-charge-layer boundaries.

The next step is to integrate the field function to produce a potential profile. (The potential of the right-hand side has arbitrarily been taken as zero.) This can be accomplished in an adequately quantitative way by simply "counting squares" under the field profile. The result is shown in Figure 3-2(c). Notice that potential ψ has been taken as increasing downward for reasons that will be explained later, and

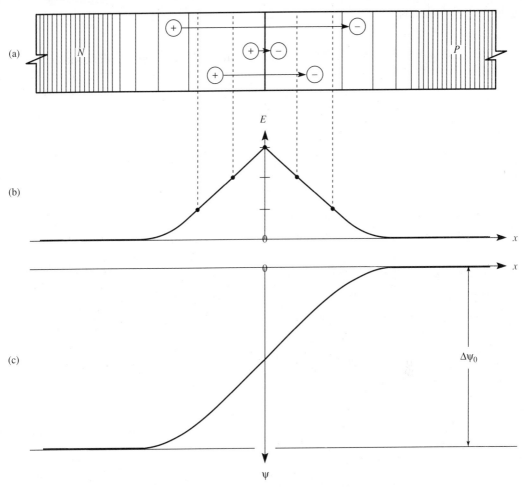

Figure 3-2 Electric-field and electrostatic-potential profiles in a *PN* junction. (a) Physical representation of the junction sample showing "exposed" ionic charges and the lines of force connecting them. (b) The electric-field profile inferred from lines-of-force density in part (a). (c) The electrostatic-potential profile obtained by integrating the electric-field profile in part (b).

that both of the end regions are *equipotential* regions because they are neutral and field-free. As a result, the height of the potential step $\Delta\psi_0$ is sharply defined. This constitutes the built-in potential step cited above. As a final point, note that the height of the potential step $\Delta\psi_0$ is *identically* the area under the field profile because, once again, $\psi = -\int E dx$.

3-1.4 Band Diagram for a Junction

This matter can be addressed in a way closely related to that employed in Figure 3-1. Let us start with the well-known band diagrams for the uniformly doped *N*-type and *P*-type blocks of Figure 3-1(a); these diagrams are shown in Figure 3-3(a).

237

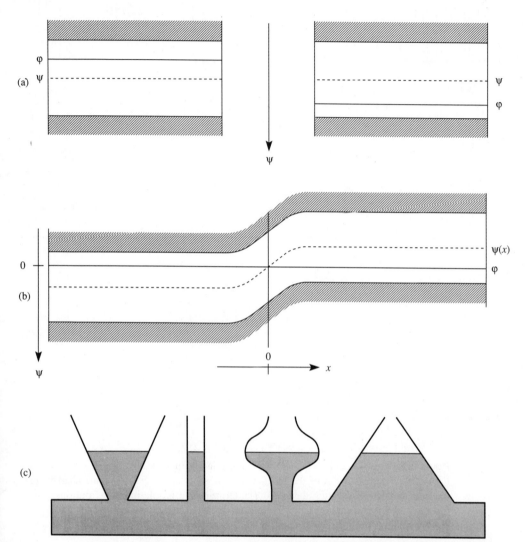

Figure 3-3 Fermi-level conditions in a junction at equilibrium. (a) Band diagrams for uniformly doped blocks of silicon before joining (in a thought experiment) to form a step junction. (b) Band diagram for the same materials after being joined and permitted to come to equilibrium. (c) Water analogy wherein the constant water level in an oddly shaped vessel is likened to the constant Fermi level in an equilibrium sample with position-dependent net doping.

After the two blocks have been joined in the thought experiment of Section 1-3.1 and have been permitted to come to equilibrium, we can gain some guidance from the water analogy used to introduce the Fermi-level concept. The Fermi level is analogous to the "top surface" of the water in the container. We dealt with a conductor there, because the bandgap of silicon complicates the application of a water analogy. But at this point, enough additional ideas have been treated that it is worthwhile performing the minor elaborations necessary to apply the water analogy to silicon.

Suppose that both of the blocks employed in the thought experiment contain deep states that are uniformly distributed in both energy and space, but that their aggregate density is too low to affect electron "supply and demand" significantly. Now the "top surface" idea can be employed even in a material having a bandgap. Two ideas are central: first, at equilibrium the Fermi level will be constant *throughout* the sample; second, a preponderance of the states below the Fermi level will be filled by electrons, and a preponderance of the states above it will be empty.

This situation is shown in Figure 3-3(b). The "top surface" idea now has meaning even in the bandgap, because states are distributed throughout it, albeit at low density. Evidently the result is that the band-diagram end regions for the junction sample have appearances essentially like those in Figure 3-3(a). But there is a gradual transition from *N*-type to *P*-type appearance in the neighborhood of the junction. The intrinsic potential has been taken as $\psi(x)$ because the Fermi level has been taken as potential reference, as explained earlier.

Next we observe that $\psi(x)$ in Figure 3-3(b) is identically $\psi(x)$ in Figure 3-2(c), apart from an arbitrary scale factor. This now makes clear the reason that the potential scale was inverted in Figure 3-2(c). We had agreed that this is an acceptable price to pay for having a band diagram in potential terms with its conduction band above, as it is in an electron-energy band diagram. Thus we reinforce the idea that a band diagram is a potential "map" of a specimen, provided the abscissa is taken as position. Furthermore, the conduction-band edge $\psi_C(x)$ or the valence-band edge $\psi_V(x)$ could serve as well as $\psi(x)$ as a potential map, with the mere adjustment of an arbitrary constant.

Now—at the risk of pushing analogy too far once more—we can recall the vessel with strangely shaped upward extensions as depicted in many general-science textbooks and in Figure 3-3(c). When water is placed in the vessel, its surface is flat, or constant throughout, regardless of shape details (neglecting second-order effects such as capillary action, of course). In an analogous way, Fermi level is constant in a semiconductor sample in spite of doping-variation details.

3-1.5 Carrier Profiles through a Junction

Our discussion of excess carriers emphasized the benefits and hazards of logarithmic and linear data presentations. The conclusion was that *both* kinds of descriptions are necessary for a complete understanding of the problem. The same observation applies to the carrier profiles existing in the neighborhood of a *PN* junction. This time we deal only with equilibrium carrier-density values, but their spatial variation makes the issue important.

In Figure 3-4(a) is shown a linear-coordinate profile of the carrier densities of the only junction considered thus far—one with uniform dopings of $10^{16}/cm^3$ on

240 PN JUNCTIONS

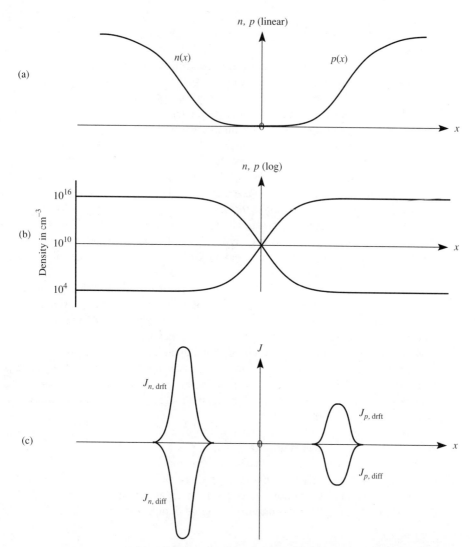

Figure 3-4 Carrier-density and current-density profiles in a symmetric step junction. (a) Linear carrier-density profiles. (b) Log-linear (semilog) carrier-density profiles in the same junction. (c) Current-density profiles in the same junction at equilibrium, showing the precise balance that exists between the electron drift and diffusion components on the one hand and the hole drift and diffusion components on the other.

either side. This profile is identical to that given in Figure 3-1(c). In Figure 3-4(b) is shown a semilog (logarithmic ordinate and linear abscissa) presentation of the same data. Both kinds of diagrams have important contributions to make, and both present misleading impressions.

Exercise 3-2. In Figure 3-4(a), estimate $n(0)$ and $p(0)$.

A literal reading of Figure 3-4(a) says that $n(0) = p(0) = 0$. This is a kind of information that the linear presentation is poorly equipped to give. Consulting Figure 3-4(b) tells us immediately that $n(0) = p(0) = 10^{10}/\text{cm}^3$, a value one million times smaller than n_{ON} and p_{OP}.

Exercise 3-3. The $p(x)$ profile in Figure 3-4(b) is identical in shape to the $\psi(x)$ profile in Figure 3-3(b). Why?

The carrier-density profile $p(x)$ is exponentially related to the potential profile $\psi(x)$. Hence the profile of the *logarithm* of $p(x)$ will have a shape identical to that of $\psi(x)$.

Exercise 3-4. In Figure 3-4(b) estimate the position at which the hole-density gradient dp/dx has a maximum value.

A superficial examination of Figure 3-4(b) might suggest that the slope of the hole density is greatest at the metallurgical junction, at $x = 0$. But then we remind ourselves that the hazard in a semilog presentation is *distortion*. It is a fully linear presentation that gives accurate information on the slope of a profile. Consulting Figure 3-4(a) shows that this occurs somewhere within the fuzzy region we have named the space-charge-layer boundary. It shows further that the hole gradient at $x = 0$ is far lower, and it is indeed several orders of magnitude lower.

Exercise 3-5. Consider the equilibrium symmetric-step-junction sample of Figure 3-2(a). Assume that it has a uniform density of one kind of recombination center throughout. This state has a fixed energy position within the bandgap relative to the band edges. Focus now on the N-type region alone. (The issues we are about to examine are applicable to the P-type region in a parallel way.) Within the transition region, evidently $\psi = \psi(x)$. From Equations 2-35 and 2-36 it follows that we therefore have a condition not met before in which $n_0 = n_0(x)$, and $p_0 = p_0(x)$, even in this sample of completely homogeneous doping. Do you expect to find $R_0 = R_0(x)$ and $G_0 = G_0(x)$? If your answer is yes, will $R_0(x) = G_0(x)$?

Under the assumed conditions, with $p_0 = p_0(x)$, the recombination rate will be $R_0(x) = p_0(x)/\tau$ insofar as τ is constant, which is a good approximation through an appreciable range of $p_0(x)$. [See Exercise 2-38, which describes lifetime constancy with varying $\psi(x)$-to-$\phi(x)$ spacing brought about by a different means, namely changing doping.]

3-1.6 Symmetric Step Junction

A *PN* junction in which there is a totally abrupt transition from uniform doping of one kind to uniform doping of the opposite kind is termed a *step junction*. It is the simplest kind of junction to analyze, and has therefore been taken as the first example. We have, to be sure, added yet another simplifying element. We have assumed equal doping values on the two sides. Hence the only junction considered thus far is the *symmetric* step junction.

Exercise 3-6. The diagrams in Figure 3-4 are symmetric from left to right for the obvious reason that equal doping values exist on the two sides. Why is Figure 3-4(b) symmetric from top to bottom?

The law of mass action asserts that

$$pn = n_i^2.$$

Because the ordinate in Figure 3-4(b) is logarithmic, the appropriate relationship there is

$$\log p + \log n = 2 \log n_i.$$

Thus the sum of p and n ordinates at any arbitrary value of x must add to a constant value. If we measure the ordinates from $p = n = 10^{10}/\text{cm}^3$ where $\log p = \log n = 0$, then the sum of the ordinates must be

$$2 \log n_i = 2 \log(10^{10}/\text{cm}^3) = 20.$$

We grant a certain lack of mathematical rigor here, because one should extract the logarithm of a pure number only, and not a quantity with dimensions. While it sometimes serves engineering purposes to do this, the "impropriety" can be avoided by normalization—with n_i being a logical constant to choose for the purpose—as explained in Section 1-2.5. Notice also that at the junction, $p(0) = n(0) = 10^{10}/\text{cm}^3$, a value one million times smaller than $n_{0N} = p_{0P} = 10^{16}/\text{cm}^3$.

3-1.7 Current-Density Profiles in the Junction

We have identified the region of the dipole layer as the space-charge region, or space-charge layer, of the junction. Another pair of terms is often used, however. Because the respective majority-carrier densities are depressed or "depleted" in the space-charge layer, the terms *depletion region* or *depletion layer* are also frequently used to identify this same junction feature. But *space-charge* and *depletion* are not synonymous adjectives. The former is more general because there are conditions other than majority-carrier depletion that can lead to the condition of space charge in a semiconductor crystal.

Yet another term that is sometimes encountered to designate this important junction attribute is *transition region*, or *transition layer*. The basic idea here is that within this portion of space, the electron-density profile changes steeply from its majority-density value n_{ON} on one side to its minority-density value n_{OP} on the other side, while the hole-density profile undergoes a comparable transition.

These steep transitions in carrier density give rise to large carrier-density gradients, so evident in Figure 3-4(a). As a result, there are inevitable diffusion currents of both carrier types, each peaking at the position where the gradient in that carrier density is a maximum. The resulting current-density profiles are shown with solid curves in Figure 3-4(c). The holes obviously diffuse leftward, constituting a negative current-density component when its direction is referred to the x axis. The electrons diffuse rightward, but that constitutes a leftward current-density component because of the electron's negative charge. The qualitative shapes of these current-density profiles are easily inferred from the carrier-density profiles, because diffusion current depends *only* on density gradient.

But there are drift-current components in the transition region as well, because of the electric field assoc..ed with ionic space charge. These current-density profiles are shown using boldface lines in Figure 3-4(c). Notice that they *precisely balance* the diffusion components at equilibrium, so that *net current vanishes* in the junction at equilibrium.

Exercise 3-7. Why does the drift component of hole current density in Figure 3-4(c) peak near the depletion-layer boundary, while electric field peaks at the metallurgical junction, $x = 0$, as can be seen in Figure 3-2(b)?

This occurs because drift transport, unlike diffusion transport, depends upon *two* variables—electric field and carrier density. Electric field at the space-charge-layer boundary is smaller than peak field by a small factor (much less than ten), while carrier density at the same (boundary) position is over five orders of magnitude *larger* than at $x = 0$. Note too that this is another case where we are served by the principle of detailed balance. Knowledge of $p(x)$ alone (and hence of dp/dx) permits a direct calculation of $J_{p,\text{diff}}$, whereas more information is needed for a direct calculation of $J_{p,\text{drft}}$.

Exercise 3-8. The current-density profiles in Figure 3-4(c) are obviously not symmetric about $x = 0$. How is it possible, then, that $E(x)$ is symmetric about $x = 0$ and the carrier densities are symmetric about $x = 0$?

The balance that exists in the junction at equilibrium is between *hole drift* and *diffusion* on the one hand, and between *electron drift* and *diffusion* on the other. In view of the Einstein relation, diffusivity and mobility are proportional, so this kind of balance is entirely consistent with the symmetries cited.

Once again it is important to point out that the four current-density components described above are a convenient mathematical fiction. That is, as noted before, we do not have carriers streaming past each other. The proof is that these components do not have power dissipation associated with them, as was pointed out in the case of the nonuniformly doped sample employed to introduce the Einstein relation. In a short period of time, a given carrier executes a three-dimensional "random walk" in a particular vicinity.

3-2 DEPLETION APPROXIMATION

It is possible to treat the step junction analytically by making an approximation [5] that seems coarse at first encounter. Nonetheless, this approximation yields results that are extremely useful, and in some respects, surprisingly accurate.

3-2.1 Assuming Total Depletion

The regions of positive and negative space charge associated with a junction must first be idealized. The space-charge-layer (or depletion-layer) *boundary,* which we have seen to be gradual or fuzzy, is now assumed to be perfectly sharp; that is, one assumes that the transition from the neutral condition of the end region to the space-charge condition on a given side of the junction occurs in an infinitesimal distance. Furthermore, one assumes that the space-charge density in the region lying between this sharp boundary and the metallurgical junction is *constant,* and is fixed by the net impurity density—$(N_D - N_A)$ or $(N_A - N_D)$, depending upon which side of the junction is being examined—with the impurity states assumed to be fully ionized. In the simplified junction sample considered up to this point, the relevant densities are simply N_D and N_A, respectively. It follows that the relevant charge densities will be $+qN_D$ and $-qN_A$ in the respective cases. Conveniently, generalization of the charge-density expression so that it will apply on either side of the present sample is unusually straightforward:

$$\text{Volumetric charge density} = q(N_D - N_A). \tag{3-1}$$

On the *N*-type side of the junction, the net charge density is appropriately positive, while on the *P*-type side, it is negative. Note in addition that Equation 3-1 is valid even for compensated samples.

Exercise 3-9. The assumption of a perfectly abrupt space-charge-layer boundary is convenient, but highly unrealistic. Why?

It is evident that a perfectly abrupt transition in carrier density must be accompanied by an infinite density gradient, and hence by an infinite diffusion current at the boundary. Pondering this issue, we see that what is really being done is to eliminate carriers from the picture. The actual semiconductor problem is being replaced by a pure-electrostatics problem involving fixed charges, with the purpose of generating electric-field and electrostatic-potential profiles that approximate those of a real junction.

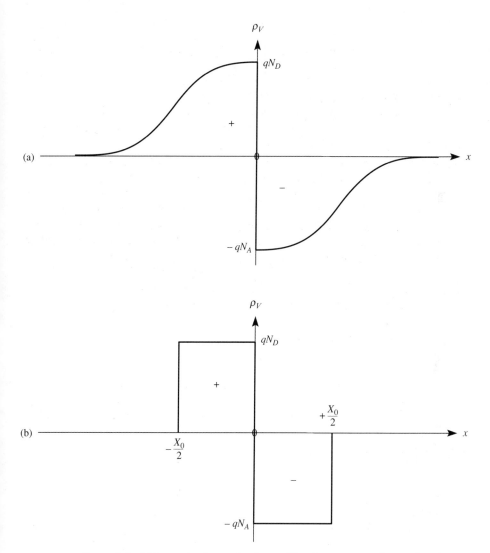

Figure 3-5 Volumetric charge-density profiles for a symmetric step junction at equilibrium. (a) Actual charge-density profile. (b) Approximate charge-density profile assumed for depletion-approximation analysis.

3-2.2 Charge-Density Profile

In Figure 3-1(c), the shaded areas represent the ionic charges "uncovered" near the junction because the majority-carrier densities are forced by the abrupt doping change of the junction to make a rapid transition through many orders of magnitude. In the specific example selected, the change amounts to twelve decades, or a factor of one trillion. It follows, then, that these impurity ions, having lost their neutralizing mobile majority carriers, constitute a fixed *net space charge*. Plotting the resulting space-charge density ρ_v in a manner consistent with Figure 3-1(c), and with due regard for algebraic sign, yields the charge-density profiles that can be seen in Figure 3-5(a). Taking advantage of symmetry, we have placed the spatial origin at the metallurgical junction. The essence of the depletion approximation is to substitute for these actual profiles, the approximate and perfectly rectangular charge-density profiles shown in Figure 3-5(b).

We assign the symbol X_0 to designate the overall thickness of the idealized double layer of charge, or the distance from sharp boundary to sharp boundary. The subscript zero is employed to denote the fact that this is an equilibrium value of depletion-layer thickness. The convenient choice of spatial origin thus makes the two boundary positions $-X_0/2$ and $+X_0/2$, as shown in Figure 3-5(b). A further exploitation of symmetry now becomes possible, because one need only analyze half the stated problem (left side or right side) and then obtain the remaining half of the solution by writing symmetric equations or constructing symmetric curves.

Exercise 3-10. The rectangle on the left-hand side of Figure 3-5(b) has a clear physical interpretation. Determine it by examining units.

The height of the rectangle is qN_D, having the dimensions C/cm^3. The base of the rectangle is $X_0/2$, having dimensions cm. Therefore the area of the rectangle represents a physical quantity having the dimensions C/cm^2. The area represents the *areal charge density* on the left-hand side of the junction. That is, it represents the number of coulombs of charge residing there per unit of junction area.

3-2.3 Electric-Field Profile

It was shown in Section 1-3.7 that the vehicle for conversion from a volumetric charge-density profile to an electric-field profile is Poisson's equation. The issue of boundary conditions raised in that discussion is unusually clear-cut in the present problem because it involves a uniform distribution of positive charges on the left-hand side of the junction, connected by imaginary lines of force to an equal number of negative charges on the right-hand side. This yields an instant qualitative description of the field profile that will emerge from the analysis: The field must vanish at both boundaries because it is purely "internal." It will have its maximum value at the metallurgical junction where the lines-of-force density has a maximum value. And the field has a positive sign because these lines have the positive-x direction. All of these features were anticipated in Figure 3-2(b). Let us choose to analyze the left-hand side of the junction problem. The appropriate form of Poisson's equation then becomes

$$\frac{dE}{dx} = \frac{qN_D}{\epsilon}, \tag{3-2}$$

where ϵ is of course the dielectric permittivity of silicon. It is approximately (within 4%) 1 pF/cm, a number that enters calculations so frequently that it is worth remembering. Next, separate the variables and integrate from the position $-X_0/2$ to an arbitrary x position (on the left-hand side), and from the zero value of electric field at the boundary to the value E that corresponds to the arbitrary value of x. Thus,

$$\int_0^E dE_1 = \frac{qN_D}{\epsilon} \int_{-\frac{X_0}{2}}^x dx_1, \tag{3-3}$$

where E_1 and x_1 are dummy variables. Carrying out the (trivial!) integration and substituting limits yields

$$E(x) = \frac{qN_D}{\epsilon} \left(x + \frac{X_0}{2} \right). \tag{3-4}$$

When plotted, this linear expression yields the solid straight line on the left-hand side of Figure 3-6(a). To be sure, the line has infinite extent, but only the portion plotted between $x = -X_0/2$ and $x = 0$ has physical meaning.

Letting $x = 0$ in Equation 3-4 yields the value of the maximum (or peak) electric field E_{OM}, already known to exist at $x = 0$:

$$E_{OM} = \frac{qN_D X_0}{2\epsilon}. \tag{3-5}$$

Once again the zero subscript indicates that this value is an equilibrium value. The right-hand side of the electric-field profile plotted in Figure 3-6(a) has been constructed on the basis of symmetry.

Exercise 3-11. Write an equation for the electric-field profile on the right-hand side of the junction under discussion.

Evidently the expression

$$E(x) = \frac{qN_D}{\epsilon} \left(\frac{X_0}{2} - x \right)$$

yields the desired straight line, as can be seen by substituting $x = 0$ and $x = X_0/2$ into it.

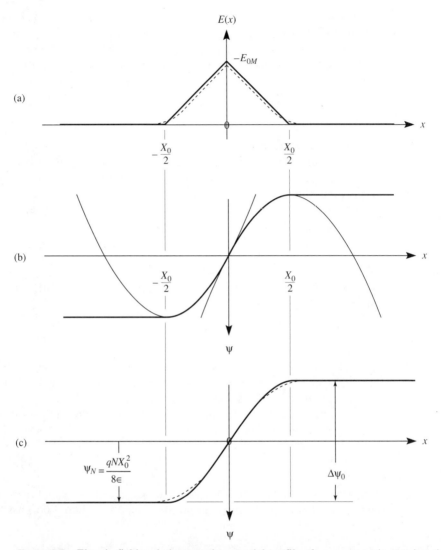

Figure 3-6 Electric-field and electrostatic-potential profiles for a symmetric step junction at equilibrium. Solid lines are predictions of the depletion approximation. (a) The electric-field profile. (b) The two parabolic components of the electrostatic-potential profile. (c) The electrostatic-potential profile.

The dashed lines in Figure 3-6(a) represent the electric-field profile obtained by more accurate analysis. It has a "skirt" at either side because the space-charge-layer boundary is in fact diffuse, rather than being sharp as assumed in the depletion approximation. But the surprising point is that the actual and approximate profiles are so similar, with actual profile exhibiting a prominent near-linear portion. The approximation is appreciably better than the starting (charge-density) approximation depicted in Figure 3-5. The reason for this is that integration is a "smoothing" operation.

Exercise 3-12. Is the area of the rectangle on the left-hand side of Figure 3-5(b) equal to the area under the curve on the left-hand side of Figure 3-5(a)? How do you know?

The two $E(x)$ curves on the left-hand side of Figure 3-6(a) represent the *integrals* of the other two curves under discussion. The fact that the two integral curves differ slightly at $x = 0$ (differing values of E_{OM}) shows that the two charge-density profiles do not have equal areas.

3-2.4 Electrostatic-Potential Profile

The next step is to substitute the expression $-d\psi/dx$ for $E(x)$ in Equation 3-4. Separating variables once more and designating the N-side potential to be ψ_N (the constant value in the neutral and field-free end region) yields

$$\int_{\psi_N}^{\psi} d\psi_1 = -\frac{qN_D}{\epsilon} \int_{-\frac{X_0}{2}}^{x} \left(x_1 + \frac{X_0}{2}\right) dx_1. \tag{3-6}$$

Carrying out the integration and substituting limits then gives

$$\psi - \psi_N = -\frac{qN_D}{\epsilon} \left[\left(\frac{x^2}{2} + \frac{X_0 x}{2}\right) - \left(\frac{X_0^2}{8} - \frac{X_0^2}{4}\right)\right], \tag{3-7}$$

or

$$\psi - \psi_N = -\frac{qN_D}{\epsilon} \left[\frac{x^2}{2} + \frac{X_0 x}{2} + \frac{X_0^2}{8}\right]. \tag{3-8}$$

It is prudent to exploit symmetry still further by letting $\psi = 0$ at $x = 0$. Thus the left-hand-side potential ψ_N can be evaluated from Equation 3-8 as

$$\psi_N = \frac{qN_D X_0^2}{8\epsilon}. \tag{3-9}$$

Substituting this expression back into Equation 3-8 then yields the desired equation for the potential profile on the left-hand side of the junction:

$$\psi(x) = -\frac{qN_D}{2\epsilon}(x^2 + X_0 x). \tag{3-10}$$

This second-order equation defines a parabola with its vertex displaced from the origin. It specifically is the parabola plotted lightly on the left-hand side of Figure 3-6(b), but once again only the portion lying between $x = -X_0/2$ and $x = 0$ has physical meaning. A companion parabola is plotted by symmetry on the right-hand side of Figure 3-6(b). The two parabolic segments with physical meaning and the constant values in the end regions have been emphasized with heavy lines. Extracting the physically relevant portions of Figure 3-6(b) yields the desired potential profile plotted in Figure 3-6(c). Notice that ψ has been shown as increasing downward in both Figure 3-6(b) and (c). This has been done for consistency with our band-diagram convention, since band diagrams are indeed potential plots. Notice that the band-diagram curves in Figure 3-3(b) differ from that in Figure 3-6(c) only by a scale factor.

In Figure 3-6(c), once again the more accurate profile is shown by means of dashed lines. The accuracy of the approximation this time is even better than for the case of electric field, because the smoothing effect of yet another integration is involved. Thus the depletion approximation, which seemed primitive at the early stage, is remarkably accurate at the later stages of the analysis.

3-2.5 Contact Potential

The potential step of height $\Delta\psi_0$ that is evident in Figure 3-6(c) is known as the *contact potential* of the junction. It is identically the *built-in potential drop* associated with a dipole layer that is obtained by integrating through its *built-in field*, as we have just done. This term contact potential is probably familiar from other studies, especially in relation to bringing dissimilar metals together. Precisely the same phenomena are involved whether a *PN* junction or dissimilar metals are being considered. In the latter case, a charge exchange occurs upon making contact, an exchange quite parallel to that invoked in the thought experiment on junction formation. Turning the argument around, we must concede that with respect to contact-potential considerations, differently doped silicon samples are "dissimilar materials."

There is a noteworthy difference, to be sure, between the cases of dissimilar metals and differently doped silicon samples. The difference is that in the semiconductor case, the problem is simple enough to analyze from first principles! This is yet another example of the simplicity inherent in our well-described single-crystal materials, the kind of simplicity that facilitates rapid progress.

In the particular junction treated here, the contact potential $\Delta\psi_0$ is readily determined. It is evident in Figure 3-6(c) that it (from symmetry once more) amounts simply to twice the value determined for ψ_N, so that

$$\Delta\psi_0 = 2\psi_N = \frac{qNX_0^2}{4\epsilon}. \tag{3-11}$$

Notice that no subscript on N is necessary here because $N_D = N_A = N$ in the symmetric case.

We are more frequently interested in finding the depletion-layer thickness X_0, knowing $\Delta\psi_0$, than the reverse, so it is useful to invert Equation 3-11. Thus,

$$X_0 = 2\sqrt{\epsilon\Delta\psi_0/qN} \qquad (3\text{-}12)$$

for the symmetric step junction. A further step makes this expression even more useful. Recall from Section 2-4.5 that the Boltzmann relation says that the ratio of the densities of a particular carrier type (holes or electrons) at two positions is an exponential function of the potential difference between the same two positions. In the present problem—where n_{ON} is the uniform majority-electron density on the left and n_{OP} is the minority-electron density on the right, with a total potential difference of $\Delta\psi_0$—we have

$$\frac{n_{ON}}{n_{OP}} = \exp\frac{q|\Delta\psi_0|}{kT}. \qquad (3\text{-}13)$$

The magnitude of $\Delta\psi_0$ has been taken here to make the result independent of arbitrary reference choice. But from the law of mass action, $n_{OP} = n_i^2/p_{OP}$, we can rewrite Equation 3-13 in terms of the two majority densities (or net-doping values, if preferred). Doing so and inverting the result yields

$$|\Delta\psi_0| = \frac{kT}{q} \ln \frac{n_{ON}p_{OP}}{n_i^2}. \qquad (3\text{-}14)$$

This result has general validity. For the symmetric case, of course, $n_{ON} = p_{OP} \approx N$, so that Equation 3-14 becomes

$$|\Delta\psi_0| = 2\frac{kT}{q} \ln \frac{N}{n_i}. \qquad (3\text{-}15)$$

Continuing to take $10^{16}/\text{cm}^3$ as a typical value of silicon doping, we find from Equation 3-15 that $\Delta\psi_0 = 0.71$ V for the junction being examined. From Equation 3-12, then, or placing Equation 3-15 in Equation 3-12, we find that $X_0 = 0.43$ μm.

The contact potential $\Delta\psi_0$ can be looked upon as the potential hill or barrier that keeps electrons on the left in the junction under study, and that keeps holes on the right. Referring back to Figure 3-3(b), we recall that holes ''fall up'' on such a diagram, so the same potential hill is thus able to do double duty. Either carrier ''slides down'' its hill—an alternate description of the drift phenomenon. And precisely balancing this tendency at equilibrium is the diffusion phenomenon, by which a carrier can surmount its hill. For this reason, contact potential, or built-in potential, is also sometimes termed *diffusion potential*.

Exercise 3-13. If we had chosen to match the areas of the charge-density-profile curves in Figures 3-5(a) and (b), then the two electric-field profiles in Figure 3-6(a) would have had the same values of E_{0M}, and their linear regions would have coincided. This appears to be a better fit of approximate to actual analyses. Why was it not done?

In most cases the usefulness of the approximation is greatest when the actual and approximate values of $\Delta\psi_0$, Figure 3-6(c), are matched. With this choice, taken here, more substantial mismatches in the other two sets of profiles are enforced. There are, however, cases of great importance where it is helpful to let the two linear regions coincide. One of these is described in Section 5-3.6 and Figure 5-48.

At this point, a question will occur to a number of readers. By applying voltmeter probes to the two ends of a junction sample, can one make a direct measurement of $\Delta\psi_0$? The answer is no. Upon a moment's reflection, we see that if the answer were otherwise, we would have a perpetual-motion apparatus in our hands! Power is required to operate an ordinary voltmeter, albeit a very small amount of power.

The explanation for the state of affairs just described resides at the dissimilar-material interfaces that are present where the voltmeter probes touch silicon. A charge exchange and a contact potential will exist at each such interface, as well as at the *PN* junction. To simplify conditions, assume that the voltmeter is constructed entirely of copper, which is a practical possibility. This situation is shown in Figure 3-7(a). As a result, there are only two additional contact potentials to deal with, beyond that at the junction. We find that these two contact potentials are different from one another because one involves *N*-type silicon, and the other, *P*-type. Also, we find that they precisely counterbalance one another, making the two probes equipotential, as shown in Figure 3-7(b). The Fermi level is constant *throughout the entire system,* including the voltmeter, and the entire system is at equilibrium.

There is another facet to this matter that is nicely introduced by the points just made. It is a fact that a voltmeter measures *Fermi-level* difference, and not potential difference. The two end regions of the junction sample, after all, are truly at different potentials because of the built-in voltage of the junction. Yet our voltmeter gave us a reading of zero. If, on the other hand, we had used a battery to alter Fermi-level positions on the two sides relative to each other, the voltmeter would faithfully indicate this difference. In what follows, we shall use the symbol ΔV to designate a difference in "voltmeter voltage" or simply a voltage difference. Further, we shall continue to use the symbol $\Delta\psi$ to designate an electrostatic-potential difference.

Exercise 3-14. The region of potential change at the probe interfaces in Figure 3-7(b) is largely confined to the silicon. Why?

This is the same as saying that the region of electric field is confined to the silicon, which occurs because a metal cannot sustain an electric field without having an accompanying net current and, under the existing equilibrium conditions, a net current is not possible. The ionic charges in silicon make extended regions of electric field possible; in the field regions, carriers diffuse in one direction and drift in the other, so that there is no net current.

DEPLETION APPROXIMATION   253

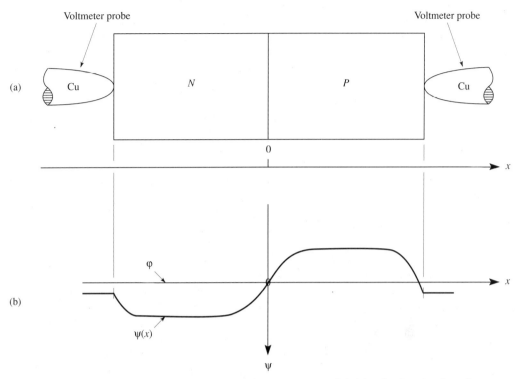

Figure 3-7 An attempt to measure the contact potential $\Delta\psi_0$ of a junction by using a voltmeter. (a) Physical representation of the junction sample. (b) Compensating contact potentials arise at the probe interfaces, rendering the two probes equipotential.

The symmetric step junction considered exclusively up to this point is in many respects the simplest case. Although it has modest practical importance, there are other cases that are encountered much more frequently. Fortunately, however, the numerous other kinds of *PN* junctions yield to depletion-approximation analysis that differs only in detail, and not in principle, from that already presented. We next retain the step-junction feature, but drop the condition of symmetry. That is, the two sides will have uniform but different net-doping values, and there will be a perfectly abrupt doping transition at the metallurgical junction.

3-2.6 Asymmetric Step Junction

In this case the *N*-type and *P*-type regions continue to have uniform doping, but have differing values of net doping, and the condition of a perfectly abrupt transition from one conductivity type to the other is retained. For the sake of a specific example, let us assume once more that each region contains but one impurity, and let $N_A = 1 \times 10^{16}/\text{cm}^3$ on the right-hand side as before. But on the left-hand side,

254 PN JUNCTIONS

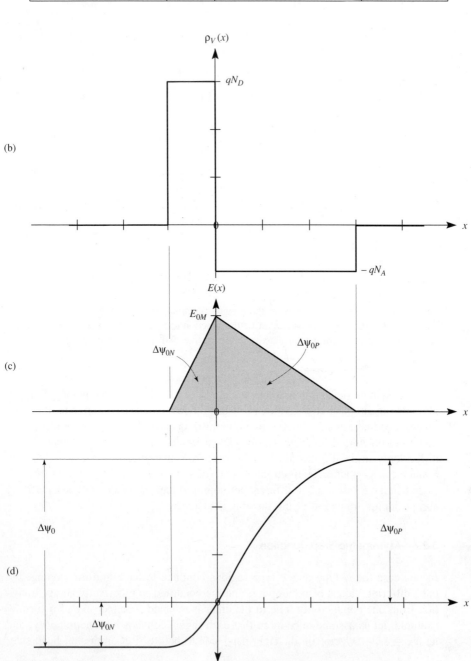

let $N_D = 3 \times 10^{16}$/cm^3. A physical representation of a small portion of this junction is depicted in Figure 3-8(a). Moving directly to the depletion approximation, we obtain the charge-density profile shown in Figure 3-8(b).

Exercise 3-15. Are the two rectangles in Figure 3-8(b) equal or unequal in area? Why?

In Exercise 3-8, it was pointed out that the area of each charge-density-profile rectangle is proportional to areal charge density at the junction. The areal densities on the two sides of the junction must be equal. If this were not so, a line of force would have to penetrate into a neutral region (that contains an ocean of mobile carriers) in search of a compensating charge. But the result would be charge motion and a readjustment of the space-charge layer until all lines of force were appropriately terminated, guaranteeing equal amounts of charge per unit area on the two sides of the junction.

This time Poisson's equation must be applied to the two sides separately. The result once more, after separating variables and integrating, is a triangular electric-field profile, but no longer an isosceles triangle. It is shown in Figure 3-8(c). We made the point earlier that the integral of field through the junction is equal to the contact potential $\Delta\psi_0$. But the same reasoning can be applied to the individual potential drops $\Delta\psi_{ON}$ and $\Delta\psi_{OP}$ on the left-hand and right-hand sides, respectively, of the junction in Figure 3-8. Inspection of Figure 3-8(c), then, shows that $\Delta\psi_{OP} = 3\Delta\psi_{ON}$, consistent with the potential profile shown in Figure 3-8(d). Expressions for these parabolic profiles can be obtained, as before, by substituting $(-d\psi/dx)$ for E, separating variables, and integrating a second time.

Exercise 3-16. The junction of Figure 3-8 has a depletion-layer thickness of $X_0 = 0.356$ μm, and a contact potential of $\Delta\psi_0 = 0.737$ V. Find the peak electric field, E_{OM}.

To solve this problem, one requires the (not very sophisticated) knowledge that the area of a triangle is half its base times its height. In the present case, X_0 is the base and $\Delta\psi_0$ is the area of the triangle in Figure 3-8(c). Hence the height, E_{OM}, can be determined from

$$E_{OM} = \frac{2\Delta\psi_0}{X_0} = 2\frac{(0.737 \text{ V})}{(3.56 \times 10^{-5} \text{ cm})} = 41.4 \text{ kV/cm}.$$

← **Figure 3-8** An asymmetric step junction. (a) Physical representation of a junction sample having $N_D = 3N_A$ on the respective sides. (b) Depletion-approximation charge-density profile for the junction. (c) Depletion-approximation electric-field profile. (d) Depletion-approximation electrostatic-potential profile.

3-2.7 One-Sided Step Junction

A grossly asymmetric step junction, or *one-sided junction*, has great practical importance, occurring in most semiconductor devices. Again, a perfectly abrupt transition (at the metallurgical junction) from one doping type to the other is assumed. The one-sided case can be regarded as the result of carrying the example just given to a practical extreme. That is, if the ratio N_D/N_A in that case had been 50 or more rather than three, the junction would qualify as one-sided. The "one side" referred to is of course the lightly doped side, because changes in the doping of the other side have little effect on junction properties once one has crossed the "one-sided" threshold, as we shall see.

While a doping ratio of 50 or more is a good criterion for gross asymmetry, let us take a more modest example for greater clarity. Figure 3-9(a) shows a physical representation of a junction for which $N_D = 20N_A$. (Previous assumptions have been retained.) The corresponding charge-density profile is shown in Figure 3-9(b). Employing Poisson's equation as before yields the electric-field profile in Figure 3-9(c). Note that the profile is *nearly a right triangle*. Imposing further asymmetry would move it even closer to that condition.

In Figure 3-9(d), then, is shown the potential profile that is inferred by employing the principles put to use in both of the preceding examples. The depletion approximation predicts in this case that $(\Delta\psi_{0P}/\Delta\psi_{0N}) = 20$. An interesting detail emerges at this point, however. More precise analysis shows that as the doping asymmetry is increased, the "high-side" potential drop ($\Delta\psi_{0N}$ in this example) does not approach zero, but rather approaches the value $(kT/q) \approx 26$ mV [8, 16]. Since the maximum value of $\Delta\psi_0$ in a junction with both sides doped heavily but not degenerately is about one volt ($\sim\psi_G$), it follows that the potential-drop ratio on the two sides of a one-sided junction will "top out" or saturate at a value less than 40. For one-sided junctions, it is common to indicate which is the heavily doped side by using a *plus* superscript, thus producing convenient and specific shorthand designations, namely, N^+P and P^+N.

3-2.8 Comparing the Step Junctions

The asymmetric step junction can be regarded as the most "general" of its class. Depletion-approximation analysis informs us that its depletion-layer thickness is given by

$$X_0 = \left[\frac{2\epsilon\Delta\psi_0}{q}\left(\frac{1}{N_1} + \frac{1}{N_2}\right)\right]^{\frac{1}{2}}, \qquad (3\text{-}16)$$

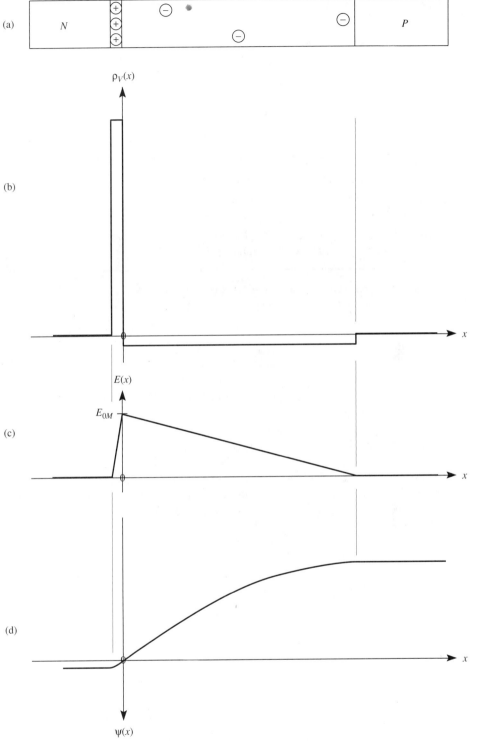

Figure 3-9 A one-sided, or grossly asymmetric, step junction. (a) Physical representation of a junction sample having $N_D = 20N_A$ on the respective sides. (b) Depletion-approximation charge-density profile for the junction. (c) Depletion-approximation electric-field profile. (d) Depletion-approximation electrostatic-potential profile.

where N_1 is the net doping on the more heavily doped left-hand side, and N_2 that on the more lightly doped right-hand side. This should be compared to the expression for the symmetric case, which we repeat for convenience:

$$X_0 = \left[\frac{4\epsilon\Delta\psi_0}{qN}\right]^{\frac{1}{2}}. \tag{3-17}$$

Exercise 3-17. How can one infer Equation 3-17 from Equation 3-16?

Since $N_1 = N_2$ in the symmetric case, the two terms in parentheses in Equation 3-16 are equal, introducing a factor of two into the numerator, and permitting us to drop the subscript on N, so that Equation 3-17 results.

In analogous fashion, the expression for the one-sided case can be inferred from Equation 3-16. As $N_1 \rightarrow \infty$, the term $1/N_1$ vanishes, and the result for the one-sided junction is

$$X_0 = \left[\frac{2\epsilon\Delta\psi_0}{qN_2}\right]^{\frac{1}{2}}, \tag{3-18}$$

where N_2 is the net doping on the more lightly doped side. For this kind of junction, the thickness of the depletion layer on the lightly doped side approximates the total depletion-layer thickness. In other words, the lightly doped side is "in command" in the sense that it single-handedly determines junction properties to a large degree.

For most purposes it is not essential (or even desirable) to memorize these expressions, but it is *extremely* important to realize that in *all* these cases, X_0 is the base of a triangle, E_{0M} is its height, and $\Delta\psi_0$ is its area. Thus, given any two of the three quantities, one can readily calculate the third. Note too that in this discussion we have adhered to the term depletion-layer *thickness*, a term usually far more descriptive than sloppy reference to depletion-layer "width"—a usage that causes completely unnecessary confusion when extended into transistor work. (Emitter-stripe *width* has a clear meaning, and it is a dimension orthogonal to emitter-region or emitter-junction *thickness*.)

Exercise 3-18. Consider a symmetric junction having $\Delta\psi_0 = 1$ V. What are the Fermi-level positions at equilibrium in the two neutral regions? (Recall that $\psi_G \approx 1.12$ V.)

The spacing of each side's Fermi level from its respective nearest band edge will be half of $\psi_G - \Delta\psi_0$, or in the present case,

$$\tfrac{1}{2}(1.12 \text{ V} - 1.0 \text{ V}) = 0.06 \text{ V}.$$

> With the Fermi level only 60 mV from each band edge, a condition of degeneracy is being approached on both sides. Sketching a band diagram for this junction at equilibrium helps to clarify and fix the concepts involved.

Before leaving this topic, we should point out that there exists a general solution for the equilibrium-step-junction problem [14], and a depletion-approximation replacement (DAR) based upon it [15], to be outlined in Section 3-7. The DAR is an approximate-analytic method that is almost as easy to use as the depletion approximation, and almost as accurate as a numerical solution. These new techniques are equally applicable to the semiconductor-surface problem (Section 5-3.1).

3-3 JUNCTION UNDER BIAS

A *biased junction* is one in which a voltage has been added to or subtracted from its contact potential. Usually this is accomplished by means of an external voltage source of some kind, such as a battery. The practical use of a junction, of course, requires its biasing. Since biasing is inevitably accompanied by a net current through the junction, it follows that biasing involves nonequilibrium conditions.

3-3.1 Algebraic-Sign Convention

In the foregoing discussion of junctions at equilibrium, it was sufficient usually to concern ourselves only with the *magnitude* of the potential difference involved, namely the contact potential $\Delta\psi_0$. But now, *algebraic sign* assumes importance also, because an externally applied voltage can add to or subtract from the height of this all-important potential hill. The convention is easy to remember because *forward bias* involves applying *positive* voltage to the *P-type* side and *negative* voltage to the *N-type* side, conditions that can be summarized thus:

$$\text{Forward Bias} \begin{cases} +P \\ -N \end{cases}.$$

A condition of forward bias diminishes the size of the potential hill that exists at the *PN* junction, causing substantial current through it.

Reverse bias, on the other hand, serves to increase the height of the potential barrier. It can obviously be summarized in this manner:

$$\text{Reverse Bias} \begin{cases} -P \\ +N \end{cases}.$$

Because of this fact, the contact potential $\Delta\psi_0$ can be legitimately regarded as a "built-in reverse bias."

To the sign convention for junctions, let us add a comment on the Ohm's-law convention, because the junction sample will soon be placed in a circuit. In Figure

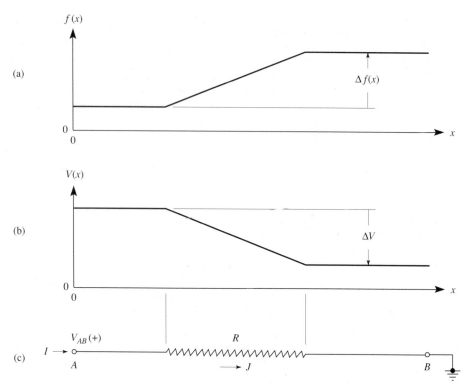

Figure 3-10 The Ohm's-law sign convention for algebraic sign. (a) The usual mathematical convention takes an *increase* in a function such as f(x) to be a positive increment. (b) The Ohm's-law convention takes a *decrease* in voltage, V(x) here, or a voltage drop to be positive. (c) This choice is made to have algebraic-sign consistency in $\Delta V = IR$, since an inward-flowing terminal current was arbitrarily chosen to be positive. Construction of this diagram additionally makes the current density J within the resistor positive as well, by reference to the positive-x axis.

3-10(a) we note the fact that an increase in a function is normally considered a positive increment. The Ohm's-law convention reverses this choice. A voltage *drop,* as illustrated in Figure 3-10(b), is taken to be a positive voltage increment, ΔV. Under such conditions, a conventional current I flows *into* terminal A of the resistor illustrated in Figure 3-10(c). Conventional current *into* a terminal is taken as positive, and this then yields sign consistency in Ohm's law, $\Delta V = IR$.

At this point it is convenient to adopt the double-subscript notation that eliminates any remaining sign ambiguity. Consistent with the Ohm's-law convention we can write

$$V_{AB} \equiv V_A - V_B; \tag{3-19}$$

thus a positive value of V_{AB} in Figure 3-10(c) ensures that a positive current I will flow *into* terminal A from outside the resistor.

Now let us adapt the same convention to the *PN*-junction sample. Arbitrarily, let it be a symmetric step junction. We choose to "ground" the *P*-type side. (Using a neutral region as potential reference is usually desirable when device currents or voltages are to be examined.) Then let us apply the voltage V_{NP} to the *N*-type side, where of course

$$V_{NP} \equiv V_N - V_P. \tag{3-20}$$

This situation is depicted in Figure 3-11(a). The equilibrium potential profile for the junction, assumed to be a symmetric step junction, is given by the solid line in Figure 3-11(b). It is essentially that of Figure 3-6(c), but with the important difference that here the constant potential in the neutral end region of the *P*-type side is now taken as $\psi = 0$.

When V_{NP} is negative, we have a case of forward bias, because negative voltage is being applied to the *N*-type side. Consequently there is a reduction of the potential barrier $\Delta\psi_0$ by an amount V_{NP}, seen to be negative in Figure 3-11(b), with a new barrier height of $\Delta\psi$ resulting. We shall see that for *both* bias polarities through wide voltage ranges it is an excellent approximation to consider that all of the externally applied voltage affects the height of the potential barrier, with little effect in the neutral end regions.

Exercise 3-19. Under what conditions would the change in barrier height *not* account for virtually all of the applied voltage?

Under extreme forward voltage, *IR* drops in the quasineutral end regions become significant and must be considered.

It was pointed out earlier that in the equilibrium junction, hole drift and diffusion are balanced, as are electron drift and diffusion. Therefore, diminishing the barrier height below the diffusion potential $\Delta\psi_0$ through forward bias causes carriers to flood through the transition region. Majority electrons travel by diffusion from the *N*-type side into the *P*-type region, where they become minority carriers. In analogous fashion, holes travel from the *P*-type region to the *N*-type region.

There is an alternative description of this phenomenon, and realizing that the two descriptions are equivalent is a matter of considerable importance: Reducing the height of the potential hill means that more carriers will have (or can acquire) sufficient energy to surmount it. Very few carriers are able thus to pass through the transition region without interruption; most that accomplish this feat have experienced several interruptions—interactions with phonons and ions that were sufficiently favorable with respect to energy and direction so that the carrier eventually arrives on the opposite side. This hill-climbing description constitutes a different way of describing diffusion.

Because carriers are able to move with relative ease over the barrier of reduced height, the junction sample exhibits relatively low resistance with bias of this polarity. It is for this reason that the diode symbol, Figure 3-11(c), is an arrowhead pointing in the direction of easy conduction under the forward-bias condition.

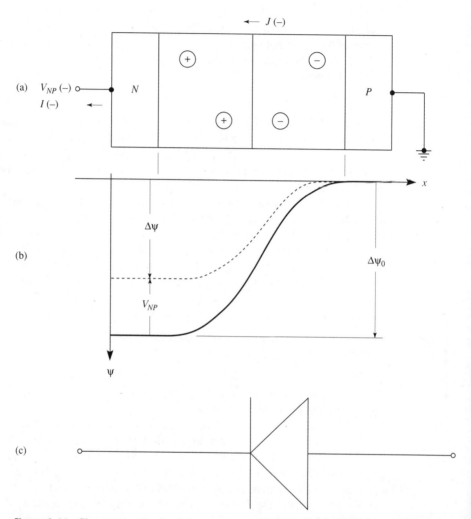

Figure 3-11 Sign convention for junction bias. (a) Physical representation of junction sample and external connections; negative current density, terminal current, and applied voltage correspond to forward bias. (b) Potential profile showing potential-barrier reduction by *forward* bias. (c) Junction-diode symbol.

3-3.2 Reverse Bias

It is evident in Figure 3-11(b) that conversion to reverse bias, which means a positive value of V_{NP} (positive applied to the N-type side), will add to the height of the junction's potential barrier, $\Delta\psi$. That is,

$$\Delta\psi = \Delta\psi_0 + V_{NP}. \tag{3-21}$$

Let us continue to use a symmetric step junction for illustration, and even more specifically, let the net doping on either side continue to be $10^{16}/cm^3$. The growing potential hill occupies a thicker region of (x-direction) space, which is another way of saying that the space-charge-region boundaries retreat from the metallurgical junction. These issues will be examined quantitatively in a moment, but for now let us see the effect as presented in the physical representation of Figure 3-12(a), and the resulting charge-density profile of Figure 3-12(b). The space-charge-layer boundaries have retreated from their equilibrium positions (solid lines) to their reverse-bias positions (dashed lines) in the process, "uncovering" additional impurity ions on both sides of the junction.

Just as the barrier-height lowering of forward bias leads to an increasing flux of carriers that penetrate the transition region via diffusion transport, the barrier-height raising of reverse bias causes the flux of such carriers to drop. Furthermore, the dependences of these numbers on barrier height are exponential. Phenomena to be described later give rise to a small "leakage current," typically one to ten picoamperes, that flows through a reverse-biased silicon junction. Consequently, a silicon junction biased in reverse approximates an open circuit, or at worst, a very high resistance. Furthermore, the fact that so little current flows in reverse bias explains why *IR* voltage drops in the end regions are negligible, so that all of the reverse applied voltage adds to the height of the barrier at the junction.

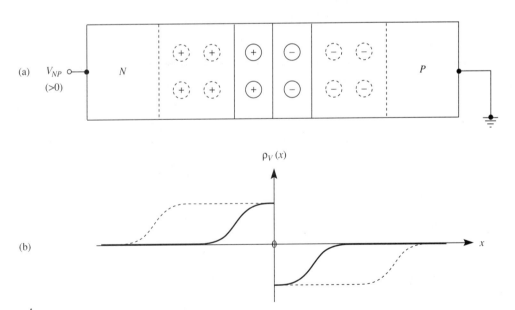

Figure 3-12 A symmetric step junction subjected to reverse bias. (a) Physical representation of the sample at equilibrium (solid lines) and under reverse bias (dashed lines). (b) Volumetric-charge-density profiles corresponding to the boundary positions shown in part (a).

Exercise 3-20. In the light of the information just given, explain the congruence of the boundary portions of the charge-density profiles in Figure 3-12(b) for the equilibrium and reverse-biased cases.

At equilibrium, net current through the junction is strictly zero. Under reverse bias, it is approximately zero. The shape of the boundary-region profile is set by the shape of the majority-carrier profile in that neighborhood, and that is in turn set by the "magic" shape that precisely balances the (very large!) drift and diffusion components of majority-carrier current in the boundary region.

Because the boundary-region profile in Figure 3-12(b) simply translates in response to reverse bias without a significant change of shape, it follows that the depletion approximation improves steadily with reverse bias. That is, substituting rectangles for the dashed curves in Figure 3-12(b) is more readily acceptable than for the solid curves, as was done before. (Yet even there, we found that after two integrations the approximation was very good.) For this reason, let us use the depletion approximation exclusively to examine the other relevant profiles in the reverse-biased symmetric step junction.

Figure 3-13(a) repeats the charge-density profiles of Figure 3-12(b), but with depletion-approximation profiles substituted for the actual profiles. The corresponding approximate electric-field profiles for equilibrium and reverse bias are shown in Figure 3-13(b).

Exercise 3-21. Why are the electric-field slopes on a given side of Figure 3-13(b) the same?

Poisson's equation informs us that dE/dx is proportional to ρ_v, and ρ_v near the junction is unaltered by reverse bias; it is fixed solely by net doping density, which is constant (independent of position) in this example.

At this point recalling the meanings of the bases, heights, and areas of the triangles in Figure 3-13(b), it is evident that the height of the potential hill in the present case of reverse bias has been increased by a factor of nine, with the result shown in Figure 3-13(c).

Exercise 3-22. Find V_{NP} in terms of $\Delta\psi_0$ in the reverse-biased junction being examined.

From Equation 3-21 we have

$$\Delta\psi = 9\Delta\psi_0 = \Delta\psi_0 + V_{NP}.$$

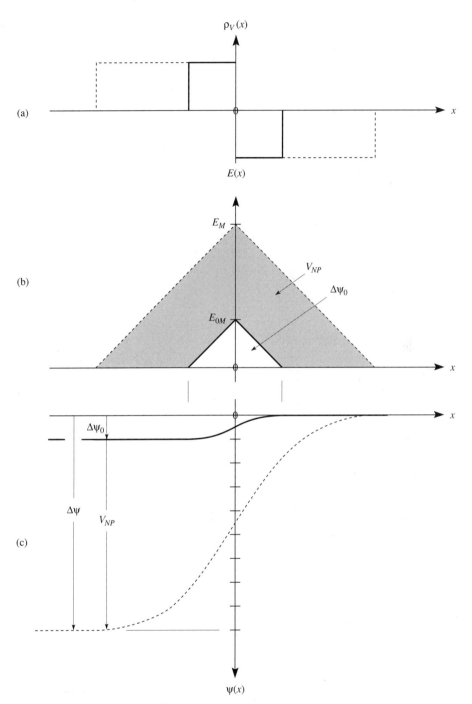

Figure 3-13 Profiles for a symmetric step junction under reverse bias as predicted by the depletion approximation, with solid lines representing equilibrium profiles and dashed lines representing a particular case of reverse bias. (a) Charge-density profiles. (b) Electric-field profiles. (c) Electrostatic-potential profiles.

From this it is evident that

$$V_{NP} = 8\Delta\psi_0.$$

Thus the shaded area in Figure 3-13(b) represents V_{NP}.

The profile curves for the other two kinds of step junctions considered earlier behave in analogous fashion under conditions of reverse bias. Further, the expressions given for equilibrium depletion-layer thickness X_0 can be converted into expressions for depletion-layer thickness X in a junction under bias simply by replacing $\Delta\psi_0$ by $\Delta\psi$; in short, simply drop the zero subscripts in Equations 3-16 through 3-18, noting carefully that $\Delta\psi = \Delta\psi_0 + V_{NP}$. The resulting equations, given here, are also found in Table 3-1, placed at the end of the chapter for convenient reference.

ASYMMETRIC STEP JUNCTION
$$X = \left[\frac{2\epsilon\Delta\psi}{q}\left(\frac{1}{N_1} + \frac{1}{N_2}\right)\right]^{1/2}; \tag{3-22}$$

SYMMETRIC STEP JUNCTION
$$X = \left[\frac{4\epsilon\Delta\psi}{qN}\right]^{1/2}; \tag{3-23}$$

ONE-SIDED STEP JUNCTION
$$X = \left[\frac{2\epsilon\Delta\psi}{qN_2}\right]^{1/2}. \tag{3-24}$$

Exercise 3-23. Examine the parabolas on the right-hand side of Figure 3-13(c). Comment on how their shapes compare.

Their shapes are *identical*. This is because both result from integrating straight lines of equal slope. In the case of the potential profile for reverse bias, a larger portion of the parabola becomes physically meaningful.

Exercise 3-24. In diagrams such as those in Figures 3-11(a) and 3-12(a), many authors prefer to reverse the junction sample, grounding the N-type region on the right, and applying external voltage to the P-type side. This choice gives them the convenience of having positive applied voltage, positive terminal current, and positive (rightward) current density all correspond to forward bias. What difficulties or problems, if any, are engendered by this choice?

The difficulty that arises can best be seen by referring to Figure 3-13(c) wherein $\Delta\psi = \Delta\psi_0 + V_{NP}$. If sample orientation is reversed, then the built-in voltage $\Delta\psi_0$ will be negative and the arrow representing $\Delta\psi_0$ will point upward. This is necessary because the tail of the arrow must be at the potential chosen as reference. But with $\Delta\psi_0$ negative, one has the awkward choice of writing, for example, $\Delta\psi_0 = -0.8$ V, or else introducing yet another explicit negative sign, writing "$-\Delta\psi_0$" consistently. There is no agreement on this issue, and appreciable confusion has occurred in the literature as a result. This problem can be avoided by using our P-type-side reference, while at the same time one has consistent algebraic signs for currents and voltages.

3-3.3 Forward Bias and Boltzmann Quasiequilibrium

The case of forward bias is depicted in Figures 3-11(a) and 3-11(b). This time V_{NP} is negative (note that its arrow has the negative-ψ direction); consequently both the barrier height and the depletion-layer thickness are diminished. To illustrate the consequences of this new condition, let us once again employ the symmetric step junction having a single-type doping of $10^{16}/\text{cm}^3$ on each side. The semilog presentation of carrier densities is unusually informative here. Therefore, in Figure 3-14 we repeat the equilibrium carrier-density profiles for this particular junction from Figure 3-4(b), employing solid lines for the purpose.

Now refer back to Section 3-2.5 where the Boltzmann relation was applied to the junction at equilibrium in Equation 3-13; the Boltzmann relation, you will recall, states that the ratio of densities for a particular carrier between two positions is exponentially related to the potential difference between those positions. Taking advantage of the fact that a specific choice of potential reference has since been made (Figure 3-11), we can be explicit on algebraic sign, with the result

$$(n_{0N}/n_{0P}) = \exp(q\Delta\psi_0/kT) = \exp \Delta U_0, \tag{3-25}$$

where a conversion has been made to *normalized* (or dimensionless) built-in voltage ΔU_0, using the thermal voltage (kT/q) for normalization. A parallel relationship holds for holes. The essential assumption in the present analysis is that an expression of the *same form as* Equation 3-25 is valid when the junction is under forward bias, or that

$$[n_{0N}/n_P(+X/2)] = \exp(\Delta U), \tag{3-26}$$

where $\Delta U = (q\Delta\psi/kT) = q(\Delta\psi_0 + V_{NP})/kT$. This is termed the assumption of *Boltzmann quasiequilibrium*. That is, we assume that the electron-density ratio at the junction boundaries is exponentially related to the now-diminished potential difference between the boundaries, ΔU, an assumption that proves to be good for wide-ranging conditions. The conditions that must be satisfied for this important assumption to be valid are treated in Section 4-4.2.

Because the new junction potential difference $\Delta U < \Delta U_0$, it follows that the electron-density ratio has been reduced, with a result that can be seen for one particular bias in Figure 3-14. Notice that the condition $n_N(-X/2) \approx n_{0N}$ has been retained, while the boundary value $n_P(+X/2)$ is now inflated, thus diminishing the

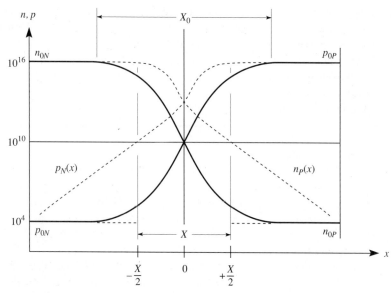

Figure 3-14 Semilogarithmic carrier-density profiles for a symmetric step junction. Solid lines show equilibrium profiles. Dashed lines show profiles for a modest forward bias, $V_{NP} = -0.35$ V.

ratio. A similar relationship holds for holes. The constancy of $n_N(-X/2)$ in the face of (moderate) forward bias is explained as follows: Neutral conditions outside the junction boundaries are assumed (again, this has proven to be a valid assumption for wide-ranging conditions). Therefore $n_N(-X/2)$ departs negligibly from n_{0N}, since the majority excess-carrier density required to preserve neutrality is negligible compared to n_{0N}. In the situation shown in Figure 3-14, the majority–minority density ratio at either boundary is one million. This qualifies as a *low-level* condition, and as a result, the minority-carrier density assumes the full burden of adjusting in order to obey the Boltzmann quasiequilibrium condition.

The boundary-neutrality condition informs us that we will have $p' \approx n'$ at each boundary of the transition region. A numerical example is in order at this point. Let us assume that U_{NP} is such that at the left-hand boundary we have $p'_N(-X/2) = n'_N(-X/2) = 10^{10}/cm^3$. Then the minority-hole density there becomes $p_N(-X/2) = 10^4/cm^3 + 10^{10}/cm^3 \approx p'_N(-X/2)$, while the majority-electron density becomes $n_N(-X/2) = 10^{16}/cm^3 + 10^{10}/cm^3 \approx n_{0N}$. Thus the equilibrium majority-carrier density at the boundary is essentially unaltered, as emphasized in the previous paragraph, a result of *low-level* forward bias.

The dashed lines in Figure 3-14 display the million-fold increases in the boundary values of the minority-carrier densities on either side. The majority-carrier densities at the boundaries, by contrast, change by one part per million. Note, too, that the boundary positions have shifted toward the metallurgical junction, again because forward bias has reduced the potential barrier (the area of the field-profile triangle), and this in turn requires shrinkage of the depletion-layer thickness.

As a result of minority-carrier-density inflation at both boundaries, it is said that minority carriers are *injected* in both directions when a junction is forward biased. The injected carriers diffuse away from the junction, recombining as they go. Suppose we focus on the right-hand side where electrons are the minority carriers. This situation can be treated by means of the constant-E continuity-transport equation, Equation 2-101. Letting the forward bias be fixed, one has a steady-state problem, so that the term $\partial n/\partial t$ will vanish. Under low-level conditions, the two regions outside the transition region are (to an excellent approximation) neutral and field-free, so that the drift term also vanishes. This leaves the diffusion and recombination terms, and the solution for this differential equation is

$$n'_P(x) = n'_P(0)\ \exp[-x/\sqrt{D_n\tau}], \qquad (3\text{-}27)$$

where the position of the right-hand transition-region boundary $(+X/2)$ in Figure 3-14, is taken temporarily as spatial origin, and the quantity $\sqrt{D_n\tau}$ is the *electron diffusion length L_n*, as was explained in Section 2-7.2.

Exercise 3-25. Verify that Equation 3-27 satisfies the specialized continuity equation that applies to the right-hand neutral region in Figure 3-14.

If we shift the spatial origin from the metallurgical junction, then the desired verification is identical to that given in Exercise 2-44, Section 2-7.2.

Because the excess-minority-carrier profiles in the neutral end regions are declining exponentials, their semilog representations are linear, as can be seen on both sides of Figure 3-14. Thus the disturbance of equilibrium carrier densities "fades away" as one departs from the junction. If the end regions are extensive, the carrier densities will have returned to equilibrium before one reaches the ohmic contact.

When the junction at equilibrium was introduced, terms such as "fuzzy" and "diffuse" were used to characterize the space-charge-layer boundaries. But in addressing the junction under bias, great emphasis is placed on conditions at those boundaries, and the analysis of the biased junction requires quantitative treatment of boundary values. Fortunately, in the biased junction, the boundaries are sharply defined, and especially so in the case of forward bias.

This fact is illustrated in Figure 3-15. The symmetric step junction under forward bias depicted in Figure 3-14 is represented in Figure 3-15(a), with boundary positions corresponding to the same amount of forward bias. In the end regions, but near the boundaries, current is carried primarily by the diffusing minority carriers. But as their densities and gradients decline with distance from the boundaries, current continuity is maintained by the drift of majority carriers in a growing (but still tiny!) electric field. The integral of this electric field through a neutral end region constitutes the voltage drop in that region, a quantity that we have explicitly assumed to be negligible compared to the voltage change in junction barrier as a consequence of applied bias. On the right-hand side of the sample in Figure 3-15(a), the majority carriers are holes, drifting leftward toward the junction (which they

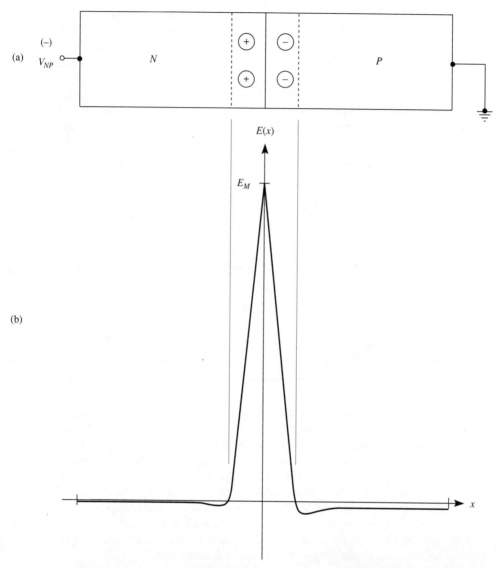

Figure 3-15 In a junction under forward bias, depletion-layer boundaries are sharply defined. (a) Physical representation of the forward-biased symmetric step junction of Figure 3-14. (b) Qualitative field profile, showing that electric field changes sign at the space-charge-region boundaries, in the process defining boundary positions.

will pass through to be injected into the N-type region). Hence the miniscule field in the P-type region is also directed leftward. Analogous reasoning shows that the field direction in the N-type region is also directed leftward. By contrast, however, the huge field in the double-layer region of the junction is positive, directed from donor ions on the left to acceptor ions on the right. The result is shown qualitatively in

Figure 3-15(b), and its importance is that *electric field vanishes at both boundaries*. This, of course, is another way of saying that electric field changes sign, or passes through zero, at these specific positions. In so doing, the field gives us *sharply defined surfaces* that can be taken as the boundaries of the space-charge layer.

Exercise 3-26. The electric field in a symmetric junction at equilibrium is strictly symmetric, a fact employed in the construction of Figure 3-6(a). But under forward bias, however small, the profile becomes asymmetric. Why?

Under steady-state conditions, the net current density J in the one-dimensional sample of Figure 3-15 must be continuous. For long end regions, J consists almost purely of majority-carrier drift near the ohmic contacts. It follows from this that the electric-field values in those two regions stand inversely as the mobilities.

Furthermore, the field profile exhibits extrema—minima in Figure 3-15(b)—near but outside the junction boundaries because of the interplay of majority-carrier drift and diffusion with minority-carrier diffusion. These features are shown qualitatively in Figure 3-15. Since low-level forward bias is intended there, the electric-field magnitudes in the end regions are represented in grossly exaggerated fashion compared to values in the junction region. One facet of this condition is treated in Problem A3-7.

3-3.4 Law of the Junction

Since the small values of current that accompany low-level conditions ensure that the voltage drops (*IR* drops) in the end regions are small enough to be neglected, as just noted, the condition

$$V_{NP} = \Delta\psi - \Delta\psi_0 \tag{3-28}$$

will hold. Thus, adopting the further definition $U_{NP} \equiv (qV_{NP}/kT)$, we can write the Boltzmann-quasiequilibrium expression for electrons as

$$[n_{0N}/n_P(+X/2)] = \exp(\Delta U_0)\exp(U_{NP}), \tag{3-29}$$

or as

$$n_P(+X/2) = [n_{0N}\exp(-\Delta U_0)]\exp(-U_{NP}). \tag{3-30}$$

But given the Boltzmann relation at equilibrium, the product

$$[n_{0N}\exp(-\Delta U_0)] = n_{0P}, \tag{3-31}$$

so,

$$n_P(+X/2) = n_{0P}\exp(-U_{NP}). \tag{3-32}$$

The corresponding expression for the left-hand boundary is

$$p_N(-X/2) = p_{0N} \exp(-U_{NP}). \tag{3-33}$$

Equations 3-32 and 3-33 jointly constitute the *law of the junction,* also known as the *Shockley boundary conditions* [5]. The information they convey is that under forward bias, a junction's minority-carrier boundary value (either side) is inflated by the exponential factor $e^{-U_{NP}}$, where U_{NP} is applied voltage normalized by means of the thermal voltage.

Useful further insight is obtained by multiplying Equation 3-32 through by p_{0P}. Evidently, then, the *pn* product at the right-hand boundary is given by

$$p_{0P}n_P(+X/2) = p_{0P}n_{0P} \exp(-U_{NP}) = n_i^2 \exp(-U_{NP}). \tag{3-34}$$

Multiplying Equation 3-33 through by n_{0P} gives an identical result. Thus the case of forward bias can be summarized by pointing out that the *pn* product *at* the space-charge-layer boundaries, and everywhere *between* them is given by

$$pn = n_i^2 \exp(-U_{NP}). \tag{3-35}$$

The case of a symmetric junction was chosen for convenience and clarity, but reviewing this section will convince the reader that all of the arguments employed are equally valid for a step junction of arbitrary asymmetry. It follows from Equation 3-35, then, that the *pn* product at *both* boundaries must be equal for the forward-biased junction. While this result is trivial in the symmetric case, it is nontrivial in the asymmetric case, and because of its generality it has considerable importance.

Thus, given the million-fold increase in the boundary-value *pn* product displayed in Figure 3-14, we find from Equation 3-35 that $U_{NP} = -\ln(pn/n_i^2) = -\ln(10^6) = -13.8$. This corresponds to a bias on the junction sample of about -0.35 V, a value of forward bias that is approximately half the typical value observed on a junction in operation in a device. Because densities (and as we shall see, current) depend exponentially on voltage, this bias value is very modest indeed.

As a final point we should note that the net current of a particular carrier (let us choose electrons) through a forward-biased junction results from the dominance of diffusion over drift. The huge values of the maximum drift and diffusion components of current density shown on the left-hand side of Figure 3-4(c) for the junction at equilibrium are perturbed by the forward bias. The drift component shrinks a bit, and the diffusion component expands. But typically, the *net* electron current remains small compared to the magnitudes of the maximum value for either component. This fact plays an important part in certain forms of junction analysis.

3-4 STATIC ANALYSIS

The next task is to develop an expression for the dc or *static* properties of a *PN* junction. These are but one class of a wide range of descriptions, outlined in Section 3-4.3. At this point we shall shift to the one-sided, grossly asymmetric, step junction for two reasons. First, such asymmetry leads to equally asymmetric carrier-injection properties, with the result that for many purposes, only the conditions on one side need to be considered. Second, such a junction is highly relevant to the bipolar junction transistor that is treated in the next chapter.

STATIC ANALYSIS 273

3-4.1 Forward Current–Voltage Characteristic

The specific junction of interest is shown in Figure 3-16(a). As was noted earlier, such a junction is often termed an N^+P junction, with the symbol N^+ denoting very heavy N-type doping. Also, the P-type region is taken to be very "long," or extensive in the x direction. The equilibrium carrier-density profiles are shown using solid lines in Figure 3-16(b).

Exercise 3-27. It was noted earlier that the high-side potential drop in a grossly asymmetric junction at equilibrium "saturates" at kT/q. What degree of density drop, such as that shown on the left side in Figure 3-16(b), corresponds to this amount of potential drop?

The electron density as a function of electrostatic potential is given by

$$n = n_i e^{q\psi/kT}.$$

Hence a potential difference of kT/q corresponds to a density change by a factor of

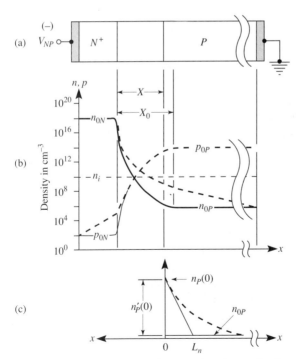

Figure 3-16 A one-sided (N^+P) step junction under forward bias. (a) Physical representation of the junction sample. (b) Log-linear carrier-density profiles. (c) A linear presentation of the injected-carrier profile in the neutral P-type region.

e, or roughly three. On a logarithmic scale, such as that in Figure 3-16(b), a factor of three is approximately half a decade.

The carrier-density profiles for a small forward bias are shown in Figure 3-16(b) with dashed lines. A very important point to notice is that the minority-carrier boundary values on the two sides exhibit equal *factors* of increase as a result of the forward bias. In this example, the factor of increase amounts to about 1000. On a logarithmic scale, equal *factors* are translated into equal *increments,* thus accounting for the appearance of Figure 3-16(b). As a result, the electron density at the right boundary is $10^9/cm^3$, while the hole density at the left boundary is only $10^5/cm^3$.

Exercise 3-28. What values of V_{NP} and U_{NP} are needed to increase minority-carrier boundary values by a factor of 1000?

The law of the junction informs us that

$$U_{NP} = -\ln \frac{n}{n_0} = -\ln(1000) = -6.91.$$

Therefore,

$$V_{NP} = \left(\frac{kT}{q}\right) U_{NP} = -0.177 \text{ V}.$$

Now focus on the right-hand boundary. To its right is an exponentially declining electron density. Because of an elementary property of the declining-exponential function, the electron-density gradient at the boundary (needed to calculate the diffusion current of electrons) is given by the value of n' at the boundary divided by the characteristic length for the exponential function, which is the electron diffusion length L_n. The next important point is that electron transport at the boundary is *only* by diffusion because electric field vanishes there. Yet another important point to appreciate is that the law of the junction gives n at the boundary and not n'—but that for the conditions assumed here, the approximation $n \approx n'$ is excellent. (It is, in fact, an excellent approximation for all but ultra-low-level situations.)

If the diffusion lengths for minority carriers on the two sides are comparable, then it follows that the electron-density gradient at the right-hand boundary will exceed the hole-density gradient at the left-hand boundary by four orders of magnitude. And it follows from this that the electron injection on the right will grossly exceed the hole injection on the left, because the injection phenomenon is driven by diffusion alone. This is the current asymmetry alluded to earlier. Asymmetric injection by a junction is a very useful property, as this chapter will show.

It is unlikely that the diffusion lengths on the two sides will be equal, or even comparable, however, because diffusion length declines with doping. A high estimate for diffusion-length asymmetry in this particular sample is a factor of 10^2. But even this situation would lead to a gradient asymmetry of a factor of $(10^4/10^2) = 10^2$, and therefore, to hole injection into the N-type region that is only about one percent of the electron injection into the P-type region. This certainly qualifies as gross asymmetry with respect to injection.

Exercise 3-29. Estimate the diffusion-length ratio to be expected on the two sides of the sample in Figure 3-16.

Carrier diffusion length L is equal to $\sqrt{D\tau}$. Figure 2-22 shows that carrier mobility declines with doping—by a factor between five and six for the doping values in the sample of Figure 3-16. But since electrons are diffusing into the lightly doped side and holes into the heavily doped side, this ratio is augmented by the mobility ratio (which is the same as the diffusivity ratio) for a combined factor of about 17. In addition, carrier lifetime declines with doping, as shown in Figure 2-34, and a reading of the data for the sample in question suggests a ratio of about 20. The diffusion-length ratio is the geometric mean of these two factors, or about 19. This amounts to one to two orders of magnitude, with two being a generously high-side estimate.

Next let us look at the injection phenomenon quantitatively. Figure 3-16(c) embodies a significant departure from all of the foregoing junction diagrams, in that the spatial origin has now been placed at the right-hand boundary of the space-charge layer. The current density J is approximately equal to $J_n(0)$ because of the injection asymmetry already examined in some detail. Because $J_n(0)$ is purely diffusive,

$$J_n(0) = qD_n \frac{dn}{dx}\bigg|_{x=0} = qD_n \left[-\frac{n'_P(0)}{L_n} \right], \tag{3-36}$$

with the minus sign introduced because the gradient is negative. But using the definition of excess-carrier density,

$$J_n(0) = -\frac{qD_n}{L_n}\left[n_P(0) - n_{0P} \right]. \tag{3-37}$$

And using the law of the junction,

$$J_n(0) = -\frac{qD_n}{L_n}(n_{0P}e^{-U_{NP}} - n_{0P}) = -\frac{qD_n n_{0P}}{L_n}(e^{-U_{NP}} - 1). \tag{3-38}$$

Letting $J_0 \equiv (qD_n n_{0P}/L_n)$, and noting again that $J \approx J_n(0)$ yields

$$J = -J_0(e^{-U_{NP}} - 1). \tag{3-39}$$

For the geometry and conventions adopted, J is directed leftward and thus is fully consistent with the outward-flowing terminal current (also negative) and the forward-bias polarity that is negative.

Equation 3-39 is the current–voltage characteristic of the junction examined. Its derivation required four very important items of information: (1) the equation for carrier transport by diffusion; (2) knowledge of the gradient of a declining exponential function at a chosen point; (3) the definition of excess-carrier density; (4) the law of the junction.

Noting further that for a junction area A, we have $I = AJ$ and $I_0 = AJ_0$, we may write the corresponding equation

$$I = -I_0(e^{-U_{NP}} - 1). \tag{3-40}$$

For $V_{NP} \approx -0.1$ V (or $U_{NP} \approx -4$), the unity term in Equation 3-40 can be dropped with only a 2% error. Thus, to an excellent approximation, the simple-theory forward current is an exponentially rising function of voltage for larger voltages. The typical range for this behavior is -0.1 V to -0.7 V, with high-level effects entering in the range beyond -0.7 V.

It is informative to examine a semilog plot of forward current versus voltage, as is presented in Figure 3-17(a). The simple-theory exponential character of the I–V characteristic is clearly displayed. For a forward-bias magnitude of less than about 0.1 V, the unity term becomes important, and as voltage is further reduced, the current heads for zero (which can never be reached on a logarithmic scale!). An extrapolation of the linear portion of the curve to $V_{NP} = 0$ defines the *saturation current* I_0. It is very useful for writing I as a function of V in simple-exponential form for use through much of the forward range.

The forward characteristic presents a very different appearance in a linear plot, Figure 3-17(b). Because of the exponential behavior of current, the forward-voltage magnitude remains in the neighborhood of 0.7 V through a wide range of current values encountered in typical junction operation. In other terms, the forward-biased silicon junction can be described as a "voltage regulator" at the approximate value of 0.7 V. This characteristic quantity for evident reasons is often termed an *offset voltage*.

Before leaving the subject of forward bias, let us examine another of its important facets. In Figure 3-18(a) we repeat the physical representation of the N^+P junction shown in Figure 3-16(a), but stress that the P-type region is extensive in the positive-x direction, or is very "long." Consequently, the injected excess electrons will all have time and space to recombine before reaching the ohmic contact at the right-hand end of the sample. This situation is accompanied by an exponential decay of the diffusion component of electron current. But for the steady-state conditions assumed, there can be no accumulation of charge at any x position. Putting the matter a different way, the total current density J must be *constant,* which is to say, independent of x in this one-dimensional problem. The current that makes total current constant is the *net* drift of majority carriers, holes. We stress the term "net" because hole drift and diffusion coexist in the region of appreciable hole-density gradient. (Recall that the excess-hole profile is nearly a replica of the excess-

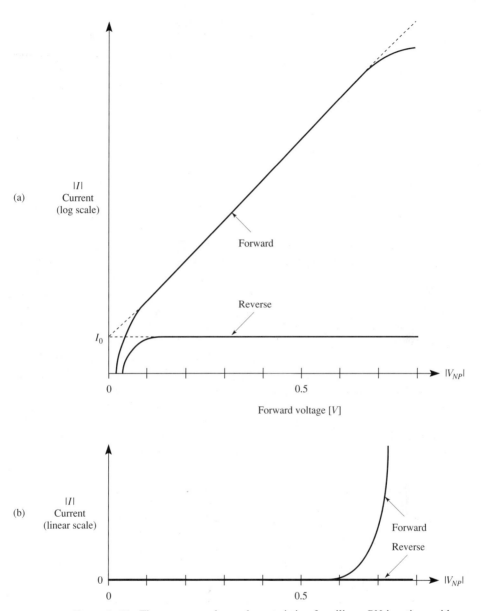

Figure 3-17 The current–voltage characteristic of a silicon PN junction, with generation–recombination effects assumed to be negligible. (a) Semilog plot. (b) Linear plot.

electron profile.) Electric field and hole drift current vanish at $x = 0$ and assume constant values near the ohmic contact. Because majority-hole density has a constant value throughout the neutral right-hand region under the low-level conditions assumed, it follows that the functional form of the hole drift current is identical to that for the electric field, a field profile discussed in Exercise 3-26.

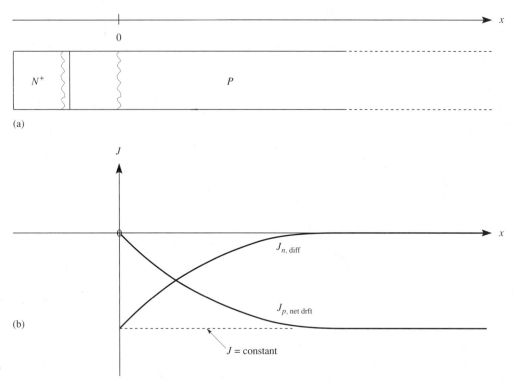

Figure 3-18 Current-density components on the (extensive) lightly doped side of an N^+P junction under forward bias. (a) Physical representation of sample. (b) The changing composition of total current density—100% electron diffusion at the boundary and 100% hole drift at the ohmic contact.

3-4.2 Reverse and Overall Characteristics

The law of the junction yields highly accurate boundary values for minority-carrier densities in typical silicon junctions under low-level forward bias. These densities, to review, are exponentially inflated by the applied voltage. Under reverse bias, carrier densities are depressed throughout the transition region, and specifically at the boundaries, but the boundary densities are not accurately predicted by the law of the junction. In short, the law has only qualitative validity for reverse bias. Interestingly, this inaccuracy has little practical effect because for a reverse bias of 0.1 V or more, the boundary value is approximately zero! (This is one of those cases where Mother Nature is curiously cooperative.) The qualitative prediction of the law of the junction is that the boundary density (of minority carriers) declines, and once it is close to zero, the exact prediction is unimportant. A fact that is important, however, is that the minimum in minority-carrier density sharply defines the junction boundaries in reverse bias.

The consequence of having a boundary value approximating zero is shown in

Figure 3-19. (We have reverted to a spatial origin placed at the metallurgical junction.) Once more the one-sided junction sample is shown physically in Figure 3-19(a), but with reverse bias applied this time. In Figure 3-19(b) is shown the electron profile at equilibrium with dashed lines. The law of the junction predicts the equilibrium electron density n_{0P} at the boundary, and this is indeed the case. With a certain value of reverse bias then, the depletion layer boundary retreats from its equilibrium position at X_0 to a new position, X_1. The result of this situation is an electron density gradient to the right of the boundary with a positive value that moves electrons leftward by diffusion.

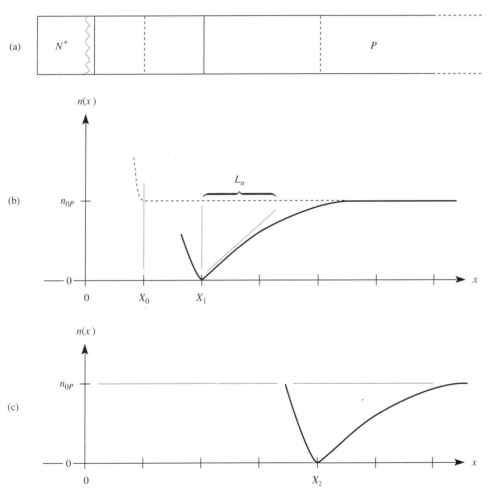

Figure 3-19 One-sided step junction under reverse bias. (a) Physical representation of an N^+P sample. (b) Minority-electron profile under reverse bias (solid curve), with equilibrium profile for comparison (dash curve). (c) The electron profile for greater reverse bias.

One can be quantitative at this point. Just to the right of the boundary, the gradient is given by n_{OP}/L_n, as can be seen in Figure 3-19(b). Electrons that diffuse to the boundary are picked up by the electric field of the double layer and are *swept* to the left-hand side, where they are now *majority* carriers. The diffusion-transport equation gives us the magnitude of this current density as

$$J_n(X_1+\delta) = qD_n(n_{OP}/L_n), \qquad (3\text{-}41)$$

where δ is a small interval on the x axis. The electrons diffusing leftward constitute a tiny conventional "leakage" current to the right.

Now let the reverse bias be increased. The result is the electron-density profile shown in Figure 3-19(c), with X_2 being the new boundary position. A repetition of the electron-density calculation just carried out yields the same result as before, because the electron-density profile has simply translated to the right, rather than changing its form.

Exercise 3-30. Determine the reverse-bias values in terms of $\Delta\psi_0$ that were needed to place X_1 at $2X_0$, and X_2 at $4X_0$.

We have noted that $\Delta\psi = KX^2$. Therefore

$$\Delta\psi_1 = K(2X_0)^2 = 4KX_0^2.$$

And thus,

$$V_{NP1} = \Delta\psi_1 - \Delta\psi_0 = KX_0^2(4 - 1) = 3KX_0^2 = 3\Delta\psi_0.$$

Similarly,

$$\Delta\psi_2 = K(4X_0)^2 = 16KX_0^2;$$

$$V_{NP2} = 15KX_0^2 = 15\Delta\psi_0.$$

Now refer back to Equation 3-38. Note that when V_{NP} is given a value greater than about $+0.1$ V (reverse bias), corresponding to $U_{NP} \approx 4$, the exponential term can be ignored. Further, note that the value of this *saturating* reverse current density is identically the value given in Equation 3-41. Hence the simple theory derived above and encapsulated in Equations 3-38 and 3-39 predicts a constant or "saturated" reverse current for a voltage in excess of 0.1 V or so. It is the differing theoretical saturation currents in junctions of various materials, related in turn to ξ_G through n_{OP} in Equation 3-38, that explain the differing offset voltages they exhibit under forward bias. The relationships are shown by means of distorted current–voltage characteristics in Figure 3-20. The 0.7-V offset in the silicon-junction characteristic is replaced by an offset of about 0.3 V in the case of germanium, a fact consistent with its much larger saturation current. For gallium arsenide, then, with a bandgap some 25% larger than that of silicon, the offset voltage is ~ 1 V.

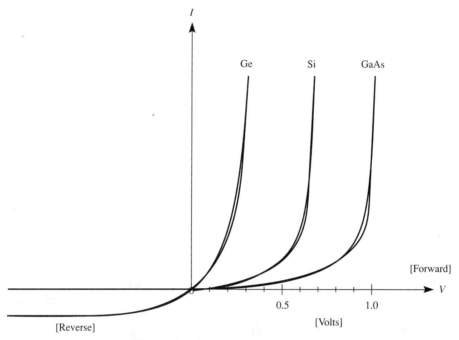

Figure 3-20 The relationships between offset voltages and reverse-saturation currents for junctions fabricated from three different semiconductor materials, represented qualitatively.

3-4.3 Defining *Models* and Related Terms

The constructs known as *models* are among the most useful in solid-state electronics. Broadly, there are two kinds of models in common use. An *analytic model* is a set of equations that interrelate the variables and parameters relevant to a particular phenomenon or device and yield a description that is accurate enough to be useful. As we shall see, there is a constant trade-off between *accurate* and *useful,* and we often sacrifice accuracy to gain a simpler model, thus enhancing its utility.

The term *parameter* is used here in the mathematical sense. It is a variable that is held constant in one fashion or another. For example, if the output port of a device is characterized for a series of constant values of an input variable, the input variable is a parameter. As another example, consider a series of N^+P junction samples that are identical except that the x-direction extent of the P-type region X_P has a different value in each. In such a case, X_P is a parameter. Typically X_P is small compared to $\sqrt{A}$, where A is junction area, so X_P is rationally described as a *thickness* dimension. But in the following analysis, extremes are to be considered, including cases where X_P becomes quite large. Such a structure has often been described as a *long diode,* terminology that we adopt. Since *diffusion length* is a highly relevant dimension measured in the x direction, this is a completely logical and consistent choice. (Regrettably, there are too many who eschew *long* and *short* and substitute "wide" and "narrow," choices devoid of plausible basis.)

The second kind of model in extensive use is an *equivalent-circuit model*. In this case one devises a set of appropriately interconnected ideal devices with combined electrical properties that approximate those of the entity being modeled. Again the object is to use as few elements as possible, consistent with the degree of accuracy that is needed. The aim of equivalent-circuit modeling is to permit one to write simple equations by inspection. This is possible because *ideal* components have been assumed. Such components include resistors, capacitors, inductors, voltage sources, and current sources. That latter two may be ac or dc. A battery constitutes an excellent approximation to an ideal dc voltage source, and an ordinary capacitor is nearly an ideal device. Solid-state electronics has given us near-ideal current sources as well, but only for restricted voltage ranges. The similarity of real and ideal devices is not very important, however, because the equivalent-circuit model is rarely implemented physically, and is instead a tool for understanding and analysis.

Both equivalent-circuit and analytic models are subject to numerous further classifications, usually binary but not always so. For example, there are static models and dynamic models, with the term *static,* as noted previously, designating dc conditions, or at most, quasistatic conditions, where the time derivatives of voltage and current are so small that reactive effects can be ignored.

The term *dynamic* designates time-varying conditions. Under the dynamic-model heading are small-signal and large-signal models. In the former case, a perfectly plausible question is, "Small compared to what?" This will be answered specifically as subsequent models are developed, and in the process, both small and large signals will be defined. The signal in turn may be sinusoidal or may involve a *switching transition*. The latter is taken to mean that applied voltage (or current) is initially constant and then undergoes a rapid transition to another constant value. This kind of problem can qualify as small-signal or large-signal, depending on the magnitude of the transition. In the large-signal case, the problem of interest may involve going from bias of one polarity to the opposite polarity. In either the large-signal or small-signal case, the situation of interest may also involve a return to the initial condition.

For sinusoidal signals, most interest attaches to the small-signal case, but the large-signal case is not ruled out. Sinusoidal signals are almost always implied by adjectives such as *low-frequency* and *high-frequency*. Because we sometimes must deal with more than one characteristic time, the classification of frequency ranges may be more than binary. For example, dielectric-relaxation time t_D and carrier lifetime τ are typically quite different and may enter a single problem; other and different time constants may enter as well.

From this brief discussion it is evident that models come in considerable variety. They also are the basis for a significant fraction of the device literature. A number of important models for the *PN* junction will be examined in the balance of Section 3-4, as well as in Sections 3-7 through 3-9.

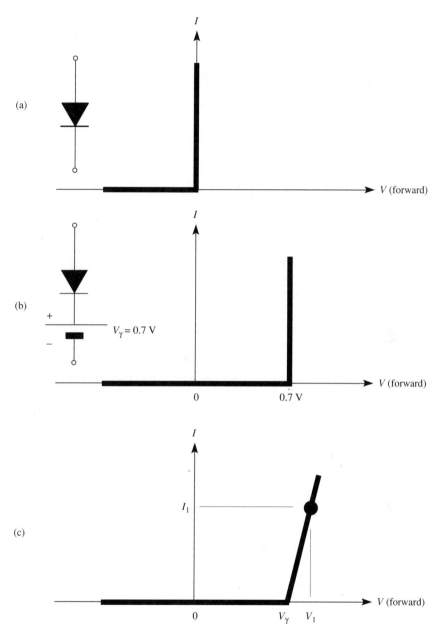

Figure 3-21 Piecewise-linear I–V characteristics. (a) Ideal-diode (ideal-rectifier) characteristic with its symbol. (b) Simplest piecewise-linear characteristic and equivalent-circuit model for a silicon junction, with the *offset voltage* $V_\gamma = 0.7$ V represented by a voltage source (battery) in series with an ideal diode. (c) Next-order refinement of a piecewise-linear model for a silicon junction, exhibiting the slope at the bias point (I_1, V_1) that matches the slope of the actual characteristic.

3-4.4 Piecewise-Linear Model

Frequently it is desirable to approximate the exponential forward characteristic of a *PN* junction by using a simpler function of some kind. When the simplest choice is made, namely a linear function, then more than one linear segment must obviously be used. An assemblage of two or more such linear segments to simulate a nonlinear function is termed a *piecewise-linear* model, or approximation.

An unusually simple piecewise-linear model is that for the *ideal rectifier*, which is a two-terminal device that permits conduction in one direction only. Figure 3-21(a) shows the current–voltage characteristic of an ideal rectifier, or ideal diode, and the symbol that we shall employ henceforth for such an ideal device. It is the same symbol that was introduced in Figure 3-11(c), except that the arrowhead is black. Note that the polarity convention in this diagram grounds the *N*-type side, so that a positive applied voltage corresponds to easy conduction. This is done to eliminate the absolute-value signs. The properties of an ideal rectifier can be summarized thus in terms of the resistance it exhibits:

$$\begin{bmatrix} R_{\text{forward}} = 0 \\ R_{\text{reverse}} = \infty \end{bmatrix}. \tag{3-42}$$

The silicon junction as a rectifier constitutes a better approximation than ever achieved before its era. Its primary departure from the ideal is the offset voltage described earlier that it exhibits under forward bias. It follows, therefore, that a useful equivalent-circuit model can be constructed by placing an ideal diode in series with an ideal voltage source (that is, a battery) as is shown in Figure 3-21(b); the resulting current–voltage characteristic shown there is a sufficiently realistic approximation of that in Figure 3-17(b) so that it is extensively used for circuit calculations. The piecewise-linear model again consists of two segments: A vertical line at $V_\gamma = 0.7$ V represents its conducting regime, and a horizontal line along the voltage axis represents its nonconducting regime.

There are other situations where a more accurate piecewise-linear model is necessary. The next stage of refinement involves determining the forward current at which the diode is to be operated, such as I_1 in Figure 3-21(c). Then a tangent to the forward characteristic is drawn at the point so determined, replacing the vertical line of Figure 3-21(b), and intersecting the voltage axis at a new voltage V_γ. Once again, the other of the two straight-line segments lies along the voltage axis. This basic piecewise-linear model and further elaborations of it are presented quantitatively in Section 3-8.1.

3-4.5 Charge-Control Model

Consider an N^+P junction incorporated in a "long" diode structure, where the term is being used in the same sense as in Section 3-4.3. Under dc conditions, the forward current through this device is proportional to the excess-carrier charge stored on the *P*-type side. This is the essence of the static *charge-control model*.

STATIC ANALYSIS 285

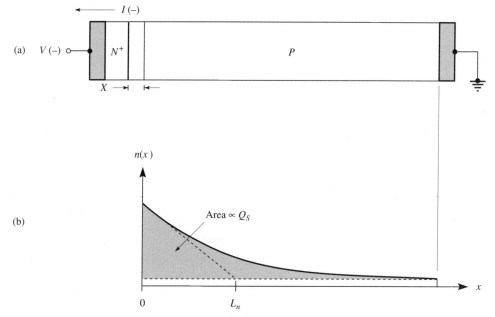

Figure 3-22 Static charge control, or the proportionality of steady-state current and stored charge. (a) Physical representation of a long N^+P diode. (b) Electron profile displaying excess-electron stored charge Q_S that resides in the P-type region.

The reason underlying this proportionality is easy to demonstrate. Figure 3-22(a) represents the structure just described, and indicates a forward-bias condition according to our convention. The shaded area in Figure 3-22(b) represents the charge Q_S of excess minority carriers stored on the right. But this entire population of carriers will recombine on average in one carrier lifetime τ, because none of the carriers are able to reach the ohmic contact at the right. The only source of the electrons needed to maintain a steady-state condition is the N^+ region, which injects electrons through the junction. Hence this injection mechanism needs to replenish the entire charge Q_S in one average lifetime, so that we may write

$$I = Q_S/\tau, \tag{3-43}$$

which is the essential charge-control equation.

Exercise 3-31. Calculate the stored charge Q_S, place it in Equation 3-43, and compare the resulting equation for I to an equation obtained by considering the current of diffusing electrons at $x = 0$.

As noted earlier, the area under a declining exponential function is equal to the value of the function at the zero of the independent variable times the characteristic value of the independent variable. Thus, here the shaded area in Figure 3-22(b) is $n'_P(0)L_n$. From Equation 3-43, then, it follows that

$$I = \frac{-qAn'_P(0)L_D}{\tau},$$

where A = junction area. But using the definition $L_n = \sqrt{D_n\tau}$ we have

$$I = \frac{-qAD_n n'_P(0)}{L_n} = qAD_n \frac{dn'_P}{dx}\bigg|_{x=0}.$$

It is evident that the last expression gives the current of diffusing electrons at $x = 0$, a negative current by virtue of the gradient's sign.

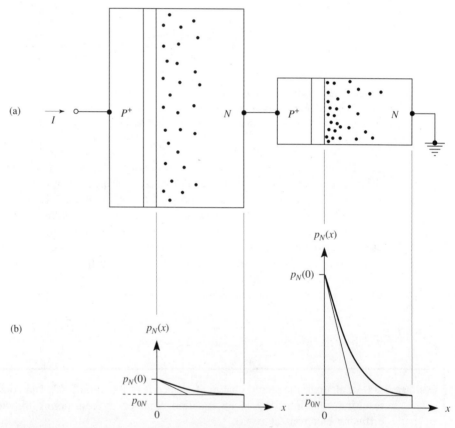

Figure 3-23 The charge-control principle requires differing voltages on diodes carrying the same current when the diodes are identical except in area. (a) Pair of identical-profile P^+N diodes, each carrying the current I. (b) Carrier profiles in the two devices, showing a higher value of $p_N(0)$, and hence higher forward voltage, in the smaller one.

Thus we have the observation that forward-current magnitude is proportional to *both* the stored charge and boundary value for excess minority carriers. The latter point permits us to explain a relationship presented earlier without explanation. Figure 3-23(a) depicts a pair of diodes connected in series and carrying a static current I. The two devices, P^+N this time for variety, are identical to each other in every respect except junction area. The principle just developed means that excess-hole *charge* must be the same in each device, a fact pictorially displayed using dots in Figure 3-23(a). Both devices have the same numbers of dots. Thus the boundary *density* of carriers must differ in the two cases, as can be seen heuristically in Figure 3-23(a), and explicitly in Figure 3-23(b). Returning next to the law of the junction, we learn that the forward voltage on the smaller device exceeds that in the larger. We shall return to this point in Section 3-8.1.

3-4.6 Characteristic of a Real Silicon Junction

Germanium junctions fit the simple theory of Sections 3-4.1 and 3-4.2 superbly, and the demonstration of this fact was one of the triumphs of early solid-state electronics [6]. Silicon, however, fits it poorly for both forward bias and reverse bias. The reasons underlying silicon's departure from simple theory were detailed in 1957 by Sah, Noyce, and Shockley [10], and can be summarized as follows: The value of n_i^2 in silicon is roughly one million times smaller than in germanium. Since minority-carrier density is proportional to n_i^2, it follows that the current of minority carriers diffusing to a junction to constitute a saturation current for a diode of given doping is a million times smaller in silicon than in germanium. Consequently, this diffusion mechanism produces only a tiny saturation current in silicon (about 0.01 pA in a typical junction) and as a result it is obscured or "swamped" by another mechanism. On the other hand, the diffusion mechanism is dominant in germanium and a "clean" saturation is observed—at about 0.1 μA in a typical germanium junction.

The additional mechanism that must be considered is a generation–recombination (g–r) mechanism. For a silicon junction, it is dominant in reverse bias, and plays an important part in shaping the low-voltage forward characteristic as well. Recall the point made in Chapter 2 that in a region of depressed carrier density, generation exceeds recombination and the natural "drive" is to restore equilibrium conditions. In a reverse-biased junction, as noted in Section 3-4.2, carrier densities are depressed throughout the transition region. Consequently, generation exceeds recombination throughout the transition region under reverse bias. But the large electric field there "separates" the carriers at birth, pushing electrons toward the N-type side and holes toward the P-type side. This *generation* current is the dominant current component in a reverse-biased silicon junction, with a typical value being some one to ten picoamperes (about 10^3 times greater than the diffusion or "saturation" component). Furthermore, the generation component does not level off, but continues to increase as the depletion region expands, and also for other reasons. The generation component exists in germanium junctions as well as in silicon, but is typically 10^3 times smaller than the diffusion component in the germanium case.

In forward bias, it is majority carriers that flood into the transition region from both sides. Again, the recombination centers in a silicon sample play an important role; opposite-type carriers from the two sides meet at these centers and recombine there. Although the recombining carriers constitute a forward current, they *are not injected* into the opposite sides of the sample. They have disappeared in the transition region. The law of the junction is still valid, dictating rising boundary values for minority carriers and the diffusion current that inevitably accompanies the rise, but this mechanism is swamped in silicon by the recombination current. To appreciate how this can occur, refer back to Figure 3-20 and note how much more rapidly diffusion current rises with forward voltage in germanium than in silicon because of germanium's high value of n_i^2, a consequence of its smaller bandgap. The mechanism underlying the extra components of forward and reverse current is sometimes labeled the *Sah-Noyce-Shockley* (SNS) effect.

Interplay of the diffusion mechanism and the recombination–generation mechanism in a silicon junction is nicely illustrated in data recorded by Barber and reported by Nussbaum [17]. Great care is required in this kind of measurement to ensure constant junction temperature. In Figure 3-24, a heavy curve has been drawn through the thirty-five closely spaced points that Barber observed for the forward characteristic of a commercial silicon device. Let us start at the top of the diagram. First note that a light line has been drawn through the linear portion (upper portion) of the semilog experimental forward characteristic. This line precisely fits simple theory, and through its extrapolation to zero volts, serves to define the theoretical saturation current for this particular sample,

$$I_0 = 2.7 \times 10^{-14} \text{ A}. \tag{3-44}$$

Clearly shown at the top of the diagram is the curving of the experimental data away from the simple-theory asymptote. This is a combined result of series resistance in the end regions and high-level effects involving the junction itself. But from about 0.35 V to 0.7 V of forward bias, agreement with diffusion theory is very good.

The Sah-Noyce-Shockley (SNS) theory in simplest form predicts a second asymptotic straight line (semilog) applying to the forward characteristic as a result of recombination, one having half the slope of the diffusion-component asymptote. This too is plotted, and consistent with their further observation, the two straight lines intersect at about $10(kT/q) \approx 0.26$ V. Thus the recombination mechanism accounts for the low-voltage "bulge" in the experimental forward characteristic. Again an intercept observation permits us to determine a key current value,

$$I_{g-r} = 4.4 \times 10^{-12} \text{ A}. \tag{3-45}$$

Accordingly, it is possible to approximate the experimental characteristics, combining diffusion and g–r effects, and the result is the following two-term expression:

$$-I = (2.7 \times 10^{-14} \text{ A}) [\exp(-qV_{NP}/kT) - 1] \tag{3-46}$$
$$+ (4.4 \times 10^{-12} \text{ A}) [\exp(-qV_{NP}/2kT) - 1].$$

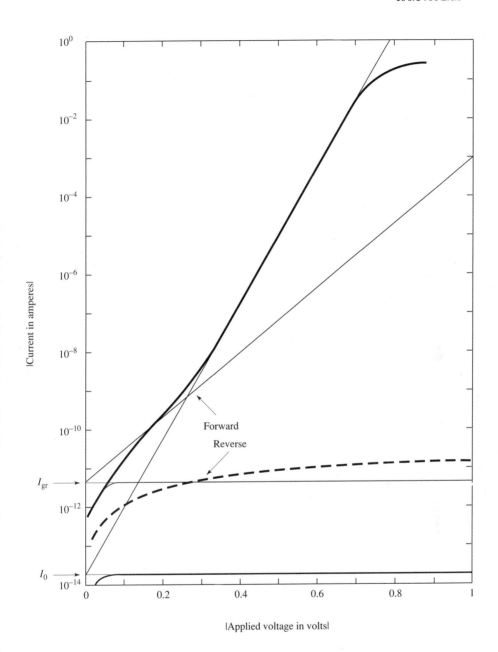

Figure 3-24 Experimental forward current–voltage characteristic of a silicon junction, from data observed by Barber and reported by Nussbaum [17].

Barber has recently supplied to the present authors a detailed set of data on his experimental conditions and on the sample he employed [18]. The data were recorded at 273 K. Also, the junction area in his device was 5.06 mm². This, it should be noted, is very large by the standards of integrated circuits.

Exercise 3-32. Determine the slope of the straight line of higher slope in Figure 3-24.

Because Figure 3-24 presents both forward and reverse characteristics in the same quadrant, it is appropriate to focus on current and voltage magnitudes. In view of this, the line in question corresponds to a pure-exponential function that can be inferred from Equation 3-40 as

$$|-I| = I_0 \exp |-U_{NP}|.$$

Hence, dropping the algebraic and absolute-value signs,

$$\ln(I/I_0) = U_{NP} = (qV_{NP}/kT),$$

so that the slope in question is given by

$$\frac{d}{dV_{NP}}[\ln(I/I_0)] = (q/kT) = 39/\text{V}.$$

Because the curve in question is linear, we can confirm the result by noting that

$$\frac{d[\ln(I/I_0)]}{dV_{NP}} = \frac{\Delta[\ln(I/I_0)]}{\Delta V_{NP}}.$$

Let us choose a current increment corresponding to change by a factor of ten and calculate the corresponding voltage increment. Then

$$\Delta V_{NP} = \frac{\Delta[\ln(I/I_0)]}{q/kT} = \frac{\ln(10)}{39/\text{V}} = 0.059 \text{ V},$$

a result that can be confirmed by a scaling measurement in Figure 3-24.

3-4.7 High-Level Forward Bias

The high-current portion of the curve at the top of Figure 3-24 has practical importance because the everyday use of silicon devices involves operation near or in that regime. There the steep rise of current with voltage declines because of *high-level* effects. The term is used here in the same sense as previously, namely, to mean that minority-carrier density in some part or parts of the device now approaches the equilibrium majority-carrier density in the same region or regions.

The problem can be readily visualized by referring back to Figure 3-14, representing a symmetric junction under forward bias. We have taken the junction boundaries as the surfaces at which electric field changes sign (or "vanishes"), as illustrated in Figure 3-15. For the low-level bias condition shown in Figure 3-14, minority densities at these boundaries are one million times smaller than the majority densities there, and hence are negligible in comparison. But when forward bias is increased to the point that the minority densities are smaller by only one to two decades, then they can be said to "approach" the majority densities. A nearly matching rise in the majority densities then will develop outside the boundaries to neutralize the minority-carrier charge. Realize that the matching is in absolute terms, and not in the relative terms presented by the logarithmic ordinate in Figure 3-14. This rearrangement of majority carriers occurs, of course, because a little charge produces a huge field, in the process causing carrier redistribution toward the condition of quasineutrality, if not strict neutrality.

It is evident in Figure 3-14 that locally rising majority density will lead to the formation of maxima or "bumps" in the majority carrier profiles. These will occur very near, but not precisely at, the boundaries, a matter that has been examined in some detail [19]. Given the fact that the minority-carrier profiles in the boundary neighborhoods are monotonically declining, as can be seen in Figure 3-14, whereas the majority-carrier profiles exhibit maxima for reasons just presented, it follows that the match of the excess majority- and minority-carrier profiles cannot be perfect. In other words, quasineutral conditions, at best, prevail in the boundary region. Quantitative treatment shows that the approach to neutrality is close when the corresponding end region of the device is long, or extensive in the x direction, and the corresponding minority-carrier diffusion length is also relatively large. The departure from neutrality increases steadily as the end region is made shorter, or as the diffusion length becomes smaller. With the shrinkage of dimensions in solid-state devices that has occurred over the decades, it follows that details such as this must be scrutinized carefully in many situations.

The problem of low-level forward bias was treated in Section 3-3.3. There two unknowns were involved, namely the minority-carrier densities at the two boundaries, since the majority-carrier densities can be treated as constant in relative terms. The two necessary equations were provided by the assumption of Boltzmann quasiequilibrium, yielding one equation for electrons and one for holes. The assumption of neutrality (or quasineutrality) outside the depletion-layer boundaries was used to show by physical argument that changes in the minority densities account for all of the change demanded by the Boltzmann-quasiequilibrium condition. But neutrality was not used explicitly for writing equations.

In high-level forward bias, by contrast, we now have four unknowns—minority *and* majority densities at both boundaries. This problem was addressed by Fletcher [20], who invoked the Boltzmann-quasiequilibrium assumption to secure two of the necessary equations. He, in fact, is the one who gave this key assumption its name and was first to use it in the manner described here and in Section 3-3.3. The two necessary additional equations were obtained by explicitly invoking the neutrality assumption at both boundaries. That is, in one he equated hole and electron excess-carrier densities at the left-hand boundary, and in the other, did the same for the right-hand boundary.

Developing four expressions for the four boundary values is straightforward, but it is a lengthy procedure and results in lengthy equations. Because the principles are the same as for the case of low-level conditions, we shall not make this investment. Shortly before Fletcher's work, a different treatment of high-level forward bias was offered by Misawa [21]. He in essence used an extension of the method used by Shockley [5] for the low-level case. In Section 3-3.3, we opted not for Shockley's approach, but for a low-level version of Fletcher's, and the two are of course fully equivalent. Furthermore, Misawa's and Fletcher's handling of the high-level problem are also fully equivalent [22], but some effort is necessary to demonstrate this fact because of the complexity of both sets of equations.

3-5 JUNCTIONS OTHER THAN PN STEP JUNCTIONS

The step junction, which incorporates a perfectly abrupt doping transition, has existed as an abstraction since the earliest days of junction theory. (See, for example, Reference 5.) Ironically, it is only in relatively recent years that we have seen technology capable of producing truly abrupt doping transitions—literally on an atomic-plane scale—with a preeminent technology being molecular-beam epitaxy (MBE) [23]. In the meantime, the closest approach to an ideally abrupt junction was the kind of one-sided junction produced by the *alloying* technique [24]. In this method a molten metal in which the semiconductor crystal could dissolve was permitted to consume a portion of the crystal. The metal was doped with a column-III or column-V element, or else was itself a dopant; when the crystal was permitted to regrow on top of the unmelted portion, the regrown layer of semiconductor would be degenerately (heavily) doped.

Other sharp-transition crystal-growth methods were proposed along the way, and some, such as sputter epitaxy [25], have recently shown significant promise. But in spite of proliferating technologies for this purpose, nonabrupt junctions have great importance, and will be examined next.

3-5.1 PIN Diode

A structure having a thin intrinsic (or near-intrinsic) layer sandwiched between heavily doped *P*-type and *N*-type regions has interesting and useful properties [26]. The label "diode" (rather than "junction") has been used here because such a structure, in a formal way, includes two step junctions—a P^+I junction and an IN^+ junction. Thus, while this structure is not a "gradual" junction, neither is it a step junction in the sense defined in Section 3-1. Using the notation adopted in Section 3-3.2, the structure is identified as P^+IN^+, although the less precise designation "*PIN* diode" is often encountered as well.

For parallelism with earlier examples, let us choose the equivalent N^+IP^+ case to examine. Its physical representation is given in Figure 3-25(a). When we once more employ the depletion approximation, the result is the charge-density profile shown in Figure 3-25(b). The central region, assumed truly intrinsic, is charge-free,

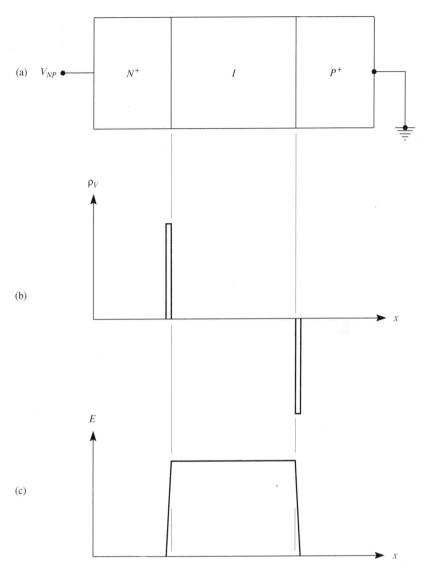

Figure 3-25 An N^+IP^+ version of the "*PIN* diode." (a) Idealized structure, with a thin intrinsic layer defined by two one-sided junctions. (b) Charge-density profile, very similar to that of an ordinary capacitor. (c) Electric-field profile, also (and necessarily) similar to that of an ordinary capacitor.

while equal and opposite sheets of charge exist in the flanking N^+ and P^+ layers, very close to their respective metallurgical junctions. The first integration of Poisson's equation, then, yields the nearly rectangular electric-field profile of Figure 3-25(c), which is virtually identical to that in a conventional parallel-plate capacitor. Correspondingly, this structure exhibits an essentially voltage-independent capacitance, simulating the conventional device. Partly because of this property, the *PIN* diode has found application as a microwave switch.

3-5.2 Linearly Graded Junction

If, in a region containing a *PN* junction, the net-doping density is proportional throughout the region to distance from the metallurgical junction, this structure constitutes a *linearly graded junction*. The problem it poses has been analyzed a number of times [5, 27–29]. The analytical description of a linearly graded junction requires but a single equation:

$$N_D - N_A = -ax. \tag{3-47}$$

For negative x, apparently, the net doping is N-type, placing the N-type side on the left once more. Also, this is again an example of a symmetric junction.

Exercise 3-33. Find the dimensions of the constant a.

Evidently,

$$a = -\frac{N_D - N_A}{x}.$$

The corresponding unit equation is

$$\frac{[1/\text{cm}^3]}{[\text{cm}]} = \left[\frac{1}{\text{cm}^4}\right].$$

Thus the quantity a is a doping-density *gradient* in precisely the sense that (dn/dx) and (dp/dx) are carrier-density gradients.

While the quantity a is usually described as the gradient of a junction, the term *grade constant* is also occasionally encountered. We previously considered in Section 2-5.8 a sample having a net-doping profile that varied linearly with distance. But in that case, $(N_D - N_A)$ was not permitted to pass through zero, and therefore the sample was of a single conductivity type. In that case we assumed that the gradient was so small that the assumption of neutrality was valid—that is, the assumption that $n(x) \approx N_D(x) - N_A(x)$ for an N-type sample. This is a good assumption for gradient values below about $10^{11}/\text{cm}^4$. With increasing gradient, however, the presence of space charge becomes increasingly pronounced. This fact is illustrated in Figure 3-26. There we show the numerically determined, or actual,

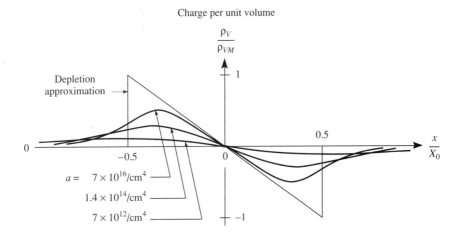

Figure 3-26 Equilibrium charge-density profiles in the linearly graded junction, comparing actual (numerically determined) profiles with the profile predicted by the depletion approximation.

charge-density profiles for several values of gradient a. At a gradient value in the neighborhood of $10^{17}/\text{cm}^4$, the charge-density profile begins to approach the purely triangular shape defined by the depletion approximation, which is also shown in the diagram. Note that the abscissa carries normalized calibration, with X_0, the total depletion-layer thickness at equilibrium, being used as the normalizing constant. This facilitates comparing the gradient-dependent actual charge-density profiles. Similarly, the ordinate normalization employs ρ_{0M}, the maximum or peak charge density predicted by the depletion approximation. Junctions with gradients ranging up to some $10^{25}/\text{cm}^4$ have practical significance, and in the upper range, the depletion approximation is extremely meaningful.

In a step junction, with its constant (zeroth-order) doping profile, we found (using the depletion approximation) a first-order field profile and a second-order potential profile through successive integrations of Poisson's equation. In the linearly graded junction, with the first-order doping profile of Equation 3-47, we have a second-order field profile, and a third-order potential profile. These functions are represented using solid lines in Figures 3-27(a), (b), and (c), respectively. Dashed lines show numerically determined profiles that are to be compared with them. In Figure 3-27(d) is shown a band diagram based upon the dashed (actual) profile in Figure 3-27(c). The main point is that this result bears a strong qualitative resemblance to that for the now-familiar symmetric step junction.

In all of the discussion thus far, it is assumed that the constant grading of doping extends well beyond the depletion-layer boundaries (whether approximate or actual). This represents one extreme possibility. In the other extreme case, deple-

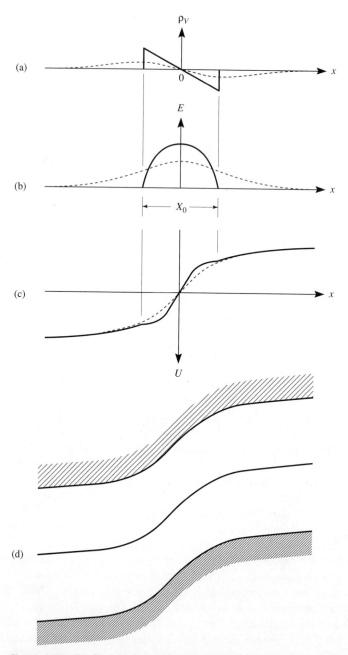

Figure 3-27 Profiles for an equilibrium linearly graded junction with a gradient of $a = 3.5 \times 10^{13}/cm^4$. (a) Charge-density, (b) electric-field, and (c) potential profiles obtained, respectively, from the depletion approximation—solid lines—and numerically (after Morgan and Smits [27])—dashed lines. (d) Band diagram consistent with the dashed curve in part (c).

tion-layer boundaries fall well outside the region of grading. This case is shown for a symmetric example in depletion-approximation terms in Figure 3-28; integration of the charge-density profile in part (a) yields the field profile in part (b), which strongly resembles that for a symmetric step junction. However, the triangular field profile this time is truncated, and has a parabolic portion smoothly substituted for the triangular top portion that was removed.

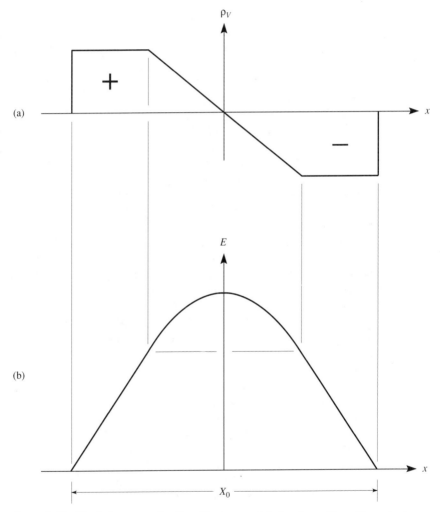

Figure 3-28 Depletion approximation for a symmetric junction with uniform doping, except for a linearly graded region at the metallurgical junction. (a) Charge-density profile. (b) Electric-field profile, consisting of a truncated triangle and fitted parabola. (b) Linear presentation of the same functions that are shown in part (a), identically positioned with respect to one another [33].

For the linearly graded junction with constant grading throughout, the depletion-layer thickness predicted by the depletion approximation is

$$X = \left[\frac{12\epsilon}{qa}(V_{NP} + \Delta\psi_0)\right]^{1/3}. \qquad (3\text{-}48)$$

Correspondingly, depletion-layer capacitance is given by

$$C_T = A\epsilon\left[\frac{12\epsilon}{qa}(V_{NP} + \Delta\psi_0)\right]^{-1/3}. \qquad (3\text{-}49)$$

Both of these equations are repeated in Table 3-1 at the end of the chapter.

3-5.3 Diffused Junctions

The technique of junction formation using *solid-phase diffusion* was developed in the 1950s [30]. From the early 1960s to the present time it has been the most widely used junction-production method in the semiconductor industry. Solid-phase diffusion is accomplished by placing a source of impurities on or near the surface of a semiconductor sample that is held at an elevated temperature in a furnace. Typically, for silicon, the temperature range is from 1100 C to 1250 C. While this range is well below the silicon melting point of 1415 C, it is high enough so that silicon atoms engage in appreciable "site hopping." They trade positions with one another because of the high level of thermal energy in the crystal. When impurity atoms are introduced at the surface, they, too, participate in the random atomic motion. Adding to this is the fact that the high density of impurities at the plane of introduction gives rise to a large gradient of impurity density (a steep density profile). Thus the two basic conditions for the diffusion phenomenon are present—random motion and a density gradient. The laws governing diffusion that were developed in Section 2-5.6 for carriers are fully applicable at these extreme temperatures to impurity atoms as well.

While the silicon sample (let us continue with this dominant example) retains its crystalline character, its atoms no longer are rigidly fixed in position in the manner stressed previously. It is important to realize, however, that even at a relatively high temperature such as 500 C, diffusion by ordinary impurities is totally negligible.

The impurity profile obtainable in a diffusion process is relatively inflexible, with a high density near the surface of introduction, smoothly declining to zero toward the interior of the sample. Fortunately, however, this qualitative situation is highly desirable for many purposes. In particular, one frequently wishes to make electrical contact to the diffused region; having a high-impurity-density layer near the surface facilitates the creation of an ohmic contact. This is examined in greater detail in Section 3-5.4. A second factor accounting for the popularity and longevity of diffusion methods for junction formation is economic. A single furnace procedure can handle many silicon slices, or a very large "batch," so that the aggregate surface area being processed is large. Competing technologies are greatly restricted in this respect, and hence more costly, although they possess various advantages not found in a diffusion process. Among these other technologies are the MBE process cited earlier [23], the growth of a semiconductor crystal from the vapor phase, known as *epitaxial growth* [31], and doping by directing energetic impurity ions at a semiconductor sample, or *ion implantation* [32].

There are two primary independent variables in a diffusion process. One is the *time* that the sample is held at the elevated temperature. The other is the *temperature* itself. The latter variable enters through the diffusion coefficient D (or diffusivity), the constant of proportionality in Fick's first law:

$$f = -D \frac{\partial N}{\partial x}. \tag{3-50}$$

(The symbol N stands for impurity density as usual, and the other features of this equation are precisely as described in Section 2-5.6.) Diffusivity increases exponentially with increasing temperature. To be sure, the values are miniscule compared to those for mobile carriers in silicon. For example, both phosphorus and boron exhibit D values of approximately 10^{-12} cm^2/s at 1150 C. Consequently, typical times involved in diffusion processes vary from a few minutes to a few hours.

Besides setting time and temperature, the process engineer must choose arrangements that fix boundary conditions. Various options are available, but two are of particular importance. In one, a constant volumetric density of impurities $N(0)$ is maintained at the surface of the sample, just *inside* the silicon. This density is usually identified simply as *surface concentration*. Diffusion under conditions of constant $N(0)$ is a classic problem in diffusion theory. (It is, in fact, the problem posed by the thought experiment involving nitrogen and oxygen molecules in Section 2-5.6). The solution for this impurity-diffusion problem is

$$N(x) = N(0)\left[1 - \text{erf}\left(\frac{x}{\sqrt{4Dt}}\right)\right] = N(0)\,\text{erfc}\left(\frac{x}{\sqrt{4Dt}}\right). \tag{3-51}$$

The symbol "erf" stands for the *error function*, and

$$\text{erf}\left(\frac{x}{\sqrt{4Dt}}\right) = \frac{2}{\sqrt{\pi}} \int_0^{x/\sqrt{4Dt}} \exp(-\lambda^2 d\lambda), \tag{3-52}$$

the same function that is the subject of Problem C1-1. In turn, the symbol "erfc" stands for *error-function complement*, or one minus the error function.

It is sometimes desirable to maximize $N(0)$. The upper limit is set by the *solid solubility* of the given impurity in the semiconductor, a number that varies rather widely. For example, in silicon at 1150 C, phosphorus has a solid solubility slightly in excess of 10^{21}/cm^3, while for boron the value is about half that. For aluminum, on the other hand (another column-III impurity), the value is only about 2×10^{19}/cm^3.

The second basic method for boundary-value setting involves two steps. In the first, a given quantity of impurity is introduced into the sample, very near the surface. This is frequently done by means of a short diffusion step. Then the source is eliminated, and in a second step, the fixed amount of impurity is caused to diffuse into the sample. In trade jargon, the two steps are known as "predep," and "drive-in." In recent years it has been commonplace to substitute ion implantation for a diffusion operation in the first step, for the reason that the "dose" of impurities placed in the sample can be precisely controlled by ion implantation. By "dose" is meant, of course, an areal charge density N_S, or number of impurities per unit area. It will not be surprising to learn that the impurity profile resulting from the two-step procedure is the Gaussian function,

$$N(x) = \frac{N_S}{\sqrt{\pi Dt}} \exp\left(-\frac{x^2}{\sqrt{4Dt}}\right), \tag{3-53}$$

which also entered Problem C1-1. Note that here the surface concentration

$$N(0)|_{\text{Gaussian}} = \frac{N_S}{\sqrt{\pi Dt}} \tag{3-54}$$

is a function of time; for each quadrupling of drive-in time t, $N(0)$ drops by a factor of two because the fixed charge of impurities spreads toward the interior of the sample.

Exercise 3-34. In a certain two-step diffusion operation, $N_S = 10^{16}/\text{cm}^2$. The impurity in question is phosphorus, for which $D = 10^{-12}$ cm²/s at the chosen temperature. Find $N(0)$ at a drive-in time of $t = 1$ h.

The surface concentration is

$$N(0) = \frac{N_S}{\sqrt{\pi Dt}} = \frac{10^{16}/\text{cm}^2}{\sqrt{\pi(10^{-12}\ \text{cm}^2/\text{s})(3600\ \text{s})}} = 9.4 \times 10^{19}/\text{cm}^3.$$

Exercise 3-35. Find the thickness of a layer uniformly doped at $N(0)$ in the preceding problem and having the same areal density N_S.

Evidently,

$$X = \frac{N_S}{N(0)} = \frac{10^{16}/\text{cm}^2}{9.4 \times 10^{19}/\text{cm}^3} = 1.1\ \mu\text{m}.$$

The constant-$N(0)$ and constant-N_S analyses just offered constitute idealizations. There are real-world constraints on our ability to hold temperature, time, and areal density constant. For example, there is the matter of cooling samples from the diffusion temperature to room temperature. Damage to the samples, even breakage, can result if cooling is done too quickly. Hence both effective temperature and effective diffusion time are perturbed by cool-down, especially in a short diffusion process. As another example, there is the problem of impurity "outdiffusion" during drive-in, and this obviously affects N_S. But in spite of such factors, the idealized analyses are often accurate enough to guide experimental work.

Both the error-function complement and the Gaussian function can be reasonably well approximated by a simple exponential function through several decades of impurity density, in a range lying below a value about three orders of magnitude below $N(0)$. This fact makes it possible to write useful exponential approximations for certain density ranges [33], and is graphically displayed in the semilog plot of normalized $N(x)$ versus normalized distance, Figure 3-29(a). The asymptotic exponential function is plotted along with the other two. The same three curves are plotted in linear fashion in Figure 3-29(b), as usual providing differing insights [33]. Note, first of all, that the Gaussian function has zero slope at the surface, while the erfc function does not. The same information is contained in the semilog plot, but with less emphasis. Second, the semilog plot displays clearly the approximate coincidence of the three functions, but the linear plot is not very helpful in this respect.

To form a PN junction using a diffusion, one introduces N-type impurities into a P-type sample, or vice versa. Where $N_D - N_A = 0$, the metallurgical junction resides. This is illustrated in Figure 3-30. In this case a uniform P-type net doping is assumed in the starting sample, at the value N_A. The profile of diffused-in N-type impurities is shown in the neighborhood of the junction. Where the two curves cross, we have the junction position x_J.

Further worthwhile information is presented in Figure 3-30. In this particular example, representing the case of a relatively "deep" junction, the diffused profile is well approximated by a linearly graded profile when the junction is at equilibrium. (The fact of equilibrium is displayed by designating X_0.) As a result, the dependence of depletion-layer thickness X on voltage goes as $(V_{NP} + \Delta\psi_0)^{1/3}$ in this voltage neighborhood. With appreciable reverse bias, however, the left-hand boundary of the depletion layer is stabilized by the steeply rising impurity density. (Remember that it is approximately exponential.) Furthermore, the right-hand boundary is in a region that is for practical purposes uniformly doped. Thus the junction exhibits the behavior of a step junction—a one-sided step junction, to be specific. There we have $X \propto (V_{NP} + \Delta\psi_0)^{1/2}$. All diffused junctions make a transition of this kind, from graded-junction behavior at very small $\Delta\psi$, to step-junction behavior at large $\Delta\psi$. There may be factors that rule out operation of a particular junction in a particular voltage range, but the foregoing statement is nonetheless valid in principle. These properties of diffused junctions were described through extensive analysis and numerical computation in 1960 [34].

In view of the differences in the Gaussian and erfc functions that are displayed dramatically in Figure 3-29(b), it may well come as a surprise that distinguishing between the two functions is unnecessary for certain purposes. In particular, when the two functions are matched at the points $N(0)$ and $N(x_J)$, their differences are

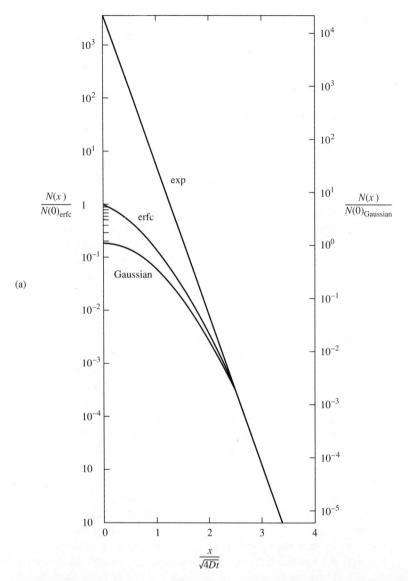

Figure 3-29 Lower portion of the Gaussian and erfc functions can be approximated through several decades by means of an exponential function. (a) Semilog presentation of superimposed "tails" [33]. (b) Linear presentation of the same functions that are shown in part (a), identically positioned with respect to one another [33].

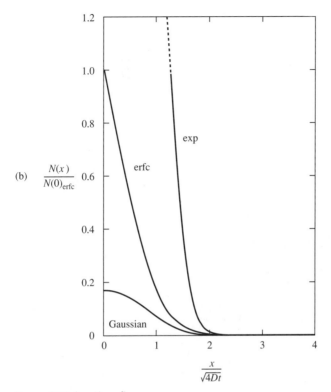

Figure 3-29 (*continued*)

surprisingly small. This is displayed in semilog fashion in Figure 3-31. When the data are plotted linearly, the near coincidence of the two curves is even more striking. For this reason, the curves presented in Reference 34 can for many purposes be used for *either* function.

In the mid-1950s, a number of techniques were combined with solid-phase diffusion to produce a result that has had enormous economic and practical significance. It was found that silicon oxide has the ability to block, at least partially, the entry of common diffusant impurities into a silicon surface. Furthermore, a silicon-oxide layer (essentially a glass) can be formed readily on a silicon surface by heating the sample in an oxidizing atmosphere—oxygen or steam. Then, selected portions of the oxide layer could be removed by chemical means—specifically, hydrofluoric acid, a well-known etchant for glass.

The area selection was accomplished by the application of *photoresist technology*, a term designating use of an acid-resisting organic film that is photosensitive. When such a film is applied over the silicon oxide and then exposed by the contact printing or projection printing onto it of an image existing on a photographic plate,

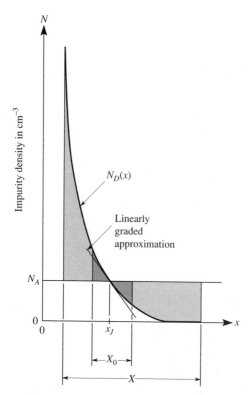

Figure 3-30 Impurity-density profile, presented linearly, for a diffused junction with junction position x_J well below the surface. The linearly graded approximation is applicable at equilibrium, and the one-sided-step approximation is applicable at large reverse voltage. For a shallow junction, the step approximation is valid at equilibrium [33].

the "information" on the plate is transferred to the organic layer on the silicon sample. Removal of the exposed portion of the photoresist layer (or, alternatively, the unexposed portion, depending on the kind of material chosen) is accomplished in a "development" process. Then the oxide layer is etched away in the resist-free areas, the rest of the photoresist is removed, and impurities are introduced by diffusion into the oxide-free areas of the sample. In this manner, lateral delineation of the diffused layer is achieved, complementing the relatively precise depth control that diffusion technology affords.

Many people contributed portions of this technology in the 1950s. Then one person assembled these ingredients, added one small but important ingredient of his own, and secured what became one of the most influential patents in solid-state electronics [35].

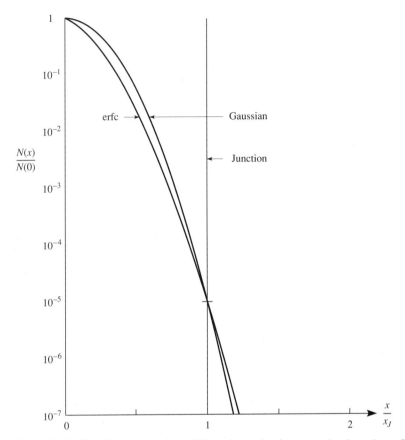

Figure 3-31 Semilog presentation of Gaussian and erfc curves that have been fitted at $x = 0$ and $x = x_J$, with $N(x_J) \ll N(0)$. The two functions approximate each other closely near the junction in this case [33].

3-5.4 High-Low Junctions and Ohmic Contacts

An ideal *high-low junction* is a surface in a semiconductor single crystal separating a region of high doping of one conductivity type from another of uniform and low or moderate doping of the *same* type. Because both regions are of the same type, the high-low junction differs in significant ways from all of the PN junctions: For one thing, the space charge on one side (the high side) is purely ionic, while that on the other side consists of *mobile carriers*. This state of affairs is illustrated in Figure 3-32(a), a physical representation of an N^+N sample at equilibrium. In Figure 3-32(b), the net-doping profile $N(x)$ is indicated by means of a solid line, and the electron profile $n(x)$, by means of a dashed line. The electron profile must, of course, be continuous and of noninfinite slope as it passes through the junction; electrons ''spill over'' from the high side, creating an electron deficiency responsi-

306 PN JUNCTIONS

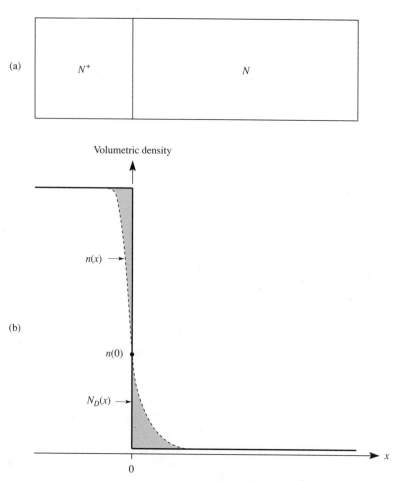

Figure 3-32 An N^+N high-low junction. (a) Physical representation. (b) Net-doping profile $N_D(x)$ and electron profile $n(x)$.

ble for ionic space charge on the left, and creating a mobile-carrier charge on the right. These two charge components are represented by the shaded areas in Figure 3-32(b), and these areas must of course be equal.

Exercise 3-36. Given that the two sides in the sample of Figure 3-32 have net-doping values of $10^{20}/cm^3$ and $10^{16}/cm^3$, respectively, estimate the values of potential drop on the two sides.

Given the Boltzmann relation, Equation 2-35, we can calculate the total equilibrium potential difference between the two neutral regions:

$$\Delta U_0 = \ln \frac{N_D^+}{N_D} = \ln 10^4 = 9.2,$$

where N_D^+ and N_D are taken to be the respective net-doping values. It was noted in Section 3-2.7 that the high-side potential drop in a grossly asymmetric junction is one normalized unit, kT/q [8, 16]. Because a high-low junction is grossly asymmetric, the same observation holds here as well. Therefore the two potential-drop values are about 1 and 8.2 normalized units, respectively, or 0.026 V and 0.21 V, respectively.

Exercise 3-37. Determine the value of $n(0)$ in Figure 3-32(b), using the data of Exercise 3-36.

Using the Boltzmann relation once more informs us that the density $n(0)$ must differ from N_D^+ by a factor of e, because the left-hand potential drop is kT/q. Hence,

$$n(0) = \frac{N_D^+}{e} = (0.37) N_D^+.$$

It makes sense to talk about forward bias and reverse bias in a high-low junction. Because the left-hand side is "more N-type" in the sample of Figure 3-32, a negative bias on that side constitutes forward bias. The response of a high-low junction to forward bias is qualitatively different from that of a *PN* junction. In the *PN* case, minority-carrier diffusion currents play an important role in the neutral regions just outside the junction boundaries. In the high-low case, by contrast, both drift and diffusion minority-carrier components of current are negligible throughout the sample. Instead, majority-carrier drift currents account for virtually all of the neutral-region current. The junction region, too, exhibits ready conductivity in the forward-bias case under consideration. The reduction in potential difference $\Delta \psi$ that results is accompanied by an increase in carrier-density gradient values through the junction, and thus electrons readily move from the high side to the low side. In the high-low case, as in the *PN* case, the mythical drift and diffusion components of equilibrium current are enormous, and typical values of net forward current are small by comparison.

There is a further and extremely important point of difference between the two kinds of junctions. It is because forward current in a *PN* junction requires minority-carrier injection that such a junction exhibits an offset voltage. The boundary value of minority-carrier density must rise sufficiently to produce the diffusion current, and the larger the energy gap (the smaller n_i), the greater the voltage required to do this, or the greater the offset voltage, as Figure 3-20 shows. But in the high-low junction, the immediate response of diffusion in the transition region and drift outside it to forward bias means that there is *no* offset voltage. Furthermore, the potential barrier in a high-low junction, small to begin with (refer back to Exercise 3-36), is only slightly altered by the passage of even a relatively large current. Combining all of these observations means that the forward characteristic of a high-low junction is linear and lies very close to the current axis. In short, the structure exhibits very low resistance.

The reverse characteristic, it turns out, is very nearly an extension of the forward characteristic. Once again majority-carrier drift accounts for neutral-region current, and a slight relative increase in the (equilibrium value of) drift current in the transition region is able to maintain current continuity. It follows, then, that the I–V characteristic of a high-low junction overall is that of a near-ideal ohmic contact.

Recall that two features must be present in an ohmic contact. First, resistance must be low, a point just made. Second, equilibrium densities must be preserved, and this condition, too, is met. Ohmic conduction by drift in the neutral regions is a nonperturbing process. And the minor adjustments of field profile and majority-carrier profile that occur in the transition region are equally nonperturbing. Minority-carrier densities undergo large relative changes, but negligible absolute changes, and contribute but little to conduction in any case.

For these reasons, high-low junctions are universally incorporated in today's devices to serve as ohmic contacts, essentially supplanting the kind of ohmic contact described in Section 2-7.5. It is not necessary that the doping transition be abrupt, as was assumed earlier for the sake of convenient description. In fact, the diffusion process is extensively used to place an N^+ region in contact with an N region, or a P^+ region in contact with a P region. We have seen that such doping profiles are smooth and gradual, but they serve well nonetheless.

An ohmic contact is needed, for example, at the terminal of a device. In many cases this terminal is to be connected electrically to another point, or other points, in a circuit, and a patterned metallic layer is introduced to accomplish this. (Sometimes a heavily doped polycrystalline-silicon patterned layer is substituted for the metal layer.) An intimate connection of a metal to the heavily doped portion of a high-low junction places an additional small resistance in series with that of the high-low junction. The resulting metal–semiconductor junction is a *Schottky junction*. By virtue of the heavy doping of the semiconductor portion in this case, however, the potential barrier that accompanies such a junction is very thin. Consequently, carriers can penetrate it by tunneling, a mechanism discussed in Section 3-6.2.

At other times a metallic layer is placed on a heavily doped semiconductor region to facilitate the attachment of another metallic element, often in the form of a wire. First, the metal–semiconductor system is heated short of melting to assure intimate contact of the two phases. Then the additional element is attached by a combination of elevated temperature, again small enough to avoid melting, and appreciable pressure between the two metal components. This technique, known as *thermocompression bonding* [36], is used to establish connections from a discrete device or integrated circuit to a terminal on the inside of its protective package.

3-6 BREAKDOWN PHENOMENA

A number of processes can occur in a sample containing a junction that cause a sharp increase in current with a small increment in applied voltage. These are collectively described as *breakdown* phenomena. The most important of these are nondestructive, provided that precautions are taken to avoid excessive current. That is, the device can be repeatedly taken into and out of the regime of steep current increase without damage. Other factors and conditions can cause permanent change in the device, an example being localized melting, but with the onward march of technology, most of these phenomena have become less important than in earlier days. For this reason, and because of the relative complexity of the destructive phenomena, they are omitted from consideration here.

3-6.1 Avalanche Breakdown

The first mechanism we examine occurs in a *PN* junction under reverse bias. It can cause the steep current increase to occur at a voltage anywhere from thousands of volts to a few volts. (Toward the lower end of this range, it first coexists with and then is superseded by the tunnel mechanism, the subject of Section 3-6.2.) The critical value of reverse voltage at which breakdown occurs is primarily fixed by certain gross features of the doping profile in the junction, with some of the profile details being relatively unimportant. Accordingly, it is possible to design and fabricate junctions to exhibit sharp, reproducible, and useful breakdown at a desired voltage. One obvious use of the resulting device is as a voltage regulator. There is a practical upper limit on the breakdown voltage in such devices, amounting to a few hundred volts, because of power-dissipation problems that lead to unstable properties, and ultimately to damage.

We saw in Section 3-4.2 that minority carriers on both sides move toward a reverse-biased junction, and that the supply of such carriers fixes a theoretical reverse leakage current. At the breakdown voltage, some of these carriers are able to acquire sufficient energy from the electric field of the junction (that is, to become "hot carriers") so that they are energetic enough to break a covalent bond by a process known as *impact ionization,* producing a hole–electron pair as a result. The minimum energy required, of course, is ξ_G. In turn, some of the newly created carriers can also create additional pairs.

A consequence of this *multiplication* process is that the steady-state population of carriers present in the space-charge region of the junction is much higher than it was at a slightly lower voltage. In turn, higher carrier density in the high-field region inevitably means higher steady-state reverse current. This is the "breakdown regime" of operation, wherein current is extremely voltage-dependent. With a further voltage increment, the probability that a particular carrier entering the high-field region will be able to produce a pair is increased; the result is a still higher steady-state carrier population and a still higher reverse current.

It should be understood that a particular carrier quickly passes out of the high-field region and no longer participates in the multiplication process, but that other carriers, both progeny and other entering carriers, are present at any subsequent instant. The essence of the multiplication process, of course, is carriers that beget carriers that beget carriers, and so on; the term *avalanche* breakdown has been applied to this mechanism because of the high degree of analogy between it and a literal avalanche involving snow on a steep slope. The essence of the avalanche process is shown in Figure 3-33(a), where a minority electron enters the high-field region of a reverse-biased N^+P junction from the right. After creating a pair it continues on its way, and the carrier population has been enhanced threefold by that event.

Extensive early measurements were made by Miller on avalanche breakdown in N^+P junctions as well as in the homologous P^+N junction [37]. These junctions were fabricated by means of the alloying process described at the beginning of Section 3-5. Subsequent theoretical investigations have described in considerable detail the underlying events; a good summary has been given by Moll [38]. Such analysis starts with a picture much like that of Figure 3-33(a), and considers the *ionization rate* of a carrier, or the number of pairs it produces per unit length of drift path [37]. These rates are steeply field-dependent. For example, for electrons in silicon, the value is about 10^5/cm at a field value of 5×10^5 V/cm, and drops roughly two orders of magnitude when the field is halved. For holes, the values are below those for electrons, being roughly three times smaller at the higher field and one hundred times smaller at the lower [39]. Knowledge of these field-dependent rates and the field profile in a particular junction permits one to calculate a *multiplication factor M*. This can be defined as the current under conditions of avalanching divided by the current that would be present if the avalanche mechanism could be "turned off." The role of multiplication is not clearly displayed in the breakdown region of the *I–V* characteristic shown in Figure 3-33(b) with a current calibration in the commonplace neighborhood of 1 mA; but with a more "sensitive" scale, as in Figure 3-33(c), the gradual nature of the dependence of multiplication on field becomes evident. The usual criterion for breakdown voltage V_B is taken as the value for which $M \rightarrow \infty$. Some multiplication can occur well below V_B because statistical variations exist in the length of the free paths during which carriers absorb energy from the electric field.

The early experiments with one-sided junctions showed equal V_B values for N^+P and P^+N samples having the same light-side doping. This might at first be surprising in view of the disparity between hole and electron ionization rates, but when we consider that *both* kinds of carriers inevitably enter into breakdown in either case, the correspondence becomes reasonable. Miller's data on alloy-junction diodes are shown in Figure 3-34, along with calculated and approximate-analytic step-junction data by other workers [39, 40]. To acquire a feeling for magnitudes, note that a net doping of 10^{15}/cm^3 on the light side of a one-sided silicon junction leads to $V_B \approx 300$ V. This diagram also gives approximate-analytic data on linearly graded junctions [39], and experimental data on diffused junctions recorded by

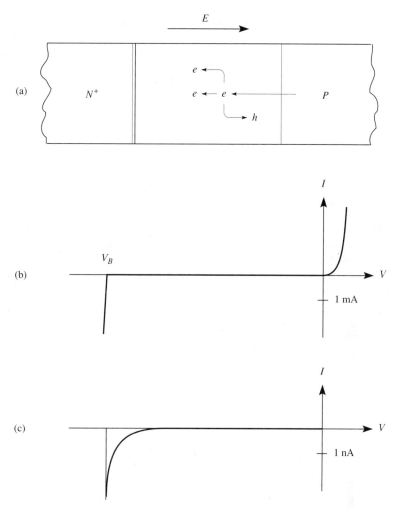

Figure 3-33 Avalanche breakdown in an N^+P junction. (a) Space-charge layer of reverse-biased junction, showing minority hole entering from the right and creating two additional carriers by impact ionization. (b) Reverse-current increase at voltage V_B as a result of avalanching. (c) The same characteristic with an expanded current scale; carrier multiplication can occur below V_B because free-path lengths exhibit statistical variations.

Carlson [41]. The latter data were taken in the 1950s, but were analyzed over a decade later [42].

Out of the potentially infinite set of possible diffused junctions, Carlson chose those fabricated by a standard diffusion process (and hence a fixed diffused-impurity profile) carried out on starting samples of various net-doping values. The choices he made caused all of his diodes to resemble linearly graded junctions, a fact consistent with the positions of his experimental points on Figure 3-34.

Sections 3-2 and 3-5 make it clear that electric-field profiles vary appreciably from one kind of junction to another. We single out the field profile because of the sensitivity of ionization rate to electric field. A number of efforts have been made to relate breakdown voltage with the peak-field value in a reverse-biased junction, but the correlation is so poor as to be meaningless. The reason is that a carrier must have "running room" to acquire the kinetic energy of about 1.1 eV that it must have to break a bond. Thus an enormous field in a very thin region does not lead to avalanche breakdown.

A more meaningful look at the avalanche phenomenon focuses jointly on the average field value through a given region and the thickness of that region. A technique for comparing various field profiles with respect to this dual criterion employs a field-distribution function [42]. When linearly graded junctions and step junctions are compared on this basis, they prove to have field distributions that are more similar than one might expect. Then, recall that diffused junctions exhibit step behavior at one extreme and linearly graded behavior at the other, and something between in the intermediate region. This property of diffused junctions was discussed in connection with Figure 3-30; because of it, we can accord diffused junctions the same treatment given to step and graded junctions, obtaining a very similar result. Because of the similarity of all possible step junctions, diffused junctions, and linearly graded junctions (but specifically not *PIN* junctions) with respect to

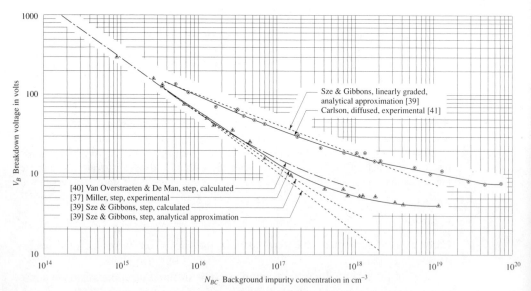

Figure 3-34 Calculated, approximated, and measured breakdown voltage V_B versus background-impurity concentration N_{BC} for step, graded, and diffused junctions, as reported by various workers [37, 39–41].

field distribution, an approximate *single* criterion for breakdown emerges: The thickness of the depletion layer *at* the breakdown voltage. Let us make this definition involving depletion-layer thickness:

$$X(V_B) \equiv X_T. \tag{3-55}$$

The subscript on X_T connotes *total* thickness *at breakdown*. On this basis, avalanche-breakdown voltage V_B is predicted within about 7% for all three categories of junctions by the empirical expression

$$V_B = (5.8 \times 10^4 \text{ V/cm}^{0.84}) X_T^{0.84}, \tag{3-56}$$

in the range ~ 15 V $< V_B <\ \sim 300$ V. This function is plotted in Figure 3-35, and is compared with various calculated and experimental data sets.

All of the foregoing data are for plane junctions, the case that is most important for our purposes. Any curvature in the junction surface leads to a reduction of V_B. The reason relates to the electric-field perturbation that accompanies such surface-shape perturbation. The effect can be visualized readily with the help of the lines-of-force picture introduced in Section 1-1.4; junction curvature leads to crowding of the lines of force, which is another way of describing a locally enhanced electric field magnitude.

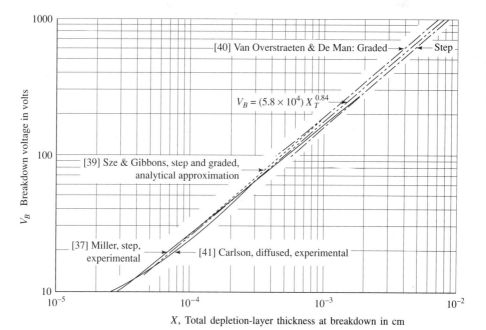

Figure 3-35 Breakdown voltage V_B as a function of depletion-layer thickness *at* V_B for step, graded, and diffused junctions. Calculated, approximated, and measured data supplied by several workers [37, 39–41] have been jointly approximated by a single function [42].

An early effort by Zener [43] to explain reverse-bias breakdown in a *PN* junction invoked the quantum-mechanical tunneling phenomenon. It was later learned that while his explanation had validity in devices exhibiting breakdown at a few volts, the avalanche phenomenon dominates in most of the range of designable and useful breakdown voltages. Nonetheless, for historical reasons, the commercial diodes that are extensively sold for voltage regulation and related applications are still known as *Zener diodes*. Now we examine the kinds of *PN* junctions in which tunneling plays a dominant role.

3-6.2 Tunneling

It was pointed out in Section 2-1.4 that valence electrons in a silicon crystal reside in energy wells and are separated by an energy barrier that is very thin. These wells are depicted in Figure 2-4; the barriers formed by squeezing such wells into a silicon crystal are only about one angstrom thick. It was further pointed out there that the motion of a valence electron from one bond to an adjacent bond is responsible for the perceived motion of a "hole" in the valence band, and the valence electron accomplishes its relatively easy penetration of the barrier by *tunneling* through it. This quantum-mechanical phenomenon forces us to drop temporarily the (otherwise useful!) particle picture of an electron. Instead it is the *wave function* of the electron that is relevant, a concept introduced in Section 1-4.6. As pointed out there, knowledge of the wave function permits one to calculate the probability that the electron resides at a particular point in space. And the wave function is typically extensive enough so that an electron "centered at" one bond location has a significant probability of showing up at an adjacent bond previously vacated (a bond that is the location of a hole). By this means valence electrons move easily *through* a barrier, rather than over it, when invited by an empty state at the destination bond. The second requirement for tunneling, besides the spatial proximity of the initial and final states, is their nearness in energy.

In a *PN* junction it is possible to have a situation that is analogous to the valence-bond situation just described, although the analogy is not initially obvious. For tunneling to occur in a junction, it is necessary for the equilibrium depletion layer to be extremely thin; for this to be true, the junction must have very heavy doping on both sides. In Figure 3-36(a) is shown a band diagram for a step P^+N^+ junction in which the Fermi level resides right at the band edge on either side. The net doping required to accomplish this is a bit over $10^{19}/cm^3$. A physical representation of the sample is given in Figure 3-36(b). Now suppose that a reverse bias is applied to the junction, causing the barrier height to increase. This causes valence-band states on the left to be (1) close to, and (2) at the same energy as conduction-band states on the right. Given empty states in the conduction band (and we know that this is the norm), and close enough physical spacing of the two band edges, then electrons can tunnel from left to right. In other words, this junction will conduct at relatively low reverse-bias values.

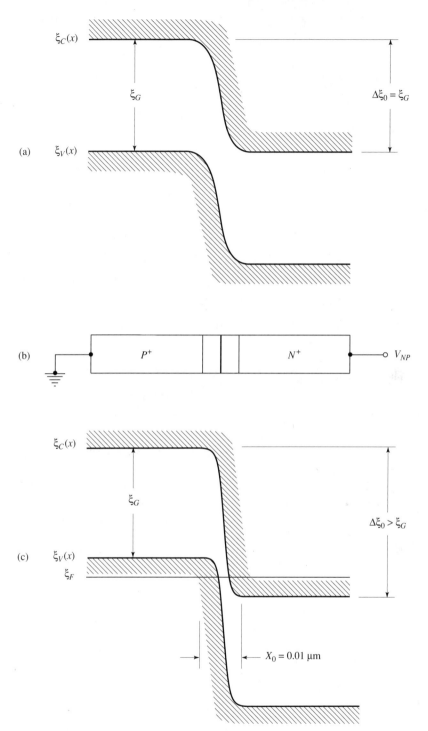

Figure 3-36 Symmetric step junction of heavy doping. (a) Band diagram for a particular P^+N^+ junction. (b) Physical representation of the same device. (c) Band diagram for more heavily doped device constituting a "backward diode."

Exercise 3-38. Calculate the depletion-layer thickness for a step P^+N^+ junction with $N = 3 \times 10^{19}/\text{cm}^3$.

From Equation 3-17,

$$X_0 = \left[\frac{4\epsilon\Delta\psi_0}{qN}\right]^{1/2};$$

$$\Delta\psi_0 = 2\frac{kT}{q}\ln\frac{N}{n_i} = 2(0.02566 \text{ V})\ln\frac{3 \times 10^{19}/\text{cm}^3}{10^{10}/\text{cm}^3} = 1.12 \text{ V}.$$

Hence,

$$X_0 = \left[\frac{4(10^{-12} \text{ F/cm})(1.12 \text{ V})}{(1.6 \times 10^{-19} \text{ C})(3 \times 10^{19}/\text{cm}^3)}\right]^{1/2} = 1.0 \times 10^{-6} \text{ cm} = 100 \text{ Å}.$$

Exercise 3-39. Comment on band-edge positions relative to the Fermi level in the sample of the previous exercise.

Because the contact potential $\Delta\psi_0 = 1.12$ V, which is approximately ψ_G, it follows that the band edges are very close to the Fermi level in the two end regions.

The probability of electron tunneling through a barrier 100 Å thick is very low. Note, however, that with increasing reverse bias, the band-edge spacing continuously decreases, a matter that has recently been analyzed quantitatively [44]. To emphasize this point, note that reverse bias by a mere three volts will cause an approximately fourfold increase in junction potential difference, and the resulting band-edge distortion is substantial. (If further insight is needed, a sketching exercise can easily provide it.) For these reasons, a junction of the kind represented in Figure 3-36(a) will exhibit "breakdown," or rapidly increasing current, at a reverse bias of a few volts, and it does so through the tunneling mechanism described by Zener [43].

Further increases in doping cause further depletion-layer shrinkage, and therefore tunneling, at even smaller reverse bias. Note, too, that the band edges are caused to move beyond the Fermi level, as shown in Figure 3-36(c), an equilibrium band diagram for such a junction. Thus, valence states on the left reside opposite (at the same energy as) conduction states on the right, even when the junction is at equilibrium. Now reverse-bias tunneling can occur at a reverse voltage that is well below the forward offset voltage in magnitude, with a result depicted in Figure 3-37. This device is known as a *backward diode* [45] because it approximates an ideal diode [in the sense of Figure 3-21(a)] when used with its two terminals interchanged. That is, in the direction of easier flow, it exhibits an offset voltage of only

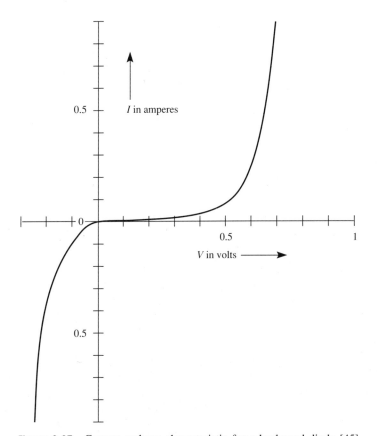

Figure 3-37   Current–voltage characteristic for a backward diode [45].

about 0.2 V. In the opposite direction, it is useful up to 0.6 V or so, after which it "breaks down" by exhibiting the offset voltage normal for that polarity.

With a still further doping increase, the band diagram at equilibrium continues to resemble that in Figure 3-36(c), except that vertical range of "overlap" increases further, and the horizontal (spatial) separation of the band edges decreases. As a result, a qualitative change occurs in the *I–V* characteristic, one that first aroused the curiosity of a researcher named Esaki [46], who explained the observation correctly. Consequently the structure he analyzed has become known as the *Esaki diode*, as well as the *tunnel diode*. It is true with the further doping increase that tunneling conditions are satisfied even with appreciable *forward* bias. Tunneling occurs (this time from right to left, to be sure) from the conduction band to the valence band. The result is an "anomalous" forward current through a limited voltage range, as is illustrated in Figure 3-38 [47]. Because forward bias diminishes barrier height in the junction, the band edges proceed in the direction of "uncrossing." That is, with a few tenths of a volt of forward bias, the diagram would resemble that in Figure 3-36(a), where there no longer remain valence and conduc-

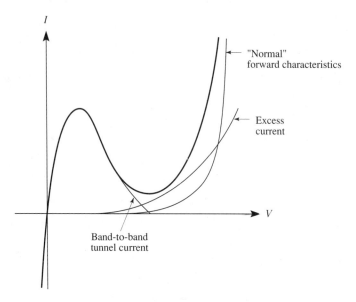

Figure 3-38 The forward characteristic of a tunnel diode and the component currents that produce it [47].

tion states at the same energy. As a result, initial resistance in forward bias is small, approximating that near the origin in reverse bias, because tunneling occurs readily with a small bias of either polarity. Forward current reaches a maximum at a bias of 0.1–0.2 V, and then declines as the uncrossing phenomenon proceeds.

Finally, ordinary carrier injection causes forward current to rise again near 0.6 V as it does in other *PN* junctions. Correspondence of the tunnel-diode forward characteristic and the normal-diode characteristic in the injection regime is not perfect, however. The degenerate doping necessarily present in the tunnel diode enforces high densities of crystalline imperfections throughout the device. These, as we have seen, introduce intragap states. Such states, in turn, introduce new conduction mechanisms in forward bias, involving, for example, carrier tunneling to and from such states within the transition region of the junction. The result of this is the presence of the *excess current* component displayed in Figure 3-38.

As noted in Section 2-6.5, silicon is an indirect-gap material. This fact has a bearing on tunneling probability in a tunnel junction [47], and as a result, tunnel diodes made of direct-gap materials such as gallium arsenide are appreciably more efficient. These matters were explored extensively because of early hopes that the tunnel diode could be exploited as a switching device. The regime wherein forward current declines with increasing forward voltage is termed, fairly obviously, a regime of *differential negative resistance,* and makes possible one type of switching application. Further motivation was supplied by the realization that a quantum-mechanical process such as tunneling occurs with no time delay, setting it apart from the more numerous kinds of switching processes that involve carrier transport,

BREAKDOWN PHENOMENA   319

and inevitably, transport delay. The promise was not fulfilled, however, because using diodes in "active" applications is at best awkward; there are great advantages in having a third electrode capable of controlling what occurs between the other two.

Nonetheless, the tunnel phenomenon is exploited extensively in today's devices in the more mundane but absolutely essential ohmic-contact role. Often the junctions are of the metal–semiconductor nature, rather than P^+N^+ nature. In the high-low type of ohmic contact described in Section 3-5.4, for example, it is common to have a metal layer in contact with the degenerate region, forming a metal–semiconductor tunnel junction. In integrated circuits, N^+P^+ tunnel junctions as ohmic contacts are fairly common. In the future, they may prove indispensable for three-dimensional all-semiconductor integrated circuits [48].

The tunnel junction discussed earlier was described as a step junction. For first-order analysis, however, a *PIN* structure (see Section 3-5.1) proved convenient [49]; although this assumption was quite unrealistic, the analysis provided useful guidance. More recently it has been shown that a linearly graded junction [Section 3-5.2] provides an appreciably more realistic and useful model for tunnel-junction analysis [44].

3-6.3 Punchthrough

A third breakdown phenomenon is somewhat apart from those just discussed in that it requires a structure with *two* closely spaced metallurgical junctions. We place it on the same footing as the other two, however, because (1) it involves a physical mechanism totally different from avalanching and tunneling, and (2) it involves a structure that is simple, important, and commonplace. (The latter point is analogous to our reason for listing the *PIN* diode (Section 3-5.1), which incorporates two metallurgical junctions, under "other junctions.") The present two-junction structure is known as a *punchthrough diode* for reasons that will now be explained.

In its classic form, this diode structure incorporates a thin, lightly doped region of one conductivity type sandwiched between two heavily doped regions of the opposite type. Figure 3-39(a) depicts an N^+PN^+ example, and the dashed line in Figure 3-39(b) shows the equilibrium potential profile of the two back-to-back junctions. With the application of voltage, one junction will be forward-biased and the other will be reverse-biased. Initially the current is limited to a value approximating the leakage current of the reverse-biased junction. With such a small current passing through the forward-biased junction, its minority-carrier boundary values depart only slightly from their equilibrium values. Since these boundary values are tied by the law of the junction to the applied voltage experienced by the forward-biased junction, it follows that only a tiny voltage appears upon that junction, and *ipso facto*, that virtually all of the applied voltage appears on the reverse-biased junction. As a result, the depletion layer of the reverse-biased junction advances toward the essentially unperturbed depletion layer of the forward-biased junction. The value of applied voltage that causes the two depletion layers to touch is known

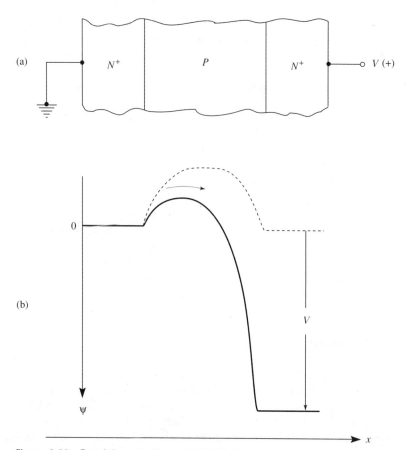

Figure 3-39 Punchthrough diode. (a) Physical representation. (b) Potential profiles at equilibrium (dashed line) and at an applied voltage greater than the reach-through voltage V_{RT} (solid line).

as the *reachthrough voltage* V_{RT}. Then, as the applied voltage begins to exceed V_{RT}, the two depletion layers begin to interact. The effect of this interaction is illustrated by the solid-line potential profile in Figure 3-39(b). For the punchthrough diode to function as intended, it is necessary to design it so that the reach-through voltage V_{RT} is well below the avalanche-breakdown voltage V_B of the reverse-biased junction.

Now visualize the potential hill on the left-hand side of Figure 3-39(b) as the barrier that retains conduction-band electrons on that side of the device. As the hill begins to shrink as a result of depletion-layer interaction, these electrons begin to spill over the barrier. Once over the barrier, they find themselves in the high-field region of the reverse-biased junction.

Exercise 3-40. How can electric field be visualized in Figure 3-39(b)?

Electric-field magnitude is proportional to the slope of the potential profile shown there, and hence is easily visualized. As an option, one can perceive the electrons as "sliding down" the potential hill, with drift velocity increasing as they move rightward because field increases there.

The next point concerns the population of available electrons in the conduction band. Figure 2-9 illustrates the fact that the density-of-states function, the electron "container," is approximately parabolic—a slowly varying power-law function. The occupancy probability for these states, on the other hand, is the nearly exponential Boltzmann "tail" of the Fermi-Dirac function. Now idealize both functions, assuming a rectangular density-of-states function and a purely exponential probability function. Under these conditions, the availability of conduction-band electrons able to spill over the barrier will increase exponentially with decreasing barrier height. As a result, current through the punchthrough diode increases steeply with applied voltage V in the voltage regime $V > V_{RT}$. The phenomena responsible for the regime of rapidly increasing current are known collectively as *punchthrough*.

The behavior just described is closely related to that occurring in a forward-biased junction. It was noted in Section 3-4 that the exponentially increasing current exhibited in Figure 3-24 is the result of carrier spillage over a declining potential barrier. Figure 3-40 compares the experimental I–V characteristic of a single diode at the extreme left (copied from Figure 3-24) with the experimental characteristic of a punchthrough diode at the extreme right [50]. For the latter device, $V_{RT} \approx 11$ V. Note that the two characteristics are comparably steep.

Exercise 3-41. What differences, if any, exist in the barrier-modulation phenomenon in a forward-biased junction and that in a punchthrough diode that might lead one to expect different steepnesses in the two I–V characteristics?

In the forward-biased junction, the increment in barrier height approximates the increment of applied voltage. In the punchthrough diode, the barrier-height increment can exceed the applied-voltage increment. The depletion-layer interaction responsible for this effect, which occurs only beyond V_{RT}, can be visualized in Figure 3-39(b). A second point of difference is that minority carriers injected through the boundary of a forward-biased junction move away from the boundary by (nearly) pure diffusion. In the punchthrough diode, carriers are injected into a region of electric field that increases with distance, so that diffusion is augmented by drift. These factors suggest faster rise in the punchthrough-diode current, but as Figure 3-40 shows, the curves are comparable in steepness.

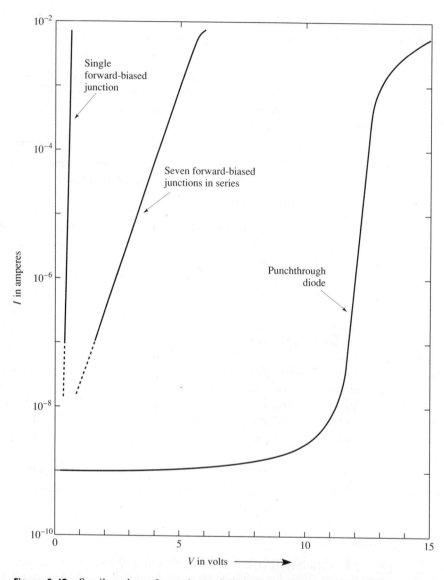

Figure 3-40 Semilog plots of experimental I–V characteristics for (left to right) a single forward-biased silicon junction, seven forward-biased junctions in series, and punchthrough diode.

Forward-biased junctions are extensively used for voltage regulation, or as voltage "level shifters," because of the steep rise in current at the offset voltage. The obvious shortcoming of a forward diode in this application, of course, is the inflexibility of offset voltage in a given material—about 0.7 V in silicon and 1.0 V in gallium arsenide. A small measure of flexibility is secured by putting two or more such diodes in series, thus achieving voltage regulation at a multiple of the offset voltage. Figure 3-40 shows the I–V characteristic of seven forward-biased junctions in series (middle curve). It is evident that one pays a penalty in this approach, however, because the curve is no longer as steep as before.

Exercise 3-42. What accounts for the differing steepnesses of the two curves at the left in Fig. 3-40?

Since these are static I–V characteristics, the relevant junction property is conductance at a particular current, a topic examined in Section 3-8.1. Since the seven junctions are in series, the seven small-signal resistances are in series, so that

$$r_7 = 7r,$$

where r_7 is the small-signal resistance of the series string, and hence

$$g_7 = g/7.$$

Exercise 3-43. It has been emphasized in several places previously that because of logarithmic distortion, it is risky, or at best difficult, to infer relative slopes from a semilog diagram. But inspection of the two left-hand curves in Figure 3-40 suggests that their slopes indeed differ by about a factor of seven, the value cited in Exercise 3-39. Why is this so?

Start with the case of a single junction forward-biased in the range where the simple theory developed in Section 3-8.1 is valid. Its current can be written

$$I = -I_0 \exp(-qV_{NP}/kT).$$

Plotting I on a logarithmic ordinate as in Figure 3-40 is the same as plotting $\ln I$ on a *linear* ordinate. Choosing the second option,

$$\ln I = (-I_0) - (qV_{NP}/kT).$$

Thus the left-hand curve has the constant slope

$$\frac{d(\ln I)}{d(-V_{NP})} = \frac{q}{kT}.$$

Again it is worth noting that the expression "ln I" is objectionable from the purist's point of view because I has dimensions. The objection could have been avoided easily by rewriting the first equation in normalized form as

$$\frac{I}{I_0} = -\exp(-qV_{NP}/kT).$$

This approach was not taken here, though, because Figure 3-40 is unnormalized.
In the case of forward junctions in series, assuming identical properties, each will experience the voltage (V_{NP}/n_d), where n_d is the number of diodes in series.

Hence in this case

$$I = -I_0 \exp(-qV_{NP}/n_d kT),$$

and

$$\ln I = (-I_0) - (qV_{NP}/n_d kT).$$

Thus

$$\frac{d(\ln I)}{d(-V_{NP})} = \frac{q}{n_d kT},$$

and the slope will again be constant, but smaller by the factor n_d. In Figure 3-40, therefore, the two left-hand curves will differ in slope by the evident factor of seven.

One can create an analog of the punchthrough diode by placing two parallel metal plates in a vacuum. When the evacuated vessel is tubular in form, as it usually is, and assuming external electrical connections to the individual plates have been established, the result is a *vacuum tube*. The analog, to be sure, predates the punchthrough diode by many decades! The qualitative similarities are strong. The metal plates, like the two N^+ regions of the punchthrough diode, have conduction bands populated by electrons that obey a Fermi-Dirac distribution. The quantitative differences, however, are significant. The conduction bands are "deeper" in the metal case. Compare Figure 2-7(c) to Figure 2-9(c), for example, acknowledging that the latter is for an intrinsic sample rather than an N^+ sample. More important, the potential barrier at the surface of a metal that prevents electrons from "leaking out" at room temperature is typically several volts high. On the other hand, the two potential barriers displayed by the dashed curve in Figure 3-39(b) are typically smaller than one volt in silicon. As a result, even at equilibrium, electrons in the N^+ regions surmount the barriers by diffusion, although countervailing electron drift maintains a zero net current everywhere.

In the case of a vacuum-tube electrode one can give the electrons sufficient energy to escape by raising its temperature, this because of the energy-distribution changes displayed in Figure 2-6(d). The resulting evolution of electrons from the hot electrode is often characterized as "boiling off" or "evaporating" electrons from the solid. The modern term for the phenomenon, however, is *thermionic emission,* which describes the escape of carriers over a barrier. When the second electrode remains cold, except for heating incidental to its proximity to the hot electrode, then it will intercept some of the emitted electrons and consequently a dc transport of electrons can be detected in its external lead. This observation was first made by Thomas Edison in 1883 in the course of his development work on the incandescent lamp, and for years thereafter was known as the *Edison effect,* until the modern term was adopted.

A significant additional variable in thermionic emission is a voltage applied externally between the electrodes. The current of transferred electrons increases steadily with applied voltage when the polarity is correct—positive on the cold electrode (anode) and negative on the hot electrode (cathode). Reversing the polarity causes the vacuum diode to approximate an open circuit because the cold plate is incapable of emitting electrons. This one-way-conduction property accounts for the British term *valve* that is equivalent to the American term *vacuum tube*.

Since the punchthrough diode as well as the vacuum diode involves the delivery of carriers over a potential barrier, the term *thermionic emission* applies equally to it. In both cases the phenomenon is described by an expression known as *Richardson's equation* that was worked out in the vacuum-tube era [51]. It is an exponential function, but not the simple-exponential function that describes the forward-biased junction, which has but two variables in the exponent, applied voltage V and absolute temperature T. These two variables also appear in the exponent of Richardson's equation, but are augmented by a parameter, namely, a critical voltage. In the vacuum diode, the critical voltage is determined mainly by material properties of the cathode, specifically surface properties. In the punchthrough diode, it is determined mainly by properties of the middle layer, its thickness and doping. It can be visualized in the punchthrough case as the voltage required to push the depletion layer of the reverse-biased junction all the way to the metallurgical-junction plane of the forward-biased junction. As we shall see shortly, this condition is not physically reachable. It is because of this critical voltage that a punchthrough diode can be realized with a current rise occurring at a few volts or at several thousand volts, but yet with a steepness approximating that in a forward-biased junction.

The regime of thermionic emission in a typical punchthrough diode extends through several decades of current, and then is terminated by a regime of lower slope (in the I–V characteristic) identified as the *space-charge-limited* regime. This occurs when the carriers in transit achieve a density of the same order as the ionic density in the lightly doped middle region. Under these conditions, the electric-field profile is perturbed in a way that terminates the near-exponential rise of current with voltage. In a typical device, this occurs at large current-density values—of the order of thousands of amperes per square centimeter.

Exercise 3-44. Give a qualitative physical explanation for the fact that field perturbation by carriers in transit slows down the current rise with voltage.

Consider first the effect of increasing the doping N of the middle region at a particular voltage on the reverse-biased junction. This is another way of introducing additional space charge. This would yield a thinner depletion layer with a larger peak field. Thus, when carrier charge density begins to be significant, a series of variables begin to change more slowly with voltage than they would otherwise, these variables being depletion-layer thickness in the reverse-biased junction, barrier-height lowering in the forward-biased junction, and current through the punchthrough diode.

The punchthrough-diode structure attracted the interest and attention of solid-state pioneers Shockley and Prim [52]. They recognized that a thermionic-emission regime would occur in such a structure. Curiously, they displayed little interest in this regime of operation, and focused instead on the space-charge-limited regime, carrying out a lengthy and detailed theoretical analysis. (Technology of the time did not afford them experimental opportunities.) Their description employed Child's law, a relevant expression also developed in the vacuum-tube era [53], because this phenomenon has extreme importance in vacuum-tube operation. The reason for Shockley's absorption with vacuum-tube-like phenomena was his desire to make an *analog transistor,* or a device wherein vacuum-tube-like structures and principles were incorporated in a solid-state device.

Exercise 3-45. Why is space-charge-limited transport more important in a vacuum tube than in the punchthrough diode described earlier?

The volume between vacuum-tube electrodes has no space charge *except* the charge of electrons in transit. Therefore, perturbation of the zero-current (or cold-cathode) field begins at very low current density. In fact, the exponential thermionic regime can be observed only through a small range of low current in a vacuum diode or triode [54].

Exercise 3-46. How could Shockley and Prim simulate vacuum-tube-like conditions in their solid-state device?

They proposed to employ an intrinsic-semiconductor middle layer in their punchthrough diode instead of the doped-semiconductor middle layer described above, in the process avoiding ionic space charge.

The exponential dependence of current upon voltage in a thermionic regime was demonstrated with experimental punchthrough diodes [55, 56], and good agreement of experiment with theory in the various regimes of operation was reported. The voltage at which the thermionic regime of a punchthrough diode occurs is obviously designable. It can be at a few volts, as in Figure 3-40, or at thousands of volts [57], provided care is taken to design for a higher value of avalanche-breakdown voltage. In extreme cases, this requirement might lead to an asymmetric punchthrough diode. As a high-voltage and high-power voltage regulator, the punchthrough diode is superior to one employing avalanche breakdown [57]. The current density in a punchthrough device is quite uniform over its area, and this often is not the case in avalanching junctions, where "hot spots" can develop, effectively limiting the practical breakdown-voltage target to a few hundred volts.

The thermionic regime is bracketed in voltage by the reachthrough voltage on the low side, just before current starts to rise steeply, and the critical voltage cited earlier on the high side. This critical voltage was termed *punchthrough voltage* V_{PT} in the early literature [52], and *flatband voltage* V_{FB} in more recent literature [56], with the latter term implying complete elimination of the voltage barrier in the forward-biased junction.

Exercise 3-47. Explain the correspondence of a "flatband condition" and "elimination of the voltage barrier" in the forward-biased junction of the punchthrough diode.

Return to the potential profile in Figure 3-39(b). If applied voltage could be increased until the left-hand voltage barrier is reduced to zero height, then the potential profile at the left-hand boundary of the middle region would start out in a horizontal or "flat" manner.

The potential barrier in a forward-biased junction cannot actually be reduced to zero, either by direct biasing (as noted earlier) or in a punchthrough diode. The reason is that either current density will rise to a destructive level, or else other phenomena enter the picture, placing a cap on current. In the punchthrough diode, it is the space-charge limit on current that does so, and therefore V_{FB}, which has only theoretical significance, occurs in that regime. This relationship can be confirmed by calculating from data given in the literature [56], but is not emphasized or even pointed out there.

Child's law predicts a current–voltage relationship that goes as V^2/X^3, where V is applied voltage and X is the thickness of the middle layer. Thus current rise in the space-charge-limited regime, while no longer exponential, is still much faster than a linear or "ohmic" rise. A simplistic treatment of the space-charge-limited flow problem yields a functional dependence in agreement with Child's law [58].

Punchthrough diodes were first applied in integrated circuits as protective devices, for "clipping" voltage spikes. But now they are finding increasing use as voltage-regulating elements. Their advantage compared to series-connected forward-biased *PN* junctions is shown clearly in Figure 3-40 with respect to voltage-regulating ability. As noted earlier, the designability of the constant-voltage level is a further significant advantage. The property of designability is one that the punchthrough diode shares in principle with an avalanche-breakdown diode (Section 3-6.1), but design variation in the punchthrough case can be achieved by altering a dimension, whereas in the avalanche case, net doping must be precisely controlled, and that is often a more awkward requirement.

3-7 APPROXIMATE-ANALYTIC MODEL FOR THE STEP JUNCTION

A closed-form analytic expression can be obtained for electric field as a function of electrostatic potential in a step junction at equilibrium. But the subsequent integration needed to obtain electrostatic potential as a function of position can be carried out only for special and asymptotic cases. As a result, computer methods have been extensively employed to complete the task, starting in the 1950s. While these solutions are precise, they fail to provide the physical insight that is afforded by an analytic treatment, even an approximate-analytic treatment. The depletion approximation is an example of the latter case. As we have seen, it is primitive in formulation, but still very useful. Its popularity four decades after its introduction [5] confirms that even a coarse analytic solution is a useful supplement to an exact numeric solution. Recent work has provided another option, however, an approximate-analytic treatment that is almost as simple as the depletion approximation and almost as accurate as a computer solution.

3-7.1 The Poisson-Boltzmann Equation

The first workers to address the problem in this way were specifically analyzing semiconductor surface regions at equilibrium [7–9], but their results are equally applicable to the equilibrium step-junction problem that is our immediate concern. Following them, we must first formulate the appropriate differential equation, and then integrate it to obtain electric field as a function of electrostatic potential.

Exercise 3-48. The transport equation involves electric field and position. Can the junction problem be attacked by writing a transport equation and equating it to zero to define the equilibrium condition?

No. At equilibrium the two terms of a transport equation must be equal in magnitude. As a result the equation constitutes an identity, and therefore cannot be used to obtain a solution.

Exercise 3-49. Derive the identity described in Exercise 3-48.

Focus arbitrarily on electrons. Their transport equation at equilibrium is

$$q\mu_n n E + qD_n \frac{dn}{dx} = 0.$$

Or,

$$\mu_n n(-d\psi/dx) = -D_n(dn/dx).$$

Using the Einstein relation and $n = n_i \exp(q\psi/kT)$, gives

$$n_i \exp(q\psi/kT)(d\psi/dx) = (kT/q)(d/dx)n_i \exp(q\psi/kT);$$

hence

$$\exp(q\psi/kT)(d\psi/dx) = \exp(q\psi/kT)(d\psi/dx),$$

constituting the expected identity.

Exercise 3-50. Is the constant-E continuity-transport equation an appropriate starting point for equilibrium junction analysis?

No. The step-junction problem is *not* a constant-E problem.

What is needed is an equation in potential and its spatial derivatives that comprehends *space charge* as well. In short, Poisson's equation is appropriate, just as it was for initiating the depletion-approximation solution in Section 3-2. In fact, the outline there is identical to the one to be followed now.

The four basic approximations of Chapter 2 are explicitly retained here—Boltzmann statistics, equivalent densities of states, band symmetry, and complete ionization. From the last one it follows that in the neutral end regions of the junction sample, well away from the metallurgical junction,

$$N_D - N_A = n_0 - p_0. \tag{3-57}$$

The symbols n_0 and p_0 are as usual the constant, neutral-equilibrium values of carrier density found in a uniformly doped sample. Thus Equation 3-57 sets net impurity density equal to the difference of the majority- and minority-carrier densities, a relationship that must hold in a neutral sample. Near the metallurgical junction, however, the sample is *not* neutral, and the density of positive space charge can be written

$$\rho_v = q\{[p_0(x) - n_0(x)] - [N_D - N_A]\}. \tag{3-58}$$

The functional notation in the first term on the right-hand side is a reminder that these carrier densities, while equilibrium values, are spatially varying because of a departure from uniform doping in the sample of interest, notation carried over from Section 3-1.5 and Exercise 3-5. The departure here is of course the *PN* junction itself. Combining Equations 3-57 and 3-58 yields

$$\rho_v = q\{[p_0(x) - n_0(x)] - [p_0 - n_0]\}, \tag{3-59}$$

the expression needed for substitution in the right-hand side of Poisson's equation, yielding:

$$\frac{d^2\psi_n(x)}{dx^2} = -\frac{q}{\epsilon}\{[p_0(x) - n_0(x)] - [p_0 - n_0]\}. \tag{3-60}$$

Once again using the convenient normalized voltage $U_0 \equiv q\psi_0/kT$, and invoking Equations 2-33 and 2-34, which are based on Boltzmann statistics, we have for the second term in brackets

$$p_0 - n_0 = n_i(e^{-U_0} - e^{U_0}). \tag{3-61}$$

But recalling that $\sinh \lambda \equiv [\exp(\lambda) - \exp(-\lambda)]/2$, we can rewrite Equation 3-61 as

$$p_0 - n_0 = -2n_i \sinh U_0. \tag{3-62}$$

For the first term in brackets on the right-hand side of Equation 3-60 it is convenient to simplify notation by dropping both subscript and functional notation, letting $p_0(x) \equiv p$, and $n_0(x) \equiv n$; similarly, let $\psi_0(x) \equiv \psi$, and $U_0(x) \equiv U$, but remember that these quantities are spatially varying equilibrium values. With this change,

$$p - n = -2n_i \sinh U. \tag{3-63}$$

Since

$$\frac{d^2 U}{dx^2} = \frac{q}{kT} \frac{d^2 \psi}{dx^2} \tag{3-64}$$

from the definition of normalized potential, Equation 3-60 (Poisson's equation), can be rewritten as

$$\frac{d^2 U}{dx^2} = \frac{2qn_i}{\epsilon} \frac{q}{kT} (\sinh U - \sinh U_0). \tag{3-65}$$

With the definition

$$\left[\frac{\epsilon}{2qn_i} \frac{kT}{q} \right]^{1/2} \equiv L_{Di}, \tag{3-66}$$

Equation 3-65 finally becomes

$$\frac{d^2 U}{dx^2} = \frac{1}{L_{Di}^2} (\sinh U - \sinh U_0). \tag{3-67}$$

This expression is often identified as the *Poisson-Boltzmann equation,* a felicitous description in view of its derivation from Equation 3-60, Poisson's equation, and Equation 3-61, which is based upon Boltzmann statistics. The quantity L_{Di} is the *intrinsic Debye length,* a special case of the characteristic length we take up next.

3-7.2 Debye Length

In Section 2-5.1 we described the tendency of a fixed charge in a semiconductor sample to produce an enhanced density of opposite-type mobile charge in its vicinity. As a result, lines of force emanating from the fixed charge are terminated in the surrounding cloud, so that the fixed charge is "invisible" when "viewed" from an appropriate distance. This phenomenon is named *screening,* as was noted there, and it determines the distance at which a fixed charge can be "felt" in such a medium (to use another anthropomorphic term). When mobile charges of both sign are present, they both participate in the screening phenomenon; in the neighborhood of the fixed charge, density is depleted for like-sign mobile charges and enhanced for opposite-sign mobile charges. This problem was first addressed in the context of electrolytes by Debye and Hückel [59, 60]; hence the characteristic length associated with the screening phenomenon is known as the *Debye-Hückel length*, usually shortened simply to *Debye length*.

Let us move now to a one-dimensional problem. Think, for example, of a sheet of positive charge buried in a block of N-type silicon. Electron density on both sides of the sheet will exceed n_0, and can be assumed to decline exponentially with departure from the sheet. The constant with length dimensions that enters into the exponent can be described as a *characteristic screening distance*, or one-dimensional Debye length. We shall confine our attention to one-dimensional cases. (For a two-dimensional or three-dimensional problem, the term *Debye radius* is sometimes encountered.)

For a sample in which both holes and electrons have significant density, it is the *general Debye length* that is relevant:

$$L_D = \left[\frac{\epsilon}{q(n_0 + p_0)} \frac{kT}{q} \right]^{1/2} \tag{3-68}$$

This expression can be rewritten as

$$L_D = \left[\frac{\epsilon}{2qn_i} \frac{kT}{q} \frac{1}{\cosh U_0} \right]^{1/2} \tag{3-69}$$

by recalling that $\cosh \lambda \equiv [\exp(\lambda) + \exp(-\lambda)]/2$. Comparing this result with Equation 3-66 makes it evident that

$$L_{Di} = L_D \sqrt{\cosh U_0}, \tag{3-70}$$

a relationship that will be used below. Evidently the conversion from L_D to L_{Di} can be made easily by noting in Equation 3-68 that $(n_0 + p_0) = 2n_i$ in the intrinsic case, or in Equation 3-69, that $\cosh U_0 = 1$ in the intrinsic case. To repeat, the quantity L_{Di} is the *intrinsic Debye length*.

A further and more important fact is that for an extrinsic sample, either n_0 or p_0 will drop out of Equation 3-68. Hence, the *extrinsic Debye length* is

$$L_{De} = \left[\frac{\epsilon}{qn_{0m}} \frac{kT}{q} \right]^{1/2} \approx \left[\frac{\epsilon}{q|N_D - N_A|} \frac{kT}{q} \right]^{1/2}, \tag{3-71}$$

where n_{0m} stands for majority-carrier density, n_0 or p_0 as appropriate.

Exercise 3-51. For what physical reason do absolute-value bars enter Equation 3-71?

Net-doping *magnitude* is relevant here because carriers of either sign can accomplish screening. They move away from a like-sign fixed charge and toward an opposite-sign fixed charge.

The relationships among L_D, L_{Di}, and L_{De} are shown for silicon in Figure 3-41.

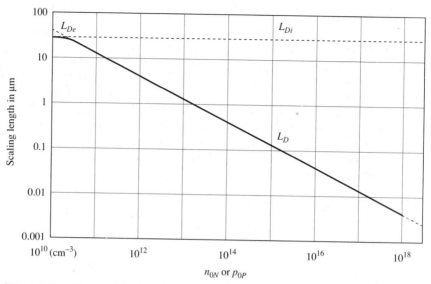

Figure 3-41 Three scaling lengths: the general (L_D), extrinsic (L_{De}), and intrinsic (L_{Di}) Debye lengths at 24.8 C as a function of majority-carrier density.

While we have not derived expressions for L_D and L_{De}, we showed in Section 3-7.1 that the quantity L_{Di} is "natural," in the sense that it appears spontaneously when the Poisson-Boltzmann equation, Equation 3-67, is derived. Notice in Figure 3-41 that the approximation $L_D \approx L_{De}$ is excellent for all but near-intrinsic conditions. For this reason, many authors ignore the distinction. However, when Debye length is used for normalization—our primary application of it—it is convenient to maintain the distinction. A physically meaningful length used as a normalizing constant is often described as a *scaling length,* for the reason that a well-chosen constant can simplify a problem dramatically.

Exercise 3-52. Carry out a full normalization of Poisson's equation. Assume the case of an extrinsic sample, and start with Poisson's equation in the form

$$\frac{dE}{dx} = \frac{qN}{\epsilon},$$

where $N \equiv |N_D - N_A|$. Use the thermal voltage and the extrinsic Debye length for normalizing potential and distance, respectively.

Multiply both sides of the given equation by L_{De} to normalize distance:

$$\frac{dE}{d(x/L_{De})} = L_{De}\frac{qN}{\epsilon}.$$

By definition,

$$E = -\frac{d\psi}{dx} = -\frac{kT}{q}\frac{dU}{dx} = -\frac{1}{L_{De}}\frac{kT}{q}\frac{dU}{d(x/L_{De})},$$

where potential is normalized in the second step and distance in the third. Substituting this result into the equation immediately preceding it yields

$$-\frac{d}{d(x/L_{De})}\left[\frac{dU}{d(x/L_{De})}\right]\frac{1}{L_{De}}\frac{kT}{q} = L_{De}\frac{qN}{\epsilon}.$$

Thus

$$\frac{d^2U}{d(x/L_{De})^2} = -L_{De}^2\frac{qN}{\epsilon}\frac{q}{kT}.$$

But in view of Equation 3-71,

$$\frac{d^2U}{d(x/L_{De})^2} = -1.$$

This result illustrates the power of normalization, and one of several interpretations of Debye-length significance. Refer back to Section 1-3.7, where it was pointed out that field-profile slope is proportional to space-charge density. In normalized form, Poisson's equation informs us that the field-profile slope is unity in a region of constant space-charge density. The electric field rises one normalized unit in each normalized unit of distance. This situation exists near a step junction, where the depletion of majority carriers leads to a constant space-charge density of magnitude qN.

Another interpretation of Debye length is that it is the distance over which carrier density changes by a factor of e in a screening situation, a point made previously. Yet another, and nonobvious, meaning is the distance that a majority carrier can travel in one dielectric-relaxation time t_D when it moves in an uninterrupted path at its thermal velocity v_t [61]. This link to dielectric-relaxation time is significant. We noted that t_D depends on permittivity ϵ and resistivity ρ, and the latter is closely related to carrier density. Debye length at a fixed temperature also depends on these two factors, permittivity and mobile-charge density.

3-7.3 First Integration of Poisson-Boltzmann Equation

Just as there are an infinite number of possible step junctions, there would seem to be an infinite number of accompanying field profiles. But several researchers noticed that redundancy among the possible solutions is vastly reduced when either the general or the extrinsic Debye length is employed for spatial normalization [12–14]. In each case, curiously, the discovery was made independently. The fact that observation was not made by the large number of additional workers who addressed the step-junction problem is partly explained by the fact that in the Poisson-Boltzmann equation (repeated below for convenient reference), it is the *intrinsic* Debye length that in essence volunteers to serve for normalization:

$$\frac{d^2U}{dx^2} = \frac{1}{L_{Di}^2} (\sinh U - \sinh U_0). \tag{3-72}$$

It is desirable to define sample geometry explicitly, as has been done in Figure 3-42; this choice, and the notation and approach that will be used, follow prior practice [14]. Note that normalized potential U (a spatially varying but equilibrium quantity) increases downward, and the Fermi level is used as reference. The junction position is x_J, and the potential there is U_J. Focus on the right-hand side of the sample. Our object is to integrate Equation 3-72 from the end region where $U = U_{02}$ to an arbitrary position x where the potential is U. The left-hand side of Equation 3-72 is rendered easily integrable by means of a straightforward manipulation:

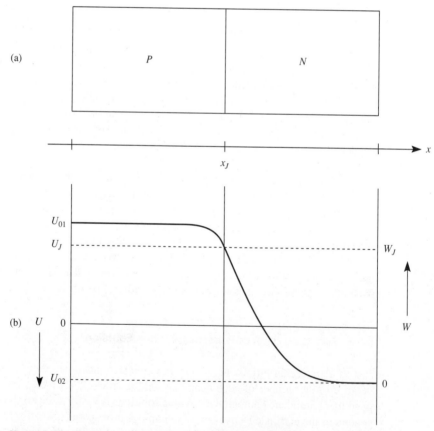

Figure 3-42 Step-junction sample of the specific orientation for explaining the approximate-analytic solution. (a) Physical representation. (b) Normalized potential as a function of position.

$$\int \frac{d^2U}{dx^2} dU = \int \frac{d}{dx}\left(\frac{dU}{dx}\right) dU = \int \left(\frac{dU}{dx}\right) d\left(\frac{dU}{dx}\right). \tag{3-73}$$

Substituting this revised expression for the left-hand side of Equation 3-72 makes it clear that consistent limits for integration are from zero to the variable value (dU/dx) on the left-hand side, and from U_{02} to the corresponding variable value U on the right-hand side. Choosing U' as a dummy variable places Equation 3-72 in shape for integration:

$$\int_0^{dU/dx} \left(\frac{dU'}{dx}\right) d\left(\frac{dU'}{dx}\right) = \frac{1}{L_{Di}^2} \int_{U_{02}}^{U} (\sinh U' - \sinh U_{02}) dU'. \tag{3-74}$$

Integrating, then, gives

$$\frac{1}{2}\left(\frac{dU}{dx}\right)^2 = \frac{1}{L_{Di}^2}[(U_{02} - U) \sinh U_{02} - (\cosh U_{02} - \cosh U)], \tag{3-75}$$

and extracting the square root of this result yields an expression for partly normalized electric field as a function of normalized potential:

$$\frac{dU}{dx} = \pm \frac{\sqrt{2}}{L_{Di}}[(U_{02} - U) \sinh U_{02} - (\cosh U_{02} - \cosh U)]^{1/2}. \tag{3-76}$$

The positive sign is appropriate for the situation shown in Figure 3-42, where dU/dx and x were made consistent in sign by choosing to have $U_{01} < U_{02}$.

Now, for reasons illustrated in Section 3-7.2, use Equation 3-70 to eliminate L_{Di} from Equation 3-76, and to introduce the general Debye length L_D, yielding

$$\frac{dU}{d(x/L_D)} = \pm \sqrt{2} \left[\frac{(U_{02} - U) \sinh U_{02} - (\cosh U_{02} - \cosh U)}{\cosh U_{02}}\right]^{1/2}. \tag{3-77}$$

Next, convert from hyperbolic-function notation to exponential notation:

$$\frac{dU}{d(x/L_D)} = \pm \sqrt{2} \left[\frac{(U_{02} - U)(e^{U_{02}} - e^{-U_{02}}) - (e^{U_{02}} + e^{-U_{02}} - e^U - e^{-U})}{e^{U_{02}} + e^{-U_{02}}}\right]^{1/2}. \tag{3-78}$$

Now it is convenient to introduce a new normalized potential,

$$W \equiv U_{02} - U. \tag{3-79}$$

The effect of this transformation is displayed in Figure 3-42. First, the potential origin is switched from the Fermi level to the equilibrium potential U_{02} in the end region, often called the *bulk potential*. The term "bulk" is meant to convey a region well away from disturbances (such as a junction) so that neutral-equilibrium conditions exist therein. Second, W and U are opposite in sign. Substituting the quantity $U = U_{02} - W$ into Equation 3-78 yields the result sought,

$$\frac{dW}{d(x/L_D)} = -\sqrt{2} \left[\frac{e^{U_{02}}(e^{-W} + W - 1) + e^{-U_{02}}(e^W - W - 1)}{e^{U_{02}} + e^{-U_{02}}}\right]^{1/2}, \tag{3-80}$$

an exact, closed-form expression for normalized electric field as a function of normalized electrostatic potential W. The negative sign is appropriate for the conditions of Figure 3-42.

More useful, to be sure, would be a relationship between electric field and position, but computer intervention is necessary to get there, as will be described in Section 3-7.4. The result of that procedure is shown in Figure 3-43 [62]. Observe the extensive region wherein normalized electric field versus normalized position exhibits a slope of unity magnitude. This is the deeply depleted region discussed in Section 3-7.2, and the field profile can legitimately be described as a "universal" curve, applicable in part to any junction. The term *general solution* has also been used to emphasize its universality [14].

An extrapolation of the linear part of the curve to the spatial axis defines a spatial origin that will be exploited below. The origin falls in the vicinity of the space-charge-layer boundary—the region of transition from neutrality to deep depletion. The utility of this particular origin choice can be illustrated by recalling

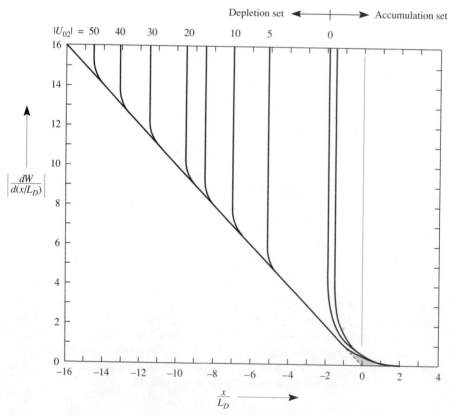

Figure 3-43 Normalized electric-field magnitude as a function of normalized distance, showing spatial-origin definition in terms of linear-profile extrapolation [after 62].

Poisson's equation in fully normalized form, derived in Section 3-7.2 (Exercise 3-49). It can be written, with separated variables, as

$$d\left[\frac{dU}{d(x/L_D)}\right] = -d(x/L_D). \tag{3-81}$$

(For increased generality, L_D has been substituted for L_{De}.) Obviously this expression can be integrated with ease to yield normalized electric field versus normalized position. If arbitrary limits are used, the result will include a constant of integration. But with the origin choice just made we have coordinated limits—extrapolated electric field vanishes at the positional zero. Under these conditions, the result of integration is

$$\frac{dU}{d(x/L_D)} = -\frac{x}{L_D}, \tag{3-82}$$

which is valid in the region of deep depletion where the field profile is linear. The negative sign is consistent with the case shown in Figure 3-43. For P-type material on the right-hand side of the junction (instead of the choice made in Figure 3-42), the sign would of course be positive.

This issue of algebraic sign is easily resolved by physical reasoning. More serious issues exist in Equation 3-80, however, where U_{01}, U_{02}, and W each can be either positive or negative. Therefore it has been shown to be worthwhile to rewrite Equation 3-80 in a way that eliminates sign ambiguity [63]:

$$\left|\frac{dW}{d(x/L_D)}\right| = \sqrt{2}\left[\frac{e^{|U_{02}|}(e^{-W} + W - 1) + e^{-|U_{02}|}(e^W - W - 1)}{e^{|U_{02}|} + e^{-|U_{02}|}}\right]^{1/2}. \tag{3-83}$$

In this form, the electric field expression describes depletion conditions (consistent with Figure 3-43), and is valid for U_{02} positive or negative (an N-type or P-type region on the right-hand side of the junction), with the further proviso that W should be taken as positive always. Absolute-value signs have been placed on the normalized-field symbol, however, as a reminder that algebraic sign for the overall expression must be established, to repeat, by physical reasoning.

A prominent feature of Figure 3-43 is the family of near-vertical curves that proceed smoothly upward from all portions of the primary curve. These are a consequence of the phenomenon of *inversion* that will be addressed in Section 3-7.6. In addition, the dashed lines shown near the spatial origin in Figure 3-44 will be explained in Section 3-7.5.

3-7.4 Second Integration of Poisson-Boltzmann Equation

The next aim is to determine normalized potential W as a function of normalized position x/L_D. Hence the first step is to separate variables in Equation 3-83, yielding

$$|d(x/L_D)| = \frac{1}{\sqrt{2}}\left[\frac{e^{|U_{02}|} + e^{-|U_{02}|}}{e^{|U_{02}|}(e^{-W} + W - 1) + e^{-|U_{02}|}(e^W - W - 1)}\right]^{1/2} dW. \tag{3-84}$$

It has been determined numerically that the value of W at the spatial origin in Figure 3-43 is

$$W(0) = 0.59944 \approx 0.6. \tag{3-85}$$

Thus consistent limits are from zero to an arbitrary value of x/L_D, and from $W(0)$ to an arbitrary (and positive) value of W. Carrying out this numerical integration yields the result shown in Figure 3-44. In the spatial interval where the field profile is linear, the potential profile is of course purely parabolic. For now we will concentrate on this portion of the solution and again defer consideration of the steeply rising portions.

To apply these findings, let us choose the simplest case first, namely the symmetric step junction. As a result, in Figure 3-42, the junction potential $U_J = 0$, or $W_J = U_{02}$. Of primary interest is the carrier profile $n_0(x)$, which for convenience will once again be labeled simply as n. Knowledge of the density profile comes immediately from the potential profile. From the definition of W, Equation 3-79, we have

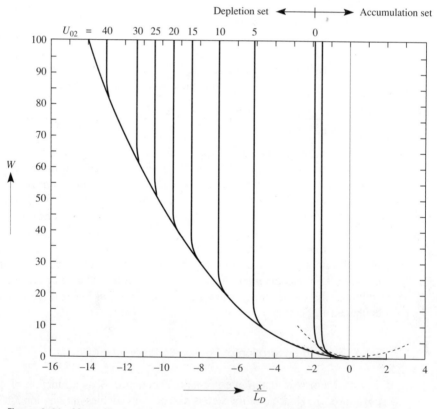

Figure 3-44 Normalized potential as a function of normalized distance for a step junction with various values of normalized bulk potential U_{02} on the right-hand side [15].

APPROXIMATE-ANALYTIC MODEL FOR THE STEP JUNCTION

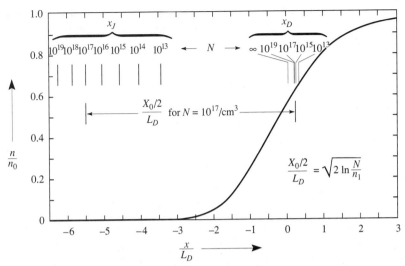

Figure 3-45 Relative majority-carrier density as a function of position in a symmetric step junction. Depletion-approximation boundary position x_D and junction position x_J are shown for several values of net doping N [15].

$$W \equiv U_{02} - U = \ln \frac{n_n}{n_i} - \ln \frac{n}{n_i} = \ln \frac{n_n}{n}. \quad (3\text{-}86)$$

Consequently,

$$\frac{n}{n_0} = e^{-W}. \quad (3\text{-}87)$$

Plotting this function with the aid of the numerically determined function $W(x/L_D)$ yields the result shown in Figure 3-45. The single curve is valid for all possible symmetric step junctions. The position of this curve in relation to the metallurgical junction, however, is dependent on doping, and the label x_J shows junction position for several net-doping values.

Exercise 3-53. Comment on a counterintuitive feature of Figure 3-45.

We observed that the depletion approximation improves with reverse bias, because depletion-layer thickness increases while the boundary-region profile is virtually unaltered. As the common net doping is diminished in a symmetric junction, the depletion layer again gets thicker, but this time the depletion approximation gets *worse*; Figure 3-45 shows the junction position x_J advancing toward the boundary region. But because distance is normalized in this diagram, what it tells us is that the depletion layer *does* get thicker but the boundary-region profile gets "fuzzier" at an even more rapid rate.

As a clincher one can use a *reductio-ad-absurdum* argument. In the limit of light doping, the junction sample becomes intrinsic throughout and space charge disappears altogether. Thus it is plausible that as this condition is approached, the region of deep depletion must gradually disappear, as Figure 3-45 indicates.

Moving in the other direction, we see the depletion approximation improving with increased doping. But in the limit, a practical fact enters. If even slight impurity diffusion has occurred during junction formation, the distance over which the doping transition occurs is no longer negligible compared to a Debye length. Consequently the structure can be more accurately modeled as a linearly graded junction than as a step junction [44]. It was shown in Figure 3-26 that in such a case, the larger the density gradient, the better the accuracy of the depletion approximation.

On the right-hand side of Figure 3-45 is indicated just where the sharp depletion-layer boundary defined by the depletion approximation will fall; in this comparison of exact and depletion-approximation results, the two have been adjusted for equal contact potential. The depletion-approximation boundary is near, but not precisely at $x = 0$, although it tends toward $x = 0$ as $|N_D - N_A| = N$ approaches infinity.

The same exact solution can be applied to the unsymmetric step junction, as is shown in Figure 3-46. This time the right-hand side of the junction is taken as having a constant and "typical" doping of $10^{16}/cm^3$. When the left-hand side is also $10^{16}/cm^3$, we have a symmetric junction and a position x_J corresponding to a doping ratio of unity. The correctness of the x_J position in this case can be verified by referring back to Figure 3-45 for the case $N = 10^{16}/cm^3$. Next let the left-hand side doping be decreased by successive factors of ten. Once again the x_J position advances toward the depletion-layer boundary. Recall the point made in Section 3-2.7 to the effect that potential drop on the high side of a grossly asymmetric junction assumes a limiting value of kT/q, or one normalized unit. But when $W = 1$, Equation 3-87 yields

$$\frac{n}{n_0} = e^{-1} = \frac{1}{e}, \quad (3\text{-}88)$$

so that

$$n = n_0/e. \quad (3\text{-}89)$$

This point is indicated in Figure 3-46, corresponding to the indicated doping ratio of infinity.

3-7.5 Depletion-Approximation Replacement

The universality of the curves in Figures 3-43 through 3-46 suggests the writing of approximate-analytic expressions to describe them. This is a task that was greatly simplified by defining the spatial origin that appears in Figures 3-43 through 3-46.

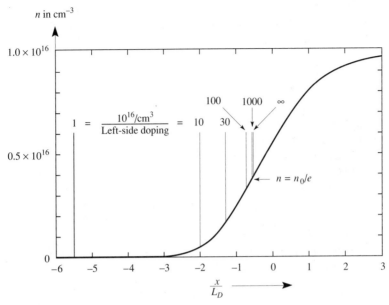

Figure 3-46 Majority-carrier density as a function of normalized position for asymmetric step junctions, showing junction position x_J for various doping ratios.

The expressions that result have been collectively named a *depletion-approximation replacement* [15], sometimes abbreviated *DAR*.

To start, examine the potential profile in Figure 3-44 and note the dashed lines near the spatial origin. The one projecting toward the right is the right-hand branch of the parabolic function that accurately gives $W(x/L_D)$ in the deep-depletion regime toward the left. The vertex of this parabola has the coordinates (0, 1), rather than (0, 0) as one might expect. That is, the vertex is shifted upward on the W axis, falling one unit above the origin of the diagram. The parabola so positioned is an accurate asymptote in the deep-depletion regime toward the left, and the function can be written

$$W = \frac{1}{2}\left(-\frac{x}{L_D}\right)^2 + 1. \tag{3-90}$$

Exercise 3-54. Examine the field profile in Figure 3-43 and make an observation that stems from the fact that $W = 1$ in Equation 3-90 when $x = 0$.

Near the spatial origin, the field profile deviates slightly from the linear asymptote, producing the familiar "skirt." The present observation tells us, recalling that $\Delta\psi = -\int E dx$, that the small shaded area shown there amounts to one normalized unit, or kT/q.

The dashed curve projecting toward the left in Figure 3-44 is asymptotically valid also, and applies in the region of near neutrality, toward the right (or where $W \ll 1$). It is an exponential function, and can be written

$$W = \exp\left(-\frac{x}{L_D} - 0.41209\right). \qquad (3\text{-}91)$$

It is evident that the spatial origin could have been chosen in a manner that eliminates the second term in the exponent, but at the expense of complicating Equation 3-90. We did not take that option, however, because of the useful and clear meaning of having extrapolated field vanish at the origin, as in Figure 3-43.

For a clearer comparison of the two asymptotes, it is helpful to expand the scale and examine them only near the spatial origin. Further, as shown in Figure 3-47, a semilog presentation displays the exponential asymptote in linear fashion. Both asymptotic curves are plotted as solid lines, and are individually valid outside the boundaries of the rectangle sketched inside the diagram. The numerical solution that smoothly joins them at opposite corners of the rectangle is also shown with a solid curve. But then it becomes evident that a further approximation is possible. The exponential expression plotted as a dashed line nearly matches the intervening numerical solution (inside the rectangle), and hence can serve as a kind of "bridge" approximation between the asymptotes. This third equation is

$$W = (0.582)\exp(-0.9x/L_D). \qquad (3\text{-}92)$$

Similar treatment can be accorded to electric field and other important variables in the problem [62]. Further, a much more detailed and extended treatment of subject matter in all of Section 3-7 is available [64].

For step junctions that are symmetric or only modestly asymmetric, the three approximate-analytic expressions, Equation 3-89 through 3-92, are sufficient. They are to be applied to one side of the junction at a time, producing an overall approximate solution. The only information lacking at this point in order to tie the two together is the value of junction potential U_J, but it can be readily calculated [14]:

$$U_J = \frac{\cosh U_{02} - \cosh U_{01} + U_{01}\sinh U_{01} - U_{02}\sinh U_{02}}{\sinh U_{01} - \sinh U_{02}}. \qquad (3\text{-}93)$$

But as asymmetry increases, it becomes necessary to deal with the additional phenomenon that we address next.

3-7.6 Inversion Layer and Accumulation Layer

Let us first take the extreme case of a grossly asymmetric step junction exhibiting the limiting high-side potential drop of kT/q. Suppose net doping in the P-type region on the left is $10^{13}/\text{cm}^3$, and in the N-type region on the right is $10^{16}/\text{cm}^3$. Insofar as the N-type region is concerned, this situation is depicted in Figure 3-46, where the indicator for a doping ratio of 1000 falls roughly at the point where

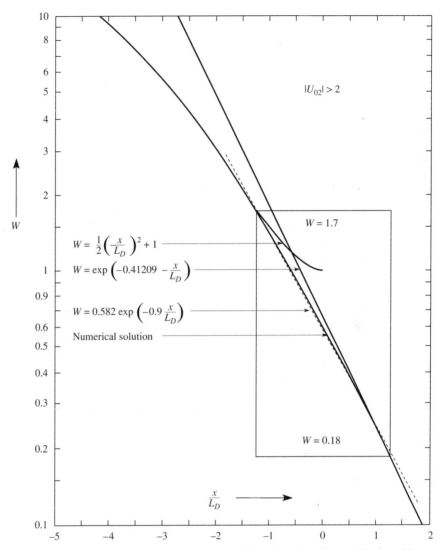

Figure 3-47 Semilog plot of normalized potential as function of normalized position showing three approximate-analytic functions. Useful and nonuseful (extrapolated) portions of asymptotic functions are shown with solid lines, as is numerical solution. Approximate ''bridge'' expression for use inside the rectangle is shown with a dashed line [15].

$n = n_0/e$. In other words, in the example chosen, the electron density right at the metallurgical junction *on the P-type side* is $n_0(\approx 0) \approx (0.37) \times 10^{16}/\text{cm}^3$, because $n_0(x)$ must be continuous. The density profile then declines in some manner toward the left, eventually reaching its minority-carrier equilibrium density of $n_{0P} = (n_i^2/p_{0P}) = 10^7/\text{cm}^3$. Thus we have here a situation not treated before. There are electrons present in a *P*-type region *at equilibrium* at densities orders of magnitude larger than $p_{0P} = 10^{13}$/cm. That is, we deal with a region having *P-type* net doping in which *electrons* make a vastly greater contribution to conductivity than do holes.

Exercise 3-55. What is the hole density where $n_0(\sim 0) \approx 0.37 \times 10^{16}$/cm^3?

Because equilibrium obtains,

$$p_0(\sim 0) = \frac{n_i^2}{n_0(\sim 0)} = 2.7 \times 10^4 \text{/cm}^3.$$

This situation is often described by saying that conductivity type is *inverted* in this region, and so the layer of electrons in the *P*-type region is termed an *inversion layer*. Also, the term *surplus carriers* has been introduced to distinguish such "extra" carriers clearly from excess carriers [22].

Exercise 3-56. What is the distinction between surplus carriers and excess carriers?

Surplus carriers exist in a region of gross space charge. They constitute a case of *equilibrium nonneutrality*. Excess carriers, in the simplest possible case, like that depicted in Figure 2-26, constitute a case of *nonequilibrium neutrality*.

An asymmetric junction is shown in Figure 3-48 that is similar to the one just considered [64]. A doping ratio of 100 has been chosen this time, as well as lower and (unrealistic) doping values, because so doing simplifies construction and scaling in the diagram. This can be seen in the fact that normalized scaling of distance on the two sides differs by a factor of ten, because $L_D \propto 1/\sqrt{N}$. The location of the inversion layer is displayed at the left of the diagram. Notice that $n_0(x)$ in the inversion layer does not follow the universal curve, but lies above it.

Exercise 3-57. Why must the inversion layer profile lie above the "universal" curve?

The latter curve is valid when electric field continues to rise linearly toward the left. But in the present case, field peaks at the metallurgical junction and then *declines* toward the left. Declining field means declining drift, which in turn requires declining diffusion in this equilibrium situation, and that in turn requires declining density gradient, which finally brings us to an electron profile that must now be higher.

The situation here is closely related to that in the high-low junction, depicted (with reverse orientation) in Figure 3-32. There, too, electrons have poured over to the lightly doped side. These surplus carriers do not, however, constitute an inversion layer, but rather an *accumulation layer,* because they are in a same-type region. This fact makes an important difference in the two situations, specifically in the contact potential.

Exercise 3-58. Compare $\Delta\psi_0$ for a *PN* junction and a high-low junction in which the respective net dopings are $10^{18}/\text{cm}^3$ and $10^{14}/\text{cm}^3$.

For the *PN* junction

$$\Delta\psi_0 = \frac{kT}{q} \ln \frac{N_D N_A}{n_i^2} = 0.71 \text{ V},$$

where N_D and N_A are net-doping values on the two sides. For the high-low junction

$$\Delta\psi_0 = \frac{kT}{q} \ln \frac{N_H}{N_L} = 0.24 \text{ V},$$

where N_H and N_L are the net-doping values on the high and low sides, respectively.

Exercise 3-59. In each case just given, the high-side potential drop will be about one thermal voltage, and the low side has a thin, dense layer of carriers concentrated adjacent to the metallurgical junction. What accounts for the difference in $\Delta\psi_0$ values?

In the *PN* junction, the low side exhibits a depletion layer that is *very* thick compared to the inversion layer. This situation is depicted in Figure 3-16; although no mention of the inversion layer was made there, it can be easily seen in the electron profile. *Most* of the charge on the P-type side is in the inversion layer, and only a small fraction is in the ionic (depletion) layer. But charge placement has a major effect on voltage, with the long lines of force reaching out to the ionic charges leading to a large value for the field integral. In the high-low junction, all the low-side charge is in the accumulation layer and there is *no* depletion layer on the lightly doped side, so the contact potential is smaller—almost three times smaller in this example.

Now refer back to Figure 3-16, which represents a one-sided *PN* junction with a low-side doping of $10^{15}/\text{cm}^3$. It can be seen that the electron density in the *P*-type region at equilibrium rises gradually as the junction is approached. But when it reaches the point at which

$$n_{OP}(x) = p_{OP}, \tag{3-94}$$

it exhibits sharply increased slope. Equation 3-94 defines a condition known as the *threshold of strong inversion*. The corresponding position is sometimes termed the threshold-of-inversion plane. It is evident in Figure 3-16, and is explicitly labeled in Figure 3-48. Other ways of stating the threshold condition are (assuming it is on the right-hand side of Figure 3-42)

$$U = -U_{02},\qquad(3\text{-}95)$$

and equivalently,

$$W = 2U_{02}.\qquad(3\text{-}96)$$

It is because an inversion layer is a thin, dense sheet (nearly) of charge that electric field departs from the universal curve in Figure 3-43 at the threshold plane. Recall that Poisson's equation dictates a field slope proportional to charge density. Refer to Figure 3-44 to observe that the departure point of the inversion-layer curve is at $W = 20$ for $U_{02} = 10$, at $W = 40$ for $U_{02} = 20$, and so forth, consistent with Equation 3-96. The inversion-layer representation in Figure 3-48 does not appear steep because the positional scale has been grossly expanded. A quantitative look at this will be provided in an analytic problem.

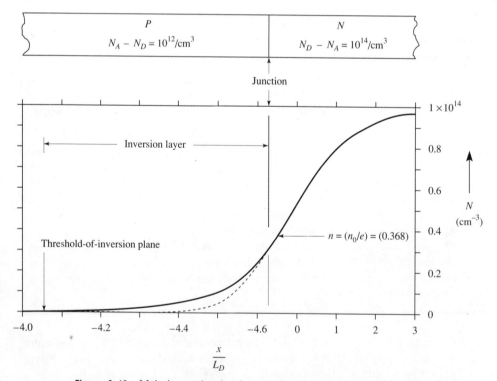

Figure 3-48 Majority-carrier density as a function of normalized position for an asymmetric step function possessing an inversion layer [64].

Inversion layers also lend themselves to approximate-analytic modeling [65, 66]. The expressions are more complex than those above, but are proving useful [67]. These approximate-analytic functions describe the family of curves in Figure 3-44 to the left of the curve for $U_{02} = 0$. This "dividing-line" case of course represents a sample that is extrinsic on the left and intrinsic on the right, a special case that can be treated exactly [68]. Moving to the right of the $U_{02} = 0$ curve brings us into the "accumulation set," or the case of a high-low junction that therefore incorporates an accumulation layer. Because the phenomenon of depletion is now no longer possible, these cases quickly converge on another universal curve. That is, the indicated curve is valid for any right-hand side bulk potential $|U_{02}| > \sim 2$.

Thus, an inversion layer that exists on the lightly doped side of a grossly asymmetric junction is a thin, dense layer of carriers of the type that constitute majority carriers on the heavily doped side. It exists within a thicker layer of ionic charge of the same polarity. In a high-low junction, there is an accumulation layer on the low side, while the ionic charge (depletion layer) is totally absent there.

In addition to the problems of potential, field, and charge-density profiles at equilibrium, certain nonequilibrium problems can be addressed by these methods as well. For example, an extension has been made to step junctions under forward bias [69]. It is likely that the approximate-analytic techniques of Section 3-7 will be applied in the future to problems of increasing range.

Finally, let us note that the existence of simple analytic expressions describing the "universal" portions of functions such as potential and electric field versus position (a universality achieved by appropriate normalization) is a consequence of the prudent choice of spatial origin. To illustrate the nontrivial nature of origin selection, we pointed out in Section 1-2.4 that the essence of the shift from an obscure Ptolemaic model of the solar system to the simple and elegant Copernican model was the matter of spatial-origin choice. We repeat the observation here, even at the risk of making too grand a comparison.

3-8 SMALL-SIGNAL DYNAMIC ANALYSIS

We have seen that the forward characteristic of a *PN* junction is grossly nonlinear; in fact, it approximates a pure-exponential function through several decades of current. Not surprisingly, the many solid-state devices that incorporate *PN* junctions exhibit comparable nonlinearity in some of their properties. The methods of small-signal analysis can be applied to nonlinear devices under certain circumstances. The term *signal* in this context means a current or voltage change or excursion. As noted earlier, a reasonable question about the adjective *small-signal* is, "How small is small?" The answer is that the signal should be small enough so that the portion of the static characteristic (or dc characteristic) that is traversed when the signal is applied can be approximated by a straight line with acceptable error. For this reason, the application of small-signal methods to a nonlinear problem is often described as *linearization*.

To the small-signal restriction, which applies throughout Section 3-8, we add the restriction of quasistatic conditions in Sections 3-8.1 through 3-8.3. Let us define the adjective *quasistatic* to mean "involving voltage and current variations so slow that a capacitor may be treated accurately as an open circuit." For greater generality one could add, "and an inductor may be treated accurately as a short circuit." (Although the solid-state devices of interest to us possess some inductive properties, their capacitive properties have much more general importance.) In a still more general way, *quasistatic* describes a condition that is infinitesimally removed from equilibrium. The same symbols will be used for quasistatic variables as for static variables, namely, capital letters with capital subscripts, if any. Examples are V_{NP}, Q_S, I, J, and J_D. This brings us to a point of some subtlety. Although reactive effects can be ignored under quasistatic conditions, the choice of such conditions is ideally suited to describing and defining dynamic capacitance, as well as dynamic resistance and conductance, which we now proceed to do.

3-8.1 Small-Signal Conductance

In a typical case, the forward bias on a silicon *PN* junction is greater than 0.35 V. Simple theory yields a good description of junction behavior in such a case, short of the high-level regime, for a junction like that represented in Figure 3-24. Hence Equation 3-40 can be realistically used to answer the following question: What small-signal (or incremental, or dynamic, or ac) resistance is exhibited by a silicon *PN* junction biased at a specific voltage lying between 0.35 V and 0.7 V? Or in algebraic terms, what is the value of $r = (dV/dI)$ for a particular point on the static *I–V* characteristic? Small-signal conductance g, the inverse of r, can be found directly from Equation 3-40. First, rewrite it in the form

$$I_0 \exp(-qV_{NP}/kT) = -I + I_0 \approx -I, \qquad (3\text{-}97)$$

where the approximate equality is valid for a forward-bias magnitude in excess of about 0.1 V. Since the range of specific interest lies above 0.35 V, this is an unimportant constraint. Because forward current and voltage are both negative for the *NP* junction under examination,

$$g = \frac{d(-I)}{d(-V_{NP})} = \frac{q}{kT} I_0 \exp(-qV_{NP}/kT) = \frac{q}{kT}(-I) = \frac{1}{r}. \qquad (3\text{-}98)$$

This is a striking result. Small-signal conductance in the forward-biased junction depends *only* on the current passing through it. It does *not* depend on junction area. Nor does it depend on whether the sample in question is made of gallium arsenide, silicon, or germanium! A qualitative appreciation for the latter fact can be derived from Figure 3-49(a). It repeats the forward-characteristic comparison of Figure 3-20, but more precisely (Figure 3-20 was intentionally distorted), and focuses on a particular value of current. The curves for all three materials exhibit the same slope at that current. In other words, the curves all have identical shape, and

SMALL-SIGNAL DYNAMIC ANALYSIS   349

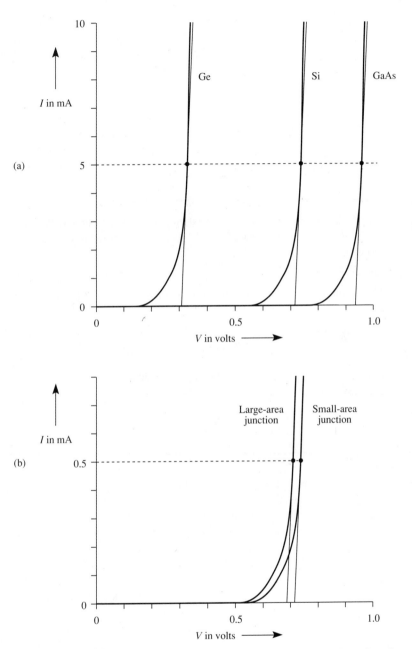

Figure 3-49 Dynamic resistance in a forward-biased *PN* junction depends only on current. (a) The effect of changing junction material. (b) The effect of changing junction area.

are translated from the origin by an amount (the "offset voltage") that increases with energy gap ξ_G. The application of a small signal at a particular current involves the manipulation of the voltage barrier, a phenomenon that is material-independent. For a current small enough to avoid parasitic-resistance effects in the junction sample, it is the behavior of the barrier alone that determines small-signal resistance.

In a similar way, changing junction area in a particular material produces a small change in offset voltage, but does not alter curve shape. This is illustrated for silicon in Figure 3-49(b). The explanation for this observation was given in Section 3-4.5; the voltage difference is dictated by the law of the junction, as was demonstrated graphically in Figure 3-23.

Exercise 3-60. Calculate the small-signal resistance of a forward-biased *PN* junction carrying 25.7 mA, and carrying 25.7 µA. Do so for germanium, silicon, and gallium arsenide.

The result will be the same for all three materials. From Equation 3-97, we will have

$$r = [kT/q(-I)] = \frac{25.7 \text{ mV}}{25.7 \text{ mA}} = 1 \text{ ohm}$$

in the first case. This clearly illustrates the steepness of the *I–V* characteristic at ordinary current values. In the second case,

$$r = \frac{25.7 \text{ mV}}{25.7 \text{ µA}} = 1000 \text{ ohm}.$$

The concept of small-signal resistance just developed is valid for both periodic and aperiodic signals. In the periodic case, the quasistatic restriction means extremely low frequency, of course. In the aperiodic case, the analogous assumption is that the voltage (or current) excursion is extremely slow. The following two sections (3-8.2 and 3-8.3) describe junction-capacitance effects that must be negligible if small-signal resistance alone is considered to characterize the junction dynamically. But in addition to this quasistatic application of the *r* and *g* concepts, there is an obvious pure-static application: We can now calculate the resistance value needed to complete the piecewise-linear model described in Section 3-4.4.

Suppose, for example, we are given the *I–V* characteristic in Figure 3-50(a) and wish to approximate it when the bias point of interest is at $I = I_1$. Let the corresponding junction voltage be $V = V_1$. Drawing a tangent to the characteristic at the bias point serves to define an additional voltage $V_{\gamma 1}$, the intercept of the tangent on the voltage axis. If, instead, we had been concerned with operation at the bias point (I_2, V_2), then a different and steeper tangent would be appropriate, as can be seen in Figure 3-50(a).

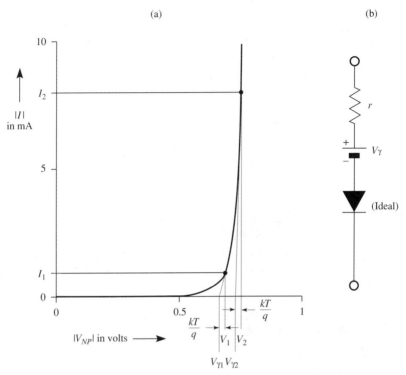

Figure 3-50 Models for the *PN*-junction diode. (a) Piecewise-linear models for operation at (I_1, V_1) and at (I_2, V_2). In each case, the voltage axis is a second element of the model. (b) Equivalent-circuit model corresponding to the piecewise-linear models in part (a). The diode shown is ideal in the sense of Figure 3-21(a).

Exercise 3-61. Derive expressions for $V_{\gamma 1}$ and $V_{\gamma 2}$.

Let r be the dynamic resistance of the I–V characteristic at a particular operating point, as just determined. Then

$$V - V_\gamma = |I|r = |I|(kT/q|I|) = (kT/q).$$

Thus, apparently $V - V_\gamma$ is independent of operating point for an ideally exponential forward characteristic. In other words, the voltage V_γ always lies one thermal voltage below the bias voltage, as is indicated in Figure 3-50(a).

The piecewise-linear model for the forward characteristic in Figure 3-50(a) can be converted into a static and low-frequency equivalent-circuit model by using the resistor r in series with a battery V_γ, the upper two elements in Figure 3-50(b). This provides two linear segments in the I–V characteristic, one lying along the voltage axis, and the other extending upward from V_γ. For smaller values of forward voltage in a silicon junction (<0.35 V), the recombination current is dominant, a fact that is evident in Figure 3-24. But the value of g (or r) derived from ideal-junction theory is still useful through part of the low-voltage range. Observe in Equation (3-46) that a recalculation of g in a regime where recombination is dominant simply introduces a small factor into its value. For reverse bias and very small forward bias, conductance computed from ideal-junction theory is not meaningful. Fortunately, in many cases this regime of junction operation can be modeled simply as an open circuit. Under these conditions, the equivalent-circuit model can be modified to handle the reverse characteristic as well as the forward by employing an ideal diode as a third series element, as is also shown in Figure 3-50(b).

If more accuracy is absolutely required in the reverse and low-voltage forward regimes, the piecewise-linear and equivalent-circuit models can be further modified. A large but finite resistance can be placed in shunt across the three components shown in series in Figure 3-50(b) without appreciably upsetting the relationships just established. This simulates the small currents that exist in those bias regimes. We repeat that this kind of elaboration should be avoided *unless* necessary to the application, with the trade-off that always exists between accuracy on the one hand and model convenience and utility on the other.

Now return to the forward-voltage regime where simple theory yields an accurate value for small-signal resistance. It is time for a quantitative look at the error that accompanies linearization. For simplicity, let us first consider a small, rapid increase in forward voltage—a switching transition.

Exercise 3-62. Determine the percentage error involved in linearly approximating the junction current increment that results from a 10-mV voltage increment. Assume in this case that the forward voltage is 0.5 V, and that the junction obeys simple theory, with $I_0 = 10^{-14}$ A.

The current will be
$$I = -I_0 e^{-qV_{NP}/kT} = -(10^{-14} \text{ A})e^{(0.5 \text{ V}/0.02566 \text{ V})} = -2.90 \text{ }\mu\text{A}.$$
Thus the small-signal conductance is
$$g = (-qI/kT) = (2.9 \times 10^{-6} \text{ A})/(0.02566 \text{ V})$$
$$= 1.13 \times 10^{-4}/\text{ohm},$$
and the resulting current increment is
$$\Delta I_\text{approx} = g\Delta V = (1.13 \times 10^{-4}/\text{ohm})(10^{-2} \text{ V}) = 1.13 \text{ }\mu\text{A}.$$
The actual current increment is
$$\Delta I_\text{actual} = I_\text{final} - I_\text{initial}$$
$$= (10^{-14} \text{ A})e^{(0.51 \text{ V}/0.02566 \text{ V})} - 2.90 \text{ }\mu\text{A} = 1.38 \text{ }\mu\text{A}.$$

Thus the approximate current increment is about 18% lower than the actual value. If we had excursions of 5 mV in each direction instead of a unidirectional 10-mV excursion, the error would be much smaller.

Inspection of the equations just presented indicates that for a small percentage error as a result of linearization, the applied-voltage increment must be small relative to the thermal voltage, kT/q. It is also worthwhile to inquire about the voltage increment for which the predicted current increment has order-of-magnitude validity. As an illustration, take the case of $\Delta V = kT/q$, and let other conditions remain the same as in Exercise 3-62. This time

$$\Delta I_{approx} = g\Delta V = (1.13 \times 10^{-4}/\text{ohm})(0.02566 \text{ V}) = 2.90 \ \mu\text{A}. \quad (3\text{-}99)$$

But from the exponential expression for forward current, it is evident that I_{final} will exceed $I_{initial}$ by the factor e, so that

$$\Delta I_{actual} = I_{final} - I_{initial} = eI_{initial} + I_{initial} \quad (3\text{-}100)$$
$$= I_{initial}(e + 1) = (2.90 \ \mu\text{A})(3.718) = 10.8 \ \mu\text{A}.$$

Inspecting Equation 3-100 makes it clear that the ratio of results is

$$\frac{\Delta I_{actual}}{\Delta I_{approx}} = 3.72 = (e + 1). \quad (3\text{-}101)$$

In summary then, even for order-of-magnitude accuracy, we must have $\Delta V < (kT/q)$.

Exercise 3-63. Why is the value of ΔI_{approx} calculated in Equation 3-99 equal to the value of $I_{initial}$?

Reference back to Figure 3-50(a) makes it evident that the linear approximation predicts zero current for a *reduction* in forward voltage of kT/q. Hence, since a linear approximation is being used, an *increase* in forward voltage by the same amount doubles the bias-point current.

Equation 3-98 for small-signal conductance was derived from Equation 3-40. This in turn was based on Figure 3-16, which represents a long diode. Let us turn now to the case of a short diode. It is represented in Figure 3-51(a), along with the corresponding long diode. The short-diode case is defined by the condition $X_P \ll L_n$, where these two dimensions are depicted in Figure 3-51(b). We have assumed that the constraint $n_P(X_P) = n_{0P}$ exists because at the position X_P we have placed an ohmic contact of the kind described in Section 2-7.5. Let us assume further that the long and short diodes are identical, except for the dimension X_P. As a final assumption, we shall consider that the boundary conditions will be identical

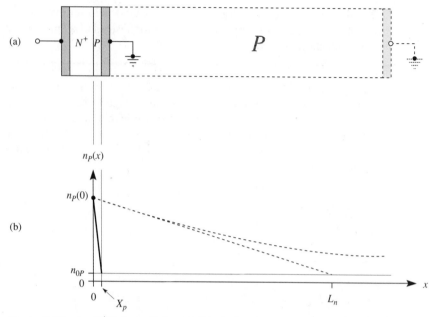

Figure 3-51 An N^+P short diode. (a) Physical representation, with that for a corresponding long diode shown with dashed lines. (b) Minority-carrier profile in the P-type region for forward bias, with that in the corresponding long diode shown with dashed lines.

in the two diodes for a given forward voltage. It has been shown that the short-diode boundary conditions diverge from those in the long diode as the device is made shorter [19], but quantitative examination of this effect is beyond our present scope.

With these several assumptions and since the injected electron current (essentially the total diode current) is proportional to $|dn_P/dx|_{x=0}$, it follows that current in the short diode will exceed that in the long diode by the factor (L_n/X_P). That is, the short-diode current can be written

$$-I_{\text{short}} = I_0(L_n/X_P)\exp(-qV_{NP}/kT). \qquad (3\text{-}102)$$

Proceeding as before gives us

$$g = (q/kT)(-I_{\text{short}}) = 1/r; \qquad (3\text{-}103)$$

that is, the dependence of g on *current* is unchanged, but the current is much larger at a given *voltage*. Now we turn to a junction property where the long–short differences are more marked.

3-8.2 Diffusion Capacitance

The static charge-control model (Section 3-4.5) focused on the excess-carrier charge Q_S stored in a long diode under forward bias, as shown in Figure 3-22. Let us continue with the long diode. The law of the junction causes the charge Q_S to be exponentially related to applied voltage through most of the low-level forward-bias range, the range wherein

$$Q_S = Aqn'_P(0)L_n \approx Aqn_P(0)L_n. \qquad (3\text{-}104)$$

The rate of change of this charge with respect to voltage is known as *diffusion capacitance*, an example of a small-signal capacitance. That is,

$$C_s \equiv dQ_S/dV_{NP}. \qquad (3\text{-}105)$$

Since Q_S is exponentially related to V_{NP}, it is evident from Equation 3-105 that C_s will be as well, so we again resort to small-signal linearization. And once again, signal voltage must be of the order of the thermal voltage (kT/q) to achieve even order-of-magnitude accuracy. The desired approximate expression for diffusion capacitance could be found readily from Equation 3-104, but can be found even more readily by combining Equation 3-97 and the basic charge-control relation, $Q_S = I\tau$. Hence,

$$Q_S = -I_0\tau \exp(-qV_{NP}/kT) = Q_{0S} \exp(-qV_{NP}/kT). \qquad (3\text{-}106)$$

From Equation 3-105, then, the diffusion capacitance is

$$C_s = (q/kT)(-Q_{0S}) \exp(-qV_{NP}/kT) = (q/kT)(-Q_S), \qquad (3\text{-}107)$$

in a long diode. Because Q_S and Q_{0S} are negative in the diode assumed, the diffusion capacitance is positive.

Exercise 3-64. A junction area of several square millimeters is large by today's standards. For such a junction, $I_0 \approx 10^{-13}$ A. Determine an order-of-magnitude value for Q_{0S}, the charge analog of theoretical saturation current I_0.

Taking $\tau \approx 5$ μs as a typical value (see Figure 2-34), we have

$$Q_{0S} = -(10^{-13} \text{ A})(5 \times 10^{-6} \text{ s}) = -5 \times 10^{-19} \text{ C}.$$

Evidently this charge is extremely tiny, amounting approximately to that of only about three electrons!

The close relationship of Equations 3-97 and 3-107 is stressed in Figures 3-52(a) and (b), and this relationship suggests an additional important equality. Applying the static charge-control equation to Equation 3-107, or Figure 3-52(b), yields

$$C_s = (q/kT)(-Q_S) = (q/kT)(-I)\tau = g\tau, \qquad (3\text{-}108)$$

where the last step comes from Equation 3-98. Thus diffusion capacitance is equal to the product of the conductance that delivers the charge and the lifetime that preserves it.

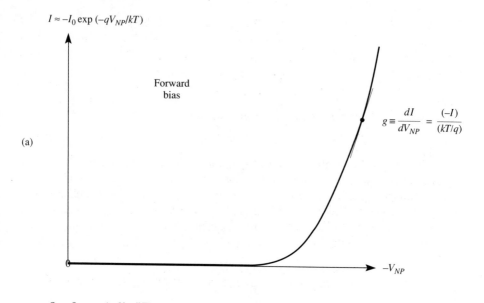

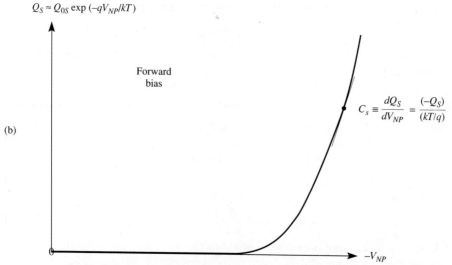

Figure 3-52 Small-signal junction properties and their close relationship. (a) Small-signal conductance, the slope of the I–V characteristic. (b) Small-signal diffusion capacitance, the slope of the Q–V characteristic.

The relationship of dQ_S to Q_S for the case just examined is shown in Figure 3-53(a). Let us move now to the other extreme, the case of the short diode. For a slow excursion in voltage, this time there exists the relationship of dQ_S and Q_S that is shown in Figure 3-53(b). As before, it is assumed that the right-hand ohmic contact located at $x = X_P$ preserves equilibrium carrier densities at its surface. Let us again assume that the short diode has the same boundary conditions as the long

(a)

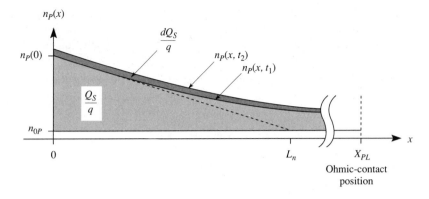

(b)

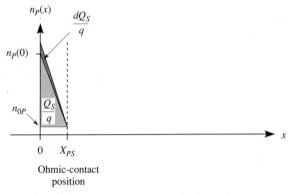

Figure 3-53 The excess-carrier storage (N^+P-diode example) that is responsible for diffusion capacitance. (a) A case of a "long" diode, wherein excess electrons disappear completely before reaching the right-hand ohmic contact. (b) A case of a "short" diode, wherein excess carriers undergo virtually no recombination before reaching the right-hand ohmic contact, which is assumed to maintain equilibrium densities at its surface.

diode at a given forward voltage and that it is identical in every physical respect except the length of the P-type region. This time, charge storage at a given voltage is reduced by the factor $(X_P/2L_n)$, and so diffusion capacitance is similarly reduced. Thus,

$$C_{s(\text{short})} = (X_P/2L_n)\, C_s. \tag{3-109}$$

On the other hand, it is evident in Figure 3-52 by inspection that for the short diode

$$I_{\text{short}} = (L_n/X_P)\, I, \tag{3-110}$$

as was noted in Equation 3-102, where I is long-diode current. In summary, current is higher in the short device and capacitance is lower.

Because recombination in the short diode's P-type region is negligible, carrier lifetime τ is no longer relevant. Therefore, an equation analogous to Equation 3-108 is not appropriate. We can, however, as a matter of curiosity, examine an "effective lifetime'' τ^* (defined in this case as the ratio of stored charge to current in the short diode). In future work we will use τ^* as a generalized or effective lifetime and will then define an effective capacitance C_s^* as $\tau^* g$, paralleling Equation 3-108.

Exercise 3-65. Given a diode with $g = 1$ mS, $\mu_n = 1400$ cm^2/V·s in the P-type region, and $X_P = 1$ μm, calculate the values of τ^* and C_s^*.

Forming the quotient of charge by current to determine τ^*,

$$\frac{Q_S}{I_S} = \frac{qAn_P'(0)(X_P/2)}{qAD_n n_P'(0)/(X_P/2)} = \frac{X_P^2}{4D_n} = \frac{X_P^2}{4(kT/q)\mu_n}$$

$$= \frac{(10^{-4} \text{ cm})^2}{4(0.02566 \text{ V})(1400 \text{ cm}^2/\text{V·s})} = 6.959 \times 10^{-11} \text{ s} \approx 70 \text{ ps} = \tau^*,$$

a value close to typical dielectric relaxation time for the same medium. Also,

$$C_s^* = \tau^* g = (70 \times 10^{-12} \text{ s})(10^{-3} \text{ S}) = 70 \text{ fF}.$$

3-8.3 Depletion-Layer Capacitance

A second charge-storage mechanism in the junction is closely related to that in an ordinary capacitor, a comparison already made in Section 3-1.2. The primary difference is that in the conventional capacitor, the charge resides in planes of roughly atomic-dimension thickness in the metallic plates, while in *depletion-layer capacitance* double-layer charge is spread through regions that are thicker by many orders of magnitude, yet still typically a fraction of a micrometer overall. The latter situation is depicted in Figures 3-1(d) and 3-5.

In the conventional capacitor governed by the law $Q = CV$, the quantity C is voltage-independent. Because C is constant, the capacitor is a *linear* device, and as a result, $(Q/V) = (dQ/dV)$. The reason that C is constant is that the planes of stored charge have a constant, voltage-independent spacing. Depletion-layer capacitance is by contrast nonconstant, and the corresponding "device" is thus nonlinear. The voltage dependence of depletion-layer thickness is the underlying physical reason for these facts.

We shall use the symbols C_t for depletion-layer capacitance, and dQ_T for the relevant charge, exploiting the convenient subscript provided by the equivalent term *transition-region capacitance*. Using the small-signal technique for linearization, we write

$$C_t = dQ_T/dV_{NP}. \tag{3-111}$$

The next step is to examine the stored charge Q_T. Because the parallel-plate capacitor is typically a symmetric device, let us first examine the symmetric step junction in order to enhance the similarity. And for simplest illustration, let us use the depletion approximation. It will be shown later that this simplification can be dropped with no alteration of principles. Figure 3-54 depicts the sample to be analyzed, showing its depletion layer of sharply defined thickness X. Now note an elementary and important fact. When the charge on a capacitor is cited, what is meant is the charge on *one capacitor plate only*. The algebraic sum of the charges on the two plates is usually zero. Let us arbitrarily consider, therefore, only the positive charge stored on the N-type side of the junction. It is evident that this charge can be written

$$Q_T = AqN(X/2). \tag{3-112}$$

Combining this expression with

$$X = \left[\frac{4\epsilon(V_{NP} + \Delta\psi_0)}{qN}\right]^{1/2}, \tag{3-113}$$

which is itself a combination of Equations 3-21 and 3-23, we have

$$Q_T = A\left[\epsilon qN(V_{NP} + \Delta\psi_0)\right]^{1/2}. \tag{3-114}$$

Comparing this expression to the corresponding law for the conventional capacitor, $Q = CV$, we see the additional complexity associated with the nonlinear junction case. Equation 3-114 is plotted in Figure 3-55(a). While we could define a total capacitance for the junction at a point such as P_1 as the quotient Q_T/V_{NP}, (the slope of the line from the point P_1 to the origin), it is a quantity that is rarely useful. On

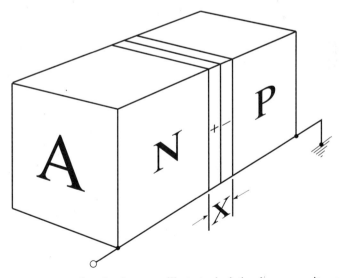

Figure 3-54 Sample chosen to illustrate depletion-layer-capacitance calculation.

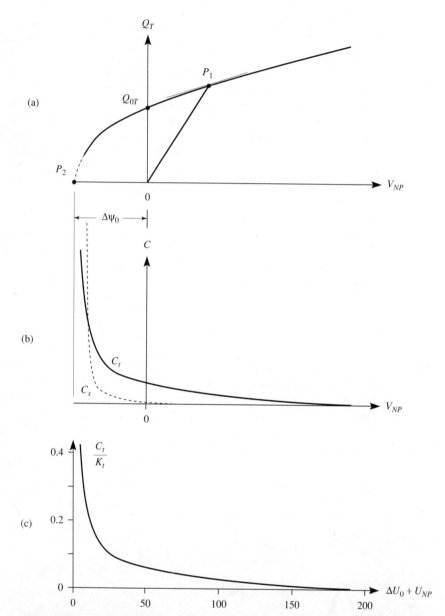

Figure 3-55 Functions related to depletion-layer capacitance C_t. (a) Total charge Q_T versus applied voltage V_{NP} for a step-junction sample. (b) Comparing depletion-layer capacitance C_t versus voltage (solid line) to diffusion capacitance C_s versus voltage (dashed line). (c) Normalized depletion-layer capacitance versus normalized potential difference through the junction, valid for any step junction.

the other hand, C_t, the small-signal or incremental capacitance at the same point (represented by the slope of the tangent at P_1), has appreciable utility.

Exercise 3-66. Locations of the equal and opposite charges for the case of depletion-layer capacitance are clearly shown in Figure 3-54, and these can be described crudely as defining the locations of "capacitor plates." Describe the locations and identities of the two analogous charges in the case of diffusion capacitance.

In Figure 3-53(a), the shaded area represents the charge Q_S of excess electrons stored in the P-type region. The equal and opposite charge is the charge of excess majority carriers stored *in the same volume*. Thus, this case does not involve physically separated charges.

Inspection of Equation 3-114 shows that the curve in Figure 3-55(a) is a parabola with its vertex at point P_2. This point corresponds to a forward voltage (negative here) equal to the contact potential $\Delta\psi_0$. This point is not physically reachable because attempting to reduce the junction barrier to zero height causes enormous currents that destroy the device. The Q_T-axis intercept Q_{0T}, by contrast, is reachable and is of course a point corresponding to the device at equilibrium. The value of Q_{0T} is readily obtained from Equation 3-114 by setting V_{NP} equal to zero, and is the total charge "stored" on the capacitor at zero bias. It will be seen in Section 3-9.6, however, that SPICE analysis chooses to place the origin of the $Q_T - V_{NP}$ plane at the point $(Q_{0T}, 0)$ in Figure 3-55(a).

The desired symmetric-junction expression for C_t as a function of voltage can be obtained by differentiating Equation 3-114:

$$C_t = \frac{d}{dV_{NP}}[A\sqrt{\epsilon q N}(V_{NP} + \Delta\psi_0)] = \frac{A\sqrt{\epsilon q N}}{2\sqrt{V_{NP} + \Delta\psi_0}}. \tag{3-115}$$

This expression is given in Table 3-1, and is plotted as the solid curve in Figure 3-55(b). It is a power-law relationship, but *not* a parabola. In fact, C_t does not increase without limit (as Equation 3-115 indicates) because the depletion approximation worsens with forward voltage, and fails for large values, a matter treated further in Section 3-9.6.

Observe that C_t is dominant for reverse bias and for small forward bias, a range wherein the depletion approximation is acceptable. The exponential dependence of C_s on forward bias causes it to sweep past C_t with its power-law dependence, and to dominate from there on, as can be seen (dashed curve) in Figure 3-55(b). For a typical junction this occurs in the upper half of the low-level forward-bias range, and well below the upper limit of that range, a matter examined in Section 3-8.4.

Interestingly, through normalization it is possible to construct a universal curve for C_t as a function of total potential difference between the two sides of the junction. It is valid for any step junction. For this purpose we choose the most general kind of step junction, the asymmetric case, whose C_t expression is given in Table 3-1 along with expressions for other cases. Multiplying numerator and denominator of this expression, inside the radical, by kT/q gives us C_t in terms of normalized voltage:

$$C_t = Aq\left[\frac{\epsilon/kT}{(1/N_1) + (1/N_2)}\right]^{1/2} (\Delta U_0 + U_{NP})^{-1/2}. \tag{3-116}$$

With the definition

$$K_t = Aq\left[\frac{\epsilon/kT}{(1/N_1) + (1/N_2)}\right]^{1/2}, \tag{3-117}$$

Equation 3-116 becomes

$$(C_t/K_t) = (\Delta U_0 + U_{NP})^{-1/2}, \tag{3-118}$$

which is plotted in Figure 3-55(c).

Exercise 3-67. Can an analogous universal curve be constructed for C_s versus V_{NP} in a step junction, and if not, why not?

The universal curve for C_t is possible because diffusion capacitance is uniquely related to total potential difference, $\Delta\psi = \Delta\psi_0 + V_{NP}$, which in effect places the voltage origin at the vertex of the parabola of Figure 3-55(a). But by contrast, C_s depends upon applied voltage, $V_{NP} = \Delta\psi - \Delta\psi_0$, which places the voltage origin at the equilibrium point, as in Figures 3-55(a) and 3-55(b). Since this voltage origin is, through $\Delta\psi_0$, a function of individual diode properties, the previous simplicity is lost.

Now return to Equation 3-115 giving C_t for the symmetric step junction. Notice that it can be rewritten with the aid of Equation 3-111 as

$$C_t = \frac{A\epsilon}{\sqrt{4\epsilon(V_{NP} + \Delta\psi_0)/qN}} = \frac{A\epsilon}{X}. \tag{3-119}$$

Comparing this expression with the analogous expression for the ordinary capacitor,

$$C = \frac{A\epsilon}{d}, \tag{3-120}$$

where d is plate spacing, leads to a seeming paradox. Why should *total* depletion-layer thickness X correspond to the plate spacing of the ordinary capacitor in view of the fact that charge in the junction is distributed through the entire volume of thickness X? The matter is resolved by noting that a small-signal voltage variation superimposed on a fixed dc bias causes the storage and recovery of charge at the *boundaries* of the depletion layer. The physical situation is depicted in Figure

3-56(a) which shows the $\rho_v(x)$ profile for a symmetric junction in depletion-approximation terms. To the charge Q_T stored on one side, an incremental charge dQ_T has been added by the application of an increment of voltage. (The incremental charge may be positive or negative, depending on the sign of the voltage increment.) The main point is that the physical locations of the equal and opposite charge increments

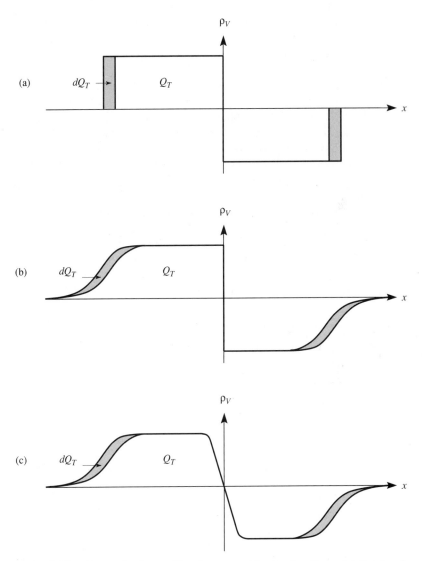

Figure 3-56 Charge-density profiles for symmetric step junctions. (a) Total and incremental charges for an ideally abrupt junction with the depletion approximation assumed. (b) Charges for an ideally abrupt junction with realistic charge profiles. (c) Charges for a realistically less-than-abrupt junction with realistic charge profiles.

are spaced apart by the dimension X, while in a conventional capacitor the spacing is d, thus explaining the congruence of Equations 3-119 and 3-120.

The mechanism of charge storage at the boundary of a depletion layer has a binary interpretation. To be specific, again focus on the positive ionic charge Q_T stored on the N-type side of the sample. A positive voltage increment dV_{NP} will cause depletion-layer expansion and the storage of a positive increment dQ_T. This positive increment can be viewed as an incremental layer of donor ions "uncovered" by the application of dV_{NP}. Alternatively, it can be viewed as a *reduction* in the *negative* charge present on the N-type side, because majority electrons are withdrawn from there when dV_{NP} is applied.

Now turn to the more realistic depiction of the $\rho_v(x)$ profile in Figure 3-56(b). The charge increment dQ_T is now distributed through an appreciably greater volume than is acknowledged in the depletion-approximation representation, Figure 3-56(a). But because the true distribution of dQ_T has its center very nearly at the position defined by the depletion approximation, Equation 3-119 constitutes an excellent approximate expression for depletion-layer capacitance. Experimental observations have shown repeatedly that the depletion-approximation expression for C_t is in error by no more than a few percent for reverse bias and low forward bias.

While today's technology permits one to fabricate step junctions of near-ideal abruptness in the doping profile, it is nonetheless commonplace to observe a measure of "grading" or gradualness in the doping transition. Figure 3-56(c) illustrates the effect of such a departure from the ideal doping profile on the charge-density profile, $\rho_v(x)$. Again, the effect on the positions of the increments dQ_T at a given bias voltage is small, and the effect on the field profile is simply to round the peak a bit. Thus the capacitance expression derived from the apparently simplistic depletion approximation is very useful in the real world through a wide voltage range.

Exercise 3-68. Explain how to derive depletion-layer capacitance expressions for any kind of junction.

In any PN junction the storage and recovery of depletion-layer charge under small-signal conditions also occurs at the boundaries of the space-charge layer. Hence to find C_t it is simply necessary to write $A\epsilon$ divided by the voltage-dependent depletion-layer thickness X. See Table 3-1 for step-junction cases.

3-8.4 Junction-Capacitance Crossover

Large-signal applications of junction diodes force one to deal with both diffusion and depletion-layer capacitance. Cases involving bias-polarity reversal are commonplace. There are other cases involving large excursions of forward voltage, and these, too, force consideration of both C_s and C_t, as Figure 3-55(b) shows. In spite of the simplifying generality displayed in Figure 3-55(c), C_t is awkward analytically because of its nonlinearity. Some degree of approximation is always a practical necessity. But C_s, by contrast, is linearly related both to current and to stored charge Q_S, as Section 3-8.2 shows, and this constitutes a significant simplification.

In much of the analysis below, C_s is assumed to dominate C_t sufficiently so that the latter can be neglected. But it is further assumed that forward voltage is confined to the low-level regime in order to preserve C_s linearity, and hence it is natural to inquire whether any realism remains in the analytic results based upon the double restriction. We examine this question now. As usual, take the case of an N^+P diode with the P-type region long compared to L_n. Assume also that $(N_D - N_A) = 10^{20}/\text{cm}^3$ in the N^+ region. Let us first determine the applied voltage value at which high-level conditions commence. By definition, this borderline condition is reached when

$$n_P(0) = p_{OP} \approx N_A, \tag{3-121}$$

where N_A is net doping in the P-type region. Substituting this expression and $n_{OP} = n_i^2/N_A$ into the law-of-the-junction equation,

$$n_P(0) = n_{OP} \exp(-qV_{NP}/kT), \tag{3-122}$$

yields

$$N_A = (n_i^2/N_A) \exp(-qV_{NP}/kT), \tag{3-123}$$

so that

$$V_{NP} = -2(kT/q) \ln(N_A/n_i). \tag{3-124}$$

This result is plotted second from the top in Figure 3-57, for N_A ranging from $10^{13}/\text{cm}^3$ to $10^{18}/\text{cm}^3$. (The top curve shows contact potential $\Delta\psi_0$, an absolute limit on forward imposed junction voltage.) Note that for the "typical" doping of $N_A = 10^{16}/\text{cm}^3$, the high-level regime is reached at the "typical" forward-bias magnitude of 0.7 V. Determining the voltage at which $C_s = C_t$ requires iteration, and the result is shown by the lower curve in Figure 3-57 for the same doping range. Thus the intermediate band of voltages below the *high-level* curve and above the *crossover* curve constitutes the zone of validity for much of the following analysis. This regime of relatively simple analysis is emphasized because our purpose here as always is to stress principles. But beyond this, it represents a practically important bias range. In terms of forward current (or stored charge) it represents variation by a factor in excess of 0.7×10^3 at the low-doping end, and a factor in excess of 45×10^3 at the high-doping end. Fortunately, it is also in this regime that the complications of space-charge-layer recombination are least important.

3-8.5 Coexisting Phenomena and Multiple Time Constants

The incremental junction properties defined quasistatically in Sections 3-8.1 through 3-8.3 are simple and precise. And as we have seen, there are bias ranges where each is unimportant. In reverse bias and low forward bias, small-signal conductance and diffusion capacitance are so small as to be negligible. At large forward

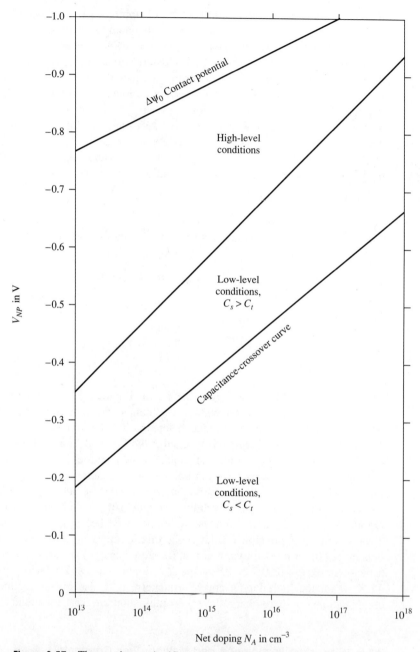

Figure 3-57 There exists a significant band of forward voltages wherein low-level conditions hold and yet $C_s > C_t$, presented here for N^+P diodes with net P-type doping ranging through a factor of 10^5.

bias, depletion-layer capacitance is small enough in comparison with diffusion capacitance so that it can be neglected. But in an intermediate and significant forward-voltage range, the three phenomena must be treated in combination. Moreover, complexity mounts steeply when rapidly varying current and voltage enter the picture. Potentially important phenomena then coexist: (1) displacement current, as well as conduction current through carrier drift and carrier diffusion; (2) carrier-density changes through generation and recombination; (3) electromagnetic-wave propagation; and (4) dielectric relaxation. Along with these phenomena come various time constants: carrier lifetime; dielectric-relaxation time; transit time of an electromagnetic wave through most of the diode's length; and durations of particular features of the current or voltage stimulus.

To develop a physical feeling for the interplay of phenomena and time constants in the general case, let us examine a short series of examples of graduated complexity, but all simpler than the junction diode itself. First, we adopt a specific voltage waveform, as shown in Figure 3-58(a). We assume that an ideal voltage supply delivers a voltage ramp of short duration that starts at $t = 0$ and that is superimposed on a static initial voltage of the order of 1 V. The duration of the ramp voltage is $t_R = 1$ ns, and its amplitude is 10 mV. The signal chosen thus qualifies as a small signal. Also, the time t_R is very short compared to a typical carrier lifetime, which we shall take here as $\tau = 1$ μs. The following examples will compare t_R and τ to two other time constants in specific circumstances.

Now assume that the sample is a silicon resistor (or capacitor, depending on point of view) of cubical form and having an edge dimension of 1 mm. Let the ohmic contacts (or capacitor plates) be applied to opposite faces. The remaining arbitrary choice that must be made is the matter of resistivity. First take the case of intrinsic silicon. For the problem at hand, we need to know dielectric-relaxation time t_D in relation to $t_R = 1$ ns.

Exercise 3-69. Determine dielectric-relaxation time for intrinsic silicon.

Given $t_D = \epsilon\rho$, we have for intrinsic silicon

$$t_D = \frac{\epsilon}{qn_i(\mu_n + \mu_p)} \approx \frac{(10^{-12} \text{ F/cm})}{(1.6 \times 10^{-19} \text{ C})(10^{10}/\text{cm}^3)[(1400 + 450) \text{ cm}^2/\text{V} \cdot \text{s}]}$$

$= 0.34$ μs.

Thus the result is half a decade (logarithmically) below 1 μs, so $t_R \ll t_D < \tau$.

Recalling from Problem A1-21 that the conduction and displacement currents would be equal at t_D, were the ramp that long in duration, we can see that conduction current is negligible relatively during the short time interval of interest. This is shown in Figure 3-58(b), where the dashed line represents an extrapolation of the conduction-current density. The displacement-current density $J_D(t)$ is constant dur-

(a)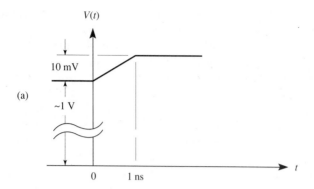

(b)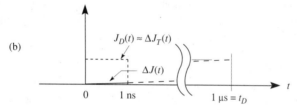

(c)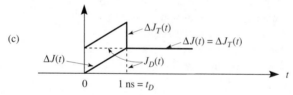

(d)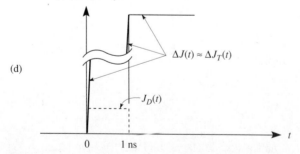

Figure 3-58 Transient waveforms in 1-mm^3 cubical silicon "resistor-capacitors" of differing resistivity. (a) Short-duration voltage-ramp waveform delivered by ideal power supply to each diode. (b) Current-density waveform for silicon sample with $\rho = \rho_i$; (c) with $\rho = 10^3$ ohm·cm; (d) with $\rho = 1$ ohm·cm.

ing the voltage rise because dV/dt is constant, and the conduction-current density averages only about one thousandth of that value. Thus $J_D(t) \approx \Delta J_T(t)$, where the latter is the incremental total-current density. Multiplying it by $A = 10^{-2}$ cm^2 will yield the incremental current observed in the external circuit.

Exercise 3-70. Calculate the value of J_D in the example just treated.

From first principles

$$J_D = \frac{dD}{dt} = \epsilon \frac{dE}{dt} = \frac{\epsilon}{L} \frac{dV}{dt} = \frac{\epsilon}{L} \frac{\Delta V}{\Delta t};$$

hence,

$$J_D \approx \frac{(10^{-12} \text{ F/cm})(10^{-2} \text{ V})}{(10^{-1} \text{ cm})(10^{-9} \text{ s})} = 0.1 \text{ mA/cm}^2.$$

Exercise 3-71. Calculate the value of ΔJ and compare it with the value just found for J_D.

First,

$$J = \frac{E}{\rho} = \frac{V}{\rho L}.$$

But from Exercise 3-69,

$$\rho = \frac{t_D}{\epsilon} = \left[\frac{0.34 \times 10^{-6} \text{ s}}{10^{-12} \text{ F/cm}}\right] = 3.4 \times 10^5 \text{ ohm·cm},$$

so that

$$J = \frac{(1 \text{ V})}{(3.4 \times 10^5 \text{ ohm·cm})(10^{-1} \text{ cm})} = 29.4 \text{ }\mu\text{A/cm}^2.$$

Since 10 mV is 1% of 1 V, it follows that $\Delta J = (0.01)J = 0.3$ μA/cm^2, which is negligible compared to $J_D = 0.1$ mA/cm^2.

Next choose the resistivity value $\rho = 1000$ ohm·cm for the silicon sample. This is a high, but technologically realizable, value. Multiplying it by $\epsilon \approx 10^{-12}$ F/cm yields a dielectric-relaxation time of $t_D = 1$ ns. According to principles just reviewed, it follows that the three current components of interest will be as shown in Figure 3-58(c). Comparing the areas of the current transients, we see readily that one third of the associated external-circuit charge transport is accounted for by the conduction current and two thirds by the displacement current.

Now turn to the case of $\rho = 1$ ohm·cm, for which $t_D \approx 1$ ps. This was described previously as a typical case in silicon. This time the displacement current makes a negligible contribution to the total, so $\Delta J(t) \approx \Delta J_T(t)$, as is illustrated in Figure 3-58(d).

Exercise 3-72. Calculate the incremental conduction-current density ΔJ at $t = 1$ ns in the last example.

From Ohm's law,

$$\Delta J = \rho \Delta E = \rho \Delta V / L = \frac{(1 \text{ ohm·cm})(10^{-2} \text{ V})}{(10^{-1} \text{ cm})} = 100 \text{ mA/cm}^2.$$

Exercise 3-73. Calculate the static current density J prior to application of the ramp voltage.

From Ohm's law,

$$J = \frac{I}{A} = \frac{V}{AR} = \frac{V}{\rho L} = \frac{(1 \text{ V})}{(1 \text{ ohm·cm})(10^{-1} \text{ cm})} = 10 \text{ A/cm}^2.$$

Next turn to the sample shown in Figure 3-59(a). The silicon dimensions are about the same as those in the preceding cases, but two thin layers have been added. At the left is an N^+ layer of negligible thickness, and next to it is an intrinsic layer with a thickness of 0.1 mm, while the P-type region remains 1 mm in thickness or "length." Let it have a resistivity $\rho = 1$ ohm·cm. This structure simulates that of an N^+P diode; the depletion-layer capacitance of such a structure was treated in Section 3-8.3. It is itself, in fact, an N^+IP diode, closely related to structures that were examined in detail in Section 3-5.1. What is important at the moment is that the N^+IP "sandwich" possesses a depletion layer, and that in turn exhibits capacitance for exactly the reasons given in Section 3-8.3. Its depletion-layer capacitance is given by $C_T = A\epsilon/X$.

Now subject this device to identically the same signal used before, Figure 3-58(a). For reasons that should be clear from the foregoing discussion, conduction current will dominate in the P-type region, and displacement current will dominate in the intrinsic region. The resulting incremental total-current density $\Delta J_T(t)$ will assume the form shown in Figure 3-58(b).

Exercise 3-74. Calculate the constant value of $\Delta J_T(t)$ that exists during the ramp.

Since $I = CdV/dt$,

$$J_T = \frac{I_T}{A} = \frac{C}{A} \frac{dV}{dt} = \frac{\epsilon}{X} \frac{\Delta V}{\Delta t} = \frac{(10^{-12} \text{ F/cm})(10^{-2} \text{ V})}{(10^{-2} \text{ cm})(10^{-9} \text{ s})} = 1.0 \text{ mA/cm}^2.$$

SMALL-SIGNAL DYNAMIC ANALYSIS   371

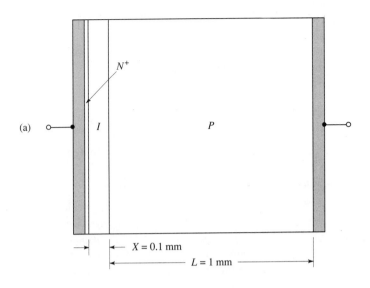

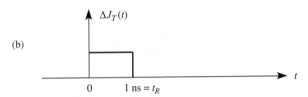

Figure 3-59 Transient experiment using a *P*-type silicon cube of $\rho = 1$ ohm·cm with added intrinsic and N^+ layers, forming an N^+IP diode. (a) Physical representation. (b) Total-current-density waveform observed upon imposing the voltage waveform shown in Figure 3-57(a).

Here we have a situation closely related to that in Section 1-3.5. Conduction and displacement currents coexist in time, but not in space, while in the examples just preceding, they coexisted in *both* time and space. Specifically, in the *P*-type region in the present example, $\Delta J_T(t) \approx \Delta J(t)$. But in the intrinsic layer, $\Delta J_T(t) \approx J_D(t)$. This is the classic situation in a depletion-layer capacitance problem, as well as in a conventional capacitor.

Exercise 3-75. In the situation shown in Figure 3-59(a), a capacitor C_t and a resistor R (the *P*-type region) are in series. Why was this fact ignored in examining current waveforms?

The answer lies in the magnitude of the RC product in relation to $t_R = 1$ ns. Evidently,

$$RC \approx \rho \frac{L}{A} \cdot \frac{\epsilon A}{X} = \rho\epsilon \frac{L}{X} = t_D \frac{L}{X},$$

where t_D is the dielectric-relaxation time for the P-type region, which is about 1 ps. Thus

$$RC \approx (10^{-12} \text{ s}) \left[\frac{10^{-1} \text{ cm}}{10^{-2} \text{ cm}} \right] = 10 \text{ ps},$$

and characteristic time for charging the capacitor is small enough to ignore in showing the current waveform in Figure 3-58(b), since it is only 1% of t_R.

The next complication is to add a new conduction mechanism—namely, the injection of excess minority carriers—that exists in a forward-biased junction. Then conduction current by this mechanism coexists with displacement current in the transition region of the junction, and is added to the (majority-carrier) conduction current already considered in the P-type region. The complicating factors are several. Minority carriers with their finite lifetime must be "tracked" in time and space. Furthermore, even signal polarity matters—whether minority carriers are being injected or extracted.

To summarize the present results, we can say that conduction and displacement currents can coexist in time and space, or in time alone. Which component dominates depends upon (1) a material property, dielectric-relaxation time, which is a measure of how long it takes carriers to "get moving"; (2) dimensional factors, which determine resistance and capacitance, and therefore determine the circuit characteristic time, usually termed *RC time*; (3) a critical time or times in the applied signal, analogous to the ramp-duration time t_R in the preceding examples; and (4) the relationship of carrier lifetime τ to the three foregoing time constants.

Although explicit time dependence in the form of a ramp voltage has been useful for the purposes of this section, the forms of more general interest are sinusoidal waveforms and switching transitions. In any of the cases, when extremely rapid changes are involved, or when a diode is very short, still further complications enter. These are described in Section 2-5.4, and in such cases the energy-transport method of analysis is useful.

3-8.6 Small-Signal Equivalent-Circuit Model

The three incremental properties described in Sections 3-8.1 through 3-8.3 can easily be combined in an equivalent circuit that models the junction diode in a valid way for the quasistatic condition, the only condition treated there. This is a step on the way to the more important and interesting problem of modeling the device for rapidly varying signals. Quantitative discussion is helped by defining a specific structure, and we will focus primarily on the simple N^+P long-diode case. For further simplification, we continue to consider only low-level forward voltage.

Figure 3-60(a) specifies the diode to be examined. The spatial origin has been placed at the right-hand boundary of the depletion layer, residing in the long and uniformly doped *P*-type region. The corresponding small-signal equivalent circuit for quasistatic conditions is shown in Figure 3-60(b). Placing the three incremental elements in parallel is justified by the fact that each is directly controlled by the applied voltage for the low-level conditions assumed. That is, there are no significant internal *IR*-drop losses that would diminish the controlling voltage, and hence virtually all of the external voltage is applied directly to the junction.

Choosing the forward voltage $V_{NP} = -0.6$ V yields a boundary value $n_P(0)$ that is nearly two orders of magnitude lower than p_{OP}, and that hence is well within the low-level regime. Using a similar one-percent criterion for small-signal magnitude, let the quasistatic increase in forward voltage be 6 mV, as is indicated in Figure 3-61(a). The resulting change in electron profile is shown in Figure 3-61(b). Note that the 1-percent change in voltage produces a 27-percent change in boundary density, and hence in current. The spatial origin is considered to move along with the boundary, an assumption that has a negligible relative effect on *P*-region length. The boundary, in turn, is taken to be one of the surfaces where electric field vanishes, always present in a forward-biased junction. The essence of the quasistatic condition is that $n_P(x, t)$ remains accurately a declining exponential function during the applied-voltage transition, as well as in the *before* and *after* cases shown in Figure 3-61(b).

Now turn to a small-signal problem that is an opposite extreme from the one just considered. Let a voltage source that we shall initially consider to be ideal

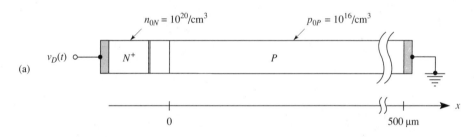

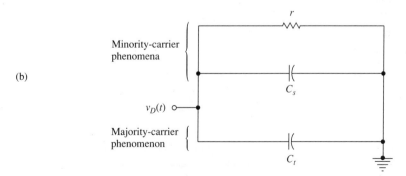

Figure 3-60 Junction-diode modeling. (a) Physical representation of a specific long N^+P diode structure for analysis. (b) Small-signal equivalent circuit.

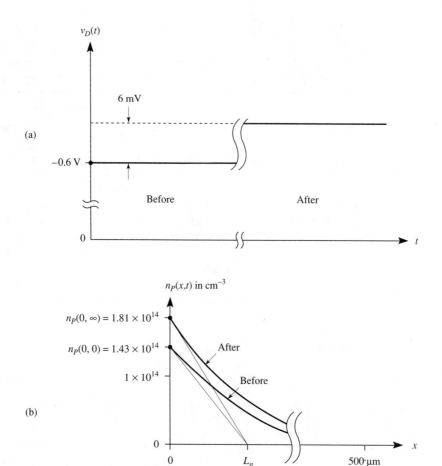

Figure 3-61 Assumed quasistatic small-signal conditions. (a) Significant but still-low-level forward bias with a slow 1% increase then imposed. (b) Corresponding before-and-after minority-electron profiles near the depletion-layer boundary.

deliver a 6-mV step-function voltage to the diode, as is shown by the solid line in Figure 3-62(a). Under these conditions, the coexisting phenomena treated in Section 3-8.5 must be considered.

Exercise 3-76. Describe the properties of the ideal voltage source just assumed and compare them to those of the quasistatic voltage source.

The source now assumed has negligible internal resistance, whereas in the quasistatic case, internal resistance is unimportant and can be large or small.

One of the incremental junction properties—namely, the transition-region capacitance C_t—is characterized by very rapid response. An electromagnetic disturb-

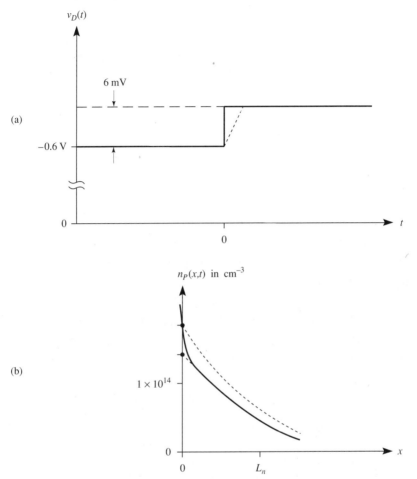

Figure 3-62 Small-signal switching. (a) Small-signal voltage waveform applied to the diode. The ideally abrupt case (solid line) is intractable analytically, but a ramp (dashed line) can be slow compared to majority-carrier response, yet fast compared to minority-carrier response. (b) Minority-electron profiles, with the dashed curves the same as the profiles in Figure 3-61(b). The solid curve obtains shortly after the voltage step is completed.

ance created by the applied-voltage step is propagated through the length of the diode. Since the P-type side is longer, the time delay on that side is greater than on the N^+ side.

Exercise 3-77. Calculate transit time for the electromagnetic disturbance through the P-type region.

The velocity of light in free space is given by

$$c = (\epsilon_0 \mu_0)^{-1/2},$$

where c depends only on the permittivity and permeability constants. The permeability of silicon, a nonmagnetic material, is essentially that of free space, but its permittivity ϵ is 11.7 times greater than ϵ_0. Using values of ϵ and μ_0 from Tables 1-3 and 2-4, we have for the disturbance velocity in silicon

$$v = (\epsilon \mu_0)^{-1/2}$$
$$= [(1.035 \times 10^{-12} \text{ F/cm})(4\pi \times 10^{-7} \text{ V·s/A·m})(10^{-2} \text{ m/cm})]^{-1/2}$$
$$= 0.88 \times 10^{10} \text{ cm/s}.$$

Elapsed time for transit through the P-type region is thus

$$\Delta t = \frac{500 \times 10^{-4} \text{ cm}}{0.88 \times 10^{10} \text{ cm/s}} = 5.7 \text{ ps},$$

a very short time.

Exercise 3-78. What, if anything, is unrealistic about the calculation in Exercise 3-77?

The calculation assumes a plane wave in an infinite (or at least large) volume of uniform material, while the diode has dimensions that are not only finite but small.

The last point deserves elaboration. A spherical electromagnetic wave can be treated as a plane wave when attention is focused on a region of dimensions that are vastly smaller than the radius of the sphere, and the expression used earlier to calculate wave velocity is valid in that case. One deals in such a case with a uniform-material volume so huge that the boundaries effectively are infinitely removed from the small region of interest, and hence boundary conditions are unimportant. In the diode problem, by contrast, the boundaries are proximate. Boundary conditions become very important. The governing differential equations are Maxwell's equations—the full set. This would mean that magnetic phenomena could not be ignored. Recall that we have simply set aside inductive effects, for example.

Aside from wave propagation, there are additional phenomena that must be factored in, because the diode contains fixed space charges and free-carrier charges. Grossly simplifying the problem outlined earlier, let us focus on just the longitudinal (x-directed) component of electric field. The phenomenon of dielectric relaxation will coexist with wave propagation. There will be a readjustment of the (longitudinal) space-charge profile, an accompanying readjustment of the corresponding electric-field profile, and also a readjustment of the current-density profile, involving both displacement and conduction currents. The conduction component is accounted for by majority-carrier drift. Thus a detailed analysis of diode response to an ideal voltage step, even a small-signal step, is truly daunting, and is perhaps impossible to solve when one considers that the majority-carrier readjustment (relaxation process) and the wave propagation coexist and must interact.

There is a way, however, to finesse this essentially unsolvable problem: We can simply let the applied-voltage step be relaxed to a ramp with a rise time that is large compared to the characteristic times of the complex events just outlined. The ramp can still be very steep when compared to minority-carrier responses that will mainly interest us. To justify such an approach, it is worthwhile to gauge the order of speed of dielectric relaxation by treating it as a phenomenon isolated from the numerous complications reviewed qualitatively above.

Exercise 3-79. Calculate dielectric-relaxation time t_D for the P-type region.

From Figure 2-22, we have for the P-type region $\mu_p = 430$ cm^2/V·s, so that

$$\rho_P = (q\mu_p p_{0P})^{-1}$$
$$= (1.60 \times 10^{-19} \text{ C})(430 \text{ cm}^2/\text{V·s})(10^{16}/\text{cm}^3)$$
$$= 1.45 \text{ ohm·cm}.$$

Hence,

$$t_D = \rho_P \epsilon = (1.45 \text{ ohm·cm})(1.035 \times 10^{-12} \text{ F/cm})$$
$$= 1.5 \text{ ps}.$$

Thus the propagation and relaxation phenomena are very fast (picosecond processes when the simple calculations are taken at face value). Hence, the small modification of the input-voltage waveform shown by the dashed line in Figure 3-62(a) will permit them to "follow" the changing applied voltage, $v_D(t)$. What we are in effect saying is that the charging of C_t is now able to follow $v_D(t)$. It is nonetheless a rapid event and involves a current "spike." This confirms that the component C_t should be positioned in shunt across the equivalent-circuit terminals in the general small-signal dynamic model, as was done in Figure 3-60(b).

Now turn to the problem of minority carriers, noting how their conditions differ from those of the majority carriers that now obediently follow the signal. First of all, the densities of minority carriers, unlike those of the majority carriers, depart widely from equilibrium values, even under low-level conditions. Second, they are governed by diffusion and recombination, which are relatively low-velocity, large-time-constant phenomena. As a result, both the space and time dependence of minority-electron density must be considered in an accurate treatment of the general problem. This can be appreciated by once again examining the function $n_P(x, t)$ near the boundary for the same *before* and *after* conditions as in the quasistatic case. This is presented in Figure 3-62(b), where the solid curve shows qualitatively the electron profile at an instant shortly after completion of the applied-voltage transition, while the dashed curves repeat the profiles from Figure 3-61(b).

Recall now that conduction current through the diode, following the current spike that charges C_t, is very nearly equal to the electron diffusion current at $x = 0$.

And because electron current is purely diffusive at that surface, it follows that $(dn/dx)|_{x=0}$ is proportional to the total current. The amount by which the density gradient at $x = 0$ in Figure 3-62(b) departs from that of the upper pure-exponential curve is a measure of the amount by which the instantaneous value of current that is charging the diffusion capacitance departs from the steady-state diode current that exists after the transient has passed. It is evident that an initial spike of current will occur in the case of diffusion capacitance as well as in the case of depletion-layer capacitance, but spike duration is appreciably greater in the diffusion case. It follows that the value of C_s that was determined quasistatically has uncertain relevance during the current transient. The initial high current delivers charge only to the region near the boundary; regions farther to the right are supplied subsequently, as the minority-carrier disturbance propagates in that direction. This description of a *distributed* problem suggests that a transmission-line model of some sort would be useful for treating the minority-carrier phenomena that are present. A plausible transmission-line option is shown in Figure 3-63(a). It contains a large number N of identical resistors, and $N + 1$ identical capacitors.

Exercise 3-80. Justify the bilateral symmetry of the transmission line, and the choice of placing a capacitor in shunt at each port.

Bilateral symmetry is elected because the P-type region has uniform properties throughout. The shunt-capacitor option is taken because storage can and does occur immediately adjacent to the depletion-layer boundary, as in Figure 3-61(b).

Now an important point must be made. A transmission line is a two-port entity, and we are often concerned about stimulating one port and observing the response at the other. But the diode to be modeled, Figure 3-63(b), is obviously a one-port entity. A voltage waveform is delivered at that port, and current at the same port is observed, or vice versa. It follows that a transmission line employed for diode modeling must be converted into a one-port entity by terminating the second port in some fashion.

One actually uses a transmission-line analogy to the minority-carrier phenomena in the lightly doped P-type region. The electron density $n_P(x, t)$ is the analog of the voltage $v(x, t)$ in an actual transmission line. In fact, one can construct analogs of the resistive and capacitive elements in terms of diffusive properties of the region, such as D_n [70]. When the transmission line is introduced into the diode equivalent circuit, note must be taken of the fact that $n_P = n_{0P}$ permanently at the right-hand end of the region being modeled, for the reason that a conventional-assumption ohmic contact was placed there. Translating back to the actual transmission line, this condition is the analog of having $v = 0$ permanently at the right-hand port, a condition easily arranged by short-circuiting that port. Hence we place the transmission line into the diode equivalent circuit with the farthest-right capacitor

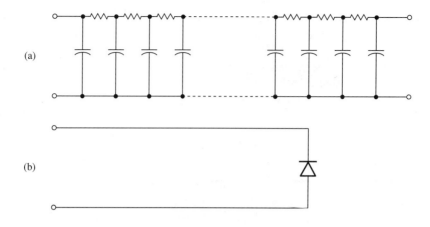

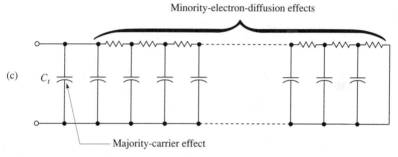

Figure 3-63 Transmission-line modeling of minority-carrier effects in the junction diode. (a) Coupled RC sections forming a two-port transmission line for modeling electron diffusion. (b) The one-port N^+P diode that is to be modeled. (c) Combined model including majority- and minority-carrier effects. This circuit predicts a linear voltage drop in the series resistors under static conditions, and hence models the short diode. Shunting each capacitor with an additional resistor to model recombination leads to the long-diode equivalent circuit.

eliminated by the short-circuit termination. The result is shown in Figure 3-63(c). Now there are N capacitive elements as well as N resistive elements.

It is evident that under static conditions, voltage will decline from left to right in the transmission line in linear fashion. Hence the diode that is modeled here is a short diode, where the analogous electron density declines linearly. In other words, recombination has been omitted from the model. The only modification to Figure 3-63(c) that is needed to model the long diode is to shunt each capacitor by another resistor, representing electron loss to recombination. When N is very large, voltage will decline exponentially in the series-resistor elements. Note that the current in each of these is proportional to dv/dx, just as the diffusion current of electrons in the actual case is proportional to dn/dx.

By adjusting the ratio of the series-element resistor to the shunt-element resistor, we can adjust the point at which voltage has fallen to $1/e$ of its value at the driven port, which is to say, we can adjust "diffusion length." Then, holding the ratio constant for the two kinds of resistors, we can adjust their absolute values so that the incremental resistance seen looking into the left-hand port is r, the incremental resistance of the diode observed quasistatically. After that, a corresponding adjustment of the capacitor values will give agreement between aggregate charge ΔQ_T stored on the capacitors for a given small-signal input-voltage increment, and hence agreement between C_s and the effective capacitance seen looking into the network.

Exercise 3-81. Why is the transmission line merely "analogous to" minority-electron effects? The pattern of charge storage in the transmission line can be made identical to that in the diode.

The voltage $v(x)$ in the transmission line is the analog of $n_P(x)$ in the diode; drift transport in the series resistors is the analog of diffusion transport in the diode; carrier "loss" through the shunt resistors is the analog of recombination in the diode; majority carriers in the network are the analog of minority carriers in the diode; and equal and opposite charges stored on spatially separate capacitor plates in the network are the analog of equal and opposite charges stored in the diode in spatially common fashion.

It is important to realize that the transmission line models the diffusion and recombination effects occurring in the P-type region. It has *nothing* to do with the bulk-resistive properties of that region, a point that has at times been obscured in the literature. A further point is that with N (the number of identical sections in the line) large, the equivalent-circuit model will predict precisely the response of the diode to a given dynamic signal. The problem, of course, is dealing with the complex circuit problem so defined. As a practical matter, numerical methods would be needed. For an exact solution, a direct analytic attack on the diffusion problem is more productive than an analytic treatment of the transmission-line problem. But for an approximate solution, a simplified form of the equivalent-circuit model yields useful results.

Figure 3-64(a) shows an ultimately simplified version of the transmission line presented in Figure 3-63(a). The number of sections has been reduced from a large value N to unity. The values of the capacitors are chosen to yield C_s exactly in the quasistatic case. Capacitor-shunted ports and bilateral symmetry are chosen for the same reason as before. When this simplified "line," with the right-hand port short-circuited, is introduced into the diode equivalent circuit, Figure 3-64(b), one of the two $C_s/2$ capacitors drops out and one remains, now in parallel with r and C_t. Hence the model corresponds to the quasistatic model with respect to resistance but not capacitance. That is, the one-section transmission-line model is not valid for quasi-

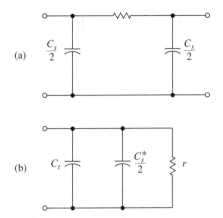

Figure 3-64 Simplified small-signal equivalent circuit for the junction diode. (a) Single-section transmission line for modeling minority-electron effects. (b) Small-signal equivalent circuit for the diode after combining C_t and a transmission line with short-circuit termination.

static conditions because diffusion capacitance is half the correct value. It does, however, model dynamic properties in a surprisingly serviceable way, and for this reason we make the definition

$$C_s^* \equiv C_s/2, \qquad (3\text{-}125)$$

where C_s^* is an effective capacitance. This important discrepancy between the quasistatic model and the dynamic one-section transmission-line model for diffusion phenomena in the long diode will be discussed more extensively later.

Although the simplistic equivalent circuit of Figure 3-64(b) predicts diode response in a remarkably accurate way, waveform details differ from those predicted by exact analysis. A useful test for displaying this relationship employs a current-step stimulus. Let the diode be biased at the same point as before carrying a current I_D, and then let the current increase abruptly at $t = 0$ by the amount $\Delta I_D = 0.01\ I_D$, as is shown in Figure 3-65(a). For further simplification, note in Figure 3-57 that C_t is much smaller than C_s^* for the chosen voltage and diode, and can be neglected. With the model thus reduced to a simple shunt combination of r and C_s^*, the result is the voltage waveform shown in Figure 3-65(b).

Exercise 3-82. Give a qualitative explanation for the voltage waveform in Figure 3-65(b).

For $r = \infty$ (and $C_t = 0$), the current step ΔI_D would charge the capacitor C_s^* at a constant rate, yielding the ramp voltage $v_d(t) = Kt$, where

$$K = \frac{\Delta V}{\Delta t} = \frac{\Delta I_D}{C_s^*}.$$

But with voltage on the capacitor rising, current through r, now finite, increases and the capacitor charges progressively more slowly.

Exact analysis of the same problem yields a voltage-versus-time curve that rises more rapidly at first, for reasons that by now should be qualitatively clear. The exact curve then crosses the approximate curve at about $(0.8)\Delta V$.

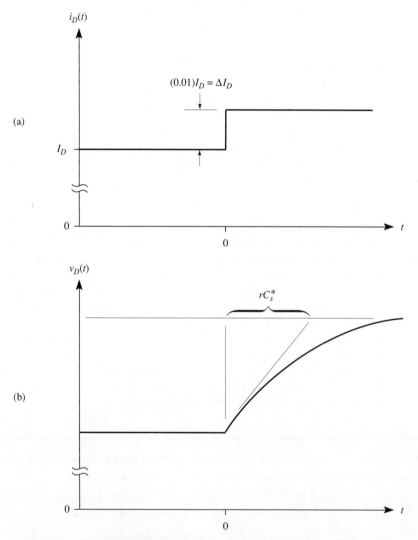

Figure 3-65 Current-step stimulus of the diode. (a) Assumed small-signal input waveform from an ideal current source. (b) Voltage response predicted by equivalent circuit of Figure 3-64(b), with $C_t = 0$.

Finally, let us summarize the difference between the two capacitances in the equivalent circuit through this set of observations: It is the behavior of *majority* carriers, driven mainly by the drift mechanism, that accounts for depletion-layer capacitance. It is the behavior of *minority* carriers, driven mainly by the diffusion mechanism, that accounts for diffusion capacitance. The latter phenomenon is slower and involves relatively complex patterns of charge in space and time. For this reason, diffusion capacitance is appreciably more complicated to analyze in a detailed way. However, as we shall see in Section 3-9.3, diffusion capacitance in combination with dynamic resistance exhibits a voltage independence that constitutes a major analytic simplification. Depletion-layer capacitance, on the other hand, either alone or in combination with other elements, is implacably voltage-dependent in the face of any but very small voltage variations. For this reason we introduce the notation $C_T(V_D)$ for the explicitly voltage-dependent large-signal value, in contrast to C_t for the constant small-signal value.

3-8.7 Effective Lifetime and Diffusion Capacitance

Section 3-8.2 emphasizes the close relationship between quasistatic I versus V_{NP}, and quasistatic Q versus V_{NP}. Both functions are accurately exponential in the upper half of the low-level forward-bias regime that is of greatest interest. Furthermore, small-signal conductance g is given by $I/(kT/q)$, and small-signal diffusion capacitance C_s, by $Q_S/(kT/q)$, as is emphasized in Figure 3-52. Solving each expression for kT/q and equating the results then gives

$$\frac{I}{g} = \frac{Q_S}{C_s}. \tag{3-126}$$

Combining this with the long-diode charge-control equation, $I = Q_S/\tau$, where τ is carrier lifetime, yields

$$g = C_s/\tau. \tag{3-127}$$

A range of diode problems can be addressed by defining an *effective diffusion capacitance* C_s^* and an *effective carrier lifetime* τ^* that are assumed to bear the same relationship to diode conductance g:

$$g = C_s^*/\tau^*. \tag{3-128}$$

Any desired value of g can be obtained by adjusting applied voltage, or by manipulating diode geometry. Once this is done, the quotient of C_s^* by τ^* is known. Hence if one of the effective quantities can be determined, the other is easily found.

In the quasistatic analysis of a short diode, for example, this approach was used. (See Exercise 3-65 and the associated discussion.) In that case, I and Q_S can be written from first principles, and applying the charge-control principle gives directly the effective lifetime as $\tau^* = X_P^2/4D_n$, so it follows that $C_s^* = gX_P^2/4D_n$. As another example, turn to the long diode under small-signal dynamic conditions, Section 3-8.6. There the one-section transmission-line analysis gave $C_s^* = C_s/2$, where C_s is quasistatic diffusion capacitance for the same device. It therefore follows, in this case, that $\tau^* = \tau/2$, where τ is actual carrier lifetime.

Once effective diffusion capacitance has been determined, either directly or through an intermediate determination of effective lifetime, then more powerful mathematical tools are applicable. One of these employs the unit step function $u(t)$. This useful abstraction is closely related to the Dirac delta function encountered earlier. (See, for example, Section 2-2.4.) In the present context, the appropriate delta-function form is $\delta(t)$. It is a dimensionless pulse or "spike" of unit area, and it maintains its unit area while pulse height approaches infinity and pulse duration approaches zero. From this description it is evident that its integral is unity, and this integral in fact defines the unit step function:

$$u(t) \equiv \int \delta(t) \, dt, \qquad (3\text{-}129)$$

where the integration limits enclose the origin. The utility of the unit step function now becomes evident. The waveform information presented graphically in Figure 3-65(a) can now be expressed mathematically as

$$i_d(t) = \Delta I_D \, u(t), \qquad (3\text{-}130)$$

where $i_d(t)$ is the small-signal component of current. For the three-branch circuit in Figure 3-64(b) that includes C_t, input current can be written simply as the sum of the currents in the three branches, or as

$$i_d(t) = \frac{v_d(t)}{r} + (C_s^* + C_t)\frac{dv_d}{dt}. \qquad (3\text{-}131)$$

Here, $v_d(t)$ is of course the small-signal component of voltage. This equation, a first-order ordinary differential equation with constant coefficients, is readily solvable for various input functions. For the unit-step-function input, in particular, its solution is

$$v_d(t) = \Delta I_D r \left[1 - \exp\left(-\frac{t}{r(C_s^* + C_t)} \right) \right]. \qquad (3\text{-}132)$$

This result matches the one plotted in Figure 3-65(b) when $(C_s^* + C_t)$ is substituted for C_s^*. Furthermore, this result is valid for any of the many diode cases where using an effective diffusion capacitance is valid.

3-8.8 Small-Signal Charge-Control Analysis

The static charge-control analysis treated in Section 3-4.5 has a dynamic counterpart that is of great importance because of the inherent linearity in the basic equations. For the small-signal problem, however, it offers little advantage over the equivalent-circuit analysis just examined because linearity is already present there. Nonetheless, it is informative to compare the two approaches as an introduction to dynamic charge-control methods.

Let us start with the equivalent-circuit formulation and derive a basic charge-control equation from it. Using a well-established convention, we can write the total charge of diffusing carriers as the sum of a static component and a small-signal time-varying component, or as $q_S(t) = Q_S + q_s(t)$. The symbol $q_s(t)$ is of course parallel to small-signal current $i_d(t)$ and small-signal voltage $v_d(t)$. Static (or quasistatic) analysis focuses upon Q_S, and small-signal dynamic analysis, upon $q_s(t)$. To avoid any possible confusion of the variable-charge symbols $q_S(t)$ and $q_s(t)$ with the electronic-charge constant, q, we shall always affix a subscript to the variable symbols. Equivalent-circuit analysis started with a differential equation for $i_d(t)$, Equation 3-131, with each of its three terms directly associated with one of the three shunt elements in Figure 3-64(b), and with $v_d(t)$ as independent variable. Charge-control analysis converts to $q_s(t)$ as independent variable, where of course

$$q_s(t) = C_s^* v_d(t),$$

and hence

$$v_d(t) = \frac{q_s(t)}{C_s^*} = \frac{rq_s}{\tau^*}. \tag{3-134}$$

Thus the first term in Equation 3-131 becomes $q_s(t)/\tau^*$. The derivative coefficient of the last two terms can be converted from dv_d/dt to $(dq_s/dt)/C_s^*$, and hence Equation 3-131 can be converted to the charge-control form as

$$i_d(t) = \frac{q_s(t)}{\tau^*} + \left[1 + \frac{C_t}{C_s^*}\right] \frac{dq_s}{dt}. \tag{3-135}$$

The solution of this equation for a step-function input is,

$$q_s(t) = \Delta I_D \tau^* \left[1 - \exp\left(-\frac{t}{\tau^*(1 + C_t/C_s^*)}\right)\right], \tag{3-136}$$

which is fully consistent with the equivalent-circuit solution that is depicted in Figure 3-65(b).

The charge-control analysis just described will be extended from small-signal to large-signal problems in Section 3-9.3. It turns out that the charge-control approach is much to be preferred for long-signal problems, because the equivalent-circuit method becomes cumbersome in such cases.

3-8.9 Linear Differential Equations

The small-signal charge-control equation, Equation 3-135, has the general mathematical form

$$\frac{dx}{dt} + ax = F(t), \tag{3-137}$$

where a is a constant and $F(t)$ is called the driving force. If we define the differential operator D as d/dy then we can rewrite Equation 3-137 as

$$(D + a)x = F(t). \tag{3-138}$$

Moreover, if we define a second differential operator L as $(D + a)$ then

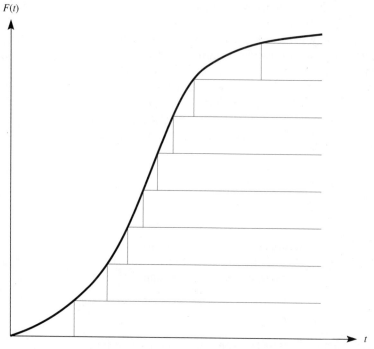

Figure 3-66 An arbitrary driving-force function $F(t)$ can be approximated by a series of unit step functions for analyzing a linear system.

$$Lx = F(t), \quad (3\text{-}139)$$

which has exactly the same meaning as the original equation, Equation 3-137. The operator L has the following properties, which are easily verified:

$$L(x + y) = Lx + Ly, \quad (3\text{-}140)$$

and

$$L(ax) = aLx. \quad (3\text{-}141)$$

The first property states that if x and y are two separate solutions, then the sum $(x + y)$ is also a solution. The second one states that if a is a solution, then the product ax is also a solution, provided that a is a constant. All differential equations that exhibit these two properties are called *linear differential equations*.

In this book we address linear differential equations almost exclusively, primarily because they have relatively simple solutions. Linearity is also important because it permits one to use the principle of superposition, defined as follows: Assume that x_a and x_b are the solutions for driving forces $F_a(t)$ and $F_b(t)$, respectively, so that

$$Lx_a = F_a(t), \quad (3\text{-}142)$$

and

$$Lx_b = F_b(t). \qquad (3\text{-}143)$$

The principle of superposition states that the sum $x_a + x_b$ is the solution for the total force $F_a(t) + F_b(t)$. Thus,

$$L(x_a + x_b) = F_a(t) + F_b(t). \qquad (3\text{-}144)$$

This result permits us to solve a complicated problem by adding together the solutions of a number of simplified problems. For example, an arbitrary driving force can be treated as a succession of step functions, as shown in Figure 3-66. If we know the solution for the step-function force, then the solution for an arbitrary force can be found by adding together a series of step responses. The principle of superposition is an extremely powerful technique that is valid only for linear differential equations.

3-9 ADVANCED DYNAMIC ANALYSIS

The modern solid-state devices used in switching circuits outnumber those used in amplifying circuits by five or ten to one. In switching applications, the input signal makes a rapid transition from one value to another. Both small-signal and large-signal transitions are possible. The small-signal problem was initially treated in Section 3.8, and will be further developed here. The large-signal problem involves some differential equations that are very difficult to solve, so we will in some cases simply outline the solution methods. In certain switching applications the input signal may "dwell" at the starting and ending values for times that are long compared to the signal's transition time. In other commonplace applications, the dwell times are short, which makes the problem yet more difficult.

In the following discussion, we will again focus on the long N^+P diode in order to be specific. In general, we will assume that (1) the N-type region is heavily doped; also we assume that (2) the P-type region is uniformly doped, (3) is long compared to minority-carrier diffusion length, and that (4) low-level conditions prevail. Further, we will consider only two kinds of input waveforms—step and sinusoidal. We do so for three reasons. First, the resulting responses are relatively easy to calculate compared to those for an arbitrary input, such as a ramp function. Second, these responses give us the most direct information about the high-speed operation of a device, our primary interest in this section. And third, the response to an arbitrary input function can be obtained most easily from the step and sinusoidal solutions by applying the principle of superposition, as was noted in the last section.

3-9.1 Survey of Analytic Techniques

Numerous methods exist for analyzing the dynamic behavior of a PN-junction diode, and we can address only a few of them here, primarily those described by linear equations. In addition to being the easiest to solve, linear problems provide the greatest insight. Matters involving the dynamic behavior of a diode can be divided into two broad categories: The first focuses on the internal physics of the device; the second, on the combined behavior of the device and the external circuit in which it has been embedded.

First, let us examine the issue of internal physics. Once again the structure chosen is that of the diode defined in Figure 3-60. Only phenomena in the P-type region will be considered. Significantly, the operating regime to be considered primarily is the intermediate regime emphasized in Figure 3-57; forward bias is large enough to cause diffusion capacitance to be dominant over depletion-layer capacitance, but small enough to yield low-level conditions. The relevant differential equation is the time-dependent diffusion equation, which is a special case of the constant-E continuity-transport equation for electrons (Equation 2-101) discussed in Section 2-7.1. Assuming that E is zero, we have the equation that must be solved with appropriate boundary conditions on x and t:

$$\frac{\partial n'_P(x, t)}{\partial t} = -\frac{n'_P(x, t)}{\tau} + D_n \frac{n'_P(x, t)}{\partial x^2}. \tag{3-145}$$

There are four primary approaches to a solution, each associated with a particular model. These are enumerated in Figure 3-67, along with the common designations of the resulting solutions. The models named in Figure 3-67 appear from top to bottom in order of increasing difficulty. The simplest of these, treated initially in Section 3-9.2, is the *charge-control* model [70, 71]. It effectively eliminates spa-

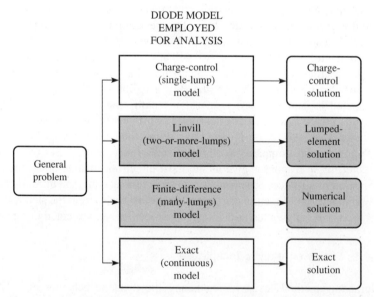

Figure 3-67 Techniques used to analyze the dynamic behavior of a junction diode in terms of its internal physics are listed here in order of increasing accuracy and difficulty.

tially dependent information, and reduces the general problem to one described by a first-order time-dependent ordinary differential equation. In other words, it treats the P-type region as a single "lump."

The first intermediate solution divides the P-type region into two or more parts. If N, the number of these parts, is relatively small (usually two or three), the general problem can still be handled analytically. The relevant model in this case is usually identified as the *Linvill* model [72, 73].

When N is large, we have the *finite-difference* case [74–76], which requires numerical methods. This kind of analysis was introduced in principle in Section 3-8.6, where a multisegment transmission line, as in Figure 3-63(c), was used to model minority-carrier phenomena.

Finally, there is the *exact* solution [77–80], which is presented in Section 3-9.4. The two intermediate models, providing intermediate levels of accuracy and simplicity, will not be presented here in detail. Instead, we concentrate on the accuracy extremes.

Turn next to the second broad issue, the circuit behavior of the diode. Figure 3-68 outlines methods for addressing this problem. In Section 3-9.3 we apply these methods using the charge-control model; in Section 3-9.5 we do so using the exact model. Linear network analysis will be used for most solutions. With a step-function input, the Laplace-transform method is usually the most convenient tool for obtaining the time-domain response, while with a sinusoidal input, the Fourier-transform method is better and yields the frequency-domain response.

Section 3-9.6 will present our first treatment of nonlinear problems encountered in describing the dynamic behavior of a diode. The large-signal switching response of a diode is inherently nonlinear because the diode is typically switched from a high-impedance to a low-impedance state (from OFF to ON), or vice versa. Moreover, we must in general include the depletion-layer capacitance that is itself a

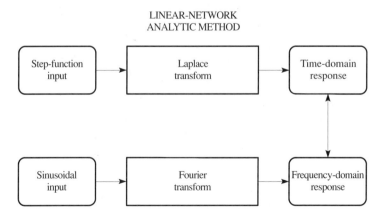

Figure 3-68 Techniques used to analyze the circuit behavior of a junction diode, with the diode represented by one of the models in Figure 3-67.

nonlinear function of applied voltage for all but very small voltage excursions. To solve such problems, one must resort to computer methods. In Section 3-9.6 is described one relevant example of such a method—the SPICE program.

3-9.2 Device-Physics Charge-Control Analysis

The static excess minority-carrier distribution in the P-type region of the long N^+P diode can be written

$$n'_P(x) = n_{0P}\{\exp[-qV_D/kT] - 1\} \exp(-x/L_n), \qquad (3\text{-}146)$$

simply by combining the law of the junction with the information presented in Figure 3-22. The key assumption of dynamic charge-control theory is that an equation having the form of Equation 3-146 is applicable to time-varying conditions, so that

$$n'_P(x, t) = n_{0P} \{\exp[-qv_D(t)/kT] - 1\} \exp(-x/L_n). \qquad (3\text{-}147)$$

Its burden is that the carrier-density profile remains accurately a declining exponential with the boundary value $n'_P(0, t)$ fixed by the instantaneous value of $v_D(t)$. Since this condition holds strictly only under quasistatic conditions, the key assumption is also known as the *quasistatic approximation.* Accepting Equation 3-147, we can proceed directly to a determination of stored charge—the "control" charge—as a function of time. We shall continue to use the symbol $q_S(t)$ to designate the total charge of diffusing excess minority carriers, static plus dynamic, so that $q_S(t) = Q_S + q_s(t)$.

It is now evident that

$$q_S(t) = \int_0^\infty [-qAn'_P(x, t)]dx. \qquad (3\text{-}148)$$

(We will not use the prime symbol on q_S even though it represents excess-carrier charge.) Substituting Equation 3-147 into Equation 3-148 and completing the integration yields

$$q_S(t) = -qAL_n n_{0P} \{\exp[-qv_D(t)/kT] - 1\}. \qquad (3\text{-}149)$$

In this step is the essence of charge-control analysis. Spatial integration has eliminated position as a variable in the problem. (In the exact analysis of Section 3-9.4, by contrast, information on the spatial distribution of carriers is retained and used to infer their transport.)

Now return to the last equation. With the definition

$$Q_{0S} \equiv -qAL_n n_{0P}, \qquad (3\text{-}150)$$

Equation 3-149 can be inverted to yield

$$v_D(t) = -\frac{kT}{q} \ln[1 + q_S(t)/Q_{0S}], \qquad (3\text{-}151)$$

which is the first of two basic charge-control relations.

ADVANCED DYNAMIC ANALYSIS

Exercise 3-83. According to the static charge-control model, there exists a positive and constant excess-carrier charge stored in the P-type region of a reverse-biased diode obeying ideal diffusion theory, and it amounts to $I_0\tau$, where I_0 is the positive saturation current. Explain the physical nature of this positive charge, and show that Q_{OS} as defined in Equation 3-150 is the same as $Q_{OS} = -I_0\tau$.

From Equations 3-38 through 3-40 and the fact that $D_n\tau = L_n^2$, it is evident that

$$I_0\tau = qAD_n(n_{0P}/L_n)\tau = qAL_n n_{0P} = -Q_{OS}.$$

The positive charge accompanying I_0 is a deficiency of electrons, or an integrated "negative" excess-electron density, visible in Figures 3-19(b) and (c). Thus Q_{OS} is a "natural" unit of excess-electron charge (negative) stored in forward operation of the N^+P junction.

Equation 3-151 relates the diode voltage and the stored charge. To relate the diode *current* and stored charge in a similar way, we begin once again with Equation 3-145, the time-dependent diffusion equation:

$$\frac{\partial n'_P(x, t)}{\partial t} = -\frac{n'_P(x, t)}{\tau} + D_n \frac{\partial^2 n'_P(x, t)}{\partial x^2}. \tag{3-152}$$

But the last term can be rewritten as

$$D_n \frac{\partial^2 n'_P(x, t)}{\partial x^2} = \frac{\partial}{\partial x}\left[D_n \frac{\partial n'_P(x, t)}{\partial x}\right] = \frac{\partial}{\partial x}\left[\frac{i(x, t)}{q}\right], \tag{3-153}$$

where $i(x, t)$ is the diffusion current of electrons as a function of time and position inside the P-type region. Now use Equation 3-153 to rewrite Equation 3-152. Holding t constant for the next two steps and multiplying Equation 3-152 through by dx, we have

$$\frac{\partial n'_P(x, t)}{\partial t} dx = -\frac{n'_P(x, t)}{\tau} dx + \frac{d}{dx}\left[\frac{i(x, t)}{q}\right] dx. \tag{3-154}$$

Next, integrate through the P-type region, dropping the functional notation on n'_P for convenience:

$$\frac{\partial}{\partial t}\int_0^\infty n'_P dx = -\frac{1}{\tau}\int_0^\infty n'_P dx + \int_0^\infty d\left[\frac{i(x, t)}{q}\right]. \tag{3-155}$$

Since the external diode current $i_D(t) \approx i(0, t)$, and since $i(\infty, t) = 0$, we then have, after multiplying through by q,

$$-\frac{\partial}{\partial t}\int_0^\infty (-qn'_P) dx = \frac{1}{\tau}\int_0^\infty (-qn'_P) dx + [0 - i_D(t)]. \tag{3-156}$$

But

$$\int_0^\infty (-qn_P')dx \equiv q_S(t), \qquad (3\text{-}157)$$

so

$$i_D(t) = \frac{q_S(t)}{\tau} + \frac{dq_S(t)}{dt}, \qquad (3\text{-}158)$$

which is the second basic relation that completes the charge-control model. The time-dependent diode current, in words, consists of two components—one that sustains the stored charge and a second that adapts it to the time rate of change of charge.

Exercise 3-84. Explain the differences between Equations 3-135 and 3-158.

Equation 3-135, derived by considering the small-signal equivalent-circuit model, confines itself to small-signal charge $q_s(t)$. Equation 3-158, on the other hand, is more general. It was derived by considering *total* charge, and hence is relevant for both small-signal and large-signal problems.

In circuit problems, it is a longstanding custom to use externally applied voltage and current variables that are the negative of the variables just employed. This provides a uniform approach in circuit theory and removes sign ambiguities between *PN* and *NP* diodes. Thus, we will define the applied diode current $i_a(t)$ as $-i_d(t)$ for an *NP* diode and as $i_d(t)$ for a *PN* diode. Note that the positive diode current is always in the same direction as that given by the arrow of the diode symbol. We will also define the applied diode voltage $v_a(t)$ to be positive when it forward biases the diode. Finally, we will define the stored charge in forward operation as the absolute value of the stored excess-minority-carrier charge. With polarity reversal, negative charge will be stored in reverse operation once all transients have decayed. (Exercise 3-83 dealt with the stored-charge polarity reversal accompanying junction-voltage polarity reversal.) To designate this diffusing charge, let us choose the subscript *U* because of its phonic link to "diffusion."

With the new variable definitions, the two charge-control relations become

$$v_A(t) = \frac{kT}{q} \ln\left[1 + \frac{q_U(t)}{Q_{0U}}\right], \qquad (3\text{-}159)$$

and

$$i_A(t) = \frac{q_U(t)}{\tau^*} + \frac{dq_U(t)}{dt}. \qquad (3\text{-}160)$$

Note that effective lifetime τ^* is being used here for greater generality. These equations are valid for either *PN* or *NP* diodes, and $q_U(t)$ must remain positive. Note also that we have omitted the depletion-layer capacitance here since its inclusion produces a nonlinear problem, one that we shall address quantitatively for the first time in Section 3-9.6 using SPICE analysis. Equation 3-159 in its present general form is nonlinear, but as an algebraic equation is tractable enough. The fact that Equation 3-160 is an ordinary, first-order, constant-coefficient, *linear* differential equation is very significant, though, and underlies its utility for large-signal as well as small-signal problems. Notice that these two basic charge-control equations are parametric in form, with the controlling charge $q_U(t)$ being the parameter.

To study the small-signal behavior of a diode, it is also a longstanding custom to write current and voltage equations describing static effects, and other current and voltage equations describing small-signal dynamic effects. To start this process, and at the same time to review the relevant symbols, we can write the currents and voltages as

$$i_A(t) = I_A + i_a(t), \qquad (3\text{-}161)$$

and

$$v_A(t) = V_A + v_a(t), \qquad (3\text{-}162)$$

while the charge parameter is written in parallel fashion as

$$q_U(t) = Q_U + q_u(t). \qquad (3\text{-}163)$$

Let us start with the second basic charge-control relation, Equation 3-160, substituting the right-hand sides of Equations 3-161 and 3-163 into it to obtain static and time-dependent terms on each side. Equating the terms separately, we get from the static terms,

$$I_A = Q_U/\tau^*, \qquad (3\text{-}164)$$

and from the dynamic terms

$$i_a(t) = (q_u/\tau^*) + [dq_u(t)/dt]. \qquad (3\text{-}165)$$

Substituting the right-hand side of Equation 3-163 into Equation 3-159, in order to treat the first charge-control equation similarly, we find

$$v_A(t) = \frac{kT}{q} \ln\left[1 + \frac{Q_U + q_u(t)}{Q_{0U}}\right]. \qquad (3\text{-}166)$$

Using a common denominator in the argument, and multiplying both numerator and denominator by the quantity $(Q_{0U} + Q_U)$, gives us

$$\begin{aligned} v_A(t) &= \frac{kT}{q} \ln\left[\left(\frac{Q_{0U} + Q_U}{Q_{0U} + Q_U}\right)\left(\frac{Q_{0U} + Q_U + q_u(t)}{Q_{0U}}\right)\right] \\ &= \frac{kT}{q} \ln\left[\frac{Q_{0U} + Q_U}{Q_{0U}}\right] + \frac{kT}{q} \ln\left[\frac{Q_{0U} + Q_U + q_u(t)}{Q_{0U} + Q_U}\right] \\ &= \frac{kT}{q} \ln\left[1 + \frac{Q_U}{Q_{0U}}\right] + \frac{kT}{q} \ln\left[1 + \frac{q_u(t)}{Q_{0U} + Q_U}\right]. \end{aligned} \qquad (3\text{-}167)$$

The definition of Q_{0U} can be inferred from Equation 3-150. Comparing Equation 3-167 with Equation 3-162 yields

$$V_A = \frac{kT}{q} \ln\left[1 + \frac{Q_U}{Q_{0U}}\right], \tag{3-168}$$

and

$$v_a(t) \approx \frac{kT}{q}\left[\frac{q_u(t)}{Q_{0U} + Q_U}\right], \tag{3-169}$$

where we have used for the latter item the fact that $\ln(1 + x) \approx x$ for small values of x. Equations 3-164 and 3-168 represent, in parametric form, the conventional static solution for the N^+P diode. These equations can be used to derive a needed expression for the diode conductance, $\Delta I_A/\Delta V_A$, in terms of Q_U. To this end we apply small but finite variations to the static expressions, yielding

$$\Delta I_A = \Delta Q_U/\tau^*, \tag{3-170}$$

and

$$\Delta V_A = (kT/q)[\Delta Q_U/(Q_{0U} + Q_U)]. \tag{3-171}$$

Now divide Equation 3-170 by Equation 3-171 to obtain the diode conductance:

$$g = (kT/q\tau^*)/(Q_{0U} + Q_U). \tag{3-172}$$

This equation in turn permits us to rewrite Equation 3-169 as

$$v_a(t) = q_u(t)/(\tau^* g), \tag{3-173}$$

the desired expression for $v_a(t)$ in terms of $q_u(t)$. The companion to this small-signal expression is Equation 3-160, repeated here for convenience:

$$i_a(t) = \frac{q_u(t)}{\tau^*} + \frac{dq_u(t)}{dt}. \tag{3-174}$$

The last two equations describe the small-signal response of the N^+P diode.

3-9.3 Circuit-Behavior Charge-Control Analysis

Let us now place the diode in a circuit that delivers a small-signal current step. This problem was already examined intuitively using the equivalent-circuit model, with Equation 3-132 resulting. But, armed with the charge-control expressions that have just been developed in a formal way and with the new and more general variable definitions, we may now profitably revisit the problem. Figure 3-69(a) presents the circuit, and Figure 3-69(b) shows graphically the current waveform it delivers. As in Equation 3-161, this current can be divided into static and small-signal dynamic parts, the latter now being written

$$i_a(t) = \Delta I_A u(t), \tag{3-175}$$

an expression parallel to Equation 3-130. Substituting the right-hand side of Equa-

ADVANCED DYNAMIC ANALYSIS 395

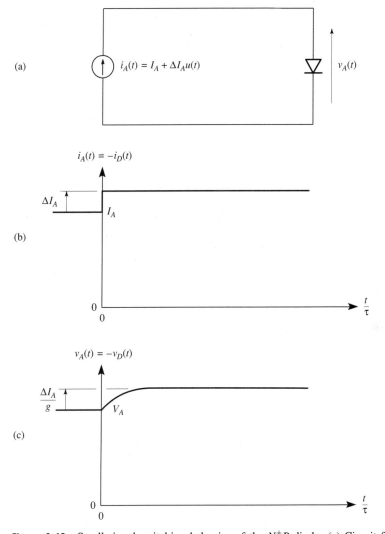

Figure 3-69 Small-signal switching behavior of the N^+P diode. (a) Circuit for the switching experiment. (b) A small, abrupt increment ΔI_A in forward current is applied to the diode. (c) The resulting voltage response of the diode.

tion 3-175 for the left-hand side of Equation 3-174 and solving the resulting differential equation yields

$$q_u(t) = \tau^* \Delta I_A [1 - \exp(-t/\tau^*)]. \tag{3-176}$$

Substituting this expression into Equation 3-173 yields

$$v_a(t) = (\Delta I_A/g)[1 - \exp(-t/\tau^*)], \tag{3-177}$$

which is the desired small-signal solution, shown graphically in Figure 3-69(c).

Equation 3-177 is consistent with the result plotted in Figure 3-65(b). A repetition of that equivalent-circuit treatment is in order, however, to complete our conversion to the more general variable symbols, adopting in addition the symbol C_u^* for effective small-signal diffusion capacitance. The small-signal equivalent circuit given in Figure 3-70 shows that the present problem is equivalent to the problem of charging a capacitance $C_u^* = g\tau^*$ in parallel with a conductance $g = qI_A/kT$, with the former relationship given in Equation 3-128. The initial small-signal current is zero and the capacitance has zero small-signal charge. At $t = 0$, then, the current through the circuit is abruptly increased by the small amount ΔI_A; it is clear that the voltage on the capacitor will increase exponentially as a function of time, and will approach a limiting small-signal value of $\Delta I_A/g$, with the relevant time constant being $\tau^* = C_u^*/g$. For completeness, let us solve the same problem explicitly by means of circuit theory. Using Figure 3-70, we can write the small-signal diode current as

$$i_a(t) = \Delta I_A u(t) = g v_a(t) + C_u^*[dv_a(t)/dt]. \tag{3-178}$$

The Laplace transform of this equation is

$$i_a(s) = (\Delta I_A/s) = g v_a(s) + s C_u^* v_a(s). \tag{3-179}$$

Solving for $v_a(s)$ yields

$$v_a(s) = \frac{\Delta I_A}{s}\left[\frac{1}{g(1 + s\tau^*)}\right], \tag{3-180}$$

where we have used the fact noted just above that C_u^*/g equals τ^*. Transforming back to the time domain yields

$$v_a(t) = (\Delta I_A/g)[1 - \exp(-t/\tau^*)], \tag{3-181}$$

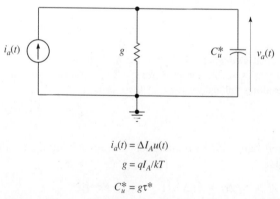

$i_a(t) = \Delta I_A u(t)$

$g = qI_A/kT$

$C_u^* = g\tau^*$

Figure 3-70 Small-signal equivalent circuit for a forward-biased diode subjected to a small-signal switching transition. Effective diffusion capacitance C_u^* is in parallel with small-signal conductance g.

which is the same as the result given by Equation 3-177. This equation is plotted in Figure 3-71 for $\tau^* = \tau$ as curve (a), and for $\tau^* = \tau/2$ as curve (b). The figure also gives the exact result (to be discussed in Section 3-9.5) as curve (c). A comparison of the curves shows that the best agreement between the exact and charge-control models is obtained when $\tau^* = \tau/2$, a result consistent with the discussion given in Section 3-8.6. The concept of voltage *rise time* t_{RISE} aids a quantitative comparison, where this is taken to be the time required to go from 10% to 90% of the final voltage. The values entered in Figure 3-71 show that using $\tau^* = \tau/2$ in the charge-control model gives best agreement with the exact model.

As a second problem, consider a small-signal sinusoidal input voltage

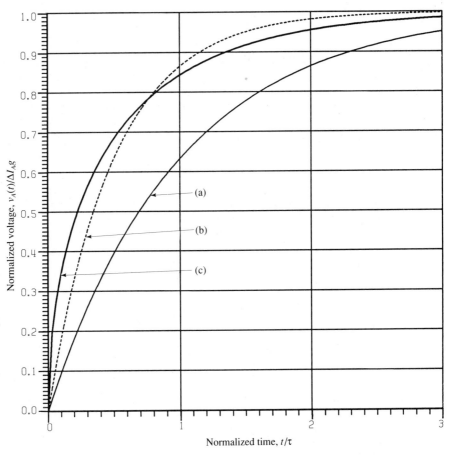

Figure 3-71 Small-signal switching behavior of the N^+P diode, normalized voltage versus normalized time, comparing charge-control and exact results. (a) Charge-control result, with $\tau^* = \tau$; $(t_{RISE}/\tau) = 2.2$. (b) Charge-control result, with $\tau^* = \tau/2$; $(t_{RISE}/\tau) = 1.1$. (c) Exact result, from Section 3-9.5; $(t_{RISE}/\tau) = 1.3$.

$$v_a(t) = V_a \exp(+j\omega t), \tag{3-182}$$

where V_a is in general a complex number. We assume that diode current and voltage have the same form, so that

$$i_a(t) = I_a \exp(+j\omega t). \tag{3-183}$$

Eliminating $q_u(t)$ from the charge-control expressions for small-signal voltage and current, Equations 3-173 and 3-174, yields

$$i_a(t) = v_a(t)g + [dv_a(t)/dt]\tau^* g. \tag{3-184}$$

Substituting Equations 3-182 and 3-183 into this expression gives, after removing the common factor $\exp(+j\omega t)$,

$$I_a = V_a g(1 + j\omega\tau^*). \tag{3-185}$$

We can define an ac admittance $y(\omega)$ as I_a/V_a so that

$$y(\omega) = g(1 + j\omega\tau^*). \tag{3-186}$$

The inverse of the diode's admittance, which is to say, its impedance $z(\omega)$, is plotted in Figure 3-72 for $\tau^* = \tau$ as curve (a) and for $\tau^* = \tau/2$ as curve (b), along with the corresponding exact result as curve (c). A comparison of the curves again shows that the best agreement is obtained when $\tau^* = \tau/2$ is used in the charge-control model.

At this point, generalization is in order. As a first item, let the stimulus applied to the diode be converted from a small-signal positive-going current step to a small-signal negative-going rectangular current pulse of duration T. Further, let $T \gg \tau^*$. This current stimulus is depicted in Figure 3-73(a), where the static component of current is now labeled $I_A = I_F$. (The new subscript denotes "forward," anticipating a subsequent generalization of the switching process that will allow reverse current and voltage.) An alternate description of this kind of current stimulus is that it amounts to a sequence of equal but opposite-polarity current steps. Not surprisingly, the charge-control model predicts that both of the resulting transitions in stored charge $q_U(t)$ will be exponentially decaying functions exhibiting the characteristic time τ^*, as is shown in Figure 3-73(b).

The next step of generalization is to alter the amplitude of the current pulse, increasing it until truly large-signal switching exists. The resulting transition in controlling charge $q_U(t)$ is depicted for this case in Figure 3-73(c). Note carefully that the gap has been eliminated from the ordinate axis, thus signifying that ordinate calibrations differ grossly in parts (b) and (c) of Figure 3-73. But the resulting charge-transition curves are qualitatively identical! This result is a consequence of the important fact that $q_U(t)$ is the solution of a *linear* differential equation, the charge-control current expression, Equation 3-160. Putting the matter a different way, the expression for small-signal current as a function of $q_u(t)$, Equation 3-174, is identical in form to the expression for total current as a function of $q_U(t)$, Equation 3-160. As a result, the small-signal solution, Equation 3-178, has the same form as the large-signal solution, except that the static charge $Q_U = \tau^* I_F$ must be added in the large-signal case:

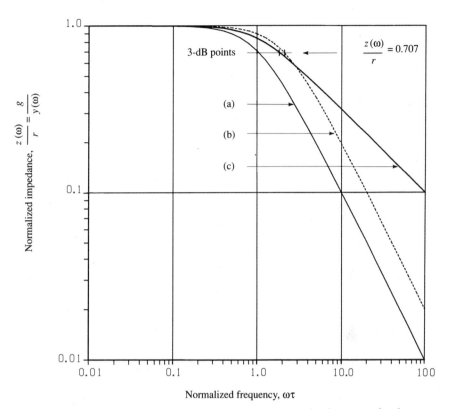

Figure 3-72 Normalized impedance versus normalized angular frequency for the junction diode, comparing charge-control and exact results, and giving the values of normalized 3-dB frequency ($\omega_{3\text{-}dB}\tau$) read from the curves. (a) Charge-control result, with $\tau^* = \tau$; $(\omega_{3\text{-}dB}\tau) = 1.0$. (b) Charge-control result, with $\tau^* = \tau/2$; $(\omega_{3\text{-}dB}\tau) = 2.0$. (c) Exact result, from Section 3-9.5; $(\omega_{3\text{-}dB}\tau) = 1.8$.

$$q_U(t) = \tau^* \Delta I_A [1 - \exp(-t/\tau)] + \tau^* I_F. \tag{3-187}$$

There is yet another way of explaining the all-important linearity in charge-control analysis, an explanation that appeals to the small-signal equivalent circuit. Equation 3-128 informs us that

$$\tau^* = (C_u^*/g) = rC_u^*. \tag{3-188}$$

That is, insofar as minority carriers are concerned, the diode's characteristic time τ^* is equal to its "*RC* time," where the dynamic resistance and diffusion capacitance in the small-signal equivalent-circuit model are the relevant elements. But in Equation 3-98 it is shown that

$$r = kT/q(-I),$$

where I is forward current, negative as defined there. Thus r varies inversely with

400 PN JUNCTIONS

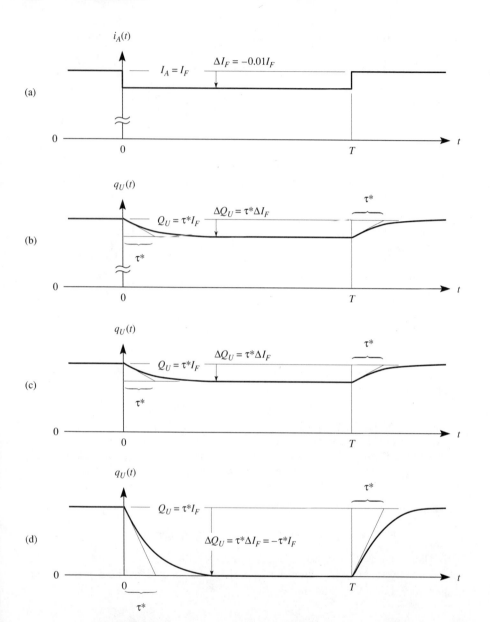

Figure 3-73 Current-pulse switching in a junction diode. (a) Small-signal current-pulse stimulus. (b) Small-signal controlling-charge response resulting from the given current pulse. (c) Large-signal charge response. (d) Large-signal charge response wherein the stored charge is reduced to zero during the pulse.

forward current. But combining Equation 3-189 with Equation 3-188 yields

$$C_u^* = (\tau^* q/kT)(-I), \quad (3\text{-}190)$$

showing that C_u^* varies directly with forward current. Thus Equation 3-188, derived strictly for small-signal conditions, holds actually for large-signal conditions as well because the rC_u^* product is independent of forward current. This fact underlies the similarity of Figures 3-73(b) and 3-73(c).

Although the case depicted in Figure 3-73(c) involves a truly large signal, the signal satisfies the constraints outlined in the second paragraph of Section 3-9.1 and falls within the regime where charge-control analysis is most meaningful. To extract a maximum of information from the charge-control model, however, it is worthwhile to examine its predictions outside that regime, even allowing polarity reversal. Let us simply set aside the junction-current mechanism of space-charge-layer recombination, letting the junction obey simple theory. We shall defer until Section 3-9.6 (on SPICE analysis) the effects of the nonlinear depletion-layer capacitance $C_T(V_A)$—see also Section 3-8.6—because it demands numerical treatment.

As the next increase in signal amplitude, assume that current is reduced to zero during the rectangular-pulse interval. The controlling-charge response would then be as illustrated in Figure 3-73(d), an obvious extrapolation from Figure 3-73(c). For any of the curves in Figures 3-73(b) through 3-73(d), one can write the accompanying voltage waveform by invoking Equation 3-159, the nonlinear though simple relation between charge and voltage:

$$v_A(t) = \frac{kT}{q} \ln\left[1 + \frac{q_U(t)}{Q_{0U}}\right]. \quad (3\text{-}191)$$

Because the current pulse employed to produce the charge response in Figure 3-73(d) takes the diode from a conducting state to a nonconducting state, and then back again, the successive current steps (or responses) are often termed *turn-off* and *turn-on* transitions, respectively.

Exercise 3-85. Determine the value of $dq_U(t)/dt$ immediately after $t = 0$ in each of Figures 3-73(b) through 3-73(d).

In the geometry of Figure 3-73(b), the slope of the curve $q_U(t)$ versus t just after $t = 0$ is seen to be

$$[dq_U(t)/dt] = (\Delta Q_U/\tau^*) = (\tau^* \Delta I_F/\tau^*) = \Delta I_F.$$

Combining this result with the information in Figure 3-73(a) then gives us

$$[dq_U(t)/dt] = -(0.01)I_F.$$

In Figure 3-73(c), we have $[dq_U(t)/dt] = \Delta I_F$, where ΔI_F is a large but unspecified negative current increment. In Figure 3-73(d),

$$[dq_U(t)/dt] = \Delta I_F = -I_F.$$

For the next degree of generalization, let us permit diode current and voltage to change sign, a problem of considerable practical importance. The cause of realism is served, however, by augmenting the idealized current source employed for the switching cases illustrated in Figure 3-73. Let us adopt the simple but useful circuit shown in Figure 3-74(a), wherein another voltage source and resistor could indeed be substituted for the current source if that were desired. Initially the N^+P diode has a constant forward current I_F and voltage V_F. Then at $t = 0$, the bias current is abruptly switched, placing the diode in series with the resistor R and the reverse voltage $-V_R$. The minus sign here comes from another longstanding tradition in the switching literature. Both forward and reverse voltages, as well as currents, are taken to be positive, so that explicit negative signs are needed for $-V_R$ and $-I_R$. Then at time T (long compared to τ^*), the bias circuit is switched back to the current I_F. Let us consider the two transitions one at a time.

The abrupt circuit-voltage reversal at $t = 0$ is accompanied by an equally abrupt current reversal. This is because extraction of charge stored in the diffusion capacitance C_u^* commences immediately. However, diode voltage—unlike circuit voltage—does not change abruptly. The diode briefly is "batterylike," or more accurately, is a charged capacitor. Departure from the small voltage $V_F \approx 0.7$ V is gradual. With $V_R \gg V_F$, the voltage on R just after $t = 0$ is approximately $-V_R$, so that the reverse current in the circuit is $-I_R \approx -V_R/R$.

Having arbitrarily specified V_R, we can adjust I_R in an equally arbitrary way through the adjustment of R. Suppose we let $I_R = I_F$. Now, since the principle illustrated in Exercise 3-85 is still valid, the value of $dq_A(t)/dt$ immediately after $t = 0$ is given by ΔI_F. But under present conditions,

$$\Delta I_F = (-I_R) - I_F = -2I_F. \tag{3-192}$$

In other words, the nature of the turn-off response of $q_U(t)$ is markedly affected by choice of I_R, a fact illustrated geometrically in Figure 3-74(b). The turn-off transition is now much steeper than in Figure 3-73(d), and could be made steeper still by choosing larger I_R. Note carefully, however, that while the construction of Figure 3-74(b) yields a charge-control assessment of the turn-off transient, the dashed portion of the curve has *no* meaningful physical interpretation. A literal reading of the model's prediction says that charge drops steeply to zero, as is shown in Figure 3-74(b), and then goes to a miniscule negative value, a subject illustrated in Exercise 3-83. The charge-versus-time curve stays there until turn-on commences.

In the case of turn-on, the charge-control model delivers the meaningful prediction shown at the right in Figure 3-74(b). Especially important is the fact that the value of I_R has *no* effect on the turn-on transient in stored charge so long as $T \gg \tau^*$. Since the turn-on transient exhibits a characteristic time of τ^*, calculation of a rise time for charge is straightforward, using a definition parallel to that used previously for voltage rise time. The time to go from 10% to 90% of the final forward charge amounts to $2.2\tau^*$.

Next consider the behavior of $i_A(t)$ versus t. We made the point earlier that the current transition is abrupt at $t = 0$, as is shown in Figure 3-74(c) (with the condition $I_R = I_F$ arbitrarily chosen). The reverse current continues as long as there is

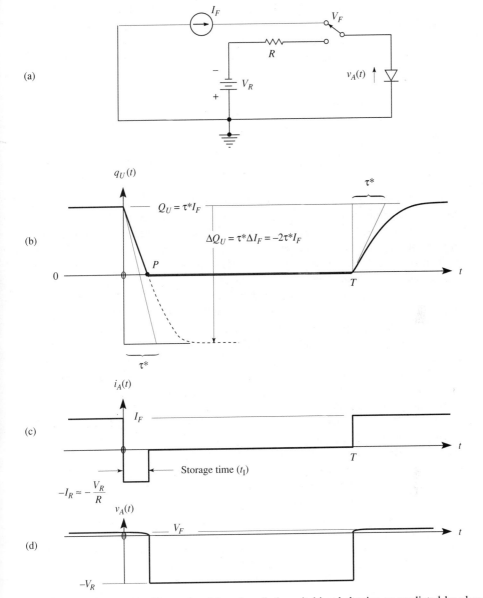

Figure 3-74 Large-signal junction-diode switching behavior as predicted by charge-control theory. (a) Circuit used to switch the diode from forward to reverse bias, and back again at a time $T \gg \tau^*$. (b) Charge response for case $I_R = I_F$, showing elimination of stored charge at point P. (c) Current response, showing storage time t_I defined by reduction of stored charge to zero, a time that depends on current ratio I_R/I_F. (d) Voltage response.

stored charge in the diffusion capacitance to be extracted. The duration of this *storage time* can be inferred from Figure 3-74(b), since $q_U(t)$ has dropped to zero at the point P. Let us use the symbol t_I for the storage time. For $t > t_I$, stored excess minority carriers go from zero density to the negative density mentioned above. As a practical matter, this negative charge has negligible consequences because the density magnitudes are so small. More important than the reversal of charge sign is the fact that the boundary of the majority-carrier "sea" retreats from the metallurgical junction, which is simply a way of describing depletion-layer capacitance $C_T(V_A)$, a subject that has been deferred until later (Section 3-9.6). Hence, another literal reading of the charge-control model says that current goes abruptly toward zero at the time t_I, as is shown in Figure 3-74(c), and becomes constant at its tiny reverse-current value.

Another subtle effect is appropriately mentioned here. There is an interaction of depletion-layer capacitance $C_T(V_A)$ and diffusion capacitance $C_U(V_A)$. Here, notation has been introduced for diffusion capacitance that is parallel to that introduced in Section 3-8.6 for depletion-layer capacitance. And since we are now considering explicit time dependence, these two quantities can also be legitimately written as $C_T(v_A)$ and $C_U(v_A)$. The interaction that enters is that as the depletion layer expands during turn-off, the moving boundary of the junction "eats into" the distribution of minority carriers in transit. While this effect could be important under some circumstances, it is small at present, because boundary motion is measured in micrometers in our diode, while diffusion length has a value of the order of hundreds of micrometers.

The abrupt current transition to zero at $t = t_I$ is the most unrealistic feature of charge-control analysis, and stems from its fundamental assumption: Charge stored in the diffusion capacitance is taken to be zero at the instant that excess minority-carrier density at the junction boundary hits zero because the shape of the density profile is taken to be qualitatively constant. In fact the more accurate ("exact") analysis of Section 3-9.4 will show that the boundary value hits zero well before all stored charge has been eliminated. As a result, the storage period of duration t_I is followed by a period of gradual return to zero charge, having a duration we shall label t_{II}, often termed *decay time*.

The last step is to plot $v_A(t)$ versus t, using Equation 3-191 for the positive-bias portions of the switching waveform. Since charge-control theory predicts that the diode approximates an open circuit in reverse bias, the reverse-voltage curve shown in Figure 3-74(d) is plausible. The diode experiences the voltage $-V_R$ from the end of the storage period until turn-on.

Now let us examine charge-control predictions quantitatively. During the storage time, the solution of the charge-control relation is

$$q_U(t) = I_F \tau^* - (I_F + I_R)\tau^*[1 - \exp(-t/\tau^*)], \tag{3-193}$$

where an arbitrary relationship of I_R to I_F is now allowed. This is a general expression that can be tailored specifically to the solutions plotted in Figures 3-73(c), 3-73(d), and 3-74(b). At $t = t_I$, $q_U(t) = 0$, so that Equation 3-193 yields

$$[I_F/(I_F + I_R)] = [1 - \exp(-t_1/\tau^*)], \qquad (3\text{-}194)$$

an expression that can be solved for t_1. It yields

$$t_1 = \tau^* \ln\left[\frac{I_R/I_F}{1 + (I_R/I_F)}\right]. \qquad (3\text{-}195)$$

The storage time is plotted in normalized fashion (using lifetime τ for normalization) in Figure 3-75, for both $\tau^* = \tau$ and $\tau^* = \tau/2$. Also given for comparison is the exact result to be developed in Section 3-9.4.

Returning now to the case wherein $I_R = I_F$ and $\tau^* = \tau$, we find that Equation 3-195 yields $t_1 = 0.65\tau$. As can be seen in Figure 3-75, this is a value nearly three times larger than the exact result, which is $t_1 = 0.23\tau$. When $\tau^* = \tau/2$, the agreement is appreciably better between the charge-control and exact models.

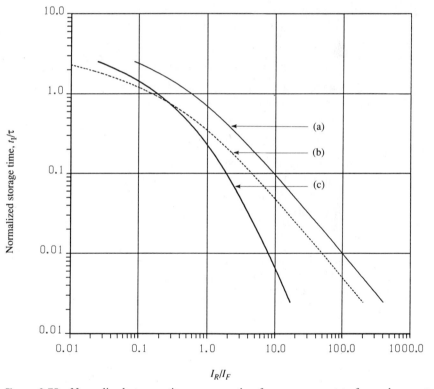

Figure 3-75 Normalized storage time versus ratio of reverse current to forward current, comparing charge-control and exact results. (a) Charge-control result, with $\tau^* = \tau$. (b) Charge-control result, with $\tau^* = \tau/2$. (c) Exact result, from Section 3-9.4.

3-9.4 Device-Physics Exact Analysis

The analytic method we have labeled *exact* differs from charge-control analysis by preserving and using information on the spatial distribution of charge. Let us continue to use $v_A(t)$ and $i_A(t)$ for the total voltage and current applied to the diode, positive when in the forward direction. In charge-control analysis, a parametric "lump" of charge was defined by eliminating spatial dependence though spatial integration. With time as the only independent variable, we then had $q_U(t)$, $q_u(t)$, and Q_U as the total charge, dynamic component of charge, and static component of charge, respectively. These symbols conveniently designated the excess minority-carrier charge stored in the diode's diffusion capacitance. It is now a further convenience to introduce simplified notation for what has previously been designated as $n'_P(x, t)$, the density of excess minority carriers as a function of position and time in the *P*-type region. Let the respective symbols, parallel to those for charge, be $n_U(x, t)$, $n_u(x, t)$, and $N_U(x)$.

The starting point for exact analysis focuses on the internal physics of the diode, uses the time-dependent diffusion equation (Equation 3-145), and also uses the diode defined in Figure 3-60(a), Section 3-8.6. During the initial turn-off period t_I, the appropriate boundary conditions on $n_U(x, t)$ are

$$n_U(\infty, t) = 0; \tag{3-196}$$

$$n_U(x, 0) = n_{0P}\{\exp[qv_A(0)/kT] - 1\} \exp(-x/L_n). \tag{3-197}$$

The meaning of the symbol n_{0P}, equilibrium minority-electron density in the *P*-type region, is clear and unaltered, so we continue to use it. The reverse current for $t > 0$ is given by a third boundary condition,

$$I_R = qAD_n[\partial n_U(x, t)/\partial x]|_{x=0}, \tag{3-198}$$

where a thicket of negative signs must be sorted out. First of all, I_R carries an explicit negative sign by definition, and this cancels the Fick's-law negative sign. Second, consistent with the treatment of q_U as positive charge, it is necessary to take electronic charge q as positive. Finally, the algebraic sign of the density gradient at $x = 0$ switches instantly from negative to positive at $t = 0$, and so I_R is properly positive.

Next, the time-dependent diffusion equation is solved. A number of authors have addressed this problem [70, 77–79] and they find that the solution of the problem is given by

$$\frac{n_U(x, t)}{n_U(0, 0)} = \exp\left[-\frac{x}{L_n}\right] \tag{3-199}$$
$$- \frac{1}{2}\left[1 + \frac{I_R}{I_F}\right] \times \left\{\exp\left[-\frac{x}{L_n}\right] \text{erfc}\left[\frac{1}{2}\frac{x/L_n}{\sqrt{t/\tau}} - \sqrt{\frac{t}{\tau}}\right]\right.$$
$$\left. - \exp\left[-\frac{x}{L_n}\right] \text{erfc}\left[\frac{1}{2}\frac{x/L_n}{\sqrt{t/\tau}} + \sqrt{\frac{t}{\tau}}\right]\right\},$$

where $n_U(0, 0) = I_F L_n/qAD_n$. It is of course the true lifetime that occurs in this expression.

Exercise 3-86. Justify the expression immediately preceding for $n_U(0, 0)$.

Letting current and carrier density be positive quantities, one sees that the forward current I_F is equal to the magnitude of the diffusion current at $x = 0$, or $I_F = -AqD_n[-n_U(0, 0)/L_n]$, which is consistent with the expression for $n_U(0, 0)$.

Figure 3-76(a) gives the Equation 3-199 function, the normalized electron profiles in the P-type region, for six values of t. At $t = 0$, the profile is given by Equation 3-197, one of the boundary conditions. Note that $n_U(\infty, t) = 0$, as is required by Equation 3-196, the first boundary condition. As time increases, the

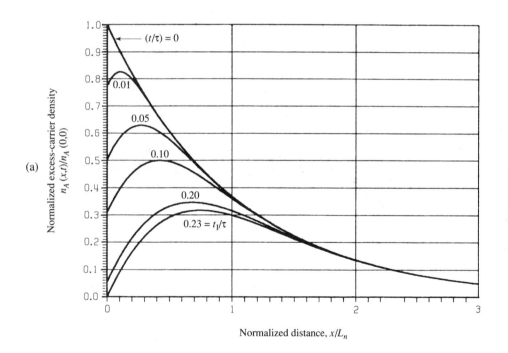

Figure 3-76 Normalized excess-electron density versus position with time as a parameter. (a) Curves calculated using diffusion-equation (exact) analysis for $I_F = I_R$, showing conditions existing during the storage period. (b) Curves calculated using charge-control analysis for $I_F = I_R$, unrealistic except under quasistatic conditions.

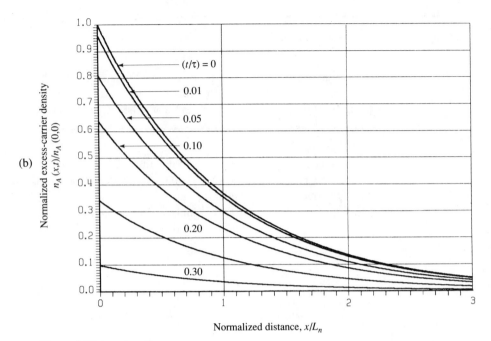

(b)

Figure 3-76 (*continued*)

electron concentration at $x = 0$ decreases until it equals zero at $t = t_1 = 0.23\tau$. The slope at $x = 0$ is always the same as is required by Equation 3-198, the third boundary condition. Figure 3-76(b) presents analogous curves that are associated with charge-control analysis, unrealistic both quantitatively and qualitatively. The qualitative inaccuracy stems from the fundamental charge-control assumption, and is dramatized by the fact that the slope at $x = 0$ has the wrong sign!

At the boundary $x = 0$, Equation 3-199 becomes

$$n_U(0, t) = \frac{I_F L_n}{qAD_n}\left[1 - \left(1 + \frac{I_R}{I_F}\right)\text{erf}(\sqrt{t/\tau})\right], \quad (3\text{-}200)$$

where we have used the identity $\text{erfc}(-x) - \text{erfc}(x) = 2\,\text{erf}(x)$. The time-dependent diode voltage is given by

$$v_A(t) = -\frac{kT}{q}\ln\left[1 + \frac{n_U(0, t)}{n_{OP}}\right]. \quad (3\text{-}201)$$

Since $n_U(0, t)$ vanishes at $t = t_I$, we can determine t_I by setting Equation 3-200 equal to zero and substituting t_I for t in the result, yielding

$$\text{erf } \sqrt{t_I/\tau} = \frac{1}{1 + (I_R/I_F)}. \tag{3-202}$$

This function is plotted in Figure 3-75 as curve (c), and shows clearly that $t_I = 0.23\tau$ when $I_R/I_F = 1$, as was already noted. Thus the constant-current phase lasts about one quarter of the minority-carrier lifetime.

When $t > t_I$, the diode enters the decay period t_{II}, wherein the diode current is no longer constant but instead declines toward zero. To solve this problem, we again use the time-dependent diffusion equation. To be as accurate as possible, one should use an initial electron profile identical to that at the end of the period t_I. But this would require some difficult mathematics, so we again resort to approximate methods. In particular, we retain the first two boundary conditions, Equations 3-196 and 3-197, and then examine the consequence of assuming that diode *voltage* is suddenly switched from V_F to zero. The sudden imposition of the voltage step from V_F to zero, however, means that a third boundary condition becomes valid at $t + \delta t$, where δt is vanishingly small:

$$n_U(0, t) = 0. \tag{3-203}$$

Using these equations, one obtains the normalized electron profiles given in Figure 3-77. The boundary value for excess-electron density drops to zero as quickly as does the applied voltage. The slope at $x = 0$ is large for small values of t and decreases rapidly as t increases. The diode current can be calculated from this slope and the result is given in Figure 3-78.

Comparing the curves in Figures 3-76 and 3-77 suggests that the curve labeled $(t/\tau) = 0.1$ in the latter can be taken as a reasonable approximation to the bottom curve in the former. The electron-profile curve labeled $(t/\tau) = 0.1$ in Figure 3-77 is very similar to that calculated for the end of the period t_I, labeled $(t/\tau) = 0.23$ in Figure 3-76. When $t_I > 0.1\tau$, the calculated profiles for the period t_{II} will approximate closely those that would have been obtained if we had used the actual electron profile at the end of the period t_I as the initial condition. Furthermore, in Figure 3-78, the point at the top of the diagram indicates that for $(t/\tau) = 0.1$, the ratio $i_A(t)/(-I_F)$ approximates unity, the initial condition desired to make the present case of voltage switching approximately congruent with the previous case of current switching, wherein current was changed abruptly to $I_R = -I_F$.

Figure 3-77 shows that $n_U(x, t)$ approaches zero everywhere as t becomes large, which is approximately the expected result. In fact, of course, the excess-electron density takes on negative values within a few diffusion lengths of the junction; but the maximum magnitude of this excess density is n_{0P}, and we have arranged to have $n_U(0, 0) \gg n_{0P}$.

If we define t_{II} as the time required to reduce the diode current $i_A(t)$ to 10% of $I_R = I_F$, minus the initial interval 0.1τ, then Figure 3-78 shows that $t_{II} = 0.59\tau$. Thus, for $(I_R/I_F) = 1$, the period t_I requires a time of 0.23τ, and the period t_{II}

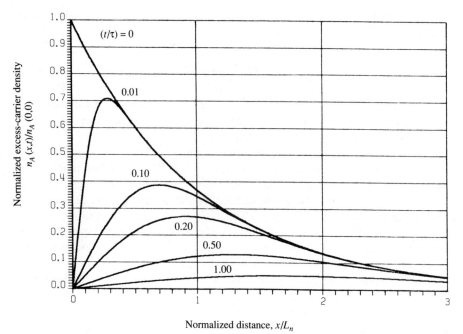

Figure 3-77 Normalized excess-electron density versus position with time as a parameter, employed for calculating decay-period response in the exact analysis.

requires a time of 0.59τ, so that the total switching time amounts to 0.82τ. Results for other values of I_R/I_F are given in Figure 3-79. In computations for these curves, however, we have omitted the initial small time interval involved in voltage switching to simplify the calculations. That is, the value of 0.82τ just calculated has been plotted as 0.92τ.

Figure 3-80 summarizes the exact results for the large-signal switching response of our N^+P diode for the condition $(I_R/I_F) = 1$. The diode current is constant and equal to I_F when $t < 0$. At $t = 0$, the current changes abruptly to $-I_R$ and remains constant until $t = t_I$, the end of the storage period. As t increases further, the current decreases and approaches zero for large values of t. At $t = t_I + t_{II} - 0.1\tau \approx t_I + t_{II}$, the diode current equals 10% of $-I_R$, where we recall that I_R is defined to be positive. This ends our discussion of the decay period and of large-signal switching behavior.

As a second example of the exact analysis, consider the internal device behav-

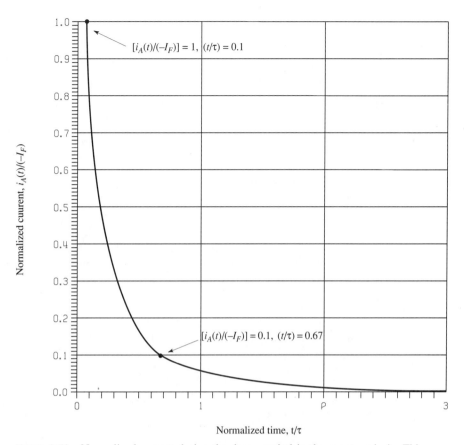

Figure 3-78 Normalized current during the decay period in the exact analysis. This result is inferred from the slope values at $x = 0$ in Figure 3-77.

ior of a diode with a small-signal sinusoidal input voltage. The boundary condition on excess electron density at $x = 0$ is given in general by

$$n_U(0, t) = n_{OP}\left[\exp\left(\frac{qv_A(t)}{kT}\right) - 1\right], \qquad (3\text{-}204)$$

where $v_A(t)$ is defined by

$$v_A(t) = V_A + v_a(t) = V_A + V_a \exp(j\omega t). \qquad (3\text{-}205)$$

Here we let V_A be the dc part of the diode voltage and V_a, the complex coefficient (phasor) of the ac term. If $|V_A| \gg kT/q$ and $|V_a| \ll kT/q$, then Equations 3-204 and 3-205 yield

$$n_U(0, t) \approx n_{OP}\left\{\exp\left(\frac{qV_A}{kT}\right)\exp\left[\frac{qV_a}{kT}\exp(j\omega t)\right] - 1\right\}$$

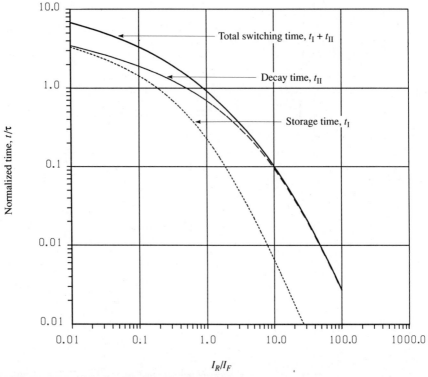

Figure 3-79 Normalized duration of (a) storage period, (b) decay period, and (c) total switching transient, versus the ratio of reverse current to forward current during the storage period, all calculated using the exact approach. Curve (a) is the same as curve (c) in Figure 3-75.

$$\approx n_{OP}\left\{\exp\left(\frac{qV_A}{kT}\right)\left[1 + \frac{qV_a}{kT}\exp(j\omega t)\right] - 1\right\}. \tag{3-206}$$

Multiplying out this expression and manipulating the second major term that results gives us

$$n_U(0, t) = n_{OP}\left[\exp\left(\frac{qV_A}{kT}\right) - 1\right]$$
$$+ n_{OP}\left[\exp\left(\frac{qV_A}{kT}\right) - 1 + 1\right]\left(\frac{qV_a}{kT}\right)\exp(j\omega t). \tag{3-207}$$

Letting

$$n_{OP}\left[\exp(qV_A/kT) - 1\right] \equiv N_U(0), \tag{3-208}$$

Equation (3-207) becomes

$$n_U(0, t) = N_U(0) + [N_U(0) + n_{OP}](qV_a/kT)\exp(j\omega t), \tag{3-209}$$

and letting
$$[N_U(0) + n_{0P}](qV_a/kT) \equiv N_u(0), \tag{3-210}$$

Equation (3-209) becomes
$$n_U(0, t) = N_U(0) + N_u(0) \exp(j\omega t). \tag{3-211}$$

Equation 3-206, and hence Equation 3-211, constitutes the required boundary con-

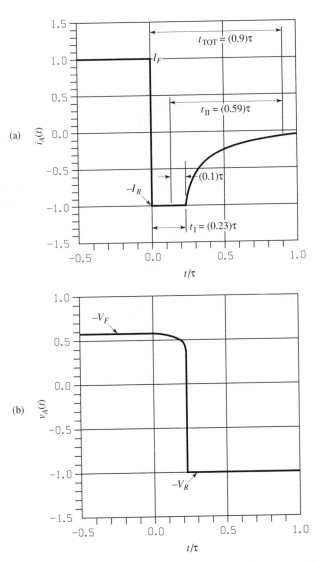

Figure 3-80 Transient waveforms predicted by the exact analysis for $(I_R/I_F) = 1$ and neglecting transition-region capacitance. (a) Current versus normalized time. (b) Voltage versus normalized time.

dition at $x = 0$. The boundary condition at $x = X_P$ is $n_U(X_P, t) = 0$, because $X_P \gg L_n$. In what follows, we shall encounter long, short, and intermediate-length diodes, so let us adhere to the general symbol X_P throughout.

Given the boundary conditions, the problem is to solve the time-dependent diffusion equation, which is in the present notation

$$\frac{\partial n_U(x, t)}{\partial t} = -\frac{n_U(x, t)}{\tau} + D_n \frac{\partial^2 n_U(x, t)}{\partial x^2}. \tag{3-212}$$

Applying the principles of perturbation theory, we can divide the excess-carrier profile $n_U(x, t)$ into dc and ac parts, which is to say, into static and dynamic parts:

$$n_U(x, t) = N_U(x) + N_u(x) \exp(j\omega t). \tag{3-213}$$

One frequently encounters heuristic graphical representations of Equation 3-213 in which $N_U(x)$ is a declining exponential function, while the time-dependent term is a multicycled sinusoidal waveform superimposed upon it, exhibiting about the same rate of decay as the static term. Such depictions are grossly inaccurate, however. The reason is that diffusion is the only transport mechanism present in the problem assumed. To appreciate the result, focus on the time-dependent small-signal component of applied voltage. The signal does not directly deliver and recover minority-carrier charge. What it actually does is manipulate the carrier density at the boundary, $x = 0$. A density perturbation at one position is "felt" for a distance of approximately one diffusion length, to use anthropomorphic language. When the signal reverses direction, the new disturbance also penetrates a diffusion length, pushing the previous disturbance deeper into the medium, at which new position it is drastically attenuated. As a result, a snapshot of the true profile exhibits a heavily damped appearance.

A qualitative appreciation of the resulting waveform's character can be gained from Figure 3-77. The curves there show the result of an abrupt change in voltage from a positive maximum to zero. Had the change in voltage been from a positive maximum to a negative maximum as in a square wave, what the qualitative nature of the new curves would be can be inferred from the diagram. The normalized densities shown there can as well be for a small-signal case as for the original large-signal case. Thus we see that even with a square wave substituted for the sinusoidal signal, the profiles will be far different from the "wiggle" waveforms often assumed, and with a return to the sinusoidal case, the profiles will have still gentler curvature.

Returning to the sinusoidal problem, we substitute Equation 3-213 into Equation 3-212 to obtain

$$j\omega N_u(x) \exp(j\omega t) = -\frac{N_U(x)}{\tau} - \frac{N_u(x)}{\tau} \exp(j\omega t)$$
$$+ D_n \frac{d^2 N_U(x)}{dx^2} + D_n \frac{d^2 N_u(x)}{dx^2} \exp(j\omega t), \tag{3-214}$$

which can be rewritten as

$$D_n \frac{d^2 N_U(x)}{dx^2} - \frac{N_U(x)}{\tau} = \exp(j\omega t)\left[-D_n \frac{d^2 N_u(x)}{dx^2} + N_u(x)\left(\frac{1 + j\omega t}{\tau}\right)\right]. \quad (3\text{-}215)$$

Since the left-hand side of this equation is time-independent, and the right-hand side is time-dependent, the equation can be valid in the general case only if both sides are separately equal to zero. Thus,

$$\frac{d^2 N_U(x)}{dx^2} - \frac{N_U(x)}{L_n^2} = 0, \quad (3\text{-}216)$$

and

$$\frac{d^2 N_u(x)}{dx^2} - \frac{N_u(x)}{L_n^2}(1 + j\omega t) = 0. \quad (3\text{-}217)$$

The solution for Equation 3-216 is the standard dc expression given in Equation 3-27, which in the present notation is

$$N_U(x) = N_U(0) \exp\left(-\frac{x}{L_n}\right). \quad (3\text{-}218)$$

By comparing Equations 3-216 and 3-217 we can guess that the solution of the second is

$$N_u(x) = N_u(0) \exp\left(-\frac{x\sqrt{1 + j\omega\tau}}{L_n}\right). \quad (3\text{-}219)$$

Exercise 3-87. Verify that Equation 3-219 is the solution for Equation 3-217.

Upon substituting the trial solution into

$$\frac{d^2 N_u(x)}{dx^2} = \frac{N_u(x)}{L_n^2}(1 + j\omega\tau),$$

we find that the left-hand side becomes

$$\frac{d}{dx}\left[N_u(0) \frac{\sqrt{1 - j\omega\tau}}{L_n} \exp\left(\frac{x\sqrt{1 + j\omega\tau}}{L_n}\right)\right],$$

or

$$N_u(0) \frac{1 + j\omega\tau}{L_n^2} \exp\left(\frac{x\sqrt{1 + j\omega\tau}}{L_n}\right) = \frac{N_u(x)}{L_n^2}(1 + j\omega\tau),$$

which is identical to the right-hand side.

Thus we have a dc spatial solution $N_U(x)$, Equation 3-218, in which the dc boundary condition $N_U(0)$ is given by Equation 3-208, and an ac spatial solution

$N_u(x)$, Equation 3-219, in which the ac boundary condition $N_u(0)$ is given by Equation 3-210. These two spatial solutions enter into the space-time solution $n_U(x, t)$ in the manner indicated in Equation 3-213. With the quantity $n_U(x, t)$ in hand, we are in a position to evaluate the current through the diode, which can be written in general form as

$$i_A(t) = I_A + i_a(t) = I_A + I_a \exp(j\omega t). \tag{3-220}$$

Let us take advantage of the fact that this current observed externally is essentially equal to the minority-carrier current at $x = 0$ for the device chosen. The dc term is thus

$$I_A = -qAD_n \frac{dN_U(x)}{dx}\bigg|_{x=0}. \tag{3-221}$$

Substituting Equation 3-218 into this expression yields

$$I_A = qAD_n N_U(0)/L_n, \tag{3-222}$$

where $N_U(0)$ is given in Equation 3-208.

The ac term of Equation 3-220 becomes

$$i_a(t) = I_a \exp(j\omega t) = -qAD_n\left(\frac{dN_u(x)}{dx}\bigg|_{x=0}\right)\exp(j\omega t), \tag{3-223}$$

where $N_u(x) \exp(j\omega\tau)$ is the second term of Equation 3-213, and $N_u(x)$ is given in Equation 3-219. Thus, substituting Equation 3-219 into Equation 3-223 yields

$$i_a(t) = (qAD_n/L_n)N_u(0)\sqrt{1 + j\omega\tau} \exp(j\omega t), \tag{3-224}$$

where

$$(qAD_n/L_n)N_u(0)\sqrt{1 + j\omega\tau} \equiv I_a(t) \tag{3-225}$$

and $N_u(0)$ is given in Equation 3-210.

Combining Equations 3-222 and 3-208 gives the conventional dc diode equation,

$$I_A = \frac{qAD_n n_{OP}}{L_n}\left[\exp\frac{qV_A}{kT} - 1\right] = I_{0A}\left[\exp\frac{qV_A}{kT} - 1\right]. \tag{3-226}$$

Since quasistatic conductance is given by $g = \partial I_A/\partial V_A$, we have from Equation 3-226,

$$g = (q/kT)I_A \exp(qV_A/kT) = (q/kT)(I_A + I_{0A}). \tag{3-227}$$

The complex ac amplitude I_a as given by Equation 3-225 is a function of $N_u(0)$. To rewrite this equation in terms of V_a we proceed as follows: Add n_{OP} to both sides of Equation 3-222 after solving it for $N_U(0)$, obtaining

$$N_U(0) + n_{OP} = \frac{L_n}{qAD_n}(I_A + I_{0A}). \tag{3-228}$$

The quantity I_{0A} is defined in Equation 3-227. Using Equation 3-227, we convert this equation into

$$N_U(0) + n_{OP} = \frac{L_n}{qAD_n} \frac{kT}{q} g. \tag{3-229}$$

Substituting this result into Equation 3-210 gives us

$$N_u(0) = \frac{L_n}{qAD_n} gV_a, \tag{3-230}$$

so that the complex ac amplitude I_a defined in Equation 3-225 becomes

$$I_a = V_a\, g\sqrt{1 + j\omega\tau}, \tag{3-231}$$

which is the desired result. As before, the ac admittance $y(\omega)$ is defined by

$$y(\omega) = (I_a/V_a) = i_a(t)/v_a(t), \tag{3-232}$$

so from Equation 3-231,

$$y(\omega) = g\sqrt{1 + j\omega\tau} = 1/z(\omega). \tag{3-233}$$

The magnitude of $z(\omega)$ from the present analysis is plotted in Figure 3-72, along with two corresponding charge-control curves, as already noted. Equation 3-233 is readily approximated for all but high-frequency conditions as

$$y(\omega) = g(1 + j\omega\tau/2). \tag{3-234}$$

Exercise 3-88. How was Equation 3-234 derived?

When $\delta \ll 1$, we have $(1 + \delta)^m \approx 1 + m\delta$. Thus it is evident that

$$g(1 + j\omega\tau)^{1/2} \approx g(1 + j\omega\tau/2).$$

Exercise 3-89. Given an N^+P diode with a P-type region having a carrier lifetime of one microsecond, what constitutes "low" frequency?

Let us use a one-percent criterion. Then $\omega\tau = 0.01$, so that $\omega = (0.01)/\tau$, or

$$\omega = [(0.01)/10^{-6}\ \text{s}] = 10^4\ \text{rad/s}.$$

Hence,

$$f = [(10^4/\text{s})/2\pi] \approx 1.6\ \text{kHz}.$$

Equation 3-234 suggests that insofar as diffusing carriers are concerned, we can represent the junction diode at low frequencies by using a parallel combination of a conductance g, and an effective diffusion capacitance $C_u{}^*$ given by $g\tau/2$, consistent with the conclusion reached in Section 3-8.6.

Equation 3-233 is general with respect to frequency. Through standard complex-number manipulations it can be rewritten as

$$y(\omega) = g\sqrt{\frac{1}{2}\sqrt{1+(\omega\tau)^2} + \frac{1}{2}} + jg\sqrt{\frac{1}{2}\sqrt{1+(\omega\tau)^2} - \frac{1}{2}}. \qquad (3\text{-}235)$$

In Table 3-2 at the end of the chapter we have repeated Equation 3-235 for the long diode, and have also given corresponding expressions for a short diode, and for the intermediate-length diode, which can be regarded as the general case.

For high-frequency conditions, the constant terms in Equation 3-234 can be dropped, yielding the approximation

$$y(\omega) = g\left[\sqrt{\frac{\omega\tau}{2}} + j\sqrt{\frac{\omega\tau}{2}}\right]. \qquad (3\text{-}236)$$

The approximate admittance expressions for the long diode are given in Table 3-3 at the end of the chapter, along with the corresponding expressions for the short diode. Since under high-frequency conditions, $\omega > 1/\tau$, the diode poses a *distributed* problem. It becomes a kind of transmission line, a matter also considered in Section 3-8.6. Under these conditions, both the real and imaginary terms of the admittance become functions of frequency. This subject is examined quantitatively in Problem A3-37.

3-9.5 Circuit-Behavior Exact Analysis

Now we examine the circuit behavior of a diode using the exact approach, and will compare and contrast the results with those obtained in Section 3-9.3 where charge-control methods were used. As before, let us consider only the case of a single diode driven by a single generator. Altogether, four binary options are involved here: (a) small-signal and large-signal sources, (b) charge-control and exact analyses, (c) step-function and sinusoidal sources, and (d) current and voltage sources. There are thus sixteen (2^4) solutions to be presented. In the case of step-function sources, only positive-going steps are treated, to avoid further proliferation; this is in spite of the fact that the diode in some cases exhibits qualitatively different "up" and "down" behaviors.

Table 3-4 lists the eight small-signal solutions. The current-step source yields the curves plotted in Figure 3-71. Charge-control analysis gives a result that is an exponential function of time, Equation 3-177. Here, however, we introduce a refinement. In offering the expression for time-dependent voltage given in Equation 3-177, we implicitly assume that it is to be applied only for $t \geq 0$, where the current step of course occurs at $t = 0$. But by multiplying this solution by the unit step function $u(t)$, we create the condition $v_a(t) = 0$ for $t < 0$. This modification is incorporated into the time-dependent solutions for step problems in Table 3-4, so that those expressions are good for all time.

The exact result for the small-signal current step is most easily obtained by assuming that we can start with Equation 3-234, the admittance expression in terms of $j\omega$, and substitute s for $j\omega$ and τ for τ^*, obtaining

$$y(s) = g\sqrt{1 + s\tau}. \tag{3-237}$$

This result suggests that the charge-control solution given in Equation 3-180 can be converted to the exact solution by substituting $\sqrt{1 + s\tau}$ for $(1 + s\tau^*)$, to obtain

$$v_a(s) = \frac{\Delta I_A}{s} \frac{1}{g\sqrt{1 + s\tau}}. \tag{3-238}$$

Then, taking the inverse transform of this expression by using standard tables, we find that

$$v_a(t) = \frac{\Delta I_A}{g} \operatorname{erf}\sqrt{\frac{t}{\tau}}, \tag{3-239}$$

which is plotted in Figure 3-71. This result can be proven rigorously in two ways. First, we could use the methods discussed in the last section for the small-signal sinusoidal input-voltage problem and modify them for the small-signal voltage step-function input. Second, we could prove that Equation 3-239 is a special case of the combined large-signal solutions given as Equations 3-200 and 3-201. We leave both of these proofs as practice problems for the reader.

Now consider the voltage-step source, second in Table 3-4. For the charge-control model, the small-signal solution is

$$i_a(t) = g\Delta V_A[\tau^* \delta(t) + u(t)]. \tag{3-240}$$

where $\delta(t)$ is the delta function. The delta function is present because the voltage step ΔV_A must deliver a charge $C_u^*\Delta V_A$ instantaneously to the small-signal diffusion capacitance C_u^*. Also, a constant component of current fixed by $g\Delta V_A$ is added for $t \geq 0$. This charge-control result is plotted in Figure 3-81(a).

The corresponding result given by the exact model is obtained as follows: The small-signal complex admittance $y(s)$ is still given by Equation 3-237. But for a voltage step, Equation 3-231 is replaced by

$$i_a(s) = (g\Delta V_A/s)\sqrt{1 + s\tau}. \tag{3-241}$$

Using standard tables, one finds the inverse transform to be

$$i_a(t) = g\Delta V_A\left[\frac{\exp(-t/\tau)}{\sqrt{\pi t/\tau}} + \operatorname{erf}\sqrt{\frac{t}{\tau}}\right]u(t). \tag{3-242}$$

This result is plotted in Figure 3-81(b) for comparison with the charge-control result. It is closely related to the curve in Figure 3-78, where a voltage step was assumed in a portion of the current-reversal problem.

Proceed now to the case of a sinusoidal current generator (next to the bottom in

420 PN JUNCTIONS

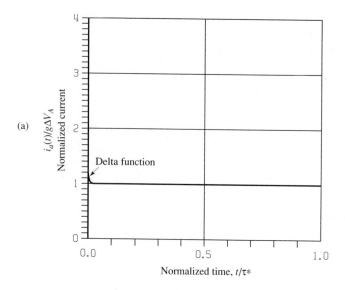

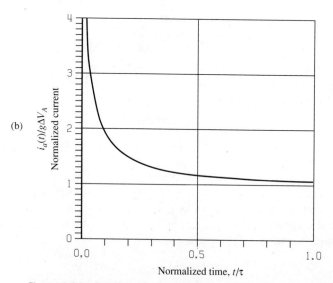

Figure 3-81 Solutions for the positive-going small-signal voltage-step problem. (a) Charge-control solution consists of a delta function $\delta(t)$ at $t = 0$, followed by constant incremental current. (b) Exact solution involves the sum of a rising error-function term and a declining exponential term having a time-dependent coefficient that diverges at $t = 0$.

Table 3-4). The desired expressions for time-dependent voltage can be written directly by forming the quotients of the given time-dependent current and the respective admittances, derived earlier. Equation 3-186 gives the charge-control expression for diode admittance $y(\omega)$, and Equation 3-233 gives the corresponding exact expression, functions that are plotted in Figure 3-72. Thus the solutions here are

$$v_a(t) = i_a(t)/g(1 + j\omega\tau^*) \tag{3-243}$$

for the charge-control model and

$$v_a(t) = i_a(t)/g\sqrt{1 + j\omega\tau} \tag{3-244}$$

for the exact model.

Finally, move to the sinusoidal voltage generator at the bottom of Table 3-4. Again, the solutions can be written directly. The expressions for time-dependent current are the product of the given time-dependent voltage source and the respective admittances.

The sinusoidal source is of course fundamentally different from the previous step source. However, since the diode is the same for both sources, these solutions must be closely related. Figure 3-82 qualitatively compares the diode's response to a current step with its response to a sinusoidal current for three values of lifetime τ. As lifetime increases, the rise time increases in the time-domain response, as is shown in Figure 3-82(a). The exact-model 10%-to-90% rise time is

$$t_{\text{RISE}} = 1.07\tau, \tag{3-245}$$

which follows directly from Equation 3-239. In the frequency domain, Figure 3-82(b), the 3-dB bandwidth f_B decreases according to

$$f_B = \frac{\sqrt{3}}{2\pi\tau} \tag{3-246}$$

as τ is increased. This follows from Equation 3-244. Forming the product $t_{\text{RISE}} f_B$, we find that

$$t_{\text{RISE}} f_B = \frac{\sqrt{3}(1.07)}{2\pi} = 0.29, \tag{3-247}$$

which is a fundamental relationship between the time-domain and frequency-domain solutions. A similar result can be obtained using the charge-control model, in which case the dimensionless constant is 0.35. Thus, the time-domain and frequency-domain responses are linked by the fact that the product $t_{\text{RISE}} f_B$ is a constant. Given the linking expression, one can approximately construct the time-domain response from the frequency-domain response, and vice versa.

Exercise 3-90. Show that Equation 3-246 follows from Equation 3-244.

Normalizing Equation 3-244 yields

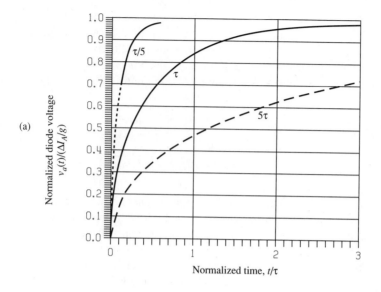

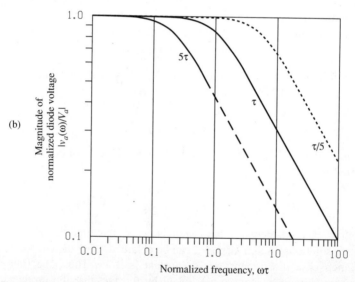

Figure 3-82 Relating exact-analysis diode responses for the small-signal current-step source and the small-signal sinusoidal-current source. (a) Time-domain (current-step) response for three lifetime values. (b) Frequency-domain (sinusoidal-current) response for the same three lifetime values. Note that for $\omega\tau = 1$, normalized voltage is $2^{-1/4} = 0.841$, rather than $2^{-1/2} = 0.707$ as predicted by the charge-control model.

$$\frac{v_a(t)}{i_a(t)/g} = \frac{1}{\sqrt{1 + j\omega\tau}}.$$

The quantity of interest is the magnitude of the voltage ratio. Hence, using the magnitude of the right-hand side, and equating it to $1/\sqrt{2}$ by the definition of the 3-dB bandwidth gives us

$$\frac{1}{\sqrt[4]{1 + (\omega_B\tau)^2}} = \frac{1}{\sqrt{2}}.$$

From this expression,

$$1 + (\omega_B\tau)^2 = 4,$$

so that $\omega_B\tau = \sqrt{3}$ and $\omega_B = \sqrt{3}/\tau$. Thus, $f_B = \sqrt{3}/2\pi\tau$.

The four source functions of Table 3-4 represent the simplest types of small-signal generators, and their corresponding solutions are also relatively simple. The solutions for more complicated source functions can be found most easily by using the principle of superposition. Let us consider one example using a periodic input function. First, we assume that the function can be resolved into a Fourier series:

$$v_a(t) = \sum_{k=1}^{n} V_k \exp(j\omega_k t). \tag{3-248}$$

Each term of this series has the same form as that of the simple source function at the bottom of Table 3-4, so that it follows from Equation 3-186 that

$$y(\omega_k) = g(1 + j\omega_k\tau^*). \tag{3-249}$$

The principle of superposition applied to this example yields a diode current $i_a(t)$ given by

$$i_a(t) = \sum_{k=1}^{n} V_k \exp(j\omega_k t)\, y(\omega_k). \tag{3-250}$$

Thus the solution is a Fourier series of exponential functions.

Now we turn from small-signal to large-signal problems, with solutions for the latter given in Table 3-5. Consideration here has been limited to the large-signal generators that are easiest to analyze. The first source is a current step, with results described in Sections 3-9.3 and 3-9.4. The charge-control solution in this case consists of two equations: Equation 3-191 for time-dependent voltage and Equation 3-193 for time-dependent charge. We have assumed here that I_A and ΔI_A are positive, as noted before. That is, the two solution equations are valid only for $q_U(t) + Q_{0U} \geq 0$, and the current step is positive-going. If the magnitude of the current step ΔI_A is very small, then each of the equations can be separated into static and dynamic parts, where the latter are identical in mathematical form to the corresponding small-signal solutions.

The exact solutions also consist of two equations. Specifically, the time-dependent-voltage and excess-carrier-density expressions were given, respectively, as Equations 3-201 and 3-200, with the imposed currents in the latter being defined as in the charge-control case. These equations are valid only for $n_U(0, t) + n_{0P} \geq 0$. When the current step ΔI_A is very small, the comment offered for the charge-control case is valid here as well.

The next source is a voltage step—described quantitatively in Section 3-9.3, for the charge-control case, and qualitatively in Section 3-9.4, for a special case of the exact model wherein $\Delta V_A = -V_A$. Here again, for a step magnitude that is small, the large-signal solutions can be used to derive the small-signal expressions.

In the last two cases, a large-signal sinusoidal variation is imposed upon a dc value, and once more the various small-signal solutions can be derived from the corresponding large-signal solution. The current-source solutions are much simpler than the voltage-source solution, even though the source functions have the same mathematical form. This is because current is linearly related to both $q_U(t)$ and $n_U(0, t)$. With the voltage source, by contrast, the corresponding relations are exponential, and as a result Fourier-series and superposition methods are needed.

3-9.6 SPICE Analysis

The SPICE program focuses on the terminal properties of a diode and avoids consideration of its internal physics. For this reason, and to take advantage of inherent linearities, it employs an empirically modified charge-control model, and makes no use of the exact model. In the SPICE model we introduce the nonlinear depletion-layer capacitance into circuit analysis for the first time. This introduces such mathematical complexities that we must now resort to numerical methods of solution. Small-signal depletion-layer capacitance C_t, discussed in Section 3-8.3, is treated as a constant; a truly static component of voltage V_A (in present notation) combined with a signal small enough to achieve linearization, caused this differential capacitance to be constant. But in the present large-signal problem, the explicitly time-dependent total voltage $v_A(t)$ causes important nonlinear changes in the depletion-layer differential capacitance, so we use the notation $C_T(v_A)$ for this case. While it is still a differential capacitance that we deal with, we use a capital subscript as a reminder that the capacitance is nonconstant. Thus the depletion-layer capacitance can be written

$$C_T(v_A) = \frac{C_{0T}}{\sqrt{1 - \dfrac{v_A(t)}{\Delta \psi_0}}}. \tag{3-251}$$

Exercise 3-91. Relate Equation 3-251 to the appropriate expression for C_t in Table 3-1, and evaluate C_{0T}.

The quantity $\Delta\psi_0$, a positive constant now and previously, is of course contact potential, a property fixed by junction structure. The voltage $v_A(t)$ is opposite in sign to that originally used. Table 3-1 gives expressions for the three kinds of step junctions. Thus the transition from C_t for the one-sided case (of interest here) with N being light-side doping, to $C_T(v_A)$ can be made as follows:

$$A\sqrt{\frac{\epsilon qN}{2(V_{NP} + \Delta\psi_0)}} = \frac{A\sqrt{\epsilon qN/2\Delta\psi_0}}{\sqrt{1 + (V_{NP}/\Delta\psi_0)}} = \frac{C_{0T}}{\sqrt{1 - [V_A(t)/\Delta\psi_0]}}.$$

It is evident that $C_{0T} \equiv A\sqrt{\epsilon qN/2\Delta\psi_0}$.

To start, we write a dynamic charge-control expression that includes depletion-layer capacitance. That is, to Equation 3-160 we add the nonlinear term $C_T(v_A)dv_A(t)/dt$ to give

$$i_A(t) = \frac{q_U(t)}{\tau^*} + \frac{dq_U(t)}{dt} + C_T(v_A)\frac{dv_A(t)}{dt}. \tag{3-252}$$

The two variables $q_U(t)$ and $v_A(t)$ that appear on the right are not independent of one another, with the latter being logarithmically related to the former. This relationship is displayed in Equation 3-191, which can be inverted to give

$$q_U(t) = Q_{0U}\{\exp[qv_A(t)/kT] - 1\}. \tag{3-253}$$

Using Equation 3-253, we can convert Equation 3-252 into a form having $q_U(t)$ as the only independent variable:

$$i_A(t) = \frac{q_U(t)}{\tau^*} + \left[1 + \frac{kT}{q}\frac{C_T(q_U)}{[Q_{0U} + q_U(t)]}\right]\frac{dq_U(t)}{dt}. \tag{3-254}$$

This expression uses $C_T(q_U)$, which can be obtained from Equation 3-251 by substituting Equation 3-191 into it. Equation 3-254, which we shall call the *nonlinear dynamic charge-control equation*, is a first-order nonlinear differential equation relating $i_A(t)$ and $q_U(t)$.

Exercise 3-92. Show that Equation 3-254 follows from Equations 3-252 and 3-253.

The two expressions differ only in their final terms. Differentiating Equation 3-253 yields

$$\frac{dq_U(t)}{dt} = \frac{q}{kT}Q_{0U}\exp\left[\frac{qv_A(t)}{kT}\right]\frac{dv_A(t)}{dt} = \frac{q}{kT}[q_U(t) + Q_{0U}]\frac{dv_A(t)}{dt}.$$

Thus,

$$\frac{dv_A(t)}{dt} = +\frac{kT}{q}\frac{1}{[Q_{0U} + q_U(t)]}\frac{dq_U(t)}{dt},$$

and multiplying this result by $C_T(q_U)$ gives us the last term of Equation 3-254. Hence we have changed the functional dependence from $v_A(t)$ and $q_U(t)$, to $q_U(t)$ alone.

A second useful form of the nonlinear dynamic charge-control equation employs $v_A(t)$ as the only independent variable, a form that results from placing Equation 3-253 into Equation 3-252. It is convenient first, however, to note that Equation 3-253 can be rewritten by means of the charge-control definition $I_{0A} \equiv Q_{0U}/\tau^*$, as

$$q_U(t) = I_{0A}\tau^*\{\exp[qv_A(t)/kT] - 1\}. \quad (3\text{-}255)$$

With this definition, Equation 3-254 becomes

$$i_A(t) = I_{0A}\left[\exp\left(\frac{qv_A(t)}{kT}\right) - 1\right]$$
$$+ \left[\tau^*\frac{q}{kT}I_{0A}\exp\left(\frac{dv_A(t)}{kT}\right) + C_T(v_A)\right]\frac{dv_A(t)}{dt}. \quad (3\text{-}256)$$

This expression is equivalent to the nonlinear dynamic charge-control equation, Equation 3-254. It is, however, an implicit expression in the sense that the independent variable—the total applied voltage $v_A(t)$—is not a charge quantity.

The actual SPICE program employs several additional generalizations of this equation. First, a series resistance R_S is added to account for ohmic drops in the diode end regions and at the contacts. This means that the applied voltage $v_A(t)$ is now an external voltage, so that we must introduce an internal (boundary-to-boundary) voltage $v_J(t)$ that fixes carrier densities, capacitances, and the like. The sign convention for $v_J(t)$ is the same as for $v_A(t)$. A diode quality factor N is also added to describe nonideal diode characteristics. Finally, a junction-grading exponent M is employed to take account of various doping profiles. The exponent has the value 1/2 for step junctions and 1/3 for linearly graded junctions. With these additions, we have

$$v_A(t) = v_J(t) + R_S i_A(t), \quad (3\text{-}257)$$

where

$$i_A(t) = I_{0A}\left[\exp\frac{qv_J(t)}{NkT} - 1\right] + \left[\tau^*\frac{qI_{0A}}{NkT}\exp\frac{qv_J(t)}{NkT} + C_T(v_J)\right]\frac{dv_J(t)}{dt}, \quad (3\text{-}258)$$

and where

$$C_T(v_J) = C_{0T}\left[1 - \frac{v_J(t)}{\Delta\psi_0}\right]^{-M}. \quad (3\text{-}259)$$

It should be noted in the last equation that $C_T(v_J)$ approaches an infinite value as $v_J(t)$ approaches $\Delta\psi_0$, a fact previously displayed in Figure 3-55(b). As noted in Section 3-8.3, this divergence is a feature of depletion-approximation analysis, an analysis that progressively worsens in forward bias. A more accurate analysis of the problem shows that $C_T(v_J)$ displays a maximum at extreme forward bias, with a value that exceeds $C_T(0)$ by only a small factor [81]. The approach used in SPICE assumes an even simpler voltage dependence for $C_T(v_J)$ in the high-forward-voltage range. It employs a linear extrapolation, in the manner indicated in Figure 3-83. The point of tangency is fixed by the adjustable parameter FC that is noted in Table 3-6. The default value given there, 0.5, places the point of tangency at the center of the forward-bias range. While the linear extrapolation yields too-small values of $C_T(v_J)$ in the high-forward-bias range, this fact is unimportant because diffusion capacitance dominates so totally in that range.

The general SPICE program incorporates a number of additional effects, some of which are illustrated in Computer Problems at the end of this chapter. Table 3-6 gives the correlation between the standard SPICE notation and our notation as used in the present analytic discussions.

3-9.7 A Numerical Example

To examine the large-signal transient response of a diode, it is helpful to define several new variables. They do not in any way modify the basic SPICE model, but

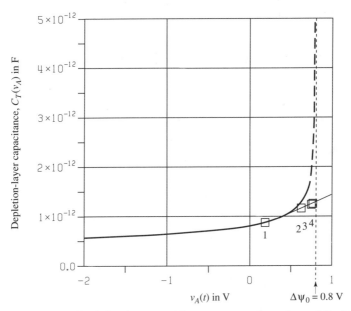

Figure 3-83 Depletion-layer capacitance versus voltage (curved line) for the device of the numerical example in Section 3-9.7, with plotted points relating to that example. The SPICE program avoids diverging capacitance near $\Delta\psi_0$ by employing a linearly extrapolated tangent at a lower voltage.

they describe various elements of the model and simplify the associated equations. First, the variables $i_R(v_J)$ and $i_C(v_J)$ represent, respectively, the "resistive" and "capacitive" components of the total current $i_A(t)$. Hence Equation 3-256 can be written

$$i_A(t) = i_R(v_J) + i_C(v_J). \tag{3-260}$$

The "resistive" current component can be written

$$i_R(v_J) = I_{0A}\left[\exp\frac{qv_J(t)}{NkT} - 1\right], \tag{3-261}$$

where $i_R(t)$ equals I_A and $v_A(t)$ equals V_A under static or quasistatic conditions. Thus, under these conditions Equation 3-261 becomes the conventional dc diode equation. The "capacitive" component $i_C(v_J)$ is the diode current necessary to charge or discharge the total capacitance, which in turn is given by

$$C_{\text{TOT}}(v_J) = C_U(v_J) + C_T(v_J). \tag{3-262}$$

The diffusion capacitance $C_U(v_J)$ is given by

$$C_U(v_J) = \frac{qI_{0A}\tau^*}{NkT}\left[\exp\frac{qv_J(t)}{NkT}\right], \tag{3-263}$$

and the depletion-layer capacitance $C_T(v_J)$, by Equation 3-259.

Let us introduce the symbol $g_R(v_J)$ for voltage-dependent (and hence time-dependent) conductance, distinct from the constant small-signal conductance g. It can be written

$$g_R(v_J) = \frac{di_R(v_J)}{dv_J} = \frac{qI_{0A}}{NkT}\exp\left[\frac{qv_J(t)}{NkT}\right]. \tag{3-264}$$

The last two equations can be combined to give

$$[C_U(v_J)/g_A(v_J)] = \tau^*, \tag{3-265}$$

an expression that is similar in form to the corresponding small-signal equation, $(C_u/g) = \tau^*$.

Using the charge-control variable $q_U(v_J)$, we can write the diffusion capacitance as $C_U(v_J) = dq_U(v_J)/dv_J$. And using an analogous charge $q_T(v_J)$, we can write the depletion-layer capacitance as $C_T(v_J) = dq_T(v_J)/dv_J$. From Equations 3-259, 3-262, and 3-263, then, it follows that a total stored-charge variable can be written

$$q_{\text{TOT}}(v_J) = q_U(v_J) + q_T(v_J), \tag{3-266}$$

where the respective individual charge variables are

$$q_U(v_J) = I_{0A}\tau^*\left[\exp\frac{qv_J(t)}{NkT} - 1\right], \tag{3-267}$$

and

$$q_T(v_J) = \frac{\Delta\psi_0 C_{0T}}{M}\left[1 - \frac{v_J(t)}{\Delta\psi_0}\right]^M. \tag{3-268}$$

The nonlinear dynamic charge-control equation can be written in terms of the total charge $q_{TOT}(v_J)$ as

$$i_A(t) = i_R(v_J) + dq_{TOT}(v_J)/dt. \qquad (3\text{-}269)$$

Thus the external current $i_A(t)$ is given at any instant by the sum of the resistive current $i_R(v_J)$ and the capacitive current, $dq_{TOT}(v_J)/dt$.

Now we turn to a problem defined in Figure 3-84. The source function indicated is applied to the circuit that is also given. Using the SPICE model, Equations 3-257 through 3-259, and the parameter values contained in Figure 3-84, we can calculate the four dc bias values given in the second row of Table 3-7, and plotted on the current–voltage characteristic given in Figure 3-85(a). This characteristic is equivalent to the curve given in Figure 3-52(a), which was plotted in that case purely for the quasistatic (or static) case. At present, however, each axis and each point on the curve carries a pair of designations. The upper symbol in each pair is a repetition of the static interpretation, but in present notation. Specifically, for the axes these symbols are I_R and V_J, and for the four points on the curve, the captions from lowest to highest are these: (1) OFF voltage, (2) crossover voltage, (3) partly

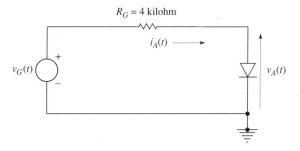

$v_G(t) = (0.2\ \text{V}) + (4.4\ \text{V})u(t)$

Name	SPICE Symbol	Analytic-Treatment Symbol	Parameter Value	Unit
Saturation current	IS	I_{0A}	10^{-16}	A
Internal series resistance	RS	R_S	0	ohm
Quality factor	N	N	1	—
Carrier lifetime	TT	τ	10^{-8}	s
Depletion-layer capacitance				
(at zero bias)	CJO	C_{0T}	8×10^{-13}	F
Contact potential	VJ	$\Delta\psi_0$	0.8	V
Grading exponent	M	M	0.3	—

Figure 3-84 Diode circuit and parameter values assumed for a numerical example.

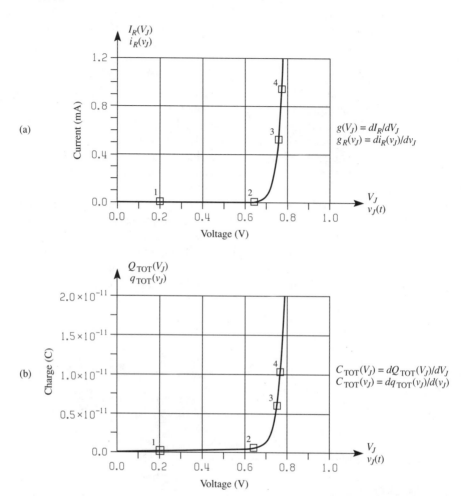

Figure 3-85 Curves supporting the SPICE analysis of the numerical example described in Table 3-7 and Figure 3-84. The four numbered points on each curve correspond to the points in Figure 3-83 and the columns in Table 3-7. (a) Current versus voltage for the static case (upper symbols) and for the resistive portion of the dynamic case (lower symbols). (b) Charge stored in both capacitances versus voltage for the static case (upper symbols) and for the dynamic case (lower symbols).

ON voltage, and (4) ON voltage. The term "crossover" refers to the voltage regime wherein the diffusion and depletion-layer capacitances approximate each other, in the sense of Figure 3-57. The four voltage values correspond to those of the plotted points on Figure 3-83.

Because of the fundamental charge-control assumption (charge distribution that is always exponential), it is identically the static charge-control *I–V* curve that is used in dynamic analysis. The lower symbols in each pair relate to the dynamic case. It is the *resistive* component of diode current $i_R(t)$ that when plotted against the junction voltage $v_J(t)$ yields the same curve as before. The times $t^{m=1}$ through $t^{m=4}$ are associated with the plotted and numbered points, and designate the instants at which the four voltages identified earlier are reached in the transient-switching process of the present numerical example. Figure 3-85(b) is like Figure 3-52(b), except that it presents, plotted versus voltage, the charge stored in both the diffusion and depletion-layer capacitances rather than diffusing charge alone. Combining this static information with the nonlinear dynamic charge-control equation, Equation 3-269, one can calculate the dynamic response of the diode.

Before addressing the problem stated in Figure 3-84, however, let us examine two simpler problems. These serve to illustrate methods that are used in SPICE computations. First, consider a voltage step applied directly to the diode, rather than to the diode and a resistor in series as in Figure 3-84. Assume that the voltage step causes $v_A(t)$ to change abruptly from 0.2 V to 0.643 V, values that appear in the second row and the first two columns of dc data given in Table 3-7. This table also shows that the resistive current will increase abruptly from 0 to 0.006 mA; the total capacitance, from 0.872 pF to 3.60 pF; and the total charge, from 0.166 pC to 0.676 pC. (The resistive-current values just given are consistent with Figure 3-85(a), although the scale of the figure does not make it evident.) The average total capacitance $\langle C_{TOT} \rangle$ for this change is

$$\langle C_{TOT} \rangle \equiv \frac{\Delta q_{TOT}}{\Delta v_A} = 1.15 \text{ pF}. \tag{3-270}$$

Now consider a current step applied to the diode that causes $i_A(t)$ to increase abruptly from 0 to 1 mA. The diode response can be roughly determined by assuming that Equation 3-269 can be approximated by

$$\Delta i_A = \Delta i_R + \langle C_{TOT} \rangle \frac{\Delta v_A}{\Delta t}, \tag{3-271}$$

so that we can estimate the time Δt required for the diode voltage to increase from 0.2 V to 0.643 V. The result is

$$\Delta t = \frac{\langle C_{TOT} \rangle \Delta v_A}{\Delta i_A - \Delta i_R} = 1.95 \text{ ns}. \tag{3-272}$$

More accurate calculation yields $\Delta t = 1.0$ ns, so even this primitive approximation is in error by less than a factor of two.

Now we turn to the more complicated problem defined in Figure 3-84. Its solution requires an understanding of two basic numerical techniques: (1) the Newton-Raphson algorithm as applied to find the dc solution of a nonlinear network, and (2) the trapezoidal formula for determining the time response of a network. Let us review these methods and use them to calculate the transient-response problem posed in Figure 3-84. (The SPICE program employs these same techniques, but with generalizations and improvements.) In the present problem, the diode current is given by

$$I_A = I_{0A}\left[\exp\frac{qV_J}{kT} - 1\right], \tag{3-273}$$

where we have assumed for simplicity that $N = 1$ and $V_J = V_A$. Using circuit theory, we can write that

$$I_A - g_G(V_G - V_J) = 0. \tag{3-274}$$

This is an implicit expression in V_J that cannot be solved directly, since I_A is exponentially related to V_J. But it can be solved by iteration as follows: Equation 3-274 has the mathematical form

$$f(x) = 0. \tag{3-275}$$

Assuming that the derivative exists, we then write

$$f'(x) = \frac{df(x)}{dx}. \tag{3-276}$$

The problem is to find by iteration the value of x that satisfies Equation 3-275. The Newton-Raphson method is based on the formulation

$$x^{n+1} = x^n - \frac{f(x^n)}{f'(x^n)}, \tag{3-277}$$

where x^n is an estimate of the final value of x, and x^{n+1} is an improved value. The basic algorithm is this:

(1) Set $n = 0$ and make an initial guess x^0.
(2) Calculate an improved value using Equation 3-277.
(3) If $|x^{n+1} - x^n|$ is smaller than a specified error value, then stop; otherwise replace n with $n+1$ and x^n with x^{n+1}, and go to step (2).

For the given problem, it is possible to use an analogous algorithm for which the iteration formula is

$$V_J^{n+1} = V_J^n + \frac{g_G V_G^n - g_G V_J^n - I_A^n}{g_G + g^n}, \tag{3-278}$$

and for which the following definitions are used:

$$I_A^n = I_{0A}\left[\exp\frac{qV_J^n}{kT} - 1\right], \tag{3-279}$$

and

$$g^n = \frac{qI_{0A}}{kT}\exp\frac{qV_J^n}{kT}. \tag{3-280}$$

Note that the dc iteration formula, Equation 3-278, involves the n value of the large-signal diode current I_A^n and the n value of the small-signal conductance g^n. Thus both large-signal and small-signal quantities must be known for application of the Newton-Raphson method.

For a transient problem, Equation 3-269 can be solved for $dq_{TOT}(v_J)/dt$ to give

$$\frac{dq_{TOT}(v_J)}{dt} = i_C(v_J), \tag{3-281}$$

where we have used the fact that $i_A(t) = i_R(v_J) + i_C(v_J)$. Equation 3-281 has the mathematical form

$$\frac{dx}{dt} = f(x, t). \tag{3-282}$$

One method for solving this differential equation is to divide the time interval into small steps, so that

$$h = t^m - t^{m-1}, \tag{3-283}$$

and to use the trapezoidial formula for integration, which is

$$x^m = x^{m-1} + \frac{h}{2}\left[f^m + f^{m-1}\right]. \tag{3-284}$$

Here x^{m-1} is the value of x at step $(m-1)$, and x^m is the subsequent value, while f^{m-1} and f^m are the corresponding values of f. For the transient problem, the corresponding formula is

$$q_{TOT}^m = q_{TOT}^{m-1} + \frac{h}{2}\left[i_C^m + i_C^{m-1}\right], \tag{3-285}$$

which can be rewritten as

$$i_C^m = -i_C^{m-1} + \frac{2(q_{TOT}^m - q_{TOT}^{m-1})}{h}. \tag{3-286}$$

This equation contains two unknown variables, i_C^m and q_{TOT}^m, which must be evaluated by iteration. To do this, we can generalize the dc iteration formula as follows:

$$v_j^{n+1,m} = v_j^{n,m} + \frac{g_G v_G^{n,m} - g_G v_j^{n,m} - i_R^{n,m} - i_C^{n,m}}{g_G + g_R^{n,m} + g_C^{n,m}}, \tag{3-287}$$

and then use the Newton-Raphson method. Here we have defined four new variables:

$$i_R^{n,m} = I_{OA}\left[\exp\frac{qv_j^{n,m}}{kT} - 1\right], \tag{3-288}$$

$$i_C^{n,m} = -i_C^{n,m-1} + \frac{2(q_{TOT}^{n,m} - g_{TOT}^{n,m-1})}{h}, \tag{3-289}$$

$$g_R^{n,m} = \frac{qI_{OA}}{kT}\left[\exp\frac{qv_j^{n,m}}{kT}\right], \tag{3-290}$$

and

$$g_C^{n,m} = \frac{2C_{TOT}^{n,m}}{h}. \tag{3-291}$$

The conductance $g_R^{n,m}$ is a generalization of the dc conductance g^n given by Equation

3-280 and it is defined by

$$g_R^{n,m} = \frac{\partial i_R^{n,m}}{\partial v_J^{n,m}},\qquad(3\text{-}292)$$

where

$$i_R^{n,m} = I_{0A}\left[\exp\frac{qv_J^{n,m}}{kT} - 1\right].\qquad(3\text{-}293)$$

The conductance $g_C^{n,m}$ is defined by

$$g_C^{n,m} = \frac{\partial i_C^{n,m}}{\partial v_J^{n,m}},\qquad(3\text{-}294)$$

where $i_C^{n,m}$ is given by Equation 3-289. Notice that Equation 3-291 follows from the fact that

$$C_{TOT}^{n,m} = \frac{\partial q_{TOT}^{n,m}}{\partial v_J^{n,m}}.\qquad(3\text{-}295)$$

Now we have all the information necessary to calculate the dynamic response of the diode in the transient problem given in Figure 3-84. The basic algorithm for the problem is this:

1. Use the Newton-Raphson method and the dc iteration formula, Equation 3-278, to find the dc values of the several variables for the OFF condition.

2. Set $m = 0$ and set the initial transient values equal to the corresponding dc values.

3. Use the Newton-Raphson method and the transient iteration formula, Equation 3-287, to find new values for the various variables at time index m from the previous values at index $(m - 1)$.

4. Repeat steps (2) and (3) for the desired sequence of m values.

The transient problem stated in Figure 3-84 must now be modified slightly because numerical methods are unable to deal with the discontinuous function $v_G(t)$ defined there. Hence we shall replace the voltage step by a steep voltage ramp. Let us assume that the generator voltage rises from 0.2 V at 10 ns to 4.6 V one nanosecond later. In preparation for the first numerical treatment of this problem, we have plotted the corresponding values of $v_G(t)$ at 0.1-ns intervals in Figure 3-86(a). Also, four particular times and the corresponding values of $v_G(t)$ are entered in the first two rows of Table 3-8.

Next we wrote a computer program using the algorithm just outlined, and used it to calculate $v_A(t)$ and $i_A(t)$ at 0.1-ns intervals, with results plotted as points in Figures 3-86(b) and 3-86(c), respectively. (We shall assume that series resistance R_S is negligible.) Again, the values of these variables, which we shall term *external* variables, at the same four times are entered in rows 3 and 4 of Table 3-8.

Next, the SPICE program was used to analyze the same transient problem, with the continuous curve in Figure 3-86(a) giving generator voltage. SPICE was instructed to calculate the external variables $v_A(t)$ and $i_A(t)$ at 1-ns intervals and did so, with the continuous-curve results plotted in Figures 3-86(b) and (c), respectively.

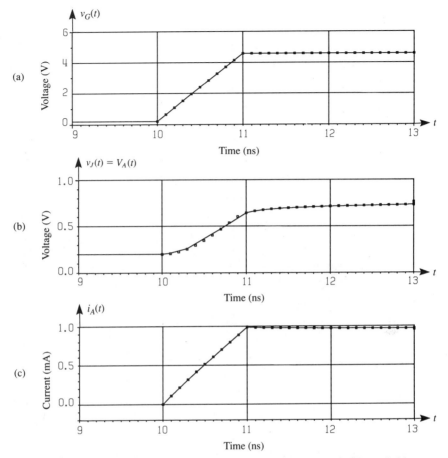

Figure 3-86 Numerical solutions of the transient problem defined in Figure 3-84. Plotted points at 0.1-ns intervals represent values computed using the algorithm given in the text. Continuous curves were calculated by the SPICE program instructed to employ 1-ns intervals except where smaller intervals were needed. (a) Voltage waveform applied to the series combination, with a ramp substituted for the step. (b) Diode-voltage waveform. (c) Diode-current waveform.

But where the given function undergoes rapid change, SPICE calculates additional points. In the present case SPICE chose to calculate, for example, transient variables at $t = 10.3$ ns and $t = 10.7$ ns, falling between the points at $t = 10$ ns and $t = 11$ ns. In Figure 3-86(b), it can be seen that the continuous curve is a broken line in this range, with its values at $t = 10.3$ ns and $t = 10.7$ ns coinciding with those of the previous numerical calculation of points at 0.1-ns intervals. This should indeed not be surprising because both calculations employ the same basic algorithms. But beyond this, even two different iteration methods should converge on the same values, provided both methods are valid.

Figure 3-87 gives further information on the SPICE simulation. Parts (a)

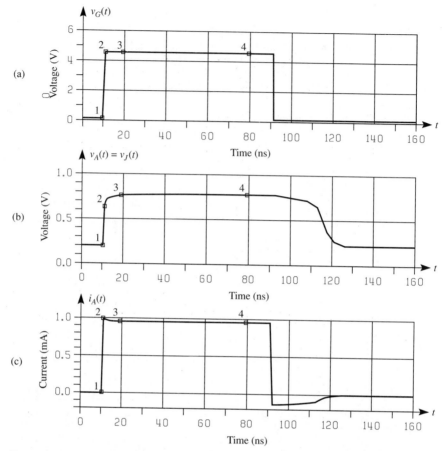

Figure 3-87 Time dependence of the "external" variables (those delivered by the normal SPICE program) in diode switching. This is the same experiment as that of Figure 3-86, but the time interval is longer and now includes turn-off. (a) Voltage waveform of the generator. (b) Diode-voltage waveform. (c) Diode-current waveform.

through (c) present all of the data in the corresponding parts of Figure 3-86, except that the time scale has now been changed to show the interval from 0 to 160 ns, instead of just from 9 to 13 ns. To illustrate turn-off behavior in addition to the turn-on behavior that has occupied us up to this point, we have assumed that the generator voltage $v_G(t)$ drops in one nanosecond to 0.2 V at a time in the neighborhood of $t = 90$ ns. Note that the diode voltage, Figure 3-87(b), changes slowly at first because appreciable stored charge must be removed to establish the OFF condition—a lower level of forward voltage. In Figure 3-86(c), reverse current is seen during this interval even without reverse voltage. But because reverse current enters, this is the same storage-time effect as described in Section 3-9.3. A subtle point of interest in Figure 3-87(c) is that the current overshoots about 3 percent

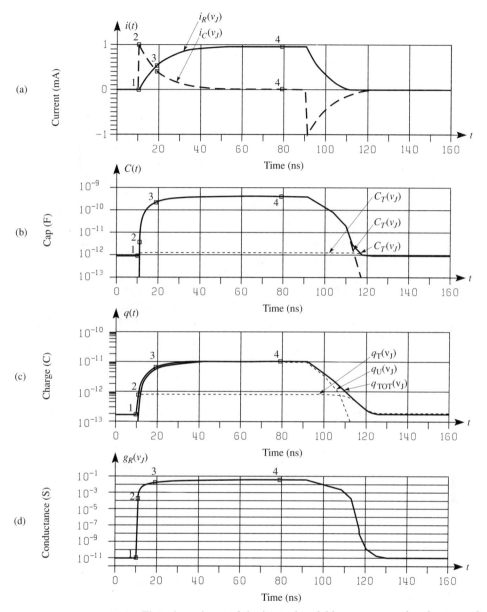

Figure 3-88 Time dependence of the internal variables, accompanying the external variables presented in Figure 3-87. (a) The resistive and capacitive current waveforms. (b) Time dependence of the total, depletion-layer, and diffusion capacitances. (c) Time dependence of the total, depletion-layer, and diffusion charge components. (d) Time dependence of the diode conductance.

during turn-on, settling back then to its ON value with a time constant equal to the effective carrier lifetime, $\tau^* = 10$ ns.

In contrast to these external variables, certain *internal* variables display more complex time dependences. The SPICE program computes these functions but does not normally deliver the data to the SPICE user. Because of the insight to be gained from examining their behaviors, however, we have inserted additional code into the program to secure the data, entered in the last nine rows of Table 3-8 for the same four times as before, and plotted in Figure 3-88 for the full time interval, from 0 to 160 ns. Figure 3-88(a) shows the time dependences of $i_C(v_J)$ and $i_R(v_J)$, the capacitive and resistive components, respectively, of the diode current $i_A(t)$. The first increases abruptly and then decreases slowly, with a time constant equal to the effective lifetime, whereas the second increases with the same lifetime. However, the resistive component of current increases only after the short delay required to charge the diffusion capacitance and establish the ON current, with the resulting waveform shift accounting for the overshoot noted earlier. During turn-off, the dominance of capacitive over resistive current accounts for the reverse current observed during the storage period.

Time dependences of the two components of capacitance, together with their sum, are given in Figure 3-88(b), while dependences for the charges they store can be seen in Figure 3-88(c). In both cases, the OFF-to-ON asymmetry results from the storage phenomenon. Notice that depletion-layer capacitance with its power-law behavior changes by much less than a factor of two, whereas diffusion capacitance changes by through nine orders of magnitude! Diode conductance, Figure 3-88(d), exhibits an increase through more than ten orders of magnitude, with diode resistance falling from 10^{12} ohms to 27 ohms. Although $g_R(v_J)$ and $C_U(v_J)$ change rapidly with time, their quotient remains constant and equal to τ^*. Similarly, the quotient of $i_R(v_J)$ and $q_U(v_J)$ is constant and equal to τ^*, but their changes are less rapid. Finally, the ratio of reverse to forward current displayed in Figure 3-87(c) is about 0.15, for which we determine a normalized storage time of two from Figure 3-75. This is evidently in good agreement with the storage time of approximately 20 ns that is clearly displayed in Figure 3-87(c).

Figures 3-87 and 3-88 contain a wealth of information, and the reader will profit by examining them in detail. (See also the computer problems at the end of the chapter.) These figures demonstrate that the internal behavior of the diode is very complex, even for the relatively simple charge-control analysis. The external behavior of the diode gives only a gross indication of the turbulent internal events.

SUMMARY

A *PN* junction is an interface that separates a *P*-type region from an *N*-type region in a semiconductor single crystal. A perfectly abrupt transition from one doping value to another, with uniform doping on each side, constitutes a step junction. The analytical problem posed by a step junction is closely related to that posed by a metal–semiconductor junction and a semiconductor surface. Ionic space charge exists on both sides of the *PN*-junction surface (metallurgical junction), with equal and opposite charge per unit of junction area on the two sides, thus forming a double layer or dipole layer. In a band diagram of a junction, either band edge can be taken as a potential map of the structure, but it is convenient and "evenhanded" to choose the Fermi level as potential origin, in which case the midgap trace or "intrinsic" potential becomes *the* potential. A constant or "flat" Fermi level throughout the sample guarantees equilibrium. Under such conditions, electron drift precisely balances electron diffusion at any position, as do hole drift and hole diffusion.

The depletion approximation for a step junction assumes that ionic space charge is uniform from the metallurgical junction to a boundary at which there is a sharp transition back to neutrality. With this assumed charge profile, integration of Poisson's equation yields a linear (triangular) electric-field profile, and a second integration yields a parabolic potential profile. The integral of the field profile, or the area of the triangle, has potential dimensions and represents the contact potential (or built-in potential drop or diffusion potential) of the junction at equilibrium. The height of the field-profile triangle is peak field E_{0M}, and its base is depletion-layer thickness X_0, so the geometrical relationships in a triangle show how these three important step-junction properties are related.

The contact potential of a junction cannot be measured with a voltmeter because compensating contact potentials exist at other interfaces in the voltmeter circuit. Forward bias means positive voltage on the *P*-type side and negative on the *N*-type side, and causes the field profile to shrink to a smaller congruent triangle, thus shrinking all three of the junction properties just cited. Reverse bias does the opposite, adding to contact potential. The depletion approximation improves with reverse bias, and worsens with forward bias. In any step junction, symmetric, asymmetric, or grossly asymmetric (one-sided), depletion-layer thickness X goes as the square root of total potential difference, $\Delta \psi = \Delta \psi_0 + V_{NP}$.

Boltzmann quasiequilibrium assumes that a Boltzmann relation holds for carrier densities at the boundaries of a junction under bias, with $\Delta \psi$ replacing $\Delta \psi_0$. It leads to the law of the junction, which says that under low-level forward bias, a boundary value of minority-carrier density (either side) is given by the equilibrium density multiplied by a Boltzmann factor $e^{-U_{NP}}$, where $U_{NP} \equiv qV_{NP}/kT$. The relevant boundaries in a forward-biased junction are well defined, as surfaces where electric field changes sign. Because minority-carrier boundary values are inflated by forward bias, minority carriers are injected into both neutral end regions. In a one-sided junction, injection into the lightly doped side grossly exceeds injection into the other side. Through a wide range of forward voltage, minority-carrier-density boundary value, density gradient at the boundary, and hence diffusion current at the boundary, where electric field and hence drift current vanish—all rise exponentially with voltage. The resulting *I–V* characteristic is linear on a semilog plot, but on a linear plot exhibits an offset voltage with a magnitude that depends on reverse saturation current, which in turn decreases as energy gap increases. This simple diffusion picture is complicated for silicon by recombination–generation current. In forward bias, carriers flood into the space-charge region and recombine there, and hence are not injected. At low voltages this current component swamps out the diffusion component (carrier-injection component) of current. In reverse bias, carriers generated in the space-charge layer swamp the "saturating," or constant, diffusion components of current. In high-level forward bias there are four boundary-value unknowns instead of the two unknowns of the low-level case. Applied bias must "operate" on majority density as well as minority density, and current increases more slowly with voltage than at lower values.

A *model* can be a set of equations or an equivalent circuit. In both cases there is a trade-off between accuracy and simplicity, and utility is greatly enhanced by simplicity. Both kinds of models permit us to understand the relationships of variables and parameters. A parameter is a "temporarily constant variable." Such models can be static (meaning for dc or near-dc conditions) or dynamic (meaning for time-varying conditions). Dynamic models can deal with sinusoidal or switching-transition variations in current and voltage. In either case, the transitions may be small-signal or large-signal in nature. In the small-signal case the I–V or Q–V characteristic can be considered linear without serious error, and such modeling is sometimes described as *linearization*. In the large-signal case the inevitable nonlinearities of semiconductor devices must be addressed. A piecewise-linear model, which simulates a nonlinear characteristic with a set of straight-line segments, is one way of attacking some features of such a problem. An ideal rectifier in the equivalent-circuit sense has a piecewise-linear characteristic with $R = 0$ in forward bias and $R = \infty$ in reverse bias, and with these lines superimposed upon the current and voltage axes, respectively. In the static charge-control model, the stored charge Q_S of diffusing minority carriers must be replaced "every τ seconds" so that the current of those carriers is Q_S/τ.

A PIN diode involves two step junctions, can be designed for high breakdown voltage, and exhibits a voltage-independent depletion-layer capacitance, thus simulating an ordinary capacitor. In a linearly graded junction, the depletion approximation predicts a triangular charge-density profile on each side, and improves in accuracy with increasing doping-density gradient. Junction formation by solid-phase diffusion involves delivering impurity atoms to the surface of a semiconductor sample while it is held at an elevated temperature ($\sim$1100 C for silicon). The diffused-impurity profile depends in detail upon boundary conditions in the process, but normally involves a high density near the surface and a vanishing density well away from the surface. The surface at which this profile intersects a preexisting opposite-type net-impurity profile is the location of the diffused *PN* junction. A high-low junction is a grossly asymmetric step junction with same-type doping on both sides. It has a small contact potential and carries large currents of either polarity with small applied voltages, thus serving as a valuable ohmic contact. Its space charge consists of ions on the high side and surplus majority carriers on the low side in the form of an accumulation layer. In the simplest cases, surplus carriers constitute an example of equilibrium nonneutrality, and excess carriers, a case of nonequilibrium neutrality.

The most important breakdown phenomena are nondestructive and repeatable, and cause a sharp increase in current through a device with a small increase in applied voltage. Avalanche breakdown is the result of impact ionization, or the breaking of a bond by a carrier that has extracted enough energy ($\sim \xi_G$) from a large electric field to eject an electron from a covalent bond, thus elevating it from the valence band to the conduction band. It requires (1) sufficient field, and (2) sufficient "running room," or distance over which that field obtains. It is *not* simply a critical-field phenomenon. The new carriers so produced also acquire energy from the electric field, and subsequent bond breaking by these carriers is the cause of the avalanche feature. For all diffused junctions, graded junctions, and step junctions of any variety, depletion-layer thickness at the breakdown voltage is about the same for *PN* junctions of a given breakdown voltage.

The phenomenon of tunneling causes reverse breakdown at a low voltage (a few volts or less) in a junction with heavy doping on both sides. This occurs because empty conduction-band states on the *N*-type side reside at about the same energy as filled valence-band states on the *P*-type side, and the two sets of states are physically very close (separated by a few tens of angstroms), with the barrier being the essentially state-free energy-gap region. Thus the prerequisites for quantum-mechanical tunneling are satisfied—a thin barrier and energy-matched states, filled on one side of the barrier and empty on the other. With even higher doping, tunneling can occur in the opposite direction under forward bias through a small voltage range, leading to the incremental negative resistance of the tunnel diode. In such a device, a linearly graded model is more accurate than a step-junction model.

Punchthrough "breakdown" occurs in a diode with a pair of parallel *PN* junctions separated by a thin layer. The depletion layers interact within the thin layer, causing that of the reverse-biased junction to "pull down" that of the forward-biased junction. As a result, current increases exponentially with voltage through several decades of current (at a voltage that is designable), constituting a case of thermionic emission.

Electric field as a function of potential in a step junction at equilibrium can be obtained exactly by analytical methods. The second integration of the Poisson-Boltzmann equation needed to obtain potential as a function of position requires numerical methods. The redundancy in either function is enormously diminished by spatial normalization using a scaling length known as the *Debye length*. At a particular temperature this is a function of sample permittivity and net doping and is a measure of screening distance, or the distance at which a charge can be "felt" or "seen" in the sample. The field and potential spatial profiles both exhibit asymptotes of simple functional form for the deep-depletion region and for the nearly neutral region. Intervening "bridge" expressions also of simple form can be written, and these expressions together with asymptotic expressions collectively constitute a depletion-approximation replacement (DAR), or an approximate-analytic description of these junction properties that is about as accurate as a numerical solution and about as easy to use as the depletion approximation. Analogous treatment can be accorded to other junction properties. A one-sided junction possesses an inversion layer made up of surplus carriers. It is a thin, dense layer of carriers adjacent to the metallurgical junction on the lightly doped side, consisting of carriers that have "spilled over" from the heavily doped side.

When a current or voltage excursion imposed on a diode (or other device) is so small that the relevant portion of its grossly nonlinear I–V characteristic can be approximated by a straight line, then small-signal analysis (sometimes described as *linearization* of the problem) is applicable. Extremely slow current or voltage excursions are termed *quasistatic*, and are ideal for defining the small-signal properties of a *PN* junction. The first of these is the small-signal conductance g of a forward-biased junction, or dI/dV_{NP}. It is given through several decades of current by q/kT times current for any semiconductor material. Also, this relation holds for both long and short diodes, but current at a given voltage is larger in the latter. Placing a resistor $r = 1/g$ in series with a voltage source yields a useful piecewise-linear model for the forward-biased junction.

In low-frequency diffusion capacitance, positive and negative charges are stored in the same physical volume in the form of excess carriers. This charge, and hence diffusion capacitance dQ_S/dV_{NP}, increases exponentially with forward voltage. Depletion-layer capacitance dQ_T/dV_{NP} involves the storage and recovery of majority carriers at the boundaries of the space-charge layer. Analysis based upon the depletion approximation yields good approximate expressions for depletion-layer capacitance. It typically exhibits a power-law dependence upon voltage, dominating in reverse bias and low forward bias, but being swamped by diffusion capacitance at larger forward bias. The two capacitances are equal, or "cross over," well below the onset of high-level conditions. Through normalization, one can write a universal relation between depletion-layer capacitance and voltage that is valid for any step junction. The same cannot be done for diffusion capacitance.

Diffusion, drift, and displacement currents all can play a role in dynamic junction phenomena. So too can carrier lifetime, dielectric-relaxation time, signal rise time, and RC product. A useful equivalent-circuit model of the junction diode consists of the conductance g, the diffusion capacitance C_s, and the depletion-layer capacitance C_t in a parallel combination.

In charge-control analysis, a dynamic extension of static charge-control analysis is assumed. The basic charge-control assumption concerning diffusion capacitance is that the spatial distribution of charge under dynamic conditions is the same as under static conditions. Since this is significantly inaccurate, one often employs an effective capacitance, accompanied by an effective carrier lifetime. Useful approximate results are often obtained by letting the effective values be half those defined quasistatically.

Much charge-control analysis is applied in the upper portion of the low-level forward-bias regime, wherein depletion-layer capacitance can be neglected. Thus one exploits the fact that the dynamic charge-control expression is a linear differential equation, and the principle of superposition can be applied, adding together the solutions of several simple linear problems in order to treat a more complicated problem. For still more general problems, the grossly nonlinear depletion-layer capacitance is sometimes added subsequently to the analysis.

There are at least four techniques or models for analyzing the junction diode in terms of its internal physics. The most accurate, termed *exact* here for brevity, starts with the time-dependent diffusion equation; this expression is the result of dropping the field-dependent term from the contant-E continuity-transport equation. A second method, finite-difference, divides up the region of minority-carrier storage into many smaller segments, or "lumps," and treats the problem numerically. In the limit of many segments, its accuracy approaches that of the exact model. In the third, usually called the *Linvill model,* only two or three lumps are used, so that analytical treatment of the problem is possible. The last and least accurate is the charge-control method, where a single lump is employed. Here we have focused on the two extreme methods—charge-control and exact.

Also, we use linear-network methods to treat a sinusoidal input signal and a step or switching input. In switching, a diode or transistor is taken quickly from one operating state to another. This binary application of such devices has enormous practical importance. It has been the basis of electronic computing since its beginning in the 1930s and 1940s, and is now encroaching steadily into applications that previously were exclusively "analog," or amplifierlike.

A device-physics analysis of the diode that starts with the diffusion equation and combines this with the charge-control model permits derivation of the linear differential equation for current as a function of time-dependent stored charge, as well as voltage and conductance as functions of charge. Solving the current equation for the case of a small-signal current step yields a circuit-behavior solution with an exponential voltage waveform involving effective lifetime as the characteristic time. Circuit-theory analysis confirms the result. Converting to a sinusoidal source permits determination of diode admittance. Generalizing to the large-signal switching problem yields a differential equation of the same form as that for the small-signal problem, and hence a solution of the same form as well. Generalizing further to the case of current turn-off, or the reversal from forward to reverse current, one can obtain an estimate of the storage time required to remove stored minority carriers, a time that is dependent on the ratio of the reverse-current to the forward-current magnitude. The model predicts unrealistically an abrupt transition to zero current at the end of the storage time. (The more realistic exact analysis predicts a decay time following storage.) In the opposite case, turn-on, effective lifetime again determines the exponential response.

Device-physics exact analysis starts with the time-dependent diffusion equation and determines the time-dependent spatial distribution of charge during large-signal switching. From this it infers storage time as the time required at constant current to reduce stored-charge density at the junction boundary to zero. Thereafter it uses an approximate method to determine current versus time during removal of the remaining stored charge, and infers (current) decay time from this function. Analysis of diode response to a small-signal sinusoidal applied voltage also starts with the time-dependent diffusion equation. The resulting carrier profile can be described as an exponential static component with an extremely heavily damped sinusoidal dynamic component superimposed upon it. Perturbation theory permits the writing of static and time-dependent current components, as well as diode conductance and admittance. The general admittance expression can be readily approximated for high and low frequencies, and for diodes of various lengths.

Using circuit-behavior exact analysis, we present four solutions corresponding to small-signal sinusoidal and positive-going step stimuli, and current and voltage sources. For comparison these are augmented by four corresponding solutions based upon charge-control analysis. It is shown that the small-signal current-source step

Table 3-1 Depletion-Approximation Equations for Junctions*

Junction Type	Depletion-Layer Thickness
1. Symmetric step	$X = \left[\dfrac{4\epsilon(V_{NP} + \Delta\psi_0)}{qN} \right]^{1/2}$
2. Asymmetric step	$X = \left[\dfrac{2\epsilon(V_{NP} + \Delta\psi_0)}{q} \left(\dfrac{1}{N_1} + \dfrac{1}{N_2} \right) \right]^{1/2}$
	$X_1 = \left[\dfrac{2\epsilon}{q} \dfrac{N_2}{N_1} \dfrac{(V_{NP} + \Delta\psi_0)}{(N_1 + N_2)} \right]^{1/2}$
	$X_2 = \left[\dfrac{2\epsilon}{q} \dfrac{N_1}{N_2} \dfrac{(V_{NP} + \Delta\psi_0)}{(N_1 + N_2)} \right]^{1/2}$
3. One-sided step (N_2 is light-side doping)	$X = \left[\dfrac{2\epsilon(V_{NP} + \Delta\psi_0)}{qN_2} \right]^{1/2}$
4. P^+IN^+	$X = $ constant
5. Linearly graded ($a \equiv$ gradient)	$X = \left[\dfrac{12\epsilon}{qa} (V_{NP} + \Delta\psi_0) \right]^{1/3}$

and sinusoidal responses are linked by the fact that the product of rise time and bandwidth is a constant—with a value of 0.29 from the exact model, and 0.35 from the charge-control model. As a result, time-domain response can be determined from the frequency-domain response, and vice versa. Eight corresponding and significantly more complicated large-signal solutions are also given, for a total of sixteen cases.

The SPICE program for junction-diode switching simulation focuses on terminal properties and is based upon an empirically modified charge-control analysis. It is capable of incorporating voltage-dependent transition-region capacitance in addition to diffusion capacitance, thus producing a nonlinear dynamic charge-control equation. Further refinements of the model include parasitic series resistance, the diode quality factor for better description of nonideal I–V characteristics, and a grading exponent for improved modeling of doping profiles. Understanding of the SPICE model is aided by dividing current into two components: one charging or discharging the capacitances, the other being "resistive" current. The advantage of so doing is that the latter involves identically the current–voltage–charge relationship examined quasistatically, while the former can be handled numerically through repeated application of the $i = CdV/dt$ relationship. The relevant capacitance now is of course the sum of the diffusion and depletion-layer capacitances, both being voltage-dependent. A numerical example tracks these two current components in time, along with the associated charges (as well as the resistive current component), with a result that is surprisingly complex in view of the simple total-current waveform. The normal SPICE program does not deliver information on these "internal" variables (current and charge components); it is necessary to insert additional code into the SPICE program to obtain such data.

Table 3-1 (*continued*)

Peak (Maximum) Field	Depletion-Layer Capacitance
$E_M = \left[\dfrac{qN}{\epsilon}(V_{NP} + \Delta\psi_0)\right]^{1/2}$	$C_T = \dfrac{A}{2}\left[\dfrac{\epsilon qN}{V_{NP} + \Delta\psi_0}\right]^{1/2}$
$E_M = \left[\dfrac{2q(V_{NP} + \Delta\psi_0)}{\epsilon[(1/N_1) + (1/N_2)]}\right]^{1/2}$	$C_T = A\left[\dfrac{\epsilon q}{2(V_{NP} + \Delta\psi_0)[(1/N_1) + (1/N_2)]}\right]^{1/2}$
$E_M = \left[\dfrac{2qN_2}{\epsilon}(V_{NP} + \Delta\psi_0)\right]^{1/2}$	$C_T = A\left[\dfrac{\epsilon qN_2}{2(V_{NP} + \Delta\psi_0)}\right]^{1/2}$
$E_M = \dfrac{V_{NP} + \Delta\psi_0}{X}$	$C_T = \dfrac{A\epsilon}{X}$
$E_M = \dfrac{qaX_0^2}{8\epsilon}$	$C_T = A\left[\dfrac{\epsilon^2 qa}{12(V_{NP} + \Delta\psi_0)}\right]^{1/3}$

*For cases 1–4, the contact potential (built-in voltage) is given by $\Delta\psi_0 = (kT/q)\ln(N_1 N_2/n_i^2)$, and for case 5, $\Delta\psi_0 = (qa/\epsilon)(X_0^3/12) = (kT/q)\ln(aX_0/2n_i)^2$, which requires iterative solution.

Table 3-2 Junction-Diode Admittance for Arbitrary Frequency

Long diode	$y(\omega) = g\sqrt{1 + j\omega\tau} = g\sqrt{\tfrac{1}{2}\sqrt{1 + (\omega\tau)^2} + \tfrac{1}{2}} + jg\sqrt{\tfrac{1}{2}\sqrt{1 + (\omega\tau)^2} - \tfrac{1}{2}}$
Short diode	$y(\omega) = g\sqrt{j\omega\dfrac{X_P^2}{D_n}}\coth\sqrt{j\omega\dfrac{X_P^2}{D_n}}$
Intermediate-length diode	$y(\omega) = g\sqrt{1 + j\omega\tau}\tanh\left(\dfrac{X_P}{L_n}\right)\coth\sqrt{\dfrac{X_P^2}{L_n^2} + j\omega\dfrac{X_P^2}{D_n}}$

Table 3-3 Junction-Diode Admittance Approximations

	Low Frequency	High Frequency
Long diode	$y(\omega) = g[1 + (j\omega\tau/2)]$	$y(\omega) = g\left[\sqrt{\dfrac{\omega\tau}{2}} + j\sqrt{\dfrac{\omega\tau}{2}}\right]$
Short diode	$y(\omega) = g\left(1 + j\omega\dfrac{X_P^2}{3D_n}\right)$	$y(\omega) = g\sqrt{j\omega\dfrac{X_P^2}{D_n}}\coth\sqrt{j\omega\dfrac{X_P^2}{D_n}}$

Table 3-4 Small-Signal Diode-Circuit Solutions

Source Function	Charge-Control-Model Solution	Exact-Model Solution
Step current $i_a(t) = \Delta I_A u(t)$	$v_a(t) = \dfrac{q_U(t)}{\tau^* g} u(t) =$ $\dfrac{\Delta I_A}{g}\left[1 - \exp\left(-\dfrac{t}{\tau^*}\right)\right] u(t)$	$v_a(t) = \dfrac{kT}{q}\dfrac{n_u(0,t)}{n_{0P}} u(t) = \dfrac{\Delta I_A}{g}\left[\mathrm{erf}\left(\sqrt{\dfrac{t}{\tau}}\right)\right] u(t)$
Step voltage $v_a(t) = \Delta V_A u(t)$	$i_a(t) = g\Delta V_A [\tau^*\delta(t) + u(t)]$	$i_a(t) = g\Delta V_A \left[\dfrac{\exp(-t/\tau)}{\sqrt{\pi t/\tau}} + \mathrm{erf}\left(\sqrt{\dfrac{t}{\tau}}\right)\right] u(t)$
Sinusoidal current $i_a(t) = I_a \exp(j\omega t)$	$y(\omega) = g(1 + j\omega t)$ $v_a(t) = \dfrac{i_a(t)}{g(1 + j\omega t)}$	$y(\omega) = g\sqrt{1 + j\omega t}$ $v_a(t) = \dfrac{i_a(t)}{g\sqrt{1 + j\omega t}}$
Sinusoidal voltage $v_a(t) = V_a \exp(j\omega t)$	$y(\omega) = g(1 + j\omega t)$ $i_a(t) = v_a(t) g(1 + j\omega t)$	$y(\omega) = g\sqrt{1 + j\omega t}$ $i_a(t) = v_a(t) g\sqrt{1 + j\omega t}$

Table 3-5 Large-Signal Diode-Circuit Solutions

Source Function	Charge-Control-Model Solution	Exact-Model Solution
Step current $i_A(t) = I_A + \Delta I_A u(t)$	$v_A(t) = \dfrac{kT}{q} \ln\left[1 + \dfrac{q_U(t)}{Q_{0U}}\right]$ $q_U(t) = \tau^*\left\{I_A + \Delta I_A\left[1 - \exp\left(\dfrac{t}{\tau^*}\right)\right]u(t)\right\}$	$v_A(t) = \dfrac{kT}{q} \ln\left[1 + \dfrac{n_U(0,t)}{n_{0P}}\right]$ $n_{U'}(0,t) = \dfrac{L_n}{qAD_n}\left\{I_A + \Delta I_A\left[\mathrm{erf}\sqrt{\dfrac{t}{\tau}}\right]u(t)\right\}$
Step voltage $v_A(t) = V_A + \Delta V_A u(t)$	$i_A(t) = I_A + \Delta I_A[\tau^*\delta(t) + u(t)]$ $I_A = \dfrac{Q_{0U}}{\tau^*}\left[\exp\left(\dfrac{qV_A}{kT}\right) - 1\right]$ $\Delta I_A = \dfrac{Q_{0U}}{\tau^*}\left[\exp\left(\dfrac{q(V_A + \Delta V_A)}{kT}\right) - \exp\left(\dfrac{qV_A}{kT}\right)\right]$	$i_A(t) = I_A + \Delta I_A\left[\dfrac{\exp(-t/\tau)}{\sqrt{\pi t/\tau}} + \mathrm{erf}\sqrt{\dfrac{t}{\tau}}\right]$ $I_A = \dfrac{qAD_n}{L_n}n_{0P}\left[\exp\left(\dfrac{qV_A}{kT}\right) - 1\right]$ $\Delta I_A = \dfrac{qAD_n}{L_n}n_{0P}\left[\exp\left(\dfrac{q(V_A + \Delta V_A)}{kT}\right) - \exp\left(\dfrac{qV_A}{kT}\right)\right]$
Sinusoidal current $i_A(t) = I_A + I_a \exp(j\omega t)$	$v_A(t) = \dfrac{kT}{q} \ln\left[1 + \dfrac{q_U(t)}{Q_{0U}}\right]$ $q_U(t) = \tau^*\left[I_A + \dfrac{I_a \exp(j\omega t)}{(1 + j\omega t)}\right]$	$v_A(t) = \dfrac{kT}{q} \ln\left[1 + \dfrac{n_U(0,t)}{n_{0P}}\right]$ $n_{U'}(0,t) = \dfrac{L_n}{qAD_n}\left[I_A + \dfrac{I_a \exp(j\omega t)}{\sqrt{1 + j\omega t}}\right]$
Sinusoidal voltage $v_A(t) = V_A + V_a \exp(j\omega t)$	$i_A(t) = I_A[1 + 2V_a \Sigma A_n(\omega) \exp(jn\omega t)]$ $I_A = \dfrac{Q_{0U}}{\tau^*}\left[\exp\left(\dfrac{qV_A}{kT}\right) - 1\right]$ $A_n(\omega) = I_n(qV_A/kT)[1 + jn\omega\tau]$†	$i_A(t) = I_A[1 + 2V_A \Sigma A_n(\omega) \exp(jn\omega t)]$ $I_A = \dfrac{qAD_n}{L_n}n_{0P}\left[\exp\left(\dfrac{qV_A}{kT}\right) - 1\right]$ $A_n(\omega) = I_n(qV_A/kT)\sqrt{1 + jn\omega\tau}$

†$I_n(qV_A/kT)$ modified Bessel function of the first kind

Table 3-6 Conversion Chart for SPICE Junction-Diode Symbols

SPICE Symbol	Description	Default Value	Units	Present Symbol	Description
IS	Saturation current	1.0 E−14	A	I_{0A} (I_0)	Saturation current
RS	Ohmic resistance	0	ohm	R_S	Series resistance
N	Emission coefficient	1	—	N	Quality factor
TT	Transit time	0	—	τ	Lifetime
CJO	Zero-bias junction capacitance	0	F	C_{0T}	Zero-bias depletion-layer capacitance
VJ	Junction potential	1	V	$\Delta\psi_0$	Contact potential
M	Grading coefficient	0.5	—	M	Grading exponent
EG	Activation energy	1.11	eV	ξ_G	Energy gap
XTI	Saturation-current temperature exponent	3.0	—	χ	Saturation-current temperature exponent
KF	Flicker-noise coefficient	0	—	[SPICE equations relating to topics in the lower portion of the table are not treated in this book.]	
AF	Flicker-noise exponent	1	—		
FC	Forward-bias depletion-capacitance coefficient	0.5	—		
BV	Reverse breakdown voltage	∞	V		
IBV	Current at breakdown	1.0 E−3	A		

Table 3-7 Calculated dc Values, Numerical Example

Symbol	Units	(1) OFF	(2) Crossover	(3) Partly ON	(4) ON
V_G	V	0.200	0.667	4.580	4.600
$V_A = V_J$	V	0.200	0.643	0.758	0.773
$I_A = I_R$	mA	4.3×10^{-10}	0.006	0.533	0.957
$i_R(V_J) = I_R$	mA	4.3×10^{-10}	0.006	0.533	0.957
$i_C(V_J) = 0$	mA	0	0	0	0
$C_{TOT} = C_u + C_t$	pF	0.872	3.604	207.2	371
$C_u = C_U(V_J)$	pF	9.8×10^{-8}	2.439	206.0	370
$C_t = C_T(V_J)$	pF	0.872	1.165	1.249	1.26
$q_{TOT} = q_u + q_t$	pC	0.166	0.676	6.078	10.34
$q_u = q_U(V_J)$	pC	4.3×10^{-9}	0.063	5.327	9.567
$q_t = q_T(V_J)$	pC	0.166	0.613	0.751	0.770
$r = r_A(V_J)$	kilohm	1.0×10^8	4.10	0.049	0.027

Table 3-8 Calculated Transient Values, Numerical Example

(Particular times) →		$t^{m=1}$	$t^{m=2}$	$t^{m=3}$	$t^{m=4}$
(Diode-voltage values) →		OFF	Crossover	Partly ON	ON
Symbol	Units				
t	ns	10	11	19.1	79.1
$v_G(t)$	V	0.200	4.600	4.600	4.600
$v_A(t) = v_J(t)$	V	0.200	0.643	0.758	0.773
$i_A(t)$	mA	1.1×10^{-8}	0.989	0.961	0.957
$i_R(v_J)$	mA	4.3×10^{-10}	0.006	0.533	0.956
$i_C(v_J)$	mA	9.8×10^{-8}	0.983	0.428	9.4×10^{-4}
$C_{TOT}(v_J)$	pF	0.872	3.604	207.2	371
$C_A(v_J)$	pF	9.8×10^{-8}	2.439	206.0	370
$C_T(v_J)$	pF	0.872	1.165	1.249	1.26
$q_{TOT}(v_J)$	pC	0.166	0.676	6.079	10.32
$q_A(v_J)$	pC	4.3×10^{-9}	0.063	5.327	9.558
$q_T(v_J)$	pC	0.166	0.613	0.751	0.770
$r_R(v_J)$	kilohm	1.0×10^8	4.10	0.049	0.027

REFERENCES

1. P. Davidov, "The Rectifying Action of Semi-Conductors," *Tech. Phys. (USSR)* **5**, 87 (1938).

2. W. Schottky, "Zur Halbleitertheorie der Sperrschnicht- und Spitzengleichrichter," *Z. Phys.* **113**, 367 (1939).

3. N. F. Mott, "The Theory of Crystal Rectifiers," *Proc. Roy. Soc. London* **171**, 27 (1939).

4. H. Y. Fan, "Contacts Between Metals and Between a Metal and a Semiconductor," *Phys. Rev.* **42**, 338 (1942).

5. W. Shockley, "The Theory of p-n Junctions in Semiconductors and p-n Junction Transistors," *Bell Syst. Tech. J.* **28**, 435 (1949).

6. F. S. Goucher, G. L. Pearson, M. Sparks, G. K. Teal, and W. Shockley, "Theory and Experiment for a Germanium p-n Junction," *Phys. Rev.* **81**, 637 (1951).

7. W. L. Brown, "n-Type Surface Conductivity on p-Type Germanium," *Phys. Rev.* **91**, 518 (1953).

8. C. G. B. Garrett and W. H. Brattain, "Physical Theory of Semiconductor Surfaces," *Phys. Rev.* **99**, 376 (1955).

9. R. H. Kingston and S. F. Neustadter, "Calculation of the Space Charge, Electric Field, and Free Carrier Concentration at the Surface of a Semiconductor," *J. Appl. Phys.* **26**, 718 (1955).

10. C. T. Sah, R. N. Noyce, and W. Shockley, "Carrier Generation and Recombination in p-n Junctions and p-n Junction Characteristics," *Proc. IRE* **45**, 1228 (1957).

11. J. L. Moll, "The Evolution of the Theory for the Voltage-Current Characteristic of P-N Junctions," *Proc. IRE* **46**, 1076 (1958).

12. Ph. Passau and M. van Styvendael, "La Jonction p-n Abrupte dans le Cas Statique," *Proc. Int. Conf. Solid State Phys. in Electron. and Telecomm.* Brussels, June 2–7, 1958, M. Desirant and J. F. Miciels (Eds.), *Int. Un. Pure and Appl. Phys.*, Vol. I, Academic Press, New York, 1960, p. 407.

13. C. Goldberg, "Space Charge Regions in Semiconductors," *Solid-State Electron.* **7**, 593 (1964).

14. R. P. Jindal and R. M. Warner, Jr., "A General Solution for Step Junctions with Infinite Extrinsic End Regions at Equilibrium," *IEEE Trans. Electron. Devices* **28**, 348 (1981).

15. R. M. Warner, Jr. and R. P. Jindal, "Replacing the Depletion Approximation," *Solid-State Electron.* **26**, 335 (1983).

16. R. M. Warner, Jr., R. D. Schrimpf, and P. D. Wang, "Explaining the Potential

Drop on the High Side of a Grossly Asymmetric Junction," *J. Appl. Phys.* **57,** 1239 (1985).

17. H. D. Barber, 1968, private communication to A. Nussbaum, "Theory of Semiconducting Junctions," *Semiconductors and Semimetals,* Vol. 15, Academic Press, New York, 1981, p. 86.

18. H. D. Barber, 1988, private communication. For further discussion of his data set, see also R. M. Warner, Jr., and B. L. Grung, *Transistors: Fundamentals for the Integrated-Circuit Engineer,* Wiley, New York, 1983, pp. 429–432; reprint edition, Krieger Publishing Company (P.O. Box 9542, Melbourne, Florida 32902-9542), 1990.

19. R. M. Warner, Jr. and K. Lee, "Modeling the Space-Charge-Layer Boundary of a Forward-Biased Junction," *J. Appl. Phys.* **53,** 5304 (1982).

20. N. H. Fletcher, "General Semiconductor Junction Relations," *J. Electronics* **2,** 609 (1957).

21. T. Misawa, "A Note on the Extended Theory of the Junction Transistor," *J. Phys. Soc. Jpn.* **11,** 728 (1956).

22. R. M. Warner, Jr., and B. L. Grung, *Transistors: Fundamentals for the Integrated-Circuit Engineer,* Wiley, New York, 1983; reprint edition, Krieger Publishing Company (P.O. Box 9542, Melbourne, Florida 32902-9542), 1990.

23. A. Y. Cho and J. R. Arthur, "Molecular Beam Epitaxy," *Prog. Solid-State Chem.* **10,** 157 (1975).

24. R. N. Hall and W. C. Dunlap, "P-N Junctions Prepared by Impurity Diffusion," *Phys. Rev.* **80,** 467 (1950).

25. G. K. Wehner, U.S. Patent 3,021,271, filed April 27, 1959, issued February 13, 1962.

26. R. N. Hall, "Power Rectifiers and Transistors," *Proc. IRE* **40,** 1512 (1952).

27. S. P. Morgan and F. M. Smits, "Potential Distribution and Capacitance of a Graded *p-n* Junction," *Bell Syst. Tech. J.* **39,** 1573 (1960).

28. J. L. Moll, *Physics of Semiconductors,* McGraw-Hill, New York, 1964, p. 121.

29. Warner and Grung, p. 322.

30. C. S. Fuller and J. A. Ditzenberger, "Diffusion of Donor and Acceptor Elements in Silicon," *J. Appl. Phys.* **27,** 544 (1956).

31. H. Christensen and G. K. Teal, U.S. Patent 2,692,839, filed April 7, 1951, and issued October 26, 1954.

32. W. Shockley, U.S. Patent 2,787,564, filed October 28, 1954, and issued April 2, 1957.

33. Warner and Grung, pp. 338–345.

34. H. Lawrence and R. M. Warner, Jr., "Diffused Junction Depletion Layer Calculations," *Bell Syst. Tech. J.* **39,** 389 (1960).

35. J. Hoerni, U.S. Patent 3,025,589, filed May 1, 1959, and issued March 20, 1962.

36. O. L. Anderson, H. Christensen, and P. Andreatch, "Techniques for Connecting Electrical Leads to Semiconductors," *J. Appl. Phys.* **28,** 923 (1957).

37. S. L. Miller, "Ionization Rates for Holes and Electrons in Silicon," *Phys. Rev.* **105,** 1246 (1957).

38. J. L. Moll, *Physics of Semiconductors,* McGraw-Hill, New York, 1964, p. 224.

39. S. M. Sze and G. Gibbons, "Avalanche Breakdown Voltages of Abrupt and Linearly Graded p-n Junctions in Ge, Si, GaAs and GaP," *Appl. Phys. Lett.* **8,** 111 (1966).

40. R. van Overstraeten and H. de Man, "Measurement of the Ionization Rates in Diffused Silicon p-n Junctions," *Solid-State Electron.* **13,** 583 (1970).

41. F. R. Carlson; these data were presented in an undated brochure titled "Diffused Silicon Diodes," printed for external distribution by U.S. Semiconductor Products, Inc., Phoenix, Arizona, in approximately 1959.

42. R. M. Warner, Jr., "Avalanche Breakdown in Silicon Diffused Junctions," *Solid-State Electron.* **15,** 1303 (1972).

43. C. Zener, *Proc. Roy. Soc. London, Ser. A* **145,** 523 (1934).

44. R. J. Gravrok, "Tunnel Junctions as Ohmic Interconnections for Three-Dimensional Integrated Circuits," MSEE Thesis, University of Minnesota, Minneapolis, Minn., August 1986; available from *Masters Abstracts International,* Vol. 23, No. 3, University Microfilms International, Ann Arbor, Mich., 1987.

45. A. G. Chynoweth, W. L. Feldmann, C. A. Lee, R. A. Logan, G. L. Pearson, and P. Aigrain, "Internal Field Emission at Narrow Silicon and Germanium p-n Junctions," *Phys. Rev.* **118,** 425 (1960).

46. L. Esaki, "New Phenomenon in Narrow Germanium p-n Junctions," *Phys. Rev.* **109,** 603 (1958).

47. S. M. Sze, *Physics of Semiconductor Devices,* Wiley, New York, 1981, p. 517.

48. Warner and Grung, p. 64.

49. Moll, p. 240.

50. R. J. Gravrok and R. D. Schrimpf, private communication.

51. O. W. Richardson, "On the Negative Radiation from Hot Platinum," *Cambridge Phil. Soc. Proc.* **11,** 286 (1901).

52. W. Shockley and R. C. Prim, "Space-Charge Limited Emission in Semiconductors," *Phys. Rev.* **90,** 753 (1953).

53. C. D. Child, "Discharge from Hot CaO," *Phys. Rev.* **32,** 498 (1911).

54. A. van der Ziel, private communication.

55. S. Denda and M. A. Nicolet, "Pure Space-Charge-Limited Electron Current in Silicon," *J. Appl. Phys.* **37,** 2417 (1966).

56. J. L. Chu, G. Persky, and S. M. Sze, "Thermionic Injection and Space-Charge-Limited Current in Reach-Through p^+np^+ Structures," *J. Appl. Phys.* **43,** 3510 (1972).

57. P. J. Kannam, "Design Concepts of High Frequency Punchthrough Structures," *IEEE Trans. Electron Devices* **ED-23,** 879 (1976).

58. Warner and Grung, p. 548.

59. P. Debye and E. Hückel, "Zur Theorie der Elektrolyte," *Z. Phys.* **24,** 185 (1923).

60. P. Debye and E. Hückel, "Zur Theorie der Elektrolyte II," *Z. Phys.* **24,** 305 (1923).

61. R. M. Warner, Jr., "Normalization in Semiconductor Problems," *Solid-State Electron.* **28,** 529 (1985).

62. R. M. Warner, Jr., R. P. Jindal, and B. L. Grung, "Field and Related Semiconductor-Surface and Equilibrium-Step-Junction Variables in Terms of the General Solution," *IEEE Trans. Electron Devices* **31,** 994 (1984).

63. Warner and Grung, p. 360.

64. Warner and Grung, p. 346 and p. 697.

65. J. R. Hauser and M. A. Littlejohn, "Approximations for Accumulation and Inversion Space-Charge Layers in Semiconductors," *Solid-State Electron.* **11,** 667 (1968).

66. D.-H. Ju and R. M. Warner, Jr., "Modeling the Inversion Layer at Equilibrium," *Solid-State Electron.* **27,** 907 (1984).

67. R. D. Schrimpf and R. M. Warner, Jr., "A Precise Scaling Length for Depleted Regions," *Solid-State Electron.* **28,** 779 (1985).

68. R. P. Jindal, Bulk and Surface Effects on Noise and Signal Behaviour of Semiconductor Devices, Ph.D. Thesis, University of Minnesota, Minneapolis, Minn. March 1981.

69. R. D. Schrimpf and R. M. Warner, Jr., "An Approximate-Analytic Solution for the Forward-Biased Step Junction," *IEEE Trans. Electron Devices* **35,** 698 (1988).

70. A. van der Ziel, *Noise in Solid State Devices and Circuits,* Wiley, New York, 1986.

71. S. K. Ghandi, *The Theory and Practice of Microelectronics*, Wiley, New York, 1968.
72. J. G. Linvill, "Lumped Models of Diodes and Transistors," *Proc. IRE* **46,** 949 (1958).
73. J. G. Linvill, *Models of Diodes and Transistors*, McGraw-Hill, New York, 1963.
74. H. K. Gummel, "A Self-Consistent Iterative Scheme for One-Dimensional Steady State Transistor Calculations," *IEEE Trans. Electron Devices* **ED-11,** 445 (1964).
75. J. W. Slotboom, "Iterative Scheme for 1- and 2-Dimensional DC Transistor Simulation," *Electron Lett.* **5,** 677 (1969).
76. C. M. Snowden, *Semiconductor Device Modeling*, Peregrinus, London, 1988.
77. R. H. Kingston, "Switching Time in Junction Diodes and Junction Transistors," *Proc. IRE* **42,** 1148 (1954).
78. B. Lax and S. F. Neustadter, "Transient Response of a p-n Junction," *J. Appl. Phys.* **25,** 1148 (1954).
79. S. Wang, *Solid-State Electronics*, McGraw-Hill, New York, 1966.
80. A. van der Ziel, *Solid State Physical Electronics*, Prentice-Hall, Englewood Cliffs, N. J., 1957, 1968, 1976 (three editions).
81. B. R. Chawla and H. K. Gummel, "Transition Region Capacitance of Diffused p-n Junctions," *IEEE Trans. Electron Devices* **18,** 178 (1971).

TOPICS FOR REVIEW

R3-1. Define a *PN* junction.
R3-2. In the thought experiment on junction formation, why does charge exchange occur?
R3-3. What is the physical identity of the charges responsible for space charge at a junction?
R3-4. What is meant by *space-charge-region boundary*?
R3-5. What is meant by *metallurgical junction*?
R3-6. What is meant by *double layer*? What is an alternate term for it?
R3-7. What is meant by *built-in field*? What is its origin?
R3-8. What is meant by *built-in potential difference*?
R3-9. State a concise criterion that assures equilibrium conditions.
R3-10. What is a *semilog* diagram? What are its advantages and limitations?
R3-11. What are the advantages and limitations of a linear diagram?
R3-12. Why does a semilog carrier profile have the same shape as the corresponding electrostatic-potential profile when algebraic sign is adjusted appropriately?

R3-13. Is a linear or a semilog diagram better for estimating the location of a density-gradient maximum?
R3-14. What is a *step junction*?
R3-15. What is meant by *symmetric step junction*?
R3-16. Distinguish between the terms *space-charge layer* and *depletion layer*.
R3-17. Describe the four current-density components that exist in a junction at equilibrium. What balances exist among them?
R3-18. Why are these four components described as a *convenient mathematical fiction*?
R3-19. What is the essence of the *depletion approximation*?
R3-20. What is meant by the "uncovering" of ionic charge near the junction and what causes it?
R3-21. What is the value of the constant volumetric charge density on either side of a step junction according to the depletion approximation?
R3-22. What is meant by *areal charge density*?
R3-23. What physical interpretation attaches to the integral of the volumetric-charge-density profile on a given side of a junction?
R3-24. Describe the electric-field profile of a step junction as predicted by the depletion approximation.
R3-25. What physical interpretation can be placed on the integral of the electric-field profile of a junction?
R3-26. In what respect is the depletion approximation most primitive? In what respect is it most precise? Why?
R3-27. Give two synonyms for *contact potential*.
R3-28. Can contact potential be measured with a voltmeter? Why?
R3-29. Why must the areal charge densities on the two sides of a junction be equal?
R3-30. Describe the relationships among the depletion-layer thickness X_0, peak electric field E_{0M}, and contact potential $\Delta\psi_0$ in a step junction at equilibrium.
R3-31. Describe the depletion-approximation electric-field profile of a step junction that is (a) symmetric; (b) asymmetric; (c) one-sided.
R3-32. What is meant by *junction bias*?
R3-33. Describe *forward bias* and *reverse bias*.
R3-34. Explain the double-subscript convention for junction bias.
R3-35. Why is it that virtually all of the reverse voltage applied to a junction appears between the boundaries of the space-charge layer?
R3-36. Explain the diode symbol.
R3-37. Why does the boundary-region majority-carrier profile translate in response to reverse bias? In what direction does it move?
R3-38. Why does the depletion approximation improve with reverse bias?
R3-39. What bias polarity reduces the height of the potential barrier in a *PN* junction? What is its effect?
R3-40. How does one convert the equilibrium depletion-approximation expressions relating X_0 and $\Delta\psi_0$ to expressions for arbitrary junction bias?
R3-41. Define *Boltzmann quasiequilibrium*.

R3-42. Define *low-level forward bias*.

R3-43. What is meant by *carrier injection*?

R3-44. What condition precisely defines the boundary of a forward-biased junction? Of a reverse-biased junction?

R3-45. Describe the minority-carrier profile in a neutral region on one side of a forward-biased junction. Describe that on one side of a reverse-biased junction.

R3-46. State the *law of the junction*.

R3-47. What carrier-density condition exists between the space-charge-layer boundaries of a forward-biased junction?

R3-48. What four facts are needed to derive the current–voltage equation for an ideal junction?

R3-49. How does one write the slope of a declining exponential function at any value of the independent variable?

R3-50. What is the probable relationship of the diffusion lengths on the two sides of a grossly asymmetric (one-sided) junction? Why?

R3-51. Describe the primary current components as a function of position on the lightly doped side of a grossly asymmetric junction under forward bias and under reverse bias.

R3-52. What is meant by *saturation current*?

R3-53. Why does the reverse current of an ideal *PN* junction saturate?

R3-54. What determines the magnitude of reverse saturation current?

R3-55. What forward-characteristic property is determined by saturation current?

R3-56. What carrier-density condition exists between the space-charge-layer boundaries of a reverse-biased junction?

R3-57. Describe the two kinds of models used extensively in solid-state electronics.

R3-58. What is meant by *parameter*?

R3-59. What is meant by *long* diode and *short* diode?

R3-60. Describe two different kinds of dynamic conditions in terms of signal character. Describe two different kinds in terms of signal magnitude.

R3-61. Define *piecewise-linear* approximation. Give an example.

R3-62. Describe the current–voltage characteristic of an ideal rectifier.

R3-63. In what respect does the *I–V* characteristic of a silicon junction primarily depart from that of the ideal rectifier?

R3-64. Explain the physical plausibility of the static charge-control model.

R3-65. What phenomenon is dominant in fixing the reverse current of a silicon junction? Of a germanium junction?

R3-66. What nonideal phenomenon dominates the low-voltage forward *I–V* characteristic of a silicon junction? Explain.

R3-67. What is *high-level forward bias*?

R3-68. How many boundary values must be determined in the high-level forward-bias problem?

R3-69. Describe a *PIN diode*. Comment on its depletion-layer-capacitance behavior.

R3-70. Describe a *linearly graded junction*. Describe its depletion-approximation charge profile.
R3-71. Describe qualitatively the impurity profile resulting from a solid-phase diffusion process.
R3-72. Why are diffused junctions so important?
R3-73. Name two functions that describe diffused-impurity profiles, and cite the boundary conditions associated with each.
R3-74. Describe a *high-low junction*.
R3-75. What is the most important electrical property of a high-low junction?
R3-76. Describe junction *breakdown*.
R3-77. What causes *avalanche* breakdown? Whence the name?
R3-78. What conditions must exist at a junction for reverse-bias tunneling to occur? For forward-bias tunneling?
R3-79. What is a *backward diode*?
R3-80. Describe a *punchthrough diode*.
R3-81. What phenomenon causes an exponential dependence of current upon applied voltage in the punchthrough diode?
R3-82. Define *space-charge-limited* current.
R3-83. What is meant by *flatband voltage*?
R3-84. What variables are related by the Poisson-Boltzmann equation?
R3-85. Give two interpretations of physical significance for *Debye length*.
R3-86. What is a desirable consequence of using general or extrinsic Debye length for spatial normalization in the semiconductor-surface or step-junction problems?
R3-87. How do the exact field and potential profiles differ from those predicted by the depletion approximation?
R3-88. In what sense does the depletion approximation improve when doping is increased in a symmetric step junction?
R3-89. What is the character of the two asymptotic functions entering the DAR?
R3-90. Distinguish between *inversion layer* and *accumulation layer*.
R3-91. Why is there so little redundancy in the general solution for accumulation?
R3-92. Define *threshold of inversion*. Give a qualitative description of its significance.
R3-93. What criterion determines the validity of a small-signal treatment?
R3-94. Explain *linearization*.
R3-95. Define *quasistatic*.
R3-96. State the expression for small-signal conductance in a diffusion-dominated forward junction.
R3-97. Tell how to determine V_γ for a piecewise-linear forward-biased-junction model employing a tangent at the bias point.
R3-98. Define *diffusion capacitance*.
R3-99. Explain why stored charge Q_S exhibits the same dependence upon applied voltage V_{NP} as does the forward current I.

R3-100. Explain depletion-layer capacitance physically.

R3-101. Why does total depletion-layer thickness correspond to plate spacing in a parallel-plate capacitor?

R3-102. In view of boundary-region diffuseness in a depletion layer, why does the depletion approximation give a good estimate of depletion-layer capacitance?

R3-103. Why is the voltage at which the diffusion and depletion-layer capacitances are equal, or "cross over," a matter of interest? In what portion of the forward-bias range does it occur?

R3-104. Name the kinds of current components that may exist in a junction sample under dynamic conditions.

R3-105. Name several time constants that may be relevant in determining the response of a junction sample to a time-varying signal.

R3-106. Identify the three elements of a useful small-signal equivalent-circuit model of a junction diode, and state how they are interconnected.

R3-107. Comment qualitatively on the characteristic times involved in charging the diffusion and depletion-layer capacitances.

R3-108. Describe the transmission-line analog applicable to the modeling of the long diode. Why is it merely an "analog"?

R3-109. Explain how the simplified transmission line leads to an effective diffusion capacitance C_s^* that is half the value defined quasistatically. What is the physical justification for using the smaller value?

R3-110. Identify the carrier populations and transport mechanisms mainly involved in the capacitances C_s and C_t.

R3-111. What justification is there for defining an effective lifetime to accompany C_s^*?

R3-112. How is the unit step function related to the Dirac delta function?

R3-113. What linear relationship is the key to static and dynamic charge-control analysis?

R3-114. Under what conditions may the principle of superposition be applied?

R3-115. How does one obtain the time-dependent diffusion equation?

R3-116. Describe and compare the charge-control, Linvill, finite-difference, and exact models.

R3-117. Describe the differences between device-physics and circuit-behavior analysis.

R3-118. In small-signal diode switching, how does the exact-model voltage-rise curve differ qualitatively from the adjusted (effective-lifetime) charge-control curve?

R3-119. Why does large-signal switching pose a more difficult problem than small-signal switching?

R3-120. In current-reversal switching, why is the character of the turn-off response current-dependent while the turn-on response is not?

R3-121. What is the fundamental assumption of charge-control theory, and in what respects is it inaccurate?

R3-122. What is the constant product that links the time-domain and frequency-domain responses of a diode?

R3-123. What model is employed for SPICE simulation of a PN-junction diode?
R3-124. What model details are included in SPICE simulation that are difficult or impossible to treat in hand calculations?
R3-125. How does SPICE deal with the diverging depletion-layer capacitance encountered at large forward bias?

ANALYTIC PROBLEMS

A3-1. At a certain point in a certain junction at equilibrium, the electric field is $+7,698$ V/cm and $n = 10^{11}/\text{cm}^3$.
  a. Calculate with due regard for algebraic sign the density gradient for electrons at that point.
  b. Do the same for holes at that point.

A3-2. A certain symmetric step PN junction has $N_D = 10^{15}/\text{cm}^3$ (donors only) on the left-hand side and $N_A = 10^{15}/\text{cm}^3$ (acceptors only) on the right-hand side. Using the depletion approximation, calculate, for equilibrium conditions,
  a. Contact potential $\Delta\psi_0$.
  b. Depletion-layer thickness X_0 in μm.
  c. Field at the junction E_{0M} in kilovolts per centimeter.

A3-3. For the junction of A3-2, and continuing to use the depletion approximation, sketch dimensioned diagrams of
  a. Charge-density profile.
  b. Field profile.
  c. Potential profile.

A3-4. Considering the spatial origin to be positioned at the metallurgical junction in the equilibrium sample of A3-2 and A3-3,
  a. Calculate, with due regard for algebraic sign, the four current-density components at $x = 0$, $J_{n,\text{drft}}$, $J_{p,\text{drft}}$, $J_{n,\text{diff}}$, $J_{p,\text{diff}}$, using realistic carrier profiles, but using E_{0M} from the depletion approximation.
  b. Calculate both density-gradient values at $x = 0$.

A3-5. A certain asymmetric step junction has a doping on the left-hand side of $N_1 = N_D - N_A = 10^{14}/\text{cm}^3$ and on the right-hand side of $N_2 = N_A - N_D = 4 \times 10^{14}/\text{cm}^3$. Using the depletion approximation, calculate, for the junction at equilibrium,
  a. Contact potential $\Delta\psi_0$.
  b. Depletion-layer thickness X_0.
  c. Electric field at the junction, E_{0M}.
  d. Thickness of the depletion-layer portion on the N-type side, X_1. [*Hint:* Use Poisson's equation.]
  e. Thickness of the depletion-layer portion on the P-type side, X_2.

A3-6. For the junction of A3-5,
  a. Calculate the ratio X_1/X_2.
  b. Explain why $(X_1/X_2) = (N_2/N_1)$.

c. Sketch a dimensioned diagram of the field profile.
d. Calculate the potential drops $\Delta\psi_1$ and $\Delta\psi_2$ on the N-type and P-type sides, respectively.

A3-7. Holes are being injected by a forward-biased junction under low-level,

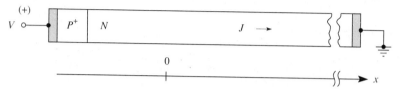

steady-state conditions at the left end of a long extrinsic-N-type silicon bar. In this problem you are only concerned with the N region for $x > 0$. The junction is several diffusion lengths to the left of the spatial origin, and $p'_N(0)$ is several times p_{0N}. The total current density in the bar is J.

a. Write an expression for $J(\infty)$, that is, current density where $x \gg L_p$. Make any reasonable approximations.
b. Given that $p'_N(x) = p'_N(0)e^{-x/L_p}$, write an expression free of primed (or excess-carrier) variables for the hole-density gradient as a function of x for $x > 0$.
c. Write an expression for $J_{p,\text{diff}}(x)$ for $x > 0$.
d. Write an expression for $J_{n,\text{diff}}(x)$ for $x > 0$.
e. Given that $J_{p,\text{diff}}(0) = 0.01\, J(\infty)$, find $E(0)$ in terms of $E(\infty)$.

A3-8. You are given an N^+P forward-biased junction.

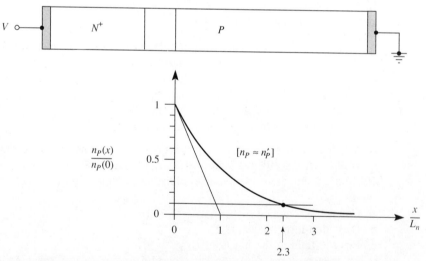

The minority electron profile on the right-hand side of the sample, $n_p(x)$ versus x, is plotted above with reasonable accuracy using normalized *linear* coordinates. In the following diagram, the minority-carrier profiles on both sides of the junction are plotted using unnormalized *semilog* coordinates, with linear abscissa and logarithmic ordinate.

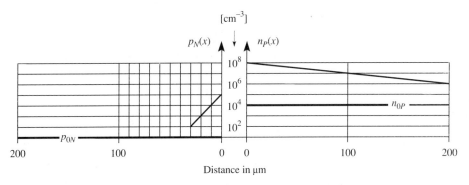

a. Determine $(dp_N/dx)|_{x=0}$ and $(dn_P/dx)|_{x=0}$. Neglect algebraic sign, but *include units*.

b. Given that $D_n = 19$ cm²/s on the right-hand side, find carrier lifetime τ on the P-type side.

A3-9. Given the N^+P junction shown here with $N_1 = 10^{19}/\text{cm}^3$, $N_2 = 10^{14}/\text{cm}^3$, and $\tau_2 = 1$ μs:

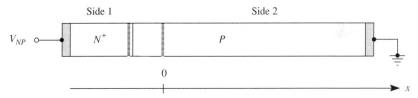

a. Find $J_n(0)$ for $V_{NP} = -0.04$ V.
b. Find $J_n(0)$ for $V_{NP} = -0.4$ V.
c. Repeat a and b with $\tau_2 = 100$ μs and all else held constant.
d. Repeat a and b with $N_2 = 10^{15}/\text{cm}^3$ and all else held constant.
e. Find $\Delta\psi$ for a and b.
f. Find an approximate value for $J_n(0)$ in b as a percentage of total current density J in the sample, given $L_p = 1$ μm and $D_p = 1.5$ cm²/s in Side 1, and assuming negligible recombination in the space-charge region.
g. Assuming the junction to be ideal, find the reverse-current density for $V_{NP} = +100$ V, $+10$ V, $+1$ V, for the sample of a and b.

A3-10. Following is a reasonably accurate sketch of the numerical solution for an equilibrium symmetric step junction having $N = 10^{17}/\text{cm}^3$.

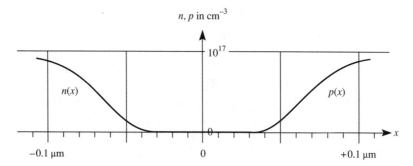

a. Estimate the maximum value of carrier-density gradient displayed in the diagram without regard for algebraic sign.

b. Calculate the maximum value of the current densities $|J_{n,\text{diff}}| = |J_{n,\text{drft}}|$ without regard for algebraic sign. Use $\mu_n = 700$ cm^2/V·s.

c. Given that $X_0 \approx 0.15$ μm (as can be confirmed on the sketch), calculate the maximum electric field in the junction, E_{OM}.

d. Calculate the value of the current densities $|J_{n,\text{diff}}(0)| = |J_{n,\text{drft}}(0)|$ without regard for algebraic sign.

e. In what approximate fraction of the space-charge-layer thickness does electric field exceed 50 kV/cm, the value at which the drift velocity of electrons saturates at 10^7 cm/s?

f. In view of your answer to part e, what is the appropriate value of μ_n to use in solving part d? Explain.

A3-11. A certain one-sided silicon step junction under bias exhibits this carrier profile.

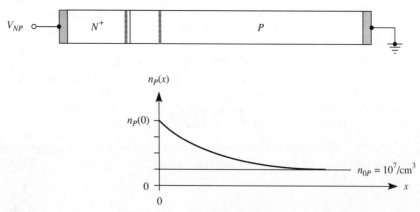

a. Find the magnitude and sign of the applied voltage V_{NP}. Use units in calculation.

b. V_{NP} is changed so that $n_P(0) = 10^8$/cm^3. Given that the total current density is $J = -5.175 \times 10^{-8}$ A/cm^2, find the minority-carrier diffusion length L_n in μm for the P-type region. Use units in calculation.

c. The sample of a is replaced by another of identical doping that exhibits a minority-carrier diffusion length of $L_n = 1$ mm. Find the carrier lifetime τ in μs for the P-type region of the new sample. Use units in calculation.

A3-12. A silicon sample at equilibrium contains an asymmetric step junction. Shown here is a portion of its field profile that is based upon the depletion approximation.

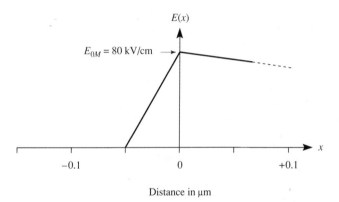

Distance in µm

a. Using Poisson's equation, derive an expression for net doping on the left-hand side. Put your final expression in terms only of symbols given in the diagram and the right-hand side of Poisson's equation.
b. Using $\epsilon \approx 1$ pF/cm, evaluate in approximate fashion the expression derived in part a. Use units.
c. Determine the potential drop on the N-type side of the junction, $\Delta\psi_{ON}$. Calculate numerical value and units.

A3-13. This N^+P junction is under a bias of $V_{NP} = -kT/q$. In the P-type region, it has the values $D_n = 25$ cm²/s and $\tau = 1$ μs. The law of the junction is not very accurate at such a low bias, but for purposes of this problem, assume that it is accurate. The value of n_{OP} is 10^6/cm³.

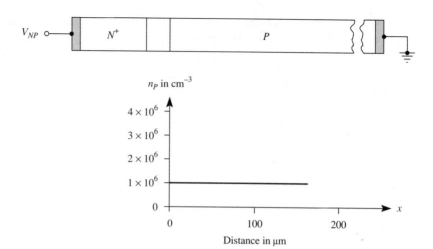

a. Calculate $(dn_p/dx)|_{x=0}$; give the correct units and algebraic sign along with the numerical value.

b. Using a solid line, sketch $n_p(x)$ for $x > 0$. The quantity $(dn_p/dx)|_{x=0}$ should be accurately presented.

c. Calculate Q', the excess-electron charge stored in the P-type region. The cross section of the sample is square, 1 mm × 1 mm. Give units.

A3-14. To analyze this one-sided step junction at equilibrium,

- Use the depletion approximation.
- Neglect potential drop on the P^+ side.
- Neglect depletion-layer penetration on the P^+ side.
- Include units in all necessary calculations.

The properties of the sample are these: On the left-hand side, $(N_D - N_A) = N_1 = 10^{15}/\text{cm}^3$; on the right-hand side, $(N_A - N_D) = N_2$; the depletion-layer dimension $X_1 \approx X_0 = 1$ μm; $\epsilon \approx 1$ pF/cm; $(kT/q) = 0.02566$ V; and finally,

$$\psi(x) = -\frac{qN_1}{\epsilon}\left[\frac{x^2}{2} + X_0 x\right].$$

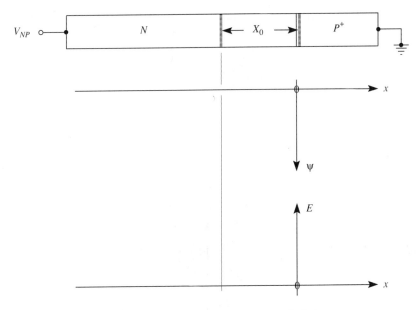

a. Sketch $\psi(x)$ versus x on the upper axis pair. Label the axes.
b. Derive an expression for $E(x)$ versus x.
c. Use your expression to calculate $E(-X_0)$, and $E(0)$ in terms of X_0.
d. Plot $E(x)$ in the space provided.
e. Derive an expression for $\Delta\psi_0$.
f. Calculate $\Delta\psi_0$.
g. Derive an expression for E_{0M} in terms of $\Delta\psi_0$.
h. Calculate E_{0M} in kV/cm.
i. Given $\Delta\psi_0 = (kT/q) \ln(N_1 N_2 / n_i^2)$, find N_2.

A3-15. Following is a one-sided junction sample having an extensive P-type region and a cross-sectional area of 10^{-2} cm². It is subjected to a steady-state forward bias. Calculate the number of excess minority carriers stored in the P-type region.

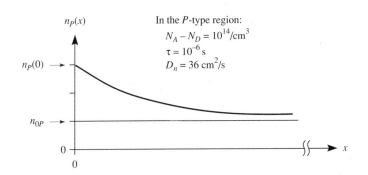

In the P-type region:
$N_A - N_D = 10^{14}/\text{cm}^3$
$\tau = 10^{-6}$ s
$D_n = 36$ cm²/s

A3-16. Given this symmetric step junction at equilibrium with a net doping on each side of $10^{16}/cm^3$, perform approximate calculations, estimating where necessary.

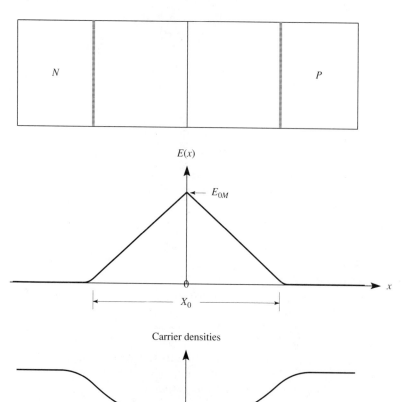

a. Calculate $\Delta\psi_0$.
b. Calculate X_0.
c. Calculate E_{0M}.
d. Estimate the value of maximum hole gradient.
e. Compute the approximate value of $J_{p,\text{diff}}$ there.
f. What is the value of $J_{p,\text{drft}}$ there?
g. Is $J_{p,\text{drft}}$ higher or lower at $x = 0$?
h. Calculate $E = E_{0M}$ there; why doesn't $J_{p,\text{drft}}$ peak at $x = 0$?
i. Calculate $J_{p,\text{drft}}$ at $x = 0$.

A3-17. A single-crystal silicon sample has a thin ideally intrinsic region flanked by heavily doped N-type and P-type regions, forming a *PIN* (or *NIP*) junction. It is *not* a step junction.

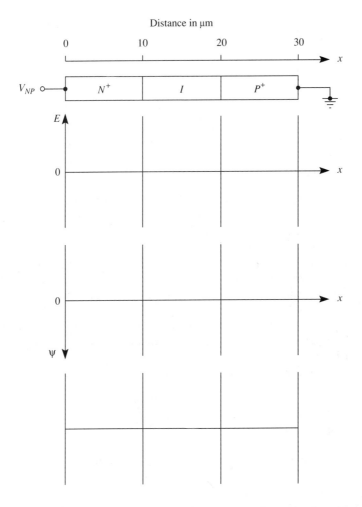

a. Considering that there is sufficient space charge in the given sample to launch and terminate four lines of force when the sample is at equilibrium, sketch these lines in the *top* diagram, using arrowheads to indicate direction.
b. Sketch *and dimension* the approximate field profile in the space provided, given $\Delta\psi_0 = 1.0$ V.
c. Sketch *and dimension* the equilibrium potential profile in the space provided, taking the potential of the P^+ region as reference. Your sketch should show potential throughout the entire device, from $x = 0$ to $x = 30$ μm.
d. Sketch the equilibrium band diagram for this sample, letting $\psi_G = 1.1$ V. Show the Fermi level ϕ.
e. Assuming the two extrinsic regions have equal net-doping magnitudes, calculate net-doping density approximately.
f. Calculate the magnitude and sign of applied voltage V_{NP} required to produce a maximum field of $E_M = 5$ kV/cm in this junction.

A3-18. Write a completely general expression for volumetric space-charge density $\rho_v(x)$ at equilibrium in a region with arbitrary doping profiles $N_D(x)$ and $N_A(x)$.

A3-19. Given the depletion approximation for an equilibrium step junction of arbitrary doping, one can specify the junction completely by specifying two of its independent variables. Given the six variables X_1, X_2, N_1, N_2, $\Delta\psi_0$, and E_{0M}, demonstrate the validity of this assertion.
a. By writing four independent equations in the six variables.
b. By writing three independent equations in the first five variables.
c. By writing two independent equations in the first four variables.
d. By writing one equation in the first three variables.

A3-20. A forward-biased long N^+P diode is acceptor-doped only on one side ($N_A = 10^{14}/\text{cm}^3$) and has $\tau = 11.1$ μs on that side. We are interested in the dominant current component.
a. Calculate the relevant diffusion length in μm.
b. Let the spatial origin be placed at the space-charge-layer boundary on the lightly doped side; calculate the equilibrium minority-carrier density at $x = 0$.
c. Calculate the magnitude of the minority-carrier gradient at $x = 0$, given that $n_P(0) = 10^7/\text{cm}^3$.
d. Calculate the magnitude of the minority-carrier current density at $x = 0$.
e. Calculate the magnitude of the excess minority-carrier charge per unit area stored in the P-type region.
f. Calculate the applied voltage V_{NP}.

A3-21. You are given an ideal silicon junction having a saturation current of $I_0 = 10^{-14}$ A, and carrying a forward current of 1 μA.
a. Assuming a reverse resistance of $R = \infty$, compute values for a piecewise-linear model of the junction.
b. Sketch and dimension the resulting I–V characteristic.
c. Repeat part a for $I = 1$ A.
d. Repeat part b for $I = 1$ A.
e. Sketch and label a static equivalent-circuit model for part c.

A3-22. A P^+N junction diode with a junction area of 1 mm^2 has a charge of 10^{-6} C of excess holes stored in it when it carries a current of 50 mA. Compute the minority-carrier diffusion length in the N-type region.

A3-23. The silicon-junction diode shown has the characteristic $I_D = (10^{-10}$ A$) \times [\exp(qV_D/kT) - 1]$. Find the amplitude of the ac output voltage.

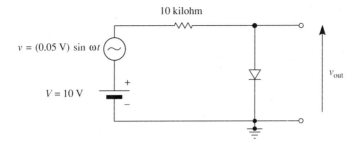

A3-24. A certain N^+P step junction receives a bias of $V_{NP} = -2.4$ V and in the process its depletion-layer thickness X as determined from capacitance measurements increases by a factor of two.
a. Find its built-in voltage $\Delta\psi_0$.
b. A certain symmetric step junction exhibits identical behavior. Determine $\Delta\psi_0$ in this case.

A3-25. A particular NP^+ step junction has a contact potential of $\Delta\psi_0 = 0.8$ V. It is being used as a voltage-controlled capacitor and exhibits an incremental depletion-layer capacitance of $C_{T1} = 100$ pF at a reverse bias of $V_{NP1} = 2$ V.
a. What bias value is required to reduce the capacitance to $C_{T2} = 25$ pF?
b. Given that the light-side net doping is $10^{14}/cm^3$, find the area of the junction.
c. Repeat part a for the case of a symmetric step junction.
d. Given that the two devices have the same junction area, find net doping N for the symmetric step junction.

A3-26. Following are electric-field profiles for four different N^+P junctions at the avalanche-breakdown condition, $V_{NP} = V_B$.

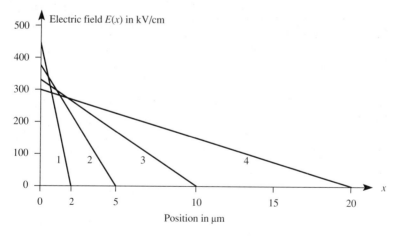

Peak-field values in the four respective cases are $E_{M1}(0) = 460$ kV/cm; $E_{M2}(0) = 380$ kV/cm; $E_{M3}(0) = 330$ kV/cm; $E_{M4}(0) = 300$ kV/cm.
a. Calculate light-side net doping N for each case.
b. Calculate the avalanche-breakdown voltage V_B for each case.
c. Calculate V_B from Equation 3-106 and compare its predictions with the results in part (b).

A3-27. You are given a symmetric step junction at equilibrium having $N = 10^{15}/\text{cm}^3$.
a. Construct a band diagram to scale with reasonable accuracy. Indicate the Fermi level, and let it be the potential reference. Let the abscissa be calibrated in micrometers, and let the ordinate be dimensionless, using $U \equiv \psi/(kT/q)$. Using arrows, indicate the values of ΔU_0 and the potential on the P-type side, U_P.
b. Using the P-type region as reference, let a voltage of $V_{NP} = -0.2$ V be applied to the junction. Using dashed lines, superimpose the new band diagram on the equilibrium diagram. Using arrows, indicate the values of ΔU and U_{NP}.

A3-28. The symbol ν is sometimes used to stand for *near-intrinsic N-type material*, and the symbol π for *near-intrinsic P-type material*. Sketch qualitative electric-field profiles for two variations on an N^+IP^+ junction, namely, for an $N^+\nu P^+$ junction and an $N^+\pi P^+$ junction. Place the spatial origin at the metallurgical junction in each case, and explain the profile shapes.

A3-29. You are to estimate the ratio of inversion-layer thickness to light-side depletion-layer thickness in a grossly asymmetric junction. (The latter dimension is usually considered to include the former.)
a. Do so first for the case depicted in Figure 3-48.
b. Do so for a PN junction with $N_1 = 10^{17}\text{cm}^3$, and $N_2 = 10^{20}/\text{cm}^3$.

A3-30. Shown here is a physical representation and field profile for a symmetric step junction at equilibrium. Assume the depletion approximation. If the permittivity of silicon is needed in calculations below, use $\epsilon = 1$ pF/cm.

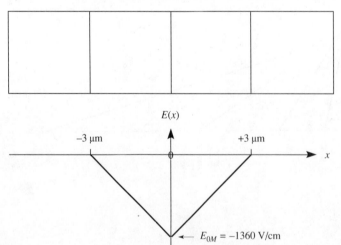

a. Using the symbols N and P, indicate the neutral N-type and P-type regions of the sample.
b. Using appropriate symbols, indicate the regions where positive and negative space charge resides in the sample.
c. Find the magnitude of $\Delta\psi_0$.
d. Calculate net doping N using Equation 3-15.
e. Calculate (dE/dx) on the right-hand side, with due regard for algebraic sign.
f. Using the result in part e, verify the result in part d with the aid of Poisson's equation.
g. Explain the physical significance in relation to the given junction of the expression $D_{0M} = \epsilon E_{0M} = qN(X_0/2)$.
h. Using the depletion-approximation values just given, combined with your knowledge of actual carrier-density values, calculate $J_{p,\text{diff}}(0)$ with due regard for algebraic sign.

A3-31. The long junction diode shown below is biased so that $I = -100 \ \mu A$. Also, $N_1 = 10^{19}/\text{cm}^3$, and on the right-hand side, $N_2 = 10^{15}/\text{cm}^3$, $D_p = 11.5 \ \text{cm}^2/\text{s}$, $D_n = 36 \ \text{cm}^2/\text{s}$, and $\tau = 1 \ \mu s$.

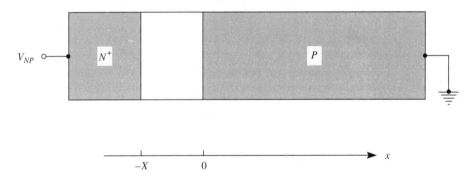

a. With due regard for algebraic sign, calculate the stored excess-carrier charge.
b. Where is this charge stored, and what charged species is involved?
c. For the given bias condition, $p_P(0) = 1.0 \times 10^{15}/\text{cm}^3$, and $n_P(0) = 1.04 \times 10^{13}/\text{cm}^3$. Calculate the junction area A in square millimeters.

A3-32. In the diagram below, the equilibrium electric-field profile is shown with a solid line, and that for the diode under bias is shown with a dashed line. The spatial origin is placed at the right-hand boundary of the junction and moves with it. You may use the depletion approximation.

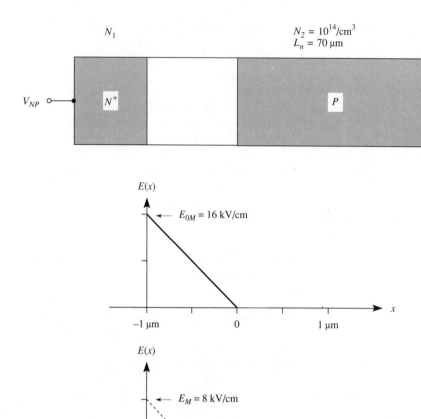

a. Calculate the bias voltage V_{NP}, with due regard for algebraic sign.
b. Calculate N_1, the net doping on the N^+ side.
c. The applied voltage is changed to a forward bias of one quarter volt. Calculate the value of $n_P(0)$.

A3-33. We have noted that the net motion of majority carriers through the transition region of a forward-biased junction is by diffusion.
a. Sketch a band diagram for the junction under forward bias.
b. Represent appropriately on the band diagram the motion of an energetic electron that moves through the transition region without interaction.
c. In an analogous way, represent the motion of an electron that starts with average energy in a passage through the transition region involving several interactions.
d. Discuss qualitatively the relative probabilities of the events described in parts b and c.
e. In fact, electron mean free path in the transition region is smaller than in the corresponding end region. Explain.

A3-34. Section 1-3.5 concentrated on steady-state aspects of the "diode" structure of Figure 1-14(a) when subjected to the ramp voltage of Figure 1-14(b), namely, $J_{OHMIC} = J_{D,CAP} = J_T$. That is, the ohmic current density in the conducting portion of the sample equaled the displacement current density in the capacitor portion, and because these were the sole components in those respective regions, each equaled the total current density. But there is an aspect of the problem that is not steady-state with respect to current density, and hence with respect to the ohmic drop in the conducting regions, one that is illuminated by the discussion of transient phenomena in this chapter.

a. On a single diagram, sketch curves of voltage versus time and indicate intervals that represent $V_{CAP}(t)$ and $V_{OHMIC}(t)$ as well as total voltage $V(t)$, all at some arbitrary time.

b. Extrapolate your curves backward to $t = 0$, and explain what accounts for the relationships existing in the neighborhood of $t = 0$.

A3-35. An N^+P diode with a small vertical dimension and an extensive x-direction dimension is subjected to steady-state penetrating radiation on one of its major surfaces, as is shown here; it constitutes a solar cell.

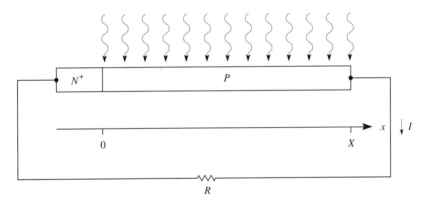

Starting with the constant-E continuity-transport equation, show that for low-level conditions the current I in the external circuit is given by

$$I = I_0[\exp(-qV_{NP}/kT) - 1] - I_L,$$

where

$$I_L = qAL_n g.$$

A3-36. Using charge-control theory, find the small-signal ac admittance of an N^+P diode. Compare your results with Equation 3-234 of Section 3-9.4. [*Hint:* Begin with Equations 3-173 and 3-174.]

A3-37. The ac admittance of an N^+P diode is given by Equation 3-233, repeated here:

$$y = g\sqrt{1 + j\omega\tau}. \tag{1}$$

This expression can be put in the following form:

$$y = g(\omega) + jb(\omega). \tag{2}$$

Using Equation 3-235 from exact analysis, we see that

$$g(\omega) = g\sqrt{(\sqrt{1 + \omega^2\tau^2} + 1)/2}; \tag{3}$$

$$b(\omega) = g\sqrt{(\sqrt{1 - \omega^2\tau^2} + 1)/2}. \tag{4}$$

Plot $g(\omega)$ and $b(\omega)$ as a function of ω.

A3-38. The ac impedance of an N^+P diode is given by $z = 1/y$. Using the information given in Problem A3-37, plot z in the complex plane. Also plot the corresponding result using charge-control theory.

A3-39. Excess-electron concentration as a function of position and time in large-signal switching of an N^+P diode from forward to reverse voltage is given by Equation 3-199 for the storage phase (phase I). Using Equation 3-201, derive an expression for $v_A(t)$ as an explicit function of time. Assume that $(I_R/I_F) = 1$. Also, note that

$$\text{erf}(x) = 1 - \text{erfc}(x)$$

and $\text{erf}(-x) = -\text{erf}(x).$

A3-40. In Section 3-9.4 it was pointed out that Equation 3-199 gives the electron density as a function of position and time. This equation is plotted in Figure 3-76 for six values of t. Prove that Equation 3-199 satisfies the boundary condition given as Equation 3-198. That is, show that the diode current is constant and equal to $-I_R$ during the storage phase. For this problem you will need the identities in Problem A3-39 as well as this one (all of which can be found in standard mathematical references):

$$\frac{d}{dx}\text{erf}(x) = -\frac{d}{dx}\text{erfc}(x) = \frac{2}{\sqrt{\pi}}\exp(-x^2).$$

A3-41. A certain symmetric step junction at equilibrium has a depletion-layer thickness of $X_0 = 1.694\ \mu\text{m}$, and a maximum electric field of $E_{OM} = 6.555$ kV/cm.
a. Calculate its contact potential, $\Delta\psi_0$.
b. Using Poisson's equation, find its net doping N.

A3-42. Given here by the heavy curve is the minority-carrier profile in the lightly doped side of a one-sided step junction under low-level forward bias. The light line is tangent to the curve at $x = 0$.

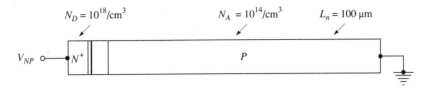

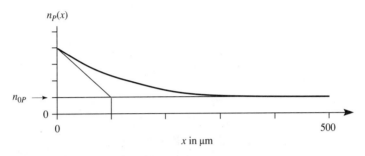

a. Write an equation for the dominant current-density component at $x = 0$.
b. Determine the value of the expression written in part (a), with due regard for algebraic sign.
c. Write an equation for the dominant current-density component at $x = 500$ μm.
d. Determine the approximate value of the expression written in part (c), with due regard for algebraic sign.
e. Calculate the carrier lifetime τ in the P-type region.
f. The applied voltage is increased so that $[n_P(0)/n_{OP}] = 10^8$. Calculate the value of the applied voltage V_{NP} with due regard for algebraic sign.

A3-43. Use the depletion approximation to complete the analysis of this asymmetric step junction. Use $\epsilon = 1$ pF/cm to simplify arithmetic. The symbols N_1 and N_2 represent net-doping values on the respective sides.

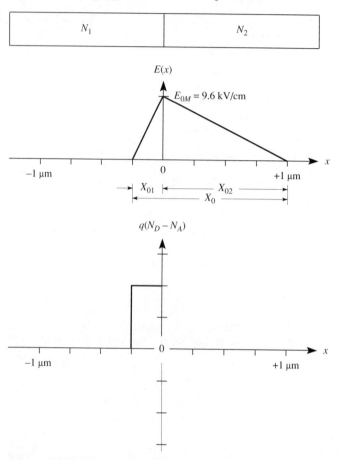

a. Calculate the contact potential $\Delta\psi_0$.
b. Complete the charge-density profile (bottom diagram); sketch the balance of the profile accurately to scale.
c. Using the symbols N and P, label the regions in the top diagram according to conductivity type.
d. Find N_1 and N_2. [*Hint:* Use Poisson's equation.]

A3-44. Following is a semilog presentation of the complete carrier profiles in an *NP* junction at equilibrium. Transfer the data to the linearly calibrated axes also given. Using a solid line for $p(x)$ and a dashed line for $n(x)$, draw the relevant portions of the profiles *as accurately as possible*.

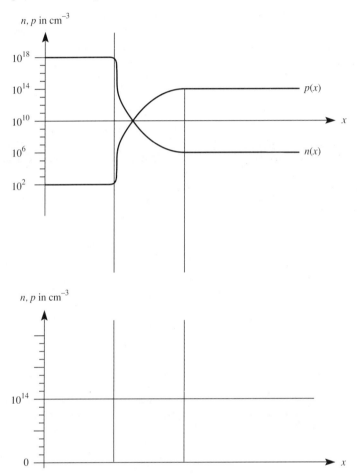

A3-45. In a certain asymmetric step junction subjected to steady-state low-level forward bias, $J_n(0) = J_p(0)$. Three pairs of axes are provided here. On each, sketch *carefully* the current-density profile indicated. The symbol J stands for total current density, whereas J_n and J_p are each a net current density. The net current density for a particular carrier is the algebraic sum of its drift and diffusion components.

PN JUNCTIONS

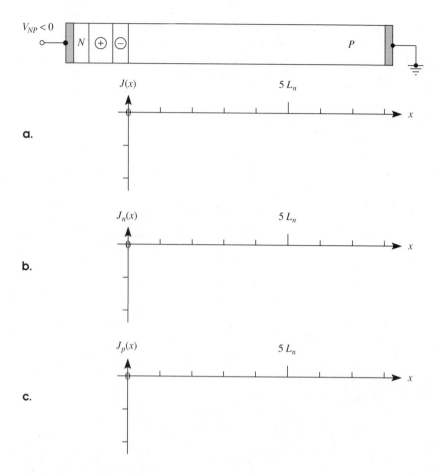

a.

b.

c.

d. In the chart provided, indicate with an X the current-density components that are significant at each of the three spatial positions stated at the left. The term *significant* is defined as *greater than 1 percent of the total*.

	$J_{n,\text{diff}}$	$J_{p,\text{diff}}$	$J_{n,\text{drft}}$	$J_{p,\text{drft}}$
$x = 0$				
$x = L_n$				
$x = 8 L_n$				

A3-46. You are given an N^+P step junction under low-level steady-state forward bias. Net doping in the P-type region is $10^{16}/cm^3$.
  a. Using symbols only (no numbers), derive an expression for the value of applied voltage that places the given device at the borderline between the low-level and high-level regimes of operation.
  b. Evaluate the expression derived in part a.

COMPUTER PROBLEMS

C3-1. Using SPICE, we wish to simulate the dc properties and certain temperature-dependent properties of a real silicon diode. Because we are interested in terminal properties this time rather than in internal relationships among such things as contact potential and applied voltage, it is convenient in this case to let voltage be applied to the P-type side, resulting in positive voltage V and terminal current I for the case of forward bias. Here we shall include the effect of parasitic series resistance R_S within the diode. Consequently the voltage V_J that actually appears at the junction from boundary to boundary (sometimes termed the imposed junction voltage) is given by

$$V_J = V - IR_S. \tag{1}$$

It is of course V_J that determines boundary values and hence is directly related to diode current. The form of the current–voltage equation that is used in SPICE is

$$I = I_0 \exp(q/NkT). \tag{2}$$

The quantity N is a *quality factor* that determines the slope of the (linear) semilog I–V characteristic. Many diodes exhibit two distinct operating regimes, with the diode whose characteristic is shown in Figure 3-24 being an example. Each regime is characterized by a particular value of saturation current I_0 and an accompanying value of N. The physical reasons underlying this behavior are explained in Section 3-4.6. To model this situation, one can connect two diodes in parallel, as was done in effect in Equation 3-46.
  a. Use SPICE to calculate the I–V characteristics of a diode having the parameter values listed in the following table for the two parallel diodes of the model, these being appropriate to the device represented in Figure 3-24:

	Unit	Diode #1	Diode #2
R_S	ohm	0.4	0.4
I_0	A	2.7×10^{-14}	4.4×10^{-12}
N	—	1	2

Plot the I–V characteristics of diodes #1 and #2 individually and in parallel combination; in each case, plot both linear and semilog curves.

b. Temperature effects in a diode are modeled through the thermal voltage that enters the exponent in Equation 2, and by empirically modeling the temperature dependence of saturation current I_0:

$$I_0(T) = I_0(T_1)[T/T_1]^{(\chi/N)} \exp(-\xi_G/kT). \qquad (3)$$

Here, T_1 is a reference temperature, usually room temperature; the quantity ξ_G is an empirical activation energy, in the present case the energy-gap value; χ is the exponent for the power-law temperature dependence, and N is again the quality factor. Equation 3 is an empirical version of the approximate theoretical expression for the temperature dependence of saturation current in a junction diode:

$$I_0(T) = KT^3 \exp(-\xi_G/kT). \qquad (4)$$

Calculate the I–V characteristic for the parallel combination of diodes #1 and #2 at the temperature values -50 C, 23 C, and 100 C, using the additional SPICE parameters given below.

	Unit	Diode #1	Diode #2
ξ_G	eV	1.11	1.11
χ	--	3	3

These particular values are *default values* in SPICE, and need not be explicitly entered. On a single linear diagram, superimpose the curves for the simulated device at all three temperatures, and do the same on a single semilog diagram.

C3-2. The switching response of a junction diode is to be modeled using SPICE. The effective diode capacitance will be modeled by means of the expression

$$C_{EFF}(t) = C_S(t) + C_T(t),$$

where

$$C_S(t) = C_{0T}(t) \{1 + [v_J(t)/\Delta\psi_0)]\}^{-M}.$$

The effective diode current will be modeled using

$$I_{EFF}(t) = I_0 \{\exp[qv_J(t)/NkT] - 1\},$$

where

$$v_A(t) = v_J(t) - i_A(t)R_S.$$

Using SPICE, calculate the switching response of a diode with the following characteristics:

SPICE Notation	Present Notation	Value
IS	I_0	2.7×10^{-14} A
RS	R_S	0.4 ohm
N	N	1
TT	τ	2×10^{-6} s
CJO	C_{0T}	50×10^{-12} F
VJ	$\Delta\psi_0$	0.6 V
M	M	0.5

(See Table 4-5 for a more complete description of these quantities.) The circuit description is shown here (a), along with the applied-voltage waveform (b). Plot the diode current $i_A(t)$ versus time.

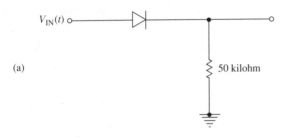

(a)

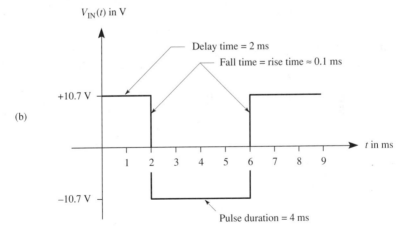

(b)

C3-3. In Section 3-9.2 we gave three expressions for the small-signal switching response of an N^+P diode, and plotted the results in Figure 3-82. Write a PASCAL program to reproduce these results. To evaluate the function erf(x), use a series expansion rather than integration because this produces a much faster program. (See, for example, A. R. Miller, *PASCAL Programs for Scientists and Engineers*, Sybex Inc., Alameda, California, 1981, pp. 326–336.)

C3-4. Here we make use of the results from the last problem. Figure 3-76 gives normalized electron density as a function of normalized position for an N^+P diode, where $(I_R/I_F) = 1$. Write a PASCAL program to generate the curve when $t = t_1$, the end of the storage period. The electron density is given as a function of x [where $X = (x/L_n)$ and $T = (t/\tau)$] by

$$N = \exp(-X) - \frac{1}{2}\left[\exp(-X)\,\mathrm{erfc}\!\left(\frac{X}{2\sqrt{T}} - \sqrt{T}\right) + \exp(X)\,\mathrm{erfc}\!\left(\frac{X}{2\sqrt{T}} - \sqrt{T}\right)\right].$$

DESIGN PROBLEMS

D3-1. Using the depletion approximation, design an asymmetric step junction having $N_2 = N_1/4$, and having $X_0 = 1.0$ μm.

D3-2. In a symmetric step junction, there is a maximum in the equilibrium depletion-layer thickness X_0 as a function doping of N. Why? Design a symmetric step junction with the largest possible value of X_0, and calculate that value.

D3-3. Design an N^+P junction for which $J = 0.1$ A/cm^2 at a forward bias of 0.5 V. The materials specialist on your design team guarantees you $\tau = 10$ μs in the lightly doped region, which is to be relatively thick.

D3-4. Design an asymmetric step junction with a net doping on one side of 10^{16}/cm^3, and an avalanche-breakdown voltage of 100 V.

D3-5. Using the results of Problem D3-4, design a voltage-controlled capacitor that is to be operated from $V_{NP} = 0$ to the maximum allowed reverse voltage. The diode in question exhibits the value $C_T = 2$ pF at $V_{NP} = 0$.
a. Determine the minimum value of capacitance in the stated range.
b. Determine the dynamic range of this device, or the ratio of maximum to minimum capacitance. The intended application of the capacitor requires a dynamic range of ten. Is this specification met?
c. Can the ratio of $C_T(0)/C_T(V_{NP})$ for a step junction be altered by means of a doping change, where V_{NP} is taken to be a fixed voltage? Assume that $V_{NP} \gg \Delta\psi_0$ in every case. Explain, using the symmetric step junction as your example.

D3-6. Following is a voltage-controlled attenuator circuit for low-frequency use.

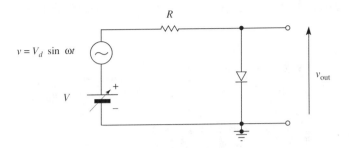

The diode that is to be employed exhibits a near-ideal I–V characteristic in the range $0.4 \text{ V} < V_D < 0.6 \text{ V}$ that is given by

$$I_D = I_0 \left[\exp(qV_D/kT) - 1\right],$$

where $I_0 = (10^{-12} \text{ A})$. The controlling voltage may not exceed $V = 5$ V. The circuit must have a dynamic range (ratio of maximum to minimum attenuation) of at least a factor of 80. The circuit must have its greatest precision at an attenuation in the neighborhood of a factor of 25. Design the circuit to meet these specifications.

4 The Bipolar Junction Transistor

The bipolar junction transistor (BJT) was invented in January 1948 [1], a few weeks after the invention of the first transistor (the point-contact transistor), and months before the point-contact device was announced [2]. The reason for the invention of the BJT is a prime object lesson in engineering practice. The point-contact transistor was fabricated by primitive and "artistic" methods, and its detailed structure and operation remain unclear to this very day. Even worse, it posed a complex three-dimensional problem. The inventor of the BJT, William Shockley, was at the time the supervisor of the two inventors (or "discoverers") of the point-contact transistor, John Bardeen and Walter Brattain. He was endeavoring, with sound engineering instinct, to conceive of a one-dimensional analog of the point-contact device as a way of opening the door to analytical investigation of its operation [3]. Suddenly he realized that the structure he was postulating was itself a transistor, and one that lent itself to analysis at a basic level! It became the bipolar junction transistor. Reduced to practice in 1951 [4], it had for all practical purposes shouldered the point-contact device aside by the mid-1950s. A genuine understanding of the point-contact transistor will probably never be achieved, because it is inferior to competing devices in so many respects that it no longer arouses even academic interest. Nonetheless, its place in history is assured as the first practical amplifying device in the era of solid-state electronics.

In its early decades, the bipolar junction transistor was known as the *junction transistor* or simply "the transistor." But it in turn suffered the fate of being pushed from its position of dominance by a field-effect transistor—specifically the MOSFET treated in Chapter 5. The decline of the bipolar junction transistor occurred only in relative terms, however. In absolute terms its use continues to grow.

Early in the 1950s, Shockley proposed the adjective *bipolar* to describe his newly conceived device because both carrier types play important roles in its operation; by contrast, he described devices in the field-effect family (which he also launched insofar as practical embodiments were concerned) as *unipolar*, because their operation depends primarily on a single carrier type. In more recent times, the need to distinguish among the proliferating varieties of transistors has caused the term *bipolar junction transistor* to come into common use, and the resulting acronym *BJT* to come into even more common use. This kind of distinction is important. Shockley's first field-effect device was a "unipolar junction transistor," but it is usually designated the *junction field-effect transistor* [5]. The term "unipolar" is not widely used.

4-1 BJT RUDIMENTS

The unfolding story of solid-state electronics can be told rather completely in terms of evolving fabrication technology, constantly expanding the number of options available to the device and integrated-circuit designer. It was for technological reasons that an early and important kind of BJT was a germanium *PNP* device [6]. The term *PNP* labels the conductivity types of the three regions within a BJT, regions separated by two *PN* junctions. In later years, and again partly for technological reasons, the dominant BJT was a silicon *NPN* device. In integrated circuits today, the combination of silicon *NPN* and *PNP* devices is a growing practice because the resulting *complementary* circuits have important power-dissipation and performance advantages. For convenience and consistency, however, and because of its continuing importance, the silicon *NPN* BJT will be the vehicle for this chapter.

4-1.1 Structure and Terminology

The essential structure of a BJT is represented in Figure 4-1(a). The very earliest such devices had structures literally of this kind. Two closely spaced junctions were created by crystal-growth methods [7], and a "bar" or parallelepiped was then cut out of the germanium crystal. Electrical leads were attached to it (an enormous challenge!) and the result was a BJT. For reasons that will be explained shortly, these electrical terminals are given the names, respectively from left to right, *emitter*, *base*, and *collector*. These names were chosen with an eye to distinctive initial letters, which are displayed in Figure 4-1(a) in association with the three terminals. The shaded regions in Figure 4-1(a) represent the space-charge regions (or depletion regions) of a pair of *PN* junctions. The *boundaries* of these regions are emphasized because conditions there assume far greater importance than conditions at the metallurgical junctions (which are not even represented in the drawing).

The term "base" may well be a puzzling choice in the context of Figure 4-1(a). Actually, the term is a vestigial remnant (and practically the only remnant!) of the point-contact transistor. In that device a small germanium crystal was mounted on a pedestal, and two point-contact wires (designated as emitter and collector) were positioned on top of the crystal, so that the crystal was literally the "base" of the

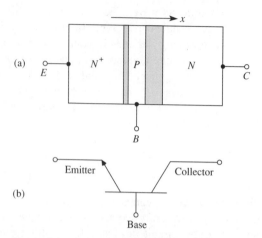

Figure 4-1 A representation of the *NPN* bipolar junction transistor (BJT) by means of (a) a physical-structure diagram and (b) a conventional symbol. The three terminal designations also apply to the corresponding regions of the device.

structure [8]. The central region in Figure 4-1(a) plays an electrical role analogous to that of the *base region* in a point-contact device.

Having given names and single-letter labels to the three terminals of the BJT, we have effectively provided names and labels to the three regions of the BJT as well, because each region is associated uniquely with one terminal. The matter of junction identification is almost as simple. The two junctions are, respectively, the *emitter–base junction* and the *collector–base junction*. However, without any ambiguity we can (and usually do) refer to them simply as the *emitter junction* and the *collector junction*.

The corresponding BJT symbol, Figure 4-1(b), was adopted very early. Once again an arrowhead is used to point from a *P*-type region to an *N*-type region, the direction of the "easy" conduction of conventional current through a *PN* junction. But only one junction—the emitter junction—is so labeled. (This is done for identification.) Once the emitter junction has been so identified and labeled, the polarity of the collector junction becomes evident.

In 1960 the BJT took on a structure that it has retained to the present day in a wide range of applications [9], the structure shown in Figure 4-2. The orientation of this diagram has been chosen to correspond to that of the BJT in Figure 4-1(a). It is evident that this structure bears little resemblance to the prototypical "bar" structure. Nonetheless, in a large number of devices having the newer structure, the currents through the junctions retain an essentially one-dimensional character, with important current fractions and current components being normal to the junctions. Consequently, a good approximate treatment of such devices can be achieved by calculating currents in the region emphasized by dotted lines in Figure 4-2 and then multiplying by an appropriate junction area to determine total current through the

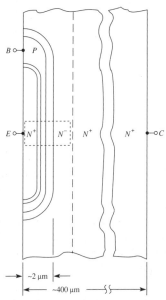

Figure 4-2 A BJT structure that has been widely used since 1960, employing the epitaxial growth of silicon and the solid-phase diffusion of impurities.

junction. Thus Shockley's original one-dimensional intent is still being honored.

Figure 4-2 specifically represents the structure of a *discrete* or separately packaged BJT. It is evident from the dimensions shown that the active portion of the device is confined to a small portion of the overall crystal volume. The bulk of the silicon crystal, in fact, is simply a mechanical support for the active portion. Any electrical resistance to the passage of current through this thick region is a *parasitic* feature of the device that one would prefer to avoid. Most BJTs today are incorporated into an *integrated circuit* (the "microchip" of the layman), defined as a useful combination of devices fabricated *in* and in some cases *on* a semiconductor single crystal. As a result, the parasitic feature of having current pass through a thick mechanical support is partly eliminated. All of the currents (in virtually all integrated circuits) are confined to a very thin region near one face of a semiconductor crystal.

The portions of the BJT structure shown in Figure 4-2 that are near the top and the bottom of the diagram also are to some degree "parasitic." (For a more usual orientation of this diagram with the silicon surface horizontal, these regions would be described as *lateral* portions of the structure.) It is a curious fact that advancing technology is now generating structures that eliminate the parasitic lateral regions, thus increasingly resembling the original bar structure! These BJTs are not freestanding parallelepipeds, but rather are defined by parallel grooves cut into a silicon crystal, and then subsequently filled by an insulating material such as silicon dioxide (SiO_2). The individual BJTs are interconnected to perform useful circuit functions by means of a variety of methods; the result is an integrated circuit that may incorporate over 100,000 devices all fabricated simultaneously in the same silicon crystal.

4-1.2 Biases and Terminal Currents

An *input port* or an *output port* in the circuit sense consists of a pair of terminals. Since the BJT is a three-terminal device, one of the three terminals is permitted to be common to the input and output ports, and the other two terminals then are each uniquely associated with each port. For reasons that we shall examine, when the base and emitter terminals are chosen for the input port with the collector and emitter terminals taken for the output port, the BJT can exhibit both current gain and voltage gain. This useful combination of properties has made the *common-emitter configuration* the most widely used of the several possibilities, a term acknowledging that the emitter terminal is common to the input and output ports. The adjective "grounded-emitter" is also sometimes applied to this connection.

Figure 4-3(a) shows the physical representation of the BJT in its common-emitter orientation, and also shows typical voltage polarities and magnitudes for the case of a silicon device. A positive voltage is applied at the input port (with reference to the common-emitter ground potential shown here), and it is evident that the emitter junction is forward-biased as a result. Emitter current is normally of such a magnitude that the near-vertical portion of the current–voltage characteristic shown

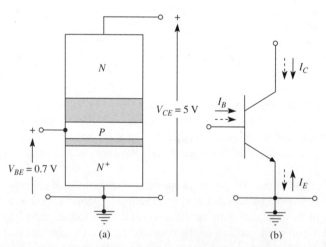

Figure 4-3 The common-emitter configuration for the *NPN* BJT. (a) Physical representation of the device, showing typical bias voltages at the input port and the output port and showing that the emitter terminal is common to both ports. (b) Using the BJT symbol, we illustrate the directions of positive terminal current (solid arrows) and positive conventional current (dashed arrows) for the given bias values.

in Figure 3-17(b) is brought into play, so that the junction voltage is near 0.7 V. Or, continuing to use the double-subscript notation introduced in Chapter 3, we have $V_{BE} \approx 0.7$ V. Collector–emitter bias, or V_{CE}, has had a typical value of 5 V in wide ranges of circuits, but for some purposes may be set as low as one volt, or as high as many hundreds of volts.

The BJT symbol in Figure 4-3(b) has associated with each terminal a pair of arrows representing current directions. The solid arrows show the directions of conventionally positive *terminal* current—always inward. The dashed arrows show the actual directions of positive conventional current at each terminal for the bias conditions shown in Figure 4-3(a). Note that in the cases of base and collector currents, the two most important to the common-emitter problem, the two current conventions are in agreement with respect to direction, a matter of significant convenience.

Exercise 4-1. Make a comparison of terminal-current versus conventional-current directions for the case of a *PNP* BJT analogous to that in Figure 4-3(b) for the case of an *NPN* device.

Conversion from an *NPN* case to the *PNP* case requires reversal of bias polarities and hence the directions of conventional current (dashed arrows). However, the convention concerning positive *terminal* current is unaltered (solid arrows). Hence in such a case, both base and collector actual-current directions make them negative terminal currents, which is to say, outward flowing.

4-1.3 Carrier Profiles

Operation of the BJT can be approached by examining the properties of its two junctions individually, considering them to be *isolated*. As we saw in Chapter 3, this is a term that describes a junction whose end regions are extensive in the x direction (since we continue to consider one-dimensional structures only). As a result, the carrier-density disturbances that accompany junction biasing have space enough to "fade away" before they reach the inevitable contacts. Thus the inherent properties of the junction under study will be seen. And since we are going to assume step junctions only, these inherent junction properties subsume those of both of its uniformly doped regions—properties such as absolute doping and carrier lifetime. Then we shall combine the junctions in order to examine their interaction. In a BJT, this *interaction* is crucial; there is no way to simulate BJT behavior using isolated junctions.

In Figure 4-4(a) is shown the BJT structure of interest. The dimensions of the various regions are shown with qualitative realism. The emitter-junction space-charge layer is typically much thinner than that of the collector junction, for two

490 THE BIPOLAR JUNCTION TRANSISTOR

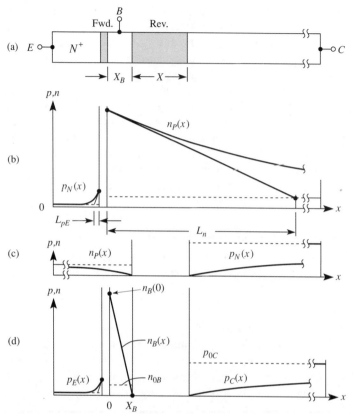

Figure 4-4 Minority-carrier profiles in a normally biased BJT. (a) Physical representation of the BJT structure. (b) Minority-carrier profiles near a forward-biased emitter-like junction that is *isolated,* meaning one that is a long distance from other junctions and contacts, using heuristic ordinate calibration. (c) Minority-carrier profiles near an isolated reverse-biased collector-like junction, again using heuristic calibration. (d) Minority-carrier profiles in the BJT structure, with the same junction biases and the same calibration as in parts (b) and (c).

reasons: First, the mean doping of the emitter and base regions (a geometric mean is appropriate here because their net-doping values have a large ratio) is typically almost two orders of magnitude larger than the mean doping of the base and collector regions; and second, the emitter junction is usually forward-biased, while the collector junction is usually reverse-biased. The thickness of the latter's space-charge region we shall designate as X. The thickness of the base region X_B is typically a few tenths of a micrometer in a modern transistor, and is a critical dimension.

Pretending that the collector junction is out of the picture, we show the minority-carrier profiles associated with the forward-biased emitter junction in Figure 4-4(b). It is the behavior of *minority* carriers that is of dominant concern. Notice also that only carrier densities *outside* the junction boundaries are considered. To elucidate certain fine points of BJT operation, it is necessary to consider carrier distributions within the junction boundaries. But explaining BJT properties in basic terms requires only knowledge of these distributions outside the boundaries, and especially *at* the boundaries. The ordinate calibration in Figure 4-4(b) is "heuristic," meaning that quasirealism is achieved by showing the profile shapes as they would appear in a linear presentation, and yet the three-order-of-magnitude net-doping ratio on the two sides of the emitter junction is temporarily submerged. (A more precise treatment would use differing ordinate calibrations on the two sides, but that complication can be omitted as long as we keep in mind that the presentation is qualitative.)

The first important bit of information in Figure 4-4(b) is that the diffusion length for electrons L_n is very large on the P-type (right-hand) side, thus requiring a very extensive region to meet the requirement set forth above. In the N^+ emitter region on the left, however, the diffusion length L_p is small for reasons discussed in Chapter 3. Therefore the N^+ region can be quite thin and still qualify as "extensive." A second important observation bears on the equilibrium minority-carrier values on the two sides of the emitter junction; the value of p_{0N} is small compared to that of n_{0P}—typically some 10^3 times smaller. Consequently, the law of the junction requires that the boundary value of p_N be smaller than the value of n_P at the other boundary *by the same factor*. This boundary-value ratio favors electron injection rightward over hole injection leftward. The actual asymmetry of the injection currents is smaller than boundary densities imply because the diffusion-length ratio is a countervailing factor.

Now examine the "collector" junction, also considered to be isolated. In Figure 4-4(c) we once again use heuristic calibration to present minority-carrier profiles near a junction, but this time a reverse-biased junction. Although the law of the junction has only qualitative validity for reverse bias, we are assured that the minority-carrier densities at both boundaries can be taken to be zero for a bias of the order of 0.1 V or more, and an appreciably larger reverse bias has been assumed here. This time the minority-carrier diffusion lengths on the two sides are large, with that on the right-hand side being the larger.

Finally, then, we let the two junctions interact by assembling them in the BJT of Figure 4-4(a). The resulting minority-carrier profiles are shown in Figure 4-4(d). The first principle to be invoked in arriving at this minority-carrier-profile picture is that in spite of the new ingredient of junction interaction, *the law of the junction still applies*. Furthermore, on the left-hand side of the emitter junction, and on the right-hand side of the collector junction, the minority-carrier profiles are essentially unchanged; the junction interaction has not affected these regions. And beyond this, applying the law of the junction at the junction boundaries that define the base region leads to the conclusion that only one possible minority-carrier profile can exist in the base region for the given conditions: A nearly straight line connects the two boundary values dictated by the law of the junction for the junctions that lie on the two sides of the base region. The reason that the line is nearly straight is that typically $X_B \ll L_n$, where L_n is the diffusion length depicted in Figure 4-4(b).

Rather than the factor-less-than-ten difference depicted in Figure 4-4, the ratio is typically of the order of one hundred. Since the curvature of the electron-density profile in Figure 4-4(b) exists because of the recombination phenomenon, it follows that very little recombination will occur as electrons diffuse through the short distance X_B, and therefore the profile can exhibit little curvature.

Notice in Figure 4-4(d) that the spatial origin has been placed at the base-region boundary of the emitter junction. This has been done because of the extreme importance of base-region conditions in BJT operation. The value of $n_P(0)$ is the same as the corresponding boundary value in Figure 4-4(b), because the same forward bias has been maintained. But notice also—and this is a point of major importance—that electron injection under the conditions of Figure 4-4(d) is *much larger* than the electron injection for the same junction bias in Figure 4-4(b).

Exercise 4-2. Determine the ratio of the electron injection in Figure 4-4(d) to that in Figure 4-4(b).

Electron transport at the junction boundary is purely diffusive, and the current ratio is therefore equal to the gradient ratio. Because the electron density is the same in both cases, it follows that the two currents stand in the ratio L_n/X_B.

Since the current density $J_n(0)$ approximates the total current density J through the emitter junction because of the junction's gross asymmetry, it follows that the *structural change* from Figure 4-4(b) to Figure 4-4(d) is responsible for a current increase through the forward-biased junction by a factor of about L_n/X_B, provided the nearby collector junction is reverse-biased.

4-1.4 Typical Dimensional and Doping Values

The range of BJT current, voltage, and power specifications is enormous. Accompanying these variations are variations in dimensions, especially in the emitter-junction area A_E required to carry the desired current. This area can be as small as a few square micrometers at one extreme, and can be in the neighborhood of square centimeters at the other extreme—a factor of 10^8! Thus there is appreciable arbitrariness in citing "typical" values. Let us choose an intermediate example, logarithmically speaking, an emitter-junction area A_E of 10^4 μm^2, or of 10^{-4} cm^2. For the sake of defining scale, this corresponds to an emitter junction measuring 100 μm by 100 μm. However, emitter junctions of this area are unlikely to be square; instead, they are more likely to be elongated rectangles, or collections of such rectangles, for reasons explained later. The junction-normal dimensions (or x-direction dimensions for our axis assignment) also vary with current, voltage, and power ratings, but not as drastically as areal dimensions.

Next, let us use the subscripts E, B, and C to replace the subscripts N and P used for single-junction analysis. The object, of course, is to eliminate the ambiguity that results otherwise, because the BJT has two regions of the same type. We shall cite *net-doping* values in the various regions, using, for example, the symbol N_{AB} to designate the net acceptor doping in the base region. With these symbol conventions adopted, we can then assign reasonable doping values as has been done in Table 4-1, given at the end of the chapter. There we also name specific dimensions in a BJT structure in order to lend concreteness to the following discussion.

The base and collector regions can be described as moderately doped, and therefore all of the usual approximations and constant values may be applied. Consulting Figure 2-22 yields the carrier-mobility values listed in Table 4-1. The emitter region, however, poses a different problem. Doping there is so heavy that three of our conventional approximations fail. These are the Boltzmann, the equivalent-density-of-states, and the 100-percent-ionization approximations. At a donor doping of $10^{20}/cm^3$, silicon is truly degenerate and the Fermi level resides well above the conduction-band edge. This is the situation treated in Computer Problem C2-1. It will be recalled that an electron density calculated from the usual collection of approximations departs significantly but not catastrophically from a density based upon the accurate Fermi-Dirac integral, even though the individual approximations are extremely inaccurate under degenerate conditions. This is a consequence of a fortuitous cancellation of errors.

Exercise 4-3. Describe the nature of the error cancellation that occurs when the conventional approximations are used in combination to analyze a degenerate sample.

When the Fermi level, to take a specific example, resides above the conduction-band edge by 59 meV, the Boltzmann approximation yields an occupancy probability at the conduction-band edge of 1,000%, which obviously is without physical meaning. Under the same conditions, the equivalent-density-of-states approximation is faulty because occupied states in the conduction band exist well above the band edge; assuming all states to be concentrated at the band edge is unrealistic. In other words, the value of N_C is too small. Hence it is the physically unrealistic occupancy probability (greater than 100%) that partly compensates for the too-small equivalent density of states.

For further complication, the energy gap shrinks in heavily doped silicon and as a result the intrinsic-carrier density is inflated. As shown in the first column in Table 4-1, the intrinsic-carrier density (n_{iE}) in an emitter region with an effective doping of $10^{20}/cm^3$ has increased by more than an order of magnitude, to approximately $1.2 \times 10^{11}/cm^3$, as a result of bandgap shrinkage to a value a bit less than one electron volt.

Exercise 4-4. Calculate a bandgap value corresponding to $n_{iE} = 1.2 \times 10^{11}/\text{cm}^3$.

The energy interval involved is of course half the gap, so that
$$n_{iE} = N_C \exp(\xi_{GE}/2kT).$$
Thus, $\xi_{GE} = 2kT \ln(n_{iE}/N_C) = 0.993$ eV.

The theory of regions as heavily doped as this is still evolving, and is somewhat controversial as well. Besides nonnegligible energy-gap changes, there is the fact that new carrier-recombination mechanisms enter, such as a three-body process known as *Auger recombination*. We do not attempt to penetrate these topics in any important way; instead, we give a set of empirically consistent values in Table 4-1 (column 1) that yield reasonable results when used in the equations valid for moderately doped silicon.

It might be thought, for example, that since μ_{pE} is given in Table 4-1 as 46 cm^2/V·s, it is known to two significant figures, but this is not the case. The most recent direct measurements of μ_p in degenerate N-type samples indicate a value somewhere in the range from 10 to 100 cm^2/V·s for a sample with a doping of 10^{20}/cm^3[10], but other, less direct, observations provide more accurate data. For example, the same workers reported significant data on bandgap, carrier lifetime, and carrier diffusion length as a function of doping in such samples. In addition, data on BJT current gain as a function of base-region thickness and doping are quite well established [11, 12]. Combining these less direct data permits us to choose the value 46 cm^2/V·s as one that provides reasonable internal consistency.

It must be conceded that this approach to treating the emitter region is at best empirical. But it (1) provides a framework for calculations, (2) gives results in agreement with experiment, and (3) displays in reasonably accurate fashion features of degenerate silicon that set it apart from moderately doped silicon—features such as smaller bandgap, larger intrinsic density, and smaller diffusion length.

Exercise 4-5. For the BJT of Table 4-1, and using the depletion approximation, calculate the equilibrium depletion-layer thicknesses of the collector junction and the emitter junction.

For an asymmetric step junction,
$$X_0 = \sqrt{\left[\frac{2\epsilon \Delta \psi_0}{q}\right]\left[\frac{1}{N_1} + \frac{1}{N_2}\right]}$$
$$= \sqrt{\left[\frac{1.292 \times 10^7}{\text{cm}\cdot\text{V}}\right]\Delta\psi_0\left[\frac{1}{N_1} + \frac{1}{N_2}\right]}.$$

Taking values from Table 4-1 for the collector junction,

$$\Delta\psi_0 = (kT/q) \ln(N_1N_2/n_i^2) = 0.750 \text{ V}$$

and $X_0 = 0.451$ μm. For the emitter junction, $\Delta\psi_0 = 1.00$ V, so that $X_0 = 0.114$ μm.

4-1.5 One-Dimensional Electron Current

The crucial minority-electron profile in the base region is shown in Figure 4-4(d). Let us now examine the electron profile throughout the entire BJT. We want the undistorted clarity of a linear plot, so let us once again employ the device of creating a gap in the ordinate scale to deal with the orders-of-magnitude changes in the transition regions of the two junctions. The result is shown in Figure 4-5(a). A light line shows the equilibrium electron-density profile, affecting mainly the base region.

The object now is to examine the dominant current components in all regions of the normally biased BJT. These regions (for present purposes) are five in number—two transition regions (of the two junctions), plus the emitter, base, and collector regions. In the N^+ emitter region, it is clearly the drift of majority electrons that will dominate.

Exercise 4-6. Offer an argument supporting the statement immediately above.

The diffusion length for holes in the emitter region provides a measure of how far injected holes proceed into the emitter region. Consulting Table 4-1, we find that

$$L_{pE} = \sqrt{D_{pE}\tau_E} = \sqrt{(kT/q)\mu_{pE}\tau_E}$$
$$= \sqrt{(0.02566 \text{ V})(46 \text{ cm}^2/\text{V·s})(10^{-9} \text{ s})} = 0.34 \text{ μm}.$$

Thus the density of injected holes and the diffusion-causing gradient will both decline by a factor of at least e^3, or about 20, through the thickness of the emitter region. Even more important is the fact that the holes diffusing into the emitter region already constitute a negligible current component because of junction asymmetry. Thus hole transport in the emitter region is a negligible factor.

A dominance of diffusion over drift exists in the emitter-junction transition region because the junction is forward-biased. And for the reason just given (asymmetry) electron diffusion dominates grossly.

In the base region, the dominant current is electron diffusion. And because of the near-constant gradient of electron density there, this current component is also constant.

In the space-charge region of the collector junction, electrons that are being

496 THE BIPOLAR JUNCTION TRANSISTOR

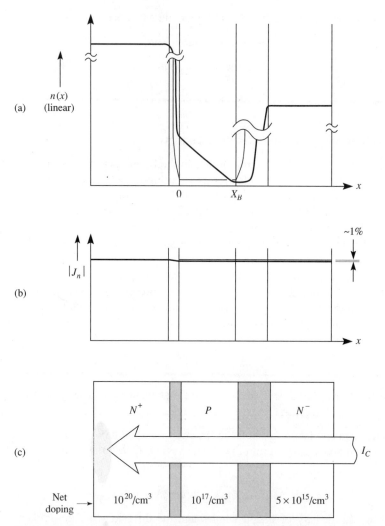

Figure 4-5 One-dimensional electron current in the BJT under bias. (a) Linear presentation of the electron-density profile, with huge and differing gaps in the ordinate scales to show electrons as both majority and minority carriers. (b) Electron current-density profile through the BJT, showing that the current density is very nearly constant in spite of the enormous fluctuation in electron density. (c) Arrow representing the conventional current that is constituted by the one-dimensional electron transport throughout the entire BJT.

delivered by diffusion through the base region are "picked up" by the large electric field and are further delivered to the collector region. There they constitute majority carriers and there it is once again the drift of majority electrons that dominates.

For reasons that will be developed more fully later, it turns out that the one-dimensional electron-current density in all five regions is *nearly constant*, and is as well the dominant current in the BJT, as is illustrated in Figure 4-5(b). Typically, the decline in J_n from emitter contact to collector contact is only about one percent! The electrons flowing out the collector terminal constitute the conventional current flowing inward, or the collector terminal current I_C. Further, this electron current *continues* to the emitter contact, as shown in Figure 4-5(c).

Exercise 4-7. For the bias conditions of Figure 4-3(a), calculate the collector-junction voltage, V_{CB}.

From the double-subscript definition we have
$$V_{CE} - V_{BE} = (V_C - V_E) - (V_B - V_E) = V_{CB}.$$
Therefore, $V_{CB} = (5 \text{ V}) - (0.7 \text{ V}) = 4.3 \text{ V}$.

Exercise 4-8. Calculate the depletion-layer thicknesses of the collector and emitter junctions of the BJT of Table 4-1 under the biases just determined.

The equations of Exercise 4-5 can be reused, simply by substituting $\Delta\psi$ for $\Delta\psi_0$. For the collector junction,
$$\Delta\psi = \Delta\psi_0 + V_{CB} = (0.75 \text{ V}) + (4.30 \text{ V}) = 5.05 \text{ V}.$$
Therefore, $X = 1.17 \ \mu\text{m}$. For the emitter junction, $\Delta\psi = \Delta\psi_0 + V_{EB} = \Delta\psi_0 - V_{BE} = (1.0 \text{ V}) - (0.7 \text{ V}) = 0.3 \text{ V}$.
Thus, $X = 0.062 \ \mu\text{m}$.

Exercise 4-9. Calculate and compare the values of peak field magnitude E_M for the two biased junctions. Comment on the result.

Because the field profiles are triangular we have
$$E_M = (2\Delta\psi/X).$$
Hence for the collector junction,
$$E_M = [2(5.05 \text{ V})/(1.17 \times 10^{-4} \text{ cm})] = 86 \text{ kV/cm}.$$
For the emitter junction,
$$E_M = [2(0.3 \text{ V})/(6.2 \times 10^{-6} \text{ cm})] = 97 \text{ kV/cm}.$$
In spite of substantial forward bias on the emitter junction, its peak field is larger

than that of the reverse-biased collector junction because the former is so heavily doped on its N-type side, and the latter so lightly doped on its N-type side. They share the same P-type doping.

Exercise 4-10. List the five regions of this biased BJT in descending order of (maximum) electric field magnitude in each, and comment on the reasons for the sequence you name. Assume low-level conditions, even though $V_{BE} = 0.7$ V approaches the boundary of high-level operation.

1. **Emitter junction.** Calculation showed E_M to be largest here in spite of forward bias, because of heavy doping on both sides of the junction.

2. **Collector junction.** This peak field is a close second.

3. **Collector region.** This region is lightly doped and requires a higher field to maintain current continuity than a more heavily doped region, but a field that is tiny compared to the peak-field values in the junction transition regions.

4. **Emitter region.** Drift-current density is approximately the same but resistivity is much lower.

5. **Base region.** Low-level transport here is by electron *diffusion*.

As a final point, note that the electron-density profile in Figure 4-5(a) exhibits a flat-bottomed minimum within the collector space-charge region. This occurs in the interval where electric field exceeds $E_s = 50$ kV/cm, the value at which velocity saturation sets in. Here (and only here) the electrons are moving at $v_s = 10^7$ cm/s. Recent work has shown that the linear portion of the base-region profile extrapolates very nearly to the collector-junction boundary defined by the depletion approximation [13]. Hence a base-region electron-density profile constructed using the simplistic condition $n_B(X_B) = 0$ yields an excellent approximation to base-region electron gradient.

4-2 ELEMENTARY DEVICE THEORY

Since the late 1950s, various techniques involving the high-temperature solid-phase diffusion of impurities have been extensively employed in BJT fabrication [14]. In particular it has been emitter regions and base regions that have been formed by these methods, yielding BJT cross sections well represented by Figure 4-2. The impurity profiles that result from impurity-diffusion procedures are not the constant-density profiles incorporated in an ideal-step-junction description. Instead they approximate the Gaussian function and the error-function complement as described in Section 3-5.3. As a result, the net-doping-density profiles for both the emitter and base regions exhibit maxima that are skewed toward the surface of the device.

Fortunately for the student and engineer, it turns out that these rather major departures from the ideal structures defined in Figure 4-4 and Table 4-1 are less consequential than one might normally assume. For a combination of reasons, calculations based upon idealized structures can provide meaningful descriptions of device properties and worthwhile guides for design modifications. Therefore, in what follows, we shall continue to adhere to the assumptions of step junctions and uniformly doped regions. It is a curious fact, once more, that modern technology is moving ever closer to the step-junction ideal employed in the simplest analyses, with uniform doping in the emitter and collector regions, and sometimes in the base region as well.

4-2.1 Internal Current Patterns

A detailed examination of current components in the biased BJT reveals a relatively complicated picture. Fortunately, there exist approximations that are simple and yet quite accurate through wide ranges of conditions. This is another way of saying that some of the most complicated features of the device are often small enough to be neglected safely.

Figure 4-6 is a physical representation of a BJT in the "common-emitter orientation" shown first in Figure 4-3(a). The large primary arrow directed from collector to emitter is identically that shown in Figure 4-5(c), representing one-dimensional electron transport [15]. The labels I_C and I_E show that it constitutes both the collector-terminal current and the emitter-terminal current, which typically differ by only about one percent. Shown also are other current "strands" that depart from the primary current or join it. The relative magnitudes of the secondary currents are represented qualitatively by their widths, but for clarity all of these widths are grossly exaggerated with respect to that of the primary arrow. The circled values give typical currents one might encounter in a small BJT (a "milliwatt device") under bias, and it is evident that they differ by many orders of magnitude.

The most important of the secondary currents is the one labeled I_B, the base-terminal current. It differs in numerous and major ways from the primary electron current that has occupied us up to this point: First, it is a majority-carrier current, or a hole current in the *NPN* device. Second, it flows *laterally* from the base contact into the active region of the BJT. Refer back to Figure 4-2, which shows the commonplace (diffused) BJT in a relatively realistic cross section. The portion of the base region contiguous with (or "under") the emitter region is the *active base region,* sometimes also termed the *intrinsic base region*. (This use of the term does not have doping connotations.) The outer portions of the base region, in analogous fashion, are sometimes designated by the adjective *extrinsic*. It is there that the base contact is made in this kind of BJT. (Base contacts, by the way, can be and often are made on both sides of the emitter.) The third major distinction of the base current from the primary electron current thus becomes evident. The current of holes traverses a relatively long path within the base region. In the example we are considering it could be several tens of micrometers, as compared to the less-than-one-micrometer base thickness traversed by the electrons.

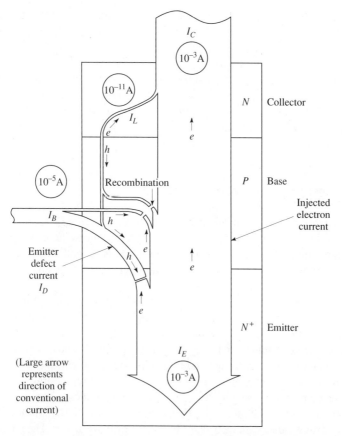

Figure 4-6 Representation of electron and hole (e and h) current components in the BJT under bias, with gaps indicating the recombination phenomenon. Typical current magnitudes in a milliwatt device are indicated by circled values and are indicated qualitatively by arrow widths. (After Warner and Grung [15].)

Exercise 4-11. The active (or intrinsic) base region is defined above as the portion lying under the emitter region. What boundaries define the active base region in the x direction?

The defining surfaces are the base boundary of the emitter-junction depletion layer and the base boundary of the collector-junction depletion layer.

Once in the intrinsic base region, most of the holes undergo a change in direction and flow into the transition region of the emitter junction. This obviously involves a complex current pattern. It is not a "one-piece" current like that shown for simplicity and convenience in Figure 4-6, but rather is somewhat like the pattern shown heuristically in Figure 4-7. In spite of the relatively small size of the base current, it is of extreme importance because, as we shall see, it is the *control current* that is supplied to the input port in the common-emitter BJT. The perturbing effects of the complicated base-current pattern on, for example, the electrostatic-potential pattern in the base region are not serious so long as the base current is small. In the typical device we are considering, base current amounts roughly to one percent of the electron current, which does qualify as "small."

The resistance encountered by the base current on its lateral path from the (ideal, let us say) base contact to the active base region, and in the active base region as well, plays a key part in fixing device performance. This *ohmic base resistance* is a parasitic element to be minimized. Its value is typically in the range of 50 to 100 ohms. It is in order to diminish base resistance that the emitter is often given the (plan-view or top-view) shape of an elongated rectangle or stripe, with the base contact, also stripelike, in close proximity to the emitter stripe. Such a structure is shown in Figure 4-8. A cross section of this device taken at A–A will approximate the structure shown in Figure 4-2. As noted earlier, the BJT designer sometimes chooses to place a base contact on both sides of the emitter stripe; doing so makes a significant reduction in ohmic base resistance.

It is the dimensions of a base or emitter stripe that are appropriately designated length and *width*. For dimensions in the x direction, into the silicon, the term

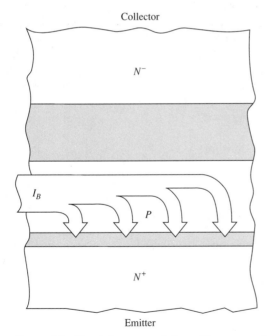

Figure 4-7 Heuristic representation of base-current pattern in the active base region.

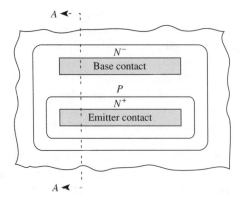

Figure 4-8 Typical plan-view geometry for a milliwatt BJT. The cross section at A–A resembles the structure depicted in Figure 4-2.

thickness is far more descriptive and appropriate, because the latter dimensions are typically much smaller. More important, clarity is served by consistency in these matters. Curiously, even engineers, who are supposed to have above-average skill in spatial visualization, sometimes seem to have difficulty in understanding and verbalizing the three-dimensional aspects of the BJT. One too frequently hears and reads tangled syntax such as, "this transistor has a very thin base width," instead of the simple declaration that the transistor has a *thin base region*.

Most of the holes shown entering the emitter junction in Figure 4-7 pass through it and recombine in the low-lifetime N^+ emitter region, so long as the BJT is not being operated with extremely low current levels. These holes are identically those represented by the profile labeled $p_N(x)$ that is shown at the left in Figure 4-4(d). In Figure 4-6, the recombination is symbolized by the gap in the I_B arrow inside the emitter region, with holes approaching it from the base side and the recombining electrons being supplied within the emitter.

4-2.2 Parasitic Internal Currents

The remaining two currents are parasitic in the sense that they vanish in the ideal BJT, and nearly vanish in the silicon BJT. First, a small current is shown in Figure 4-6 as splitting from I_B to a recombination gap within the base region. A small but finite portion of the flood of electrons diffusing from emitter to collector fails to reach the collector, recombining in transit, and it is this occurrence that the second gap symbolizes. In the early BJTs, *most* of the base current was consumed by this mechanism because the base thickness X_B amounted to tens of micrometers and base-region lifetime values τ_B were sometimes low. As a result, the crucial electron profile in the base region departed significantly from the modern-device linear profile shown in Figure 4-4(d). Rather, it took on the character represented in Figure 4-9. The *base-transport efficiency* γ_B took on considerable importance in those

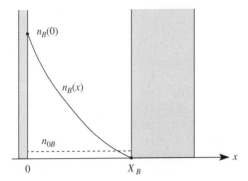

Figure 4-9 Electron-density profile in the base region of an early BJT. Significant amounts of base-region recombination caused the profile to be curved rather than straight.

days. (Today it is so close to unity that it is rarely mentioned.) The factor γ_B can be defined as the ratio of electron current departing from the active base region at its right-hand surface (see Figure 4-9) to the electron current entering the active base region at its left-hand surface.

Exercise 4-12. Explain how γ_B can be inferred quantitatively from a diagram such as Figure 4-9.

Because electron current is purely diffusive in the base region, we may write

$$\gamma_B = \frac{(dn/dx)|_{x=X_B}}{(dn/dx)|_{x=0}}.$$

The difference between the departing current and the entering current is identically the base-recombination current, which in the old BJTs dominated I_B.

Returning now to the modern transistor, we stress that I_B is very nearly equal to the current of holes being injected into the emitter. This brings us to a second efficiency factor, γ_E, one that is of great importance in the modern BJT. It can be defined as

$$\gamma_E \equiv (I_{nE}/I_E) = (I_E - I_{pE})/I_E = 1 - (I_{pE}/I_E), \tag{4-1}$$

where I_{nE} and I_{pE} are, respectively, the electron current being injected into the base region by the emitter junction and the hole current being injected into the emitter region by the same junction. The total emitter-junction current is inevitably equal to the emitter-terminal current I_E. In a "perfect" emitter, all of the emitter current would consist of electrons injected into the base region. From this point of view I_{pE} (the current of holes injected into the emitter region) is a *defect current,* a description sometimes used. The adjective "perfect" was placed in quotation marks, however, because the injected current of holes plays the critical role of the control current, and as such is a vital factor in BJT design and operation, rather than a factor to be eliminated as the potentially misleading term "defect current" may suggest.

This brings us to the last and smallest current, namely the *leakage current I_L* arising at the reverse-biased collector junction. It provides one of the sharpest contrasts between the modern silicon transistor and the early germanium transistor. As we noted in Chapter 3, the behavior of a reverse-biased silicon junction fits simple theory quite poorly, while the germanium case fits it very well. At the same time, the silicon-junction features that make it nonideal from a theoretical point of view serve to create near-ideal BJT properties! These details are worth examining because of the insight they provide into certain BJT subtleties. A lesser reason is that part of the older literature on the BJT and its applications dwells on the *nonideal* BJT properties that arise from the "ideal" germanium-junction behavior, and this literature is puzzling to a newcomer without a measure of appreciation of the responsible phenomena.

We saw in Chapter 3 that the dominant reverse-current mechanism in a silicon junction is the excess of generation over recombination in the space-charge region, where reverse bias has caused a depression of both carrier densities below their equilibrium values. (In such conditions, generation exceeds recombination, just as the reverse is true in the presence of excess carriers.) The carriers so generated are separated by the electric field and are swept to the regions where they become majority carriers. A representative value for this leakage current is of the order of ten picoamperes. The description just given shows that it consists equally of holes and electrons. The electrons become a negligible part of the collector current, departing the BJT through the collector terminal. The holes entering the base region mingle with those coming from the base contact. As a result, they, too, participate in emitter-region recombination and, to a much lesser degree, base-region recombination.

Conditions in the old germanium BJT were quite different. First, the diffusion of minority carriers to the reverse-biased collector junction provided most of the collector leakage current. Because the collector region is more lightly doped than the base region by one to two orders of magnitude, minority holes in the collector region dominate this diffusion mechanism. This density is roughly four million times larger in germanium than in silicon for given net-doping values, because the values of n_i^2 stand in about that ratio for the two materials, and their mobility ratios introduce another factor of about three. The resulting current of holes into the base region might be of the order of 0.1 microampere, amounting to one percent of the 10-μA base current cited in the example of Figure 4-6. When I_B is further diminished (and in particular, when it is reduced to zero), this leakage current is like an irreducible "parasitic base current." Further, this parasitic current is then multiplied by the gain mechanism of the BJT, typically by 100, with very undesirable effects. The details of this topic are best treated after we have introduced and described the gain mechanism itself, which we will do shortly. And after that, we shall point out why the silicon BJT does *not* multiply its collector-leakage current (already three orders of magnitude smaller!) by its current-gain factor.

4-2.3 Common-Emitter Current Gain

In the silicon BJT we may safely neglect for a wide range of purposes the two parasitic internal currents, namely, the base-recombination current and the collector-leakage current. Thus the admittedly complex pattern of Figure 4-6 reduces to the much simpler pattern of Figure 4-10. Even here, however, there remains the complexity of the base current's two-dimensional pattern, a complication that may be neglected for a small ratio of base current to electron current, as was noted earlier.

Return briefly to Figure 4-4(d). Focus on the carrier-density profiles in the emitter and base regions, and ignore that in the collector region because it affects only collector-leakage current, now dropped from consideration. Elementary diffusion theory permits us to write the current of electrons in the base region. This constitutes the *collector current* I_C, which is the output current in the common-

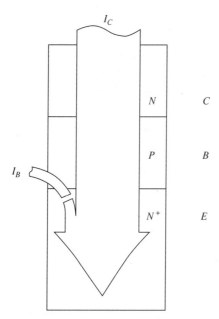

Figure 4-10 The terminal base and collector currents under typical conditions in a silicon BJT are large enough so that the internal parasitic currents may be neglected, yielding this simple current pattern.

emitter situation under study. Having already established that it is a positive terminal current, let us employ absolute-value signs on the diffusion-current expression. Given the emitter-junction area A_E,

$$I_C = |A_E q D_{nB}(dn_B/dx)|, \qquad (4\text{-}2)$$

where the gradient can be treated as constant throughout the base region, and is given simply by $n_B(0)/X_B$. It is important to stress once more at this point that X_B is taken to be the distance *from boundary to boundary*, and not from (metallurgical) junction to junction. It is clearly the base region as defined by the respective depletion-layer boundaries that enters into the gradient calculation. Thus X_B is the *effective base thickness*. To repeat, it is the thickness dimension of the active (or intrinsic) base region, whose lateral dimensions are fixed by emitter plan-view shape.

Employing the law of the junction now yields $n_B(0)$ as a function of applied voltage, in turn giving us I_C as a function of base-emitter voltage. The voltage $-V_{NP}$ in our law-of-the-junction formulation is equivalent to $-V_{EB}$, and hence to $+V_{BE}$. Therefore

$$I_C = (A_E q D_{nB} n_{0B}/X_B) \exp(qV_{BE}/kT). \qquad (4\text{-}3)$$

Identical principles permit us to write the current of holes diffusing into the emitter region, except that the critical length is L_{pE} rather than a device dimension. Accordingly,

$$I_B \approx |I_D| = |-A_E q D_{pE}(dp_E/dx)|, \qquad (4\text{-}4)$$

from which

$$I_B = (A_E q D_{pE} p_{0E}/L_{pE}) \exp(qV_{BE}/kT). \qquad (4\text{-}5)$$

It is of the greatest importance that the *same* voltage V_{BE} enters into *both* the input and output currents. Because this is true, we can define a *common-emitter static* (or dc, or steady-state) *current gain* that is a function of *structural constants* in the device:

$$\beta \equiv (I_C/I_B). \qquad (4\text{-}6)$$

Making use of Equations 4-3 and 4-5 gives us

$$\beta = \frac{D_{nB} n_{0B}/X_B}{D_{pE} p_{0E}/L_{pE}}. \qquad (4\text{-}7)$$

Thus we have from Equation 4-6

$$I_C = \beta I_B, \qquad (4\text{-}8)$$

where β from Equation 4-7 can be adjusted through wide ranges by the device designer and fabricator. Because the output current I_C is proportional to the input current I_B, the common-emitter BJT is a *linear current amplifier*.

Exercise 4-13. Calculate β for the device defined in Table 4-1.

Values taken from Table 4-1 yield

$$D_{nB} = (700 \text{ cm}^2/\text{V·s})(0.02566 \text{ V}) = 18.0 \text{ cm}^2/\text{s};$$

$$D_{pE} = (46 \text{ cm}^2/\text{V·s})(0.02566 \text{ V}) = 1.18 \text{ cm}^2/\text{s};$$

$$p_{0E} = \frac{n_{iE}^2}{n_{0E}} = \frac{(1.2 \times 10^{11}/\text{cm}^3)^2}{(10^{20}/\text{cm}^3)} = 144/\text{cm}^3;$$

$$L_{pE} = \sqrt{D_{pE}\tau} = \sqrt{(1.18 \text{ cm}^2/\text{s})(10^{-9}\text{s})} = 3.44 \times 10^{-5} \text{ cm}.$$

Thus from Equation 4-7,

$$\beta = \frac{(18.0 \text{ cm}^2/\text{s})(10^3/\text{cm}^3)/(4 \times 10^{-5} \text{ cm})}{(1.18 \text{ cm}^2/\text{s})(144/\text{cm}^3)/(3.44 \times 10^{-5} \text{ cm})} = 91.1.$$

This is a typical value for a milliwatt small-signal transistor.

Exercise 4-14. What can the BJT designer do to increase β?

The answer resides in Equation 4-7. The two diffusivities and the hole diffusion length in the emitter region are largely given by nature, but the remaining three factors are directly under the designer's control. It is clear that high β is favored by diminishing X_B and by diminishing the ratio of p_{0E} to n_{0B}, which means increasing the doping ratio, N_{DE}/N_{AB}.

The range of common-emitter current-gain values, or β values, found in practical BJTs extends through about three orders of magnitude. In power BJTs (ranging up to hundreds of watts) a combination of factors forces the designer to choose large values of X_B—several micrometers. As a result, such devices have low β—typically about ten. At the other extreme are *superbeta* transistors, such as those at the input stages of operational amplifiers. Here the base regions are made so thin that β approaches 10,000! The price that must be paid for this kind of gain performance is a drastic loss of operating-voltage range. As the collector space-charge layer expands under its normal reverse bias, its encroachment in an ultrathin base region can permit it to reach the emitter-junction space-charge region. Under these conditions, electrons are injected directly into the collector junction so that current increases precipitously and with no further control being exerted by the input current I_B. This condition is identically the punchthrough condition treated in Section 3-6.3.

4-2.4 The Gain Mechanism

The BJT places a number of conceptual barriers in the path of a student. With respect to basic device operation, there is one that is often difficult to surmount, sometimes articulated in this way: How can a defect current be a control current? Our opening description treated the electron current as "primary," so that the hole current flowing inward from the base contact may have seemed to be more of a result than a cause of the phenomena sometimes collectively termed *transistor action*. But the hole current (or base current) is indeed the *cause*, and this fact can be illuminated by examining once again the reason for the base current that is emphasized in Figure 4-10. Most of the holes flowing inward from the base contact are injected into the emitter region, to recombine. As a result, the charge-control principle of Section 3-4.5 can be applied to the base current, as we do now. For the structure of Table 4-1, L_{pE} is about one third of the emitter thickness. Therefore the injected-hole density has dropped to a few percent of its boundary value ($1/e^3 = 0.05$ or 5%) in traversing the emitter, which constitutes a fair engineering approximation of zero. As a result, the excess-hole charge stored in the emitter region can be written

$$Q_E = qA_E L_{pE} p_N(x_{\text{BDRY}}), \qquad (4\text{-}9)$$

where x_{BDRY} is the location of the emitter boundary of the emitter junction. Also it follows that

$$I_B = \frac{Q_E}{\tau_B}. \qquad (4\text{-}10)$$

Note the algebraic-sign consistency in the charge-control equation. An inward flowing, and therefore positive, I_B is controlled by the positive hole charge Q_E.

Operating the BJT involves the external setting of the input current I_B. Through the charge-control principle, this is tantamount to setting Q_E. This in turn fixes the value of V_{BE}. But V_{BE} fixes minority-carrier boundary values on *both* sides of the emitter junction. The boundary value on the emitter side is directly proportional to Q_E and I_B, as just described in some detail; the boundary value on the base side is directly proportional to dn_B/dx in the base region, and hence to I_C. From this point of view, the "controlling charge" Q_E makes the BJT a linear current amplifier.

Exercise 4-15. For a given I_B, how does V_{BE} depend upon emitter area A_E?

In Equation 4-9 it is evident that $p_N(x_{\text{BDRY}})$ will vary inversely with A_E for a fixed Q_E and hence a fixed I_B. Since $p_N(x_{\text{BDRY}})$ is exponentially related to V_{BE}, it follows that V_{BE} will vary logarithmically with the inverse of A_E.

Exercise 4-16. State in the most succinct way possible why the common-emitter BJT is a linear current amplifier.

The proportionality of output current I_C to input current I_B is a consequence of the fact that both currents are proportional to boundary values at *one* junction, the emitter junction, and *both* boundary values are proportional to $\exp(qV_{BE}/kT)$.

The points just made can be summarized in this fashion: The BJT user sets I_B. The BJT sets V_{BE} accordingly. The value of V_{BE} sets the emitter-junction boundary values. These boundary values are uniquely related to I_B and I_C.

Let us now return to the old germanium BJT with base-region recombination accounting for virtually all of I_B. Once more, a charge-control analysis is possible. This time base-region lifetime τ_B is the critical quantity, and the stored charge of excess carriers must be totally replaced every "τ_B seconds."

Exercise 4-17. The old germanium BJT with base-region recombination dominant was *also* a linear current amplifier in the common-emitter configuration. Why? In addition, how can the steady-state excess-carrier population of the base region be related to a terminal current when the electrons enter through the emitter junction and most of them exit through the collector junction?

The stored charge in the base region Q_B was *also* proportional to $n_B(0)$, but it was the *majority-hole* excess-carrier population that had charge-control relevance, and not the minority-electron excess-carrier population. The holes enter solely from the base contact (neglecting collector leakage current). They cannot depart through the collector junction because it is reverse-biased. They can only recombine in the base region, or in the emitter region after passing through the forward-biased emitter junction. Hence,

$$I_B = \frac{Q_E}{\tau_E} + \frac{Q_B}{\tau_B}.$$

The quantities I_B, Q_E, and Q_B are all positive. In the modern BJT, the first term on the right dominates. In the original BJT, the last term dominated. Both Q_E and Q_B are exponentially related to V_{BE}; thus I_B is as well. And because I_C is also exponentially related to V_{BE}, the BJT is a linear current amplifier for any "mix" of the two terms. In the original BJT we had the curious situation wherein *both* I_C and I_B were determined by a boundary value on *one* side of the emitter junction, namely, the base side.

Now we can return to the collector-leakage-current problems of the germanium BJT. A nonnegligible current of holes is delivered to the base region from the minority-carrier diffusion to the junction on the collector side. These holes are indistinguishable from base-current holes insofar as the BJT is concerned. Thus the *gain mechanism* just treated in considerable detail multiplies the collector-leakage current by β. Taking the typical values of $I_L = 0.1$ μA and $\beta = 100$ yields the appreciable primary current (collector current) of 10 μA even when *no* current is being delivered through the base contact. This effective leakage current is a major departure from the ideal value, zero.

To make matters worse, this irreducible leakage current exhibited steep temperature dependence. It arose from diffusing minority carriers whose equilibrium density follows the exponential temperature dependence of n_i^2. As a result it was a serious factor, and considerable ingenuity was applied to circuit techniques for compensating it. For silicon, the problem is far less serious.

Finally we examine why it is that this last phenomenon is avoided in the silicon BJT. In Section 3-4.6 we described the Sah-Noyce-Shockley (or SNS) effect. One component of this effect is the space-charge-generation mechanism in a reverse-biased silicon junction. The other component is that in forward bias, carriers diffusing into the space-charge region from both sides recombine in the space-charge region, and carriers thus recombining are *not injected*. At relatively low voltages this component of forward current dominates. In the silicon BJT, the absence of carrier injection at low values of V_{BE} means that the low-current-level β is very low. Another way of putting the matter is that the gain mechanism is not functioning at low current. How fortunate! The 10-pA leakage current supplied by the collector junction qualifies as a low-level current, and does not become multiplied. As a result, the silicon BJT that is "OFF" (meaning $I_B \approx 0$) exhibits a collector current that approximates the amount of the inherent leakage of the reverse-biased collector junction, a very small current indeed.

4-3 BIASING AND USING THE BJT

We have seen that the common-emitter BJT is inherently a current amplifier. Its combination of properties permit it to be simultaneously an efficient voltage amplifier, properties that we shall examine. Other properties of the common-emitter BJT make it valuable as a switching device. About 80% of the BJTs in the world are employed as switches in one of several configurations, of which the common-emitter configuration is but one. Because of the priority and simplicity of this configuration, however, we shall emphasize it here. Whether the device is to be used as switch or amplifier, the first task is to devise a biasing arrangement that permits us to exploit its useful properties.

4-3.1 Basic Bias Circuit

In order to forward-bias the emitter junction in a satisfactory manner, the distinctive forward characteristic of a *PN* junction must be clearly appreciated. Figure 4-11 illustrates once more the nature of the characteristic, and the reason that direct voltage biasing is inadvisable. Applying a battery voltage to select a point on the steeply rising curve is not effective because the "ON voltage," or *offset voltage,* of the silicon junction declines with increasing temperature. The act of passing forward current through a junction inevitably involves power dissipation. Unless elaborate precautions are taken, a temperature rise will follow, and the current so carefully established will drift upward.

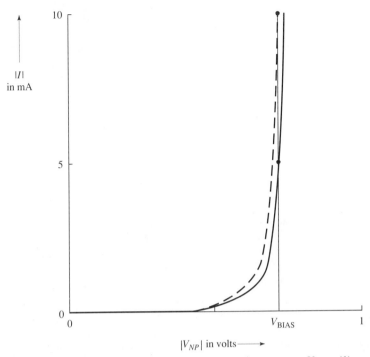

Figure 4-11 An attempt to set current with a voltage source V_{BIAS} (1) would require great voltage precision and (2) would result in major current changes with thermally induced changes in the characteristic, so this approach is normally avoided.

It is obvious that a *current-setting* arrangement is needed to forward bias a junction satisfactorily, and hence to bias the emitter junction of the BJT. There are several ways of accomplishing this. In today's integrated electronics, one normally uses one or more silicon devices to bias another silicon device. In the discrete-transistor era, resistors were usually used. Our primary aim is to illustrate and clarify principles, which can best be done in this case by using an early technique, so let us turn in that direction.

The simplest current-regulating circuit consists of a battery and a resistor. As is shown in Figure 4-12, where the current–voltage characteristic of a resistor is plotted, current changes but little when the applied voltage is changed from V_{BIAS} to the lower value indicated by the dashed line. The degree of current regulation at a particular desired current can be improved to an arbitarary extent. The cost of the improvement is a voltage source of higher value to combine with a proportionately larger resistance value.

The present problem is somewhat more involved, however, because the current so regulated is to be delivered to another component, the emitter junction. When one encounters a static problem involving two linear components (that is, resistors), very simple Ohm's-law equations permit determination of the voltage division and

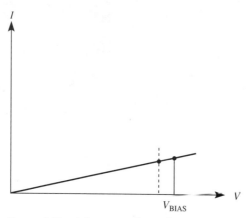

Figure 4-12 A battery and a resistor constitute the simplest current-regulating circuit.

the resulting current. But the emitter junction has a grossly nonlinear $I-V$ characteristic. It is true that we have a relatively simple equation for the junction's $I-V$ characteristic, an equation that is serviceable through at least part of the forward-voltage range, as Figure 3-24 shows. But this transcendental expression in combination with the linear expression for a resistor yields a result that is at best awkward. To solve a voltage-division and current-setting problem involving one or more nonlinear devices, there exists a graphical technique of great utility. It involves a construction known as a *loadline diagram*.

We shall illustrate the loadline technique for the case of a resistor and an emitter-junction forward characteristic, the case of immediate interest. One starts by constructing the $I-V$ characteristics of the two devices in the manner shown with solid lines in Figure 4-13, where V_{BB} is the *magnitude* of the battery voltage that will be applied to the two components in series. The next step is to rotate the characteristic of one of the devices about its current axis. In Figure 4-13 it is the resistor characteristic that has been given this treatment, with the result shown by the dashed line. Choosing the characteristic to be rotated is an arbitrary matter, but if one device is a resistor, convention usually favors its choice. The point of this construction is as follows: Because the two components are in series, they must have a common current, the current I_1 that is uniquely determined in Figure 4-13. Further, it is evident that the voltage-division problem has been solved. The diode voltage is V_1, and the voltage drop in the resistor is $V_{BB} - V_1$; these two drops add to give the total voltage V_{BB} as they must.

It is worth noting that the loadline method can often be extended to combinations of more than two components. Let us illustrate with two-terminal components. Their current–voltage characteristic may possess arbitrary nonlinearities, but these must be well-known. A pair of components in parallel can be resolved effectively into a single component by adding their currents at each voltage. A pair in series may similarly be resolved by adding their voltages at each current. Proceeding in

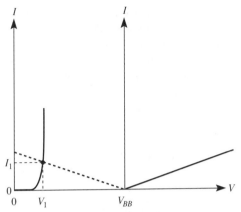

Figure 4-13 A loadline diagram for determining the voltage division and the resulting current when devices are biased in series, with one or more of the devices having a nonlinear I–V characteristic.

this manner one can reduce the number of "components" until the method of Figure 4-13 may be applied to determine a particular current and voltage.

Next, the battery-and-resistor method will be used to bias the input port of a common-emitter BJT amplifier. Biasing the output port is a less critical matter, as will be seen below. This is because the act of setting the input current I_B automatically sets the output current I_C through the relationship $I_C = \beta I_B$, Equation 4-8.

4-3.2 Static Equivalent-Circuit Model

Our aim is to devise for the common-emitter BJT the simplest meaningful assemblage of ideal components that will exhibit dc terminal characteristics simulating those of the transistor. The resulting equivalent-circuit model falls in one of the two broad categories discussed in Section 3-4.3. For many purposes, the approximate equations that are generated by the equivalent-circuit-model approach are more appropriate or more useful than the more exact determinations of static currents and voltages that can be derived from the loadline analysis of Section 4-3.1. This is true at the input port of the common-emitter BJT amplifier. Thus it constitutes a *third* approach to the input-port problem, following the fully analytic method that was mentioned and set aside, and the loadline method. Effort invested in describing the latter was not wasted, however, because loadline principles will be applied later in a context that is often more difficult to comprehend than in the diode-resistor context.

Let us start with the emitter junction, since dc equivalent circuits for junctions have already been developed. In particular, the circuit of Figure 3-21(b) is an appropriate compromise, consisting simply of an ideal diode in series with a battery. (An ordinary battery constitutes a superb real-life approximation to an ideal *independent voltage source*; its terminal voltage is sharply defined and its internal resistance is very low.) But at present we are *only* interested in the case of forward bias for which the ideal diode in that model can in turn be modeled as a short circuit, so we drop the diode and retain only the 0.7-V battery. We now shall label it V_{BE} for obvious reasons. What remains is to model the gain mechanism of the common-emitter BJT. We emphasized the fact that the BJT is a linear current amplifier, and this description points to modeling by means of a *current-controlled current source* placed in series in the collector arm of the equivalent circuit. The result is shown in Figure 4-14.

The battery V_{BE} shunting the input port provides the offset voltage needed in a faithful model. Its presence in no way hampers the component labeled βI_B because the independent voltage source presents zero resistance. But even if it possessed a finite internal resistance, the current source would prevail and would "push" the current βI_B through the resistor. (There is a finite resistance in the input portion of the BJT that sometimes requires modeling, a requirement easily met by adding a resistor to the model.) This model also informs us that the emitter terminal experiences an *outward* (and therefore, negative) current in the amount

$$I_E = -(I_B + \beta I_B) = -I_B(1 + \beta). \tag{4-11}$$

Exercise 4-18. Why doesn't the current βI_B flow out the base terminal?

In order to make the BJT operate properly, we have gone to a certain amount of trouble to insure that the current I_B flows *into* the base terminal. Use of the model in Figure 4-14 assumes that this condition has been preserved. If an inward-flowing I_B does *not* exist, this particular model is inapplicable and a different one must be used.

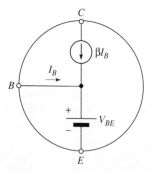

Figure 4-14 A two-component equivalent-circuit model for the normally biased common-emitter BJT. The offset voltage of the forward-biased emitter junction is modeled by the voltage source V_{BE}, and the BJT's gain mechanism is modeled by the current-controlled current source βI_B.

Finally, the equivalent-circuit model yields the very important output current. It is the *positive* collector current in the amount

$$I_C = \beta I_B, \qquad (4\text{-}12)$$

consistent with Equation 4-8.

4-3.3 Rudimentary BJT Amplifier

The circuit shown in Figure 4-15(a) converts the BJT, inherently a current amplifier, into a voltage amplifier. The bias battery supplying I_B has been replaced by a low-impedance source, V_{IN}, that is capable of relatively slow voltage variations. Associating a voltage symbol such as V_{IN} with a terminal is a shorthand way of designating that the voltage-source "box" (whatever it may be) has one terminal connected to the input terminal of the amplifier circuit and the other connected to circuit ground, which in this case is at the common-emitter node. For the desired base current, of course, it is the positive terminal of V_{IN} that must be connected to the input node.

At the output port of the amplifier circuit an element or a circuit of extremely high impedance, such as the input probe of an oscilloscope, is connected. This proviso is made so that the circuit being driven by the amplifier will "draw" negligible current. Under these conditions, the current in the resistor R_C that has been placed in series with the collector terminal will be I_C. Therefore the output voltage, V_{OUT}, will be determined by the voltage drop $I_C R_C$ and the bias voltage $+V_{CC}$. Also, it is evident that V_{OUT} is identically V_{CE}. Once again, associating the symbol $+V_{CC}$ with the top node of R_C implies that a battery (or other voltage source) is connected from that node to circuit ground. It is customary to designate the *magnitude* of the power-supply voltage associated with a particular terminal by using a corresponding double subscript, and to use an explicit algebraic sign, as has been done in Figure 4-15.

The next step is to remove the BJT symbol in Figure 4-15(a) and to "plug in" the BJT equivalent-circuit model from Figure 4-14 as a replacement, with the result shown in Figure 4-15(b). Now approximate equations can be written with ease. It is evident by applying Kirchhoff's voltage law to the "input loop" that

$$I_B = \frac{V_{IN} - V_{BE}}{R_B}. \qquad (4\text{-}13)$$

Placing this expression in Equation 4-12 yields

$$I_C = \beta I_B = \beta(V_{IN} - V_{BE})/R_B. \qquad (4\text{-}14)$$

Now examine the output loop (which is completed through the battery V_{CC} that is not explicitly shown). A loop equation here (Kirchhoff's voltage law again) yields

$$V_{OUT} = V_{CC} - \beta I_B R_C = V_{CC} - \beta(R_C/R_B)(V_{IN} - V_{BE}). \qquad (4\text{-}15)$$

Suppose that now we wish to find the *small-signal* voltage gain A_V that can be

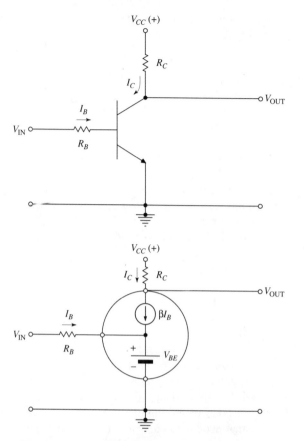

Figure 4-15 A rudimentary common-emitter-BJT voltage amplifier. (a) Biasing arrangement employing a base series resistor R_B and a collector series resistor R_C. (b) The same amplifier circuit with the BJT equivalent-circuit model substituted for the BJT symbol, for approximate analysis.

defined as the rate of change of V_{OUT} with respect to V_{IN}. Because Equation 4-15 states V_{OUT} as a function of V_{IN}, it is evident that we need merely differentiate to obtain the desired expression. That is,

$$A_V \equiv \frac{dV_{OUT}}{dV_{IN}} = -\beta \frac{R_C}{R_B}. \tag{4-16}$$

The significance of the negative sign in the gain expression is that we have here an *inverting* amplifier. That is, a positive ΔV_{IN} causes a positive ΔI_B, which causes a positive ΔI_C, which finally causes a *negative* ΔV_{OUT}. This is because the voltage drop $I_C R_C$ has increased, pushing the output voltage "down." Because of this property, the configuration in Figure 4-15(a) consisting of R_C and the BJT is often termed an *inverter*.

Equation 4-16 gives the plausible information that the voltage gain A_V of the amplifier circuit is proportional to the common-emitter current gain β of the BJT. Equally plausible (in view of the next-to-last sentence of the last paragraph) is the observation that A_V is proportional to R_C. This final bit of information can be nicely illustrated by reinvoking a loadline diagram, but this time in connection with the output port.

A diagram presenting the BJT's output current I_C as a function of its output voltage V_{CE} is termed the *output plane* for the common-emitter BJT. On this plane one plots a "family" of *output characteristics*. What is meant by this is that the controlling variable I_B is given a constant value and the resulting output curve (I_C versus V_{CE}) is plotted. Then I_B is given a different constant value, and so on. In mathematical language, I_B is being employed as a *parameter*. (This term, regrettably, has been overused and abused to such a point that many people erroneously assume that it is a synonym for "property" or "feature." We shall endeavor to confine our use of the term to cases involving the more specific mathematical meaning of a "temporarily constant variable.") A collection of *I–V* characteristics generated in this manner is termed the family of characteristics.

Figure 4-16 shows the output plane of a common-emitter BJT having $\beta = 100$ in the circuit of Figure 4-15. Our model, Figure 4-14, predicts a strictly constant value of I_C for each constant value of I_B. This condition holds up to a value of V_{CE} at which one of several possible "breakdown" phenomena cause current to increase sharply. The punchthrough phenomenon described in Section 3-7.3 can occur when the base region is extremely thin, and can be regarded as falling in this category. Also, the constant-I_C characteristics are valid down to a small but nonzero value of

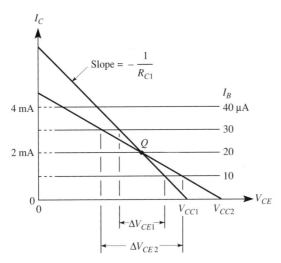

Figure 4-16 The output plane for an ideal *NPN* BJT. Loadlines corresponding to two different values of R_C are shown, with V_{CC} adjusted to maintain the same quiescent operating point Q.

V_{CE}—typically a few tenths of a volt. In that regime of the output plane, each of the constant-I_B characteristics heads precipitously for a point near the origin. Consequently these output-plane characteristics, the various members of the family, are grossly nonlinear, each exhibiting a near-ninety-degree angle. They are, to be sure, *piecewise linear* in the sense of Section 3-4.4; but since each is so nonlinear overall, we have here another case where a loadline construction is useful and informative.

Even though the BJT is a three-terminal device, we are entitled to exploit the loadline technique in dealing with the nonlinear properties of a *pair* of terminals, in this case those of the output port. Since input current I_B has been taken as a parameter, no problems arise in relation to the input port. The output loop involves the BJT, the resistor R_C and the voltage source V_{CC}. Following the previous example, let us mark off V_{CC} on the V_{CE} axis, and then apply the necessary rotation to the characteristic of the resistor, this time R_C. This has been done in Figure 4-16, with the resistor characteristic prior to rotation having been omitted. Focus on the line of labeled slope. Because voltage is the independent variable, the slope magnitude is reciprocal resistance, and because of the rotation, the sign of the slope is negative.

Let us now assume that the input voltage V_{IN} is set at such a value that I_B is 20 μA. Consequently, the loadline construction tells us that the two characteristics in question intersect at the point labeled Q, which stands for *quiescent operating point*. The term quiescent implies static, or without an ac signal applied. The voltage-division result of this situation is readily determined from the approximate Equation 4-15. Now let us assume that V_{IN} is varied through a range that causes input current I_B to vary or "swing" from 10 μA to 30 μA. The voltage-division solution provided by the loadline construction this time informs us that as a result, V_{CE} (the same as V_{OUT}) swings through the interval labeled ΔV_{CE1}. The next step then is to increase the collector resistor, changing it from the value R_{C1} to the value R_{C2}. In order to maintain the same quiescent operating point Q (which may or may not be desired), it is necessary to make a coordinated change of V_{CC} from V_{CC1} to V_{CC2}. Now note the output swing V_{CE2}. It has increased precisely in the ratio of the two R_C values, thus validating Equation 4-16.

When the collector junction is reverse-biased and the emitter junction is forward-biased, the BJT is said to be in the *forward-active* regime of operation. This condition prevails in most of the output-plane area represented in Figure 4-16. The term *active* connotes the ability of the BJT under these conditions to operate as an amplifier—and specifically, as an approximately linear amplifier. Finally, the term *forward* connotes the "normal" biasing arrangement, with the emitter junction emitting and the collector junction collecting. Next we move outside the forward-active regime into that near the collector-current axis, unexplored heretofore.

4-3.4 Saturation

To be specific, let us move the quiescent operating point Q shown in Figure 4-16 to a different position in the output plane, retaining the same ideal BJT as before. The schematic diagram in Figure 4-17(a) defines the new position. Input current I_B is fixed this time at 10 μA by an unspecified current source. The power-supply volt-

BIASING AND USING THE BJT   519

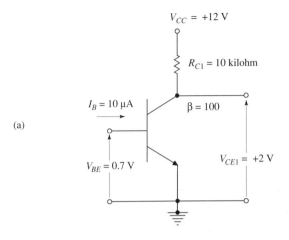

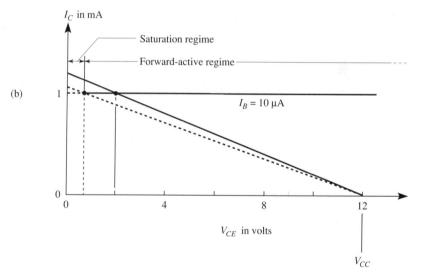

Figure 4-17 Inducing saturation in a BJT. (a) Inverter circuit with variable R_C. (b) Output plane, with the characteristic for which $I_B = 10$ μA, a loadline (solid) for $R_{C1} = 10$ kilohm, and a quiescent operating point at $V_{CE} = 2$ V, in the forward-active regime. Increasing R_C rotates the loadline to a new position (dashed) that places the operating point at the boundary of the forward-saturation regime.

age V_{CC} is set at 12 V, and R_C is a variable resistor. With β remaining at 100, the static output voltage $V_{CE1} = 2$ V because we have set $R_{C1} = 10$ kilohm. The resulting output-plane and loadline diagram is shown in Figure 4-17(b), using a solid loadline.

Now let us adjust R_C, increasing it slowly. For the assumed conditions, V_{CE} will decline but I_C will remain virtually constant at 1 mA. When we reach $V_{CE} =$

$V_{BE} \approx 0.7$ V, the loadline is represented by the dashed line in Figure 4-17(b), and the BJT has arrived at a critical condition.

Exercise 4-19. What is special about the bias condition just described, and what physical information does it communicate?

As noted in Exercise 4-7,

$$V_{CB} = V_{CE} - V_{BE}.$$

Thus, when $V_{CE} = V_{BE}$, we have the condition $V_{CB} = 0$. The important physical message is that a junction at zero bias is virtually as efficient a collector of minority carriers as a reverse-biased junction is. This is plausible upon reflection, because the collection mechanism is based upon the electric field in the junction, and even a junction at zero bias typically exhibits field values of many kilovolts per centimeter.

The effect of increasing R_C, then, is to cause the collector junction to go from its initial condition of reverse bias into *forward* bias. As Figure 4-17(b) indicates, the condition $V_{CE} = V_{BE}$ defines a boundary between the forward-active regime and a new one designated the *saturation* regime. When the boundary is crossed we have *both* junctions in the condition of forward bias; this constitutes a succinct definition of the saturation condition. Here the term *saturation* no longer means "approaching a constant value" (as in "saturation current"); rather, it communicates the message that the base region is becoming loaded up or "saturated" with excess carriers. Recall that the law-of-the-junction relationship is inviolable and unique in forward bias. With both junctions forward-biased, both boundary values in the base region rise above equilibrium values, and excess carriers exist everywhere within the intrinsic base region.

It is evident in Figure 4-17(b) that even under low-level forward bias, the collector junction *still* performs efficiently as a collector of minority electrons arriving from the emitter. The same holds until forward bias on the collector junction amounts to some 0.3 V or 0.4 V. Then the collector current starts to drop, and drops precipitously thereafter. The result is an output characteristic much like that drawn in Figure 4-18(a). A reasonably sharp "corner" in the characteristic exists at about $V_{CE} = 0.25$ V, and below this resides another roughly linear portion of the I–V characteristic. Thus the piecewise-linear label is justified for the output I–V characteristic of the ideal BJT, with the angle formed by the two line segments approximating 90°. This "corner" constitutes a dividing point between two subportions of the saturation regime, known as *weak* saturation where I_C is approximately constant, and as *deep* saturation where V_{CE} is approximately constant. The

BIASING AND USING THE BJT   521

(a)

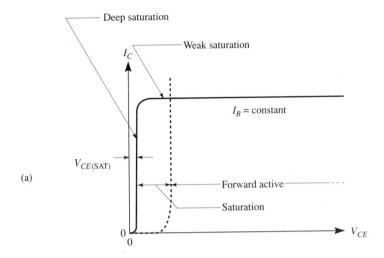

(b)

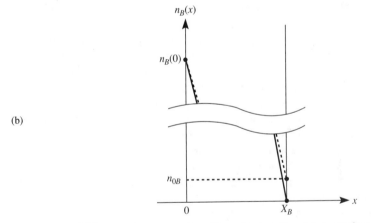

Figure 4-18 BJT saturation seen externally and internally. (a) A nearly piecewise-linear output characteristic displaying a forward-active regime at the right, and the near-vertical and near-horizontal portions of the forward-saturation regime at the left, corresponding to deep and weak saturation, respectively. (b) Minority-carrier profiles in the base region for (solid) forward-active operation, and (dashed) operation at the boundary of forward saturation, showing the negligible difference.

roughly constant latter voltage is important enough in modeling to have its own symbol: $V_{CE(SAT)} \approx 0.2$ V for the case of a typical silicon BJT at modest current, as is indicated in Figure 4-18(a).

To say that $V_{CE} = 0.7$ V constitutes the boundary between the forward-active and saturation regimes is of course an approximation. As I_B approaches zero (and as I_C does so as well), V_{BE} declines. And since $V_{CE} = V_{BE}$ is an accurate statement of where the boundary between the two regimes lies, it follows that the actual boundary has a character much like that represented by the dashed line in Figure 4-18(a).

A firm qualitative grasp of saturation behavior can be had from the base-region minority-carrier profiles corresponding to various bias points. To begin, let us compare the profiles for the two quiescent operating points shown in Figure 4-17(b). The solid profile in Figure 4-18(b) corresponds to the forward-active solid loadline in Figure 4-17(b). With $V_{BE} \approx 0.7$ V, the Boltzmann factor in the law of the junction amounts roughly to a factor of one trillion. Noting this, it is evident that Figure 4-18(b) involves appreciable distortion in spite of the fact that a gap has been introduced into the ordinate to help deal with such an enormous range of values in a linear diagram. (Without distortion, the profile would be effectively superimposed on the density axis in the upper portion of the diagram, and superimposed on the vertical line at $x = X_B$ in the lower portion.) Note, too, that the solid line shown applies for any permissible value of reverse bias on the collector junction, and not just to the value corresponding to $V_{CE} = 2$ V.

Next, let the quiescent operating point be changed to fall on the saturation-active boundary. As a result, $n_B(X_B) = n_{0B}$ because an equilibrium boundary value exists at zero bias. The result is shown by the dashed profile in Figure 4-18(b). Even with the distortion introduced there for the sake of clarity, it is evident that the slope dn_B/dx has changed very little with the change in operating point; if the diagram were rendered to scale, the change in slope would be imperceptible. And since the profile slope determines the diffusion current of electrons through the base region, and hence collector current, it follows that an imperceptible change in I_C will result from the quiescent-point change displayed in Figure 4-17(b).

Thus far, the parameter I_B has remained constant. Now examine the effect of increasing it, often described as increasing the "drive" on the BJT. In Figure 4-19(a) we present the output characteristics for three values of I_B, all of which cause the BJT to reside in deep saturation for the given bias conditions. The deep-saturation portions of the output characteristics are nearly superimposed. (Indeed, distortion has again been introduced to separate them a bit for the sake of clarity.) As a result, the change in "drive" has almost no effect on I_C and V_{CE}. In its deep-saturation condition, the BJT is "passive" in the sense that its collector current is fixed by external-circuit conditions, specifically V_{CC} and R_C. In its "active" regime, it was the BJT that fixed the collector current through the relationship $I_C = \beta I_B$.

While the base-current changes have little effect on I_C and V_{CE}, they do have a major effect on the minority-carrier distribution in the base region. This is illustrated in Figure 3-19(b), with curves labeled 1, 2, and 3, corresponding to the three levels of base current with quiescent operating points similarly labeled in Figure 3-19(a). Because the external circuit is now "in charge," fixing I_C, it follows that the BJT must now conform by adjusting its internal conditions to keep I_C constant. This means that dn_B/dx in the base region must be essentially constant, the condition

BIASING AND USING THE BJT 523

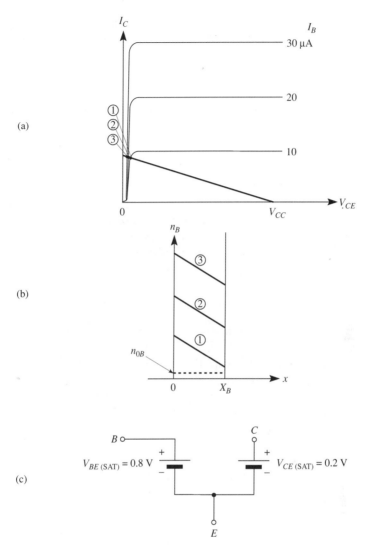

Figure 4-19 Operation in deep saturation. (a) Output plane, showing three quiescent operation points that nearly coincide as base current is changed. (b) Base region profiles corresponding to the points shown in part (a). (c) BJT equivalent-circuit model for deep saturation.

presented in Figure 4-19(b). By increasing I_B, of course, we also increase $n_B(0)$ and $n_B(X_B)$. Thus, driving a BJT deeper into saturation by increasing I_B results in rising values of V_{BE} and V_{BC}. For this reason, in the equivalent-circuit model of a BJT in deep saturation, the emitter junction is modeled by means of a voltage source $V_{BE(SAT)} = 0.8$ V, rather than the $V_{BE} = 0.7$ V of the forward-active model, Figure

4-14. To complete the deep-saturation model, all that is needed is a voltage source in shunt at the output port with $V_{CE(SAT)} = 0.2$ V, with the final result shown in Figure 4-19(c).

Exercise 4-20. According to the model of Figure 4-19(c), what value is assumed for voltage on the collector junction, V_{BC}?

Since $V_{BC} = V_{BE} - V_{CE}$, we have $V_{BC} = (0.8 \text{ V}) - (0.2 \text{ V}) = 0.6$ V.

Because the model of Figure 4-19(c) is valid only in deep saturation, while that in Figure 4-14 is valid in the forward-saturation *and* weak saturation conditions, some authors prefer to lump weak saturation into the active regime, letting deep saturation alone represent "saturation." We prefer, however, to let the boundary between the saturation and forward-active regimes be the unequivocal condition $V_{CE} = V_{BE}$.

4-3.5 Other Operating Regimes

In Figure 4-17(b) it is shown that a BJT can be moved from the forward-active to the saturation regime and back again by adjustment of the resistor R_C. Figure 4-19(a) makes it equally evident that changing the base current can accomplish the same thing. Yet another way of doing so is by manipulating the power-supply voltage V_{CC}, the case we examine next.

Exercise 4-21. Write an expression for the value of V_{CC} that places a BJT at the edge of saturation in terms of I_B, R_C, β, and other necessary quantities.

The boundary is reached when $V_{CE} = V_{BE}$. Therefore

$$V_{CC} = I_C R_C + V_{BE} = \beta I_B R_C + V_{BE}.$$

In Figure 4-20(a) is shown a series of four loadlines in the output plane of a BJT. With I_B held constant, only one output characteristic is needed. And since all the loadlines have the same slope, it is evident that R_C is being held constant and only V_{CC} is being changed.

Exercise 4-22. Estimate the value of R_C used in the construction of Figure 4-20(a), assuming that $I_B = 10$ μA and $\beta = 100$.

Since the constant value of collector current is $I_C = \beta I_B = 1$ mA, and $I_C R_C$ can be estimated at about 0.65 V, it follows that

$$R_C \approx \frac{0.65 \text{ V}}{1 \text{ mA}} = 0.65 \text{ kilohm}.$$

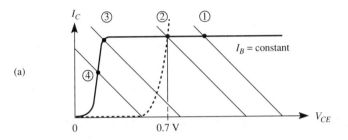

(a)

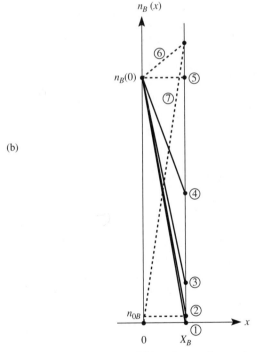

(b)

Figure 4-20 Illustrating additional operating regimes. (a) Loadlines corresponding to (1) forward-active, (2) saturation-boundary, (3) weak-saturation, and (4) deep-saturation operation. (b) Base-region profiles for the four conditions of part (a), plus (5) zero-collector-current, (6) reverse-saturation, and (7) reverse-active conditions.

As V_{CC} is changed from case 1 to 2 to 3 in Figure 4-20(a), the base-region minority-carrier profile will pass through a sequence corresponding to profiles 1, 2, and 3 in Figure 4-20(b), where once again distortion has been employed for clarity to an even greater degree than before. For simplest description, let us assume that V_{BE} and $n_B(0)$ as well as I_B remain constant in the transition from case 2 to 3, and beyond. We conclude that the slopes (and hence I_C) in cases 1 and 2 will be virtually identical; in case 3, the slope and current will be only slightly diminished. In case 4, however, both parts (a) and (b) of Figure 4-20 inform us that collector current will drop by about a factor of two.

Exercise 4-23. We assumed V_{BE} to be constant in the transition from case 2 to 3 and beyond. What would cause V_{BE} to be other than constant?

When the BJT enters saturation, the base current I_B that was assumed constant must supply holes to the forward-biased emitter junction *and* to the now forward-biased collector junction. Since hole current through the emitter junction falls, so too do V_{BE} and $n_B(0)$.

By continuing to reduce the value of V_{CC}, we come ultimately to the condition $V_{CE} = 0$. But another way of stating this condition is $V_{BE} = V_{BC}$. Thus the minority-carrier values at *both* boundaries of the base region will be the same in this device of uniform base-region doping. It follows that the electron-density profile will exhibit zero slope, as profile 5 shows in Figure 4-20(b). Consequently the collector current will be (almost) zero.

Now, keeping everything else constant, let us reverse the polarity of bias applied to the collector terminal, causing the collector junction to be *more* forward-biased than the emitter junction. The result will be case 6, wherein minority electrons in the base region diffuse from collector to emitter. Because of this reversal of primary current, the resulting regime is described as a *reverse* regime, and because both junctions clearly remain forward-biased, this regime is specifically the *reverse-saturation* regime of operation. The sequence of cases can be concluded now by (for the first time) placing reverse bias on the emitter junction. With the two junctions now in the reverse of their normal roles, we have the *reverse-active* regime of operation. To complete the terminology assignments we should point out that the regime designated simply as *saturation* previously is specifically the *forward-saturation* regime of operation.

These four operating regimes associated with the common-emitter configuration can be displayed in a single diagram. Figure 4-21 is such a diagram, with the active regimes shaded. The first quadrant is identically the output plane presented previously. Notice that the second quadrant is empty. The reason for this is that, as noted above, when V_{CE} becomes negative, I_C also changes sign, with conventional current flowing *out* of the collector terminal.

BIASING AND USING THE BJT   527

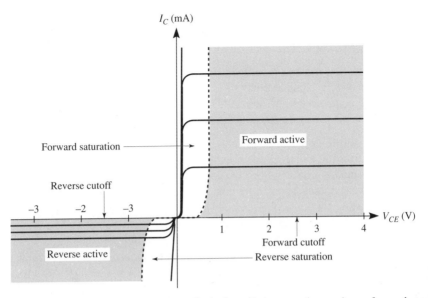

Figure 4-21 Complete output plane displaying all six operating regimes: forward- and reverse-active (crosshatched), forward- and reverse-saturation, and forward- and reverse-cutoff regimes.

We have seen before that the spacing of the horizontal portions of the output characteristics constitute a measure of current gain β. Thus it is evident that current gain in reverse operation (third quadrant) is lower than in forward operation, a fact that is typically a result of BJT geometry. (This matter will be examined further in Section 4-4.) For general static analysis of the BJT, therefore, one must employ a forward current gain β_F and a reverse current gain β_R. The current gain β of previous analysis is of course identically β_F.

Exercise 4-24. Assuming that the base-drive step in Figure 4-21 amounts to an increment of 10 μA throughout Figure 4-21, assign parameter labels with correct magnitude and sign to all of the output characteristics presented.

Start with forward operation as presented in the first quadrant. Beginning with the characteristic that lies along the V_{CE} axis, appropriate labels are $I_B = 0$, $+10$ μA, $+20$ μA, and $+30$ μA. In reverse operation with characteristics lying in the third quadrant, one *still* requires positive base current. Thus the analogous labels will be the same as in the first quadrant.

Exercise 4-25. The two current gains β_F and β_R bear a subtly different relationship to the curves in quadrants one and three, respectively, in Figure 4-21. Display this difference by writing expressions for ΔI_C in the two cases in terms of ΔI_B.

In the first quadrant, forward operation, we have $\Delta I_C = \beta_F \Delta I_B$. But in reverse operation the collector has become a "quasiemitter." Hence the desired current increment must be written $\Delta I_C = -(1 + \beta_R)\Delta I_B$.

A still more subtle difference between reverse and forward operation resides in the differing values of the roughly constant saturation voltages (with the one in the first quadrant having been labeled $V_{CF(SAT)}$). That in the third quadrant is correctly shown as being appreciably smaller than $V_{CE(SAT)}$. To explain the phenomena responsible for this difference, we need the more powerful modeling tools that will be developed in Section 4-6, and so the matter will be set aside for now.

Figure 4-21 also displays features of such importance that they can be regarded as fifth and sixth operating regimes, known as *cutoff* regimes. These features are of course the output characteristics lying along the voltage axis, those for $I_B = 0$. One can ensure the cutoff condition by reverse biasing both junctions, or at least, reducing both to zero bias. As a practical matter, however, it is not necessary in a silicon BJT to set $V_{BE} = 0$ (or $V_{BC} = 0$ in reverse operation) to achieve $I_B \approx 0$. In fact, there are typical applications in which $V_{BE} \approx 0.2$ V in the case of approximately zero input and output currents. The explanation for this convenient fact resides in the small leakage current of a silicon junction, and the small current gain exhibited by a silicon BJT at low current. These matters were treated in some detail in Section 4-2.4.

The cutoff regime (especially in forward operation) is important because it enters the simplest switching application of a BJT. This application can be described with the help of Figure 4-19(a). The output port of the common-emitter BJT is a switch that can be controlled by means of base current. With $I_B \approx 0$, the switch is open, and only a leakage current will flow through the output port. But when I_B is raised to a value sufficient to put the BJT in deep saturation, the switch is closed and exhibits low resistance. The closed or ON condition is less ideal than the open or OFF condition; there exists an ON voltage drop of $V_{CE(SAT)} = 0.2$ V. Nonetheless, this BJT switch is very serviceable indeed.

Exercise 4-26. Assuming that the BJT switching device is in a circuit similar to that shown in Figure 4-17(a), but with fixed R_C, write an approximate expression for the value of I_B required to ensure deep saturation for the sake of an acceptable ON condition.

The loadline intersects the axis at the current $I_C = V_{CC}/R_C$. Therefore the minimum acceptable value of I_B will require that $\beta I_B = V_{CC}/R_C$, or $I_B = V_{CC}/\beta R_C$.

As a practical matter, again, one usually provides a margin of safety, driving the BJT appreciably beyond the "corner" current just calculated. This is to ensure that the BJT will not "drop out of saturation" (to use the jargon)—because of a change in β with temperature, for example. Therein lies a problem. The base region is "saturated" with excess carriers when the BJT is ON; in order to return it to the OFF condition, those carriers must be extracted, or must be permitted to recombine. Either option means time delay, which is to be avoided in most switching applications. Hence, recent years have seen other switching methods rising rapidly in popularity. Especially important are those wherein the BJT remains either in the active regime or in cutoff, thus virtually eliminating the problems and delays that accompany carrier storage.

The various operating regimes can be conveniently displayed and summarized by means of a "voltage–voltage" diagram. The two voltages involved are V_{BE} and V_{BC}, with positive values meaning forward bias and vice versa. This has been done in Figure 4-22; a diagram showing one or more base-region profiles has been placed in each quadrant as a reminder of the physical conditions inside the BJT that accompany each regime. There are two saturation regimes, with the relative magnitudes of V_{BE} and V_{BC} determining which applies. A parallel comment can be made for the two cutoff regimes, although it should be remembered that requiring both junctions to be reverse-biased (as this diagram does) is an overstatement of the engineering requirement. The voltage signs are unambiguous for the cases of forward- and reverse-active operation, and so a total of six operating regimes are represented here.

Exercise 4-27. Devise an equivalent-circuit model for the BJT in forward or reverse cutoff.

In cutoff, all terminals are isolated from one another, approximately. In other words, draw a circle with three terminals on it and show nothing inside the circle, thus indicating open circuits for all three terminal pairs.

4-3.6 Other Circuit Configurations

A good reason underlies the dominant use of the common-emitter configuration—coexisting current and voltage gain. The other two possibilities have important specialized applications, however. The *common-base* configuration provides substantial voltage gain and near-unity current gain, while the *common-collector* case offers substantial current gain and near-unity (dynamic) voltage gain. Although the small-signal models described in Section 4-7 display these properties in elegant fashion, even static analysis is illuminating and worth a brief discussion.

The schematic diagram in Figure 4-23 shows a typical common-collector BJT stage. Only the collector terminal of the BJT is at a fixed voltage (V_{CC}). In dynamic

530 THE BIPOLAR JUNCTION TRANSISTOR

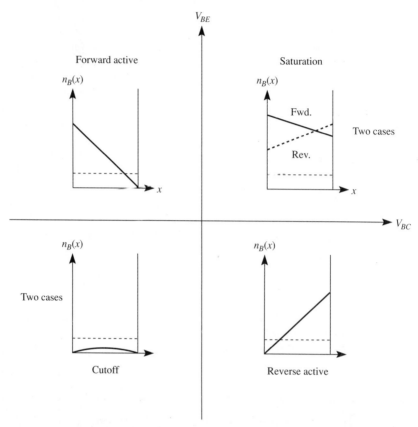

Figure 4-22 Using the V_{BE}–V_{BC} plane to display the six operating regimes, each illustrated by means of a base-region minority-carrier profile.

terms it is a "grounded" terminal and is common to the input and output ports. For a static examination of the stage, however, let us refer V_{IN} and V_{OUT} to the circuit ground, as is shown in Figure 4-23.

Our aim is to deliver substantial current and power to the low-resistance load, $R_L = 500$ ohms. Let us start with the arbitrary assumption that $I_E = -11$ mA. The downward-flowing conventional current will divide between R_E and R_L inversely as the resistances, so that $I_R = 1$ mA and $I_L = 10$ mA. Thus it is evident that $V_E = 5$ V, $V_B = V_{IN} \approx 5.7$ V, and $V_{CE} = 5$ V. Applying an input-voltage increment sufficient to change V_E to 6 V will yield $\Delta V_E = +1$ V. It is clear that the 20% increment in V_E will be accompanied by a 20% increment in I_R, I_L, I_B, and I_C, although I_C does not concern us. We see, too, that there was an increment ΔV_{IN} that approximately equals ΔV_E. To estimate ΔV_{IN} more accurately, let us make the plausible assumption that

$$I_B = I_{0B} \exp(qV_{BE}/kT), \qquad (4\text{-}17)$$

BIASING AND USING THE BJT   531

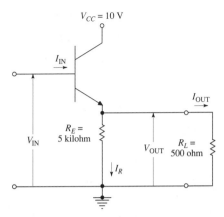

Figure 4-23 A typical common-collector (or emitter-follower) stage, providing a resistance transformation from high to low.

a matter not previously discussed. Making the equally plausible assumption that $V_{BE} = 0.7$ V in the initial condition, and once again taking β (or β_F) to be 100, we have for that condition

$$I_B = \frac{-I_E}{(1+\beta)} = \frac{0.011 \text{ A}}{101} = 1.089 \times 10^{-4} \text{ A}, \quad (4\text{-}18)$$

and

$$I_{OB} = \frac{I_B}{\exp(qV_{BE}/kT)} = \frac{1.089 \times 10^{-4} \text{ A}}{\exp(0.7 \text{ V}/0.02566 \text{ V})} = 1.547 \times 10^{-16} \text{ A}. \quad (4\text{-}19)$$

Since the new value of I_B is $(1.2)(1.089 \times 10^{-4} \text{ A}) = 1.307 \times 10^{-4}$ A, the new value of V_{BE} is

$$V_{BE} = (kT/q)\ln\left(\frac{I_B}{I_{OB}}\right) = (0.02566 \text{ V}) \ln\left(\frac{1.307 \times 10^{-4} \text{ A}}{1.547 \times 10^{-16} \text{ A}}\right) = 0.7047 \text{ V}. \quad (4\text{-}20)$$

Thus the incremental input voltage is

$$\Delta V_{IN} = (6 + 0.7047) \text{ V} - (5.7) \text{ V} = 1.0047 \text{ V}, \quad (4\text{-}21)$$

and the incremental voltage gain is

$$A_V \approx \frac{\Delta V_{OUT}}{\Delta V_{IN}} = \frac{1 \text{ V}}{1.0047 \text{ V}} = 0.995, \quad (4\text{-}22)$$

which is very close to unity. Because the output voltage excursion so nearly tracks the input excursion (to better than 1% in this example), the common-collector stage is usually termed an *emitter-follower* stage. That is, V_E "follows" V_B with considerable fidelity.

The data in the example just given can be easily used to calculate incremental current gain as well:

$$A_I \approx \frac{\Delta I_{\text{OUT}}}{\Delta I_{\text{IN}}} = \frac{(0.2)(10^{-2}\text{ A})}{(0.2)(1.089 \times 10^{-4}\text{ A})} = 91.8. \quad (4\text{-}23)$$

Exercise 4-28. The incremental current gain just calculated is less than the assumed value of β. Instead of using an emitter-follower stage, why not simply use a common-emitter stage with R_L substituted for R_C? Any awkwardness involving voltage-level shifting could be countered by substituting a *PNP* device for the *NPN* BJT. Then one could achieve $A_I = \beta$.

The common-collector stage is a resistance transformer, with a high input resistance and low output resistance; because it achieves a good impedance match to a low-impedance (or low-resistance) load it delivers more power to it. This, too, is lucidly revealed in small-signal analysis, but the insight from the present static analysis is nonetheless valuable. In the emitter follower, $\Delta V_{\text{OUT}} \approx \Delta V_{\text{IN}}$. Recall that ΔV_{OUT} in the analysis of the common-emitter case of Figure 4-16 was proportional to βR_C, and is appreciably smaller than ΔV_{IN} for very small R_C. Delivering maximum power, accomplished by an impedance match, involves maximizing the product $\Delta V_{\text{OUT}}\,\Delta I_{\text{OUT}}$, and this is achieved by the emitter-follower circuit.

Exercise 4-29. Estimate the incremental input resistance of the emitter-follower stage in the example above.

The incremental input resistance can be written

$$r_{\text{IN}} \approx \frac{\Delta V_{\text{IN}}}{\Delta I_{\text{IN}}} = \frac{1.0047\text{ V}}{(0.2)(1.089 \times 10^{-4}\text{ A})} = 46 \text{ kilohm}.$$

A common-base stage is shown in Figure 4-24(a). It is evident that since $I_C = \beta I_B$, we have

$$I_E = -(1 + \beta)I_B. \quad (4\text{-}24)$$

Static current gain in this configuration, designated α, is thus

$$\alpha \equiv -\frac{I_{\text{OUT}}}{I_{\text{IN}}} = -\frac{I_C}{I_E} = \frac{\beta}{1 + \beta}. \quad (4\text{-}25)$$

Continuing to let $\beta = 100$, we thus have $\alpha = 0.99$, which, as was pointed out initially, is very near unity. Thus, both dynamic and static current gain are given by α for the common-base stage, just as both were given by β for the common-emitter stage.

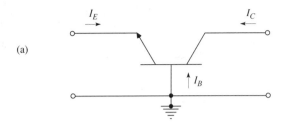

(a)

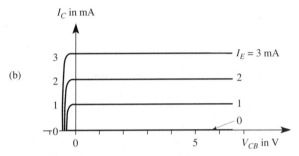

(b)

Figure 4-24 The common-base configuration. (a) Schematic diagram of the circuit. (b) The output plane and output characteristics.

Exercise 4-30. Verify the last statement.

Because $I_C = \alpha I_E$, in the common-base configuration, we can write $dI_C = \alpha dI_E$, or $\Delta I_C \approx \alpha \Delta I_E$, so that

$$\alpha \equiv \frac{dI_C}{dI_E} \approx \frac{\Delta I_C}{\Delta I_E}.$$

The common-emitter case involving β can be treated in the same way.

Exercise 4-31. Determine β as a function of α.

Since $\alpha = \beta/(1 + \beta)$, we have $\alpha(1 + \beta) = \beta$, or $\alpha = \beta(1 - \alpha)$. Hence, $\beta = \alpha/(1 - \alpha)$.

The output plane appears as shown in Figure 4-24(b) because the input current I_E is the parameter in this configuration. Since I_E is externally regulated and because $I_C \approx I_E$, we find that the output characteristics are very nearly horizontal. That is, dynamic output resistance is extremely high. Although the output characteristics closely resemble the common-emitter characteristics shown in Figures 4-16 and 4-21, except for a current gain of β in that case and a current gain of α in the present case, we shall see in Section 4-4 that the two cases are in fact quite different; Figures 4-16 and 4-21 are significantly idealized with respect to output resistance.

Exercise 4-32. Why do the output characteristics in Figure 4-24(b) have intercepts on the I_C axis very nearly equal to their corresponding values of I_E?

When $V_{CB} = 0$, the collector junction is obviously at zero bias, but as we have seen before, it is still a very effective collector. It continues to be effective until a forward bias (displayed in the second quadrant) of some 0.3 V or 0.4 V is reached.

Thus the common-base output characteristics are well matched to a high-resistance load. The loadline principles explained through Figure 4-16 are equally applicable here. When they are applied, it is evident that substantial voltage gain is achievable, in combination with near-unity current gain.

Exercise 4-33. Comment on the common-base configuration as a resistance transformer.

The extremely high dynamic output resistance just noted is accompanied by a low dynamic input resistance, which is essentially that of a forward-biased junction (Section 3-5.1). For example, at the modest current $I_E = -2.5$ mA, dynamic input resistance is about 10 ohm. Thus this configuration is a very effective low-to-high resistance transformer. This property and the gain properties already described make it almost a perfect complement to the common-collector configuration.

4-4 STRUCTURES AND PROPERTIES OF REAL BJTS

Actual BJTs have structural features that depart from the idealized versions employed for analysis thus far. Also, the real device exhibits parasitic properties, such as nonzero collector leakage current, nonzero ohmic base resistance, and nonzero $V_{CE(SAT)}$. For engineering calculations we can often safely ignore one or more of these parasitic properties, but in other cases they are crucial. Knowing which case one faces is probably part of the art of engineering, learned mainly through experience.

The significance of the adjective *real* in the present context is clear enough, but finding an unequivocal antonym is more difficult. The word "ideal" has at least two meanings in device work. It can mean adherence to a simple theoretical model, or it can signify the absence of parasitic effects and the existence of simple structures. To maintain a distinction, let us use the term *idealized* to convey the latter meaning. We can exemplify the two terms by pointing to the reverse characteristic of a germanium junction as *ideal;* it saturates cleanly and is well described by simple diffusion theory. On the other hand, the reverse characteristic of a silicon junction with a leakage current several orders of magnitude smaller constitutes a closer approach to that of an *idealized* junction, in which reverse current is small, negligible, or absent altogether.

The evolution and refinement of fabrication technology has improved some features of real devices, with controlled surface properties being one example, along with several surface-related phenomena. Also, circuit ingenuity permitted designers to live with other vexing parasitic properties. An example here is the large, temperature-dependent, and β-multiplied collector current exhibited by a germanium BJT in "cutoff." Resistor-biasing circuits for discrete BJTs dealt with the problem effectively. Those solutions are essentially obsolete, however, in spite of their cleverness. The advent of silicon technology eliminated most of the problem and the advent of integrated circuits brought alternatives to resistor biasing.

At least one other response on the part of the designer has been evoked by parasitic realities. Device structures have been modified to diminish one parasitic effect or another. In most cases, however, a price must be paid for the improvement; one or more other properties are worsened as a result. Nonetheless, such device modifications are instructive, and we shall look at several of them.

4-4.1 Electrochemical Potential

To address the first real-BJT problem it is convenient (though not obligatory) to introduce an additional tool, a concept related to that of the Fermi level. As Chapter 2 stresses, Fermi level is an equilibrium concept. Equilibrium carrier densities can be expressed succinctly by choosing the Fermi level as reference (or zero) for potential, and letting the midgap level be the "indicator" of electrostatic potential ψ, with Equations 2-33 and 2-34 being the result. But these equations can be rewritten as

$$n = n_i \exp \frac{q(\psi - \phi)}{kT}, \qquad (4\text{-}26)$$

and

$$p = n_i \exp \frac{q(\phi - \psi)}{kT}. \qquad (4\text{-}27)$$

By retaining the term ϕ in each exponent we have achieved a more general result; that is, the potential reference can be placed arbitrarily at some other level, so that ϕ as well as ψ is finite. Setting $\phi = 0$ causes these equations to revert to Equations 2-33 and 2-34.

We wish now to treat a nonequilibrium problem wherein excess carriers are present. Our approach will be to *define* a potential known as a *quasi* Fermi level. If the true Fermi level were positioned at this new potential, a conventional equilibrium computation would yield a carrier density, let us say, for electrons, equal to the nonequilibrium density of electrons actually present. Let us call the new potential ϕ_n, the *quasi Fermi level for electrons*. In a similar way, it is possible to define an additional new potential ϕ_p, the *quasi Fermi level for holes,* to be used in a completely analogous fashion. Using these definitions, then, we can write the *nonequilibrium* carrier densities as

$$n = n_i \exp \frac{q(\psi - \phi_n)}{kT}, \tag{4-28}$$

and

$$p = n_i \exp \frac{q(\phi_p - \psi)}{kT}. \tag{4-29}$$

These two expressions are algebraic statements of the verbal descriptions given earlier for ϕ_n and ϕ_p. Solving each for the quasi Fermi level then provides the desired definitions:

$$\phi_n \equiv \psi - \frac{kT}{q} \ln \frac{n}{n_i}; \tag{4-30}$$

$$\phi_p \equiv \psi + \frac{kT}{q} \ln \frac{p}{n_i}. \tag{4-31}$$

The two quantities so defined are sometimes labeled more briefly as the "Fermi level for electrons," and the "Fermi level for holes," respectively.

In each of Equations 4-30 and 4-31, the first term on the right-hand side is of course electrostatic potential. The last term, concerning a particle density, is known to the physical chemist as a *chemical potential*. Consequently, a quasi Fermi level is identically an entity known as an *electrochemical potential*. A graphical interpretation of Equations 4-30 and 4-31 in a band diagram aids understanding. Let us consider a sample not at equilibrium, and to simplify the problem, let us once again eliminate time as a variable by taking a steady-state example, and eliminate position as a variable by considering a thin sample under penetrating irradiation. This device was used repeatedly in Chapter 2. Figure 4-25(a) represents the nonequilibrium sample, arbitrarily taken to be *N*-type, and Figure 4-25(b) is its band diagram. We shall use the true (or equilibrium) Fermi level ϕ as reference and continue to use the midgap potential as the indicator of electrostatic potential ψ. Thus the arrow nearest the center of Figure 4-25(b) is the graphical representation of ψ.

Next, we introduce the left and right arrows in Figure 4-25(b), representing the electron and hole chemical potentials, respectively. Direction is assigned according to the corresponding algebraic signs in Equations 4-30 and 4-31, recalling that the positive direction is downward. The condition represented in Figure 4-25(b) is a high-level condition.

STRUCTURES AND PROPERTIES OF REAL BJTS   537

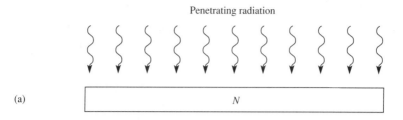

(a)

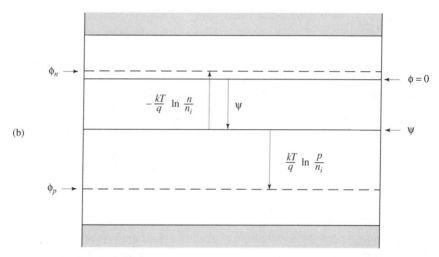

(b)

Figure 4-25 Quasi Fermi levels. (a) Physical representation of a thin N-type sample subjected to penetrating irradiation. (b) Band diagram for the sample in a uniform, high-level, nonequilibrium condition; the majority-carrier quasi Fermi level ϕ_n departs only slightly from the true Fermi level ϕ, while ϕ_p departs substantially from it.

Exercise 4-34. Why are the chemical-potential arrows in Figure 4-25(b) roughly equal in magnitude and opposite in sign?

In high-level conditions, $n \approx p$, or at most the two quantities differ by a small factor. Since the chemical potential involves the logarithm of a density, the two corresponding arrows approximate each other in length. The arrows have opposite senses, first, because holes and electrons have opposite charges, and second, because $(p/n_i) > 1$ in this high-level case. For $(p/n_i) < 1$, the logarithm would be negative, and the arrow representing hole chemical potential would point upward.

Exercise 4-35. Why is the electrostatic potential of the sample in Figure 4-25(a) not altered by its state of nonequilibrium?

The sample was neutral in its equilibrium state and retains its neutrality in its nonequilibrium state; excess carriers are produced in pairs. If we had charged the sample by directing an electron beam at it, then its potential would have changed.

Notice in Figure 4-25(b) that even for this high-level case, the majority-carrier quasi Fermi level departs only slightly from the equilibrium (true) Fermi level ϕ. The minority-carrier quasi Fermi level ϕ_p, however, departs widely from ϕ, a consequence of the hole-density change through many orders of magnitude in transition from the equilibrium state to the nonequilibrium state. From these observations it follows that inspection of ϕ_n position in this sample informs us immediately on the matter of whether we have a high-level or a low-level disturbance. In the low-level case, ϕ_n deviates negligibly from ϕ.

The power and significance of the electrochemical-potential concept can be illustrated as follows: Recall that there are two carrier-transport mechanisms, one driven by a gradient in electrostatic potential, and the other, by a gradient in carrier density. But a *single* expression involving the gradient of electrochemical potential comprehends *both* transport mechanisms. To illustrate this point, consider a general one-dimensional case involving holes. Let p, ψ, and ϕ_p all be functions of x. (We shall avoid functional notation, however, for the sake of simplicity.) First, write an expression having the form of the drift-transport equation. Second, remove $E = -d\psi/dx$ from the equation and replace it by $(-d\phi_p/dx)$. This yields

$$J_p = q\mu_p p(-d\phi_p/dx). \tag{4-32}$$

Then use Equation 4-31 to evaluate the derivative, which gives us

$$J_p = q\mu_p p\left(-\frac{d\psi}{dx}\right) - q\mu_p p \frac{kT}{q} \frac{n_i}{p} \frac{1}{n_i} \frac{dp}{dx}. \tag{4-33}$$

Simplifying and using the Einstein relation in the last term then produces

$$J_p = q\mu_p p E - qD_p(dp/dx), \tag{4-34}$$

which is identical to Equation 2-56, the hole-transport equation.

It was Shockley who introduced the electrochemical potential into the analysis of semiconductor problems. In his classic treatment of the *PN* junction [16], he renamed it a "nonequilibrium quasi Fermi level." Subsequently he adopted the shorter label "imref," which is *Fermi* spelled backwards—a bit of whimsy suggested by Fermi. While the quasi Fermi level is a powerful tool in treating certain junction problems, one can derive the equation for a junction's *I–V* characteristic without it, as Section 3-4.1 demonstrates. The assumption of Boltzmann quasiequilibrium that was employed there, however, has quasi-Fermi-level implications, as Section 4-4.2 explains.

Exercise 4-36. Devise a graphical means involving the Fermi-Dirac function for displaying the meaning of the quasi Fermi levels.

"Split" the function at its center of symmetry, attaching its upper branch to ϕ_n and its lower branch to ϕ_p.

4-4.2 Nonuniform Base-Region Doping

Base-region conditions are central to BJT operation, literally and figuratively. A consistent assumption in the foregoing analyses has been uniform doping in the base region, a condition that is rarely met in a real BJT. Figure 4-26(a) shows the net-

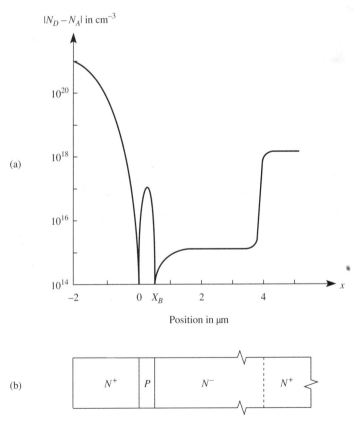

Figure 4-26 Net-doping profile in a representative BJT. (a) Semilog presentation of net-doping magnitude as a function of position. (b) Physical representation of BJT structure that relates to the profile diagram.

doping profile of a reasonably representative small BJT, whether discrete or integrated, and Figure 4-26(b) gives the corresponding physical representation. The distinctive profiles seen in the emitter and base regions are a consequence of using the solid-phase-diffusion methods discussed in Section 3-5.3 to dope these two regions. Thus, the base region in particular is far from uniformly doped. Curiously, the impact of this seemingly major departure from the idealized base regions treated heretofore is smaller than one might expect.

Exercise 4-37. Inspecting the semilog base-region profile in Figure 4-26(a) in the neighborhood of $x = X_B/2$, comment on how it would appear in a linear plot.

Insofar as this portion of the semilog profile can be approximated by a straight line, the actual profile is a declining exponential function, concave up.

Moll and Ross were first to examine the case of doping variations in the base region [17]. They treated a *PNP* device, but we shall convert to the *NPN* case for consistency with foregoing analysis. Their assumptions were as follows: (1) The BJT geometry is one-dimensional. This is highly realistic with respect to doping variations and continues to be realistic with respect to the minority-carrier current on which their analysis focused. (2) The net-doping profile $N_{AB}(x)$ is arbitrary, but is sufficiently free of abrupt changes to justify the relationship $p_B(x) \approx N_{AB}(x)$. The base region is thus quasineutral. (A quasineutral situation similar to this was previously encountered in Problem D2-2.) (3) Significant electron drift is permitted. Recall that so little charge is required to produce a significant electric field that this assumption does not violate the assumption of quasineutrality. (4) Diffusion transport of electrons occurs in the base region. In view of the statistical nature of the diffusion mechanism, an electron must on average experience several interactions with ions and (or) phonons in traversing the base region. (5) Base-region electron mobility (or diffusivity) can be treated as constant. Literal application of this assumption would enforce a maximum base-region doping in the neighborhood of $10^{15}/\text{cm}^3$, which would be a severe restriction of the analysis. However, adopting a constant average mobility is a justifiable approximation. (6) The entire base region is nondegenerate. Boltzmann statistics can be validly used. (7) The condition $n_B(x) \ll p_B(x)$ holds throughout the base region. An alternate statement is $n_B(x) \ll N_{AB}(x)$ everywhere. (8) Carrier recombination in the base region is negligible. We have taken the last condition as a hallmark of a "modern" BJT.

Exercise 4-38. Make quantitative observations on the minimum allowed base thickness X_B imposed by assumption (4).

Assume that the base region is uncompensated and take a large but nondegenerate doping value, $N_{AB} = 10^{18}/cm^3$. Ionic scattering dominates here. The mean spacing of impurity atoms can be approximated as

$$(N_{AB})^{-1/3} = (10^{-18} \text{ cm}^3)^{1/3} = 10^{-6} \text{ cm} = 0.01 \ \mu m = 100 \text{ angstroms}.$$

If we further assume that electron mean free path approximates mean impurity-atom spacing, then an electron passing through a base region with $X_B = 0.5 \ \mu m$ would experience fifty interactions if it traveled in a nearly straight line. For the case of $N_{AB} = 10^{15}/cm^3$, the corresponding number is five interactions. But the electron's path is erratic, so the number of interactions is much larger in each case. Let us estimate that the base region can be as thin as a few tenths of a micrometer.

Exercise 4-39. What is another way of stating assumption (7)?

Low-level conditions in the base region are assumed.

Now, modify Equation 4-29 to allow variations as a function of x, and write for the majority carriers in the base region

$$p_B(x) = n_i \exp \frac{q[\phi_p(x) - \psi(x)]}{kT} \approx N_{AB}(x). \tag{4-35}$$

But since low-level conditions have been assumed,

$$\phi_p(x) \approx \phi = \text{constant} \tag{4-36}$$

in the base region. Hence

$$\phi - \psi(x) = \frac{kT}{q} \ln \frac{N_{AB}(x)}{n_i} = \frac{kT}{q} \ln N_{AB}(x) - \frac{kT}{q} \ln n_i. \tag{4-37}$$

Dropping functional notation for simplicity once again, we obtain by differentiating Equation 4-37

$$E = -\frac{d\psi}{dx} = \frac{kT}{q} \frac{1}{N_{AB}} \frac{dN_{AB}}{dx}. \tag{4-38}$$

The transport equation for minority carriers is

$$J_n = q\mu_{nB}n_B E + qD_{nB} \frac{dn_B}{dx}. \tag{4-39}$$

Replacing E in this equation by the right-hand expression in Equation (4-38) then gives us

$$J_n = qn_B\mu_{nB} \frac{kT}{q} \frac{1}{N_{AB}} \frac{dN_{AB}}{dx} + qD_{nB} \frac{dn_B}{dx}, \tag{4-40}$$

and this becomes after applying the Einstein relation,

$$J_n = qD_{nB}\left[\frac{n_B}{N_{AB}}\frac{dN_{AB}}{dx} + \frac{dn_B}{dx}\right]. \tag{4-41}$$

But the right-hand side can be converted into a perfect differential by means of a rearrangement:

$$\frac{J_n N_{AB} dx}{qD_{nB}} = n_B dN_{AB} + N_{AB} dn_B = d(n_B N_{AB}). \tag{4-42}$$

Now restore the functional notation and integrate from the left-hand boundary of the base region at $x = 0$, to its right-hand boundary at $x = X_B$:

$$\frac{J_n}{qD_{nB}}\int_0^{X_B} N_{AB}(x)dx = \Big[n_B(x)N_{AB}(x)\Big]_0^{X_B}. \tag{4-43}$$

Assuming forward-active operation, with the collector reverse-biased, we can take $n(X_B) \approx 0$, so that Equation 4-43 becomes

$$\frac{J_n}{qD_{nB}}\int_0^{X_B} N_{AB}(x)dx = -n_B(0)N_{AB}(0). \tag{4-44}$$

Thus,

$$\frac{J_n}{qD_{nB}N_{AB}(0)}\int_0^{X_B} N_{AB}(x)dx = -n_B(0), \tag{4-45}$$

and hence,

$$\frac{J_n}{qD_{nB}N_{AB}(0)}\int_0^{X_B} N_{AB}(x)dx = -\frac{n_i^2}{N_{AB}(0)}\exp\left(\frac{qV_{BE}}{kT}\right). \tag{4-46}$$

Evidently $N_{AB}(0)$ can be dropped from both sides. Solving the resulting expression for J_n yields finally

$$J_n = -\frac{qD_{nB}n_i^2 \exp(qV_{BE}/kT)}{\int_0^{X_B} N_{AB}(x)dx}. \tag{4-47}$$

Here we have an expression for electron current density in the base region that is a function of areal net-doping density (the integral factor that is the denominator). But the expression is *independent of the base-region boundary value*, which has cancelled out. In view of the emphasis we have placed on such boundary values in preceding analyses, this is a striking result.

Exercise 4-40. Equation 4-36 states that

$$\phi_p(x) \approx \text{constant}.$$

What does that say about the value of the net one-dimensional majority-carrier current density $J_p(x)$?

Assuming $\phi_p(x) \approx$ constant is the same as assuming

$$\frac{d\phi_p(x)}{dx} \approx 0,$$

and according to Equation 4-32, this amounts to assuming that $J_p(x) \approx 0$.

Exercise 4-41. If $J_p(x) \approx 0$ in the base region, the majority holes must somehow be "confined." What causes their confinement?

The huge negative electric field in the collector junction pushes holes back into the base region, where they are majority carriers. There is a current of holes from the base region into the emitter region, accounting for base current. But this current (1) is small compared to one-dimensional electron current, and (2) constitutes an x-directed one-dimensional hole current only near the emitter boundary of the base region; that is, through most of the base region, the y-directed component (lateral component) of this small current exceeds the x-directed component.

Exercise 4-42. Accepting the statement that $J_p(x) \approx 0$ provides an alternative way to arrive at Equation 4-38. What is it?

Assuming that $J_p(x) = 0$, then

$$q\mu_{pB} p_B E = qD_{pB}(dp_B/dx).$$

Or,

$$E = \frac{D_{pB}}{\mu_{pB}} \frac{1}{p_B} \frac{dp_B}{dx} = \frac{kT}{q} \frac{1}{p_B} \frac{dp_B}{dx}.$$

But according to assumption (2), $p_B(x) \approx N_{AB}(x)$, so that

$$E = \frac{kT}{q} \frac{1}{N_{AB}} \frac{dN_{AB}}{dx}.$$

The case of majority holes in the BJT base region provides a good opportunity to illustrate certain principles related to the Boltzmann-quasiequilibrium assumption. This assumption was introduced in Section 3-3.3 for junction analysis, and was mentioned again in the same connection in Section 4-4.1. Under equilibrium conditions, the Fermi level ϕ is strictly constant, which is to say, x-independent. For the valid assumption of Boltzmann quasiequilibrium, the relevant quasi Fermi level (ϕ_p in the present case) must be approximately x-independent. As is shown by Equations 4-32 through 4-34, and in Exercise 4-40, this requirement is typically met in the base region of the one-dimensional BJT under bias.

In the case of a *PN* junction, the condition of nearly constant quasi Fermi levels holds typically for *both* carriers under all forward biases except extremely small values (a few times the thermal voltage kT/q). The near constancy needed for valid use of this important assumption need only exist from boundary to boundary in the junction, often a very short distance in a forward-biased junction. Another description of the necessary condition is that the net current of the carrier in question should be small. The equivalence of this statement to the previous statement concerning quasi Fermi level was the point of Exercise 4-40 for the case of majority holes in the base region. The adjective "small," however, requires us once again to answer the question, "Small compared to what?" In the present case, the answer is that net current must be small compared to the countervailing drift and diffusion components that add algebraically to yield the net current. For hole current in the base region, this point was emphasized in Equation 4-36 and in Exercises 4-40 and 4-41. For the net hole and electron currents passing through a forward-biased junction, the same point was made in Section 3-3.3. Also, Figure 3-4(c) shows the nature of the four enormous drift and diffusion components of current that exist in a *PN* junction.

4-4.3 The Gummel Number

It was noticed by Gummel [18] that Equation 4-47 permits one to evaluate its integral factor, the areal net-doping density in the base region, through a straightforward electrical measurement. Multiplying Equation 4-47 through by emitter area A_E yields

$$I_C = \frac{A_E q D_{nB} n_i^2}{\int_0^{X_B} N_{AB}(x)dx} \exp\left(\frac{qV_{BE}}{kT}\right) = I_S \exp\left(\frac{qV_{BE}}{kT}\right). \tag{4-48}$$

He wished to evaluate I_S experimentally, but at low values of current so that complications such as series resistance (ohmic base resistance, for example) could be ignored. But under low-current conditions, phenomena such as current generation in the space-charge region of the reverse-biased collector junction become nonnegligible, and so he made an empirical adjustment of Equation 4-48:

$$I_C = I_S \exp\left(\frac{qV_{BE}}{kT}\right) + I_L. \tag{4-49}$$

The added term I_L is a leakage term. By plotting experimental values of $(I_C - I_L)$ versus V_{BE} semilogarithmically, he obtained a straight line, as is shown in Figure 4-27. Consequently, to evaluate I_S, he simply extrapolated the straight line to $V_{BE} = 0$, accounting for the designation of I_S as *intercept current*, a tiny current that typically amounts to about 10^{-14} A. Figure 4-27 is sometimes described as a *Gummel plot*. Since Equation 4-48 defines this current as

$$I_S = \frac{A_E q D_{nB} n_i^2}{\int_0^{X_B} N_{AB}(x)dx}, \tag{4-50}$$

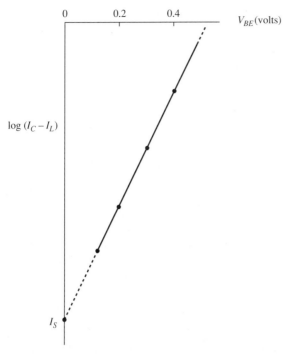

Figure 4-27 A Gummel plot. The intercept current I_S, determined from low-voltage measurements of injected-electron current, leads to evaluation of the areal net-doping density in the base region.

he was then able to evaluate the integral using the experimental value of I_S by writing

$$\frac{qA_E D_{nB} n_i^2}{I_S} = \int_0^{X_B} N_{AB}(x)dx \equiv G_B. \quad (4\text{-}51)$$

The areal net-doping density of a BJT base region has become known as the *Gummel number*, G_B.

Exercise 4-43. The leakage current of a reverse-biased silicon junction is voltage-dependent. Doesn't that complicate the "I_L correction" described after Equation 4-49?

No. Collector voltage is held constant while Gummel-plot data are observed, so I_L is constant.

Exercise 4-44. In executing a Gummel plot, how can one afford to ignore the current of recombining carriers in the emitter space-charge region, the SNS phenomenon responsible for the "bump" at low voltage that appears in the diode characteristic shown in Figure 3-24? As was noted, low-voltage data are desired for such a plot.

Recombination in the emitter space-charge region affects I_B, but does not affect I_C, which is fixed only by the *injected* electrons. For a Gummel plot, one requires data on I_C versus V_{BE}, not I_B versus V_{BE}.

Exercise 4-45. Since a BJT structure in effect permits one to "discard" the nonideal SNS component of current passing through the emitter junction, it affords a way to make a direct observation of the diffusion or "ideal" component of emitter-junction current. Hence, doesn't I_S have the same significance as I_0 for a single-junction characteristic, such as that in Figure 3-24?

The current I_S has for the emitter junction identically the meaning assigned to the current I_0 for the junction characterized in Figure 3-24. Both are ideal-theory saturation currents (as the symbol I_S suggests). Precedent exists for using the symbol I_S for the BJT intercept current, however, and the symbol I_0 for the saturation current in a sample containing a single junction.

Exercise 4-46. If the emitter junction were truly isolated, how would the ideal component of its I–V characteristic differ from that in the Gummel plot?

The straight line in the Gummel plot would be appreciably less steep, because $L_n(\gg X_B)$ enters the denominator of the expression for the current at a particular voltage in that case. However, both straight lines would extrapolate to $I_S = I_0$.

Typical values of the Gummel number range from about 10^{12} to 10^{13} net impurity atoms per square centimeter of base region, viewing the base region in a direction normal to the junctions. Current gain β varies inversely with Gummel number. We shall see in Section 4-6.1 that the Gummel number facilitates an economical BJT model.

4-4.4 Breakdown Voltage

The common-base configuration of Figure 4-24 is simplest to analyze with respect to breakdown voltage. The reason is that the bias voltages V_{EB} and V_{CB} each affect a single junction, with little complication caused by the presence of the other in many cases. That is, in this circuit configuration one can consider a given junction to be *isolated* with respect to its breakdown-voltage properties. Let us take the emitter junction first. The symbol for its breakdown voltage is BV_{EBO}, which is meant to convey *breakdown voltage from emitter terminal to base terminal with the collector terminal open*. The point of specifying an open (floating) collector terminal is to minimize any effect the nearby collector junction could have on emitter-junction properties.

Let us consider the most common BJT structure, one having emitter and base regions formed by solid-phase diffusion. Figures 4-2 and 4-26 depict such devices. Applying a reverse voltage ($V_{EB} > 0$) to the emitter junction, we typically observe $BV_{EBO} \approx 6$ V. This low value, of course, is primarily a consequence of the heavy doping on both sides of the emitter junction.

Exercise 4-47. Check the consistency of the observation $BV_{EBO} \approx 6$ V with the prediction of Figure 3-35 on breakdown voltage, and the calculation of Exercise 4-5 on depletion-layer thickness in the emitter junction. Figure 3-30 points out that a diffused junction makes a transition from linearly graded behavior to step behavior as reverse voltage is increased. For a typical emitter junction, the entire allowed range of reverse-bias voltages is well within the range of linearly graded behavior [19].

A linearly graded junction is symmetric. Therefore the magnitudes of the net-doping densities at the two depletion-layer boundaries are the same. (In a diffused junction they are not the same, but their geometric mean very nearly equals the value in a truly symmetric linearly graded junction of equivalent properties [20].) Using Figure 4-26, let us estimate this common value of net doping as $N = 10^{18}/cm^3$. This yields for the contact potential

$$\Delta\psi_0 = 2(kT/q) \ln(N/n_i) = 0.95 \text{ V}.$$

This is in reasonable agreement with the value $\Delta\psi_0 = 1.0$ V calculated in Exercise 4-5, which employed an idealized step-junction model of the BJT designed to give properties approximating those of a real BJT. According to Equation 3-48, for a linearly graded junction,

$$X = X_0(\Delta\psi/\Delta\psi_0)^{1/3}.$$

At the breakdown voltage observed for this junction,

$$\Delta\psi = V_B + \Delta\psi_0 \approx (6 \text{ V}) + (1 \text{ V}) = 7 \text{ V}.$$

If we accept the value of depletion-layer thickness at equilibrium also calculated in Exercise 4-5, namely, $X = 0.11$ μm, then we can calculate the value of emitter-junction depletion-layer thickness at breakdown to be

$$X = (0.11 \text{ } \mu\text{m})(7 \text{ V}/1 \text{ V})^{1/3} = 0.21 \text{ } \mu\text{m}.$$

Now refer to Figure 3-35, which relates breakdown voltage to depletion-layer thickness at breakdown. While a measure of extrapolation is required, it is evident that the values just calculated display reasonable internal consistency.

This relatively low value of BV_{EBO} (~6 V) is common to integrated BJTs and to small discrete BJTs as well. It has sometimes been exploited for voltage regulation, especially in integrated circuits, complementing the series assembly of forward-biased diodes. Interestingly, it is the emitter junction of an integrated BJT that shows the best overall properties as a general-purpose diode in an integrated circuit [21]. For this application the collector junction is short-circuited. A general-purpose diode must be isolated from other components, and that is true of the emitter-base diode. Besides, when other BJTs are needed in the circuit, using a BJT as a diode costs less than creating an *ad hoc* structure for the purpose. Further, significant excess-carrier storage occurs only in the base region when the emitter junction is forward-biased, and complications connected with the junction that isolates the BJT are avoided as well. Even if it is to be used as a "blocking" or reverse-biased diode, its low value of BV_{EBO} is often not of concern because the maximum voltage encountered in many bipolar integrated circuits is $V_{CC} = 5$ V.

Now turn to the collector junction of a BJT in a common-base configuration. Its breakdown voltage is designated BV_{CBO}, to be interpreted as *breakdown voltage from collector terminal to base terminal with emitter terminal open*. Let us assume that the Gummel number of the BJT in question is large enough so that

$$BV_{CBO} < V_{RT} \tag{4-52}$$

for the collector junction, where V_{RT} is the reachthrough voltage from collector to emitter. That is, the voltage for collector-junction breakdown (usually avalanche breakdown) is smaller than that at which the collector-junction depletion layer will expand to touch the equilibrium depletion layer of the emitter junction. This phenomenon is treated in Section 3-6.3. The condition of Equation 4-52 is assured by a sufficient areal net doping in the base region, identically the Gummel number.

Exercise 4-48. How is the same condition stated for the idealized step-junction BJTs of Sections 1-1 through 1-3?

The reach-through condition can be avoided by a sufficient product of $N_{AB}X_B = G_B$.

With reachthrough ruled out, observation of BV_{CBO} will inform us of the inherent breakdown voltage of the collector junction, unaffected by the presence of the emitter junction. However, the value observed will be smaller than that inferred from diagrams such as Figures 3-34 and 3-35. As is pointed out in Section 3-6.1, those data were collected for *plane* junctions. The diffused-base BJT normally has a curved region in its collector junction, emphasized by a dashed circle in Figure 4-28 (and also evident in Figure 4-2). As also noted earlier, local electric-field enhancement exists in the curved region, readily visualized in terms of the "crowding" of lines of force, and hence the breakdown voltage of such a region is lower than that of the plane region. These two distinct breakdown-voltage values are associated uniquely with the two regions of the BJT often designated *intrinsic* (near the emitter junction) and *extrinsic* (peripheral, or away from the emitter junction). The two regions are so labeled in Figure 4-28. Hence, BV_{CBO} in the device being considered characterizes the extrinsic portion of the device. Let us designate the higher breakdown voltage associated with the plane or intrinsic portion of the collector junction as BV'_{CBO}. It enters into analytical considerations later.

Exercise 4-49. Why is the effect of junction curvature on BV_{EBO} appreciably less significant than its effect on BV_{CBO}? (We did not even consider curvature in Exercise 4-47.)

From Figures 4-2 and 4-28 in combination with Exercise 4-47, we see that for the emitter junction, depletion-layer thickness at breakdown is roughly one tenth of the radius of curvature of the cylindrical portion of the junction. Hence a plane approximation to the emitter junction is justified. On the other hand, from Table 4-1 or Figure 4-26 in combination with Figure 3-45, we infer a value of BV'_{CBO} to be in the neighborhood of 100 V. Then from Figure 3-46 we learn that the depletion-layer thickness of such a junction at breakdown amounts to several micrometers, exceeding the radius of curvature of the collector junction. Consequently the perturbing effect of junction curvature upon electric-field values is substantial in the case of the collector junction.

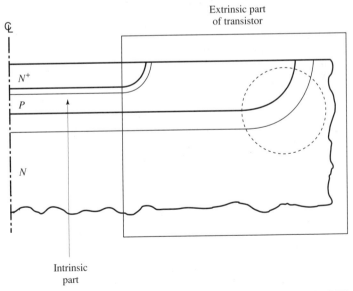

Figure 4-28 Identifying the intrinsic and extrinsic portions of the BJT, for treating breakdown properties.

There is, unfortunately, a significant opportunity for confusion over terminology at this point. The masked-diffusion procedure used to produce junctions like the two depicted in Figure 4-28, a procedure developed in the mid-1950s [22], was in the late 1950s designated the "planar process" after a small but crucial addition [23]. The intention of this choice of words was to convey the fact that the top surface of the device remains approximately a plane, or *planar,* through the fabrication process. By extension, junctions of this type have come to be called *planar junctions,* and the BJT embodying them, a *planar BJT.* This leaves us with the inconvenient common usage wherein a "planar" junction is not *planar.* As an interesting aside, we may note that the developers of the process decided not to seek trademark registration for the process name they had chosen [24]; they wanted the name to be universally used, a wish that was met to a degree far beyond expectations as their process (and patent licensing therefor) became the keystone of microelectronics. Had they sought trademark registration, they could not have used an accepted English word to designate it. They would have been required by statute to misspell it or otherwise "invent" a term; hence, wisdom can indeed reside in statutes.

Next let us turn to the case wherein

$$BV_{CBO} > V_{RT}.\tag{4-53}$$

Under such circumstances the emitter junction participates in the breakdown phenomenon, but the result is only slightly more complicated. Under such conditions,

$$BV_{CBO} \approx V_{RT} + BV_{EBO}.\tag{4-54}$$

Exercise 4-50. Explain Equation 4-54.

With the emitter terminal floating, the emitter-region voltage V_E will "float" at a voltage that is spaced away from the collector voltage V_C by an amount that is approximately V_{RT}. Continuing to increase V_C causes V_E eventually to reach a value that is BV_{EBO} above V_B, the reference voltage. At this point, breakdown in the sidewalls of the emitter junction effectively establishes a conducting path to the base terminal or "ground," and the punchthrough phenomenon establishes a path from emitter to collector. These two paths are in series, so that V_{CB} cannot be further increased without large currents. This fixes the value of BV_{CBO}.

Now we shall turn to the breakdown-voltage properties of the BJT in other circuit configurations. Start with the common-emitter case, Figure 4-3. The emitter junction is of course forward-biased for forward-active operation. But if it is reverse-biased (for extreme cutoff, for example), then BV_{EBO} is the maximum allowed reverse voltage at the input terminals of the BJT. The breakdown voltage at the output terminals, or BV_{CEO}, is both more complicated and more interesting. The applied voltage this time appears across the two junctions in series—as reverse bias on the collector junction and as forward bias on the emitter junction. Consequently, the situation inevitably is complicated by what we have called transistor action.

In the emitter-follower circuit, one directly controls V_{CB} through choice of the base-terminal voltage. But since the emitter region is inescapably lower in voltage than the base region by the amount $V_{BE} \approx 0.7$ V, then our choice of V_{CB} makes a *de facto* choice of V_{CE} in this case also. Thus BV_{CEO} is as relevant to common-collector operation as it is to common-emitter operation, provided operating limits are being approached. Having made this point, let us return to the common-emitter circuit for the balance of the discussion.

Now consider the intrinsic portion of the BJT shown in Figure 4-28. Let us define a common-base current gain α_{0F} as the value that would be observed with the effect of avalanche multiplication in the reverse-biased collector junction either absent or subtracted from the collector current. That is, α_{0F} characterizes a hypothetical device in which the multiplication process has been "turned off." Hence, for such a device (or for a real device with V_{CE} small), we may write

$$I_C = -\alpha_{0F} I_E. \tag{4-55}$$

Thus, to *include* the effect of avalanche multiplication, we may write for a real device

$$I_C = -\alpha_{0F} I_E M, \tag{4-56}$$

where M is the avalanche multiplication factor. Since all of the collector current passes through the collector junction, it is all subjected to avalanche multiplication. For the isolated junction treated in Section 3-6.1, we similarly defined M as the ratio of current actually passing through the junction to the current that would exist if the multiplication process were "turned off."

Since the sum of currents directed toward a node must vanish (according to Kirchhoff), we can write for the BJT

$$I_E = -(I_B + I_C), \tag{4-57}$$

and substituting this into Equation 4-56 yields

$$I_C = \frac{\alpha_{0F} I_B M}{1 - \alpha_{0F} M}. \tag{4-58}$$

Thus it follows that I_C increases without limit in the common-emitter BJT when

$$\alpha_{0F} M \to 1, \tag{4-59}$$

or when

$$M \to (1/\alpha_{0F}). \tag{4-60}$$

For a BJT of $\beta_F = 100$ and $\alpha_{0F} = 0.99$, we shall therefore observe breakdown when $M \to 1.01$. Contrast this with the case of the isolated junction, wherein avalanche breakdown occurs when $M \to$ infinity! (See Section 3-6.1.) It follows, therefore, that

$$BV_{CEO} < BV'_{CBO} \qquad (4\text{-}61)$$

by a significant amount. An empirical treatment of the ratio of these two voltages yields the expression

$$\frac{BV_{CEO}}{BV'_{CBO}} \approx (1 - \alpha_{0F})^{1/n}, \qquad (4\text{-}62)$$

where the value of n ranges from about 1 to 5 for silicon BJTs. Letting β_{0F} be the common-emitter current gain corresponding to α_{0F}, and since

$$(1 - \alpha_{0F}) = \frac{1}{1 + \beta_{0F}} \approx \frac{1}{\beta_{0F}}, \qquad (4\text{-}63)$$

we can also write

$$\frac{BV_{CEO}}{BV'_{CBO}} \approx \frac{1}{\beta_{0F}^{1/n}}. \qquad (4\text{-}64)$$

Letting $n = 5$, $\beta_{0F} = 100$, and $BV'_{CBO} = 100$ V, we have $BV_{CEO} \approx 40$ V.

Exercise 4-51. Why is BV'_{CBO} relevant to Equation 4-64, rather than BV_{CBO} (characterizing the extrinsic collector junction)?

Virtually all of the collector current is carried through the intrinsic portion of the collector junction, characterized by BV'_{CBO}. Hence this component of current is dominant, even though the value of M in the curved region is somewhat larger.

The I–V breakdown characteristics observed in a common-emitter BJT are illuminating. Three examples are shown in Figure 4-29. Look first at the left-hand curve, for $I_B > 0$. With base current of normal polarity being supplied, the emitter junction is forward-biased, supplying collector current. When the multiplication M in the collector junction reaches the modest value $1/\alpha_{0F}$, as given in Equation 4-60, collector current increases without limit. In this example, $M \to 1/\alpha_{0F}$ at a voltage substantially less than that required to cause $M \to \infty$ in the curved portion of the collector junction, or BV_{CBO}.

For the middle curve, $I_B = 0$. In other words, the base terminal is open. The emitter junction is forward-biased, but at low current it injects inefficiently; most of its forward current consists of carriers that recombine in its space-charge layer, thus not being injected. But the small current of injected carriers becomes multiplied by M starting well below BV_{CEO}, and as BV_{CEO} is approached, $M \to 1/\alpha_{0F}$ once more, and breakdown is observed.

In the last case, the emitter junction is slightly reverse-biased by imposing a

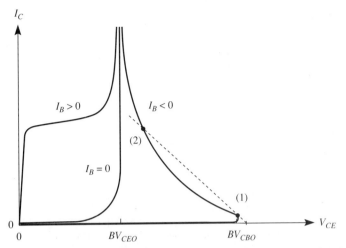

Figure 4-29 I–V characteristics of a common-emitter BJT entering breakdown. The dashed line is a portion of a loadline.

small negative base current. Hence the emitter junction does not initially inject electrons into the base region at all. Thus, the initial breakdown that will be observed occurs when $M \to \infty$ in the region of highest multiplication, the cylindrical region of the collector junction. The voltage so observed is $BV_{CBO} < BV'_{CBO}$. When multiplication increases the current sufficiently to supply the base current of holes flowing to the base terminal, the additional flow of holes into the base region is available to flow into the emitter region. In other words, in spite of the small negative base current, the emitter junction becomes forward-biased as the arriving holes make the base region more positive. When hole injection into the emitter junction has increased sufficiently to cause significant electron injection into the base region, transistor action comes into play. The concave-up right-hand curve in Figure 4-29, therefore, is the locus of points where (using functional notation),

$$\alpha_{0F}(I_C)M(V_{CB}) = 1. \tag{4-65}$$

That is, α_{0F} increases with I_C as diffusion current takes over from recombination current in the emitter junction. This forces a lower value of V_{CB} (and hence V_{CE}) in order to reduce M and maintain the product at unity. This curve of negative differential resistance is indeed a static characteristic that can be observed point by point in a loadline situation by using a load resistance that is sufficiently high. That is, it is not in any sense a "transient" characteristic. But if a lower value of load resistance is used, such as that represented by the dashed line in Figure 4-29, and V_{CE} is gradually increased, then the operating point can be seen to "snap" or switch from point 1 to point 2 as BV_{CBO} is exceeded slightly. The common value of voltage to which the several I–V characteristics in Figure 4-29 converge is sometimes termed the *sustaining voltage*.

4-4.5 Output Conductance

Idealized models for the BJT predict near-zero incremental output conductance g_o for the device in a common-emitter circuit and in forward-active operation. Figures 4-16, 4-17(b), 4-18(a), 4-19(a), 4-20(a), and 4-21 all imply this kind of model. It was just shown that breakdown effects cause a departure from this behavior—indeed, an extreme departure. Figure 4-29 displays $g_o \to \infty$ in the breakdown regime. And for $I_B < 0$ and $BV_{CEO} < V_{CE} < BV_{CBO}$, it even displays $g_o < 0$ and finite, because of the interplay of $\alpha_{0F}(I_C)$ and $M(V_{CB})$. For $I_B > 0$, g_o can be described as small only at low values of V_{CE}. (A quantitative examination of the M-to-g_o relationship is the subject of Problem A4-23.) Important as the breakdown phenomenon is, it is not the only phenomenon that is significant in fixing g_o.

It was stressed in Section 4-2.3 that the thickness of the effective (or active) base region is defined by space-charge-layer boundaries, one for the emitter junction and one for the collector junction. But the position of such a boundary is voltage-dependent. For the real BJT, this dependence is greater in the collector case than the emitter case. Additionally, the collector junction is normally subjected to much more widely varying voltage than the emitter junction, and so dominates in the effect now under scrutiny. The base thickness X_B is modulated by V_{CE} (or V_{CB}). And because β_F is inversely proportional to X_B, it follows that β_F is similarly dependent upon V_{CE}. The relationships involved here are illustrated in Figure 4-30. In Figure 4-30(a), the BJT is represented physically at a low collector-junction voltage, while Figure 4-30(b) portrays the accompanying base-region profile. To exaggerate the effect, we have chosen an N^+PN^+ structure, which causes depletion-layer expansion to occur mainly in the base region. A twentyfold increase in V_{CB}, then, reduces X_B by approximately a factor of two, as is shown in Figure 4-30(c). As a result, β_F when $V_{CB} = 20$ V is about twice β_F when $V_{CB} = 1$ V, as Figure 4-30(d) illustrates. Although the focus here is on V_{CB}, we note that V_{CE} differs from it by only a small and nearly constant V_{BE}. The explanation for this effect was first published by Early in 1952 and has become known as the *Early effect* in deference to his contribution [25].

Exercise 4-52. When breakdown effects are factored into Figure 4-30, what qualitative and quantitative changes occur in the I–V characteristics?

Consider the first quadrant first, and output voltage only in the range $0 < V_{CE} < BV_{CEO}$. All of the characteristics will curve upward as BV_{CEO} is approached, except those for very small I_B. The conductance at a given voltage increases with I_B because ΔI_C is proportional to MI_B.

In the third quadrant, the emitter junction is functioning as a quasi collector junction. Hence the low value, BV_{EBO}, becomes relevant. But β_R is so small in reverse operation that transistor action is minor. Thus breakdown will be observed at a voltage magnitude near that of BV_{EBO}, but typically this is still much smaller than that of BV_{CEO}.

The physical mechanism displayed in Figure 4-30 makes it clear that g_o should increase in proportion to I_B. That is,

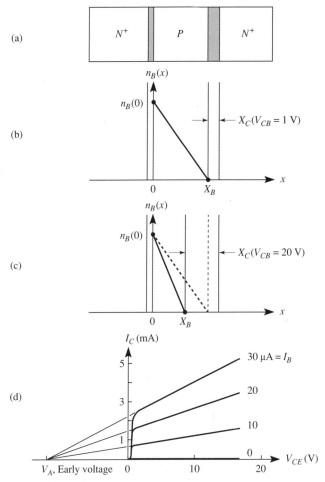

Figure 4-30 The consequence of bias-dependent base thickness, known as the Early effect. (a) Physical representation of an N^+PN^+ BJT with $V_{CB} = 1$ V. (b) Minority-carrier profile in the base region for $V_{CB} = 1$ V. (c) Minority-carrier profile in the base region for $V_{CB} = 20$ V, showing reduction of base thickness X_B by roughly a factor of 2. (d) Output characteristics showing the degraded output conductance that results, and extrapolation nearly to a single point.

$$\frac{g_2}{g_1} = \frac{\Delta I_{C2}(V_{CE}, I_{B2})/\Delta V_{CE}}{\Delta I_{C1}(V_{CE}, I_{B1})/\Delta V_{CE}} = \frac{I_{C2}(V_{CE})}{I_{C1}(V_{CE})} = \frac{I_{B2}}{I_{B1}}. \tag{4-66}$$

As a result, the near-linear portions of the I–V characteristics in Figure 4-30(d) extrapolate approximately to a common point on the V_{CE} axis, labeled V_A. (The analogous quantity for reverse operation of the BJT we shall call V_R.) When this fact was noted in 1970 [26], the voltage so defined was named the *Early voltage*. The smaller its magnitude, the larger the Early effect. The geometrical insight embodied in Figure 4-30(d) inspired substantial analytic examination of the Early effect in the ensuing years, involving about a dozen authors. Typical of these papers is a note by Van der Ziel [27].

The effects of breakdown, considered to be negligible in Figure 4-30(d), are of course superimposed on the Early effect. The left-hand characteristic in Figure 4-29 displays with reasonable fidelity the result of combining the two phenomena. The combination is a departure from idealized behavior—or a parasitic property—of considerable importance. Visualize a loadline in the output plane of a real BJT in the common-emitter configuration. There will be significant departures from the constant current gain β_F of the idealized device. And the effect of reducing incremental output resistance $r_o = 1/g_o$ is the same as the effect of reducing the loadline resistance R_C, in that voltage gain declines. Switching behavior is also affected, because the value of incremental output conductance g_o is a factor in switching speed. It is an interesting fact that base thickness, too, is modulated by V_{CB} in the common-base configuration, but there is *no* Early effect in this case. Consequently, Figure 4-24(b) is quite accurate with respect to g_o. The reasons for this behavior are examined in Problem A4-28. The Early effect by itself and in combination with the breakdown effect has inspired a number of BJT structural innovations designed to diminish the impact of these parasitic problems. Some will be examined in the next section.

4-4.6 Structural Variations

In the 1950s, semiconductor-device engineers were "technique-limited." A number of devices that were conceived and even analyzed in that era had to wait decades for realization. As new methods came along—particularly junction-forming methods—their new characteristics led to BJT structural variations. In later periods, on the other hand, with a more abundant range of process options available, a number of variations were introduced to diminish certain parasitic properties of the BJT.

The first practical BJT [4] strongly resembled the physical representation in Figure 4-1, as was noted in Section 4-1.1. These were germanium grown-junction devices, but the first practical silicon BJTs also employed grown junctions and assumed a similar form [28, 29]. In these early products, the parallelepiped that became the BJT was literally cut out of a single-crystal ingot in which the emitter and collector junctions had been created during the crystal-growth process itself.

By the mid-1950s, the dominant BJT was a germanium alloyed-junction device [6], based upon a junction-forming method developed two years earlier [30]. A "milliwatt" embodiment of the alloyed-junction BJT is represented in Figure 4-31. The essential departure here was the use of indium, a *P*-type impurity with a high solid solubility in germanium, to dope the emitter and collector regions. Pellets of metallic indium were placed in contact with an *N*-type germanium "die" (singular of *dice*), or thin parallelepiped, and this assembly, contained in a suitable "jig," was carried through a carefully designed temperature–time program. The low-melting indium formed "puddles" that dissolved germanium in a *eutectic* phase during the rising-temperature portion of the program, a phase that exhibits a lower melting point than that of either constituent. During a slow cooling interval, germanium would regrow on both sides of the parent crystal, now the base region of the eventual BJT. The regrown germanium on the two faces of the die contained heavy indium doping, and thus constituted P^+ regions, as are indicated in Figure 4-31.

A major advance in this technology occurred when it was noted that crystallographically plane junctions could be achieved by using a starting crystal of (111)

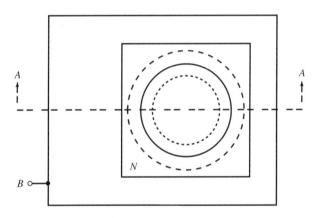

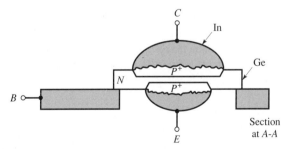

Figure 4-31 Top and cross-sectional views of a germanium alloyed-junction milliwatt BJT of the 1950s.

orientation [31]. This preferential penetration into the single crystal takes place for reasons emphasized in Section 1-5.6, and the crucial planarity of these junctions is emphasized in Figure 4-31. With this refinement, and in a high-power copper package, the alloyed-junction *PNP* BJT survived for decades as a low-cost power BJT of impressive performance. An advantage of the alloyed-junction BJT, shared by no others before or since, is that the emitter terminal can easily be attached directly to the package part that serves as heat sink, thus adapting the resulting BJT perfectly to the popular common-emitter configuration.

Another feature of the alloyed-junction technology was that true bilateral (emitter–collector) symmetry could be achieved by using indium pellets of equal area, a feature that was advantageous for some purposes. Another, and negative, feature was the fact that collector-junction depletion-layer expansion with increasing V_{CB} occurs almost totally in the base region. Thus, this P^+NP^+ BJT shares the exaggerated Early-effect properties of the hypothetical N^+PN^+ device depicted in Figure 4-30(a).

After elucidating the phenomenon responsible for degraded output conductance, James Early proposed a remedy. His structure was the P^+NIP^+ BJT [32], influenced by the technology of the time, and subsequently designated the *intrinsic-barrier transistor*. His thought was that giving the collector-junction depletion layer ample room to expand in the direction of the collector region would diminish its effect on the active base thickness X_B. Prototypes were realized with the scarce technological options of the time, and these comparatively primitive devices verified the concept [33]. Arrival of the diffused-base BJT [14, 15], with its relatively thin, heavily doped base region, brought similar advantages, and diffusion quickly became preferred technology.

The next major step in the same direction was the *epitaxial BJT* of 1960 [9], strategically combining technologies available by that time. This device (see Figure 4-2), incorporated an optimized collector structure that increased breakdown voltage, diminished Early effect, and most important for switching applications, reduced $V_{CE(SAT)}$. Figure 4-32 illustrates the improvement in $V_{CE(SAT)}$ that this innovation brought, even in its first application.

We shall return to the "planar-epitaxial" BJT in a moment, but now turn to another structural variation, the *lock-layer transistor* [34, 35]. It placed a buried layer in the collector body to inhibit expansion of the collector-junction space-charge layer. This led to a further variation [36], and generalization of the principle in a range of structures termed generically *channel-collector transistors* [37]. These, in effect, are a BJT and a junction field-effect transistor (JFET) in a *cascode* configuration, as is shown schematically in Figure 4-33(a). It does indeed virtually eliminate the Early effect, and yields $BV_{CEO} \approx BV_{CBO}$, as the experimental output characteristics in Figure 4-33(b) show, where they are superimposed on the inherent characteristics of the BJT portion alone [38]. One price of these advantages, in the inevitable engineering trade-off, is a worsened $V_{CE(SAT)}$. Also, there is an upper limit on operating current that can be seen in Figure 4-33(b), fixed by a characteristic current associated with the JFET portion of the structure.

The adjective *channel-collector* was intended to convey the idea that a single

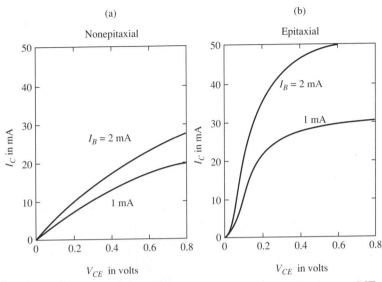

Figure 4-32 Low-voltage portions of the output characteristics for two BJT structures, showing the benefit of epitaxial structure. (a) Device with a uniformly doped collector body, showing the effect of its parasitic series resistance. (b) Epitaxial BJT, with diminished saturation resistance and $V_{CE(SAT)}$. (After Theuerer et al. [9], with permission, copyright 1960, IEEE.)

region served two distinct functions. Subsequently, the apt term *merging* was introduced to communicate this concept [39]. It is an idea as old as the transistor itself. Indeed, it is older. The pentode vacuum tube can be regarded as a cascode combination of two vacuum-tube triodes, and it produces properties very like those of the channel-collector transistor [35].

In Figure 4-34 is a cross-sectional diagram of the epitaxial transistor, drawn to scale. The dimensions indicated are representative of a small device of this type at the height of its popularity. While a discrete device is shown, the integrated version was not greatly different, except that a collector contact was made on top. The underlying N^+ region in the latter case was formed in a P-type substrate before the epitaxial-growth step by the introduction of a slowly diffusing impurity. Its role was to conduct collector current laterally to the vicinity of the collector contact. The PN junction below the N^+ layer so formed was part of the isolating junction. The sidewall portion of this junction was created by deeply diffusing boron "moats" on all four sides of the BJT through the lightly doped N-type epitaxial layer, and into the P-type substrate beneath.

The surface of this device is covered everywhere, either by a metallic contact, or by a layer of SiO_2, grown *in situ*. The steps in this oxide layer are the result of etching openings in the layer for the introduction of junction-forming impurities by solid-phase diffusion and the regrowth of oxide during the diffusion steps. The impurities used to form the base and emitter regions were boron and phosphorus,

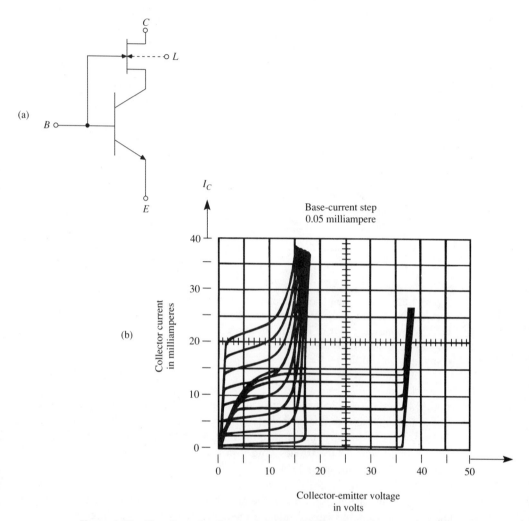

Figure 4-33 The channel-collector transistor (CCT). (a) Schematic configuration of the principal merged devices that compose the CCT. The terminal labeled L, standing for *lock*, exists in some embodiments and not in others. (b) Output characteristics of the CCT (lower right-hand curves) superimposed on those of a stand-alone BJT identical to the BJT portion of the CCT. Note that in the former, $g \approx 0$ and $BV_{CEO} \approx BV_{CBO}$, but that the price for these advantages is a worsened value of $V_{CE(SAT)}$ and an upper limit on operating current.

respectively, elements with nearly identical diffusivities in silicon. Boron was introduced first and phosphorus second, but during the second operation, both impurity populations moved significantly. As a result, the time–temperature schedules required to produce the desired emitter- and base-region profiles (Figure 4-26) were relatively inflexible. Furthermore, as Figure 4-34 shows, the base thickness is a

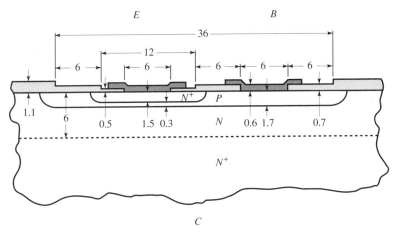

Figure 4-34 Cross-sectional diagram drawn to scale for a small discrete epitaxial BJT of the 1970s. All dimensions are in micrometers.

small difference between two relatively larger junction depths, which underscores the critical nature of this part of the fabrication process.

In the mid-1970s an apparently effective profile variation was put forward that was quite nonobvious initially. Called the low-emitter-concentration transistor, or *LEC transistor,* it placed a thin N^+ region at the surface of an appreciably thicker N^- region [40]. The emitter region, in other words, contained a high-low junction. Surprisingly, the presence of a lightly doped region adjacent to the emitter junction did not degrade emitter efficiency γ_E (a quantity discussed in Section 4-2.2); in fact, the LEC transistor exhibited unusually high values of γ_E, and hence, of β_F. Speculative explanations for these observations were advanced, but the correct answer proved to be technological [41]. Because the N^+ region was so shallow, the result of a very short diffusion process, its volumetric phosphorus density was much higher than that of the emitter in the relatively inflexible standard process (with the standard profile result depicted in Figure 4-26). It was this unusually high emitter doping that led to high values of γ_E and β_F. Once again, however, the improvement was not "free." While static current gain was high, the storage of carriers in the N^- portion of the emitter region significantly degraded dynamic performance in the LEC device.

Now note that as far as profile is concerned, we deal approximately with an LEC transistor when the device of Figure 4-34 is operated in the reverse (or "upward") direction. The comparison would be literally valid if the substrate doping were very high. Reference to Figure 4-26 shows, though, that this is not the case. Substrate doping (unlike that of the emitter region) is typically high, but not degenerate. Thus it is that an appreciation of the LEC transistor aids understanding of BJT reverse operation, a topic addressed in the following section.

Exercise 4-53. Why is the substrate not doped as heavily as an emitter, thus reducing parasitic resistance in series with the collector terminal to a bare minimum?

We noted at the end of Section 2-6.5 that crystalline quality declines monotonically with doping in the degenerate doping range. A low-quality crystal is unsatisfactory as a substrate for epitaxial growth, because crystal defects are propagated into the growing layer. As a result, a BJT fabricated therein would exhibit below-standard properties, especially with respect to leakage current, breakdown voltage, and carrier lifetime. A high, but not degenerate, doping density is used in the substrate as a matter of compromise.

The planar-epitaxial BJT was essentially unchallenged through the 1960s and 1970s. But the 1980s have seen a rush to apply more varied technologies to BJT fabrication as a way of shrinking dimensions, eliminating parasitic features, and boosting performance. Some of these techniques are new, some are old but only recently appreciated. As a result, the design and process teams have increasing numbers of increasingly versatile process options from which to choose at any given step in BJT fabrication. For example, the manipulation of the excellent and conveniently grown insulator SiO_2 has continued for three decades. Through contributions by a large number of workers, this technology has evolved into one wherein a trench cut on a silicon surface can be filled with oxide, restoring a top surface that is nearly flat.

The versatility of previously named individual process options is illustrated by the fact that several of these options can single-handedly create a BJT. Solid-phase diffusion, for example, was used exclusively in the original "planar" devices and their immediate progenitors. Epitaxial growth of single-crystal silicon from the vapor [42], applied with such significant effect to the "planar" BJT [9], was used to fabricate an entire integrated logic gate that contained a BJT [43]. Ion implantation, the doping of a semiconductor by projecting high-energy impurity ions into the surface of the crystal [44], was used similarly. All-implanted BJTs were reported by Payne et al. in 1974 [11].

Another feature that is now widely used is the *polysilicon emitter* [45]. In this process, polysilicon (polycrystalline silicon) is deposited over the emitter region, often over an extremely narrow opening in the oxide. Lateral dimensions can thus be small and tightly controlled. Phosphorus is diffused into the polysilicon, through which it diffuses very readily. Reaching the single-crystal silicon below, it then diffuses more slowly. In this way a thin and very heavily doped emitter region can be formed.

An idealized example of a modern BJT incorporating a polysilicon emitter [46] is shown in Figure 4-35. In addition to facilitating very small lateral dimensions for

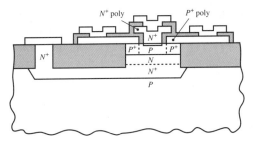

Figure 4-35 Idealized cross-sectional diagram of a modern silicon BJT incorporating (1) isolation by oxide-filled trenches, (2) self-alignment, (3) implanted base region, (4) polysilicon emitter, and (5) polysilicon ohmic contacts. (After Ashburn et al. [46], with permission, copyright 1987, IEEE.)

the emitter, this procedure opens the way to the *self-alignment* of BJT regions, a feature that grows in importance as dimensions shrink. In the example of Figure 4-35, the high-low contacts to the base region are positionally defined by the same opening that subsequently defines emitter width (its lateral dimension, we hasten to stress). Emitter widths in such devices are now approaching one micrometer. The miniscule difference in the widths of the N^+ emitter region and the P (or P^-) base region is fixed by two layers of grown oxide, the shaded regions. (The larger shaded areas represent the oxide-filled trenches that are also included in the structure of this BJT.) Both the P^- and P^+ portions of the base region are formed by ion implantation, with superb control of Gummel number and base thickness. Notice that the need for an overall isolating junction has been eliminated, and that this structure evokes the "bar" geometry of the original BJT.

4-4.7 Forward and Reverse Current Gain

The Gummel number is the sole determinant of minority-carrier current through a BJT base region, given specified values of V_{BE} and D_{nB}, as was shown in Sections 4-4.2 and 4-4.3. Section 4-4.6, in addition, described work demonstrating that forward and reverse emitter efficiencies would be the same in an intrinsic $N^+PN^-N^+$ BJT wherein the true emitter and the substrate had equivalent doping, and wherein the P and N^- regions possessed high minority-carrier lifetime. Emitter efficiency is defined in Equation 4-1. Accepting these propositions, we shall now inquire why the typical "planar-epitaxial" BJT that exhibits $\beta_F \approx 100$ also exhibits $\beta_R = 1$ to 3.

The answer resides partly in the unequal N^+ doping values, as noted previously, and the balance resides in device geometry. This subject took on much more than academic interest when a new digital-logic technology that became known as *integrated-injection logic,* or I²L, gained currency in the 1970s [39, 47]. For bipolar integrated circuits it significantly increased functional density (the number of logic gates per cm²) by exploiting *merging* [39]. Also, and significantly, it used epitaxial

BJTs in the reverse or "upward" direction. The advantage of so doing is that the grounding of the emitter regions of all of those *NPN* BJTs now occurs naturally. Still further, multiple collector regions could be formed on top within a single base region, and these are isolated from one another by their natural reverse bias. This new and different logic technology accomplished logical functions by using multiple output terminals, rather than the multiple input terminals that were previously standard. Through its radical acceptance of an *NPN* device operated in reverse, combined with its extensive merging, I^{2}L was able to eliminate then-standard junction isolation, and in so doing, achieved its high functional density.

The inevitable fee to be paid for these advantages, in the form of an engineering trade-off, was that a reverse-operated BJT of this type is an inferior device. But with evolutionary adjustments, the N^- region virtually disappeared from the intrinsic BJT in order to improve dynamic performance. With heavy substrate doping, values of β_R (or β_{UP} as it became known) were pushed into the 5-to-10 range. This was sufficient.

This brings us, then, to the geometrical issue cited above. Figure 4-36 shows somewhat idealized electron paths in an epitaxial BJT. The fundamentals of forward and reverse operation in the device represented here are identical to those in the I^{2}L BJT. In forward operation, shown in Figure 4-36(a), electron transport is confined to the intrinsic portion of the BJT; there is nothing to cause electron injection elsewhere. The slight spreading of the flow paths that would be seen in a more accurate representation would have little effect on this conclusion. In reverse or upward operation, however, the situation is quite different, as Figure 4-36(b) shows. Now the collector N^-P junction is forward-biased. In the lateral regions beyond the base region, very little electron injection occurs. The reason is that a high-low junction creates only minor carrier-density disturbances when biased, as was stressed in Section 3-6.4. Hence, little electron transport occurs in the outlying regions. Furthermore, the surface recombination velocity (see Section 2-7.4) is very low at the interface between the oxide and the high-resistivity epitaxial layer, so that this interface is not an effective sink for carriers. In the intrinsic and extrinsic regions of the collector junction, on the other hand, electron injection into the base region occurs freely. Electrons pass easily through the associated portion of the high-low junction, to "feed" the density gradient in the base region established by forward bias on the collector junction. According to the "LEC" principle [41], reverse operation of the intrinsic BJT is almost as efficient as forward operation, with electrons being collected by the reverse-biased emitter junction. But electrons entering the extrinsic base region head for the heavily doped and relatively low-lifetime region near the surface, to recombine. The doping profile, recall, is that associated with a diffused region. See Figure 4-26 for the overall picture, and refer to the undistorted representation of a diffused profile shown with linear scales in Figure 3-41 for a more accurate understanding of the degree of doping variation in the *P*-type region.

Next, turn to the analytical consequences of the graphic information in Figure 4-36. Applying Equation 4-25 to forward operation gives us

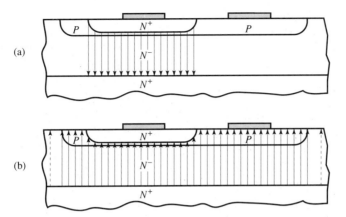

Figure 4-36 Approximate electron paths in the epitaxial BJT for (a) forward ("downward") operation, and (b) reverse ("upward") operation.

$$\beta_F = \frac{\alpha_F}{1 - \alpha_F}, \tag{4-67}$$

where α_F is common-base current gain (Section 4-3.6). Assuming no loss of electrons in the base region, then

$$\alpha_F \approx \gamma_E, \tag{4-68}$$

where γ_E is the emitter efficiency defined in Equation 4-1. It is the ratio of injected-electron current to the total current passing through the emitter junction. Combining Equations 4-67 and 4-68 gives us

$$\beta_F \approx \frac{\gamma_E}{1 - \gamma_E}. \tag{4-69}$$

Now examine the case of reverse or upward operation, Figure 4-36(b). Let γ_C represent the efficiency of injection into the *intrinsic* base region by the now-forward-biased collector junction. Evidently

$$\gamma_C = \frac{A_E}{A_C}, \tag{4-70}$$

where A_E and A_C are the emitter and collector junction areas, respectively. Thus

$$\beta_R = \frac{\gamma_C}{1 - \gamma_C} = \frac{A_E/A_C}{1 - A_E/A_C} = \frac{A_E}{A_C - A_E} \approx \frac{A_E}{A_C}, \tag{4-71}$$

where the final approximation assumes a relatively large ratio of the two junction areas. In the I^2L BJT this was usually true, because the base region had to be large enough to accommodate several "collectors" (N^+ regions at the surface). The number of collectors was usually from three to six.

Detailed measurements were made for a particular I²L BJT design where the area ratio was approximately twenty [48]. Processing variations, some of them intentionally introduced for the sake of the experiment, yielded values of β_{UP} (or β_R) ranging from about two to eleven. The corresponding values of β_{DOWN} were carefully measured as well, and data were plotted as β_{DOWN} versus β_{UP} with the result shown in Figure 4-37. The factor of about twenty is clearly evident in the near-linear portion of the curve. The "saturation" or leveling off of β_{DOWN} values is a consequence of high-level effects that we address in the next section. Thus this experiment supports the geometrical interpretation of forward and reverse current gain given in Figure 4-36.

4-5 HIGH-LEVEL EFFECTS

The term *high-level* in the present context has the same meaning that it had in Section 3-4.7. Forward bias on a junction is increased to the point that the minority-carrier density at one (or both) of its boundaries approaches or exceeds the equilibrium *majority*-carrier density there. The subject is important because BJTs are routinely used under high-level conditions. Our purpose here is to impart a good qualitative grasp of the phenomena that occur when the high-level regime of operation is entered. Analytical details on these phenomena are readily available in the references cited.

4-5.1 Rittner Effect

When minority-carrier density at a junction boundary exceeds equilibrium majority-carrier density there, the result is a failure of the law of the junction. This problem was successfully analyzed by Misawa [49] and Fletcher [50]. (See Section 3-4.7.) Not surprisingly, high-level junction conditions in the BJT are frequently accompanied by large values of current density in the junctions and adjacent regions. These in turn give rise to values of electric field in the neutral or quasineutral regions that are nonnegligible with respect to their effects on even minority carriers. The electric field in question, often described as a *longitudinal* electric field, is x-directed, or normal to the junctions.

Exercise 4-54. How do the electric-field effects of interest here differ from those in the Moll-Ross analysis, Section 4-4.2?

In the former case, substantial electric-field values were "built in" because of doping nonuniformity. Here, the fields develop only when biases and current densities become large.

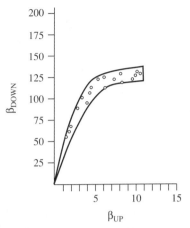

Figure 4-37 Comparing β_{DOWN} and β_{UP} for I²L BJTs all having identical plain-view geometry and $I_B = 0.1$ mA, but differing Gummel number. Each point represents measurements on one transistor. Before high-level effects enter in downward operation, the beta ratio approximately equals the collector-emitter area ratio of 20. (After Wisted et al. [48], with permission, copyright 1982, Pergamon Press.)

A pair of solid-state-device pioneers explored high-level effects in the emitter and base regions of a BJT. They did so independently for the most part, and published their findings simultaneously. They approached the problem with different points of view, and arrived at compatible conclusions. Two phenomena are mainly involved. One affects junction theory, the other, the longitudinal electric field in the base region. Although both workers dealt with both phenomena, we have chosen to treat the two separately, and to associate one of the names with each, in the cause of equity. The collective phenomena are usually identified as the *Webster effect* [51]. But let us instead begin with the *Rittner effect* [52], following our recent practice [53].

The emitter junction can justifiably be regarded as a one-sided junction, with the base side being the more lightly doped. Therefore, when V_{BE} is gradually increased, high-level effects will occur on the base side well before they occur on the emitter side. This was the situation that was analyzed by Rittner, providing an almost perfect anticipation of Fletcher's analysis. Rittner's, however, was for the special case of the one-sided junction, while Fletcher's analysis was more general.

Exercise 4-55. What value of V_{BE} is required to reach high-level conditions in a uniformly doped base region with $N_{AB} = 10^{17}/\text{cm}^3$?

The question can be rephrased to read, "At what value of V_{BE} does $n_B(0)$ approach the order of magnitude of $p_{OB} \approx N_{AB}$? Let us take $n_B(0) = 0.1 \, N_{AB}$ as a quantitative criterion for "approach the order of magnitude." The answer, since $n_{OB} = 10^3/\text{cm}^3$, is provided by the law of the junction as

$$V_{BE} = (kT/q) \ln[n_B(0)/n_{OB}] = (0.02566)\ln[(10^{16}/\text{cm}^3)/(10^3/\text{cm}^3)] = 0.768 \text{ V}.$$

Many BJTs are routinely operated under high-level conditions, so the associated analysis has more than academic interest.

Under high-level conditions, the voltage applied to the terminals of the emitter junction in general exceeds the voltage that would be measured from boundary to boundary of the junction. (This matter is examined further in Section 4-5.2.) The latter we have named the *imposed junction voltage* V_{JE}. It is important because it is the voltage that normally enters into boundary-value equations. The rising value of $n_B(0)$ is accompanied by a matching rise in $p_B(0)$, and in the high-level regime, the latter density is large compared to p_{OB}. But this is the density value that fixes hole injection into the transition region of the junction, and ultimately into the emitter region. Descriptive language that is sometimes used at this point is that the value of hole density at the left-hand boundary of the emitter junction exhibits a "faster-than-Boltzmann" rise. The allusion here, of course, is to the Boltzmann factor in the law of the junction, the factor having V_{JE} in its argument. By the opposite token, the majority-carrier density at the left-hand boundary is virtually unchanged on a relative basis because the emitter region is so heavily doped. Consequently $n_B(0)$ at the right-hand boundary obeys the simple law-of-the-junction prescription in terms of V_{JE}, as does the collector current that it determines. These behaviors can be regarded as the consequence of the Boltzmann quasiequilibrium condition (Section 4-4.2) that relates electron densities at the two boundaries on the one hand, and hole densities, on the other. Curiously, with respect to the *terminal* voltage on the emitter junction, the emitter-side boundary value exhibits simple law-of-the-junction behavior. These matters are examined further in Section 4-5.2.

Recall that for practical purposes the current of holes injected into the emitter region accounts for *all* of the base current I_B. Thus the faster-than-Boltzmann rise of minority density at the emitter boundary, which we have labeled the *Rittner effect*, when taken by itself is responsible for a *beta fall-off*, or a decline in current gain, because it inflates I_B.

4-5.2 Webster Effect

Now turn attention from the emitter junction itself to the base region. Although most aspects of high-level theory are more complicated than the corresponding features of low-level theory, there is one respect in which conditions become simpler: It is now possible to draw accurate *linear* diagrams that show both minority and majority populations. Figure 4-38 presents such a diagram for a particular case of high-level forward-active operation. Notice that these profiles, instead of being concave up like that of Figure 4-9, or linear like those of Figures 4-19(b), 4-20(b), and 4-22, are concave *down*. This curious fact was noted by Rittner in his classic paper. The longitudinal field in the base region, for which the majority carriers are

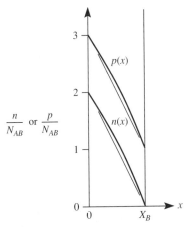

Figure 4-38 Carrier-density profiles in the base region of a forward-active BJT under modestly high-level conditions. Axis calibrations are linear.

responsible, affects the minority carriers under high-level conditions; their densities now have the same order of magnitude as the majority-carrier densities. The leftward-directed field *aids* the transport of minority electrons to the right. Thus minority-carrier transport in the base region is now through a combination of diffusion and drift.

Exercise 4-56. What accounts for the concave-down high-level profiles in Figure 4-38? Assume a uniformly doped base region.

Since the BJT is in forward-active operation, electron density is very small at the right-hand boundary. There the electric field has little effect on minority carriers, because drift transport is density-dependent. But $J_n(x)$ must be constant in the base region. Consequently, a steeper gradient is needed at the right to maintain current continuity; there the diffusion mechanism must "do it all." The electron population in Figure 4-38 can be regarded as 100% excess carriers. The matching population of excess holes sits atop the equilibrium hole profile, assumed x-independent. Thus the total-hole profile will show the same concave-down character as that for electrons.

Exercise 4-57. Describe qualitatively the longitudinal electric-field profile accompanying the conditions of Figure 4-38.

The electric field is negative, monotonically rising in magnitude toward the right. It increases slowly at first, and then faster, exhibiting maxima in slope magnitude and field magnitude at the right.

In short, the Webster and Rittner analyses of the base region showed that electron transport through the base region receives a boost from drift under high-level conditions. Under low-level conditions, only diffusion is involved. Their concluding equation, applying in the limit of high-level operation, is

$$I_C = (qA_E 2D_{nB}/X_B)[n_B(0) - n_B(X_B)]. \tag{4-72}$$

The equation in this form is general in the sense that it permits $n_B(X_B)$ to be other than zero. That is, the equation applies in saturation as well as in forward-active operation. Examination of Equation 4-72 shows that it is a perfectly conventional diffusion-current expression except for one thing: An extra factor of two appears in the coefficient. It is there because drift and diffusion contribute equally to electron transport in the high-level limit. This effect, known as the *Webster effect,* is often described as "an apparent doubling of the diffusivity."

The nature of the longitudinal electric field responsible for the Webster effect was not examined in either of the closely related pioneering papers [51, 52], although Webster offered partial and qualitative comments on its origin. It might seem at first that the Webster effect is a "favorable" phenomenon because the high-level electric field assists base-region electron transport, and hence I_C. But such is not the case. The voltage required to create the electric field is subtracted from terminal voltage, and hence the emitter junction loses "working voltage." That is, using V_{EB} for sign consistency, we may write

$$V_{EB} = V_{JE} + \Delta V, \tag{4-73}$$

where V_{JE} is the boundary-to-boundary voltage felt by the emitter junction as a result of external bias, or the *imposed junction voltage,* while ΔV is the longitudinal voltage drop in the base region. (This relationship is not totally obvious and will be treated further in the next section.) Since the aiding result of the Webster effect can at most be a factor of two, while V_{JE} enters into an exponent in the expression governing collector current, it follows that the latter fact is dominant; loss of V_{JE} hurts more than the aiding field helps.

One reason for the small amount of attention paid to electric field by Webster and Rittner was that they eliminated it as a variable early in their analyses. This they did by solving each of the transport equations for electric field, and then equating the resulting expressions. Thus the quantity E did not appear in the final result or in any of the intervening equations.

When the longitudinal electric field in the base region is integrated from $x = 0$ to $x = X_B$, the result is the longitudinal potential difference ΔV. But, as pointed out in Section 4-4.2, we can estimate the value of this potential difference simply yet accurately by the valid assumption of Boltzmann quasiequilibrium for the base region. A key point here is that the region between the two positions of interest can be a region of gross space charge, as in a junction, or a region of quasineutrality, as in the case of the base region—in either case the Boltzmann quasiequilibrium assumption can be used. One still has carrier densities that have approximately an exponential relationship to the potential difference between the two positions, provided the necessary conditions are met. In particular, those conditions are that the net current of the carriers involved must be small compared to the individual drift and diffusion components of that current, and equivalently, the quasi Fermi level for the same carriers must be approximately constant in the interval.

The origin and composition of the voltage drop ΔV will be examined in the next section. But before leaving the Webster effect, let us comment on a curious fact concerning the early analyses. As we have just noted, the longitudinal majority-hole current in the base region is typically negligible in forward-active operation. Webster and Rittner, in fact, set it equal to zero as a matter of assumption. There was a bit of good luck in the fact that this brought them to a meaningful result, however. When a similar approach is used for the collector region (neglecting hole current because it is very small), one is led to an absurd result [54].

4-5.3 Ambipolar Effect

In high-level operation, the longitudinal voltage drop in the base region of a BJT can amount to several tenths of a volt. Since this voltage is derived from V_{BE}, as just indicated, its effect is significant; hence we have labeled the resulting phenomenon (somewhat unilaterally) the *ambipolar effect*. Ambipolar conditions in general are those wherein extreme conductivity modulation exists, in combination with quasineutrality. The first study of such conditions was done in relation to plasma physics, and the first detailed examination of ambipolar phenomena in semiconductors was carried out by Van Roosbroeck [55]. In turn, the first application of the ambipolar principles to the BJT base region in high-level operation was done by Ebner and Gray [56]. Subsequently, Van Vliet and Min [57] repeated the Ebner and Gray investigation of base-region conditions, correcting errors in the previous analysis, and substantially extending the investigation.

The first step in such a study is an examination of the longitudinal electric field. The total current density J in the base region can be written simply by adding the two transport equations, $J = J_n + J_p$. From Equations 2-56 and 2-57, it is evident that the result is

$$J = q(\mu_n n + \mu_p p)E + q\left(D_n \frac{dn}{dx} - D_p \frac{dp}{dx}\right). \tag{4-74}$$

The coefficient of E is of course the conductivity σ. Also, a succinct statement of the quasineutrality condition is $(dn/dx) \approx (dp/dx)$. Making the changes thus indicated in Equation 4-74 gives us

$$J = \sigma E + q(D_n - D_p)\frac{dp}{dx}, \tag{4-75}$$

and solving for E yields

$$E = \frac{J}{\sigma} - \frac{q}{\sigma}(D_n - D_p)\frac{dn}{dx}. \tag{4-76}$$

It is often convenient to introduce a symbol for the diffusivity (or mobility) ratio. Let $b_B \equiv D_n/D_p$, where the subscript denotes "base region." Introducing b_B into Equation 4-76 yields

$$E = \frac{J}{\sigma} - \frac{q}{\sigma}\left(\frac{b_B - 1}{b_B}\right)D_n\frac{dn}{dx}. \tag{4-77}$$

Notice that this straightforward manipulation has resolved the electric field into two components. The first term is often described as the *ohmic* component of electric field, in view of the fact that $E = J/\sigma$ is Ohm's law. Its physical significance, simply, is that the existence of a current J in a medium of conductivity σ is inevitably accompanied by an electric-field component. But a second mechanism coexists with the current passage, and it, too, contributes a field component. At every position in the base region (excluding thin regions very near its boundaries) the electron and hole density gradients are almost precisely equal. But the tendency of electrons to "outrun" holes by faster diffusion gives rise to a tiny space charge and associated electric field. This component of field speeds up the holes and slows down the electrons, so the nonequilibrium condition in the base region can remain a steady-state condition. The second term of Equation 4-77 is often called the *bulk-diffusion* component of electric field.

Exercise 4-58. Consider a BJT in low-level forward active operation. How do the two terms of Equation 4-77 compare in magnitude?

For the low-level case, we can write $J \approx qD_n dn/dx$. Hence the ratio of the ohmic field magnitude to the bulk-diffusion field magnitude is

$$\frac{E_{\text{OHMIC}}}{E_{\text{BULK DIFFUSION}}} = \frac{b_B}{b_B - 1},$$

and letting $b_B = 3$ for simple computation yields the ratio 1.5. Realize, however, that both field components are very small under low-level conditions.

For simplicity, let us continue with steady-state cases only. The main point is this: Carrier-density gradients are inevitably and spontaneously accompanied by longitudinal electric fields that are required to maintain the steady-state conditions. The BJT base region provides an especially lucid illustration of these principles because we know precisely how to create any desired carrier-density gradient. If we further choose the uniformly doped case, this will eliminate longitudinal electric fields from the equilibrium problem, so that fields existing in the nonequilibrium problem are then exclusively those attributable to carrier-transport phenomena.

Now let us inquire into how one might alter the ratio computed in Exercise 4-58. It is evident that the last or bulk-diffusion term in Equation 4-77 vanishes when $b_B = 1$. Although this is not readily arranged in silicon, we do see by inspection of Figures 2-22 and 4-26 that values ranging from 2 to 3 are realistic, and that the value tends to decline as doping increases.

Even for a fixed value of b_B, the ratio of the two field-component magnitudes is variable. The reason is that the relationship of J in the ohmic term of Equation 4-77 to dn/dx in the bulk-diffusion term is a function of n and p. That is, it changes as one goes from low-level to high-level conditions. This relationship is far from transparent, and so a contribution by Van Roosbroeck [55] is particularly noteworthy here. He defined an *effective* diffusivity known as the *ambipolar diffusivity* D^*. It is an effective diffusivity in that it takes into account the effect of the net field that is present, but treats carrier transport as though it were purely diffusive. His expression for ambipolar diffusivity is

$$D^* = \frac{n+p}{\left(\dfrac{p}{D_n}\right) + \left(\dfrac{n}{D_p}\right)}. \tag{4-78}$$

Exercise 4-59. Determine the value of D^* in the limit of extreme conductivity modulation, or when $n \approx p$. Assume once more that $b_B = 3$.

For the given assumptions,

$$D^* \approx \frac{2D_p D_n}{D_p + D_n} = \frac{6D_p^2}{4D_p} = 1.5\, D_p.$$

That is, the effective diffusivity has a value intermediate between D_n and D_p. This is a reflection of the fact that part of the reason for the existence of a longitudinal field is to speed up holes and slow down electrons, this being necessary to maintain steady-state conditions.

Exercise 4-57 dealt qualitatively with the character of the *net* electric-field profile for a particular base-region condition. The net field was clearly x-dependent. Generalizing on the x dependence under high-level conditions is difficult because of the number of variables that enter into it. However, a valid generalization concerning the transition from low-level to high-level conditions is this: Under low-level conditions, the minority carriers are negligibly affected by the field, while under high-level conditions, they are affected about as significantly as the majority carriers. Notice, too, by inspection of Equation 4-77, that it is entirely possible for both components of longitudinal electric field, and the net field as well, to remain small as high-level conditions are entered. That is, small values of J and dn/dx, in combination with the large values of σ that accompany high-level conditions would combine to keep E, E_{OHMIC}, and $E_{\text{BULK DIFFUSION}}$ small.

Exercise 4-60. As a practical matter, how can one achieve small J in combination with large σ?

This is precisely the condition of deep saturation.

Exercise 4-61. Compare the algebraic signs of the ohmic and bulk-diffusion components of electric field in the base regions of *NPN* and *PNP* BJTs.

In both devices, both holes and electrons are diffusing to the right. Thus a positive (rightward) bulk-diffusion component is needed in each case to speed holes and slow electrons. In the *NPN* case, longitudinal current in the base region is negative (leftward) and must be accompanied by a negative ohmic field component, so that the two components are subtractive. In the *PNP* case, both current and field are positive, so that the two components are additive.

Corresponding to each electric-field component in Equation 4-77 there exists a voltage increment that results from integrating that component from $x = 0$ to $x = X_B$. The details of these integrations are found in the literature [53, 57]. In the event that one is interested only in the net longitudinal voltage drop, however, integration is unnecessary. As pointed out in Section 4-4.2, applying the Boltzmann-quasiequilibrium concept yields the answer directly. Let us outline the procedure. For simplicity, continue to assume that the base region is uniformly doped, so that $\psi_0(0) - \psi_0(X_B) = \Delta\psi_0 = 0$. Next, repeat Equation 2-36, the Boltzmann relation for holes. Changing it to notation that is relevant to the BJT base region, Equation 2-36 becomes

$$\frac{p_{OB}(0)}{p_{OB}(X_B)} = \exp\left[\frac{\psi_0(X_B) - \psi_0(0)}{kT/q}\right]. \tag{4-79}$$

Applying the Boltzmann-quasiequilibrium concept means assuming that a parallel relationship holds under nonequilibrium conditions, a valid assumption for reasons discussed at some length in Section 4-4.2:

$$\frac{p_B(0)}{p_B(X_B)} = \exp\left[\frac{\psi(X_B) - \psi(0)}{kT/q}\right] = \exp\left[\frac{\Delta\psi}{kT/q}\right]. \tag{4-80}$$

The increment $\Delta\psi$ is positive because the base terminal is more positive than the emitter terminal in forward operation of the BJT. Hence, inspection of Equation 4-80 verifies its validity. But because of the Ohm's-law convention discussed in Section 3-3.1, we have $\Delta\psi = -\Delta V$, where ΔV is the net longitudinal voltage drop in the base region. Thus Equation 4-80 can be rewritten as

$$\Delta V = -\frac{kT}{q} \ln\left[\frac{p_B(0)}{p_B(X_B)}\right]. \tag{4-81}$$

In the limit of extreme high-level forward operation we have $\Delta V \approx V_{JE}$, the emitter imposed junction voltage [53, 57]. This is a finding consistent with the ultimate diffusivity doubling associated with the Webster effect.

Exercise 4-62. Consider *NPN* and *PNP* BJTs of identical dimensions, net-doping magnitudes, and biases. Comment on the magnitudes of their net longitudinal voltage drops ΔV. In view of your answer to this, and in view of Exercise 4-61, comment on the magnitudes of the four field components at a particular position in the base regions.

In view of Equations 2-35 and 2-36, both devices will exhibit a ΔV of the same magnitude. Because the two field components are additive in the *PNP* device, and subtractive in the *NPN* device, at an arbitrarily chosen position no two field components will have the same magnitudes!

Equation 4-73 displays the fact that terminal voltage V_{EB} is the sum of V_{JE} and ΔV, with all three being negative for forward operation. Turn now to the promised explanation for this relationship. The density $p_B(X_B)$ is typically small in forward operation of the BJT. Insofar as it approximates p_{OB}, there will be no voltage drop (as dictated by the Boltzmann-quasiequilibrium principle) from the base contact in the extrinsic base region to the position $x = X_B$ in the intrinsic base region. A lateral voltage drop accompanying large base current I_B would alter this situation, but we shall rule that out for now. In a wide range of practical operating conditions, this is a valid assumption. Thus it is that the main contributions to the terminal voltage V_{EB} are the imposed junction voltage V_{JE} on the emitter junction and the longitudinal voltage drop ΔV in the base region, these being in series and consistent in sign (all negative).

It follows that growing ΔV is responsible for major losses of V_{JE} as high levels are approached, the phenomenon we have named the ambipolar effect. But detailed analysis reveals that the ambipolar effect precisely cancels the Rittner effect [53, 54]! That is, the boundary value on hole density that determines I_B obeys the law of the junction when V_{EB} is used in the exponential argument in lieu of V_{JE}. The result of this relationship is that a plot of log I_B versus V_{BE} is linear through the low-level, intermediate-level, and high-level regimes. Only in the high-level extreme does a deviation occur, and that is because of the lateral voltage drop associated with I_B that was briefly mentioned earlier.

A physical-geometrical argument can also be used to explain this important point [53]. In forward operation of the BJT, the Fermi level for electrons ϕ_n is constant from the emitter terminal, through the heavily doped emitter region, through the emitter junction, and well into the base region with its charge of excess carriers. The Fermi level for holes ϕ_p is also constant through the emitter junction and base region, for reasons that have been emphasized repeatedly. If the emitter junction were isolated, this would not be true, expecially near the terminal of the P-type region. But for the peculiar BJT situation, the constancy of ϕ_p holds, even laterally to the base terminal. As a result of these conditions, the magnitude of the Fermi-level splitting $\Delta\phi_J$ in the emitter junction is virtually identical to that of the terminal voltage V_{EB}. Hence the boundary value that fixes I_B, the hole density on the emitter side of the junction, obeys the law of the junction with terminal voltage in the exponential argument.

The ambipolar effect in a BJT base region in the sense of this section is a specific instance of a more general high-level problem. It is worthwhile to broaden the discussion here in order to provide at least a qualitative appreciation of other high-level phenomena, also treated by Van Roosbroeck [55]. One significant topic was introduced in Section 2-7.3, and Exercise 2-47. It is nicely illustrated, as before, using the simple geometry of the Haynes-Shockley experiment. Ohmic conduction is caused by applying a steady-state voltage to the ends of a long, uniformly doped silicon barlike sample. Let it be N-type as before. Then a focused-light flash creates a localized pulse of excess carriers, and because minority carriers are "in charge," this disturbance drifts in the electric-field direction. Under low-level conditions, the electric field is unperturbed by the excess-carrier pulse. But under high-level conditions there is by definition a local perturbation of conductivity σ. Current density J will be constant (independent of both time and position) for the case of a long sample; hence there must exist a local depression of electric field E, as required by Ohm's law, $J = \sigma E$.

Exercise 4-63. Assume that at a particular instant the hole-density maximum at the pulse center is $p = 3n_0$. Assuming that $\mu_n = 3\mu_p$, calculate electric field at that position as a percentage of the value of E in the unperturbed portions of the bar.

For the assumed conditions, $\sigma_0 \approx q(3\mu_p)n_0$. But at the pulse maximum,

$$\sigma = q(\mu_p p + \mu_n n) = q[\mu_p(3n_0) + (3\mu_p)(n_0 + 3n_0)] = (15)q\mu_p n_0.$$

Hence, $(\sigma_0/\sigma) = (1/5)$, and so the local field must be 20% of its value elsewhere.

Pulse behavior is described in terms of its *drift mobility*, which is the pulse velocity divided by the unperturbed value of electric field. But because field is locally depressed, drift mobility would be appreciably smaller than the conductivity mobility of the holes, μ_p. Again, Van Roosbroeck provided a quantitative analysis of the problem. His *ambipolar mobility* gives the drift mobility of such a pulse as

$$\mu^* \equiv \frac{p - n}{\left(\dfrac{p}{\mu_n}\right) + \left(\dfrac{n}{\mu_p}\right)}. \tag{4-82}$$

Notice that in the limit of extreme conductivity modulation, $p \approx n$, the pulse is stationary. As it decays, it begins to move with the field in accelerating fashion.

Exercise 4-64. How can one account physically for the locally depressed electric field?

The excess-carrier populations are displaced slightly by the applied field, which we will consider to be positive. The right flank of the disturbance has a tiny hole space charge, and the left flank, electron space charge. The resulting new component of field is thus negative, subtracting from the applied field.

Thus we see that ambipolar mobility μ^* is also an *effective* quantity; it is used in a calculation that employs the unperturbed applied field. Similarly, ambipolar diffusivity is an effective quantity because it comprehends but does not acknowledge the presence of an electric field. It is important to realize, therefore, that μ^* and D^* are *not* connected by an Einstein relation. Note that each of these quantities relaxes smoothly to the minority-carrier value when a transition is made from high-level to low-level conditions.

There are special applications where the two coefficients can be used in combination. As an illustration, let us refer back to the constant-E continuity-transport equation, given for minority electrons and holes, respectively, in Equations 2-101 and 2-102. These equations in unmodified form are confined to low-level applications. To see why, we need only refer to the high-level base-region problem and the high-level Haynes-Shockley experiment. In both cases, nonconstant E enters, by definition violating a restriction on the constant-E continuity-transport equation. But if D^* and μ^* are respectively substituted for D_p and μ_p in Equation 2-102 (to choose the second continuity equation arbitrarily), and also into its solution, then the low-level equation and solution are converted into an equation and solution also valid at high levels—and everywhere in between. The "constant" E in this case is the unperturbed value of E, taking the Haynes-Shockley problem for simplicity. The modified mobility correctly describes the drift velocity of the carrier pulse, and the modified diffusivity correctly takes account of electric-field effects at the flanks of the disturbance that affect the rate at which the pulse spreads and decays. This modified form of the constant-E continuity-transport equation is sometimes called the Van Roosbroeck equation. Although it is elegantly parallel to the low-level equation, it is not very useful for problem solving. The density-dependent mobility and diffusivity enforce the use of approximate or numerical methods.

4-5.4 Kirk Effect and Quasisaturation

Now let us turn to collector-region phenomena. Our discussion of the epitaxial BJT in Section 4-4.2 and before may have implied that *this* structural change, at least, is one that brought numerous benefits and exacted no price. The rule of the engineering trade-off had been temporarily suspended, in other words. But in fact, the presence of the N^- layer in the collector structure introduced a high-level effect that was finally unraveled only some fifteen years after the epitaxial BJT made its appearance. A great many people contributed to the sorting-out process; a detailed bibliography is available [53].

The phenomenon in question has become known as the *Kirk effect,* after the person who first called attention to it [58]. Kirk focused on a decline in cutoff frequency, a measure of high-frequency performance, when an epitaxial BJT is pushed into high-current operation. He sketched a qualitative picture wherein the collector junction "moved into" the N^- region at high current. The subsequent literature, accepting his description, often used the phrase "base push-out." The resulting expansion of base thickness would account for his observations, because frequency response and base thickness are inversely related.

In fact, there is no way for a junction to "move." What does happen is that the parasitic series resistance of the N^- region can cause the collector junction to be *forward*-biased in spite of apparent reverse bias measured at the collector-base terminal pair. The forward bias inevitably means inflated boundary values for minority-carrier densities, and hence additional storage of excess carriers. Some occurs in the base region, as in a case of conventional saturation; but because the BJT *appears* to be in forward-active operation in terms of the external V_{CB}, this new condition has been named *quasisaturation.*

A specific case of near quasisaturation is illustrated quantitatively in Figure 4-39. The collector terminal is more positive than the base terminal by 1 V. Under low-level conditions this would be a clear case of reverse bias on the collector junction. But in the present example, the high value of current and the parasitic resistance R_0 cause a one-volt drop to appear within the N^- layer. Consequently, the actual, or imposed, junction voltage is zero. The BJT is at the boundary of the quasisaturation regime. The N^- epitaxial region was of course introduced into the BJT to accommodate the depletion layer of the collector junction. But since quasisaturation is a low-voltage phenomenon, the N^- region is undepleted and so R_0 represents its end-to-end resistance.

Exercise 4-65. How does one cause a transition *into* the quasisaturation regime from the borderline case of Figure 4-39?

It is merely necessary to increase I_C, causing internal forward bias on the collector junction. Collector current can be increased by increasing V_{BE} or I_B, depending upon the specific circuit involved.

More important than additional carrier storage in the base region is storage in the N^- collector region. The two carrier populations there behave in a way that is curiously parallel to their high-level behavior in the base region, although with an

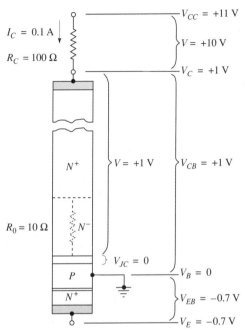

Figure 4-39 Example of BJT biasing at the boundary between the forward-active and quasisaturation regimes. Voltage drop in the N^- layer brings the collector junction to zero bias in spite of the apparent reverse bias observed externally.

important difference. Specifically, the majority and minority carriers interchange roles. Hence the conductivity-modulated portion of the N^- region (the portion nearest the collector junction) is sometimes described as an *extended base region*. This description evokes Kirk's original suggestion; even though his idea was incorrect in detail, it did indeed involve the presence of excess carriers in the N^- region.

The carrier profiles in a BJT operating in quasisaturation are shown for a specific bias condition in Figure 4-40(a). This device, shown physically in Figure 4-40(b), is identical to that defined in Table 4-1, except that the N^- region has been made more extensive in the x direction and more lightly doped in order to emphasize the quasisaturation phenomenon. These carrier profiles were calculated from a detailed high-level BJT model [54].

Exercise 4-66. How do these structural changes emphasize quasisaturation?

Both changes increase the parasitic resistance R_0 that is responsible for the effect.

In Figure 4-40(a), the region of extensive carrier storage is the extended-base

580   THE BIPOLAR JUNCTION TRANSISTOR

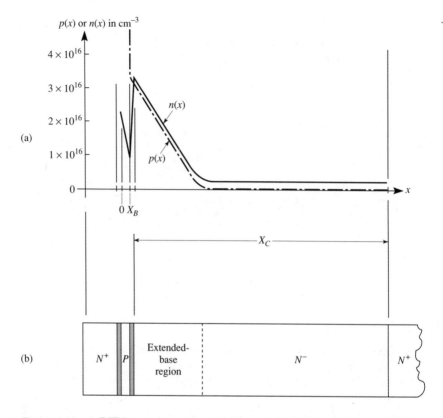

Figure 4-40 A BJT in quasisaturation. (a) Linear presentation of electron and hole profiles in the base and collector regions. (b) Physical representation of the device, indicating the conductivity-modulated "extended-base" region in the N^- layer.

zone. Carriers in this zone move by a combination of diffusion and drift, as they do in a high-level base region. For the particular operating point chosen, however, the N^- region is in a high-level condition and the base region is not, because of its heavier doping. This accounts for the factor-of-two slope difference between profiles in the two regions.

Exercise 4-67. How does the fact just cited account for the slope differences apparent in Figure 4-40(a)?

The base region presents a case of low-level operation, so the electrons are moving by diffusion only. In the N^- collector region, the electrons move by combined diffusion and drift, so lower slope is needed. The continuity of $J_n(x)$ in this one-dimensional BJT is a necessity.

Exercise 4-68. Comment on the signs and magnitudes of the various current components in the N^- region.

Consideration of the voltage drop in Figure 4-39 shows that the associated electric field in the extended-base region is negative, aiding the diffusive electron transport. The net longitudinal hole current approximately vanishes in the base region and must do the same in the collector region (no sources or sinks are present). Thus the two components of hole current cancel each other.

Last, on the subject of quasisaturation, I–V output characteristics are plotted in Figure 4-41. The point A corresponds to the conditions of Figure 4-40(a). The same model was employed to calculate both diagrams [54]. A departure in Figure 4-41 from a usual output-plane diagram is that V_{CB} is plotted on the abscissa in place of V_{CE}. This was done so that the I–V characteristic of the resistor R_0 (the dashed line at roughly 45°) would be undistorted. Thus, the triangular region between the I_C axis and the R_0 characteristic is the regime of quasisaturation. The regime of true saturation (the only kind considered previously) lies to the *left* of the I_C axis. The I–V characteristics have not been extended into that regime, but we know them to be approximately vertical. Hence, when current is pushed high enough to cause quasisaturation, an inescapable accompaniment is a relatively steep rise in $V_{CE(SAT)}$, as can be inferred readily from Figure 4-41. It is not possible to obviate this fact by

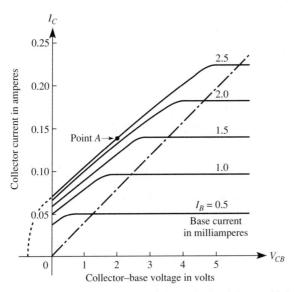

Figure 4-41 Output characteristics calculated from a high-level BJT model, showing the quasisaturation regime of operation. Point A corresponds to the carrier profiles shown in Figure 4-42(a).

means of harder "drive" (that is, by increasing I_B), because *all* of the curves exhibit this feature.

In constructing Figure 4-41 we assumed that the resistor R_0 was a linear device. Actually, we know that electron velocity saturation (Figure 2-23 and Section 2-5.4) makes an N-type silicon resistor nonlinear. In fact, the resistor's I–V characteristic exhibits precisely the shape of the curve in Figure 2-23. This complication, too, has been examined in some detail [53].

4-5.5 Lateral Voltage Drops in the Base Region

For the first time in this discussion of high-level phenomena, let us look at an effect involving lateral voltage variations. The base region presents a resistance to the lateral flow of majority holes that was noted in connection with Figure 4-7, Section 4-2.1. This ohmic base resistance, as it was there identified, is sometimes called *base spreading resistance*. The adjective "spreading" in such a context connotes nonparallel flow lines, or a diverging of flow lines—a significant departure from one-dimensional geometry. As Figure 4-7 itself emphasizes, that condition certainly holds for majority-carrier current in the base region, especially in the intrinsic base region. The issue of base spreading resistance has been examined by approximate-analytic methods for a variety of BJT geometries [59].

Further, a consideration of the IR-drop polarity in Figure 4-7 makes it clear that the portions of the emitter junction most remote from the base contact will experience "debiasing," or a degradation of forward voltage. In extreme cases, this means that electron injection into the base region will be similarly variable. Emission will be greatest at the emitter edge closest to the base contact, even if the emitter junction has the stripelike character depicted in Figure 4-8. If two base contacts are used, then preferential injection will occur at both emitter-junction edges. The term *emitter-current crowding* is often used to label such preferential injection.

Another feature of the base region must now be considered. Figures 4-2 and 4-8 provide cross-sectional and plan views, respectively, of the same BJT structure. The extrinsic base region, lying between the base contact and the emitter edge, also exhibits parasitic resistance, R_{OB}. Although it is typically only a few ohms or tens of ohms, it, too, can contribute an IR drop that becomes significant for extreme values of base current I_B. That is, the quantity $I_B R_{OB}$ is also subtracted from V_{BE}, representing a significant effect when this quantity amounts to merely a few tens of millivolts. The lateral effects in the base region under high-level conditions involve a complicated combination of emitter-junction debiasing and an IR drop in the extrinsic base region.

4-5.6 High-Level Effects in Combination

A good way to assess the aggregate impact of the five high-level effects described in the sections immediately preceding is, obviously, to examine experimental data. An appropriate data set has been given by Jespers [60]. In Figure 4-42 are shown curves

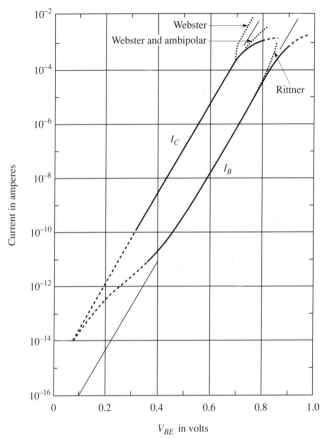

Figure 4-42 Semilog presentation of collector and base currents in a typical BJT as a function of base-emitter voltage. The solid portions of the curves are based upon data reported by Jespers [60]. The dotted portions are explained in the text.

having solid portions based directly upon his measured points. He recorded I_C and I_B as functions of V_{BE} for wide-ranging current values. The dashed lines are based upon his extrapolations of the experimental data.

Exercise 4-69. In the regime where the I_C and I_B curves are linear and parallel, estimate the common-emitter current gain for this particular BJT.

The vertical spacing of the curves amounts to about two and one-half decades, or approximately 300.

Let us begin at the low-current end of the curve of log I_B versus V_{BE}. The low-current departure from linearity is of course a result of the SNS effect. Since this effect fades away at about $V_{BE} = 0.45$ V and $I_B = 0.1$ nA, it can be regarded as a truly low-level phenomenon. (Recall that in a previous data set shown in Figure 3-24, involving a much larger junction, the corresponding values were about $V_{BE} = 0.35$ V and $I_B = 10$ nA.) The linear segment immediately above this point, extending from about $I_B = 0.1$ nA to 10 nA, is the low-level linear regime. From $I_B = 10$ nA to 1 μA can be regarded as the intermediate-level regime in this tiny device, and above $I_B = 1$ μA can be regarded as the high-level regime. The last point corresponds to a V_{BE} value a bit over 0.7 V, in rough agreement with the estimate made in Exercise 4-55.

Now let us pretend that some of the high-level effects can be observed in isolation from all the others. Start with the Rittner effect. It would cause a rising departure from linearity in the I_B curve in the high-level regime, as is indicated by a dotted curve in Figure 4-42. Note, however, that this interpretation requires us to relabel the abscissa temporarily as $|V_{JE}|$ instead of V_{BE}, but doing so prohibits display of the ambipolar effect, which involves the difference between V_{BE} and $|V_{JE}|$. Therefore, let us simply say that reintroducing the ambipolar effect in combination with the Rittner effect causes the I_B curve to revert to the linear solid line observed experimentally. The cancellation of these two phenomena, discussed at length in Section 4-5.3, is so effective that the I_B curve retains its linearity up to about 1 mA and almost 0.9 V! Beyond this point, it is the $I_B R_{OB}$ loss in the extrinsic base region combined with emitter debiasing that causes an ultimate drooping of the I_B curve.

Now turn to the I_C curve. It is identically a Gummel plot, as in Figure 4-27. Because the BJT collector current consists almost exclusively of electrons injected into the base region (with no SNS complications), the I_C curve retains its linearity down to very low current levels. Let us once again pretend that a single effect can be observed in the absence of all the others, this time the Webster effect. Again it is necessary to reinterpret the V_{BE} axis as $|V_{JE}|$. Under these conditions, I_C would actually rise above the linear curve in the high-level regime, but "top out" as shown at a factor-of-two increase, which is about one third of a decade. But reverting from $|V_{JE}|$ to V_{BE} eliminates the rise, because the ambipolar effect sets in at precisely the same point as the Webster effect; both relate to the longitudinal electric field in the base region.

A further point is worth making in relation to the ambipolar effect. In a sense, it affects I_C and I_B separately, because each of these currents is tied to but one of the emitter-junction boundary values, and the two behave differently. Their behaviors, to review, are these: The I_B-connected boundary value exhibits "faster-than-Boltzmann" behavior with respect to $|V_{JE}|$, and "Boltzmann" behavior with respect to V_{BE}. The I_C-connected boundary value exhibits "Boltzmann" behavior with respect to $|V_{JE}|$, and "slower-than-Boltzmann" behavior with respect to V_{BE}. Hence the combined Webster and ambipolar effects cause the downward curvature in the I_C curve, labeled "Webster and ambipolar effects." (The Rittner effect has no impact on I_C.) But the experimental solid curve shows even more marked downward curvature. This is caused by the Kirk effect, which is appreciably "stronger" than the others.

Virtually all of the discussion of high-level phenomena to this point has dealt with a uniformly doped base region. Both clarity and simplicity were served by this choice. Now we shall permit significant one-dimensional doping variations, with the most common nonconstant profiles being included. Again it turns out that the consequences are surprisingly minor, especially in the high-level limit.

4-5.7 General Base-Region High-Level Analysis

In Section 4-4.2 we discussed the low-level behavior of nonuniformly doped base regions, with certain restrictions on the abruptness of variations in net-doping profile. Section 4-4.3, in turn, indicated how the Gummel number can characterize base regions with widely varying profiles. In previous portions of Section 4-5, we have described high-level behavior in a uniformly doped base region. Now we address a combination of these subjects in a comparatively general way, following the work of Das and Boothroyd [61], Lindmayer and Wrigley [62], and Huang [63].

To begin, however, let us continue with the case of a uniformly doped base region, and generalize with respect to the degree of high-level operation. Minority-carrier profiles have been calculated that can be considered generalizations of the lower curve in Figure 4-38. Figure 4-43(a) shows several profiles in normalized coordinates. The parameter is a measure of the level of injection, $\delta \equiv n_B(0)/N_{AB}$. Notice the concave-down character of the intermediate curves, and the linear character of the curves for low-level and high-level extrema.

Exercise 4-70. Confirm that the quantity plotted on the ordinate in Figure 4-43(a) is dimensionless.

The normalizing quantity is $I_C X_B/qAD_n$. Its dimensions are

$$\frac{[A][cm]}{[C][cm^2][cm^2/s]} = \frac{[A \cdot s]}{[C \cdot cm^3]} = \frac{1}{[cm^3]},$$

which are the same as those of $n_B(x)$, the quantity being normalized.

Now, turn to the case of nonuniform base-region doping. A negative net-doping gradient in the base region (density declining from emitter to collector) gives rise to a built-in longitudinal electric field. A BJT having this feature is often described as a *graded-base* transistor. The electric field enhances high-frequency performance in low-level operation. Here we develop a foundation for understanding the performance-enhancing effect and show why the effect declines with current; under high-level conditions, the enhancement vanishes. The reason for this, briefly, is that mobile-carrier behavior also causes longitudinal electric fields under high-level conditions (Sections 4-5.2 and 4-5.3), and these field components overwhelm those arising from net-doping variations.

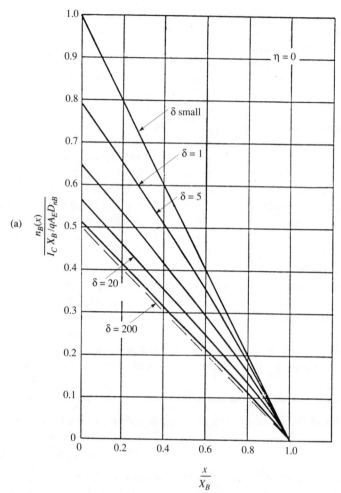

(a)

Figure 4-43 Normalized minority-carrier profiles in the base region of an *NPN* BJT. (a) The case of a uniformly doped base region, with injection level as a parameter, $\delta \equiv n_B(0)/N_{AB}$. (b) The case of an exponentially declining net-doping profile (a common graded-base condition), where the parameter η is a measure of the grading steepness, with $\eta = 0$ corresponding to uniform doping. All carrier profiles are for low-level operation. (c) The case of high-level and low-level injection into a steep ($\eta = 8$) exponentially graded base region. This case requires the more general injection-level parameter $\delta \equiv I_C X_B/qA_E N_{AB}(0) D_{nB}$. (After Lindmayer and Wrigley [62] with permission.)

General analysis of the graded-base transistor begins with the assumption that the majority-carrier current is zero everywhere in the base region. The validity of this assumption was shown for low-level conditions in Section 4-4.2. It is also valid for high-level conditions, provided the BJT continues to operate in a high-beta regime. Thus we can write for majority holes

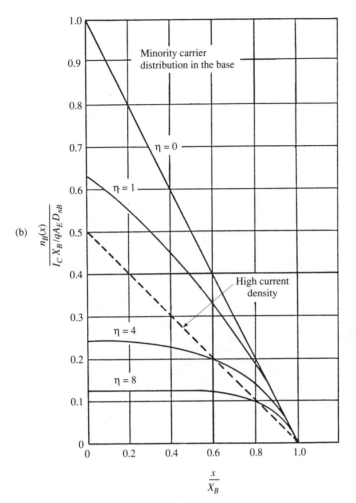

Figure 4-43 (*Continued*)

$$I_p(x) = 0 = qA_E\mu_{pB}p_B(x)E(x) - qA_E D_{pB}(dp_B/dx). \tag{4-83}$$

Using the fact that $D_{pB} = (kT/q)\mu_{pB}$, we can then solve Equation 4-83 for $E(x)$, obtaining

$$E(x) = \frac{kT}{q}\frac{1}{p_B(x)}\frac{dp_B}{dx}. \tag{4-84}$$

Assuming charge neutrality in the base region, $p_B(x) + N_{AB}(x) = n_B(x)$, we then arrive at

$$E(x) = \frac{kT}{q}\frac{1}{n_B(x) + N_{AB}(x)}\left[\frac{dn_B(x)}{dx} + \frac{dN_{AB}(x)}{dx}\right]. \tag{4-85}$$

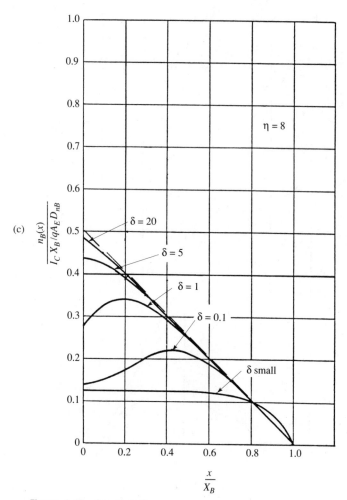

Figure 4-43 (*continued*)

From the initial assumption, it is clear that the minority-carrier current in the base region equals the collector current; thus

$$I_n(x) = I_C = qA_E\mu_{nB}n_B(x)E(x) + qA_E D_{nB}(dn_B/dx). \tag{4-86}$$

Substituting Equation 4-85 for $E(x)$ into this equation and using the fact that $D_{nB} = (kT/q)\mu_{nB}$, we find that

$$I_C = qA_E D_{nB}\left\{\left[\frac{dn_B(x)}{dx} + \frac{dN_{AB}(x)}{dx}\right]\left[\frac{n_B(x)}{n_B(x) + N_{AB}(x)}\right] + \frac{dn_B}{dx}\right\}, \tag{4-87}$$

which is the general expression for minority-carrier transport in the base region. The solution of this equation for all injection levels is possible only by numerical means.

Now let us consider special cases. Under low-level conditions, $n_B(x) \ll N_{AB}$, and hence Equation 4-87 becomes

$$I_C = qA_E D_{nB}\left[\left(\frac{n_B(x)}{N_{AB}(x)}\right)\frac{dN_{AB}(x)}{dx} + \frac{dn_B}{dx}\right], \quad (4\text{-}88)$$

which is consistent with Equation 4-41 in Section 4-4.2. Following the same procedures employed in that section, we find that

$$n_B(x) = \frac{I_C}{qA_E D_{nB} N_{AB}(x)} \int_x^{X_B} N_{AB}(x') \, dx', \quad (4\text{-}89)$$

and

$$I_C = \frac{qD_{nB} n_i^2 \exp(qV_{BE}/kT)}{\int_0^{X_B} N_{AB}(x) dx}, \quad (4\text{-}90)$$

which are consistent with Equations 4-45 and 4-47. For uniform doping these equations of course reduce to

$$n_B(x) = \frac{I_C X_B}{qA_E D_{nB}}\left[1 - \frac{x}{X_B}\right], \quad (4\text{-}91)$$

and

$$I_C = \frac{qD_{nB} n_i^2}{N_{AB}} \exp\left[\frac{qV_{BE}}{kT}\right]. \quad (4\text{-}92)$$

Now consider three possible graded-base doping profiles: exponential, error-function complement (erfc), and "double-diffused." The last term is sometimes used to identify the profile in a base region like that in Figure 4-26, a device shown in cross section in Figure 4-34; here the base region is defined by junctions formed in two solid-phase diffusion procedures, with base-type impurities introduced first, and emitter-type impurities introduced second. All three of these functions are good representations of actual and practical net-doping profiles. The simplest of the three is the exponential profile,

$$N_{AB}(x) = N_{AB}(0) \exp[-\eta(x/X_B)], \quad (4\text{-}93)$$

where η is a measure of the steepness of the profile. Typical values of η range from zero, representing uniform doping, to about eight, representing a rather extreme high-gradient case. Substituting Equation 4-93 into Equations 4-89 and 4-90 yields

$$n_B(x) = \frac{I_C X_B}{qA_E D_{nB}} \frac{1 - \exp\left[-\eta\left(1 - \frac{x}{X_B}\right)\right]}{\eta}, \quad (4\text{-}94)$$

and

$$I_C = \frac{qA_E D_{nB} n_i^2}{N_{AB} X_B}\left[\exp\left(\frac{qV_{BE}}{kT}\right) - 1\right]\left[\frac{1 - \exp(-\eta)}{\eta}\right]. \quad (4\text{-}95)$$

Equation 4-94 for the minority-carrier density profile is plotted in normalized fashion in Figure 4-43(b) for four values of η under low-level conditions, all using solid lines. The abscissa presents the normalized quantity x/X_B, and the ordinate, the quantity $n_B(x)/(I_C X_B/qAD_n)$.

For $\eta = 0$, the uniformly doped case in Figure 4-43(b), the profile is a straight line, as expected. For $\eta = 8$ (the extreme case), the profile is approximately horizontal from $x = 0$ to $x = 0.6 X_B$, and then decreases to zero at $x = X_B$. A horizontal profile implies that all minority-carrier transport is by drift and that none is by diffusion. Under low-level conditions, it is also possible to find analytical solutions for the cases of net-doping profiles of the erfc [62] and double-diffused [63] varieties. These solutions are qualitatively similar to those for the exponential case and are not presented here.

Now consider high-level conditions in the exponential graded-base region, so that $n_B(x) \gg N_{AB}(x)$. Of course, this inequality is not valid near the position $x = X_B$, but for simplicity we shall ignore this complication. Let us use the extreme case, $\eta = 8$, to dramatize the differences from the curves in Figure 4-43(a). This time it is necessary to generalize the parameter δ, because N_{AB} is not constant in the graded-base case. It can be written

$$\delta \equiv \frac{I_C X_B}{qA_E N_{AB}(0) D_{nB}}, \quad (4\text{-}96)$$

and once again $\delta = 1$ is the value separating low-level from high-level conditions; in the low-level limit, δ approaches zero.

To compare the curves in Figure 4-43(c), first note that for low-level conditions, the curve shown for $\eta = 8$ in Figure 4-43(b) is repeated. Then as injection level is increased, the profiles progressively approach the straight-line dashed curve for the limit of high-level operation. Thus the dashed curves in all three parts of Figure 4-43 are identical, indicating that doping-pattern differences have been "washed out." The intermediate profiles, however, show major qualitative differences from those for either extreme.

Under high-level conditions, the general equation becomes

$$I_C = qA_E 2D_{nB}\left(\frac{dn_B}{dx}\right). \quad (4\text{-}97)$$

This expression is identical to the low-level equation for the uniform-doping case except for the factor of two—the doubling of the diffusivity that was discussed in Sections 4-5.2 and 4-5.6. Thus, the electron-density profile is given by

$$n_B(x) = n_B(0)\left[1 - \frac{x}{X_B}\right], \quad (4\text{-}98)$$

which is plotted in Figure 4-43(c) as the dashed curve discussed previously. Thus, for high-level operations, only one curve applies for all base-doping profiles since Equation 4-97 is independent of $N_{AB}(x)$. This is a striking and unexpected result which again shows a case wherein the base-doping profile has little influence on device behavior.

4-6 EBERS-MOLL STATIC MODEL

In the era of germanium grown-junction [4] and alloyed-junction [6] BJTs, a large-signal model of fundamental importance was advanced by Ebers and Moll [64]. In spite of extensive BJT structural evolution since that time (Section 4-4.6), their model remains valid. Its primary use has been for static modeling. The original paper extended the model to deal with large-signal time-varying problems as well. But their dynamic model did not find wide acceptance, because it introduced for the extension somewhat awkward frequency-dependent current generators. For static modeling, however, the Ebers-Moll model is compact and economical. Furthermore, it is directly rooted in device physics. The physical properties stressed originally were common-base current gain, α_F and α_R, and the saturation currents exhibited by the two junctions in their BJT environment. The latter choice was a logical one in the germanium era, because as we have seen, germanium junctions are dominated by diffusion current in reverse bias and saturate cleanly. While the same approach still worked for silicon BJTs, a measure of physical relevance was lost. We now describe how that physical relevance was regained.

4-6.1 Gummel-Poon Reformulation

In 1970, a significant restatement of the Ebers-Moll model was contributed by Gummel and Poon [26]. It avoided the use of junction saturation currents, and common-base current gains, using instead β_F and β_R and the intercept current I_S that Gummel had previously introduced [18]. (See Section 4-4.3.) The reformulated Ebers-Moll model contributed by Gummel and Poon has become known in its rudimentary or unelaborated embodiment as the *transport form* of the Ebers-Moll model. In addition to using device properties relevant to modern silicon transistors, it is appreciably more transparent than the model in original form. For that reason, we shall develop the transport form of the model equations first and move to the original form later, showing how the two are related.

While the transport-form reformulation was a significant contribution, the Gummel-Poon paper went well beyond it. Their primary interests were (1) dynamic modeling for (2) computer analysis of BJTs, (3) incorporating realistic properties. Setting aside idealized devices, they wanted their model to deal with such items as emitter crowding, the Early effect, the Sah-Noyce-Shockley effect (space-charge-region recombination), the Webster-Rittner effect, and the Kirk effect.

Their approach to this challenge employed a variation and extension of charge-control techniques. The original methods, described and applied to the isolated junction in Sections 3-4.5 and 3-9, had been used extensively for the dynamic modeling of semiconductor devices. They were applied to the BJT by Beaufoy and Sparkes [65], with a result that is treated at some length in Section 4-7.6. Gummel himself contributed a different kind of charge-control model and applied it to the BJT in an earlier 1970 paper [66]. The essence of this charge-control model "of the second kind" is illustrated in Section 4-7.7. The further extension contained in the Gummel-Poon paper used a "phenomenological" or semiempirical approach to the complicated problem of the real BJT. That is, their model does not concern itself with the details of carrier distributions and physical mechanisms responsible for charge storage. For example, in the case of the Kirk effect it simply associates a certain packet of charge with the collector junction, rather than becoming involved with information such as that in Figure 4-40(a) (which, to be sure, was not available at that time). Their computer-oriented model employed twenty-one parameters that were to be assigned and then empirically adjusted by comparing model predictions to experimental data. It has been extensively used in the years since its introduction. The book by Getreu [67] gives a good account of related modeling developments in the 1970s.

Finally, significant success was achieved during the late 1970s and early 1980s in generating a fully analytic high-level model for the BJT [53, 54]. It can be regarded as a high-level version of the Ebers-Moll model. It combines previously separate analyses of high-level junction theory and ambipolar diffusion in the BJT context. Further, it yields accurate closed-form descriptions of certain effects (especially the Kirk effect) that previously had only been treated approximately, empirically, or numerically; examples of data calculated from this model are discussed in Section 4-5.4.

4-6.2 Assumptions and Problem Definition

A highly idealized device will be specified for present purposes. To leave no doubt about the device being analyzed we offer a set of assumptions more detailed and explicit than those of Ebers and Moll:

1. The emitter, base, and collector regions individually are uniformly doped. This requires that the junctions be step junctions.
2. In the base and collector regions the doping is extrinsic but nondegenerate.
3. If the emitter-region doping is degenerate, the region can nonetheless be described by the conventional semiconductor equations by employing empirically adjusted values of quantities such as diffusivity and intrinsic density.
4. In the base and collector regions, carrier lifetime is high.
5. Injected-carrier current densities are small.
6. The I–V characteristic of each junction is of the form

$$I = -I_{K0}[\exp(-qV_{NP}/kT) - 1], \quad (4\text{-}99)$$

where I_{K0} is a constant and is valid for that junction in the presence of the opposite junction when the opposite junction is open-circuited. (Gummel and Poon chose to employ a short-circuited condition on the opposite junction, which requires a different coefficient in the equation for the junction I–V characteristic.)

7. The positional dependence of the junction boundaries upon bias voltage may be neglected.
8. Breakdown phenomena may be neglected.

9. The problem is one-dimensional.
10. The ohmic contacts are ideal and of the high-recombination-velocity variety.
11. The emitter and collector regions are thick, or extensive in the x direction, relative to their respective minority-carrier diffusion lengths.
12. The current gains β_F and β_R are independent of current and voltage.

These assumptions constitute a measure of "overkill" intended to simplify and clarify the model's insights and construction. The Ebers-Moll model, in fact, achieved remarkable generality. In an appendix to the original paper, the authors showed that the model was practically geometry-independent. Also, we saw in Section 4-4.2 that base-region doping nonuniformities have a smaller-than-expected impact upon low-level BJT properties. And in Sections 3-6 and 3-7 it was pointed out that junctions departing significantly from step-junction doping profiles exhibit properties with great qualitative similarity to those of step junctions.

The aim of the present exercise is to write expressions for the currents at two of the BJT terminals in terms of the voltages at those terminals, with the third terminal being a common reference for defining the voltages. If current at the third terminal is of interest, it can be readily determined from the other two currents by applying Kirchhoff's current law. To define the terminal currents and voltages, it is especially convenient to use the common-base configuration; each voltage is thus applied directly to one junction only, as can be seen in Figure 4-44. For forward bias, the respective voltages V_{EB} and V_{CB} must obviously be negative. Ebers and Moll chose to open-circuit one junction while biasing the other; on the other hand, Gummel and Poon, as noted earlier, chose the short-circuit option. This was a key step in achieving a more economical and transparent model.

Exercise 4-71. Does choosing the common-base configuration as a vehicle for derivation mean that the resulting BJT model will be useful only for devices in that circuit?

No. The model will be based on device physics and will not depend on external-circuit details used to adjust currents or voltages.

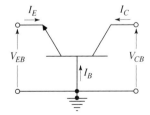

Figure 4-44 Common-base BJT configuration chosen for static analysis.

4-6.3 Equations in Transport Form

As a first step in the Gummel-Poon reformulation, we want to create the base-region condition depicted in Figure 4-45(a). This calls for a forward-biased emitter. Furthermore, the law of the junction combined with assumption 6 will guarantee that $n_B(X_B) = n_{OB}$ for the condition $V_{CB} = 0$. In other words, it is finally only necessary to short-circuit the collector junction, as in Figure 4-45(b), to assure existence of the desired base-region profile. Collector current can be written by inspection of Figure 4-45(a) as

$$I_C = A_E q D_n \left[\frac{n_B(0) - n_B(X_B)}{X_B} \right]. \tag{4-100}$$

Substituting boundary values given by the law of the junction under the specific conditions assumed yields

$$I_C = (A_E q D_n n_{OB}/X_B) \left[\exp(-qV_{EB}/kT) - 1 \right], \tag{4-101}$$

or

$$I_C = I_S \left[\exp(-qV_{EB}/kT) - 1 \right]. \tag{4-102}$$

The intercept current I_S is identically that defined in Equation 4-50.

(a)
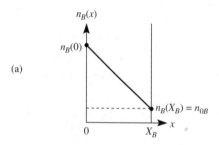

(b)
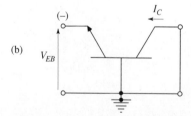

Figure 4-45 Operation at the boundary of the forward-active and forward-saturation regimes for easy analysis. (a) Profile of minority-carrier density in the base region. (b) Bias condition establishing profile in part (a).

Exercise 4-72. Show the equivalence of the two expressions for intercept current.

For the idealized BJT under consideration here,

$$\int_0^{X_B} N_{AB}(x)dx = N_{AB}X_B.$$

Substituting this result into Equation 4-50 yields

$$I_S = \frac{A_E q D_n n_i^2}{N_{AB} X_B} = \frac{A_E q D_n n_{0B}}{X_B}.$$

Exercise 4-73. Using Figure 4-45(a), justify the assertion that I_S is fixed by properties of the base region only.

The collector region is in quasiequilibrium; although it can receive a current, it cannot "cause" one. Total current through the emitter junction is determined by both the emitter and base regions, but at present we are only interested in the current of electrons into the base region, fixed by the base boundary of the emitter junction.

Since the intercept current for the present simplified case can be rewritten as

$$I_S = A_E q D_n \left[\frac{n_{0B}}{X_B} \right], \tag{4-103}$$

we are led directly to a physical interpretation. Equation 4-103 has the form of a diffusion current, wherein the responsible gradient is defined by base-region properties. In fact, the current value can in principle be experimentally observed in the circuit of Figure 4-46(a). It will be a tiny "leakage current" of electrons, but one associated with the base region rather than a junction. Because the density gradient involved is defined by n_{0B} and X_B, it is evident that changing both quantities by the same factor preserves the gradient, as is illustrated in Figure 4-46(b).

Exercise 4-74. Interpret Figure 4-46(b) in terms of Gummel number.

Changing n_{0B} and X_B by the same factor is the same as changing X_B and N_{AB} by inverse factors. In other words, the "coordinated change" in Figure 4-46(b) preserves the Gummel number. For the present simple structure, this shows why I_S is a function of Gummel number alone.

596 THE BIPOLAR JUNCTION TRANSISTOR

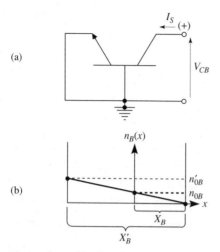

Figure 4-46 Physical interpretation of intercept current I_S. (a) Bias arrangement that in principle permits direct observation of I_S. (b) Coordinated change of base-region properties leaves I_S (and Gummel number) unchanged.

Exercise 4-75. Although the carrier profile in Figure 4-46(b) has sometimes been used to illustrate an observation of I_S in principle, a preferred illustration would short circuit the collector junction, and reverse bias the emitter junction. Why?

It makes the "in-principle" measurement easier. Because the emitter region is so heavily doped, it will contribute a negligible diffusion current of minority holes to the current observation. The collector region on the other hand, will contribute a significant diffusion current of minority holes in the situation shown in Figure 4-46(b).

The next step is to write an expression for emitter current in the circuit of Figure 4-45(b). Clearly,

$$I_E = -(I_C + I_B) = -\left(1 + \frac{1}{\beta_F}\right)I_C \qquad (4\text{-}104)$$

from Kirchhoff's current law and the definition of β_F. Combining this expression with Equation 4-102 yields

$$I_E = -I_S\left(1 + \frac{1}{\beta_F}\right)\left[\exp\left(-\frac{qV_{EB}}{kT}\right) - 1\right]. \qquad (4\text{-}105)$$

For the second part of the derivation we create a reverse-operation analog of conditions defined in Figure 4-45. That is, we will short-circuit the emitter junction and apply forward bias to the collector junction. Going through a series of steps completely parallel to those above then yields for emitter current in reverse operation,

$$I_E = I_S\left[\exp\left(-\frac{qV_{CB}}{kT}\right) - 1\right], \qquad (4\text{-}106)$$

and for collector current in reverse operation,

$$I_C = -I_S\left(1 + \frac{1}{\beta_R}\right)\left[\exp\left(-\frac{qV_{CB}}{kT}\right) - 1\right]. \qquad (4\text{-}107)$$

The next step is to note that Equation 4-105 states emitter current for forward operation, while Equation 4-106 states emitter current for reverse operation of the common-base BJT, with the opposite port short-circuited in each case. For *arbitrary* bias at the two ports, we are entitled (as explained later) to write emitter current as the *sum* of the expressions obtained for the two special cases:

$$I_E = -I_S\left(1 + \frac{1}{\beta_F}\right)\left[\exp\left(-\frac{qV_{EB}}{kT}\right) - 1\right] + I_S\left[\exp\left(-\frac{qV_{CB}}{kT}\right) - 1\right]. \qquad (4\text{-}108)$$

Proceeding in fully analogous fashion to obtain a companion expression for Equation 4-82, and combining those two yields the collector current for arbitrary biases:

$$I_C = I_S\left[\exp\left(-\frac{qV_{EB}}{kT}\right) - 1\right] - I_S\left(1 + \frac{1}{\beta_R}\right)\left[\exp\left(-\frac{qV_{CB}}{kT}\right) - 1\right]. \qquad (4\text{-}109)$$

Equations 4-108 and 4-109 are the Ebers-Moll equations in transport form. They yield the characteristics plotted in Figure 4-21.

Justification for the key step of adding two special-case expressions to obtain each general-case expression deserves emphasis and explanation. The principle of *superposition* is involved. Seeing its applicability to the BJT required considerable insight. In a familiar application of superposition, one deals with a two-port network of linear components. To find the current flowing in any given component of the network with arbitrary voltages V_1 and V_2 applied at the two respective ports, one follows this procedure: Observe the current in the component of interest with V_1 applied to port 1, and with port 2 short-circuited. Then observe the current at the same point with V_2 applied to port 2, and with port 1 short-circuited. Add the two currents algebraically, and the result will be the current in the component of interest when V_1 and V_2 are simultaneously applied to the respective ports. The nonobviousness of applying superposition in a BJT problem resides in the fact that the BJT is grossly nonlinear in terms of current–voltage relationships, while a *linear* system is necessary for the valid use of superposition.

The necessary linearity is internal to the BJT. Figure 4-45(a) and Equation 4-100 show that I_C is linearly related to $n_B(0)$. Similarly, for the conditions assumed to write Equation 4-106, I_E is linearly related to $n_B(X_B)$. But when forward voltages are applied simultaneously to the emitter and collector ports, yielding the simultaneous "application" of the boundary values $n_B(0)$ and $n_B(X_B)$, then the resulting current is the algebraic sum of the two special-case currents. Figure 4-47 displays the resulting superposition of density gradients, and hence of diffusion currents.

598 THE BIPOLAR JUNCTION TRANSISTOR

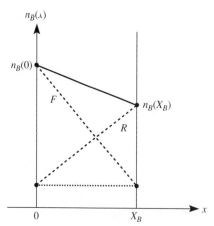

Figure 4-47 Applying the superposition principle in the base region. Addition of dashed profiles F and R gives the solid profile. By the same token, algebraic addition of the F and R gradients gives the net gradient, and algebraic addition of the F and R diffusion currents gives the net diffusion current.

Exercise 4-76. Figure 4-47 depicts a case of forward saturation. Given quantitative information, we could infer the magnitude of electron diffusion current through the base region. But in saturation, base current flows to both the emitter and collector regions and terminals. Doesn't this disrupt the simplistic superposition of electron currents? Assume that $\beta_F \gg \beta_R$, as in a typical real BJT.

No. Base current is comprehended in the Ebers-Moll equations. Look at Equation 4-108 for emitter current. The first major term is dominant, and base current enters through the small term $(1/\beta_F)$ in the coefficient of the first major term. In Equation 4-109 for collector current, the much larger term $(1/\beta_R)$ in the coefficient of the second major term accounts for the base-current fraction that enters into collector current. In this case, however, it affects the smaller of the two major terms.

Next, compare Equation 4-102 for I_C with Equation 4-106 for I_E. The only difference in the two right-hand sides is that V_{EB} occurs in the first, and V_{CB} in the second. In other words, the *same* voltage-to-current transfer function (current out for a voltage in) describes the common-base BJT in forward and reverse operation. Any two-port device, system, or network having identical forward and reverse voltage-to-current transfer functions is said to be *reciprocal*. The significantly nonobvious reciprocity of the common-base BJT was noted and proven by Shockley et al. in the very first publication reporting a working device [4]. Using a different method, Ebers and Moll also demonstrated this reciprocity in the same appendix that proved the geometry independence (or near-independence) of their analysis

[64]. The subject has continued to fascinate researchers, with a proof for the most general assumptions of all having appeared quite recently [68], using the original approach of Shockley et al. The recent work also summarizes treatments in the intervening years, each of which contributed another level of generality.

Exercise 4-77. In spite of the surprising generality in the reciprocal properties of a common-base BJT, there is a parasitic feature explicitly neglected by Ebers and Moll that degrades BJT reciprocity. What is the feature and why does it have this consequence in most real BJTs?

The Early effect degrades reciprocity. The depletion-layer encroachment caused by a given voltage increment is a function of doping-distribution details near the depletion-layer boundaries. In a typical BJT, these are very different for the emitter and collector junctions. Hence the Early-effect perturbation of base dimensions and Gummel number is typically different for forward and reverse common-base operation at given voltages.

Exercise 4-78. In Problem A4-28 we make the point that the common-base output characteristics are free of Early-effect distortion, while in Exercise 4-77 we blame the Early effect for common-base nonreciprocity. How can we have it both ways?

When the Early effect is defined in terms of output-characteristic slope, or conductance, it is true that the common-emitter BJT displays it and the common-base BJT does not. Another way of describing the Early effect, however, is to say that it causes a "voltage-dependent Gummel number." In these terms it affects both the common-emitter and common-base cases.

The issue of reciprocity is quite apart from the issue of superposition validity. Linearity of the system involved is necessary *and* sufficient for the valid application of the superposition principle. But linearity is *neither* necessary *nor* sufficient in a two-port system to guarantee its reciprocity. Two examples prove the point. The idealized common-base BJT itself is a preeminent example of a two-port entity with grossly nonlinear current-versus-voltage properties that exhibits reciprocity. On the other hand, a fundamental two-port circuit-building block known as the *gyrator* [69] is linear in its ideal form and is also explicitly nonreciprocal. (In a loose description, the gyrator transmits a signal in one direction with no phase change, and imposes a phase change of π radians on a signal passing in the opposite direction.)

Writing I_C and I_E in terms of the intercept current I_S, as is done in Equations 4-102 and 4-106, respectively, makes BJT reciprocity evident. Using the Gummel-number insight, as the transport formulation does, is a major factor in the simplicity of this derivation. (As we shall see, in the original case the reciprocity condition was applied after the fact to reduce what appeared initially to be four independently adjustable parameters to three.) Furthermore, the three BJT-characterizing parameters in the transport formulation are fixed by conditions that are comparatively localized within the device. That is, the intercept current I_S is fixed by base-region properties alone. In forward operation, the collector current is fixed by the parameter I_S and the variable V_{EB}. But the *emitter* current under these conditions is fixed by emitter-region as well as base-region properties, and by the same functional dependence on the *same* voltage variable V_{EB}. It follows, therefore, that α_F is fixed by the combination of emitter-region and base-region properties. Since $\beta_F = \alpha_F/(1 - \alpha_F)$, it further follows that the second transport-form parameter β_F is fixed by emitter-region and base-region properties. In a completely parallel way we can show that the third parameter β_R is fixed by collector-region and base-region properties.

4-6.4 Equations in Original Form

To facilitate comparisons of the transport and original forms of the Ebers-Moll equations, it is convenient to write both in compact matrix notation. Equations 4-108 and 4-109 thus become

$$\begin{bmatrix} I_E \\ I_C \end{bmatrix} = \begin{bmatrix} -I_S\left(1 + \dfrac{1}{\beta_F}\right) & I_S \\ I_S & -I_S\left(1 + \dfrac{1}{\beta_R}\right) \end{bmatrix} \begin{bmatrix} \left[\exp\left(\dfrac{-qV_{EB}}{kT}\right) - 1\right] \\ \left[\exp\left(\dfrac{-qV_{CB}}{kT}\right) - 1\right] \end{bmatrix} \quad (4\text{-}110)$$

This treatment of course emphasizes reciprocity once again, displaying the equality of the off-diagonal terms in the two-by-two matrix.

In Section 4-6.2 it was noted that for good reason, the common-base configuration was chosen as a basis for deriving the desired equations. Thus the original choice of common-base current gains α_F and α_R for two of the four parameters was logical. (It would have been nonobvious to choose β_F and β_R.) The other two parameters were the saturation currents, I_{EO} and I_{CO}. Going back to assumption 6, I_{EO} (for example) is intended to characterize the emitter junction in the BJT context, with the opposite port *open*. (The O subscript designates "open" and is distinguishable from the subscript designating equilibrium, because the latter symbol is a zero and is usually written first.) It is reasonable to speculate that Ebers and Moll chose the open option to diminish the effect of the collector junction on this emitter-junction observation. But with the benefit of hindsight we will assert that this choice complicated the analysis significantly and, in view of Section 4-6.3, unnecessarily.

Starting with Equation 4-109 and assumption 6, then, we can write for the two junctions

$$I_E = -I_{EO}\,[\exp(-qV_{EB}/kT) - 1], \quad (4\text{-}111)$$

and

$$I_C = -I_{CO}\,[\exp(-qV_{CB}/kT) - 1]. \quad (4\text{-}112)$$

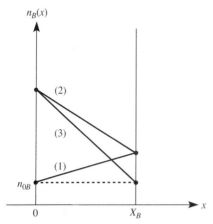

Figure 4-48 Minority-carrier density profiles in the base region of an arbitrarily assumed BJT, for (curve 2) forward bias at the emitter port and an open circuit at the collector port, (curve 3) the same emitter condition and a short-circuited collector, and (curve 1) the "difference" case. Treating these currents analytically using original-form parameters, along with three other currents for reverse operation, leads to the original Ebers-Moll equations.

Now apply a forward voltage at the emitter port, with the collector port open. The result will be a base-region electron profile that can be described as "forward-saturation-like." Such a profile is shown as curve 2 at the top in Figure 4-48. One explanation for this kind of profile is that the collector junction goes into forward bias through the "proximity effect." That is, an elevated boundary density is "applied" to the collector junction by the forward-biased emitter junction. The law of the junction demands a one-to-one relationship between voltage and boundary value without being concerned about which variable is independent and which is dependent. Another, and more physically detailed, explanation is that under the given conditions, the collector junction collects electrons, causing the collector region to assume a negative floating potential, thus forward biasing the collector junction.

Exercise 4-79. Given assumptions 1 and 10, describe the current components in the collector region that accompany base-region profile 2.

The arbitrarily assumed profile 2 shows approximately a factor-of-two increase in $n_B(X_B)$. Hence there will be the same factor of increase in the minority-hole density at the collector boundary of the junction (though a much larger absolute increase). Equilibrium densities will be maintained at the ideal ohmic contact. Thus holes will diffuse from junction to contact. The matching profile of excess (majority) electrons in the collector region will cause (net) diffusion of electrons to

> the ohmic contact at precisely the same rate, and there the carriers will recombine. In view of these equal and opposite hole and electron components of (conventional) current, the condition $I_C = 0$ is maintained. As usual, the diffusivity mismatch is compensated by the presence of a tiny electric field that slows electrons but has no effect on minority holes.

Equation 4-111, by definition, gives the emitter current associated with the open-collector profile 2 in Figure 4-48. If we now short-circuit the collector, emitter-current magnitude will increase, to a value associated with profile 3. We can immediately write down the emitter-current increment, because there is now a positive collector current in the amount I_C, causing the emitter-current increment of $-\alpha_R I_C$. The magnitude of this increment can be inferred from profile 1. (In the example arbitrarily selected, the current-component magnitudes increase in the sequence of numbering.) Thus, the emitter current with short-circuited collector can be written as

$$I_E = -\alpha_R I_C - I_{EO}[\exp(-qV_{EB}/kT) - 1]. \tag{4-113}$$

Applying similar arguments to a case of forward-biased collector junction and short-circuited emitter junction would yield

$$I_C = -\alpha_F I_E - I_{CO}[\exp(-qV_{CB}/kT) - 1]. \tag{4-114}$$

Recall that when analogous equations were written using the transport formulation, I_E was independent of I_C, and vice versa. Hence we could simply add (superpose) two special-case currents to get the general current. Here, however, we must solve the two special-case equations for a single current. To eliminate I_C, place Equation 4-114 in Equation 4-113, yielding

$$I_E = \alpha_F \alpha_R I_E + \alpha_R I_{CO}[\exp(-qV_{CB}/kT) - 1] - I_{EO}[\exp(-qV_{EB}/kT) - 1], \tag{4-115}$$

or

$$I_E(1 - \alpha_F \alpha_R) = -I_{EO}[\exp(-qV_{EB}/kT) - 1] + \alpha_R I_{CO}[\exp(-qV_{CB}/kT) - 1]. \tag{4-116}$$

Similarly, placing Equation 4-113 in Equation 4-114 yields

$$I_C(1 - \alpha_F \alpha_R) = -I_{CO}[\exp(-qV_{CB}/kT) - 1] + \alpha_F I_{EO}[\exp(-qV_{EB}/kT) - 1], \tag{4-117}$$

so the resulting equation pair in matrix form is

$$\begin{bmatrix} I_E \\ I_C \end{bmatrix} = \begin{bmatrix} \dfrac{-I_{EO}}{1 - \alpha_F \alpha_R} & \dfrac{\alpha_R I_{CO}}{1 - \alpha_F \alpha_R} \\ \dfrac{\alpha_F I_{EO}}{1 - \alpha_F \alpha_R} & \dfrac{-I_{CO}}{1 - \alpha_F \alpha_R} \end{bmatrix} \begin{bmatrix} \exp\left(-\dfrac{qV_{EB}}{kT}\right) - 1 \\ \exp\left(-\dfrac{qV_{CB}}{kT}\right) - 1 \end{bmatrix} \tag{4-118}$$

Invoking the known reciprocity of the device permits us to equate the off-diagonal elements, with the result

$$\alpha_F I_{EO} = \alpha_R I_{CO}. \qquad (4\text{-}119)$$

Thus only three of the four parameters that enter into this equation can be independently adjusted.

It is informative to write these four parameters in terms of the three transport-formulation parameters. Equating the corresponding matrix elements in Equations 4-110 and 4-118, and canceling pairs of minus signs where they occur, gives us

$$I_S\left(1 + \frac{1}{\beta_F}\right) = \frac{I_{EO}}{1 - \alpha_F \alpha_R}; \qquad (4\text{-}120)$$

$$I_S = \frac{\alpha_R I_{CO}}{1 - \alpha_F \alpha_R}; \qquad (4\text{-}121)$$

$$I_S = \frac{\alpha_F I_{EO}}{1 - \alpha_F \alpha_R}; \qquad (4\text{-}122)$$

$$I_S\left(1 + \frac{1}{\beta_R}\right) = \frac{I_{CO}}{1 - \alpha_F \alpha_R}. \qquad (4\text{-}123)$$

Eliminating I_{EO} from Equations 4-120 and 4-122 yields

$$\alpha_F = \frac{\beta_F}{1 + \beta_F}, \qquad (4\text{-}124)$$

which is of course the expected relationship of the two current gains. Similarly, Equations 4-121 and 4-123 yield

$$\alpha_R = \frac{\beta_R}{1 + \beta_R}. \qquad (4\text{-}125)$$

Placing the last two results in Equation 4-120 yields

$$I_{EO} = I_S\left(\frac{1 + \beta_F}{\beta_F} - \frac{\beta_R}{1 + \beta_R}\right). \qquad (4\text{-}126)$$

Similarly, Equation 4-123 yields

$$I_{CO} = I_S\left(\frac{1 + \beta_R}{\beta_R} - \frac{\beta_F}{1 + \beta_F}\right). \qquad (4\text{-}127)$$

Exercise 4-80. We noted earlier that one of the three parameters in the transport formulation is fixed by the properties of a single BJT region, while the other two are fixed by two regions apiece. Make parallel observations about the present four parameters.

The common-base current gain α_F is fixed by the properties of two regions, the emitter and base regions. This is shown by Equation 4-124, since β_F is fixed by the properties of the emitter and base regions. Similarly, β_R is fixed by the properties of collector and base regions, as Equation 4-124 shows. However, the saturation current I_{EO} is seen in Equation 4-126 to be a function of both β_F and β_R, so that properties of all three BJT regions are involved, and the same is true of I_{CO} [Equation 4-127].

4-6.5 Applications

The Ebers-Moll equations provide a wealth of detail concerning current-voltage relationships in the BJT. Suppose, for example, we are given a device having $\beta_F = 100$, $\beta_R = 2$, and $I_S = 10^{-14}$ A, and suppose we are then informed that $V_{EB} = -0.7$ V and $V_{CB} = -0.5$ V. This makes it clear that the BJT is in saturation, because both junctions are forward-biased. Further, with greater forward bias on the emitter than on the collector, we know that forward saturation specifically exists. But let us determine how deeply saturated it is by comparing the actual value of I_C with the value $\beta_F I_B$ that would be observed in the forward-active regime. From Equation 4-108,

$$I_E = -(10^{-11} \text{ mA}) \left\{ \frac{101}{100} [\exp(0.7 \text{ V}/0.02566 \text{ V}) - 1] - [\exp(0.5 \text{ V}/0.02566 \text{ V}) - 1] \right\} = -7.1058 \text{ mA}. \qquad (4\text{-}128)$$

From Equation 4-109,

$$I_C = (10^{-11} \text{ mA}) \left\{ [\exp(0.7 \text{ V}/0.02566 \text{ V}) - 1] - \frac{3}{2} [\exp(0.5 \text{ V}/0.02566 \text{ V}) - 1] \right\} = 7.0340 \text{ mA}. \qquad (4\text{-}129)$$

From these results

$$I_B = -(I_C + I_E) = -[(7.0340 \text{ mA}) + (-7.1058 \text{ mA})] = 0.0718 \text{ mA}, \qquad (4\text{-}130)$$

so that

$$\beta_F I_B = (100)(0.0718 \text{ mA}) = 7.18 \text{ mA}. \qquad (4\text{-}131)$$

Thus we learn that I_C falls below $\beta_F I_B$ by about 2%, so the BJT is indeed close to the forward-saturation "corner" point.

It was pointed out by Ebers and Moll that for voltage determinations, it is advantageous to solve the two equations simultaneously to obtain expressions for each junction voltage as a function of the currents. Doing so yields

$$V_{EB} = -\frac{kT}{q} \ln\left[1 - \frac{I_E + \left(\frac{\beta_R}{1+\beta_R}\right)I_C}{I_S\left(\frac{1+\beta_F}{\beta_F} - \frac{\beta_R}{1+\beta_R}\right)}\right] = -\frac{kT}{q} \ln\left[1 - \frac{I_E + \alpha_R I_C}{I_{EO}}\right]. \qquad (4\text{-}132)$$

It is curious that the original formulation gives a simpler result in this case than does the transport formulation! In a similar way we obtain

$$V_{CB} = -\frac{kT}{q} \ln\left[1 - \frac{I_C + \left(\frac{\beta_F}{1+\beta_F}\right)I_E}{I_S\left(\frac{1+\beta_R}{\beta_R} - \frac{\beta_F}{1+\beta_F}\right)}\right] = -\frac{kT}{q}\ln\left[1 - \frac{I_C + \alpha_F I_E}{I_{CO}}\right].$$

(4-133)

The equivalence of each pair of expressions can be confirmed by consulting Equations 4-124 through 4-127.

To illustrate the use of these equations, let us return to a point raised in Section 4-3.5. It was noted that in Figure 4-21, the value of $V_{EC(SAT)}$ (third quadrant) is significantly smaller than that of $V_{CE(SAT)}$. The discovery was made in the early 1970s by engineers developing I^{2}L [39, 47] that their upward-operated BJTs displayed gratifyingly small values of saturation voltage. Since these devices were employed as saturated switches, this feature was favorable. The saturation-voltage asymmetry is a consequence of $\beta_F - \beta_R$ asymmetry.

To examine this relationship let us determine the prediction of the Ebers-Moll equations, again for the device defined at the beginning of this section. With $I_B = 0.1$ mA, calculate $V_{CE(SAT)}$ at $I_C = \beta_F I_C/2$, and $V_{EC(SAT)}$ at $I_E = \beta_R I_B/2$.

Consider the first quadrant—forward operation—first. With $I_B = 0.1$ mA, and

$$I_C = (\beta_F I_B/2) = [(100)(0.1 \text{ mA})/2] = 5 \text{ mA}, \quad (4\text{-}134)$$

then

$$I_E = -(I_C + I_B) = -(5 \text{ mA} + 0.1 \text{ mA}) = -5.1 \text{ mA}. \quad (4\text{-}135)$$

From Equation (4-132),

$$V_{EB} = -(0.02566 \text{ V})\ln\left[1 - \frac{(-5.1 \text{ mA}) + (2/3)(5 \text{ mA})}{(10^{-11} \text{ mA})(1.01 - 2/3)}\right] = -0.6920 \text{ V}.$$

(4-136)

From Equation 4-133,

$$V_{CB} = -(0.02566 \text{ V})\ln\left[1 - \frac{(5 \text{ mA}) + \frac{100}{101}(-5.1 \text{ mA})}{(10^{-11} \text{ mA})\left(\frac{3}{2} - \frac{100}{101}\right)}\right] = -0.5901 \text{ V}.$$

(4-137)

Thus,

$$V_{CE(SAT)} = V_{CB} - V_{EB} = -0.102 \text{ V}. \quad (4\text{-}138)$$

For reverse operation, on the other hand, given $I_B = 0.1$ mA, and

$$I_E = (\beta_R I_B/2) = [2(0.1 \text{ mA})/2] = 0.1 \text{ mA}, \quad (4\text{-}139)$$

then

$$I_C = -(I_E + I_B) = -(0.1 \text{ mA} + 0.1 \text{ mA}) = -0.2 \text{ mA}. \quad (4\text{-}140)$$

From Equation 4-132,

$$V_{EB} = -(0.02566 \text{ V}) \ln\left[1 - \frac{(0.1 \text{ mA}) + \frac{2}{3}(-0.2 \text{ mA})}{(10^{-11} \text{ mA})\left(1.01 - \frac{2}{3}\right)}\right] = -0.5901. \quad (4\text{-}141)$$

From Equation 4-134,

$$V_{CB} = -(0.02566 \text{ V}) \ln\left[1 - \frac{(-0.2 \text{ mA}) + \left(\frac{100}{101}\right)(0.1 \text{ mA})}{(10^{-11} \text{ mA})\left(\frac{3}{2} - \frac{100}{101}\right)}\right] = -0.6084 \text{ V}. \quad (4\text{-}142)$$

Hence,

$$V_{EC(SAT)} = V_{EB} - V_{CB} = 0.018 \text{ V}, \quad (4\text{-}143)$$

and saturation voltage is about six times smaller in reverse operation than in forward operation of this particular BJT.

4-6.6 Equivalent-Circuit Model

The Ebers-Moll equivalent-circuit model for a BJT exhibits a symmetry that is a reflection of its reciprocity property. Each junction is modeled in piecewise-linear fashion with the result shown in Figure 3-21(b). As also shown there, the two components in each case are a voltage source V_γ to provide the offset voltage, and a diode that is ideal in the sense of an ideal rectifier. The leakage of each junction is modeled by an independent current source and is given the value that the present idealized BJT would exhibit in a common-emitter configuration (or its reverse), because of the importance of that configuration. The physical origin of this leakage current will be discussed below. Finally, forward and reverse current gains are modeled by means of current-controlled current sources.

With the resulting collection of eight elements in the model, predicted common-emitter output characteristics take on the highly idealized form shown in Figure 4-49. Specifically, $V_{CE(SAT)} = V_{EC(SAT)} = 0$, and output conductance $g_o = 0$ as well. The leakage currents I_{CEO} and I_{ECO} are best explained in the context of the model itself. As usual, the fidelity of the model could be improved by adding components. As a result, the idealization seen in Figure 4-49 could be moderated. The price of such an "improvement," however, would be a model that is harder to

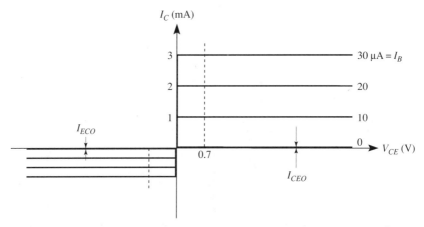

Figure 4-49 Idealized I–V output curves for forward and reverse BJT operation predicted by a general equivalent-circuit model. Compare these curves with those shown in Figure 4-21 that are predicted by the analytic Ebers-Moll model.

use. A good rule is to start with a simple model, and elaborate it only when necessary.

Figure 4-50 shows a series of six bias conditions imposed upon the equivalent-circuit model. The base terminal is open in Figure 4-50(a) (and in none of the others). With the condition $I_B = 0$ thus imposed, the two current-controlled current sources are deactivated. That condition is indicated by using light interconnecting lines to those model elements. Elements that are active are distinguished by heavy interconnecting lines. First, current from the independent source I_{ECO} simply circulates in the loop that also contains V_γ and a conducting ideal diode. Its current plays no external role.

Exercise 4-81. Why cannot some of the current from the source I_{ECO} "sneak" into the ground node?

If that did occur, the source I_{ECO} would extract an equal and opposite current *from* the ground node to satisfy its unvarying requirement, and the two currents would obviously cancel.

The only current that is seen at the terminals is the "enhanced" leakage current I_{CEO}. Its physical basis is that some holes from the collector region enter the base region. Unable to distinguish the source of holes entering the base region, the BJT treats these as base current and multiplies the hole current by β_F. In other words, the holes from the collector region continue on into the emitter region, and an enhanced current of electrons is injected by the emitter region, to constitute most of I_{CEO}.

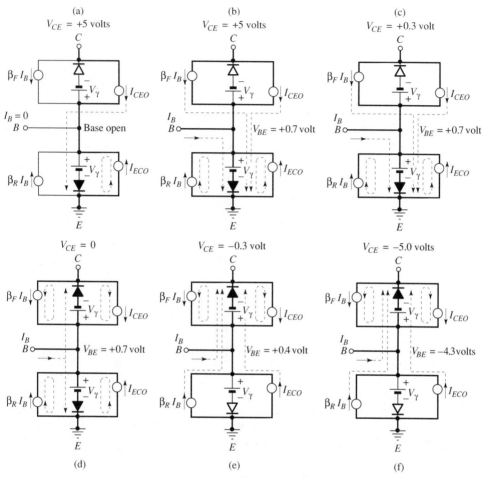

Figure 4-50 Using the equivalent-circuit model to predict currents under various bias conditions. (a) Common-emitter collector leakage current I_{CEO} is the only current seen at the terminals when V_{CE} is positive and $I_B = 0$. (b) Forward-active condition. (c) Forward-saturation condition. (d) Ambiguous case resulting from use of idealized elements in model. (e) Reverse-saturation condition. (f) Reverse-active condition.

Exercise 4-82. Determine I_{CEO} quantitatively as a multiple of I_S for the BJT defined at the beginning of Section 4-6.5.

For the condition $I_B = 0$, we have $-I_E = I_C$. Hence, since the collector junction is significantly reverse-biased, from Equations 4-108 and 4-109,

$$\left(1 + \frac{1}{\beta_F}\right)\left[\exp\left(-\frac{qV_{EB}}{kT}\right) - 1\right] + 1 = \left[\exp\left(-\frac{qV_{EB}}{kT}\right) - 1\right] + \left(1 + \frac{1}{\beta_R}\right),$$

or

$$-V_{EB} = \frac{kT}{q} \ln\left(\frac{\beta_F}{\beta_R} + 1\right).$$

Substituting this expression into Equation 4-109 gives us

$$I_C = I_{CEO} = \left(\frac{\beta_F + 1}{\beta_R} + 1\right)I_S = (51.5)\, I_S.$$

Exercise 4-83. Compare I_{CEO} as just calculated with I_{CO}, the open-emitter saturation current associated with the collector junction.

From Equation 4-127,

$$I_{CO} = I_S\left(\frac{1 + \beta_R}{\beta_R} - \frac{\beta_F}{1 + \beta_F}\right) = I_S\left(\frac{3}{2} - \frac{100}{101}\right) = (0.510)\, I_S.$$

Hence,

$$\frac{I_{CEO}}{I_{CO}} = \left(\frac{51.5}{0.510}\right) = 101 = (1 + \beta_F).$$

This reinforces the picture of a collector saturation current I_{CO} that is multiplied by the β_F gain mechanism to produce the leakage current

$$I_{CEO} = (1 + \beta_F)\, I_{CO}.$$

Only after some reflection, however, is it obvious that a saturation current defined with emitter open (I_{CO}) is the appropriate one to use in such a calculation. (This I_{CEO} calculation is examined further in Problem A4-39.)

Note that the idealized analysis above is not valid for a silicon BJT, wherein $I_{CEO} \approx I_{CO}$. As noted previously, the reason is that assumption 12 in Section 4-6.2 is not valid in the silicon case.

In Figure 4-50(b), base current is supplied, thus activating the sources $\beta_F I_B$ and $\beta_R I_B$. However, current from the latter merely circulates. The model informs us that the collector current is $\beta_F I_B + I_{CEO}$, and the voltages indicated show this to be a case of forward-active operation. Note that the two elements most essential to describing forward-active operation are the current-controlled current source $\beta_F I_B$ and the (lower) voltage source V_γ, needed to supply the offset voltage associated with the emitter junction. When these two components are preserved and the others are dropped, the result is the two-element equivalent-circuit model originally introduced in Figure 4-14, except that there V_γ is identified as "V_{BE}."

Exercise 4-84. Justify dropping six of the eight elements in the general equivalent-circuit model for the forward-active case.

The upper ideal diode is reverse-biased and can be replaced by an open circuit, which effectively eliminates the upper voltage source V_γ as well. The independent current source I_{CEO} is small and can be dropped, especially for a silicon device having $I_{CEO} \approx I_{CO}$. The currents of the lower controlled source $\beta_R I_B$ and independent source I_{ECO} simply circulate, so those two elements can be dropped. Finally, the lower ideal diode is forward-biased and can be replaced by a short circuit.

In Figure 4-50(c), the base terminal is now more positive than the collector terminal, so both junctions are now forward-biased and this is a case of saturation. The relative bias magnitudes further inform us that it is forward saturation.

Figure 4-50(d) illustrates a problem associated with an equivalent-circuit model incorporating idealized components. The model has no feature to determine how the current I_B will divide, so this case remains ambiguous. It is, however, significant that this shortcoming of the equivalent-circuit model is not shared by the analytic model.

In Figure 4-50(e), we have caused V_{CE} to change sign, thus creating a case of reverse saturation. For the first time the current sources $\beta_R I_B$ and I_{ECO} come into play. Both junctions are clearly forward-biased, but the base current goes unequivocally to the collector terminal.

Finally, in Figure 4-50(f) reverse-active operation is established. Notice that (neglecting leakage current) the *collector* current is now $(\beta_R + 1)I_B$ rather than just $\beta_R I_B$. This is a point that was made in connection with Figure 4-21.

4-7 SMALL-SIGNAL DYNAMIC MODELS

We have seen that the BJT is grossly nonlinear overall. But when it is to be used in an amplifier application for sufficiently small signals, simplification is possible. In particular, the linearization methods described in Section 3-8 for the case of a *PN* junction are equally applicable to the bipolar junction transistor. As a result, the powerful and extensive methods of linear circuit analysis can be brought into play for analyzing amplifier circuits.

4-7.1 Low-Frequency Hybrid Model

We wish to develop a small-signal equivalent-circuit model for the BJT under low-frequency conditions. Let us confine attention to the common-emitter configuration because of its importance. The first step is to represent the BJT in this configuration, as has been done in Figure 4-51, indicating the quasistatic current and voltage variables at each port; then we apply basic knowledge developed in foregoing portions of this chapter.

Examine the input port first. In Section 4-3.1 we made the point that it is unprofitable to attempt external control of V_{BE} under normal conditions. The emitter junction's steeply rising current and the temperature dependence of its offset voltage

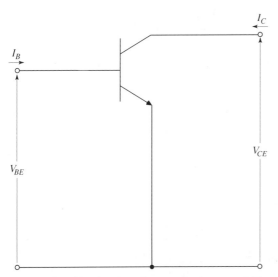

Figure 4-51 Quasistatic voltage and current variables for the common-emitter configuration, restated for formal generation of the hybrid model.

enforce this conclusion. Base current I_B, on the other hand, is readily controlled externally, using a circuit as simple as a resistor (R_B) in series with a voltage source (V_{BB}). Hence, I_B is the logical choice as independent variable at the input port. Once I_B has been adjusted, the BJT will "set V_{BE}," which is to say, V_{BE} is a dependent variable.

Now turn to the output port. Assume that the BJT is operating in the forward-active regime. This choice is appropriate if the BJT is to be used as an amplifying device. It is in amplifier applications, in turn, that small-signal models are most useful. Given these circumstances, the output current I_C is clearly a dependent variable, since $I_C = \beta_F I_B$, and I_B has already been fixed. The variable V_{CE}, on the other hand, can once again be set easily by means of a resistor (R_C) in series with a voltage (V_{CC}). Once this has been done, adjustment of *either* R_C or V_{CC} will alter V_{CE}. It is of course true that (as noted earlier), a voltage source alone can be applied at the output port to set V_{CE} without hazard to the device. But the combination of a resistor and voltage source is more generally useful. This done, input and output dependent variables, respectively, can then be written as functions of the two independent variables:

$$V_{BE} = V_{BE}(I_B, V_{CE}); \tag{4-144}$$

$$I_C = I_C(I_B, V_{CE}). \tag{4-145}$$

As the next step, write the total differential of each of these quasistatic variables:

$$dV_{BE} = \frac{\partial V_{BE}}{\partial I_B} dI_B + \frac{\partial V_{BE}}{\partial V_{CE}} dV_{CE}; \tag{4-146}$$

$$dI_C = \frac{\partial I_C}{\partial I_B} dI_B + \frac{\partial I_C}{\partial V_{CE}} dV_{CE}. \qquad (4\text{-}147)$$

The four differentials appearing in these equations are of course infinitesimal in the mathematician's rigorous sense. But as engineers we convert each differential into a finite though small excursion, with the criterion for what constitutes "small" having been discussed in the introduction to Section 3-8, and in Section 3-8.1. Using the symbols ΔV_{BE}, ΔI_B, ΔI_C, and ΔV_{CE} to represent such small but finite differences, we can then rewrite Equations 4-146 and 4-147 as

$$\Delta V_{BE} = \frac{\partial V_{BE}}{\partial I_B} \Delta I_B + \frac{\partial V_{BE}}{\partial V_{CE}} \Delta V_{CE}; \qquad (4\text{-}148)$$

$$\Delta I_C = \frac{\partial I_C}{\partial I_B} \Delta I_B + \frac{\partial I_C}{\partial V_{CE}} \Delta V_{CE}. \qquad (4\text{-}149)$$

With these equations a shift in point of view becomes possible. We can say that the difference Equations 4-148 and 4-149 are the result of expanding each of the functions given in Equations 4-144 and 4-145 in a Taylor's series about the quiescent point, and then discarding all terms of order higher than the first.

It is of course conventional in circuit and device analysis to use yet another symbolism for a small signal—a lowercase letter for the variable and lowercase subscripts. In the present case these are v_{be}, i_b, i_c, and v_{ce}. The voltage and current excursions represented by these symbols may be repetitive or nonrepetitive, unidirectional or bidirectional. For many purposes it is useful to choose a specific bidirectional and repetitive case, namely, a sinusoidal signal of small amplitude; this choice is particularly advantageous because the mathematics of the case are so convenient and highly developed. But choosing the sinusoidal case is by no means obligatory.

At this point some comments on notation are in order, augmenting and extending those made for the diode in Section 3-9.2. We use the widely accepted standard wherein, for example,

$$v_{BE} = V_{BE} + v_{be}. \qquad (4\text{-}150)$$

This states that the instantaneous total voltage v_{BE} is the sum of a static component V_{BE} and a signal component v_{be}. The corresponding small difference, appearing on the left-hand side of Equation 4-148, and equivalent to v_{be}, can thus be written

$$\Delta V_{BE} = v_{BE} - V_{BE}. \qquad (4\text{-}151)$$

A point that needs to be emphasized is this: The four continuously variable quantities in Equations 4-144 and 4-145 are *quasistatic* in the sense of Section 3-8. But sometimes a particular fixed value is desired—the static component of a time-dependent function. In a simpler situation, one easily distinguishes between the two cases with subscripts. For example, x and y are the continuous variables of the x–y plane. But if we wish to specify fixed values, such as the coordinates of point 1, we write (x_1, y_1).

We could employ a similar approach for distinguishing between the quasistatic (but variable) value of V_{BE} in Equation 4-144, and the fixed value of V_{BE} in Equation 4-151, but like most authors, we prefer not to complicate the notation to that degree. Which case is intended is usually clear from context. The reason for making this point is that some authors attempt to sidestep the issue by writing "dv_{BE}" instead of dV_{BE} and "Δv_{BE}" instead of ΔV_{BE}. This is at best illogical because it describes "an increment upon an increment"—see Equations 4-150 and 4-151—and that is not at all what is intended.

The last step, then, in generating the desired equations is to replace each partial derivative by the symbol h with distinguishing subscripts. These four quantities are known as the low-frequency *hybrid parameters*. For convenient subsequent reference, let us state these definitions explicitly:

$$\frac{\partial V_{BE}}{\partial I_B} \equiv h_{ie}; \tag{4-152}$$

$$\frac{\partial V_{BE}}{\partial V_{CE}} \equiv h_{re}; \tag{4-153}$$

$$\frac{\partial I_C}{\partial I_B} \equiv h_{fe}; \tag{4-154}$$

$$\frac{\partial I_C}{\partial V_{CE}} \equiv h_{oe}. \tag{4-155}$$

Finally, introducing the new symbols defined in Equations 4-152 through 4-155 and the lowercase symbols for small signals into Equations 4-148 and 4-149 gives us the small-signal low-frequency *hybrid equations*.

$$v_{be} = h_{ie}i_b + h_{re}v_{ce}; \tag{4-156}$$

$$i_c = h_{fe}i_b + h_{oe}v_{ce}. \tag{4-157}$$

Exercise 4-85. In what sense are the two equations immediately above "hybrid?"

The pair of equations are appropriately described as hybrid because they present an input voltage and an output current. Consequently the coefficients (the hybrid parameters) possess mixed dimensions. We shall see later that choosing two currents as a function of two voltages, or vice versa, yields coefficients with consistent units.

The symbol h with letter subscripts will be used exclusively here to stand for a real quantity. Hence only low-frequency characterization will be done with the hybrid parameters. The corresponding symbols for use at higher frequencies, where reactive properties must be considered, we shall distinguish by means of number subscripts. (Not all authors maintain this distinction, but it is a simple one and should be quite clear.) This is consistent with circuit-analysis practice, and it is in that case that the more general analysis is important.

We have seen that a hybrid parameter is fundamentally a partial derivative. When we write $h_{ie} = \partial V_{BE}/\partial I_B$, it is not necessary to add (as is sometimes done) "with $v_{be} = 0$" or "with V_{BE} = constant," because that information is embodied in the definition of a partial derivative. In Equation 4-154, however, it is evident that another valid expression for the same hybrid parameter is $h_{ie} = (v_{be}/i_b)$ with $v_{ce} = 0$. This time the specification on v_{ce} is needed.

Now let us examine the meaning of the hybrid parameters. The subscript e appearing on each of the four hybrid parameters informs us that they apply to the common-emitter configuration. Similar pairs of equations can be written for the common-base and common-collector configurations as well. When this is done, the twelve resulting hybrid parameters can be interrelated. That is, one set can be written in terms of the others. The procedure is straightforward but tedious, and will not be done here because the common-emitter case significantly overshadows the other two in importance.

The remaining four subscripts on the hybrid parameters can be explained as follows: Return to Equations 4-152 through 4-155, where the definitions are stated. The subscript i on h_{ie} stands for *input*. The independent variable in question is I_B, the independent variable at the input port. Because h_{ie} is the rate of change of input voltage with respect to input current, h_{ie} is an input resistance.

In an analogous way, the subscript o on h_{oe} stands for *output*. Inspection of Equation 4-155 shows that h_{oe} is an output conductance, the quantity we have labeled g_o previously. The parameter h_{oe} (also by inspection) is a *reverse voltage gain*, and is dimensionless. It is typically quite small in a BJT. Finally, h_{fe} is a *forward current gain*. It is the small-signal analog of the static parameter β_F. In fact, for parallelism in notation we sometimes employ the definition

$$\beta_F \equiv h_{FE}. \tag{4-158}$$

In a typical BJT there is an appreciable collector-current range in which

$$h_{fe} \approx h_{FE}. \tag{4-159}$$

Exercise 4-86. From the data presented in Figure 4-42, plot a curve of I_C versus I_B, using linear calibration on both axes. For any point on the curve you have constructed, what feature is proportional to h_{FE}? to h_{fe}?

At any point on the curve, h_{FE} is proportional to the slope of a line drawn from the origin to that point, while h_{fe} is proportional to the slope of the curve at that point.

Exercise 4-87. Describe qualitatively the curve of Exercise 4-86, commenting on the h_{fe} peak and its location as compared to that for the h_{FE} peak. Refer these peak locations to the I_B axis.

When a low value of current is chosen as the full-scale value, the curve starts with near-zero slope, and is markedly concave-up. In Figure 4-42, the interval separating the two curves is a measure of h_{FE}. Thus it is evident that h_{FE} for the device described there is approximately constant in the range 1 nA $< I_B < 100$ nA. When a current in this range is chosen as the full-scale value, the curve is nearly linear, but possesses a subtle S shape, with one point of inflection. Below this point the curve is concave-up, and above it, concave-down, especially in the high-level regime. From the answer to Exercise 4-86, it is evident that the maximum value of h_{fe} occurs at the point of inflection, and that h_{fe} exceeds h_{FE} by a small factor in the regime where both current gains are rising. The peak value of h_{fe} is typically a few percent higher than that of h_{FE} and occurs at a current usually less than ten times lower than that of the h_{FE} peak. (In Figure 4-64, very closely related curves of gain versus current, calculated with the aid of SPICE modeling, are presented, and the qualitative observations just offered are confirmed.)

4-7.2 Hybrid Model and Device Physics

It is sometimes convenient to write Equations 4-156 and 4-157, the hybrid equations, in matrix form. When this is done, it becomes evident from the discussion immediately preceding that the diagonal elements of the two-by-two matrix are an input resistance and output conductance, respectively, while the off-diagonal elements are a reverse voltage gain and a forward current gain, respectively. The first step in a physical interpretation of the model is to interrelate the formally generated coefficients, the hybrid parameters, in an equivalent-circuit model. This has been done in Figure 4-52. Let us interpret the hybrid parameters one by one, starting with h_{ie} as defined in Equation 4-152.

Exercise 4-88. The input resistance h_{ie} is to be measured with a short-circuited output port, which is to say, $v_{ce} = 0$, or V_{CE} = constant. As a practical matter, how is the measurement made?

The condition V_{CE} = constant (which is if course the same as $dV_{CE} = 0$), can be created by shunting the output port with a battery of voltage V_{CE}, or with a large capacitor.

The resistance h_{ie} effectively shunts the input port because the ideal voltage source $h_{re}v_{ce}$ has zero resistance. Furthermore, the circuit element incorporating h_{re} is a voltage-controlled voltage source, and with $v_{ce} = 0$, the voltage-source voltage is zero. The controlling voltage v_{ce} we identify with the independent variable chosen for the output port, V_{CE}.

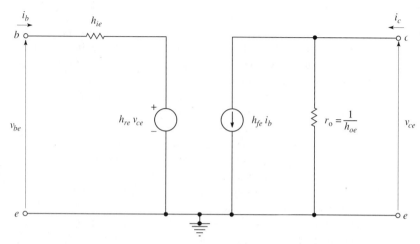

Figure 4-52 Equivalent-circuit model incorporating the four hybrid parameters.

Likewise, the current-controlled current source at the output port is

$$i_c = h_{fe}i_b. \tag{4-160}$$

This equation is the small-signal analog of the familiar static relationship $I_C = \beta_F I_B$. And finally, the resistor r_o shunting the output port in Figure 4-52 is the reciprocal of h_{oe}, the output conductance.

Having thus interconnected circuit elements that can be described in terms of the four hybrid parameters, let us turn to the physical basis of each, starting "from the top" with h_{ie}. Return to the planar BJT shown in cross section in Figure 4-34. It is evident that the conducting base-to-emitter path poses a complicated problem, at best two-dimensional and involving distributed elements (as opposed to lumped elements). Even the emitter region and contact involve some parasitic series resistance. Typically the contact resistance is so small (of the order of an ohm) compared to others in the path that we shall neglect it. It is, however, included in the computer-oriented SPICE model of Section 4-8.

A key part of the base–emitter path is the emitter junction, which we shall treat separately. Next in sequence comes the intrinsic base region, and then the extrinsic base region and the base contact. (See Figure 4-28 for identification of these two base-region portions.) The distributed nature of the intrinsic base-region problem is the main point of Figure 4-7; the base current passes through it in a two-dimensional pattern. Treating this geometrically complicated resistance as a lumped element, we observe a value that is typically in the range from 50 to 100 ohms. The extrinsic base region that is in series with it also has a distributed nature, because the base-contact stripe has a finite width. (We consistently use the term *width* to designate a dimension parallel to the surface of the device.) Base-contact resistance is of course in series with the distributed extrinsic resistance. The lumped-resistor equivalent to the combined base-contact and extrinsic-base resistance has a typical value of a few ohms, designated R_{OB} in the discussion of high-level effects. Even though this value is appreciably smaller than the intrinsic lumped resistance, it must be considered for large base current, because an *IR* drop in R_{OB} debiases the *entire* emitter junction.

The simplified equivalent circuit almost always used to represent the base–emitter path consists of a single lumped resistor representing all of the base-resistance components in series with a *PN* junction. The node between these two major components is often designated b' for dynamic purposes (or B' for static purposes.) It is an "effective" and physically inaccessible internal node. Thus the actual voltage on the emitter junction for the present assumptions is $V_{B'E}$, essentially the same in magnitude as the imposed junction voltage V_{JE}. In the current range where simple junction theory prevails, emitter current is proportional to $\exp(qV_{B'E}/kT)$. Differentiating the emitter-current expression with respect to $V_{B'E}$ to find emitter-junction conductance then yields

$$g_{\text{EMITTER JUNCTION}} = -qI_E/kT, \quad (4\text{-}161)$$

where the negative sign enters because I_E is negative when the emitter junction is forward-biased. But since the base–emitter port is the input port, the relevant current is

$$I_B = -I_E/(1 + \beta_F); \quad (4\text{-}162)$$

hence, the emitter junction when viewed from the base terminal exhibits a conductance of

$$g_{b'e} = qI_B/kT, \quad (4\text{-}163)$$

or a dynamic resistance of

$$r_{b'e} = kT/qI_B. \quad (4\text{-}164)$$

This resistance is of course in series with the series combination of the extrinsic and intrinsic base-region resistances just discussed, which can be identified as the resistance from the external node b to the internal node b'. For this reason the (combined) ohmic base resistance is often identified as $r_{bb'}$. Finally, then, we express the input-resistance hybrid parameter as

$$h_{ie} = r_{bb'} + r_{b'e} = r_{bb'} + kT/qI_B. \quad (4\text{-}165)$$

Exercise 4-89. Calculate the value of $r_{b'e}$ in a BJT with $I_C = 1$ mA and $\beta_F = 100$.

For the given conditions, $I_B = 10$ μA, and

$$r_{b'e} = \frac{kT/q}{I_B} = \left(\frac{0.02566 \text{ V}}{10^{-5} \text{ A}}\right) = 2{,}566 \text{ ohm.}$$

This value of a few kilohms is typical of a small device, and appreciably larger than the value of $r_{bb'}$ (50 to 100 ohms) that is in series with it.

The parameter h_{re} is often omitted from the equivalent-circuit model because it is small—usually smaller than 10^{-3}. Unilateral gain is an important property in an amplifying device, and an ideal BJT would have $h_{re} = 0$. The small amount of voltage feedback that does exist arises from voltage modulation of the effective base thickness X_B. This is obviously an important occurrence in the intrinsic base region only, and is identically the kind of base-thickness modulation that causes the Early effect.

Exercise 4-90. Verify the algebraic sign on the voltage-source symbol $h_{re}v_{ce}$ in Figure 4-52.

A positive voltage increment at the output port causes the intrinsic base region to become thinner, and therefore causes the associated lateral base resistance to increase. As is evident in Figure 4-52, with I_B held constant, the resulting increase in h_{ie} will cause an increase in V_{BE}, thus verifying a positive feedback voltage.

Because h_{re} represents a positive voltage feedback, there is a simple alternative to a voltage source in the input path for those cases where voltage feedback is important. In the small-signal hybrid equivalent-circuit model, we can simply connect a feedback resistor $r_{b'c}$ from the internal node b' to the output node c. This option is illustrated in Figure 4-53, along with changes in the input circuit that follow from the preceding analysis.

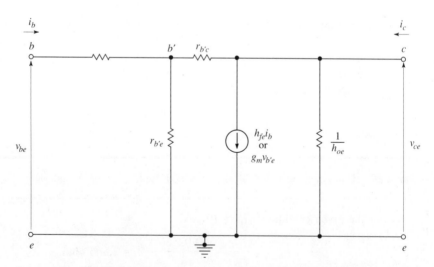

Figure 4-53 Modified hybrid equivalent-circuit model displaying the resolution of h_{ie} into $r_{bb'} + r_{b'e}$ (where $r_{b'e}$ is the dynamic resistance of the emitter junction), and substituting the feedback resistor $r_{b'c}$ for the feedback-voltage source $h_{re}v_{ce}$.

Exercise 4-91. For the BJT and bias condition of Exercise 4-89, calculate the value of $r_{b'c}$ needed to provide the functional equivalent of $h_{re} = 10^{-3}$. You may neglect the effect of $r_{bb'}$ for this calculation.

For the two kinds of voltage feedback to be equivalent, we must have, for $r_{bb'} \approx 0$,

$$\frac{v_{be}}{v_{ce}} = \frac{v_{b'e}}{v_{ce}} = 10^{-3} = \frac{r_{b'e}}{r_{b'e} + r_{b'c}}.$$

Hence,

$$r_{b'c} = \left[\frac{r_{b'e}(1 - 10^{-3})}{10^{-3}}\right] = 2.56 \text{ M ohm}.$$

This very large feedback resistor corresponds to small voltage feedback.

Physical explanations for the remaining two hybrid parameters are self-evident. The current source $h_{fe}i_b$ is associated with transistor action, its dynamic manifestation, as introduced in static form in Section 4-2.4. Lastly, the output conductance h_{oe} is a consequence of the Early effect (Section 4-4.5), combined with the breakdown phenomenon (Section 4-4.4), and when applicable, with the reachthrough phenomenon (Sections 4-4.4 and 3-6.3).

4-7.3 BJT Transconductance

The rate of change of output current with respect to input voltage is termed the *transconductance* of an amplifying device. For the BJT,

$$g_m \equiv \frac{\partial I_C}{\partial V_{BE}}. \tag{4-166}$$

The prefix *trans* (across) enters because the definition shows the dependence of an output variable upon an input variable; the fact that this partial derivative has conductance dimensions accounts for the balance of the term. Transconductance is of crucial importance because it describes the ability of the device to "drive" a load. For example, it determines how quickly the device can charge a shunt capacitance (often parasitic) at the output port. Because of the basic importance of transconductance, it is often used as a figure of merit for an amplifying device—sometimes by itself and sometimes in combination with other device-characterizing quantities.

Among the distinguishing features of the BJT are remarkable transconductance properties. Equation 4-48 in Section 4-4.3 informs us that output current for the common-emitter BJT can be written

$$I_C = I_S \exp\left(\frac{qV_{BE}}{kT}\right). \tag{4-167}$$

Thus it is immediately apparent that

$$g_m = \frac{q}{kT} I_S \exp\left(\frac{qV_{BE}}{kT}\right) = \frac{qI_C}{kT}. \tag{4-168}$$

With g_m proportional to I_C, we can in principle boost g_m to any desired level simply by raising current.

Exercise 4-92. Find the output current I_C for which transconductance equals one Siemens.

From Equation 4-168,

$$I_C = (kT/q)g_m = (26.7 \text{ mV}) (1 \text{ S}) = 26.7 \text{ mA}.$$

The value $g_m = 1$ S constitutes an enormous transconductance, as can be seen by comparing the BJT with competing devices. (This matter is examined in some detail in Section 5-6.1, where the BJT and MOSFET are directly compared with respect to transconductance.) The fact that $g_m = 1$ S is reached at the modest current $I_C = 26.7$ mA illustrates the first important feature of BJT g_m—sheer magnitude. The second point is that at a given current and temperature, BJT transconductance will be the same whether the device is made of germanium, silicon, gallium arsenide, or gallium phosphide! Finally, the same value will be observed in a device with a junction area of 1 cm², or 10^{-8} cm² (one square micrometer), provided the current chosen permits one to overlook parasitic effects. The reason underlying all of these surprising facts is that the essence of BJT operation is the modulation of a potential barrier that controls the "spillage" or flow of carriers over it. The incremental spillage is proportional to the steady-state flow.

For some device-modeling purposes, especially at high frequency, it is advantageous to introduce g_m into the equivalent-circuit model. This has been done for the BJT in Figure 4-53, by indicating that the current-controlled current source $h_{fe}i_b$ can be converted to a voltage-controlled current source, $g_m v_{b'e}$. The advantages of this change will be indicated below.

4-7.4 Hybrid-Pi and Other Models

The low-frequency hybrid-parameter model can be extended for use in the medium-frequency range by adding two capacitances. The resulting equivalent circuit is shown in Figure 4-54, and is known primarily as the *hybrid-pi model*, but sometimes as the Giacoletto model [70]. The designation "pi" arises because the configuration of its components resembles the Greek letter. It has proven to be the most useful and versatile BJT model because of two important features: (1) Its components, both real and imaginary, are fixed, frequency-independent elements; and (2) these elements can be realistically related to device physics. On the latter point, the

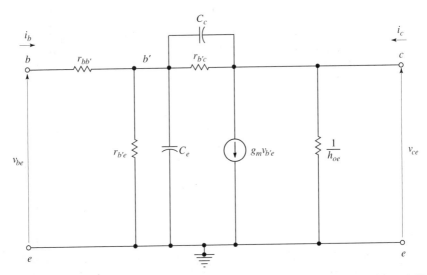

Figure 4-54 The hybrid-pi model. It is identical to the hybrid model of Figure 4-53 except for the addition of the capacitances C_e and C_c.

elements carried over from the hybrid model have the physical significance described in Sections 4-7.2 and 4-7.3. And the two new elements (capacitances) relate directly to the emitter and collector junctions. That is, because forward-active operation is assumed, C_e is the sum of the emitter-base diffusion capacitance C_{SE}, and the emitter-junction depletion-layer capacitance C_{TE}, while $C_c = C_{TC}$ is the collector-junction depletion-layer capacitance. Table 4-2 gives typical values for the hybrid-pi-model parameters of a small BJT of about 1970 [71], and also for a smaller device more recently described [72] whose parameters will be calculated in Section 4-8.7.

One could of course create a hybrid-pi model for reverse-active operation. While the resulting equivalent-circuit model would resemble that in Figure 4-54, some of the terminals and elements would require relabeling, and most of the element values would be quite different. The overriding importance of the common-emitter configuration, especially in forward-active operation, explains why it is the usual subject of small-signal modeling. The greater symmetry of the common-base configuration would favor its choice, but practical considerations usually outweigh that advantage.

In addition to the hybrid and the hybrid-pi models, three other small-signal models are advantageous in certain circumstances. All five models are summarized in Table 4-3. Notice first that the hybrid and hybrid-pi models share the advantage of having fixed elements. The latter model, at the cost of but two additional elements, is appreciably more powerful than the former. The remaining three models employ complex elements. Since both the h-parameter model and the hybrid-parameter model employ the symbol h as coefficient, we shall distinguish between the two sets of symbols (as noted earlier) by using numbers as subscripts in the h-parameter case. The numbers 1 and 2 refer to port 1 and port 2. This formalism is appropriate because derivation of the h-parameter model proceeds formally in the manner of network theory. Its purpose is to reduce the number of elements to four, with the price of this economy being the relinquishing of easy physical interpretation of some of the parameters. Nonetheless, the h-parameter elements can be related back to elements in the previous two without extreme difficulty because parallel formulations are used for all three. The reader should be aware that our method of distinguishing between the hybrid-parameter and h-parameter models is not universally used, and that considerable confusion exists in the literature because of inconsistent use of both words and symbols.

The admittance model, using y parameters, is also formal and economical. It has certain advantages for computer simulation, and is favored in SPICE analysis. Finally, the scattering-parameter, or s-parameter, model finds greatest utility in the practical extreme of high-frequency analysis, where the others cease to be useful. As an illustration of the problems that arise, consider this: Measuring the parameters h_{oe} and h_{22} involves making an observation at the output port with the input port short-circuited. At low and moderate frequencies, achieving a satisfactory short circuit is a simple matter, but in the microwave range, it is difficult. Detailed descriptions of the various models can be found in the literature—for example, see the book by Ghausi [73]. Now let us take a look at the models individually and at some of their interrelationships.

Two-port theory is used in formal BJT modeling. That is, the transistor is treated as a black box having two terminal pairs, or ports, an approach that can be used for any three-terminal or four-terminal device. In this way it is possible to describe the circuit properties of such a device in any configuration. Two-port theory has great generality. It is powerful in device modeling because the same language and methods can be applied to almost any device. For example, it can be used to compare the properties of seemingly different devices, such as an amplifier and a delay line. Such generalized comparisons provide new insights, and new ways of understanding device behavior in a circuit environment.

Let us examine the h-parameter small-signal model first. Its equivalent circuit is shown in Figure 4-55, along with the definitions of its coefficients. Consider, as an example, the short-circuit forward-current-gain parameter h_{21}. It is defined as the partial derivative of the quasistatic output current I_2 with respect to the quasistatic input current I_1. Implicit in its definition as a partial derivative is the requirement that V_2 = constant or $v_2 = 0$, which makes it the "short-circuit current gain." Figure 4-55 also gives equations in terms of the complex h parameters for the small-signal input voltage v_1 and output current i_2, written in both algebraic and matrix form. From these equations it is once again evident that we can also write, for example, $h_{21} = (i_2/i_1)$ with $v_2 = 0$.

As a first step in explaining model conversions, let us relate the hybrid-pi and h parameters. The practical result of such an exercise will be an ability to convert measured data into device information, and, in reverse, device information into expected results from experimental measurements. The exercise is made easier by

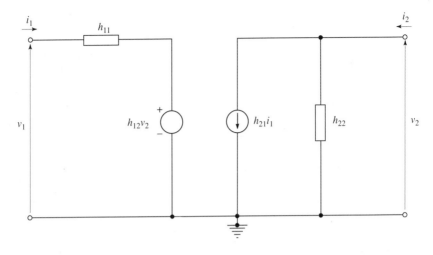

$$v_1 = h_{11} i_1 + h_{12} v_2$$
$$i_2 = h_{21} i_1 + h_{22} v_2$$

or

$$\begin{bmatrix} v_1 \\ i_2 \end{bmatrix} = \begin{bmatrix} h_{11} & h_{12} \\ h_{21} & h_{22} \end{bmatrix} \begin{bmatrix} i_1 \\ v_2 \end{bmatrix}$$

$$h_{11} = \frac{\partial V_1}{\partial I_1} \qquad h_{12} = \frac{\partial V_1}{\partial V_2}$$

$$h_{21} = \frac{\partial I_2}{\partial I_1} \qquad h_{22} = \frac{\partial I_2}{\partial V_2}$$

$$\Delta h = h_{11} h_{22} - h_{12} h_{21}$$

Figure 4-55 Equivalent-circuit model and definitions for the complex h parameters. Uppercase symbols indicate quasistatic (but continuously variable) currents and voltages, while lowercase symbols represent small-signal currents and voltages.

first simplifying the hybrid-pi model, as has been done in Figure 4-56(a). The collector-to-internal-node feedback elements C_c and $r_{b'c}$ have been dropped. A corresponding equivalent circuit for the h-parameter model is shown in Figure 4-56(b). The missing element is h_{12}.

Exercise 4-93. How does the omission of these three feedback elements appreciably simplify the model comparison?

While C_c and $r_{b'c}$ cause feedback to the internal node b', the element h_{12} involves port-to-port feedback. This difference would greatly complicate the comparison if these elements were included.

Exercise 4-94. Aren't feedback-free models unrealistic?

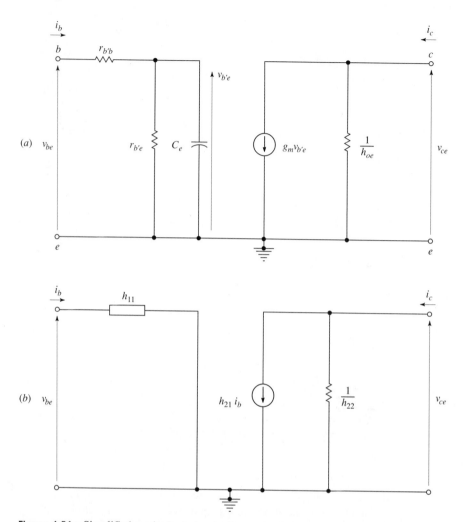

Figure 4-56 Simplified equivalent circuits for comparing two small-signal models. The simplification employed is the omission of certain feedback elements. (a) The hybrid-pi model. (b) The h-parameter model.

The resulting models are not feedback-free. The element $r_{bb'}$, ohmic base resistance, is an important feedback element and is dominant in some frequency ranges. The complex h parameters will incorporate corresponding feedback.

As a result of the simplifying changes, the hybrid-pi equivalent circuit consists of four fixed real elements and one fixed capacitance. The h-parameter equivalent circuit, on the other hand, consists of three complex elements. Also, in order to enhance the correspondence of the two models, we have renamed certain variables. For example, the input current i_b is indicated in Figure 4-56(b) in lieu of i_1.

As an example of the kind of parameter conversion intended here, let us derive h_{21}, the complex short-circuit forward current gain, in terms of hybrid-pi elements. Short-circuiting the output port of the hybrid-pi small-signal equivalent circuit in Figure 4-56(a) produces a positive (inward-flowing) small-signal output current of

$$i_c = g_m v_{b'e}. \tag{4-169}$$

For a small-signal input current i_b, the voltage $v_{b'e}$ is given by

$$v_{b'e} = \frac{i_b}{[(1/r_{b'e}) + j\omega C_e]}. \tag{4-170}$$

Combining Equations 4-169 and 4-170 to determine i_c yields an expression that is linear in i_b, so that h_{21} can be written

$$h_{21} = \left.\frac{i_c}{i_b}\right|_{V_{CE}=\text{constant}} = \frac{g_m}{[(1/r_{b'e}) + j\omega C_e]} = \frac{g_m r_{b'e}}{(1 + j\omega r_{b'e} C_e)}. \tag{4-171}$$

This result for h_{21} is given in Table 4-4, along with two other forms of the same expression. Notice that *both* the real and imaginary terms of h_{21} are frequency-dependent, a fact emphasizing the advantage of having fixed elements in the hybrid-pi model. Table 4-4 also gives conversion equations for the other three h parameters for the simplified equivalent-circuit models of Figure 4-56. Thus with these equations, one can calculate h parameters from a knowledge of values for the hybrid-pi elements. Transformation from measured h parameters to hybrid-pi elements is also a straightforward process, and one that is well documented [72]. The rather voluminous detail in this transformation is beyond our present needs, however, and the simplified examples will suffice. With the more detailed models, these transformations become very complicated, although still straightforward. What is more to the point, considerable effort and insight is required to extract meaningful device-property information from measured results.

For circuit analysis and simulation programs such as SPICE, admittance parameters, or y parameters, are favored. One minor advantage in such a case can be appreciated by noting the convenience (and greater ease in calculations) of writing $g + j\omega C$ in place of $r + j/\omega C$. The y-parameter model is summarized in Figure 4-57 by means of its equivalent-circuit model, the coefficient definitions, and equations for i_1 and i_2 in both algebraic and matrix form. Table 4-5 gives the complex y parameters in terms of the complex h parameters.

Let us examine the "transadmittance" y_{21} in terms of hybrid-pi elements. Table 4-5 informs us that

$$y_{21} = h_{21}/h_{11}. \tag{4-172}$$

But from Table 4-4,

$$h_{21} = h_{fe}/[1 + j(\omega/\omega_\beta)], \tag{4-173}$$

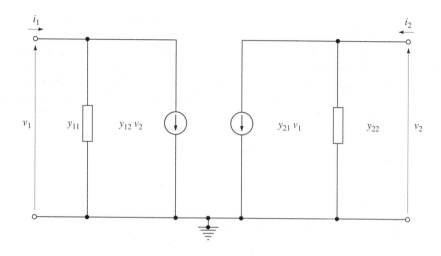

$$i_1 = y_{11} i_1 + y_{12} v_2$$
$$i_2 = y_{21} v_1 + y_{22} v_2$$ or $$\begin{bmatrix} i_1 \\ i_2 \end{bmatrix} = \begin{bmatrix} y_{11} & y_{12} \\ y_{21} & y_{22} \end{bmatrix} \begin{bmatrix} v_1 \\ v_2 \end{bmatrix}$$

$$y_{11} = \frac{\partial I_1}{\partial V_1} \qquad y_{12} = \frac{\partial I_1}{\partial V_2}$$

$$y_{21} = \frac{\partial I_2}{\partial V_1} \qquad y_{22} = \frac{\partial I_2}{\partial V_2}$$

$$\Delta y = y_{11} y_{22} - y_{12} y_{21}$$

Figure 4-57 Equivalent-circuit model and definitions for the complex y parameters. Uppercase symbols indicate quasistatic (but continuously variable) currents and voltages, while lowercase symbols represent small-signal currents and voltages.

and

$$h_{11} = r_{bb'} + r_{b'e}/[1 + j(\omega/\omega_\beta)], \qquad (4\text{-}174)$$

where ω_β is the angular frequency at which the magnitude of short-circuit current gain has fallen by the factor $1/\sqrt{2}$. Substituting Equations 4-173 and 4-174 into Equation 4-172 and dividing numerator and denominator by $r_{b'e}$ yields

$$y_{21} = g_m/\{1 + (r_{bb'}/r_{b'e})[1 + j(\omega/\omega_\beta)]\}. \qquad (4\text{-}175)$$

This result was obtained by using an important relation:

$$g_m = h_{fe}/r_{b'e}. \qquad (4\text{-}176)$$

Equation 4-176 results from combining Equations 4-164 and 4-168, and from the definition of $\beta_F \approx h_{fe}$. From Equation 4-175 it is evident that for $r_{bb'} \ll r_{b'e}$, as is often the case, $y_{21} \approx g_m$ under low-frequency conditions. The frequency-dependent behavior of y_{21} will be examined in more detail in Section 4-7.6, and it will be used to define the cutoff frequency f_T.

Thus we have illustrated the transformation from one model to another with a few examples. It should be clear at this point that many different models are possible, and that certain circumstances favor the use of one or another. Typically we work with the model that is most convenient for a particular application, and then transform to other models as needed. The examples given have included derivation of some of the complex high-frequency parameters in terms of hybrid-pi elements that are device-physics oriented. A logical next step, carrying this idea further, would be to calculate quantities such as transadmittance y_{21} in terms of basic physical properties of the BJT, such as base-region thickness and doping. Techniques that can be used to accomplish this have already been illustrated for the case of the diode in Section 3-5.5, on the subject of diode admittance. Unfortunately, such calculations are often long and tedious, and not very rewarding in terms of increased understanding. Hence, our approach will be merely to outline the method. Considerably more detail can be found in the literature [73, 74].

Let us set out to determine the common-emitter short-circuit y parameters for an *NPN* BJT. The starting point is the time-dependent diffusion equation. As noted earlier, this is another name for the constant-E continuity-transport equation with electric field assumed negligible. We have seen that this assumption is valid for the base region under low-level conditions. Further base-region assumptions, treated at length earlier in this chapter, are that majority hole current can be neglected, and that junction boundaries are fixed in position. With such complications as avalanche multiplication and punchthrough also omitted from the picture, our device thus has zero output conductance.

The BJT problem is sufficiently more complicated than the diode problem that it is convenient to assign a single symbol to a frequently recurring complex argument, letting

$$\theta \equiv \frac{X_B}{L_{nB}}\sqrt{1 - j\omega\tau}, \qquad (4\text{-}177)$$

where τ is carrier lifetime in the base region, and the other symbols have their usual meanings. From Equation 4-168, we also know that

$$g_m = qI_C/kT. \qquad (4\text{-}178)$$

Using Equations 4-177 and 4-178 we find that the y parameters can be written thus [73, 74]:

$$y_{11} = g_m\theta(\coth\theta - \cosh\theta); \qquad (4\text{-}179)$$

$$y_{12} = 0; \qquad (4\text{-}180)$$

$$y_{21} = g_m\theta\,\mathrm{csch}\,\theta; \qquad (4\text{-}181)$$

$$y_{22} = 0. \qquad (4\text{-}182)$$

Although simplifying assumptions have reduced the number of finite y parameters to two, and both can be written compactly, nonetheless the fact that they involve hyperbolic functions of complex numbers makes them cumbersome for circuit analysis. Moreover, these transcendental functions cannot be exactly represented (or synthesized) by using a finite number of R, L, and C elements. As a result, further approximations are needed in what is already a simplified model. The necessary approximations are equivalent to substituting a network of lumped elements for a distributed element, such as a transmission line. Lumped-element replacements are useful in some problems, but in others they yield results that are misleading or even false.

In BJT analysis, a major problem encountered with lumped-element models is the introduction of phase errors at high frequencies, even though the description of signal magnitude versus frequency may be acceptable. This is understandable. A signal requires a finite time to pass through a base region of finite thickness. A lumped-element network cannot simulate the resulting phase shift in the high-frequency extreme, although at lower frequencies the fit can be adequate.

For a low-to-medium-frequency model, there are two useful ways to approximate hyperbolic functions such as those in Equations 4-179 and 4-181. The more accurate of the two uses a product expansion of the hyperbolic functions. A less accurate but simpler approach uses a truncated Taylor's-series expansion. For example, we replace $\cosh \theta$ by

$$\cosh \theta \approx 1 + \theta^2/2. \quad (4\text{-}183)$$

This example can be used to illustrate some important results. From Table 4-5 it is evident that

$$h_{21} = y_{21}/y_{11}. \quad (4\text{-}184)$$

Using Equations 4-179, 4-181, 4-183, and hyperbolic-function identities we find that

$$h_{21} = \frac{\operatorname{csch} \theta}{\coth \theta - \operatorname{csch} \theta} = \frac{1}{\cosh \theta - 1} \approx \frac{1}{\theta^2/2}. \quad (4\text{-}185)$$

Then, squaring Equation 4-177 and substituting the result into Equation 4-185 gives us

$$h_{21} = \frac{1}{\dfrac{1}{2}\left(\dfrac{X_B}{L_{nB}}\right)^2 + j\left(\dfrac{\omega}{2D_{nB}/X_B^2}\right)}, \quad (4\text{-}186)$$

where we have also used the definition $L_{nB}^2 \equiv D_{nB}\tau_B$. Comparing Equation 4-186 with Equation 4-171 shows that

$$g_m r_{b'e} = 2\left(\frac{L_{nB}}{X_B}\right)^2 = h_{fe}, \quad (4\text{-}187)$$

where we have once again used the relation that $g_m r_{b'e} = h_{fe}$. Also,

$$C_e r_{b'e} = \frac{X_B^2}{2D_{nB}} = \frac{1}{\omega_T}, \quad (4\text{-}188)$$

where we will demonstrate that ω_T is the angular frequency at which the magnitude of the common-emitter current gain has fallen to unity. By making use of the y parameters we have thus established a direct link among hybrid-pi parameters C_e, $r_{b'e}$, and g_m, the BJT properties X_B, L_{nB}, and D_{nB}, and the measured quantities h_{fe} and ω_T. In view of the number of simplifying assumptions and approximations involved, of course, the relations thus established are useful only for first-order calculations.

4-7.5 Improving Model Accuracy

The small-signal models we have described can be improved substantially by incorporating into them such real-life phenomena as defect current and Early effect. Such additions are parallel to those introduced in Section 4-8 for large-signal SPICE modeling. We shall focus here on improvements such as an extra phase-shift provision, a current-dependent base resistance, and the addition of extrinsic (or "external") parasitic elements such as interwire capacitances.

The last-named refinement is comparatively obvious, and has been incorporated into the extended hybrid-pi equivalent circuit shown in Figure 4-58. Specifically, three parasitic capacitances (with obvious symbol designations) have been added, connecting each of the three terminal pairs. For precise work it is essential to

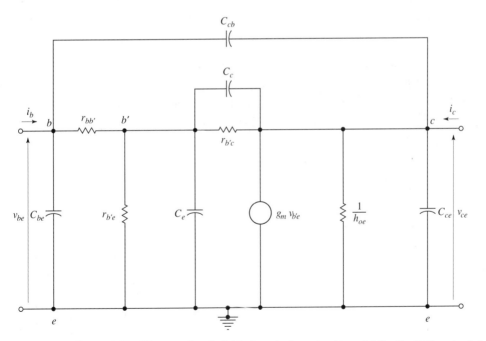

Figure 4-58 The complete hybrid-pi equivalent-circuit model for the BJT, extended to include extrinsic parasitic capacitances.

introduce the extrinsic elements before transforming from this model to another small-signal model. Significant errors can result if the intrinsic model is transformed first and the extrinsic elements are then introduced after the fact.

The effect of "current crowding" was discussed in Section 4-5.5. Two features were emphasized: (1) The complicated majority-carrier current patterns in both the intrinsic and extrinsic base introduce difficult "spreading-resistance" problems that are usually handled semiempirically. (2) Lateral voltage drops in the intrinsic base region cause "debiasing" of parts of the emitter junction, with the edge regions injecting preferentially. A third effect, alluded to in several parts of Section 4-5, is the conductivity modulation of the ohmic base resistance that occurs under certain operating conditions, especially affecting the intrinsic base region. It is clear that the last two effects (debiasing and modulation) will coexist under high-current conditions, and will interact in a complicated way. A semiempirical treatment of this problem has been incorporated into SPICE and has been described by Antognetti and Massobrio [75]. This treatment assumes that the ohmic base resistance $r_{bb'}$ can be expressed in terms of two resistance values. One is r_b, the zero-current or low-level-limit value of $r_{bb'}$, and the other is r_{bm}, the minimum value $r_{bb'}$ can assume at high current levels. Also, a base current I_{RB} is defined as the current at which $r_{bb'}$ falls to $r_b/2$. Then, dependency of $r_{bb'}$ in SPICE is taken as

$$r_{bb'} = r_{bm} + 3(r_b - r_{bm})\left[\frac{\tan z - z}{z \tan^2 z}\right], \tag{4-189}$$

where

$$z = \frac{-1 + \sqrt{1 + (144)I_B/\pi^2 I_{RB}}}{[(24)/\pi^2]\sqrt{I_B/I_{RB}}}. \tag{4-190}$$

These equations are employed in Problem C4-1.

Finally, the addition of extra phase shift to the model of Figure 4-58 is essential for high-frequency analog circuits. A semiempirical modification also described in Reference 75 replaces the output current i_c by

$$i_c = i_c \exp(-j\alpha), \tag{4-191}$$

where

$$\alpha \equiv \tanh^{-1}\frac{3\omega_0\omega}{3\omega_0^2 - \omega^2}, \tag{4-192}$$

and where the frequency ω_0 is an empirical constant.

4-7.6 Charge-Control Model

Analysis of the BJT by charge-control methods proceeds on a path parallel to that used for the junction diode (Section 3-9.1). One starts with an expression for stored excess-carrier charge that is a function of time and space. A spatial integration is then carried out to obtain total stored charge as a function of time. From this point forward, therefore, all information on excess-carrier distribution in the device is absent from the model. In particular, the final equations produced by the analysis involve time constants or frequency constants rather than information that is more device-physics oriented. For this reason, charge-control models are sometimes described as "phenomenological."

The accuracy of a charge-control model depends upon the validity of its key assumption, which is that a spatial distribution of charge known to be correct under static conditions is at least approximately correct under dynamic conditions. For the diode, this assumption is fair for small-signal cases, but very poor for large-signal cases, as is illustrated dramatically in Section 3-9.4 through a comparison of the actual carrier-density profiles in Figure 3-76(a) with the assumed profiles in Figure 3-76(b). The same generalization holds for the BJT, and in this section we confine ourselves to small-signal analysis.

Charge-control examination of the BJT was first developed in the 1950s [65], at a time when base current was dominated by the recombination of carriers in the base region. Nonetheless, the model's equations are equally applicable to the modern BJT, wherein base current is dominated by recombination in the emitter region. This observation becomes plausible in light of the point just made, namely that the physical location of the relevant charges has been "integrated out" of the model. A further compelling point, however, is that the early BJT and the modern BJT (as well as intervening devices) are linear current amplifiers in the common-emitter configuration. The reasons for this are discussed in Exercise 4-17, Section 4-2.4. Consequently, we can justify developing the BJT charge-control model in its original form, which is especially transparent because *both* the base current and the collector current can be written readily in terms of charge stored in the base region. In view of the complexity of the BJT, amply documented in the preceding sections, a model that offers clarity as well as a good description of small-signal properties is worthy of inclusion; a good qualitative understanding of these properties comes from this model, and the model is sufficiently accurate for many applications.

Let us confine ourselves to low-level as well as small-signal conditions. The distribution of base-region charge is well known. Although the carrier profile is somewhat concave up in the early-BJT case, a linear approximation to the profile is good in a wide range of cases. Hence we can move directly to consideration of excess-carrier charge stored in the base region. It can be written in the usual way as the sum of dc and ac components, or $q_B(t) = Q_B + q_b(t)$. Even though excess-carrier charge is being described, we omit the prime symbol (as was done before in Section 3-9.1) to simplify notation.

In addition to charge storage in the base region, we must consider storage on both junction capacitances. We shall defer discussion of these latter two mechanisms, but nonetheless will move directly to the charge-control equation for the total and time-dependent base current, an equation that comprehends all relevant mechanisms:

$$i_B(t) = -\frac{q_B(t)}{\tau} - \frac{dq_B(t)}{dt} + C_{TE}\frac{dv_{BE}(t)}{dt} + C_{TC}\frac{dv_{BC}(t)}{dt}. \tag{4-193}$$

Obviously the last two terms relate to the contributions to base current made by the depletion-layer capacitances, while the first two terms on the right-hand side concern base-region charge.

We first examine base-region phenomena. The minus signs on the first two terms enter because it is customary to apply an explicit sign to charge, and we have chosen to focus on minority carriers, excess electrons—a charge that can readily be related to both base current and collector current. (Recall that when sign consistency was desired, we related positive base current to *hole* storage—in the emitter region for a modern device and in the base region for the 1950s device.) The term $-q_B(t)/\tau$ gives the total (and time-varying) base-current component resulting from the classic static (or quasistatic) charge-control picture developed in Section 3-4.5, with τ being carrier lifetime in the base region. But inasmuch as diffusion capacitance is a true capacitance, there is a second component of base current that depends upon the rate of change of stored charge, or $-dq_B(t)/dt$, which is the second term on the right-hand side of Equation 4-193.

Exercise 4-95. It is evident that all terms in Equation 4-193 are time-dependent. How can the right-hand side be separated into a dc portion and an ac portion?

The dc portion will obviously be $-Q_B/\tau$, where Q_B is the static component of base-region charge. Adding to this the time-varying portion of the first term on the right-hand side of Equation 4-195 gives us

$$i_B(t) = -\frac{Q_B}{\tau} + \left[-\frac{q_B(t) - Q_B}{\tau} - \frac{dq_B(t)}{dt} + C_{TE}\frac{dv_{BE}(t)}{dt} + C_{TC}\frac{dv_{BC}(t)}{dt} \right].$$

This equation obviously reduces to Equation 4-193, but consists on the right-hand side of a dc first term, and an ac term in square brackets.

Now turn to the last two terms of Equation 4-193. The symbols C_{TE} and C_{TC} are the transition-region (depletion-layer) capacitances of the respective junctions. The appropriate values to assign to these voltage-dependent capacitances are those corresponding to the static junction voltages V_{BE} and V_{CB}.

Exercise 4-96. Why did we use different symbols, namely C_e and C_c, in the hybrid-pi model?

The symbol C_e stands for the sum of emitter-junction depletion-layer capacitance C_{TE} and diffusion capacitance. But the charge associated with diffusion capacitance is at the heart of the charge-control model, and is treated separately in the term $-dq_B(t)/dt$. The other hybrid-pi capacitance is identically C_{TC}, but was named C_c for parallelism in the hybrid-pi model.

Exercise 4-97. In what respect is the term $C_{TE}dv_{BE}(t)/dt$ approximate? Is the last term also approximate?

> The transition-region capacitance of the emitter junction is actually charged by the voltage $v_{b'e}$ rather than the voltage v_{be}. Similarly, the capacitance C_{TC} is charged by the voltage $v_{b'c}$ rather than the voltage v_{bc}, so the last term is also approximate.

At this point a note on variable choice is in order. The reader will recall that in developing the hybrid equations in Section 4-7.1, we insisted upon using the quasistatic variables I_B, I_C, V_{BE}, and V_{CE}. The reason for that was that the hybrid-equation development involved derivatives of voltage with respect to current, current with respect to voltage, current with respect to current, and voltage with respect to voltage, as well as differentials of current and voltage. Under those conditions, using time-dependent variables introduces irrelevancy and confusion that can easily be avoided by using quasistatic variables. But the present development involves time derivatives of charge, current, and voltage. And since $q_B(t)$ and $q_b(t)$, as well as the corresponding current and voltage variables, are explicitly functions of time, their time derivatives are readily interpreted. Furthermore, the essence of charge-control analysis is a proportionality between a charge and its corresponding current. Hence the quotient of a corresponding charge and current yields a time constant, or in the inverse case, a frequency constant. Similarly, derivatives involving these variables, whether the variables are static or dynamic, yield the same constants as the corresponding quotients. Thus the variables present in Equation 4-193 are logically chosen and easily interpreted.

Now we wish to rewrite the right-hand side of Equation 4-193 in terms of total collector current, $i_C(t)$. The first term we rewrite as

$$-\frac{q_B(t)}{\tau} = \frac{i_C(t)}{\beta_F}. \tag{4-194}$$

Treating β_F as a real constant in this quasistatic term implies that the three capacitances (associated with diffusion and the two junctions) will contribute the reactive portions of a model that must ultimately be complex. By appealing to rudimentary charge-control concepts, we can write

$$i_C(t) = -\frac{\beta_F}{\tau}q_B(t) = -\frac{q_B(t)}{\tau_B}. \tag{4-195}$$

The symbol

$$\tau_B \equiv \tau/\beta_F \tag{4-196}$$

is known as the *base-charging time* because it is a constant of proportionality linking collector current and (the negative of) stored base-region charge. Since quasistatic current and charge are proportional, it follows that their time derivatives have the same proportionality, or

$$\frac{di_C(t)}{dt} = -\frac{1}{\tau_B} \frac{dq_B(t)}{dt}. \tag{4-197}$$

(Notice that we could equally well write i_c in place of i_C on the left-hand side, and q_b in place of q_B on the right-hand side.) Hence the second term on the right-hand side of Equation 4-193 can be written

$$-\frac{dq_B(t)}{dt} = \tau_B \frac{di_C(t)}{dt}. \tag{4-198}$$

Using the label "base-charging time" for the symbol τ_B makes sense for the 1950s BJT but not for the modern BJT. Nonetheless we shall adhere to that designation because it is well entrenched.

Now we turn to the last two terms of Equation 4-193, converting them to expressions in total collector current. The emitter-capacitance term can be modified readily. Let us drop functional notation at this point for brevity:

$$C_{TE} \frac{dv_{BE}}{dt} = C_{TE} \frac{dv_{BE}}{di_C} \frac{di_C}{dt} = \frac{C_{TE}}{g_m} \frac{di_C}{dt} \equiv \tau_E \frac{di_C}{dt}. \tag{4-199}$$

The last step defines a time constant τ_E associated with the emitter junction. Similarly, for the collector junction there is a time constant τ_C that is developed in this way:

$$C_{TC} \frac{dv_{CE}}{dt} = C_{TC} \frac{dv_{CE}}{di_c} \frac{di_C}{dt} = \frac{C_{TC}}{g_o} \frac{di_C}{dt} \equiv \tau_C \frac{di_C}{dt}. \tag{4-200}$$

The quantity g_o is of course the output conductance of the BJT at the quiescent point of interest. (One might protest that the second expression in each of Equations 4-199 and 4-200 violates the variable-choice dictum offered just above, but viewing the step as an isolated formal manipulation avoids any problem.)

Substituting the results from Equations 4-194 and 4-198 through 4-200, into Equation 4-193 yields

$$i_B(t) = \frac{i_C(t)}{\beta_F} + (\tau_B + \tau_E + \tau_C) \frac{di_C(t)}{dt}. \tag{4-201}$$

With the definition

$$\tau_B + \tau_E + \tau_C \equiv 1/\omega_T, \tag{4-202}$$

Equation 4-201 becomes

$$i_B(t) = \frac{i_C(t)}{\beta_F} + \frac{1}{\omega_T} \frac{di_C(t)}{dt}. \tag{4-203}$$

The quantities ω_T and

$$f_T \equiv \omega_T/2\pi \tag{4-204}$$

are important figures of merit.

Next let us apply the results so far obtained to a simple but important problem, namely, determination of the complex current gain h_{21}. For this purpose let us specifically assume the sinusoidal small-signal functions

$$i_b(t) = I_b \exp(j\omega t), \qquad (4\text{-}205)$$

and

$$i_c(t) = I_c \exp(j\omega t), \qquad (4\text{-}206)$$

where the coefficients I_b and I_c are complex. Substituting total-current expressions incorporating these small-signal terms into Equation 4-203 and carrying out appropriate cancellations gives us

$$I_b = \frac{I_c}{\beta_F} + \frac{j\omega I_c}{\omega_T} = I_c\left(\frac{\omega_T + j\beta_F\omega}{\beta_F\omega_T}\right) = I_c\left(\frac{1 + j\beta_F\omega/\omega_T}{\beta_F}\right). \qquad (4\text{-}207)$$

Thus the complex current gain is

$$h_{21} \equiv \frac{I_c}{I_b} = \frac{\beta_F}{1 + j\beta_F\left(\dfrac{\omega}{\omega_T}\right)} = \frac{\beta_F}{1 + j\beta_F\left(\dfrac{f}{f_T}\right)} = \beta_F \frac{1 - j\left(\dfrac{f}{f_T/\beta_F}\right)}{1 + \left(\dfrac{f}{f_T/\beta_F}\right)^2}$$

$$= \beta_F\{[1 - j(f/f_\beta)]/[1 + (f/f_\beta)^2]\}, \qquad (4\text{-}208)$$

where $f_\beta \equiv f_T/\beta_F$.

Comparing this expression with the result given in Table 4-4 for h_{21} shows that the charge-control model is in gratifying agreement with small-signal models treated earlier. In Section 4-7.4 we related somewhat simplified hybrid-pi and h-parameter models to each other and then, with the aid of the y-parameter model, developed the Table-4-4 expression for complex h_{21}.

Multiplying the numerator of Equation 4-208 by its complex conjugate then yields the magnitude of the current gain,

$$|h_{21}| = \frac{\beta_F}{\sqrt{1 + \left(\dfrac{f}{f_T/\beta_F}\right)^2}}. \qquad (4\text{-}209)$$

The frequency dependence of $|h_{21}|$ is conveniently displayed by plotting $\log |h_{21}|$ versus $\log f$, as has been done in Figure 4-59. In the low-frequency range, the gain is constant at β_F up to about f_T/β_F. But it is evident by inspection of Equation 4-209 that for $f = f_T/\beta_F$, $|h_{21}|$ has dropped to $\beta_F/\sqrt{2}$, or 0.707 times its low-frequency value. Then the gain declines hyperbolically, translating into a linear decline at 45° in the symmetric log–log plot.

Exercise 4-98. Show that $|h_{21}|$ is very nearly unity at $f = f_T$.

From Equation 4-209 at $f = f_T$ we have

$$|h_{21}| = \frac{\beta_F}{\sqrt{1 + \beta_F^2}} \approx 1$$

for typical values of β_F.

To close the loop, let us use the hybrid-pi model and actual parameter values to generate a curve analogous to that in Figure 4-59 (which is essentially a product of the charge-control model). For this purpose, use the second set of data in Table 4-2. Applying appropriate circuit theory yields the computer-plotted curve of $|h_{21}|$ versus frequency in Figure 4-60(a). Notice that the low-frequency value of $|h_{21}|$ is approximately 100, and that f_T is about 1 GHz. In Section 4-8.7 we shall demonstrate how the same curve can be derived from the SPICE model.

Figure 4-60(b) shows the phase of h_{21} as a function of frequency. It first departs downward from zero because of the pole associated with emitter capacitance C_e, and emitter dynamic resistance $r_{b'e}$. (The Miller effect is absent here because the output port is short-circuited.) The second prominent downward change in phase versus frequency is caused by current feeding forward through the capacitance C_c, and hence *out* the collector terminal. Recall that positive collector current is *inward*.

Next we apply circuit theory to the same data set to determine the four complex

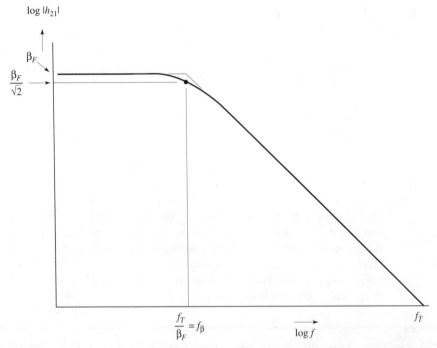

Figure 4-59 Magnitude of the short-circuit common-emitter current gain as a function of frequency.

SMALL-SIGNAL DYNAMIC MODELS **637**

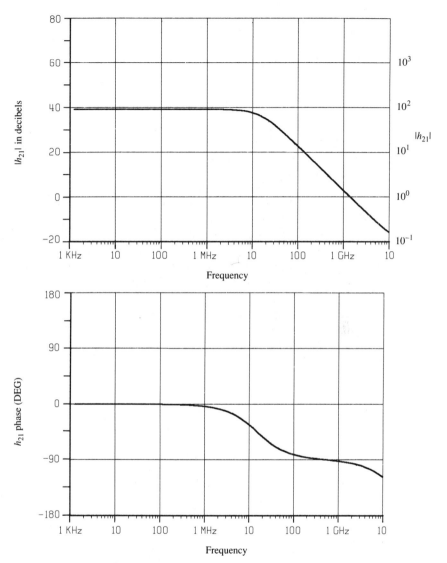

Figure 4-60 Frequency dependence of current-gain magnitude and phase, calculated by computer from experimentally determined hybrid-pi parameters. (a) The magnitude of h_{21} expressed in terms of decibels (20 log $|h_{21}|$) and as a dimensionless factor, as a function of frequency. (b) The phase of h_{21} (the arc tangent of the imaginary part divided by the real part) as a function of frequency.

h parameters as a function of frequency. These are plotted in Figure 4-61, where a solid line presents the real part of each function, and a dashed line, the imaginary part. Notice in particular the curves for h_{21}, shown in Figure 4-61(c). At low frequencies, h_{21} is real and approximately equal to h_{FE}. At frequencies above f_β, the

638 THE BIPOLAR JUNCTION TRANSISTOR

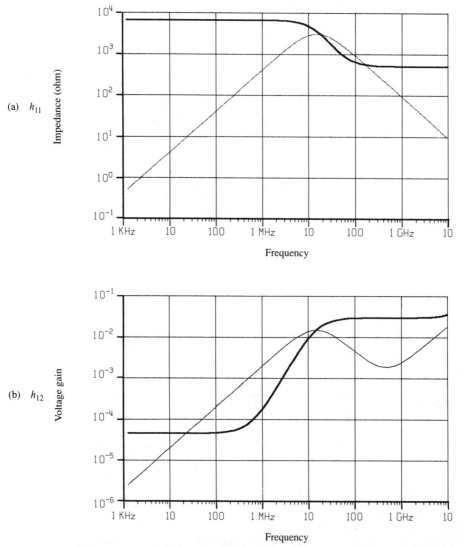

Figure 4-61 The four complex h parameters, using circuit theory and the same data set used for Figure 4-60. Solid lines show real parts and dashed lines show imaginary parts. (a) Input impedance, h_{11}. (b) Inverse voltage gain, h_{12}. (c) Forward current gain, h_{21}. (d) Output admittance, h_{22}.

susceptance of $C_{b'e}$ is greater than the conductance of $r_{b'e}$. Thus the high-frequency gain is determined by reactance. Finally, notice how Figure 4-60(a) amounts to a composite of the two curves in Figure 4-61(c). In Computer Problems C4-1 through C4-3, we shall calculate the h parameters using SPICE, starting with the hybrid-pi model and the same parameter set just noted. In addition, we will calculate the z and y parameters.

(c) h_{21}

(d) h_{22}

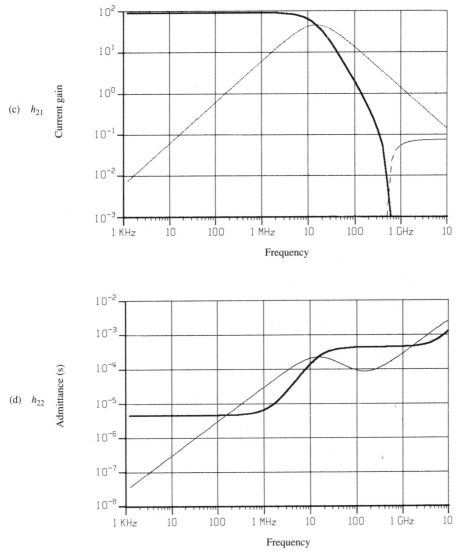

Figure 4-61 (*Continued*)

4-7.7 Base-Charging Time

The quantity τ_B to which we have assigned the traditional label of base-charging time is a fundamental device-characterizing time. Not surprisingly, it can be expressed in terms of fundamental BJT properties, such as base thickness X_B. To do so, let us first write collector current using elementary static BJT theory for an early device having a uniformly doped base region:

$$I_C \approx A_E q D_{nB} n_B(0)/X_B. \tag{4-210}$$

Also

$$Q_B \approx -A_E q n_B(0) X_B/2, \quad (4\text{-}211)$$

so that

$$\frac{I_C}{Q_B} \approx -\frac{2D_{nB}}{X_B^2}. \quad (4\text{-}212)$$

But from the static charge-control equation we have

$$\frac{I_C}{Q_B} = -\frac{1}{\tau_B}; \quad (4\text{-}213)$$

from Equations 4-212 and 4-213 it is thus evident that

$$\tau_B \approx \frac{X_B^2}{2D_{nB}}. \quad (4\text{-}214)$$

Exercise 4-99. Why are approximate equalities indicated in Equations 4-210 and 4-211?

These equations are based upon the assumption of a linear minority-carrier profile in the base region. But for the 1950s BJT, the profile was concave up, as shown for an extreme case in Figure 4-9.

To summarize, for the low-level uniform-base case we have

$$\tau_B = -\frac{Q_B}{I_C} = -\frac{q_B(t)}{i_C(t)} = -\frac{q_b(t)}{i_c(t)} = \frac{X_B^2}{2D_{nB}}. \quad (4\text{-}215)$$

Our purpose now is to generalize both to high-level conditions and to the graded-base BJT. The total time-dependent excess-carrier charge stored in the base region can be written

$$q_B(t) = qA_E \int_0^{X_B} n_B'(x, t)\, dx. \quad (4\text{-}216)$$

Gummel (and later Gummel and Poon) showed a method for writing the desired high-level collector-current equation for a BJT with a nonuniformly doped base region [26, 66]. The method constitutes a generalization of his low-level analysis employing the Gummel number [18], treated in Section 4-4.3. The Gummel number is of course the integral of the arbitrary function $N_{AB}(x)$ through the base region, and gives the areal density of net-doping impurities there, or equivalently, gives the areal density of majority carriers there at equilibrium. The generalized formulation substitutes for the Gummel number the *nonequilibrium* (high-level as well) areal density of majority carriers in the base region. The function to be integrated can be written $n_B'(x,t) + N_{AB}(x)$, on grounds that majority and minority excess densities match, and that the term $N_{AB}(x)$ will account for the equilibrium hole population.

His expression for current, then, is

$$i_C(t) = A_E q D_{nB} \frac{[n'_B(0, t) + N_{AB}(0)]n'_B(0, t)}{\int_0^{X_B} [n'_B(x, t) + N_{AB}(x)]dx}. \tag{4-217}$$

Exercise 4-100. How does Equation 4-217 reduce to the familiar expression for low-level, uniform-base conditions?

Under low-level conditions the terms $n'_B(0, t)$ and $n'_B(x, t)$ are negligible compared to $N_{AB}(0)$ and $N_{AB}(x)$, respectively, and may be dropped. Integrating $N_{AB}(x)$ yields $N_{AB}X_B$, causing $N_{AB}(0) = N_{AB}$ to cancel from the numerator, and leaving the familiar expression.

Substituting Equations 4-216 and 4-217 into Equation 4-215 and multiplying numerator and denominator by $2X_B^2$ then gives us

$$\tau_B = \frac{X_B^2}{2D_{nB}} \left[\frac{\int_0^{X_B} [n'_B(x, t) + N_{AB}(x)]dx}{[n'_B(0, t) + N_{AB}(0)]X_B} \right] \left[\frac{2\int_0^{X_B} n'_B(x, t)dx}{n'_B(0, t)X_B} \right]. \tag{4-218}$$

It is instructive to see how this generalized expression for τ_B simplifies in special cases. First, again examine the low-level, uniform-base limit. Under these conditions the terms $N_{AB}(0) = N_{AB}(x) = N_{AB}$ are again dominant. Also,

$$n'_B(x, t) \approx n_B(x, t) \approx n_B(0, t)[1 - x/X_B]. \tag{4-219}$$

Substituting these expressions into Equation 4-218 causes τ_B to revert to the simple form of Equation 4-215,

$$\tau_B = X_B^2/2D_{nB}. \tag{4-220}$$

Another simple case is the high-level limit, wherein

$$n'_B(x, t) + N_{AB}(x) \approx n'_B(x, t) \approx n_B(x, t), \tag{4-221}$$

because equilibrium carriers are swamped by excess carriers. Once again the carrier profile assumes the form given in Equation 4-219, and when these expressions are substituted into Equation 4-218, the result is

$$\tau_B = X_B^2/4D_{nB}, \tag{4-222}$$

showing the now-familiar doubling of the diffusivity.

Exercise 4-101. Verify the result in Equation 4-222.

We have

$$n_B(0, t)\int_0^{X_B}\left[1 - \frac{x}{X_B}\right]dx = n_B(0, t)\left[x - \frac{x^2}{2X_B}\right]_0^{X_B}$$
$$= n_B(0, t)\left[X_B - \frac{X_B}{2}\right] = n_B(0, t)\frac{X_B}{2}.$$

Substituting this result for both integrals in Equation 4-218 yields

$$\tau_B = \frac{X_B^2}{2D_{nB}}\left[\frac{n_B(0, t)X_B/2}{n_B(0, t)X_B}\right]\left[\frac{2n_B(0, t)X_B/2}{n_B(0, t)X_B}\right] = \frac{X_B^2}{4D_{nB}}.$$

As a third special case, assume an exponential net-doping profile and low-level conditions. Notice, now, that the generality of Equation 4-218 is not needed simply to calculate τ_B, as Equation 4-213 verifies; that is, we may substitute static functions into Equation 4-218. The necessary expressions were already developed in Section 4-5.7. Equation 4-93 shows that the exponential net-doping profile can be written as

$$N_{AB}(x) = N_{AB}(0)\exp[-\eta(x/X_B)]. \tag{4-223}$$

Also, as noted in Equation 4-94, the resulting minority-carrier profile can be written as

$$n_B(x) = \frac{I_C X_B}{qA_E D_{nB}}\frac{1 - \exp\left[-\eta\left(1 - \frac{x}{X_B}\right)\right]}{\eta}. \tag{4-224}$$

Substituting Equations 4-223 and 4-224 into Equation 4-218 yields the result

$$\tau_B = \frac{X_B^2}{2D_{nB}}\left[2\frac{\eta - 1 + \exp(-\eta)}{\eta^2}\right], \tag{4-225}$$

which is the subject of Problem A4-40. As η approaches zero, corresponding to uniform doping, the bracketed factor approaches unity, thus yielding the expected value of τ_B given in Equation 4-215.

Exercise 4-102. Verify the last statement.

Replace the exponential term by the truncated series appropriate for small values of the argument,

$$\exp(-\eta) = 1 - \eta + \eta^2/2.$$

As a result, the bracketed factor in Equation 4-225 becomes

$$\{2[\eta - 1 + (1 - \eta + \eta^2/2)]/\eta^2\} = 1,$$

thus demonstrating the desired point.

Using computer methods one can generalize the base-charging-time results, thus connecting the special cases considered just above. The results are shown in Figure 4-62, which presents normalized base-charging time as a function of normalized injection level. On the right-hand side of the diagram a *diffusion-enhancement factor* that is simply the inverse of normalized base-charging time is indicated. We shall identify it as v [Greek upsilon]:

$$v \equiv X_B^2/2D_{nB}\tau_B. \tag{4-226}$$

This factor gauges the amount of transport aid caused by the longitudinal built-in field of a graded-base device, and in combination with it, the longitudinal field accompanying high-level injection. Evidently the factor v is unity for low-level conditions in the uniform-base case ($\eta = 0$), and goes to two at high levels. For the graded-base device with $\eta = 8$, the factor decreases from six to two as injection level is increased. These overall results remind us again that the base-region doping profile has a negligible effect on high-level BJT operation, which is the norm.

The condition $\eta = 2.2$ is a special case, in that the diffusion-enhancement factor remains at approximately two for all injection levels. The field caused by high-current conditions gradually takes over from the built-in field as injection level is increased.

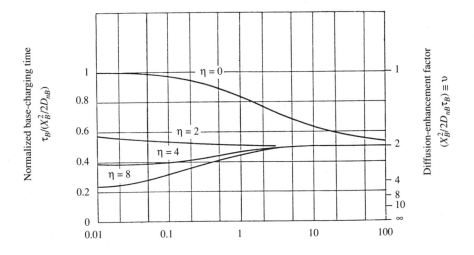

Figure 4-62 Normalized base-charging time as a function of normalized injection level for graded-base transistors. The parameter η is a measure of grading steepness, with $\eta = 0$ corresponding to uniform doping. The diffusion-enhancing factor is a measure of transport assistance by the base-region electric field. (After Lindmayer and Wrigley [62], with permission.)

4-7.8 Figures of Merit

Several critical parameters, called *figures of merit*, permit one to evaluate and compare high-performance devices. A good figure of merit must be independent of circuit variables, such as load resistance, and must be expressible in terms of fundamental device properties, such as transconductance. One widely used figure of merit is the cutoff frequency defined in Equation 4-204, $f_T = \omega_T/2\pi$. The angular cutoff frequency ω_T is defined in Equation 4-202 in terms of three time constants that were in turn defined in Equations 4-196, 4-199, and 4-200. It is worthwhile to restate these cutoff frequencies in terms of fundamental device properties, however, because doing so tells the designer how this figure of merit can be improved:

$$\frac{1}{\omega_T} = \frac{2\pi}{f_T} = \frac{X_B^2}{2vD_{nB}} + \frac{C_{TE}}{g_m} + \frac{C_{TC}}{g_o}. \qquad (4\text{-}227)$$

All three of the preceding terms exhibit current dependence. Figure 4-60 displays the I_C dependence of v, while g_m is proportional to I_C. Output conductance g_o varies with I_C in a manner that can be inferred readily from Figure 4-30(d). Furthermore, X_B, C_{TE}, and C_{TC} exhibit voltage dependence.

Cutoff frequency is especially meaningful in wide-band and high-frequency amplifier applications. It is not a universal figure of merit, however, because it is independent of $r_{bb'}$, a feedback element that sometimes has great importance. Its independence of $r_{bb'}$ can be understood by examining the usual circuit arrangements for such amplifier applications, treated at the beginning of Section 4-7 on small-signal dynamic models.

Exercise 4-103. Explain the last statement.

With I_B and i_b controlled externally, the voltages $V_{BB'}$ and $v_{bb'}$ are also controlled externally, so that $r_{bb'}$ does not play a role in determining f_T.

In devices for narrow-band amplifiers at the extreme of high-frequency operation, which is to say, for microwave transistors, the parasitic element $r_{bb'}$ is extremely important. The major figure of merit for such application is the maximum frequency of oscillation, f_{MAX}. Its derivation is relatively involved, so we simply give the expression that results:

$$f_{MAX} = \sqrt{\frac{f_T}{8\pi r_{bb'} C_{CT}}}. \qquad (4\text{-}228)$$

Note that both f_T and $r_{bb'}$ enter into the definition of f_{MAX}, thus improving its value as an overall figure of merit for the BJT.

4-8 SPICE MODELS

In the approach to be used here, we present the complete model and then derive specific known results, such as the Ebers-Moll model. This reverses our usual practice, wherein we offer specific examples and then generalize. But sufficient background has been established at this point to appreciate both the general and specific SPICE BJT models. If additional information is sought, Reference 75 is a good source. The model will be used here for both small-signal and large-signal analyses. Expressions for the small-signal hybrid-pi parameters will be derived and evaluated for the bias conditions given in Table 4-8.

4-8.1 Model Equations

The SPICE BJT model consists of twelve equations in fourteen variables. The terminal quantities $i_C(t)$, $i_B(t)$, $v_{BE}(t)$, and $v_{BC}(t)$, have definitions consistent with the symbol convention established in Section 3-9.2, but with obvious changes for application to the BJT. The quantities $v_{JE}(t)$ and $v_{JC}(t)$ are the *imposed junction voltages*, or the "voltmeter" voltages from boundary to boundary of their respective junctions. The dimensionless charge-control quantities $q_B(t)$, $q_1(t)$, and $q_2(t)$ are defined in Equations 4-233 through 4-235; the capacitances $C_{BE}(t)$, $C_{BC}(t)$, and $C_{CS}(t)$, in Equations 4-236 through 4-238; and the transconductances $g_{mF}(t)$ and $g_{mR}(t)$, in Equations 4-239 and 4-240. The collector-substrate voltage $v_{JS}(t)$ and the associated capacitance $C_{CS}(t)$ are included here for completeness but will be neglected in the following discussion. Note that we now use a simplified notation—for example, $C_{BE}(t)$ in place of $C_{BE}(v_{JE})$—rather than the more explicit notation used previously. The model involves a set of twenty-seven parameters, which are listed in Table 4-6. The first sixteen parameters are required for dc analysis, and the remaining eleven parameters, for ac and transient analysis. The table also gives dimensions and default values. There are various extensions of the SPICE BJT model involving an additional set of thirteen parameters, extensions dealing with matters such as temperature effects and noise, topics that will not be discussed here.

Let us now set down all twelve equations. The two current equations are

$$i_C(t) = \frac{I_S}{q_B(t)}\left[\exp\left(\frac{qv_{JE}(t)}{N_F kT}\right) - \exp\left(\frac{qv_{JC}(t)}{N_F kT}\right)\right]$$
$$- \frac{I_S}{\beta_R}\left[\exp\left(\frac{qv_{JC}(t)}{N_C kT}\right) - 1\right] - I_{SC}\left[\exp\left(\frac{qv_{JC}(t)}{N_C kT}\right) - 1\right]$$
$$- C_{BC}(t)\frac{dv_{JC}(t)}{dt} + C_{CS}(t)\frac{v_{JS}(t)}{dt}, \qquad (4\text{-}229)$$

and

$$i_B(t) = \frac{I_S}{\beta_F}\left[\exp\left(\frac{qv_{JE}(t)}{N_F kT}\right) - 1\right] + I_{SE}\left[\exp\left(\frac{qv_{JE}(t)}{N_E kT}\right) - 1\right] + C_{BE}(t)\frac{dv_{JE}(t)}{dt}$$

$$+ \frac{I_S}{\beta_R}\left[\exp\left(\frac{qv_{JC}(t)}{N_R kT}\right) - 1\right] + I_{SC}\left[\exp\left(\frac{qv_{JC}(t)}{N_C kT}\right) - 1\right] + C_{BC}(t)\frac{v_{JC}(t)}{dt}. \quad (4\text{-}230)$$

The two voltage equations are

$$v_{JC}(t) = v_{BC}(t) + i_B(t)R_{SB} - i_C(t)R_{SC}, \quad (4\text{-}231)$$

and

$$v_{JE}(t) = v_{BE}(t) + i_B(t)R_{SB} - [i_B(t) + i_C(t)]R_{SE}. \quad (4\text{-}232)$$

The three charge-control equations are

$$q_B(t) = q_1(t)\left[\frac{1}{2} + \sqrt{\frac{1}{4} + q_2(t)}\right], \quad (4\text{-}233)$$

$$q_1(t) = \left[1 - \frac{v_{JC}(t)}{V_A} - \frac{v_{JE}(t)}{V_B}\right]^{-1}, \quad (4\text{-}234)$$

and

$$q_2(t) = \frac{I_S}{I_k}\left[\exp\left(\frac{qv_{JE}(t)}{N_F kT}\right) - 1\right] + \frac{I_S}{I_{kr}}\left[\exp\left(\frac{qv_{JC}(t)}{N_R kT}\right) - 1\right]. \quad (4\text{-}235)$$

The three capacitance equations are

$$C_{BE}(t) = C_{OTE}\left[1 - \frac{v_{JE}(t)}{\Delta\psi_{OE}}\right]^{-M_E} + \tau_F g_{mF}(t), \quad (4\text{-}236)$$

$$C_{BC}(t) = C_{OTC}\left[1 - \frac{v_{JC}(t)}{\Delta\psi_{OC}}\right]^{-M_C} + \tau_R g_{mR}(t), \quad (4\text{-}237)$$

and

$$C_{CS}(t) = C_{OTS}\left[1 - \frac{v_{JS}(t)}{\Delta\psi_{OS}}\right]^{-M_S}. \quad (4\text{-}238)$$

Completing the basic twelve equations, the two transconductance equations are

$$g_{mF}(t) = \frac{qI_S}{kT}\left[\exp\left(\frac{qv_{JE}(t)}{N_F kT}\right) - 1\right] - \frac{qI_S}{q_B^2}\left[\exp\left(\frac{qv_{JE}(t)}{N_F kT}\right) - 1\right]\frac{\partial q_B}{\partial v_{JE}}, \quad (4\text{-}239)$$

and

$$g_{mR}(t) = \frac{qI_S}{kT}\left[\exp\left(\frac{qv_{JC}(t)}{N_R kT}\right) - 1\right]. \quad (4\text{-}240)$$

An initial understanding of the SPICE BJT model can be gained by assuming static conditions and by using the default values. Substituting these values into the preceding equations causes the SPICE model to reduce to two equations in four variables:

$$I_C = I_S\left[\exp\left(\frac{qV_{BE}}{kT}\right) - \exp\left(\frac{qV_{BC}}{kT}\right)\right] - \frac{I_S}{\beta_R}\left[\exp\left(\frac{qV_{BC}}{kT}\right) - 1\right], \quad \text{(4-241)}$$

and

$$I_B = \frac{I_S}{\beta_F}\left[\exp\left(\frac{qV_{BE}}{kT}\right) - 1\right] + \frac{I_S}{\beta_R}\left[\exp\left(\frac{qV_{BC}}{kT}\right) - 1\right]. \quad \text{(4-242)}$$

Consider the last equation as an example. It can be derived from Equation 4-230 by using the default values $I_{SE} = 0$ and $I_{SC} = 0$. From Equations 4-241 and 4-242, we can derive the following two expressions for I_C and I_E, since $I_C + I_B + I_E = 0$:

$$I_E = -I_S\left[1 + \frac{1}{\beta_F}\right]\left[\exp\left(\frac{qV_{BE}}{kT}\right) - 1\right] + I_S\left[\exp\left(\frac{qV_{BC}}{kT}\right) - 1\right], \quad \text{(4-243)}$$

and

$$I_C = +I_S\left[\exp\left(\frac{qV_{BE}}{kT}\right) - 1\right] - I_S\left[1 + \frac{1}{\beta_R}\right]\left[\exp\left(\frac{qV_{BC}}{kT}\right) - 1\right]. \quad \text{(4-244)}$$

These equations are identically the Ebers-Moll equations in transport form. Thus, low-frequency analysis requires the usual three parameters that are familiar from the static Ebers-Moll model of Section 4-6, namely I_S, β_F, and β_R. These parameters are listed at the top of Table 4-6 and are identified as the basic set. The remaining twenty-four parameters can be put into five other sets identified as (a) series resistance, R_{SE}, R_{SC}, and R_{SB}; (b) Early effect, V_A and V_R; (c) high-current effects, I_k and I_{kr}; (d) nonideal-diode effects, I_{SE}, I_{SC}, N_E, N_C, N_F, and N_R; and (e) capacitance effects, C_{OTE}, $\Delta\psi_{OE}$, M_E, C_{OTC}, $\Delta\psi_{OC}$, M_C, C_{OTS}, $\Delta\psi_{OS}$, M_S, τ_F, and τ_R. We now consider each set of parameters in turn. In Table 4-7 we give a typical set of SPICE values that are used throughout this chapter in the examples and problems.

4-8.2 Series-Resistance Effects

The imposed junction voltages $v_{JE}(t)$ and $v_{JC}(t)$ are used to include series-resistance effects in SPICE, along with Equations 4-231 and 4-232, which indicate that series resistance decreases the available junction voltage. In the forward regime of operation, the major effects of series resistance are these: Collector resistance R_{SC} increases the ON resistance of the transistor. The emitter resistance R_{SE} also increases the ON resistance, but in addition decreases the transistor gain. The base resistance R_{SB} has a major impact on high-frequency behavior and, for accurate ac or transient analysis, requires special treatment. This topic will be considered later. As a further complication, the R_{SB} value is a function of the degree of conductivity modulation in the base region. But for now, we shall assume that R_{SB} is constant.

4-8.3 Early Effect

Here we assume low-level static conditions and use default values for all SPICE parameters except the forward Early voltage V_A and the reverse Early voltage V_R. Under these conditions, the SPICE equations become

$$I_C = \frac{I_S}{q_B}\left[\exp\left(\frac{qV_{BE}}{kT}\right) - \exp\left(\frac{qV_{BC}}{kT}\right)\right] - \frac{I_S}{\beta_R}\left[\exp\left(\frac{qV_{BC}}{kT}\right) - 1\right], \quad (4\text{-}245)$$

$$I_B = \frac{I_S}{\beta_F}\left[\exp\left(\frac{qV_{BE}}{kT}\right) - 1\right] + \frac{I_S}{\beta_R}\left[\exp\left(\frac{qV_{BC}}{kT}\right) - 1\right], \quad (4\text{-}246)$$

and

$$q_B = \left[1 - \frac{V_{BC}}{V_A} - \frac{V_{BE}}{V_B}\right]^{-1}. \quad (4\text{-}247)$$

For forward-bias conditions, these equations reduce to

$$I_C = I_S \exp\left(\frac{qV_{BE}}{kT}\right)\left[1 - \frac{V_{BE}}{V_A}\right], \quad (4\text{-}248)$$

and

$$I_B = \frac{I_S}{\beta_F}\left[\exp\left(\frac{qV_{BE}}{kT}\right)\right], \quad (4\text{-}249)$$

which in turn can be combined to give

$$I_C = \beta_F I_B \left[1 - \frac{V_{BE}}{V_A}\right]. \quad (4\text{-}250)$$

The Early effect, discussed in Section 4-4.5, is described in the SPICE model by means of this equation. Using the parameter values given in Table 4-7, we can calculate the output characteristics given in Figure 4-63. At point B, the calculated slope gives an output resistance of 219 kilohm. From Equation 4-250, the output resistance should be approximately V_A/I_C, which equals 200 kilohms, in agreement with the value calculated from the slope.

4-8.4 High-Current Effects

Now let us use the default values for all SPICE parameters except the forward and reverse knee currents, I_k and I_{kr}. The forward knee current is defined as the value of collector current above which high-level conditions exist in the base region, a matter discussed in Section 4-5, and the reverse knee current has similar significance for reverse operation. Assuming conditions for which the default values apply, the SPICE equations reduce once again to Equations 4-245 and 4-246. Also, the charge-control relations become

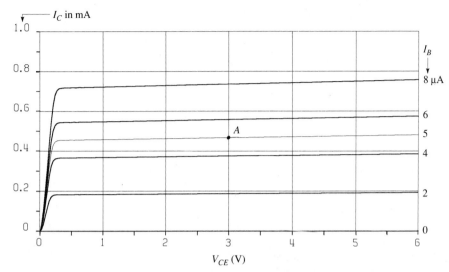

Figure 4-63 Output characteristics calculated using SPICE-parameter values given in Table 4-7. Operating-point values given in Table 4-8 correspond to point A. (See Sections 4-8.3 and 4-8.7.)

$$q_B = \left[\frac{1}{2} + \sqrt{\frac{1}{4} + q_2}\right], \qquad (4\text{-}251)$$

and

$$q_2 = \frac{I_S}{I_k}\left[\exp\!\left(\frac{qV_{BE}}{kT}\right) - 1\right] + \frac{I_S}{I_{kr}}\left[\exp\!\left(\frac{qV_{BC}}{kT}\right) - 1\right]. \qquad (4\text{-}252)$$

Under forward bias, Equations 4-245, 4-246, 4-251, and 4-252 reduce to

$$I_C = \frac{I_S \exp\!\left(\dfrac{qV_{BE}}{kT}\right)}{\left[1 + 2\sqrt{\dfrac{1}{4} + \dfrac{I_S}{I_k}\exp\!\left(\dfrac{qV_{BE}}{kT}\right)}\right]}, \qquad (4\text{-}253)$$

and

$$I_B = \frac{I_S}{\beta_F}\left[\exp\!\left(\frac{qV_{BE}}{kT}\right)\right]. \qquad (4\text{-}254)$$

Combining these equations, we find that (for all values of collector current)

$$I_C = \frac{2\beta_F I_B}{\left[\dfrac{1}{2} + \sqrt{\dfrac{1}{4} + \dfrac{\beta_F I_B}{I_k}}\right]}. \qquad (4\text{-}255)$$

For low-level conditions, the denominator equals two and Equation 4-255 becomes

$$I_C = \beta_F I_B. \tag{4-256}$$

For high-level conditions, Equation 4-255 becomes

$$I_C = \sqrt{\beta_F I_k I_B}. \tag{4-257}$$

Using Equation 4-254 again, we can rewrite Equation 4-257 as

$$I_C = \sqrt{I_k I_S}\left[\exp\left(\frac{qV_{BE}}{2kT}\right)\right]. \tag{4-258}$$

This equation shows that effective junction voltage (imposed junction voltage) is one-half the applied terminal voltage V_{BE}; thus in the limit of high-level operation, only half of the applied base-emitter voltage is available to control the emitter junction. These are two of the major effects of high-level conditions in the modern BJT, and were discussed in Sections 4-5.2, 4-5.3, and 4-5.6.

Using the data of Table 4-7, we can calculate β_F and h_{fe} as functions of I_C, with the result shown in Figure 4-64. Both β_F and h_{fe} fall off, as expected from Equation 4-258. As I_C increases, β_F approaches asymptotically the line labeled *upper current limit*. This line has a slope of minus one, and is drawn through the point $\beta_F = 100$,

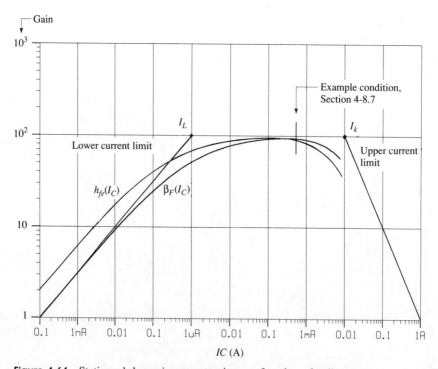

Figure 4-64 Static and dynamic current gain as a function of collector current, calculated using data of Table 4-7. (See Sections 4-8.4 and 4-8.7.)

and $I_C = I_k = 10$ mA. The two curves plotted here are closely related to the qualitative observation offered in Exercise 4-87.

4-8.5 Nonideal-Diode Effects

At very low current levels, transistor gain decreases because of a number of nonideal-diode effects. These are modeled in SPICE using I_{SC}, I_{SE}, N_E, N_C, N_F, and N_R. (See Table 4-6 for descriptions of these quantities.) Assuming low-level conditions and the default values of all of the SPICE parameters except for these six, we find that the SPICE equations reduce to

$$I_C = I_S\left[\exp\left(\frac{qV_{BE}}{N_F kT}\right) - \exp\left(\frac{qV_{BE}}{N_R kT}\right)\right]$$
$$- \frac{I_S}{\beta_R}\left[\exp\left(\frac{qV_{BE}}{N_C kT}\right) - 1\right] - I_{SC}\left[\exp\left(\frac{qV_{BC}}{N_C kT}\right) - 1\right], \quad (4\text{-}259)$$

and

$$I_B = \frac{I_S}{\beta_F}\left[\exp\left(\frac{qV_{BE}}{N_F kT}\right) - 1\right] + I_{SE}\left[\exp\left(\frac{qV_{BE}}{N_E kT}\right) - 1\right]$$
$$+ \frac{I_S}{\beta_R}\left[\exp\left(\frac{qV_{BC}}{N_R kT}\right) - 1\right] + I_{SC}\left[\exp\left(\frac{qV_{BC}}{N_C kT}\right) - 1\right]. \quad (4\text{-}260)$$

For forward-bias conditions, these equations become

$$I_C = I_S \exp\left(\frac{qV_{BE}}{N_F kT}\right) - \frac{I_S}{\beta_F} \exp\left(\frac{qV_{BC}}{N_C kT}\right), \quad (4\text{-}261)$$

and

$$I_B = \frac{I_S}{\beta_F} \exp\left(\frac{qV_{BE}}{N_F kT}\right) + I_{SE} \exp\left(\frac{qV_{BE}}{N_E kT}\right). \quad (4\text{-}262)$$

The empirical parameters N_F and N_R permit a generalization of the basic Ebers-Moll equations. Usually, we let $N_F = N_R = 1$. The major low-level effect is described by the second term in Equation 4-262. It introduces an additional base current described by a saturation current I_{SE} and a quality factor N_E, which is usually taken as equal to two. The responsible Sah-Noyce-Shockley phenomena have been described in Section 3-4.6.

To see the consequences of nonideal-diode effects, examine Figure 4-65, plotted using the second data set in Table 4-2. Here we plot log I_C and log I_B versus the base-emitter voltage. At low currents the I_C curve exhibits unity slope and extrapolates to $I_C = 10^{-16}$ A at $V_{BE} = 0$, as expected. On the other hand, the I_B curve exhibits a slope of 0.5, and extrapolates to 10^{-13} A at $V_{BE} = 0$. Referring back to Figure 4-64, we see that it is possible to define a parameter I_L for low-level behavior that is analogous to I_k for high-level behavior. Thus, β_F is roughly equal to 100 only for $I_L < I_C < I_k$.

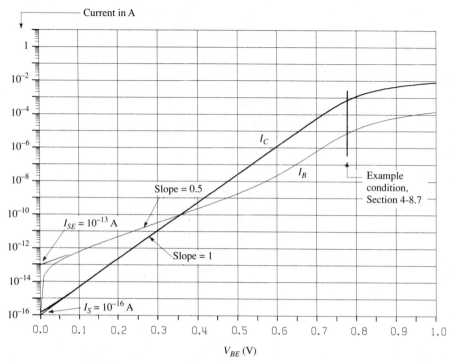

Figure 4-65 Collector and base currents as functions of V_{BE}, calculated using data of Table 4-7. (See Sections 4-8.5 and 4-8.7.)

4-8.6 Capacitance Effects

For ac and transient analysis, ten additional parameters, given in Table 4-6, are required. The first three of these parameters are the emitter–base zero-bias depletion-layer capacitance C_{OTE}, the contact potential $\Delta\psi_{OE}$ and the grading coefficient M_E. The next three, C_{OTC}, $\Delta\psi_{OC}$, and M_C, are the corresponding parameters for the collector–base junction; and the next three, C_{OTS}, $\Delta\psi_{OS}$, and M_S, are for the collector–substrate junction. The final two parameters are the forward transit time τ_F and the reverse transit time τ_R. Under low-level conditions and using default values, Equations 4-229 through 4-235 of the SPICE model reduce to the Ebers-Moll model, as was noted previously. Assuming other than default values for the capacitance parameters, Equations 4-236 through 4-240 become

$$C_{BE}(t) = C_{OTE}\left[1 - \frac{v_{BE}(t)}{\Delta\psi_{OE}}\right]^{-M_E} + \tau_F g_{mF}(t), \tag{4-263}$$

$$C_{BC}(t) = C_{OTC}\left[1 - \frac{v_{BC}(t)}{\Delta\psi_{OC}}\right]^{-M_C} + \tau_R g_{mR}(t), \tag{4-264}$$

$$C_{CS}(t) = C_{OTS}\left[1 - \frac{v_{CS}(t)}{\Delta\psi_{OS}}\right]^{-M_S}, \tag{4-265}$$

$$g_{mF}(t) = \frac{qI_S}{kT} \exp\left(\frac{qv_{BE}(t)}{kT}\right), \qquad (4\text{-}266)$$

and

$$g_{mR}(t) = \frac{qI_S}{kT} \exp\left(\frac{qv_{BC}(t)}{kT}\right). \qquad (4\text{-}267)$$

These equations are ideal in form, incorporating the character of those developed for the diode model in Section 3-9. Examples of diode capacitive effects are given in Section 3-9.6 on large-signal SPICE analysis.

4-8.7 Example of Small-Signal Analysis

Now we want to arrive at the hybrid-pi parameter values given in Table 4-2 (second set) and the Bias column of Table 4-9. For this purpose we can use the large-signal SPICE equations and the SPICE values in Table 4-7 (used previously to produce Figures 4-63 through 4-65). We shall focus on the single bias condition of Table 4-8, with $I_B = 0.005$ mA and $V_{CE} = 3$ V. Under these conditions, SPICE yields the operating-point information also given in Table 4-8. In the present section, we will use all this information to generate the small-signal hybrid-pi parameter values. In Section 4-8.8, we shall address large-signal SPICE analysis.

Here we will refer to Equations 4-229 through 4-240 as the SPICE model. Because we will consider only dc bias values, we will replace—without additional notice—a time-dependent variable such as $v_{BE}(t)$ with the dc bias value V_{BE}. Moreover, we will refer to the static values given in Tables 4-7 and 4-8 as the "dc data."

Using Equations 4-231 and 4-232 of the SPICE model and the dc data we find that

$$\begin{aligned} V_{JE} &= V_{BE} - I_B R_{SB} - (I_B + I_C) R_{SE} \\ &= (0.762 \text{ V}) - (0.0025 \text{ V}) - (0.0047 \text{ V}) = 0.755 \text{ V}, \end{aligned} \qquad (4\text{-}268)$$

and

$$\begin{aligned} V_{JC} &= V_{BC} + I_B R_{SB} + I_C R_{SC} \\ &= (-2.238 \text{ V}) - (0.0025 \text{ V}) + (0.0467 \text{ V}) = -2.194 \text{ V}. \end{aligned} \qquad (4\text{-}269)$$

Notice that the inward-flowing base current in R_{SB} actually *increases* the magnitude of the collector imposed junction voltage V_{JC}. From these equations, we conclude that R_{SB}, R_{SC}, and R_{SE} have only a small effect for the given conditions. At higher current levels, these resistances can dominate the device behavior.

Using Equations 4-233 through 4-235 of the SPICE model and the dc data we find that

$$q_1 = \left[1 - \frac{V_{JC}}{V_A}\right]^{-1} = 0.98, \qquad (4\text{-}270)$$

$$q_2 = \frac{I_S}{I_k}\left[\exp\left(\frac{qV_{JE}}{N_F kT}\right)\right] = 0.0478, \qquad (4\text{-}271)$$

and

$$q_B = q_1\left[\frac{1}{2} + \sqrt{\frac{1}{4} + q_2}\right] = 0.978(1.045) = 1.023. \qquad (4\text{-}272)$$

From these equations, we observe that the Early effect causes a 2% decrease in q_B, and that the Kirk effect causes a 4% increase, producing a net increase of 2%. Both effects become more important at higher current levels.

To calculate I_B and I_C, we start with Equations 4-228 and 4-230 of the SPICE model and use the dc data to obtain

$$I_C = \frac{I_S}{q_B}\left[\exp\left(\frac{qV_{JE}}{N_F kT}\right)\right] = 4.67 \times 10^{-4} \text{ A}, \qquad (4\text{-}273)$$

and

$$\begin{aligned}I_B &= \frac{I_S}{\beta_F}\left[\exp\left(\frac{qV_{JE}}{N_F kT}\right)\right] + I_{SE}\left[\exp\left(\frac{qV_{JE}}{N_E kT}\right)\right]\\ &= (4.8 \times 10^{-6} \text{ A}) + (0.2 \times 10^{-6} \text{ A}) = 5 \times 10^{-6} \text{ A}.\end{aligned} \qquad (4\text{-}274)$$

Thus, q_B, R_{SB}, R_{SE}, and R_{SC} decrease the collector current I_C to 33% of the first-order value calculated from $I_S \exp(qV_{BE}/kT)$. The ohmic resistances have the major impact on the results, since q_B by itself accounts for only a 2% reduction. Series resistance also reduces I_B from the first-order value calculated using $(I_S/\beta_F)\exp(qV_{BE}/kT)$. Note that q_B has no effect on I_B.

So far, we have calculated the dc bias currents and voltages for the SPICE model and have proven that they are the same as those given in Table 4-8, as expected. Now we will calculate the hybrid-pi parameters given in Table 4-6, from the dc data given in Tables 4-7 and 4-8. To begin, we will start by calculating the y-parameters, and then will transform these parameters into the hybrid-pi parameters.

Using the definitions of the y-parameters and Equations 4-273 and 4-274 we find that

$$y_{11} = \frac{\partial I_B}{\partial V_{JE}} = \frac{qI_S}{\beta_F N_F kT}\left[\exp\left(\frac{qV_{JE}}{N_F kT}\right)\right] + \frac{qI_{SE}}{N_E kT}\left[\exp\left(\frac{qV_{JE}}{N_E kT}\right)\right], \qquad (4\text{-}275)$$

$$y_{12} = \frac{\partial I_B}{\partial V_{JC}} = 0, \qquad (4\text{-}276)$$

$$y_{21} = \frac{\partial I_C}{\partial V_{JE}} = \frac{qI_S}{q_B N_F kT}\left[\exp\left(\frac{qV_{JE}}{N_F kT}\right)\right] - \frac{qI_S}{q_B^2}\left[\exp\left(\frac{qV_{JE}}{N_F kT}\right)\right]\frac{\partial q_B}{\partial V_{JE}}, \qquad (4\text{-}277)$$

and

$$y_{22} = \frac{\partial I_C}{\partial V_{JE}} = -\frac{qI_S}{q_B^2}\left[\exp\left(\frac{qV_{JE}}{N_F kT}\right)\right]\frac{\partial q_B}{\partial V_{JC}}. \qquad (4\text{-}278)$$

To evaluate the y-parameters numerically, we need to find expressions for $\partial q_B/\partial V_{JE}$ and $\partial q_B/\partial V_{JC}$. Using Equations 4-270 through 4-272, we find that

$$\frac{\partial q_B}{\partial V_{JE}} = q_1 \left[-\frac{1}{2}\left(\frac{1}{\sqrt{0.25+q_2}}\right) \frac{\partial q_2}{\partial V_{JE}} \right]$$

$$= \frac{q_1}{2\sqrt{0.25+q_2}} \frac{qI_S}{I_k N_F kT} \left[\exp\left(\frac{qV_{JE}}{N_F kT}\right) \right]. \tag{4-279}$$

Moreover,

$$\frac{\partial q_B}{\partial V_{JC}} = \frac{\partial q_1}{\partial V_{JC}} \left[\frac{1}{2} + \sqrt{\frac{1}{4}+q_2} \right] = \frac{q_B q_1}{V_A}. \tag{4-280}$$

Now we can evaluate the hybrid-pi parameters. First, we note $r_{b'e} = 1/g_{b'e} = 1/y_{11}$, so

$$r_{b'e} = \frac{1}{\dfrac{qI_S}{\beta_F N_F kT}\left[\exp\left(\dfrac{qV_{JE}}{N_F kT}\right)\right] + \dfrac{qI_{SE}}{N_E kT}\left[\exp\left(\dfrac{qV_{JE}}{N_E kT}\right)\right]} = 5.29 \text{ kilohm}, \tag{4-281}$$

where we have used Equation 4-275 and the dc data. In a first-order calculation, we expect $r_{b'e}$ to be given by $\beta_F kT/qI_B = 5.5$ kilohm, in agreement with the result just given. The feedback resistance $r_{b'c} = 1/g_{b'c} = 1/y_{12}$, giving

$$r_{b'c} = \infty. \tag{4-282}$$

The transconductance is given by $g_m = (y_{12} + y_{21}) \approx y_{12}$ so that

$$g_m = \frac{qI_S}{q_B N_F kT}\left[\exp\left(\frac{qV_{JE}}{N_F kT}\right)\right] - \frac{qI_S}{q_B^2}\left[\exp\left(\frac{qV_{JE}}{N_F kT}\right)\right]\frac{\partial q_B}{\partial V_{JE}}$$

$$= (1.805 \times 10^{-2} \text{ S}) - (7.555 \times 10^{-4} \text{ S}) = 17.3 \text{ mS}. \tag{4-283}$$

To first order, g_m is given by $qI_E/kT = 18.2$ mS, which agrees reasonably well with the value just given. The output resistance $r_{ce} = y_{22} + y_{12} \approx y_{22}$, so that

$$r_{ce} = \frac{-1}{\dfrac{qI_S}{q_B^2}\left[\exp\left(\dfrac{qV_{JE}}{N_F kT}\right)\right]\dfrac{\partial q_B}{\partial V_{JE}}} = 2.19 \times 10^5 \text{ ohm}. \tag{4-284}$$

As noted above, r_{ce} is approximately $(V_A/I_D) = 2.12 \times 10^5$ ohms in agreement with the value just calculated. Using Equations 4-237 and 4-240 and the dc data, we can find the small-signal capacitance $C_c = C_{BC}$ as follows:

$$C_c = C_{OTC}\left[1 - \frac{V_{JC}}{\Delta\psi_{OC}}\right]^{-M_C} + \tau_R g_{mR}$$

$$= (5.10 \times 10^{-14} \text{ F}) + (2 \times 10^{-25} \text{ F}) = 51.0 \text{ fF}. \tag{4-285}$$

Thus it is evident that the second term, the diffusion capacitance, is utterly negligible, a result of the fact that g_{mR} is so small:

$$g_{mR} = \frac{qI_S}{N_R kT}\left[\exp\left(\frac{qV_{JC}}{N_R kT}\right)\right] = 4.65 \times 10^{-17} \text{ S}. \quad (4\text{-}286)$$

Furthermore, the first term and its depletion-approximation basis are fully valid for the reverse-biased collector junction.

By contrast, the small-signal conductance g_{mF} associated with the forward-biased emitter junction given by Equation 4-239 is large when evaluated with the help of the dc data:

$$\begin{aligned}g_{mF} &= \frac{qI_S}{q_B N_F kT}\left[\exp\left(\frac{qV_{JE}}{N_F kT}\right)\right] - \frac{qI_S}{q_B^2}\left[\exp\left(\frac{qV_{JE}}{N_F kT}\right)\right]\frac{\partial q_B}{\partial V_{JE}}\\ &= (1.805 \times 10^{-2} \text{ S}) - (7.555 \times 10^{-4} \text{ S}) = 17.3 \text{ mS}.\end{aligned} \quad (4\text{-}287)$$

Thus diffusion capacitance dominates in the emitter-junction case, amounting to 1.73 pF, the second term of the small-signal emitter-junction capacitance given by Equation 4-236, which is repeated here for convenience:

$$C_e = C_{OTE}\left[1 - \frac{V_{JE}}{\Delta\psi_{OE}}\right]^{-M_E} + \tau_F g_{mF}. \quad (4\text{-}288)$$

The first term here is of course the depletion-layer capacitance of the emitter junction, and the present example again raises a matter treated previously in Sections 3-8.3 and 3-9.6. To repeat, the first term of Equation 4-288 is based upon the depletion approximation, which is poor for large forward bias. More accurate analysis [76] has shown that depletion-layer capacitance does not diverge, but passes through a maximum at large bias. SPICE automatically makes a still simpler adjustment. It employs a linear function of voltage that is tangent to the diverging function. The point of tangency is usually at or near the middle of the forward-bias range. This approximation errs on the low side, and is in fact a significantly poorer approximation than the diverging function through much of the forward-bias range! But it does eliminate the divergence that is very troublesome in numerical analysis, and its poor quality is relatively unimportant because diffusion capacitance is normally dominant in this range. The present example is a good case in point, because the approximate depletion-layer capacitance is 0.12 pF. This added to the value 1.73 pF cited before yields the total value of 1.85 pF that is given in Table 4-2.

Now let us consider a small-signal problem that will employ the hybrid-pi parameters just derived, and will provide a further understanding of the large-signal problem to be considered in the next section. Figure 4-66 shows two transistors connected as two inverters in cascade. For small-signal analysis of this circuit, we bias the inverters so that Q1 has the same internal currents and voltages as those used for calculating the hybrid-pi parameters. For example, the voltage across R_B is 0.2 V, which gives a bias current of 5 μA; further, the output voltage is 3 V, and both bias values are consistent with the dc data given in Table 4-8.

If we replace the dc supply voltage V_{IN} by a voltage pulse whose waveform is shown in Figure 4-67(a), SPICE produces the result given in Figure 4-67(b). Now we will calculate the same result using approximate methods. For small-signal

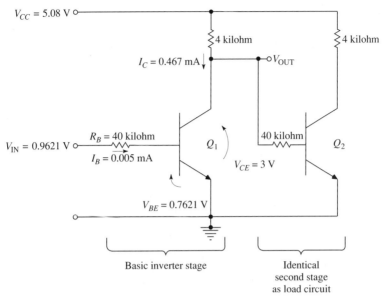

Figure 4-66 Schematic diagram of a circuit subjected to small-signal analysis using hybrid-pi parameter values derived from SPICE equations.

analysis, the circuit of Figure 4-66 can be replaced by the one of Figure 4-68 where we have used the hybrid-pi model. This circuit in turn can be approximated by that in Figure 4-69 by using circuit analysis. The effective base–emitter capacitance is given by

$$C_{\text{eff}} = [C_{b'e} + (1 + A_V) C_{b'c}]$$
$$= (1.85 \times 10^{-12} \text{ F}) + (1 + 61.9)(0.051 \times 10^{-12} \text{ F}) \quad \text{(4-289)}$$
$$= 5.06 \text{ pF}.$$

Thus Miller feedback accounts for about 63% of C_{eff}, as compared to the 37% contributed by $C_{b'e}$. The rise and fall times of the circuit are approximately $r_{b'e}C_{b'e} = 27$ ns, in agreement with the SPICE result given in Figure 4-67.

4-8.8 Example of Large-Signal Analysis

For the large-signal transient problem, let us use the same circuit as before and choose a static loadline passing through the bias point used in the small-signal problem. This loadline is plotted on the output plane in Figure 4-70, with a square labeled B marking the previous bias point. The large-signal pulse waveform to be applied to the circuit is shown with a solid line in Figure 4-71(a). At $t = 10$ ns it ramps up from 0.2 V to 4.6 V in 1 ns, returning in a similar fashion to 0.2 V after an interval of 80 ns.

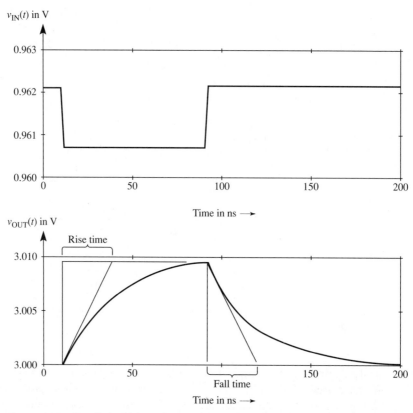

Figure 4-67 Small-signal waveforms calculated by SPICE for the circuit of Figure 4-66. (a) Voltage pulse to be substituted for V_{IN}. (b) Computer-determined output waveform.

Table 4-9 gives static values of important device variables under three input conditions. Columns A and C correspond to the OFF and ON conditions indicated in Figure 4-70, while column B of course applies to the bias point of the small-signal case. Squares are used in Figure 4-70 to designate all three of the points, A, B, and C. The dashed line in Figure 4-71(a) gives the output voltage waveform determined by SPICE simulation. Figure 4-71(b) gives the base-emitter voltage waveform, and the positive (saturation-regime) portion of the base-collector voltage waveform. Figures 4-71(c) and (d) give, respectively, the collector and base terminal currents.

The transient data given in Table 4-10 are calculated for five particular times occurring during transistor turn on, and are presented in five corresponding columns. The corresponding pairs of values for the output variables are indicated by the numbered triangles plotted along the loadline in Figure 4-70. In addition to external variables such as $v_{OUT}(t)$ and $i_C(t)$, certain "internal variables" such as $v_{JE}(t)$ and $q_{BC}(t)$ are included in Table 4-10. These have been extracted by inserting

SPICE MODELS

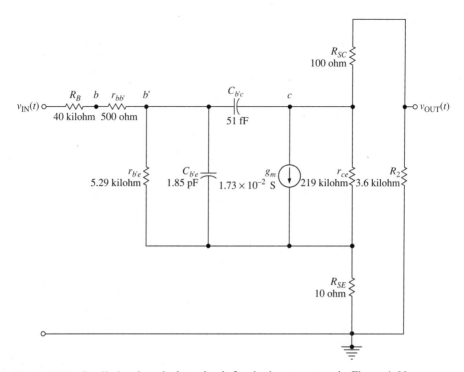

Figure 4-68 Small-signal equivalent circuit for the inverter stage in Figure 4-66, incorporating the hybrid-pi model for Q_1.

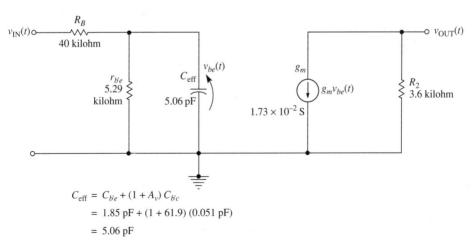

$$C_{\text{eff}} = C_{b'e} + (1 + A_v) C_{b'c}$$
$$= 1.85 \text{ pF} + (1 + 61.9)(0.051 \text{ pF})$$
$$= 5.06 \text{ pF}$$

Figure 4-69 Approximate small-signal equivalent circuit derived by circuit analysis from that in Figure 4-68.

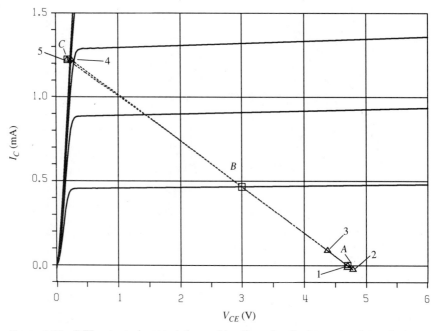

Figure 4-70 BJT output characteristics and loadlines for the large-signal transient problem, where the squares marked A, C, and B are the off, the on, and the bias conditions, respectively, with the last used previously for the small-signal problem (Figure 4-63). The five numbered triangles correspond to points plotted in Figures 4-71 and 4-72, and also to the five columns of data in Table 4-10. The upper of the two closely spaced loadlines shows the turn-on path, and the lower, the turn-off path.

additional code into the SPICE program (just as was done for the diode problem in Section 3-9.7), and are presented graphically in Figure 4-72.

The internal variables $r_{b'e}$, g_m, r_{ce}, and $r_{b'c}$ are the instantaneous values of the small-signal hybrid-pi parameters as a function of time, and they are plotted in Figures 4-72(a) and 4-72(e). The internal junction voltages $v_{JE}(t)$ and $v_{JC}(t)$ are given in Figure 4-72(b). The capacitances $C_{BE}(t)$ and $C_{BC}(t)$ and the corresponding total-charge variables $q_{BE}(t)$ and $q_{BC}(t)$ are given in Figures 4-72(f) and 4-72(g). The remaining five internal variables are suggested in Figure 4-73, and defined as will now be described. The base current as given by Equation 4-230 is divided into resistive and capacitive components and then is further divided into emitter and collector components as

$$i_B(t) = [i_B^{RE}(t) + i_B^{RC}(t)] + [i_B^{CE}(t) + i_B^{CC}(t)], \quad (4\text{-}290)$$

where

$$i_B^{RE}(t) = \frac{I_S}{\beta_F}\left[\exp\left(\frac{qv_{JE}(t)}{N_F kT}\right) - 1\right] + I_{SE}\left[\exp\left(\frac{qv_{JE}(t)}{N_E kT}\right) - 1\right], \quad (4\text{-}291)$$

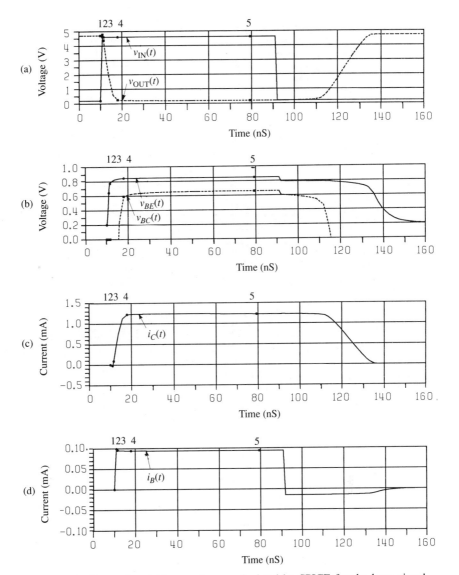

Figure 4-71 External-variable waveforms calculated by SPICE for the large-signal transient problem. The five numbered points on each waveform correspond to the five numbered triangles in Figure 4-70. (a) Input and output voltage versus time. (b) Emitter and collector terminal voltages versus time. (c) Collector current versus time. (d) Base current versus time.

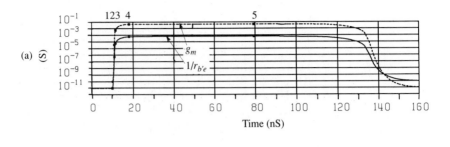
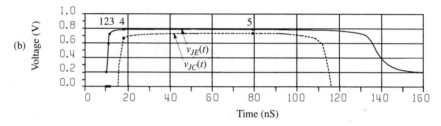
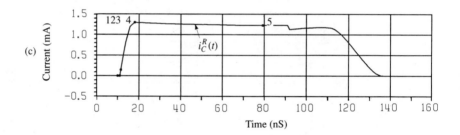
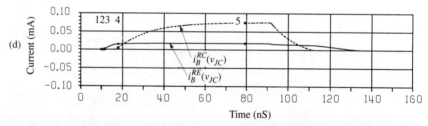

Figure 4-72 Internal-variable waveforms calculated by SPICE for the large-signal transient problem. The five numbered points on each waveform correspond to the five numbered triangles in Figure 4-70. The waveforms describe the behavior of the following variables. (a) Transconductance and the reciprocal of $r_{b'e}$ (displayed in Figure 4-68). (b) Emitter- and collector-imposed junction voltages. (c) Resistive current flowing between collector and emitter internal nodes. (d) Resistive components of base current flowing to collector and emitter junctions. (e) Two conductances displayed (inversely) in Figure 4-68. (f) Total capacitances associated with the collector and emitter junctions. (g) Charges stored in the capacitances of part (f). (h) The capacitive components of base current flowing to the collector and emitter junctions.

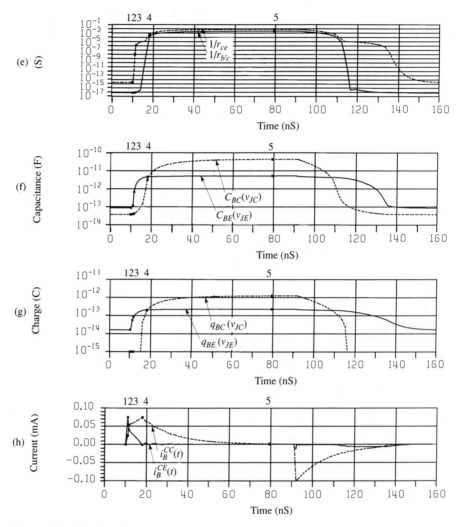

Figure 4-72 (*Continued*)

$$i_B^{RC}(t) = \frac{I_S}{\beta_R}\left[\exp\left(\frac{qv_{JC}(t)}{N_R kT}\right) - 1\right] + I_{SC}\left[\exp\left(\frac{qv_{JC}(t)}{N_C kT}\right) - 1\right], \quad (4\text{-}292)$$

$$i_B^{CE}(t) = C_{BE}(t)\frac{dv_{JE}}{dt}, \quad (4\text{-}293)$$

and

$$i_B^{CC}(t) = C_{BC}(t)\frac{dv_{JC}}{dt}. \quad (4\text{-}294)$$

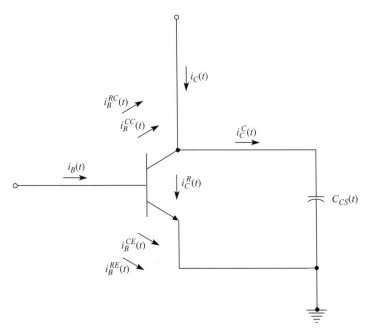

Figure 4-73 For large-signal transient analysis, base current and collector current can each be divided into four time-dependent components, indicated on the BJT symbol above.

Equations 4-291 and 4-292 could be further divided into Ebers-Moll and nonideal-diode components, but we will not do this since the nonideal components can often be neglected. Also, for simplicity we shall lump the diffusion and depletion-layer components of capacitance.

The collector current as given by Equation 4-229 can also be divided into four components that are displayed in Figure 4-73, so that

$$i_C(t) = [i_C^R(t) + i_C^C(t)] - [i_B^{RC}(t) + i_B^{CC}(t)], \quad (4\text{-}295)$$

where

$$i_C^R(t) = \frac{I_S}{q_B(t)}\left[\exp\left(\frac{qv_{JE}(t)}{N_F kT}\right) - \exp\left(\frac{qv_{JC}(t)}{N_R kT}\right)\right], \quad (4\text{-}296)$$

and

$$i_C^C(t) = C_{CS}(t)\frac{dv_{JC}}{dt}. \quad (4\text{-}297)$$

Here the resistive term $i_C^R(t)$ represents the current flowing from the internal collector node to the internal emitter node. The capacitive term $i_C^C(t)$ flows through the collector-substrate capacitance $C_{CS}(t)$, which, for simplicity, we are neglecting.

The various waveform curves associated with the emitter–base junction resemble the corresponding curves for the diode, as given in Figures 3-84 and 3-85. This resemblance is not surprising since we have selected circuit conditions and device parameters to emphasize the correspondence. In particular, we used the same input waveform and we used the same values for five of the diode parameters. However, a number of modifications are necessary to obtain a closer match between the diode response and emitter–base transistor response. For the diode circuit, we selected an external series resistance R_G that is a factor of ten smaller than that in the transistor circuit. This was done so that the diode ON conditions would approximate the ON conditions of the base–emitter diode, whose current is fixed by transistor action. As a result, the junction voltages are nearly equal and the diode current approximates the transistor emitter current. For simplicity in the diode problem, we neglected the internal series resistance R_S. The zero-bias diode-junction capacitance C_{0T} was chosen to be ten times C_{0BE}, to compensate for the lowering of R_G by the same factor. The effective lifetime τ^* for the diode equals the base-charging time τ_B, which is 100 times τ. In Section 4-7.6, we showed that the base-charging time was given by $\beta_F \tau$ for active-regime operation of the transistor.

Now we examine the turn-on response of the BJT in detail. The five points plotted in each part of Figures 4-71 and 4-72 divide the period from 10 ns to 80 ns into five phases. Phase 1 is the time interval from 10 ns to 11 ns, during which the resistive components of current can be neglected and the average base current of 0.05 mA is divided between the base–collector and the base–emitter capacitances. The average value of $C_{BE}(t)$ is 0.038 pF, and of $C_{BC}(t)$ is 0.099 pF. Thus some 60% of the base current goes to the collector, causing a small positive increase in collector-emitter voltage, $v_{OUT}(t)$, as can be seen in Figure 4-71(a), and a small negative increase in collector current, $i_C(t)$, as can be seen in Figure 4-71(c). After phase 1, collector current increases and collector-emitter voltage decreases, both monotonically. During phase 2, from 11 ns to 11.54 ns, the BJT has a small average voltage gain of about -3, and this combined with the Miller effect causes a rapid increase in i_B^{CE}, and a corresponding decrease in i_B^{CC}. At the end of phase 2 these currents are 0.040 mA and 0.054 mA, respectively, while the resistive components are still much smaller but increasing in importance. In phase 3, the BJT is in its forward-active regime with an average voltage gain of -69.3, and C_{BE} is near its maximum value, while i_B^{CE} approaches zero and i_B^{RE} approaches the fully ON value. In phase 4 the BJT enters saturation, C_{BC} rises to 43.7 pF, and average voltage gain drops to -17.6. At the end of phase 4, the capacitive currents have decayed, and the "forced beta" (in saturation) is about 4.

4-8.9 Thermal Resistance

Most of the analyses in this book are isothermal. But there is one problem involving temperature differences that has appreciable practical importance, and also is relatively simple because it is governed by a linear equation. The rate of transport of thermal energy through a medium from one position to another is proportional to the temperature difference ΔT between the two positions. The linear law applies whether we are concerned with temperature rise in a BJT or heat loss through a building wall.

An analogy to Ohm's law aids both understanding and memory. At several points earlier we have noted that the voltage variable in Ohm's law is actually a voltage difference, ΔV; this is taken to be the analog of ΔT. The analog of current I, in turn, is taken to be the rate of thermal-energy delivery to a particular region within the BJT, which is to say, the power dissipation. All that is left, then, is to define the analog of resistance R, and that is the *thermal resistance* θ. Finally, the desired equation, by analogy to $\Delta V = IR$, is simply

$$\Delta T = P\theta. \tag{4-298}$$

Exercise 4-104. What are the units of thermal resistance θ?

Since $\theta = \Delta T/P$, it follows that the dimensions of θ are [C°/W]. Thermal resistance is a function of material properties and geometry, and tells the amount of the temperature difference between two locations that will accompany a certain level of steady-state dissipation at one of the locations.

Professional bodies concerned with standardizing and rationalizing systems of units have in recent times encouraged the citing of temperature in *centigrades* or *Kelvins,* dropping the term degree or degrees. Regrettably, this change sacrifices a distinction that often serves the cause of clarity. For example, we can maintain a useful separation between *temperature in degrees centigrade* on the one hand, and *temperature difference in centigrade degrees* on the other hand. We shall employ this distinction in the present discussion. To the skeptic, let us point out that there is indeed a distinction worth maintaining between 1 C° = 1 K°, and 1°C = 274°K. The logical rejoinder is that such a distinction is not necessary between the units for voltage V and voltage difference ΔV. However, the two situations differ in that references are well known and immutable in the case of temperature scales.

Exercise 4-105. Where is the primary location or "seat" of power dissipation in a BJT operating in the forward-active regime?

The terminal current I_C passes through the portion of the collector junction associated with the intrinsic base region, and I_C approximates I_E, the largest terminal current in the BJT. Furthermore, the collector-junction voltage V_{CB} typically approximates V_{CE}, the largest voltage applied to the BJT. Hence, a localized power dissipation in the amount

$$P = I_C V_{CB}$$

occurs in the collector junction portion adjacent to the intrinsic base region. A closer approximation to total power dissipation in the transistor is

$$P = I_C V_{CE},$$

taking into account the emitter-junction contribution too.

It is convenient and customary to use subscripts on the thermal-resistance symbol to designate the two locations involved in the definition. For example, the thermal resistance most commonly cited is θ_{jc}, the thermal resistance from *junction to case*. Let us interpret "junction" to mean both junctions, so that the power in question approximates the total dissipation in the transistor. Thus the site of the dissipation is a pair of regions flanking the intrinsic base region. The term *case* designates a position (usually standardized) on the exterior of a discrete-device package. In copper power-transistor packages (such as the TO-3), one observes θ_{jc} values of a few centigrade degrees per watt. In integrated-circuit packages, on the other hand, values of a few tens or hundreds of centigrade degrees per watt are more typical.

When the BJT is in close thermal contact with a heat sink of some kind, the latter thermal resistance is obviously in series with the former. Let us assume liquid cooling, so that there is a thermal resistance θ_{cl} from the BJT case to the liquid. Hence,

$$\theta_{jl} = \theta_{jc} + \theta_{cl}. \tag{4-299}$$

More complex networks of thermal resistances are obviously possible, but the series combination serves to illustrate the point.

In many early circuits, the BJT had no heat-sink arrangement, and had to transfer its heat by radiation and convection to the environment, or the surrounding *atmosphere*. In such a case it was the case-to-atmosphere thermal resistance θ_{ca} that was placed in series with θ_{jc}. This notation employs the same subscripts as a variant, "case-to-ambient" thermal resistance, that is frequently encountered, but is mildly corrupt because *ambient* is an adjective and not a noun, and the term is being used as shorthand for *ambient* (or surrounding) *atmosphere*.

In circuit-analysis programs such as SPICE, temperature is regarded as an operating condition, and not as a model parameter. Thus, the same temperature T is used for the whole circuit. Given a temperature T, SPICE will recalculate all device parameters that are functions of temperature, and then will begin simulation. Time-dependent heating problems are disallowed.

SUMMARY

The BJT (bipolar junction transistor) yields to one-dimensional analysis. The adjective *bipolar* in its name is intended to convey the information that both holes and electrons play an important part in its operation. It was invented in 1948, reduced to practice in 1951, and became the dominant solid-state amplifying device by the mid-1950s. The basic device incorporates three regions named *emitter, base,* and *collector,* each having an electrical terminal with the same name. The three regions are separated by two *PN* junctions. Consequently, there are *NPN* and *PNP* devices. The *NPN* is more common and somewhat superior, but complementary circuits employing both types achieve performance and power-dissipation advantages. Conditions at the (space-charge-layer) boundaries of the two junctions are particularly important in BJT analysis. Minority-carrier densities there are uniquely related to terminal voltages.

Most BJTs today are incorporated into integrated circuits. An integrated circuit is defined as a useful combination of components fabricated in (or on) a semiconductor single crystal. However, specialized BJTs such as power transistors are still being manufactured as discrete (separately packaged) devices. The BJT is most frequently used in the common-emitter configuration, which means that the input signal is applied to the base terminal, the output signal is taken from the collector terminal, and the emitter terminal is common to the input and output ports. In this arrangement the BJT is capable of delivering both current gain and voltage gain. For normal bias in this configuration, both input (base) and output (collector) static conventional currents are positive, or inward, for the *NPN* device. All currents and voltages are reversed for the *PNP* device. The two closely spaced junctions of the BJT interact in the sense that their current values are very different from the values that would exist if the junctions were far apart and subjected to the same voltages. However, the law of the junction may still be used to find boundary values for carriers, just as though the junctions were isolated from one another. The gross asymmetry of the emitter junction causes asymmetric injection of minority carriers in its normal condition of forward bias; injection into the base region accounts for most of the junction current. The normally reverse-biased collector junction receives the injected carriers and delivers them to the collector region, where they are once more majority carriers. Transport of the minority carriers through the base region is the result of diffusion, with a nearly constant density gradient from boundary to boundary being the norm.

The area that enters in the most important way into fixing current density is emitter-junction area A_E. BJTs in common use have A_E values ranging through some eight orders of magnitude. On the other hand, the most critical thickness dimension, base thickness X_B, ranges in ordinary devices only a bit over a factor of ten. The primary one-dimensional current through a BJT consisting of majority carriers in emitter and collector regions typically exhibits a current density that varies by only a few percent from emitter body to collector body. The density of these carriers, by contrast, typically varies by a factor of a billion or more in this same spatial interval.

The usual semiconductor approximations and assumptions are valid in the moderately doped base and collector regions that define the asymmetric collector junction. But they are not valid in the heavily doped emitter region, so empirical methods must usually be used there. The five BJT regions in the sequence of decreasing equilibrium peak-field-magnitude values are typically emitter junction, collector junction, collector region, emitter region, base region. The dominant transport mechanisms in the five regions during operation are diffusion in the emitter junction and base region and drift in the other three.

The most common BJT has emitter and base regions formed by solid-phase impurity diffusion, and hence these regions are not uniformly doped. Nonetheless, idealized step-junction analyses yield good approximations of actual BJT behavior. Although minority carriers diffusing from emitter to collector through the base region constitute a nearly one-dimensional problem, majority carriers in the base region constitute a two- or three-dimensional problem because they travel laterally from the base contact and thence into the emitter region, where nearly all of them recombine in a modern BJT. The fact that they also constitute the normally small base current I_B, however, means that their current can be neglected for some purposes. Because the base current is typically a small fraction of the total current passing through the emitter junction, it is sometimes described as a defect current. But inasmuch as the base current is the control current in the common-emitter BJT, it plays an extremely important role. The intrinsic or active base region is defined in the thickness direction by the junction boundaries on the base sides of the two junctions and laterally by the periphery of the emitter junction (this junction usually having a smaller area than the collector junction). The balance of the base region is defined as extrinsic. In the path of the base current is the parasitic ohmic base resistance (typically ~ 50 ohms), contributed by both the extrinsic and intrinsic base regions. Emitter and base electrical contacts having the form of closely spaced stripes tend to diminish base resistance. A small fraction of the majority carriers that constitute base current recombine in the base region with minority carriers that fail to complete the short trip from emitter junction to collector junction. In a silicon BJT, leakage current associated with the normally reverse-biased collector junction arises mainly from carrier generation in its space-charge layer.

Minority carriers diffusing through the base region constitute the collector current I_C, which is the output current in the common-emitter BJT. The ratio of I_C to the input current I_B is the static (dc) current gain β of the device. It yields to elementary analysis, except for high-level-doping effects in the emitter region that play a part in fixing I_B; since β is determined by structural constants of the device, it is itself a constant characterizing the BJT. The relationship $I_C = \beta I_B$, stating that the BJT is a linear current amplifier, exists because both currents are fixed by emitter-junction boundary values for minority carriers, fixed in turn by a single voltage, the base-emitter voltage V_{BE}.

The input port of the common-emitter BJT is the normally forward-biased base-emitter junction, and therefore requires current biasing, or an arrangement to set static input current. Voltage division and current setting by two or more linear elements is best handled analytically. When one or more of the devices involved has a nonlinear I–V characteristic, as does a forward-biased junction, a graphical technique is useful, namely the loadline diagram. It exploits the fact that (for two devices in series) the two devices carry the same current and have individual voltage drops that add to equal the applied voltage. The use of resistors for biasing in the base, collector, and (sometimes) emitter legs of a BJT circuit constitutes a practical approach to biasing. But with the advent of the integrated circuit, the use of another solid-state device (often a BJT) in lieu of one or more resistors is favored because such a device requires less silicon area than a resistor and is therefore less costly. The equivalent-circuit model constitutes a third way to handle voltage-division and current-determination problems. It is an approximate-analytic technique employing ideal devices in a configuration that yields terminal properties similar to those of the device being modeled. For the common-emitter BJT one employs a 0.7-V battery (voltage source) to model the offset voltage V_{BE} of the input-port junction, and a current-controlled current source βI_B in series with the collector terminal. The output plane is a diagram of output current I_C versus output voltage V_{CE}, with input current I_B usually employed as a parameter. For equal-I_B steps, the BJT's output characteristics in the normal-bias regime are approximately a set of equally spaced horizontal straight lines. Nonetheless, these characteristics are grossly nonlinear overall because each "turns a corner" near the I_C axis and heads for the origin of the output plane. Consequently, the loadline construction is once again useful. When one places in series with the collector terminal an external resistor R_C, the resistor's I–V characteristic becomes the "loadline," extending from a point V_{CC} on the voltage axis, where V_{CC} is the voltage applied to R_C and the BJT in series, to the point V_{CC}/R_C on the current axis. The resulting diagram is useful for understanding the amplifying properties of the common-emitter BJT, which is to say, the BJT in a configuration having the emitter terminal common to the input and output ports. A positive voltage increment at the input port causes a negative increment at the output port, so the amplifier is inverting. Its small-signal voltage is fixed by component values in the biasing circuit and is proportional to current gain β.

The quiescent operating point in the common-emitter output plane of a BJT can be set by adjusting V_{CC}, R_C, and I_C (which is to say, I_B). When the emitter junction is forward-biased and the collector junction is reverse-biased, the BJT is in the forward-active regime of operation. When $V_{CE} = V_{BE}$, the collector junction has a bias voltage of zero. A still smaller value of V_{CE} means that the collector junction is forward-biased. The BJT with both junctions forward-biased is said to be in the saturation regime. The collector junction still collects efficiently, however (and collector current remains nearly constant), until its forward voltage amounts to several tenths of a volt. This is termed the *regime of weak saturation*. With still greater forward bias on the collector junction, collector current falls steeply, the condition known as *deep saturation*. Here the BJT can be modeled as two batteries—$V_{CE(SAT)} \approx 0.2$ V between the collector and emitter terminals, and $V_{BE(SAT)} \approx 0.8$ V between the base and emitter terminals. The output-plane operating point under deep-saturation conditions is fixed by the external circuit.

A BJT in forward saturation has its emitter junction more forward-biased than its collector junction. Reversing the relationship places the BJT in reverse saturation. If a common-emitter BJT in the forward-active regime has bias polarities at its input and output terminals reversed, it is in the reverse-active regime of operation. When both junctions are reverse-biased (or zero-biased), the BJT is in cutoff. The simplest switching application of the BJT uses forward (deep) saturation as the ON (or closed) condition, and cutoff as the OFF (or open) condition. While a BJT in the common-emitter configuration can deliver both current gain and voltage gain, the common-base configuration can provide near-unity current gain and large voltage gain. The common-collector (or emitter-follower) configuration, on the other hand, can provide large current gain and near-unity voltage gain.

Actual BJTs exhibit parasitic properties. Examples are nonzero collector leakage current, saturation voltage, ohmic base resistance, and the Early effect. The last is a modulation of effective base thickness X_B that is caused by variations in collector-junction voltage, causing collector current and current gain β to increase with collector voltage V_{CE}. Some of the evolutionary changes in BJT structure were made to ameliorate the performance penalties associated with parasitic effects. Notable among these changes was the collector-region structure incorporating a heavily doped substrate region upon which a lightly doped layer was created by epitaxial growth. This structural modification diminished the Early effect and improved breakdown voltage and collector saturation resistance.

Solid-phase-diffusion methods were introduced for creating the base- and emitter-region net-doping profiles. These are attractive because of relatively easy and accurate doping control and large-batch capabilities, which translate into low cost. However, the base-region net-doping profile resulting from diffusion procedures is notably nonuniform. It turns out, as Gummel demonstrated, that the details of doping profile are relatively unimportant, and that what is important is the areal density of net impurities in the base region—the net number of impurity atoms per unit area. This quantity, known as the Gummel number, ranges typically from $10^{12}/cm^2$ to $10^{13}/cm^2$.

Output characteristics (I_C versus V_{CE}) for the ideal common-emitter BJT exhibit zero conductance. In a real device, however, departures from the ideal condition are caused by the Early effect, breakdown phenomena, and the punchthrough phenomenon. Early-effect output conductance is approximately independent of voltage (V_{CE}). Breakdown and punchthrough effects, however, cause output conductance to increase with voltage. Avalanche multiplication in the collector junction causes collector–emitter breakdown (with significant collector current) to occur when the multiplication factor M approaches $1/\alpha$, where α is the common-base current gain. Since α is normally close to but a bit smaller than unity, BJT breakdown voltage BV_{CEO} (base terminal open) is appreciably lower than that of an isolated junction equivalent to the collector junction. (In the latter case, breakdown accompanies the condition wherein M approaches infinity.) In the most basic terms, the reason for this is that holes resulting from avalanche multiplication enter the base region as a "quasi base current," and this current is multiplied by the gain mechanism of the BJT. Curvature of the collector junction at its periphery has little effect on BV_{CEO} because little current passes through that portion of the junction. The negative differential resistance observable in an output characteristic for a low value of I_B occurs because α increases with current and M increases with voltage; the product of the two must equal unity for a static condition.

The ratio of forward to reverse β is significantly dependent upon the ratio of collector-junction area to emitter-junction area. Thus this BJT feature can be largely adjusted by the designer when it is important to do so. Also, possible BJT trade-offs are numerous. For example, one can achieve elimination of the Early effect and achieve a BV_{CEO} value that is essentially that of the isolated collector junction. The price of such an improvement, however, is an upper limit on collector current, and worsened (increased) saturation voltage.

The BJT is routinely used under high-level conditions, meaning that minority-carrier densities in the base region and (or) collector region approach or exceed the equilibrium majority-carrier densities there. This fact results in failure of the law of the junction, so high-level formulations of the junction problem should be used. Also, longitudinal (x-directional or junction-normal) voltage drops develop that cause imposed junction voltages (boundary-to-boundary voltages) to differ from terminal voltages. The net of these effects causes the linear curve of log I_C versus V_{BE} to deviate sharply toward lower slope in the range 0.7 V < V_{BE} < 0.8 V. The curve of log I_B versus V_{BE}, however, retains its linearity up to nearly V_{BE} = 0.9 V, so that it is clear that collector current is the cause of the common-emitter current-gain (β) decline that occurs in the high-level regime. At the lower ends of these same two curves, the I_C curve retains its linearity to very low voltage, while the I_B curve deviates toward lower slope, starting in the range 0.3 V < V_{BE} < 0.5 V. This is a consequence of the SNS (Sah-Noyce-Shockley) effect, and causes I_B to be responsible for the β decline that occurs in the low-level regime.

The longitudinal high-level voltage drops in the base and collector regions are accompanied by an electric field. Such a field can be resolved into an ohmic component that inevitably accompanies the passage of a net current, and a bulk-diffusion component that arises spontaneously to equalize the (apparent) diffusivities of holes and electrons in regions where density gradients exist. The base and collector regions involved remain in a quasineutral condition, for which a succinct criterion is $(dn/dx) \approx (dp/dx)$. Under such conditions, ambipolar analysis is applicable. Such analysis sometimes makes use of an (effective) ambipolar diffusivity that takes into account the net longitudinal electric field that is present, while pretending that it is absent, and an (effective) ambipolar drift mobility that takes into account the perturbation of a uniform applied electric field caused by a localized high-level condition, while pretending that the perturbation is absent. Ambipolar diffusivity in the high-level limit approaches a value intermediate between the hole and electron values. Ambipolar (drift) mobility in the high-level limit approaches zero. Drift mobility refers to the motion of a density disturbance, while conductivity mobility refers to the net motion of individual carriers. A graded-base BJT has a base region that is nonuniformly doped, with the nonuniformities sometimes being very marked. Longitudinal electric fields, even at equilibrium, are the result of such doping variations. In graded-base devices, low-level minority-carrier profiles are very different from those in the uniformly doped case at the same current. But under high-level conditions, the doping variations become progressively less important, and nearly linear minority-carrier profiles develop for all types of variations.

The Ebers-Moll low-level large-signal model, used primarily for static analysis, is directly based upon device physics. It yields terminal current as a function of terminal voltages. It is extremely general with respect to BJT geometrical variations, but does not comprehend certain effects, such as the bias dependence of junction-boundary positions, that are sometimes important. Gummel and Poon reformulated the model (resulting in what is called the transport form) in a manner that makes it clearer and simpler for many purposes, and that adapts it more readily to the modern silicon BJT. When the original model was first presented, only germanium BJTs had reached a practical stage of development. The Ebers-Moll model exploits the linear relationship between a base-region boundary value and the current at the opposite port that exists for the common-base configuration. This inherent linearity permits the application of the superposition principle to obtain the currents existing for arbitrary boundary values and hence, via the law of the junction, for arbitrary terminal voltages. In a BJT with negligible Early effect, common-base reciprocity holds to an excellent approximation. An eight-component equivalent-circuit Ebers-Moll model yields terminal characteristics that approximate those predicted by the two-equation analytic model.

The hybrid parameters are four partial derivatives involving terminal currents and voltages that can be treated as constant at and near a particular operating point. They are useful for linear, small-signal, dynamic characterization of the device. Because they are real numbers (and not complex numbers) they are valid only for low-frequency analysis. They lend themselves readily to equivalent-circuit application and physical interpretation. Transconductance g_m is the ratio of an incremental current delivered by an amplifying device to the incremental voltage at the input port required to produce it. In the common-emitter BJT, g_m depends on two variables only, output current I_C and absolute temperature T, being directly proportional to I_C and inversely to T. The BJT is distinguished from competing devices by remarkably large transconductance. This is because its input voltage directly modulates the voltage barrier critical to its operation.

The low-frequency hybrid-parameter model can be extended to medium-frequency use by adding two capacitances, each associated with a junction. The resulting hybrid-pi (or Giacoletto) model is useful and versatile because its elements are frequency-independent, and can be related directly to device physics. The four elements of the h-parameter model for medium-frequency use are complex numbers, and thus are distinct from those of the hybrid-parameter model, which are real. Transformation from the hybrid-pi form to the h-parameter form, or the reverse, is straightforward, but sometimes tedious. The admittance model uses four complex y parameters and has advantages for analytical calculation and computer simulation. The scattering-parameter (s-parameter) model is used for high-frequency analysis. Usually one works with the model that is most convenient for a particular purpose, and then makes transformations to other models as necessary.

A problem common to the lumped-element models is phase errors at high frequencies, even though the predictions of signal magnitudes may be adequately accurate; a distributed network is needed to model signal delay realistically. Empirical correction of phase errors is sometimes employed. A similar approach is used to introduce corrections for current-dependent base resistance, extrinsic capacitances, the Early effect, and defect current.

The charge-control model involves time constants (or frequency constants), and hence is not closely related to device physics. One starts (as in the diode case) with knowledge of stored excess-carrier charge, and then performs a spatial integration, thus giving up information on charge location. Its key assumption, that the charge distributions known for static conditions have relevance under dynamic conditions as well, is usually acceptable for small-signal cases and poor for large-signal cases. The charge-control model yields small-signal gain-versus-frequency predictions quite consistent with those of the hybrid-pi and h-parameter models.

Gummel (and later, Gummel and Poon) contributed an extended and modified charge-control model for high-level conditions. It focuses on total majority-carrier charge in the base region and writes a collector-current equation analogous to the low-level expression involving the Gummel number, but substituting the areal density of majority carriers for that of majority impurities. This approach is central in the Gummel-Poon model.

Figures of merit are useful for evaluating and comparing devices. A useful figure of merit is independent of circuit variables and is expressible in terms of device properties. The cutoff frequency for common-emitter small-signal current gain (the frequency at which it has dropped to unity) is one example, and is useful for wide-band-amplifier applications. But because it is independent of ohmic base resistance, which plays a major role in high-frequency applications, the maximum frequency of oscillation is a more useful figure of merit for high-frequency applications.

The SPICE program for numerical analysis of the BJT can be used for both large-signal and small-signal modeling. The basic large-signal model, employing twelve equations, fourteen variables, and twenty-seven parameters, comprehends such real-life factors as parasitic series resistance, high-current effects, nonideal-diode effects, and the Early effect. From the large-signal model, one can generate hybrid-pi parameters for use in small-signal modeling. In a typical small-signal common-emitter result, one finds that Miller capacitance dominates input capacitance.

The temperature rise of a BJT in steady-state operation depends linearly upon the power dissipation involved. The constant of proportionality, known as *thermal resistance,* is a collective property of the materials and structures lying in the thermal path between the locus of power dissipation and the ultimate heat sink. It is measured in degrees per watt. In SPICE analysis, temperature is treated as an operating condition that affects SPICE parameters, and only steady-state temperatures are allowed.

Table 4-1 A Specific BJT Structure

Emitter	Base	Collector
$N_{DE} = 10^{20}/cm^3$	$N_{AB} = 10^{17}/cm^3$	$N_{DC} = 5 \times 10^{15}/cm^3$
$\mu_{nE} = 100$ cm^2/V·s	$\mu_{nB} = 700$ cm^2/V·s	$\mu_{nC} = 1250$ cm^2/V·s
$\mu_{pE} = 46$ cm^2/V·s	$\mu_{pB} = 300$ cm^2/V·s	$\mu_{pC} = 430$ cm^2/V·s

Table 4-1 (continued)

Emitter	Base	Collector
$X_E = 1~\mu\text{m}$	$X_B = 0.4~\mu\text{m}$	$X_C = 1.5~\mu\text{m}$
$\tau_E = 10^{-9}$ s	$\tau_B = 10^{-5}$ s	$\tau_C = 3 \times 10^{-5}$ s
$n_{iE} = 1.2 \times 10^{11}/\text{cm}^3$	$n_i = 10^{10}/\text{cm}^3$	$n_i = 10^{10}/\text{cm}^3$
$A_E = 10^{-4}~\text{cm}^2$		

Table 4-2 Typical Hybrid-Pi Parameter Values for Room Temperature

Parameter	Value for small BJT of about 1970 at $I_C = 1.3$ mA [71]	Value for smaller BJT of about 1980 at $I_C = 0.467$ mA [72]
g_m	50 mS	17.3 mS
$r_{bb'}$	100 ohm	500 ohm
$r_{b'e}$	1 kilohm	5.29 kilohm
$r_{b'c}$	4 Megohm	6.44×10^{16} ohm
r_{ce}	80 kilohm	2.19 Megohm
C_e	100 pF	1.85 pF
C_c	3 pF	0.05 pF

Table 4-3 Important Models for Small-Signal Analysis of the BJT

Name	Parameter Types	Primary Application
Hybrid-parameter model	Fixed real (resistive, dimensionless, or controlled-source) elements	Low-frequency device-characterizing measurements
Hybrid-pi model	Fixed real (resistive or controlled-source) and capacitive elements	Low-to-medium-frequency device characterization
h-parameter model	Variable (frequency-dependent) complex elements	Low-to-medium-frequency device-characterizing measurements

Table 4-3 (*continued*)

Name	Parameter Types	Primary Application
y-parameter model (admittance-parameter model)	Variable (frequency-dependent) complex elements	Low-to-medium-frequency circuit analysis
s-parameter model (scattering-parameter model)	Variable (frequency-dependent) complex elements	High-frequency device-characterizing measurements

Table 4-4 The Complex h Parameters as Functions of the Hybrid-Pi Parameters, for the Simplified Circuits of Figure 4-56

$$h_{11} = r_{bb'} + r_{b'e}/(1 + j\omega r_{b'e}C_e)$$
$$= r_{bb'} + r_{b'e}/[1 + j(\omega/\omega_\beta)]$$
$$= r_{bb'} + r_{b'e}\{[1 - j(\omega/\omega_\beta)]/[1 + (\omega/\omega_\beta)^2]\}$$
$$h_{12} = 0$$

$$h_{21} = g_m r_{b'e}/(1 + j\omega r_{b'e}C_e)$$
$$= h_{fe}/[1 + j(\omega/\omega_\beta)]$$
$$= h_{fe}\{[1 - j(\omega/\omega_\beta)]/[1 + (\omega/\omega_\beta)^2]\}$$
$$h_{22} = h_{oe}$$

Table 4-5 The y Parameters (Admittance Parameters) as Functions of the Complex h Parameters

$$y_{11} = \frac{1}{h_{11}}$$
$$y_{12} = -\frac{h_{12}}{h_{11}}$$
$$y_{21} = \frac{h_{21}}{h_{11}}$$
$$y_{22} = \frac{h_{22}h_{11} - h_{12}h_{21}}{h_{11}}$$

Table 4-6 Conversion Chart for SPICE BJT Symbols

SPICE Symbol	Description	Default Value	Unit	Present Symbol	Description
Basic Set: I_S, β_F and β_R					
BF	Ideal forward current gain	100	—	β_F	Maximum forward current gain
BR	Ideal reverse current gain	1.0	—	β_R	Maximum reverse current gain
IS	Saturation current	1.0E–14	A	I_S	Intercept current
Series Resistance: R_{SB}, R_{SC}, and R_{SE}					
RE	Emitter ohmic resistance	0.0	ohm	R_{SE}	Emitter series resistance
RC	Collector ohmic resistance	0.0	ohm	R_{SC}	Collector series resistance
RB	Base ohmic resistance	0.0	ohm	R_{SB}	Base series resistance
Early Effect: V_A and V_R					
VA	Forward Early voltage	∞	V	V_A	Forward Early voltage
VB	Reverse Early voltage	∞	V	V_R	Reverse Early voltage
High-Current Effects: I_k and I_{kr}					
IK	Forward high-current knee current	∞	A	I_k	Forward high-level knee current
IKR	Reverse high-current knee current	∞	A	I_{kr}	Reverse high-level knee current

THE BIPOLAR JUNCTION TRANSISTOR

Table 4-6 (continued)

SPICE Symbol	Description	Default Value	Unit	Present Symbol	Description
		Nonideal-Diode Effects: I_{SC}, I_{SE}, N_E, N_C, N_F, and N_R			
ISC	Forward low-current nonideal base current	1.0E–14	A	I_{SE}	Forward saturation current for SNS (recombination) component of base current
ISR	Reverse low-current nonideal base current	1.0E–14	A	I_{SC}	Reverse saturation current for SNS (recombination) component of base current
NE	Nonideal low-current BE emission coefficient	2.0	—	N_E	BE low-current recombination quality factor
NC	Nonideal low-current BC emission coefficient	2.0	—	N_C	BC low-current recombination quality factor
NF	Nonideal low-current BE emission coefficient	1.0	—	N_F	BE forward intermediate-current quality factor
NR	Nonideal low-current BC emission coefficient	1.0	—	N_R	BC reverse intermediate-current quality factor

Table 4-6 (continued)

Capacitance Effects: C_{OTE}, $\Delta\psi_{0E}$, M_E, C_{OTC}, $\Delta\psi_{0C}$, M_C, C_{OTS}, $\Delta\psi_{0S}$, M_S, τ_F, and τ_R

CJE	Zero-bias BE junction capacitance	0.0	F	C_{OTE}	BE zero-bias depletion-layer capacitance
VJE	BE junction potential	1.0	V	$\Delta\psi_{0E}$	BE contact potential
MJE	BE grading coefficient	0.5	—	M_E	BE grading exponent
CJC	Zero-bias BC junction capacitance	0.0	F	C_{OTC}	BC zero-bias depletion-layer capacitance
VJC	BC junction potential	1.0	V	$\Delta\psi_{0C}$	BC contact potential
MJC	BC grading coefficient	0.5	—	M_C	BC grading exponent
CJS	Zero-bias CS junction capacitance	0.0	F	C_{OTS}	CS zero-bias depletion-layer capacitance
VJS	CS junction potential	1.0	V	$\Delta\psi_{0S}$	CS contact potential
MJS	CS grading coefficient	0.5	—	M_S	CS grading exponent
TF	Forward transit time	0.0	s	τ_F	Forward-active charge-control lifetime
TR	Reverse transit time	0.0	s	τ_R	Reverse-active charge-control lifetime

THE BIPOLAR JUNCTION TRANSISTOR

Table 4-7 A Representative Set of SPICE Parameter Values

Parameter	Value	Parameter	Value
β_F	100	C_{OTE}	8×10^{-14} F
β_R	3	$\Delta\psi_{OE}$	0.8 V
I_S	10^{-16} A	M_E	0.3 V
R_{SE}	10 ohm	C_{OTC}	1.1×10^{-13} F
R_{SC}	100 ohm	$\Delta\psi_{OC}$	0.6 V
R_{SB}	500 ohm	M_C	0.5
I_{SE}	10^{-13} A	C_{OTS}	0
I_{SC}	10^{-14} A	$\Delta\psi_{OS}$	0.75 V
N_E	2	M_S	0
N_C	2	τ_F	10^{-10} s
N_F	1	τ_R	5×10^{-9} s
N_R	1	I_k	0.01 A
V_A	100	I_{kr}	∞
V_R	∞		

Table 4-8 SPICE Operating-Point Information Corresponding to the Parameter Set in the Preceding Table

$I_B = 0.005$ mA	$\beta_F = 93.4$	$V_{CE} = +3$ V	$f_T = 1.16$ GHz
$I_C = 0.467$ mA	$h_{fe} = 91.5$	$V_{BE} = +0.762$ V	$V_{JE} = +0.755$ V
		$V_{BC} = -2.238$ V	$V_{JC} = -2.194$ V

Table 4-9 Calculated dc and Hybrid-Pi-Parameter Values, Numerical Example

Symbol	Unit	A OFF	B Bias	C ON
V_{IN}	V	0.200	0.962	4.700
V_{OUT}	V	4.700	3.000	0.188
V_{BE}	V	0.200	0.762	0.849
V_{BC}	V	-4.500	-2.238	0.660
I_C	mA	2.38×10^{-10}	0.467	1.224
I_B	mA	4.68×10^{-9}	5.00×10^{-3}	9.38×10^{-2}
V_{JE}	V	0.200	0.755	0.7885
V_{JC}	V	-4.500	-2.194	0.736
I_C^R	mA	2.38×10^{-10}	0.467	1.224
I_C^C	mA	0.0	0.0	0.0
I_B^{RE}	mA	4.69×10^{-9}	5.00×10^{-3}	0.018
I_B^{RC}	mA	0.0	0.0	0.076

Table 4-9 (continued)

Symbol	Unit	A OFF	B Bias	C ON
$I_B^{CE} = 0$	mA	0.0	0.0	0.0
$I_B^{CC} = 0$	mA	0.0	0.0	0.0
C_{be}	pF	0.087	2.32	5.24
q_{be}	pC	0.017	0.054	0.229
C_{bc}	pF	0.038	0.051	44.4
q_{bc}	pC	−0.253	−0.158	1.271
$1/r_{b'e}$	S	9.24×10^{-11}	1.89×10^{-4}	6.79×10^{-4}
g_m	S	9.22×10^{-12}	1.73×10^{-2}	4.44×10^{-2}
$1/r_{b'c}$	S	7.74×10^{-18}	1.55×10^{-17}	2.94×10^{-3}
$1/r_{ce}$	S	2.31×10^{-15}	4.57×10^{-6}	7.62×10^{-3}
β_F	—	0.051	93.45	13.05
h_{fe}	—	0.100	91.52	65.44
f_T	GHz	0.0	1.16	0.014

Table 4-10 Calculated Transient Values, Numerical Example

Symbol	Unit	Particular Times				
		$t^{m=1}$	$t^{m=2}$	$t^{m=3}$	$t^{m=4}$	$t^{m=5}$
t	ns	10.0	11.0	11.54	17.97	79.24
$v_{IN}(t)$	V	0.200	4.600	4.600	4.600	4.600
$v_{OUT}(t)$	V	4.700	4.783	4.372	0.253	0.188
$v_{BE}(t)$	V	0.200	0.643	0.773	0.845	0.848
$v_{BC}(t)$	V	−4.5006	−4.140	−3.599	0.591	0.660
$i_C(t)$	mA	-2.6×10^{-6}	−0.0229	0.0900	1.221	1.224
$i_B(t)$	mA	2.49×10^{-4}	0.0989	0.0957	0.0939	0.0938
$v_{JE}(t)$	V	0.200	0.593	0.724	0.785	0.788
$v_{JC}(t)$	V	−4.500	−4.192	−3.638	0.667	0.736
$i_C^R(t)$	mA	2.38×10^{-10}	9.48×10^{-4}	0.144	1.296	1.225
$i_C^C(t)$	mA	0.0	0.0	0.0	0.0	0.0
$i_B^{RE}(v_{JE})$	mA	4.67×10^{-9}	1.86×10^{-6}	1.53×10^{-3}	1.54×10^{-2}	1.77×10^{-2}
$i_B^{RC}(v_{JC})$	mA	-3.5×10^{-14}	-3.5×10^{-14}	-3.5×10^{-14}	5.18×10^{-3}	7.48×10^{-2}
$i_B^{CE}(v_{JE})$	mA	2.46×10^{-4}	7.50×10^{-2}	4.00×10^{-2}	-1.26×10^{-3}	2.00×10^{-5}
$i_B^{CC}(v_{JC})$	mA	2.60×10^{-6}	2.39×10^{-2}	5.41×10^{-2}	7.46×10^{-2}	1.19×10^{-3}
$C_{BE}(v_{JE})$	pF	0.087	0.116	0.672	4.68	5.23
$q_{BE}(v_{JE})$	pC	0.017	0.056	0.085	0.21	0.228
$C_{BE}(v_{JC})$	pF	0.038	0.039	0.041	3.26	43.67
$q_{BC}(v_{JC})$	pC	−0.253	−0.241	−0.219	0.191	1.254
$1/r_{b'e}$	S	9.24×10^{-11}	5.36×10^{-7}	5.68×10^{-5}	5.87×10^{-4}	6.78×10^{-4}
g_m	S	9.21×10^{-12}	3.66×10^{-5}	5.49×10^{-3}	4.50×10^{-2}	4.45×10^{-2}
$1/r_{b'c}$	S	7.74×10^{-18}	8.28×10^{-18}	9.50×10^{-18}	2.00×10^{-4}	2.89×10^{-2}
$1/r_{ce}$	S	2.31×10^{-15}	9.10×10^{-9}	1.39×10^{-6}	5.41×10^{-4}	7.50×10^{-4}

REFERENCES

1. W. Shockley, U.S. Patent 2,569,347, filed June 26, 1948, and issued September 25, 1951.

2. J. Bardeen and W. H. Brattain, "The Transistor—A Semiconductor Triode," *Phys. Rev.* **74**, 230 (1948).

3. W. Shockley, "How We Invented the Transistor," *New Scientist,* 689, 21 December (1972).

4. W. Shockley, M. Sparks, and G. K. Teal, "$p-n$ Junction Transistors," *Phys. Rev.* **83**, 151 (1951).

5. W. Shockley, "A Unipolar Field-Effect Transistor," *Proc. IRE* **40**, 1365 (1952).

6. J. S. Saby, "Fused Impurity P-N-P Junction Transistor," *Proc. IRE* **40**, 1358 (1952).

7. G. K. Teal and J. B. Little, "Growth of Germanium Single Crystals," *Phys. Rev.* **78**, 647 (1950).

8. W. Shockley, *Electrons and Holes in Semiconductors,* Van Nostrand, New York, 1950, pp. 34, 35.

9. H. C. Theuerer, J. J. Kleimack, H. H. Loar, and H. Christensen, "Epitaxial Diffused Transistors," *Proc. IRE* **48**, 1642 (1960).

10. J. del Alamo, S. Swirhun, and R. M. Swanson, "Measuring and Modeling Minority Carrier Transport in Heavily Doped Silicon," *Solid-State Electron.* **28**, 47 (1985).

11. R. S. Payne, R. J. Scavuzzo, K. H. Olson, J. M. Nacci, and R. A. Moline, "Fully Ion-Implanted Bipolar Transistors," *IEEE Trans. Electron Devices* **ED-21**, 273 (1974).

12. S. M. Sze, *Physics of Semiconductor Devices,* Wiley, New York, 1981, pp. 140–144.

13. R. M. Warner, Jr., D. H. Ju, and B. L. Grung, "Electron-Velocity Saturation at a BJT Collector Junction Under Low-Level Conditions," *IEEE Trans. Electron Devices* **30**, 230 (1983).

14. M. Tanenbaum and D. E. Thomas, "Diffused Emitter and Base Silicon Transistors," *Bell Syst. Tech. J.* **35**, 1 (1956); C. A. Lee, "A High-Frequency Diffused Base Germanium Transistor," *Bell Syst. Tech. J.* **35**, 23 (1956).

15. R. M. Warner, Jr. and B. L. Grung, *Transistors: Fundamentals for the Integrated-Circuit Engineer,* Wiley, New York, 1983; reprint edition, Krieger Publishing Company (P.O. Box 9542, Melbourne, Florida 32902-9542), 1990, p. 491.

16. W. Shockley, "The Theory of *p–n* Junctions in Semiconductors and *p–n* Junction Transistors," *Bell Syst. Tech. J.* **28**, 435 (1949).

17. J. L. Moll and I. M. Ross, "The Dependence of Transistor Parameters on the Distribution of Base Layer Resistivity," *Proc. IRE* **44**, 72 (1956).

18. H. K. Gummel, "Measurement of the Number of Impurities in the Base Layer of a Transistor," *Proc. IRE* **49**, 834 (1961).

19. H. Lawrence and R. M. Warner, Jr., "Diffused Junction Depletion Layer Calculations," *Bell Syst. Tech. J.* **39**, 389 (1960).

20. Warner and Grung, p. 342.

21. R. M. Warner, Jr. and J. N. Fordemwalt (Eds.), *Integrated Circuits: Design Principles and Fabrication,* McGraw-Hill, New York, 1965, pp. 195–207.

22. C. J. Frosch and L. Derick, "Surface Protection and Selective Masking during Diffusion in Silicon," *J. Electrochem. Soc.* **104**, 547 (1957).

23. J. Hoerni, U.S. Patent 3,025,589, filed May 1, 1959, and issued March 20, 1962.

24. Patent Counsel, Fairchild Camera and Instrument, Inc., private communication (~ 1963).

25. J. M. Early, "Effects of Space-Charge Layer Widening in Junction Transistors," *Proc. IRE* **40**, 1401 (1952).

26. H. K. Gummel and H. C. Poon, "An Integral Charge Control Model of Bipolar Transistors," *Bell Syst. Tech. J.* **49**, 827 (1970).

27. A. van der Ziel, "A Sufficient Condition for an Early Voltage," *Solid-State Electron.* **17**, 108 (1970).

28. W. A. Adcock, M. E. Jones, J. W. Thornhill, and E. D. Jackson, "Silicon Transistor," *Proc. IRE* **42**, 1192 (1954).

29. G. K. Teal, "Crystals of Germanium and Silicon—Basic to the Transistor and Integrated Circuit," *IEEE Trans. Electron Devices* **ED-23**, 621 (1976).

30. R. N. Hall and W. C. Dunlap, "*P–N* Junctions Prepared by Impurity Diffusion," *Phys. Rev.* **80**, 467 (1950).

31. W. E. Taylor, U.S. Patent 2,971,869, filed February 10, 1954, and issued February 14, 1961.

32. J. M. Early, "*p-n-i-p* and *n-p-i-n* Junction Transistor Triodes," *Bell Syst. Tech. J.* **33**, 517 (1954).

33. R. M. Warner, Jr. and W. C. Hittinger, "A Developmental Intrinsic-Barrier Transistor," *IRE Trans. Electron Devices* **3**, 157 (1956).

34. R. M. Warner, Jr., U. S. Patent 3,404,295, filed November 30, 1964, and issued October 1, 1968.

35. R. M. Warner, Jr. and B. L. Grung, "A Bipolar Lock-Layer Transistor," *Solid-State Electron.* **18**, 323 (1975).

36. G. R. Wilson, U.S. Patent 3,564,356, filed October 24, 1968, and issued February 16, 1971.

37. T. E. Zipperian, R. M. Warner, Jr., and B. L. Grung, "Channel-Collector Transistors," *IEEE Trans. Electron Devices* **29**, 341 (1982).

38. B. L. Grung, "Investigation of the Lock-Layer Transistor," Ph.D. Thesis, University of Minnesota, 1976.

39. H. H. Berger and S. K. Wiedmann, "Merged Transistor Logic (MTL)—a Low Cost Bipolar Logic Concept," *IEEE J. Solid-State Circuits* **7**, 340 (1972).

40. H. Yagi and T. Tsuyuki, "A Novel and High Performance Bipolar Device of Low Emitter Impurity Concentration Structure," *Proceedings of the Sixth Conference on Solid-State Devices,* Tokyo, Japan, 1974, p. 279.

41. B. L. Grung, "An Analytical Model for the Low-Emitter-Concentration Transistor," *Solid-State Electron.* **21**, 821 (1978).

42. H. Christensen and G. K. Teal, U.S. Patent 2,692,839, filed April 7, 1951, and issued October 26, 1954.

43. O. P. Frazee, with work reported by R. M. Warner, Jr., "Epitaxial Growth and Devices," *Electronics* **35**, 49 (1962).

44. W. Shockley, U.S. Patent 2,787,564, filed October 28, 1954, and issued April 2, 1957.

45. M. Takagi, K. Nakayama, C. Terada, and H. Kamioka, "Improvement of Shallow Base Transistor Technology by using a Doped Poly-Silicon Diffusion Source," *J. Jpn. Soc. Appl. Phys. (Suppl.)* **42**, 101 (1973).

46. P. Ashburn, A. A. Rezazadeh, E. F. Chor, and A. Brunnschweiler, *Digest of the 1987 Bipolar Circuits and Technology Meeting,* Minneapolis, Minnesota, September 1987, p. 61.

47. K. Hart and A. Slob, "Integrated Injection Logic; A New Approach to LSI," *IEEE J. Solid-State Circuits* **7**, 346 (1972).

48. J. M. Wisted and R. M. Warner, Jr., "A Note on β_{UP} and β_{DOWN} in I^{2}L Transistors," *Solid-State Electron.* **25**, 251 (1982).

49. T. Misawa, "A Note on the Extended Theory of the Junction Transistor," *J. Phys. Soc. Jpn.* **11**, 728 (1956).

50. N. H. Fletcher, "General Semiconductor Relations," *J. Electronics* **2**, 609 (1957).

51. W. M. Webster, "On the Variation of Junction-Transistor Current-Amplification Factor with Emitter Current," *Proc. IRE* **42**, 914 (1954).

52. E. S. Rittner, "Extension of the Theory of the Junction Transistor," *Phys. Rev.* **94**, 1161 (1954).

53. Warner and Grung, pp. 587–686.

54. B. L. Grung and R. M. Warner, Jr., "An Analytical Model for the Epitaxial Bipolar Transistor," *Solid-State Electron.* **20**, 753 (1977).

55. W. van Roosbroeck, "The Transport of Added Current Carriers in a Homogeneous Semiconductor," *Phys. Rev.* **91**, 282 (1953).

56. G. C. Ebner and P. E. Gray, "Static V–I Relationships in Transistors at High Injection Levels," *IEEE Trans. Electron Devices* **13**, 692 (1966).

57. K. M. van Vliet and H. S. Min, "Current–Voltage Relationships and Equivalent Circuits of Transistors at High Injection Levels," *Solid-State Electron.* **17**, 267 (1974).

58. C. T. Kirk, Jr., "A Theory of Transistor Cutoff Frequency (f_t) Falloff at High Current Densities," *IRE Trans. Electron Devices* **9**, 164 (1962).

59. Warner and Fordemwalt, pp. 107–109.

60. P. G. A. Jespers, "Measurements for Bipolar Devices," in F. van de Wiele, W. L. Engl, and P. G. A. Jespers (Eds.), *Process and Device Modeling for Integrated Circuit Design,* Noordhoff, Leyden, 1977.

61. H. Das and A. R. Boothroyd, "Determination of Physical Parameters of Diffusion and Drift Transistors," *IRE Trans. Electron Devices* **ED-8**, 15 (1961).

62. J. Lindmayer and C. Wrigley, "The High Injection Level Operation of Drift Transistors," *Solid-State Electron.* **2**, 79 (1961).

63. J. S. T. Huang, "Low to High Injection in Double-Diffused Transistors," *IEEE Trans. Electron Devices* **ED-15**, 940 (1968).

64. J. J. Ebers and J. L. Moll, "Large-Signal Behavior of Junction Transistors," *Proc. IRE* **42**, 1761 (1954).

65. R. Beaufoy and J. J. Sparkes, "The Junction Transistor as a Charge-Controlled Device," *ATE J. (London)* **13**, 310 (1957).

66. H. K. Gummel, "A Charge Control Relation for Bipolar Transistors," *Bell Syst. Tech. J.* **49**, 115 (1970).

67. I. E. Getreu, *Modeling the Bipolar Transistor,* Elsevier, New York, 1978.

68. Y. H. Kwark and R. M. Swanson, "A Generalized Proof of the Reciprocity Theorem," *IEEE Trans. Electron Devices,* **ED-33**, 865 (1986).

69. B. D. H. Tellegen, "The Gyrator, a New Electric Network Element," *Philips Res. Repts.* **3**, 80 (1948).

70. L. J. Giacoletto, "Study of *p-n-p* Alloy Junction Transistor from dc through Medium Frequencies," *RCA Review*, **15**, 506 (1959).

71. J. Millman and C. C. Halkias, *Integrated Electronics: Analog and Digital Circuits and Systems,* McGraw-Hill, New York, 1972.

72. M. I. Elmasry, *Digital Bipolar Integrated Circuits,* John Wiley, New York, 1983.

73. M. G. Ghausi, *Principles and Design of Linear Active Circuits,* McGraw-Hill, New York, 1965.

74. R. D. Middlebrook, *An Introduction to Junction Transistor Theory,* John Wiley, New York, 1957.

75. P. Antognetti and G. Massobrio, *Semiconductor Device Modeling with SPICE,* McGraw-Hill, New York, 1988.

76. B. R. Chawla and H. K. Gummel, "Transition Region Capacitance of Diffused *p–n* Junctions," *IEEE Trans. Electron Devices* **18**, 178 (1971).

TOPICS FOR REVIEW

R4-1. Explain the terms *unipolar* and *bipolar*. Give a device example described by each term.

R4-2. What is meant by *PNP* and *NPN*?

R4-3. Explain what is meant by *complementary* circuit.

R4-4. What are the advantages of complementary circuits?

R4-5. What is meant by BJT?

R4-6. How many *PN* junctions are there in a BJT?

R4-7. What terminals (and regions) are identified by the letters *E*, *B*, and *C*?

R4-8. Which surfaces assume greater importance in BJT theory—metallurgical junctions or space-charge-layer boundaries?

R4-9. What is the origin of the term *base*?

R4-10. Name the two junctions of the BJT.

R4-11. Draw the BJT symbol and explain it.

R4-12. What is meant by a *discrete* device?

R4-13. What is an *integrated circuit*?

R4-14. Explain the sense in which integrated circuits are two-dimensional.

R4-15. How can a *three-terminal* device be a *two-port* device?

R4-16. What is the most widely used BJT circuit configuration? Why?

R4-17. Explain the symbol V_{BE}.

R4-18. State the algebraic signs of the terminal currents in a normally biased *NPN* BJT. State those in a normally biased *PNP* BJT.

R4-19. What is meant by *isolated junction*?

R4-20. What is a typical value of X_B in a BJT?

R4-21. To understand the basic operation of a BJT, which carrier profiles (or "distributions") are more important—those outside the junction space-charge layers, or those inside?

R4-22. What is the meaning of *heuristic*?

R4-23. What is the relationship of electron injection to hole injection in an N^+P junction, and why?

R4-24. How does the junction interaction that exists in a BJT affect the law of the junction?

R4-25. Why is the minority-carrier-density profile in the base region of a BJT very nearly linear?

R4-26. Which dimensions change most significantly as BJT current rating is changed or "scaled," junction-area dimensions or region-thickness dimensions?

R4-27. State the symbols employed to designate the net-doping values in the various regions of a BJT.

R4-28. Explain the flat-bottomed minimum that exists in the electron-density profile for an *NPN* BJT.

R4-29. What widely used approximations fail in a degenerately doped BJT emitter region? Why?

R4-30. What is the Fermi-level location in a BJT emitter region?

R4-31. Explain the error cancellation that occurs among the conventional approximations in degenerate silicon.

R4-32. By what factor does the electron density vary throughout a one-dimensional BJT in normal operation?

R4-33. By what factor does electron *current* density vary throughout a one-dimensional BJT in normal operation?

R4-34. What causes the dominant current in each of the five regions (junction regions included) of a normally biased BJT? Give your answer in terms of the transport mechanism and carrier population (majority or minority) involved.

R4-35. What has been the dominant technology employed in BJT fabrication in the past 25 years?

R4-36. Identify two major current constituents in BJT operation and two parasitic current constituents.

R4-37. Explain the terms *intrinsic* base region and *extrinsic* base region.

R4-38. What is meant by *active base region*? Define all six boundaries of an active base region that is a rectangular parallelepiped.

R4-39. What is meant by *effective base thickness*?

R4-40. Describe the pattern of majority-carrier flow in the BJT base region.

R4-41. Explain *ohmic base resistance*. What is a typical value?

R4-42. Why are plan-view emitter-region geometries often stripelike? Assuming stripelike emitter and base contacts, identify the length, width, and thickness dimensions of the active base region.

R4-43. Explain *control current*.

R4-44. What is meant by *base-transport efficiency* γ_B? Why was it relatively low in the early germanium BJTs?

R4-45. What is meant by a "perfect" emitter?

R4-46. Define *defect current*.

R4-47. Describe the phenomena responsible for collector *leakage current* in the silicon BJT and the germanium BJT.

R4-48. Why is collector leakage current multiplied by a significant factor in the germanium BJT and not in the silicon BJT?

R4-49. Explain the connection between I_C and electron density gradient in the BJT base region.

R4-50. Where in the BJT are electrons most likely to be moving at their saturated velocity v_s? How does this complication affect the construction of a meaningful base-region minority-carrier profile?

R4-51. What is meant by *static* current gain? Define β.

R4-52. What is meant by the statement that the common-emitter BJT is a *linear current amplifier*?

R4-53. How can a designer increase β? What engineering trade-offs are there in doing so?

R4-54. What is a typical β value in a power BJT? Why is it so low?

R4-55. Where are "superbeta" BJTs employed and what gain values can they reach?

R4-56. What is *punchthrough*?

R4-57. What is the area under (that is, the integral of) a declining exponential function?

R4-58. Explain *charge control*. What is meant by sign consistency in a charge-control equation?

R4-59. What "stored charge" determines base current in the modern BJT? In the early germanium BJT?

R4-60. Why is the common-emitter modern BJT a linear current amplifier? Why is the early germanium common-emitter transistor a linear current amplifier?

R4-61. How does the value of n_i^2 for germanium compare with that for silicon?

R4-62. How does the low value of low-current β in a silicon BJT enhance its practical performance?

R4-63. Why is the voltage biasing of a common-emitter BJT input port usually infeasible?

R4-64. Describe the temperature dependence of offset voltage for the emitter junction of a silicon BJT.

R4-65. Describe the simplest current-regulating circuit.

R4-66. Describe the loadline construction. What does it accomplish? Why does it work? When is its use indicated?

R4-67. What elements make up the static equivalent-circuit model for a normally biased common-emitter BJT?

R4-68. What is meant by V_{CC} and by V_{BB}?

R4-69. Express small-signal voltage gain as a derivative.
R4-70. What is meant by *output plane* and *output characteristics*?
R4-71. What is the meaning of *parameter*?
R4-72. In an output-plane loadline diagram for the common-emitter BJT, the loadline itself constitutes a characteristic for which component? What fixes its V_{CE} intercept? Its I_C intercept?
R4-73. Describe the elements and their interconnection in a useful static equivalent-circuit model for the common-emitter BJT that is useful for arbitrary bias values (except for one ambiguous case).
R4-74. What is meant by *quiescent operating point*? Name three ways to alter its position, and tell how its position depends on each variable.
R4-75. How is small-signal voltage gain A_V in a rudimentary common-emitter BJT affected by β? By R_C? By R_B?
R4-76. Define BJT *saturation*. In what way is the term descriptive?
R4-77. Distinguish between *weak* and *deep* saturation.
R4-78. Describe the equivalent-circuit models for weak and deep saturation.
R4-79. Why is V_{BE} for the deep-saturation regime greater than for the forward-active regime?
R4-80. What is cutoff? What is the reverse-active regime?
R4-81. Describe the simplest use of a BJT as a switch. What operating regimes are involved?
R4-82. State the properties of the common-base configuration and of the common-collector configuration. State an alternative term for the latter.
R4-83. How are common-base and common-emitter current gain related?
R4-84. What is meant by electrochemical potential?
R4-85. Which quasi Fermi level deviates from the Fermi level more markedly when equilibrium is disturbed? Why?
R4-86. How can one employ knowledge of $\phi_n(x)$ to write a transport equation for electrons?
R4-87. Explain Gummel plot and Gummel number.
R4-88. How can you explain the de facto outcome of the Moll-Ross analysis, namely that BJTs of the same Gummel number have the same current gain?
R4-89. Explain the meaning of BV_{CEO}.
R4-90. Explain *reachthrough* in a BJT.
R4-91. What is a *planar* BJT?
R4-92. Explain qualitatively why BV_{CEO} is appreciably smaller than BV_{CBO}.
R4-93. Describe the Early effect and its cause.
R4-94. Why was the Early effect so marked in the alloyed-junction BJT?
R4-95. Why was the alloyed-junction BJT a *PNP* device?
R4-96. What improvements over older structures were achieved in the epitaxial BJT?
R4-97. Why are junction areas a factor in determining the ratio of forward to reverse current gain?
R4-98. What is meant by high-level operation of a BJT?

R4-99. What analytic complications accompany high-level conditions?

R4-100. What is the Rittner effect? The Webster effect? The ambipolar effect? Quasisaturation (the Kirk effect)? Emitter debiasing?

R4-101. What causes deviation(s) from linearity in a curve of log I_C versus V_{BE}? Of log I_B versus V_{BE}?

R4-102. What is meant by resolving an electric field into an ohmic component and a bulk-diffusion component?

R4-103. Under what conditions is ambipolar analysis appropriate?

R4-104. What limiting values are approached by ambipolar diffusivity and ambipolar mobility?

R4-105. What is a graded-base BJT? How does grading affect base-region carrier distributions as a function of current level?

R4-106. What relationships are given by the Ebers-Moll model? What primary assumptions does it employ?

R4-107. What is meant by the transport form of the Ebers-Moll model, and how does its derivation primarily differ from that of the original model? What device-characterizing quantities are employed in the two cases?

R4-108. What is the Gummel-Poon model?

R4-109. What is meant by superposition?

R4-110. What quality is needed in a circuit or device for the valid application of superposition, and how does the BJT meet the condition?

R4-111. For what conditions and assumptions does a BJT exhibit reciprocity?

R4-112. What are the hybrid parameters, why are they so called, how many are there, and what is the physical interpretation of each?

R4-113. What is transconductance?

R4-114. Name several noteworthy aspects of BJT transconductance properties.

R4-115. How does the hybrid-pi model differ from the hybrid model, and what is the consequence of the difference?

R4-116. Name two advantages of the hybrid-pi model.

R4-117. How does the *h*-parameter model differ from the hybrid-parameter model?

R4-118. For what kind of measurements is the scattering-parameter model used?

R4-119. Why is the charge-control model "phenomenological," or less related to device physics than, for example, the hybrid-pi model?

R4-120. What is the key charge-control assumption and how valid is it?

R4-121. What kind of charge is the focus of the Gummel-Poon model? In the case of the base region, how is the application of this charge similar to the application of the Gummel number, and how does the result differ?

R4-122. What is a figure of merit? Give an example.

R4-123. What attributes should a figure of merit possess?

R4-124. Name at least four parasitic effects present in the BJT that are comprehended in the SPICE model.

R4-125. How is small-signal modeling approached in SPICE?

R4-126. Define thermal resistance.

R4-127. How are temperature effects treated in SPICE modeling?

ANALYTIC PROBLEMS

A4-1. Given this silicon BJT in forward-active operation,

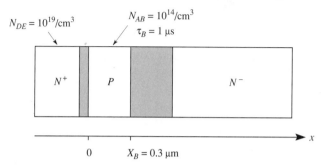

a. Find $J_n(0)$ for $V_{NP} = V_{EB} = -0.4$ V.
b. Repeat part a for $\tau_B = 100$ μs.
c. Repeat part a for $N_{AB} = 10^{15}/\text{cm}^3$.

A4-2. Given this qualitative electron-density profile throughout the BJT in forward-active (forward-biased emitter and reverse-biased collector junctions) operation, and the fact that electron current density J_n throughout the device is constant within 1%,

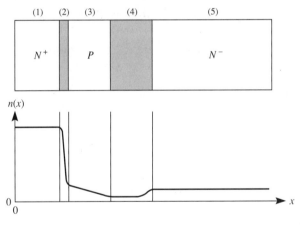

a. Cite the dominant electron-transport mechanism in each of the five regions.
b. Cite the electric-field sign (+ or −) in regions 1, 2, 4, and 5.
c. Rank regions 1, 2, 4, and 5 in the sequence of decreasing electric-field magnitude, assuming that $V_{CE} \approx 20$ V.
d. Judging from the base-region profile, is this diagram linear or semilog?
e. Given that the base region has a uniform net doping of $N_{AB} = 10^{16}/\text{cm}^3$ and that $V_{BE} = 0.6$ V, show that $n_B(x) = n'_B(x)$ is a good approximation in the base region.

A4-3. Given a silicon BJT with $\beta = 100$, $V_{BE} = 0.7$ V, and $V_{CE} = 5$ V, calculate, for this bias circuit,

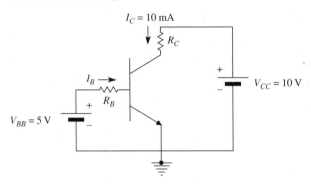

a. R_B. [*Hint:* Apply Kirchhoff's voltage law to base loop.]
b R_C. [*Hint:* Treat collector loop in similar fashion.]
c. Inserting a resistor $R_E = 0.1$ kilohm from the emitter terminal to the ground terminal, but keeping V_{BB}, V_{CC}, V_{CE}, I_B, and I_C the same, recalculate R_B.
d. Recalculate R_C for conditions of part c.

A4-4. A certain silicon BJT has $(\mu_{nB}/\mu_{pE}) = 4$, has an emitter–base doping ratio of $(N_{DE}/N_{AB}) = 10^2$, and has $L_{pE} = 0.25$ μm.
a. Design its base thickness X_B to yield a common-emitter current gain of $\beta = 200$.
b. This BJT is used in a rudimentary amplifier circuit having $R_C = 1$ kilohm, and a low-frequency-ac voltage gain $A_V = (\Delta V_{CE}/\Delta V_{IN}) = -10$. Find R_B.

A4-5. In the circuit shown here, $V_{CC} = 24$ V, $R_C = 10$ kilohm, and $R_E = 270$ ohm. If a silicon transistor is used with $\beta = 45$ and if $V_{CE} = 5$ V, find R. Neglect the reverse saturation current.

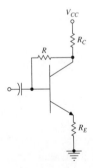

A4-6. When the transistor in the amplifier circuit shown next is represented by the model also shown, the voltage gain of the circuit $A_V = (\Delta V_{OUT}/\Delta V_{IN})$ is completely determined by the two quantities $(V_{CC} - V_{CE})$ and $(V_{BB} - V_{BE})$, where V_{CE} is the quiescent collector voltage. These quantities are the quiescent voltage drops across R_C and R_B.

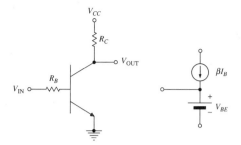

a. Substituting the BJT model for the BJT symbol, derive an expression for the voltage gain in terms of the two voltage drops stated above.

b. An amplifier of this kind using a silicon transistor ($V_{BE} = 0.7$ V) is to be designed for a voltage gain of 15. The collector supply voltage available is $V_{CC} = 20$ V, and the quiescent point can be chosen at $I_C = 1$ mA and $V_{CE} = 2$ V. Specify the values of V_{BB} and R_C required. Specify the value of R_B as a function of the transistor current gain β.

A4-7. An *NPN* silicon transistor is used in the amplifier of Problem 4-6 with $V_{CC} = 25$ V, $V_{BB} = 1.7$ V, $R_B = 100$ kilohm, and $R_C = 20$ kilohm. The base-to-emitter voltage drop can be treated as a constant, $V_{BE} = 0.7$ V. The output characteristic curves for the transistor are approximately horizontal lines corresponding to $I_C = 100\, I_B$.

a. Sketch and dimension a family of output characteristic curves for $I_B = 0$, 5 μA, 10 μA, 15 μA, 20 μA, 25 μA, and 30 μA.

b. Construct a loadline for the amplifier on the characteristics of part a, and indicate the current and voltage at which the loadline intersects the axes.

c. Show the quiescent operating point on the loadline of part b, and give its coordinates.

A4-8. You are given the bias circuit below and a silicon BJT with $\beta = 99$.

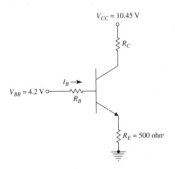

a. Substitute an appropriate large-signal equivalent-circuit model for the BJT in the given circuit. Derive an expression for R_B as a function of I_B.

b. Find R_B for $I_B = 10$ μA.

c. Derive an expression for R_C.
d. Find R_C for $V_{CE} = 5$ V.

A4-9. Following is a sketch of a rudimentary amplifier circuit and its accompanying output plane, showing the loadline and the quiescent operating point Q. Assume that $V_{BE} = 0.7$ V.

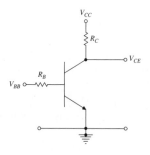

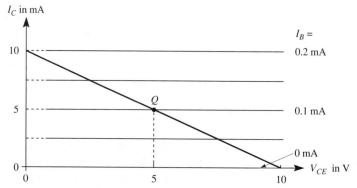

a. Find the value of V_{CC}.
b. Find the value of R_C.
c. Find the value of V_{CE}.
d. Find the value of I_C.
e. Find the value of β.
f. Assuming that $V_{BB} = 1.7$ V, calculate R_B.
g. From first principles, and without using formulas (*use Ohm's law*), calculate $A_V = (\Delta V_{OUT}/\Delta V_{IN})$, with due regard for algebraic sign.

A4-10. A certain *NPN* BJT has a net doping in the base region of $N_{AB} = 10^{16}/\text{cm}^3$, an emitter-junction area of 10^{-3} cm^2, and an active base thickness of 0.5 μm when $V_{BE} = 0.5$ V and $V_{CE} = 5$ V.
a. Calculate I_C, given $\mu_{nB} = 1{,}200$ cm^2/V·s.
b. The collector junction is an asymmetric step junction. Under the given bias conditions, $X = 2.8\ X_0$ for this junction. Using the depletion approximation, find its (equilibrium) contact potential $\Delta\psi_0$.

c. The net doping in the emitter region is $N_{DE} = 10^{19}/cm^3$, and $D_{pE} = 4\ cm^2/s$. For the given bias conditions, the "emitter-defect" current is $I_D = 5\ \mu A$. Assuming that our four basic approximations are valid for this emitter region, calculate hole lifetime τ in the emitter region.

d. Calculate common-emitter current gain, β_F.

A4-11.

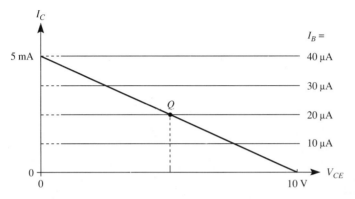

Shown here is the output plane and load line for a silicon BJT in the following bias circuit, with $V_{BB} = 2.7$ V:

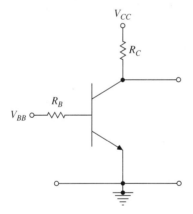

a. Calculate β.
b. Calculate R_C.
c. Calculate R_B.
d. The quiescent operating point is to be moved from $I_B = 20\ \mu A$ to $I_B = 10\ \mu A$, with V_{BB}, V_{CC}, and V_{CE} all kept the same. What changes are needed in the circuit?
e. What effect does the change of part d have on small-signal voltage gain?

A4-12. Shown next is the base-region electron-density profile for a particular silicon BJT under normal bias.

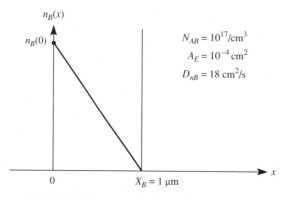

$N_{AB} = 10^{17}/\text{cm}^3$
$A_E = 10^{-4} \text{ cm}^2$
$D_{nB} = 18 \text{ cm}^2/\text{s}$

a. Calculate the magnitude of the density gradient (dn_B/dx), given that $V_{BE} = 0.5$ V.
b. Calculate the collector current I_C.
c. At the emitter boundary of the emitter junction, the hole current-density magnitude is 2.4×10^{-4} A/cm^2. Find β.

A4-13. Shown here is the minority-electron profile in the base region of a BJT fabricated in 1952. Estimate the value of the base transport factor γ_B, which is the ratio of the electron current entering the collector junction to the electron current leaving the emitter junction.

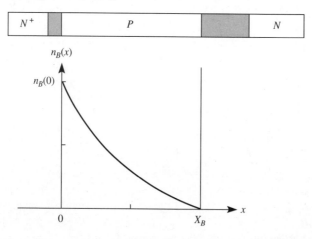

A4-14. Following is a partially completed cross-sectional drawing for a common kind of BJT.

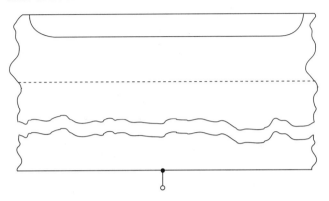

a. Complete the cross-sectional diagram, labeling each region with a symbol such as N, N^+, P, or P^+, etc.
b. Indicate the remaining terminals and label each terminal with an appropriate *word*, spelled correctly.
c. Using a ground symbol and + or − symbols, indicate at each terminal its bias condition in a common-emitter configuration.

A4-15. A certain BJT has a base-region doping of $N_{AB} = 10^{17}/\text{cm}^3$, a collector-region doping of $N_{DC} = 2 \times 10^{15}/\text{cm}^3$, and an emitter-junction area of $A_E = 1.5 \times 10^{-4}$ cm^2. This device receives a collector-junction bias of $V_{CB} = 10$ V, and has $I_C = 1$ mA. Calculate n_c, the electron density that exists at the flat-bottomed minimum in the electron-density profile that occurs in the collector space-charge region as a result of velocity saturation. [This feature is displayed in Figure 4-5(a).] Calculate the thickness of the region wherein $n = n_c$. You may assume the depletion approximation, and also that the electric-field profile is negligibly perturbed by the charge of carriers in transit.

A4-16. This power-supply problem leads into Problem A4-17, which uses a BJT to improve power-supply performance. A certain power supply delivers the voltage V_O to a load R_L and exhibits an output resistance R_O in the neighborhood of a particular operating point.
a. Display the meaning of this information graphically by showing how a change in output voltage $dV_O \approx \Delta V_O$ results when a change in load resistance $dR_L \approx \Delta R_L$ occurs. For the sake of specific construction, let $V_O = 5$ V, $R_L = 1$ kilohm, and $R_O = 0.5$ kilohm, while ΔR_L amounts to a 10% increase.
b. Derive an expression for dV_O having the form

$$dV_O = f(R_O, R_L, \ldots)dR_L .$$

c. Calculate ΔV_O for the given increase in R_L.

A4-17. Reducing the output resistance of a power supply renders output voltage less sensitive to changes in R_L. This can be achieved by placing a common-base BJT circuit between the high-output-impedance power supply and the load. In a rudimentary form, such a circuit is shown below, where the power supply of Problem A4-16 is in the box.

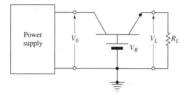

a. Choose an appropriate value of V_R for $V_L = 5$ V.
b. Derive an expression of the form $dV_L = f(R_L, I_C, \beta_F) dR_L$.
c. Using $I_L = 5$ mA, $\beta = 50$, $R_L = 1000$ ohms, and $(\Delta R_L/R_L) = 0.1$, calculate ΔV_L.
d. Derive an approximate form of the final expression in part b.
e. Evaluate the expression in part d for the given conditions, and find its percentage difference from the result in part c.
f. Explain why the load R_L no longer "feels" the voltage-changing effects of the power-supply output resistance R_O.
g. What is the output resistance of the BJT-modified power supply?

A4-18. The forward-active output characteristics of the common-emitter BJT exhibit a finite resistance because of the Early effect. The resulting collector current as a function of I_C and V_{CE} can be approximated as

$$I_C = \beta I_B (1 + V_{CE}/V_A),$$

where V_A is the Early voltage. Given the amplifier circuit shown below, with $V_{IN} \gg V_{BE}$, calculate small-signal voltage gain A_V.

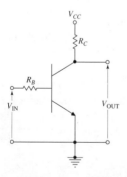

A4-19. Given here is the simplest common-collector, or emitter-follower, circuit. In Section 4-3.6 and Exercise 4-29, the input resistance R_{IN} for this circuit was determined. Using the total-differential method illustrated in Problems A4-16 through A4-18, derive a general expression for input resistance.

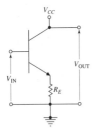

A4-20. A thin, neutral silicon sample having $N_D - N_A = 10^{15}/\text{cm}^3$ is subjected to intense penetrating radiation that causes ϕ_n to reside one normalized unit (one thermal voltage) above the Fermi level on the band diagram. Calculate the position of ϕ_p. Use ϕ as reference for computing all potentials.

A4-21. Given Equations 4-30 and 4-31, show algebraically that the assumption of constant quasi Fermi levels through the space-charge layer of a junction is equivalent to the assumption of Boltzmann quasiequilibrium. Assume that the respective boundaries are located at $x = x_{\mu N}$ on the N-type side and at $x = x_{\mu P}$ on the P-type side.

A4-22. It is pointed out in Figure 3-40 (Section 3-6.3) that the "tails" of impurity distributions formed by solid-phase diffusion are approximately exponential. Let us assume that the net-doping profile in a particular BJT is given approximately by

$$N_{AB}(x) = N_{AB}(0)\, \exp(-2x/X_B),$$

with $N_{AB}(0) = 10^{18}/\text{cm}^3$, and $X_B = 0.4\ \mu\text{m}$.

a. Derive an expression for the electric-field profile $E(x)$ in the base region and sketch the resulting profile.
b. Evaluate your expression at $x = X_B/2$.
c. Calculate the Gummel number for the given device.

A4-23. Shown here are base-region doping profiles for two BJTs.

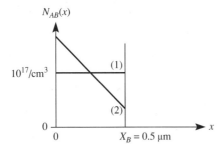

a. Calculate their Gummel numbers.
b. Offer a qualitative explanation for how it is possible that the current density J_n in the base region is the same in the two devices for a given value of V_{BE}, in view of the great emphasis on the importance of the boundary value $n_B(0)$ in Sections 1-1 through 1-3. At present, it is inescapable that $n_{0B1}(0) \neq n_{0B2}(0)$.

c. In Figures 2-37 (Section 2-7.2) and 3-16(c) (Section 3-4.1), it was stressed that the electric field affecting majority carriers and arising from their behavior in those situations had a negligible effect on minority carriers under low-level conditions. But in analyzing the BJT base region (Section 4-4.2) we have treated low-level conditions and now claim that the electric field causes significant minority-carrier transport. How can this apparent inconsistency be resolved?

A4-24. A certain BJT has a base region with linearly graded doping, from $N_{AB}(0) = 2 \times 10^{17}/cm^3$ to $N_{AB}(X_B) = 2 \times 10^{14}/cm^3$. Also, $X_B = 0.4$ μm.
a. With due regard for sign, calculate the electric field at $x = (X_B/2)$ under equilibrium conditions.
b. Calculate the Gummel number for this structure.

A4-25. The right-hand characteristic curve in Figure 4-29 can be observed in the circuit shown here:

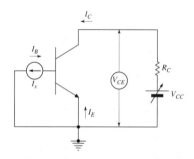

The resistor R_C is large enough so that the loadline has but a single intersection with the I–V characteristic in spite of its incremental negative resistance, and hence point-by-point observation of the curve is possible. The current-source current I_X is small and constant in the negative-resistance regime of the curve. The current of holes that sustains the negative base current I_X is provided by avalanching in the collector junction. The current of holes to the base region in excess of I_X flows to the emitter junction, to recombine in the space-charge layer or the emitter region, and thus constitutes the "emitter defect current" $\equiv I_D$. You may assume that the current increment resulting from the avalanching phenomenon is equally divided between holes flowing to the base region and electrons flowing to the collector region. You may further assume that collector leakage current and base-region recombination are negligible. The negative-resistance regime in the range $I_C \gg I_X$ can be regarded approximately as the locus of points for which the following criterion holds:

$$f(M, \alpha_{0F}) = \text{constant}.$$

a. Determine the function $f(M, \alpha_{0F})$ and the constant. [The solution being sought is relatively simplistic and requires little algebra.]
b. Give a qualitative explanation for the negative-resistance behavior.
c. Suggest an extension to this problem.

A4-26. For high values of V_{CE}, the dc behavior of a certain BJT can be described by the expression

$$I_C = I_B \alpha_F M/(1 - \alpha_F M). \qquad (1)$$

We have taken α_F to be a low-voltage value so that $\alpha_F \approx \alpha_{0F}$ in Equation 4-58 for present conditions, to arrive at Equation 1. The avalanche-multiplication factor M can be related empirically to V_{CE} by

$$M = \frac{1}{1 - (V_{CE}/BV'_{CBO})^2}, \qquad (2)$$

where BV'_{CBO} is the breakdown voltage of the intrinsic (plane) collector–base junction.

a. Assuming that $\beta_F = 50$ and $BV'_{CBO} = 50$ V, calculate the collector–emitter breakdown voltage BV_{CEO}. That is, calculate the value of V_{CE} when $I_C \to \infty$.

b. Using Equations 1 and 2, derive an expression for the transistor's output conductance g_o as a function of I_B, BV'_{CBO}, and α_F. Neglect the Early effect and assume $\alpha_F = $ constant.

c. Using the results of part b, calculate g_o for a collector–emitter voltage equal to half the breakdown voltage BV_{CEO}, and for $I_B = 1$ mA.

A4-27. A BJT base region is formed in a uniformly doped N-type sample using the usual two-step diffusion process.

a. Using *linear* axes, sketch the impurity profile qualitatively.

b. How does this impurity-profile function differ geometrically from the profile obtained when the volumetric density of impurities at the surface is held constant? Name the impurity-profile function.

c. Derive an expression for the junction depth x_J.

d. The second diffusion step in this process is carried out at $T = 1150$ C, at which the three following common doping impurities exhibit the diffusivities indicated:

Antimony	6×10^{-14} cm^2/s
Phosphorus	8×10^{-13} cm^2/s
Boron	8×10^{-13} cm^2/s

Also, in the particular process employed,

$$Q = 10^{14}/\text{cm}^2;$$
$$t = 30 \text{ min};$$
$$N_D = 10^{15}/\text{cm}^3.$$

Calculate x_J, the collector-junction depth in micrometers.

A4-28. Output characteristics for a BJT are shown with equal realism in Figure 4-24(b) for common-base operation, and in Figure 4-31(d) for common-emitter operation. In each case, an output-voltage increase causes the collector-junction depletion region to encroach upon the base region. Why is the Early-effect degradation of output conductance that is so evident in the common-emitter case not seen at all in the common-base case? Use a physical argument involving conditions in the base region for your answer.

A4-29. You are given a pair of BJTs that are precisely complementary in terms of dimensions and net-doping magnitudes. They are subjected to precisely complementary forward-active low-level bias conditions. Comment quantitatively on $I_{C(NPN)}/I_{C(PNP)}$, $I_{B(NPN)}/I_{B(PNP)}$, and $\beta_{F(NPN)}/\beta_{F(PNP)}$.

A4-30. You are given the *one-dimensional* BJT shown here.

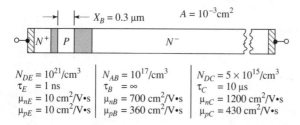

a. Calculate the Gummel intercept current I_S.
b. Write an expression for the "defect current" I_{DR} for reverse operation (current of holes into the collector region).
c. Calculate β_R.
d. Calculate the Gummel number, G_B.

A4-31. You are given a BJT identical to that of Problem A4-30, except that in the present device, $X_B = 0.5$ μm. It is in forward deep saturation, with $n(X_B) = 10^{14}/\text{cm}^3$, and with $|J_n| = 2.88$ A/cm² in the base region.
a. Calculate $|dn_B/dx|$, giving units.
b. Calculate $n_B(0)$.
c. Is this an example of high-level or low-level operation? Justify your answer.
d. Using the axes given here, calibrate the ordinate and sketch the $n_B(x)$ profile accurately.

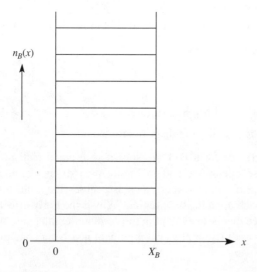

e. Calculate V_{BE}.

A4-32. You are given the *NP*-junction sample shown here under steady-state bias and no other external stimuli.

| N | ⊕ ⊖ | P |

The matrix shown here involves two bias conditions and two functions of position. Each of the latter gives a partial description of conditions in the sample.

	Forward	Reverse
$\varphi_n(x)$	(a)	(b)
$\psi(x)$	(c)	(d)

State whether each of the indicated functions (a) through (d) is monotonic or nonmonotonic under the indicated bias condition, and give the physical reason for which your answer is correct. [Monotonic ≡ free of slope reversals.]

A4-33. Consider an *NPN* BJT with uniformly doped base region, operating in the forward-active regime under low-level conditions. It is well known that the net current in the base region can be described accurately as almost 100% electron-diffusion current, so that $J \approx J_{n,\text{diff}}$. But the conductivity entering into $E = J/\sigma$ is almost exclusively a hole conductivity, $q\mu_p p$. Hence, by what kind of logic can one describe $E = J/\sigma$ as an "ohmic" field component?

A4-34. In Exercise 4-59 it was found that for $b_B = 3$ and $n \approx p$, the ambipolar diffusivity becomes $D^* = 1.5$. Carry out a direct analysis of the same situation, using the actual diffusivities D_n and D_p. Calculate an *effective* diffusivity in terms of D_p by including the effect of the field E that is present.

A4-35. Exercise 4-57 in Section 4-5.2 deals with the longitudinal electric-field profile accompanying the high-level carrier profiles shown in Figure 4-38. Find the true ratio

$$E(\sim X_B)/E(\sim 0)$$

and the true ratio

$$\frac{(dn_B/dx)|_{x=\sim X_B}}{(dn_B/dx)|_{x=\sim 0}}$$

for the relative carrier-density values in Figure 4-38, assuming $J_p = 0$. The notation $x = \sim 0$ means "a position near the emitter-junction boundary but far enough from it to avoid the slope reversal of majority-carrier density illustrated in Figure 6-16." The notation $x = \sim X_B$ means "a position near the collector-junction boundary but far enough from it so that minority-carrier transport is purely diffusive."

A4-36. Refer to Figure 4-42.
a. The Jespers data show that the SNS component of the base current is dominant up to about 0.4 V. The low-current extrapolation of the same curve indicates that the SNS component exceeds the diffusion component (the light, linear extension of the linear portion of the I_B curve) by a factor of over 100 near $V_{BE} = 0.1$ V. In Figure 3-24, by contrast, the SNS component dominates only up to about 0.35 V, and exceeds the diffusion component by about a factor of ten at the same voltage. Offer a qualitative explanation for these relationships.
b. Assume that the low-current extrapolations of the two curves are correct. They indicate that for V_{BE} a bit less than 0.1 V, we have $I_C = I_B \approx 10^{-14}$ A in this particular BJT. At this equal-current point, determine the magnitude and sign of each terminal current.
c. At the same point, state for each boundary of each junction (a total of four surfaces) the direction and approximate magnitude of the electron and hole components of current. If one component is small compared to the other, the magnitude of the smaller can be described as "≈ 0."
d. Discuss the current components existing in a grossly asymmetric junction subjected to a low forward bias at which SNS current accounts for virtually all of the conduction.

A4-37. Because of asymmetry in a real BJT, the I–V characteristics shown in Figure 4-21 pass almost but not quite through the origin. Assuming a BJT having $\beta_F = 100$, $\beta_R = 2$, and $I_S = 10^{-14}$ A, and using the Ebers-Moll equations, calculate
a. V_{CE} at $I_C = 0$
b. I_C at $V_{CE} = 0$

A4-38. For the device of Problem A4-31, calculate
a. α_F
b. α_R
c. I_{EO}/I_S
d. I_{CO}/I_S

A4-39. The BJT shown has $\beta_F = 100$, $\beta_R = 2$, and $I_S = 10^{-14}$ A. Its base terminal is open so that $I_B = 0$.

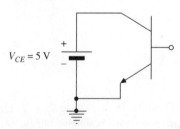

a. Determine V_{BE}.
b. Show analytically that $I_{CEO} = (1 + \beta_F)I_{CO}$.
c. Determine I_{CEO} in terms of transport-form parameters.
d. In the transport-form derivation of an expression for I_E (for example), there is a

single structure-determined current involved (I_S), fixed by the properties of a *single* region, the base region. In the original-form derivation of the I_E expression, there are *two* structure-determined currents involved (I_{EO} and I_{CO}), each fixed by the properties of *all three* regions. What feature of the transport-form derivation is responsible for its improved economy and clarity? Does it make more simplifying assumptions than were made in the original derivation, such as the neglect of a leakage current?

A4-40. Carry out the integration indicated between Equations 4-224 and 4-225.

A4-41. Using the hybrid model of Figure 4-52, determine the small-signal resistance observed when one "looks into" a particular BJT port for the conditions indicated. Do so by inspection or by means of a circuit sketch and its analysis. In all cases the BJT is within or at the boundary of the forward-active regime.
  a. Look into the *BE* port with the *CE* port short-circuited.
  b. Look into the *BC* port with the *CE* port short-circuited.
  c. Look into the *CE* port with the *B* and *C* terminals common, and connected to a bias source of infinite impedance.

A4-42. A sample-and-hold circuit is schematically represented here.

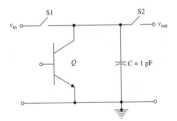

In a typical application, a signal $v_{in}(t)$ is present. The switch S1 is closed very briefly, permitting C to charge to the instantaneous value of $v_{in}(t)$. At a time Δt later, the switch S2 is closed, permitting the voltage on C to be measured and the measured value to be converted into digital form. After that, a brief positive pulse at the base terminal of Q causes the BJT to conduct, discharging C and resetting its voltage to zero, in preparation for a repetition of the three events. When the BJT is OFF, its leakage current tends to discharge C. For acceptable operation of the circuit, the voltage v on C must decline no more than one percent in the time Δt. For $v \approx 1$ V, what is the maximum tolerable leakage current of the BJT from collector to emitter, given $\Delta t = 1$ ns?

A4-43. A certain BJT is switched from a steady-state OFF condition to a steady-state ON condition. Collector current increases as the device heats up. At the conclusion of the thermal transient, the bias values are $V_{CE} = 1$ V and $I_C = 1$ mA. The case temperature of the BJT is held constant at room temperature. The device and its package are characterized by a thermal resistance $\theta_{jc} = 10$ C°/mW, and a temperature coefficient of collector current in the neighborhood of room temperature of $K_{TC} = d(\ln I_{CN})/dT = 0.075/$C°, where I_{CN} is normalized collector current. Calculate the percentage change in collector current that results from power-dissipation heating.

A4-44. The biases $V_{BE} = 0.5$ V and $V_{CE} = 5$ V are applied to the BJT specified here:

Emitter	Base	Collector
$N_{DE} = 10^{20}/\text{cm}^3$	$N_{AB} = 5 \times 10^{16}/\text{cm}^3$	$N_{DC} = 5 \times 10^{15}/\text{cm}^3$
$\mu_{nE} = 100$ cm^2/V·s	$\mu_{nB} = 900$ cm^2/V·s	$\mu_{nc} = 1{,}250$ cm^2/V·s
$\mu_{pE} = 46$ cm^2/V·s	$\mu_{pB} = 380$ cm^2/V·s	$\mu_{pc} = 430$ cm^2/V·s
$X_E = 1$ μm	$X_B = 0.5$ μm	$X_C = 1.5$ μm
$\tau_E = 10^{-9}$ s	$\tau_B = 10^{-5}$ s	$\tau_C = 3 \times 10^{-5}$ s
$n_{iE} = 1.2 \times 10^{11}/\text{cm}^3$	$n_i = 10^{10}/\text{cm}^3$	$n_i = 10^{10}/\text{cm}^3$
$A_E = 10^{-4}$ cm^2		

a. Calculate $n_B(0)$, where the spatial origin is placed at the base boundary of the emitter junction.
b. Calculate I_C.

A4-45. You are given this rudimentary amplifier circuit using a silicon BJT.

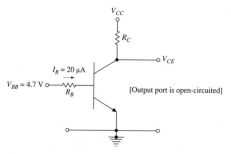

The loadline diagram for this circuit is shown here, having a quiescent operating point Q_1.

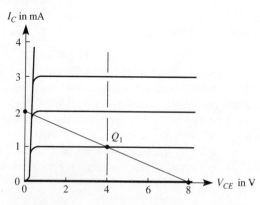

a. Calculate R_B.
b. Calculate β.
c. Calculate R_C.
d. The value of R_C is reduced by a factor of two, while all other circuit elements and power-supply voltages are held constant. Draw the new loadline on the output plane.
e. On the loadline diagram, indicate the new quiescent operating point Q_2, and calculate its coordinates.

A4-46.
a. Given the component values and voltages in the rudimentary BJT amplifier circuit shown schematically, calculate the static current gain β for the BJT.

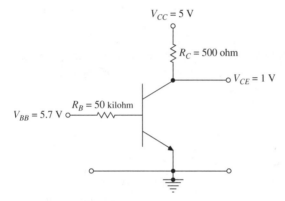

b. On the axes provided below, draw the output characteristic of the BJT for the given bias conditions. Label the characteristic with the appropriate parameter symbol and value. Plot the loadline and indicate the quiescent operating point Q. Label and calibrate the axes accurately.

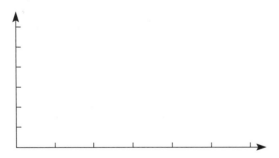

A4-47. Following is a BJT of static current gain $\beta = 100$, and emitter-junction area $A_E = 1$ mm². It is biased so that $I_C = 8.2$ mA. Heavy doping in the emitter region causes gap narrowing there that in turn causes intrinsic density there to be $n_{iE} = 1.2 \times 10^{11}$/cm³.

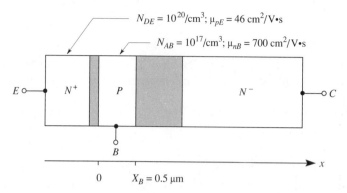

a. Using symbols only (no numbers), derive an expression for base–emitter voltage V_{BE}.
b. Evaluate the expression derived in part a.
c. Using symbols only (no numbers), and assuming that simple theory is valid for the emitter region, derive an expression for minority-hole diffusion length in the emitter (L_{pE}) in terms of collector current I_C.
d. Evaluate the expression in part c.

COMPUTER PROBLEMS

C4-1. Consider the circuit shown here. SPICE parameter values for this transistor are given in Table 4-7, and dc operating-point information is given in Table 4-8.
a. Using the SPICE program, calculate the ac voltage gain in dB as a function of frequency for values of frequency from 1 KHz to 10 GHz. Plot the results along with the phase shift in degrees. Use the .OP command in SPICE to print out the values of the small-signal hybrid-pi parameters and compare them with those given in Figure 4-68.
b. Replace the transistor in part a with the hybrid-pi circuit that uses resistors, capacitors, and a voltage-controlled current source. Repeat the calculation of ac voltage gain and show that the results are identical to those already obtained in part a.

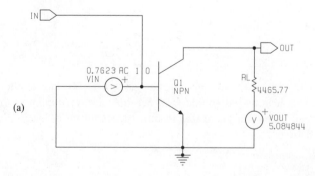

(a)

C4-2. In circuit design, it is important to know the delay time of an inverter—the time required for a signal to pass through the inverter. More specifically, delay time is defined as the time measured from the halfway point on the rising edge of an input voltage pulse to the corresponding halfway point on the falling edge of the output waveform. Unfortunately, this time depends strongly on the shape of the input pulse and the characteristics of the load device. To remove these complications and to establish a unique definition, one usually employs a circuit like the one shown here. The first inverter acts as an input buffer, and the last inverter as a load device. The delay time is then taken as average delay time for the center two inverters.

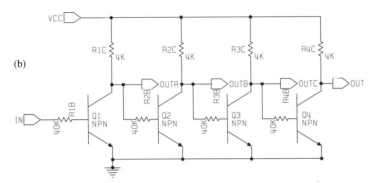

(b)

a. Using SPICE and the transistor-parameter values given in Table 4-7, calculate the transient response of the given circuit for an input pulse that is identical to the solid line given in Figure 4-71(a). Show that the output waveform at OUTA is identical to the dotted line in Figure 4-71(a).

b. Plot the waveforms at OUTA and OUTC on the same graph and show that the average delay time per stage is about 17 ns. What causes the major portion of this delay and what circuit techniques can be used to reduce this delay?

DESIGN PROBLEMS

D4-1. The following rudimentary amplifier schematic employs a silicon BJT and specifies V_{CC}, V_{CE}, R_B, and β. Complete the amplifier design by determining the remaining values needed to yield a voltage gain of $A_V = -\beta(R_C/R_B) = -10$.

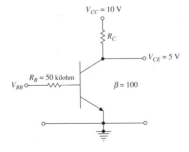

a. $V_{BB} = $ _____ V
b. $I_B = $ _____ mA
c. $I_C = $ _____ mA
d. $R_C = $ _____ kilohm

D4-2. You are to design a BJT base region by specifying its Gummel number G_B. You may assume that a mobility value corresponding to average doping density is appropriate, and that the emitter region is as defined in Table 4-1. Also you may neglect collector-junction leakage current.
a. Design a BJT with $\beta_F = 100$ and $X_B = 3.0$ μm.
b. Design a BJT with $\beta_F = 100$ and $X_B = 0.3$ μm.
c. Comment on the difference in the two values of Gummel number.

D4-3. In the early 1950s, the outstanding challenge in solid-state electronics was to develop a method for achieving a BJT base region sufficiently uniform in thickness and sufficiently thin. You were a design engineer at that time, and this curiosity concerning semiconductor etching came to your attention: An etching procedure exists that removes the quasineutral crystal at a rapid rate. But when the advancing surface reaches the boundary of a depletion layer, the rate of etching slows drastically. (In certain electrolytic etching procedures, the rate is sensitive to carrier availability; a depleted zone acts like an insulator.)
a. Outline a fabrication procedure that exploits this effect in a way that achieves a BJT base region of unprecedented uniformity and thinness. Think in terms of an alloyed-junction device with the collector junction already formed. Etch from the opposite face of the crystal to reach a surface where the emitter junction will be placed.
b. The proposed fabrication procedure is tried and works as expected. But the properties of the resulting BJTs are disappointing. Why?
c. It has been observed in electrolytic etching that the presence of a copious supply of holes speeds up the process significantly. Offer a possible explanation.

D4-4. Current-regulating ("current-source") circuits are used in a large number of linear applications and a growing number of digital applications. Shown below are two widely used options for achieving current that is approximately voltage-independent. You wish to design for the smallest possible variations in I_C in the face of manufacturing variations that cause changes in BJT properties. Assume that V_{BB} is constant and that the variation in R can be held to 1%.

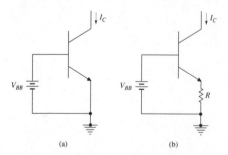

(a)   (b)

a. State which circuit, a or b, you expect to be superior, and give a qualitative reason for your choice.

b. Assume the Gummel equation $I_C = I_S \exp(qV_{BE}/kT)$, where I_S is subject to 30% manufacturing variations, $|\Delta I_S/I_S| = 0.3$. Determine the worst-case I_C variation in circuit A.

c. Find the value of $g_m R$ needed in circuit B to restrict worst-case I_C variations to 2%. [As was suggested in Problem D2-6c, start by writing the total differential of the relevant function, in this case I_C.]

d. Determine the necessary value of R if your design calls for $I_C = 1$ mA.

D4-5. Design a geometrical representation of the meaning of Equation 4-147, which was the starting point for deriving one of the hybrid equations, by following this procedure:

a. Sketch a surface representing the function $I_C = I_C(I_B, V_{CE})$, Equation 4-145, using information developed in Exercises 4-86 and 4-87 (Section 4-7.1). Interpret its major features.

b. With a second sketch of a localized part of the surface represented in part a, show the relationship of a plane surface employed to interpret the linear Equation 4-147 to the actual curved surface of the function $I_C = I_C(I_B, V_{CE})$.

D4-6. You are to design a *derating curve* for a power transistor in which the maximum permissible collector-junction temperature is $T_j = 200$ C. A derating curve is a plot of steady-state power dissipation P versus case temperature T_c. The junction-to-case thermal resistance is $\theta_{jc} = 2$ C°/W.

a. Write an equation relating P to T_c.

b. Plot the derating curve, using the equation written in part a.

c. Explain the value of the derating curve and give an example of its use.

5 The MOSFET

A thin and tightly adherent layer of glass, mainly SiO_2, can be formed on the surface of a single-crystal silicon sample by heating it in an oxidizing atmosphere. The most common choices are oxygen and steam. Starting in the late 1950s, these procedures were moved along vigorously in the hope that they might provide at least a partial solution to the vexing problem of stabilizing surface conditions on silicon devices. Ultimately they permitted not only stabilization but also control of surface conditions; to accomplish this, however, enormous investments of time and resources were required to achieve the necessary understanding of the oxide–silicon system, and especially of the oxide–silicon interface.

At a relatively early stage, these procedures benefited the BJT, because its crucial phenomena occur mainly inside the silicon crystal. Further, even a limited grasp of oxide technology opened a whole new approach to BJT fabrication. But then came the MOSFET, which we shall describe in detail. By contrast with the BJT, the decisive phenomenon in the MOSFET occurs *at* the oxide–silicon interface. Consequently, the art and science of controlling interfacial properties required an order-of-magnitude improvement to make the MOSFET a truly practical device, a process that took approximately an additional ten years. The central portion of the MOSFET is an *MOS capacitor,* a structure that received intensive study in its own right. When isolated, it is a one-dimensional "sandwich" consisting of the oxide with silicon on one side, and with an adherent field-plate electrode on the opposite side that is capable of manipulating potential at the oxide–silicon interface.

In Section 5-1 we carry out a quantitative but simplistic analysis of the MOSFET that yields a surprisingly useful result. Its primary simplifications reside in the properties assumed for the MOS capacitor. These are addressed in more realistic fashion in Sections 5-2 and 5-3. In Section 5-4, a conversion is made from the realistic but still one-dimensional MOS capacitor to the MOSFET problem, which is two-dimensional at best. Further real-life factors are also introduced to produce accurate static models. Dynamic properties are treated in the remaining sections.

5-1 BASIC MOSFET THEORY

As we saw in Chapter 4, the essence of BJT operation is applying a voltage to modulate the potential barrier in a *PN* junction. The modulated current of carriers spilling over the barrier as a result becomes an output current, and these combined phenomena are responsible for the remarkable gain properties of this important bipolar device. The operating principle in a field-effect transistor, or FET, is quite different. An FET is essentially a voltage-controlled resistor. Hence it is "unipolar" in the sense that carriers that are in the majority in the region providing the resistance are dominant in determining device properties. As the result of a voltage applied from one end of the resistor to the other, these majority carriers flow ohmically in a thin region known as the *channel*. The population of carriers in the channel is modulated by a separate *control* voltage that creates a modulated electric field essentially normal to the thin channel. As carrier population is changed, so, too, is channel resistance. The control electrode is known as the *gate* electrode.

There are several FET families, each with many members, and in each family there are differences of detail in how the control voltage interacts with the channel. Nonetheless, all of the FETs have similar gain properties, properties that are substantially inferior to those of the BJT through a wide current range. But, we should add quickly, the various families have offsetting advantages. In one case, that of the MOSFET, the sum of these advantages is so imposing that MOSFETs are the dominant solid-state devices in the world today.

5-1.1 Field-Effect Transistors

Well before the present solid-state-device era, intuitive thoughts were advanced on how to realize a solid-state amplifier; the structures proposed were essentially field-effect transistors [1, 2]. The serious postwar effort that led to the transistor also began with field-effect transistors [3, 4]. While that path was being explored, an unintended experimental event led to the discovery of the point-contact transistor [5], which found enough applications to be characterized as the first practical solid-state amplifier. But the point-contact transistor had serious shortcomings (and operating principles that remain obscure to this day). Consequently it was supplanted by the BJT within just a few years [6]. The events connected with these two devices (occurring from the mid-1940s to the mid-1950s) display a fascinating blend of scientific insight, art, engineering intuition, and happenstance [7].

A practical FET structure was described and analyzed in detail by Shockley in 1952 [8], a year *before* its reduction to practice by Foy and Wiegmann, working under the direction of Ross and Dacey [9]. This was a case of having metaphorical lightning strike twice in the same place, because as we noted in Chapter 4, the same inventor had accomplished the same feat just a few years earlier with the BJT! Shockley's FET employed a pair of *PN* junctions for defining the channel, and hence has become known as the junction FET, or JFET. His innovation placed the region of crucial properties well inside the semiconductor crystal and away from the troublesome and poorly understood surfaces. Varying degrees of reverse bias on these channel-defining junctions caused varying depletion-layer encroachment upon the channel, thus achieving the aim of altering the areal density (number per unit area) of carriers in the thin channel.

While the laboratory JFETs reported in 1953 [9] constituted a feasibility demonstration, they were not practical devices. The JFET requires a degree of control over areal impurity density in its channel that is about ten times better than that required, for example, in the base region of a BJT. Thus the JFET was a device awaiting new and more refined fabrication options. The first practical JFETs, made in 1960, incorporated channels produced by the epitaxial growth of silicon from a vapor [10, 11]. Today the technology of choice is usually ion implantation [12], which achieves unprecedented control of areal impurity density in thin layers.

In the discrete-device arena, JFETs have played a relatively small part. Unique properties, especially low noise, have earned for them a "niche market." At first JFETs did not play a role in integrated circuits either, because their peculiar requirements made them essentially incompatible with the most popular devices. Then it was found that a kind of hybrid JFET–MOSFET device could play an advantageous part in circuits dominated by the latter [13]. Today, JFETs are being incorporated into even bipolar integrated circuits, thanks to the almost commonplace availability of ion-implantation technology. In addition, the JFET may find application in monocrystalline three-dimensional integrated circuits [14].

The MOSFET [15] was the second important field-effect device to come along. The heart of its original embodiment was a metal–oxide–silicon sandwich, with initial letters responsible for the acronym MOS, as in MOSFET and MOS technology. The sandwich was in fact a parallel-plate capacitor, with one metal plate and one silicon plate. Thus the structure indeed evokes early (and unsuccessful) experiments that endeavored to modulate the majority-carrier population in a thin semiconductor layer as a way of realizing a solid-state amplifier. But important new features were present in the MOSFET, and these spelled success. First, the new device exploited the infant oxide-growth technology that started with a polished single-crystal silicon sample. Significantly, one of the MOSFET inventors was also a pioneering developer of the oxide-growth technology and student of oxide-silicon interfaces [16]. Second, the MOSFET exploited the creation of an *inversion layer* in the silicon at the interface, rather than attempting to achieve a significant modulation of majority-carrier areal density in a thin sample. More specifically, the metal plate of the capacitor was used to create a potential well at the surface of the silicon, a well capable of retaining the desired carriers—electrons in a P-doped sample or holes in an N-doped sample. This constitutes an alternative description of the inversion layer at an insulator–semiconductor interface. As an isolated entity, this inversion layer is identical to that in a grossly asymmetric junction at equilibrium, as was described in Section 3-7.6. But the two cases are different in context: In the present case the carriers are confined within the potential well because they cannot penetrate the insulator on the other side of the interface, while in the *PN*-junction case, the inversion-layer carriers are in direct communication with the identical majority carriers on the heavily doped side of the junction. Because the MOSFET has overwhelming practical importance today, and because all field-effect devices involve similar principles, our focus in the balance of this chapter will be on the MOSFET.

The FET chronicle by no means ended with the MOSFET, however. A new family was proposed in 1966 wherein control was exercised through a metal–semiconductor junction, or *Schottky junction* [17], and reduction to practice was reported the following year [18]. The channel carriers in this device are confined between the depletion region associated with a Schottky junction and that associated with a *PN* junction. (Sometimes the *PN* junction is replaced by a junction between an extrinsic region and a semiinsulating region.) This structure was given the name MESFET, for metal–semiconductor field-effect transistor. It has been realized using a number of semiconductor materials, but GaAs is by far the most common. The MESFET, in fact, is the dominant compound-semiconductor transistor today.

A still more recent FET is the MODFET, which designates the modulation-doped FET. It creates a potential well for electrons in a region of extremely light doping, and as a result, the ionic scattering of drifting electrons is reduced to near zero, leading to extreme values of electron mobility, well above 10^4 cm^2/V·s at room temperature. (Hole mobilities are typically small in the GaAs family of materials.) By cooling the sample to reduce phonon activity, one can achieve electron mobilities exceeding 10^5 cm^2/V·s. Such enhanced electron mobilities were first observed in 1978 [19], and were realized in a field-effect transistor in 1980 [20]. This feature of the MODFET has given it a second name, HEMT, the acronym for high-electron-mobility transistor. The significance of high carrier mobility is high-speed operation of the resulting device. Still another name for the MODFET is TEGFET, for two-dimensional-electron-gas FET.

The feat of separating channel electrons in the MODFET from the impurity atoms that contribute them is accomplished by means of a *heterojunction,* which is an interface between two semiconductor materials of differing energy gap. Two materials often used to form a heterojunction are the binary compound GaAs and the ternary compound AlGaAs. Heterojunctions are growing in importance in semiconductor technology, making possible proliferating device innovations. One of these is yet another FET, the HIGFET [21, 22], which stands for heterojunction insulated-gate field-effect transistor. Among its advantages is the ready achievement of complementary—or *N*-channel and *P*-channel—devices, which make possible superior performance.

A still further advance in the FET art uses heterojunctions to create a series of *quantum wells* or a *superlattice,* terms that describe a set of very thin, parallel layers of differing energy gap in a semiconductor crystal, usually involving binary and ternary compound semiconductors. Layer thickness must be appreciably less than the de Broglie wavelength (see Section 1-4.6), and typically amounts to a few tens of angstroms. (A superlattice can also be created by doping variations, and even by light or sound waves!)

One kind of FET-like structure exploiting quantum-well layers uses them as multiple channels that are under the control of a gate electrode. Because carriers are able to move in the planes of such layers at great velocity, the resulting device is capable of high gain and fast switching [23]. An even more recent proposal would replace the two-dimensional quantum-well layers by one-dimensional quantum-well "wires" [24]. The transition from classical to quantum principles of operation opens new realms of device possibilities. A substantial number of variations on these and related themes already exist, and it is safe to predict that the future holds many more.

5-1.2 MOSFET Definitions

The essential structure of a MOSFET is shown in Figure 5-1; for purposes of explanation one can imagine that this structure had been cut out of an integrated circuit. The capacitorlike arrangement formed on and by the silicon substrate accounts for the acronym *MOS,* as was noted above, and as is also noted in the diagram. The metal top plate was made of aluminum in most early devices. But since about 1970, heavily doped polycrystalline silicon (*polysil*) has been favored. (See the end of Section 1-5.6.) Many different insulating materials have been employed through the years, but thermally grown oxide formed *in situ* is currently and historically by far the most important option. This description of the silicon oxide means that the necessary silicon was supplied by the single-crystal substrate, and its growth was induced, as was noted before, by heating the sample in an oxidizing atmosphere.

The metal top plate, or field plate, becomes the control electrode of the MOS-

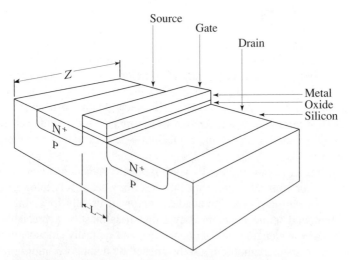

Figure 5-1 Essential structure of a MOSFET.

FET and is known as the *gate*. Applying a sufficient voltage to the gate with respect to the substrate causes electrons (normally in the minority in the *P*-type substrate) to be attracted to the interface, thus creating an inversion layer. This thin layer constitutes the resistor cited earlier, and is termed the MOSFET *channel*. Ohmic contact is made to the ends of the channel by means of heavily doped *N*-type regions, usually formed by ion implantation or solid-phase diffusion, or some combination of the two.

It is evident in Figure 5-1 that the MOSFET exhibits bilateral symmetry. That is, channel carriers can be made to flow in either direction, depending on bias polarity. This symmetry is important in certain MOSFET applications. The two channel terminals have been named according to which one supplies carriers to the channel and which one receives carriers from it. These two terms are *source* and *drain*, respectively, and it is evident that bias polarity determines which is which. This choice of terms has been carried over from the JFET [8], and was made there to provide distinctive initial letters for subscript purposes. (The physicist's classically favored terms *source* and *sink* are obviously wanting with respect to this requirement.) To summarize, for the *N*-channel device under consideration, the more positive of the two channel terminals is the drain, the other is the source.

It is customary to refer all voltages in the MOSFET to the source voltage (just as in an earlier era, terminal voltages in a vacuum tube were referred to that of the cathode). Thus, two important bias values are the drain–source voltage V_{DS} and the gate–source voltage V_{GS}, continuing to use the double-subscript notation introduced in Chapter 3. But the substrate constitutes a fourth terminal. And a bias from substrate to source indeed alters device properties; this is a situation that sometimes cannot be avoided, especially in integrated circuits. To deal with it, we need at the most simplistic level a distinctive subscript for "substrate" too, and so the homely term *bulk* has been adopted as an approximate synonym. It is intended to connote the interior of the substrate. In Section 5-4 we shall examine the effect of a bulk–source voltage V_{BS}. But for the initial description, let us assume that the source and substrate are electrically common, to serve as voltage reference. Further, for the moment let $V_{DS} = 0$.

There is a certain voltage V_{GS} at which the hole population immediately under the oxide–silicon interface will be identical to that deep within the bulk of the crystal. Under such conditions, a band diagram representing conditions from the interface to the interior of the *P*-type silicon crystal would have perfectly horizontal band edges. This magic gate-bias value is termed the *flatband voltage*, $V_{GS} = V_{FB}$, and is valuable as a conceptual starting point. When the gate terminal is now made more positive with respect to source (and substrate), the band edges begin to bend near the oxide–silicon interface. This matter will be examined in detail. At the moment, what is important is that potential ψ_S at the surface of the silicon crystal (at the oxide–silicon interface) is becoming more positive through the influence of the field plate. Hole density at the interface declines and electron density rises. This is dictated by the law of mass action, $pn = n_i^2$, because the silicon remains (for engineering purposes) at equilibrium. The reason for the caveat is that any leakage current through the oxide is accompanied by a net current in the silicon that violates equilibrium. But the superb insulating qualities of SiO$_2$—resistivity in excess of 10^{15} ohm·cm—make this a vanishing concern at present.

There exists another critical voltage at which the *electron* density right at the interface is equal to the hole density in the bulk. This condition is termed the *threshold of strong inversion* and is specified by the statement $V_{GS} = V_T$. The chosen term is apt, because channel conductance increases steeply with gate voltage for $V_{GS} > V_T$. At the threshold voltage and just below it, the electrons that form the incipient channel contribute a very small amount of drain–source conduction, leading to *subthreshold current*. For still smaller values of V_{GS}, there exists essentially an open circuit from drain to source because the applied voltage V_{DS} reverse biases the drain junction. Only a tiny fraction of V_{DS} appears as forward bias on the source junction. Thus to a good approximation, one can say that the channel conductance is "turned on" by V_{GS} at the threshold voltage V_T.

5-1.3 Rudimentary Analysis

Figure 5-2 shows explicitly the electrical commonality of source and substrate and their joint use as voltage reference. This time, however, we choose a value of V_{GS} well above V_T, as indicated by the symbol "++". But drain-source bias is small ("+"). Thus we have conditions in which there is only a small current in the channel, I_{CH}, consisting of course of electrons drifting from source to drain. The *IR* drop in the channel from drain to source amounts to V_{DS}, and by requiring the condition $V_{DS} \ll V_{GS}$ we can arrange to have the channel be nearly uniform in conductivity from one end to the other. That is, its areal density of electrons will be about the same at the source and drain ends.

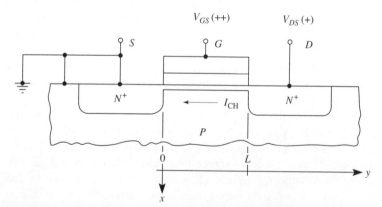

Figure 5-2 Cross-sectional diagram of a MOSFET under bias, showing axis assignments. Gate–source voltage is sufficient to produce an inversion layer; drain–source voltage is very small. The *x* direction is into the sample, with spatial origin at the silicon surface. The *y* direction is from source to drain, with origin at the source end of the channel. The *z* direction is out of the paper.

To deal with the conducting properties of a thin and uniform sheet, it is useful to introduce a concept known as *sheet resistance*. This is especially so when the layer or sheet is nonuniform in the thickness direction, which, as Figure 5-2 shows, we take to be the x direction. Such is the case with the inversion layer that serves here as channel. Electron density peaks just under the oxide and then declines monotonically with increasing x. Having said this, let us assume temporarily that the sheet of interest *is* uniform in the x direction, a restriction that will be relaxed after just two equations. Further, let us use the symbol Z for the extent of the channel in the z direction, a common practice in MOSFET work.

Consider a uniform layer like that depicted in Figure 5-3(a) that has a resistivity ρ, a thickness X, and two other equal dimensions L. Since the cross-sectional area presented to the entering current is $A = XL$, it is evident from Equation 1-26 that the resistance of this thin sample is

$$R = \rho \frac{L}{A} = \frac{\rho L}{XL} = \frac{\rho}{X}. \tag{5-1}$$

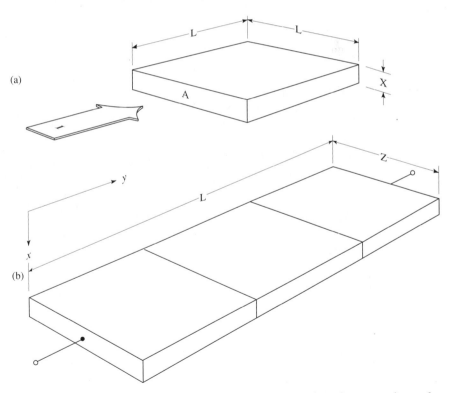

Figure 5-3 The sheet-resistance concept. (a) Current passes through a square layer of thickness X, entering through one edge of area $A = LX$. (b) The resistance of a rectangular sample is found by multiplying sheet resistance by the aspect ratio of the sample, L/Z.

Thus it is evident that the resistance of a uniform square sample is independent of its lateral dimensions, depending only on resistivity and thickness. (It does not matter whether L is one micrometer or one mile!) This invariant quantity is designated sheet resistance, R_S. It is measured in ohms. We sometimes specify it in "ohms per square" to label the quantity cited as a sheet resistance, but "square" has no dimensional-analysis significance. Because R_S is an intensive quantity, like bulk resistivity ρ, it is sometimes termed sheet *resistivity*. Either designation is correct.

Exercise 5-1. Explain qualitatively how it can be possible to have a one-micrometer square exhibit the same side-to-side resistance as a one-mile square having the same resistivity and thickness.

Increasing square size places increasing amounts of material in the series path, but also in parallel paths in a way that holds resistance constant.

Now turn to the rectangular resistor in Figure 5-3(b). Given $R_S \equiv \rho/X$, it is evident that this time we may write

$$R = R_S(L/Z). \qquad (5\text{-}2)$$

In this example, three squares are connected in series, a number provided by the ratio L/Z, and the resistance of each square is R_S so the resulting resistance is $R = 3R_S$. Had the squares been in parallel, we would have had $(L/Z) = 1/3$, and thus the resistance this time would have been $R = R_S/3$. The convenience of working with R_S in lieu of resistivity and dimensions is that *now* we need only be concerned about layer uniformity in the lateral directions. The function $\rho(x)$ is completely arbitrary; the properties of the layer need only be independent of y and z.

Now let us apply the sheet-resistance formulation to the problem posed by the MOSFET channel in Figure 5-2. Once again it is convenient to assume that we deal with a conducting layer of fully uniform properties, and then to relax the requirement later. Given a channel with a well-defined thickness X and a volumetric electron density n, we can write its areal charge density with due regard for algebraic sign as

$$Q_n = -qnX. \qquad (5\text{-}3)$$

Then using $R_S \equiv \rho/X$ to eliminate X, we have

$$Q_n = -qn\rho/R_S. \qquad (5\text{-}4)$$

But recalling that $\rho = 1/q\mu_n n$, we can rewrite Equation 5-4 as

$$Q_n = -qn/(qn\mu_n R_S) = -1/\mu_n R_S. \qquad (5\text{-}5)$$

This relationship (admittedly nonobvious) eliminates all reference to channel resistivity and thickness; thus, realistic inversion-layer properties are now fully permitted. That is, the steep variation of $n(x)$ and hence $\rho(x)$ poses no problems.

Next we write a different expression for areal charge density in the inversion layer, one that takes advantage of the fact that the MOS sandwich structure in Figures 5-1 and 5-2 is very nearly a conventional parallel-plate capacitor. Its only modest departure from such description is that the inversion layer has a thickness of the order of a few hundred angstroms rather than the few angstroms that characterize the charge-layer thickness in a metal capacitor plate. Let us assume that inversion-layer formation commences abruptly at $V_{GS} = V_T$, which proves to be an excellent working approximation. Then the resulting expression is

$$Q_n = -C_{OX}(V_{GS} - V_T), \tag{5-6}$$

where C_{OX} is capacitance per unit area for the MOS capacitor. Finally, combining Equations 5-3 and 5-6 gives us an expression for channel sheet resistance in terms of MOSFET real physical properties and one terminal voltage, and independent of artificial quantities such as ρ and X that were temporarily assumed for the channel:

$$R_S = \frac{1}{\mu_n C_{OX}(V_{GS} - V_T)}. \tag{5-7}$$

Exercise 5-2. Why is there a minus sign on the right-hand side of Equation 5-6? The quantities C_{OX} and $(V_{GS} - V_T)$ are both positive, and the minus sign that accounts for negative electron charge is subsumed in Q_n, as Equation 5-3 shows.

It was stressed in Section 1-3.3 that for algebraic sign consistency in the capacitor law, $Q = CV$, it is necessary to examine the chart Q on the plate *to which* the voltage V is applied. But in Figure 5.2, we apply the voltage to the top plate, or gate, while the charge of interest is the inversion layer on the bottom plate. Thus the extra minus sign is needed for algebraic-sign consistency.

Next let us acknowledge that in the general case there *will* be a variation in channel properties from source to drain because of the unavoidable *IR* drop accompanying I_{CH}. This situation is represented in Figure 5-4. Note the polarities involved. Both V_{GS} and V_{DS} are positive, and $V_{DS} < V_{GS}$ but V_{DS} is no longer negligible. Because the drain end of the channel is more positive than the source end, and because the conducting channel plays the role of a capacitor plate, it follows that there is less voltage drop from gate to channel at the drain end than there is at the source end. However, in this instance the values of V_{GS} and V_{DS} are adjusted to ensure that above-threshold conditions exist even at the drain end of the channel. That is, the inversion layer is continuous from source to drain. Furthermore, the channel has been made long enough so that its properties (especially areal charge density) change slowly from source to drain. This structural choice is made so that the channel will exhibit near-one-dimensional properties—so that its variation in the y direction will be negligible compared to that in the x direction. (No variation exists in the z direction.) The analogous specification was made by Shock-

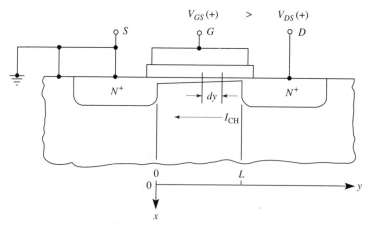

Figure 5-4 Cross-sectional representation of a long-channel MOSFET with significant gate–source voltage and drain–source voltage, but with $V_{GS} > V_{DS}$ sufficiently to ensure that an inversion layer exists at the drain end of the channel.

ley in his approximate analysis of the JFET [8], and is known as his *gradual-channel* approximation. In recent times, the same approximation or assumption is often identified through the adjective *long-channel*.

Now let the symbol $V(y)$ represent the "voltmeter voltage" in the channel, varying from $V(0) = 0$ to $V(L) = V_{DS}$ for the biasing conditions represented in Figure 5-4. Note, too, that the channel "thickness" variation shown there heuristically is to represent the monotonically declining charge density in the channel as the drain end is approached, a variation directly related to $V(y)$, which in turn stems from the aforementioned *IR* drop. The next step is to focus on a small channel element of length dy located at the arbitrary position y, as also shown in Figure 5-4. This element exhibits a sheet resistance $R_S(y)$, which also increases monotonically from source to drain. At the source end $R_S(y)$ is fixed by the voltage $(V_{GS} - V_T)$ and is given by Equation 5-7. At the drain end it is fixed by the smaller voltage $(V_{GS} - V_T - V_{DS})$. And at the arbitrary position y, $R_S(y)$ is fixed by the intermediate voltage $[V_{GS} - V_T - V(y)]$, so the appropriate modification of Equation 5-7 is

$$R_S(y) = \frac{1}{\mu_n C_{OX}[V_{GS} - V_T - V(y)]}. \tag{5-8}$$

The channel element of length dy exhibits a resistance dR that can be written immediately using the sheet-resistance formulation as

$$dR = \frac{dy}{Z} R_S(y) = \frac{dy}{Z\mu_n C_{OX}[V_{GS} - V_T - V(y)]}, \tag{5-9}$$

where Z, once more, is the dimension of the channel in the z or width direction. Now simplify the notation by letting $V(y) \equiv V$. Using Ohm's law we may write

$$I_{CH} = -\frac{dV}{dR}, \tag{5-10}$$

thus bringing us to the relationship sought between current and voltage.

Exercise 5-3. Why is there a minus sign in Equation 5-10?

The Ohm's-law convention, as noted in Section 3-3.1, defines a voltage *drop* as positive. This is evident in the present example, wherein I_{CH} is negative (leftward) while V increases toward the right.

Exercise 5-4. Is I_{CH} a function of y, as are V and R_S?

No. The inversion-layer electrons that compose the channel are confined in a potential well from which they cannot escape except at a channel end.

5-1.4 Current–Voltage Equations

The desired expression for output current as a function of input and output voltages can now be derived by combining Equations 5-9 and 5-10, which yields

$$I_{CH} = -\frac{Z\mu_n C_{OX}(V_{GS} - V_T - V)dV}{dy}. \tag{5-11}$$

Separating variables and applying appropriate limits gives

$$I_{CH}\int_0^L dy = -Z\mu_n C_{OX}\int_0^{V_{DS}}(V_{GS} - V_T - V)dV, \tag{5-12}$$

and completing the integration yields

$$I_{CH} = -\frac{\mu_n C_{OX}}{2}\frac{Z}{L}[2(V_{GS} - V_T)V_{DS} - V_{DS}^2]. \tag{5-13}$$

Finally, then, after noting that $I_D = -I_{CH}$, we have

$$I_D = \frac{\mu_n C_{OX}}{2}\frac{Z}{L}[2(V_{GS} - V_T)V_{DS} - V_{DS}^2]. \tag{5-14}$$

Exercise 5-5. Justify the last change of algebraic sign.

The leftward (and hence negative) current I_{CH} results in an *inward,* and hence positive, terminal current I_D.

Letting V_{GS} be a parameter, we can use Equation 5-14 to plot in the MOSFET *output-plane* curves of drain current I_D versus drain-source voltage V_{DS}. The result is shown in Figure 5-5(a). Inspection of Equation 5-14 shows that each curve is a parabola with vertex displaced from the origin. Only the solid portions of the parabolas have physical meaning, and consequently only the shaded portion of the output plane is described by Equation 5-14. The relevant portion can of course be expanded by assigning still larger values to the parameter V_{GS}.

Exercise 5-6. Compare the three parabolas shown in Figure 5-5 with respect to their proportions.

Each is the *same* parabola, as Figure 5-5(a) suggests and Equation 5-14 confirms. The vertex is translated when V_{GS} is changed, so that more or less of the parabola has meaning.

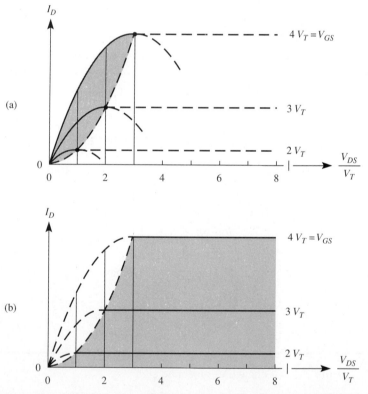

Figure 5-5 MOSFET output-plane characteristics predicted by rudimentary analysis. The input-voltage parameter V_{GS} and the output voltage are normalized using V_T. (a) The *curved* regime of operation (shaded). (b) The *saturation* regime of operation (shaded).

The physical quantities in the coefficient of Equation 5-14 deserve further description. The electron mobility μ_n in an inversion layer is typically two or three times smaller than the bulk mobility of Section 2-5.3 for a sample having the same doping as the substrate. At least a partial explanation for this fact is the diminished mean free path of the electron when it is confined in the very thin potential well. Its random motion is converted from free three-dimensional motion to a nearly two-dimensional confinement, and it is intuitively obvious that this kind of restriction will diminish average free-path length, with a consequent reduction of mobility. The oxide capacitance per unit area is given by

$$C_{OX} = \epsilon_{OX}/X_{OX}, \tag{5-15}$$

where

$$\epsilon_{OX} = \epsilon/3, \tag{5-16}$$

with ϵ being the permittivity of silicon. The oxide thickness X_{OX} is typically about 0.05 μm. Finally, as noted before, Z and L are the lateral dimensions of the channel, with L being in the current direction.

In much of the early literature, the symbol β was used to stand for the expression $\mu_n C_{OX} Z/L$, so that the coefficient of Equation 5-14 became $\beta/2$. But because MOSFETs and BJTs are being combined today with increasing frequency in the same integrated circuit, that choice is poor. Confusion with the BJT β is likely. Therefore, let us use the symbol K to stand for the entire coefficient:

$$K \equiv \frac{\mu_n C_{OX}}{2} \frac{Z}{L}. \tag{5-17}$$

Thus Equation 5-14 becomes

$$I_D = K[2(V_{GS} - V_T)V_{DS} - V_{DS}^2]. \tag{5-18}$$

Exercise 5-7. Evaluate the coefficient K for a MOSFET in which $\mu_n = 580$ cm^2/V·s, $X_{OX} = 500$ angstroms, and $(Z/L) = 1$.

Evidently,

$$C_{OX} = \left(\epsilon/3X_{OX}\right) = \frac{(1.035 \times 10^{-12} \text{ F/cm})}{3(500 \times 10^{-8} \text{ cm})} = 6.9 \times 10^{-8} \text{ F/cm}^2.$$

Thus

$$K = \frac{\mu_n C_{OX}}{2} \frac{Z}{L} = \frac{1}{2}(580 \text{ cm}^2/\text{V·s})(6.9 \times 10^{-8} \text{ F/cm}^2) = 20 \ \mu\text{A/V}^2.$$

We shall identify the operating regime indicated in Figure 5-5(a) as the *curved* regime. Portions of the literature use the term "linear" to identify the same region of the output plane, but such a characterization is patently a poor way to describe a parabola. On the other hand, for operation extremely near the origin, linear approximations are meaningful and justified. Such operation is sometimes advantageous.

Physical reasoning was applied to the FET by Shockley (before any working devices had been fabricated), with the prediction that current would be essentially independent of V_{DS} after the value corresponding to the parabolic vertex was exceeded. This prediction was soon confirmed experimentally. The resulting operating regime for the MOSFET is shown in Figure 5-5(b), and is usually designated the *saturation* regime. The term is used here in the sense of *leveling off*, or becoming constant. It is both more important than the curved regime and easier to describe analytically. Given that each constant-current curve of the saturation regime corresponds to the vertex or maximum of a parabola, one can simply differentiate Equation 5-18 with respect to V_{DS} to obtain the key relation. The result is

$$\frac{\partial I_D}{\partial V_{DS}} = -K[2(V_{GS} - V_T) - 2V_{DS}]. \tag{5-19}$$

Setting the right-hand side equal to zero gives

$$V_{DS} = (V_{GS} - V_T). \tag{5-20}$$

Thus the curve passing through the parabolic vertices in Figure 5-5 is the locus of points for which Equation 5-20 holds. This curve, plotted in Figure 5-6(a), is itself a parabola, and is in fact identical to the other parabolas, but with different orientation; it is rotated 180° in the plane of the paper with respect to the *I–V* curves.

Exercise 5-8. Why is the curve passing through the parabolic vertices of the output characteristics described as a "locus of points" for which Equation 5-20 is valid? Why do we not simply describe Equation 5-20 instead as "the equation for" the curve in Figure 5-6(a)?

Equation 5-20 incorporates the variables V_{DS} and V_{GS}. But the equation for the curve in Figure 5-6(a) must involve the variables I_D and V_{DS}. (See Problem A5-6.)

Substituting Equation 5-20 into Equation 5-18 yields

$$I_D = K[2(V_{GS} - V_T)(V_{GS} - V_T) - (V_{GS} - V_T)^2], \tag{5-21}$$

or

$$I_D = K(V_{GS} - V_T)^2. \tag{5-22}$$

As noted earlier, operation of the MOSFET in the saturation regime is the most

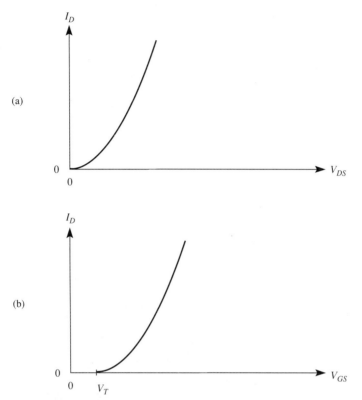

Figure 5-6 Parabolas in the $I_D - V_{DS}$ and $I_D - V_{GS}$ planes that match those in Figure 5-5. (a) The locus of the parabolic vertices in Figure 5-5, conforming to $V_{DS} = V_{GS} - V_T$, as Figure 5-5 confirms. (b) The current-to-voltage transfer characteristic of the MOSFET.

common and the most important situation. Equation 5-22 indicates that this operation is governed by an extremely simple relation, and also explains why the MOSFET is often described as a *square-law device*. Equation 5-22 is plotted in Figure 5-6(b), and is the current-to-voltage *transfer characteristic* for the MOSFET operating in the saturation regime. This transfer characteristic is identical in shape to the curve plotted in Figure 5-6(a) for equivalent calibration of the voltage axes.

5-1.5 Universal Transfer Characteristics

Treating the MOSFET simplistically as a parallel-plate capacitor with a lateral current and a resulting *IR* drop in one plate neglects important considerations, as we shall see. Nonetheless, Equation 5-22 is extremely useful. Figure 5-7 presents experimental data from long-channel MOSFETs of the early 1960s. Plotting $\sqrt{|I_D|}$

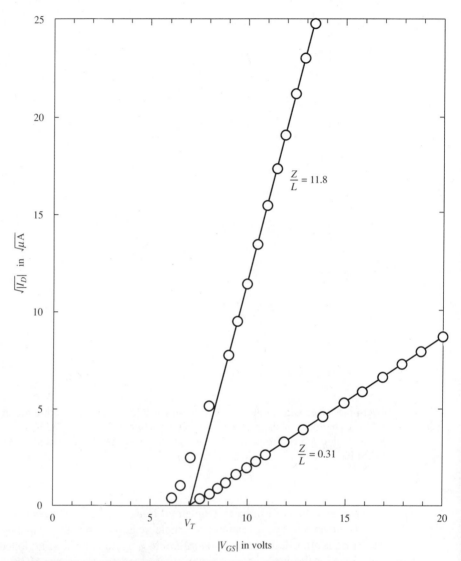

Figure 5-7 Experimental data from the long-channel MOSFETs of the early 1960s demonstrating the qualitative correctness of Equation 5-22.

versus V_{GS} should yield a straight line if Equation 5-22 is valid. This treatment has been accorded to two MOSFETs fabricated simultaneously, but having very different aspect ratios Z/L. Both curves have extensive linear regions. The curvature near the bottom, most evident in the higher-current device, exists because some conduction begins below the threshold voltage V_T, the phenomenon labeled *subthreshold conduction*. Nonetheless, because the two devices are identical in all respects except aspect ratio, extrapolating the linear portions of their *I–V* curves leads to a consistent and meaningful theoretical value of V_T. The value of seven volts observed for these early devices is extremely high by today's standards, and was the result of fabrication problems. A typical value today is of the order of one volt.

The device described up to this point is known as the *enhancement-mode,* or *E-mode,* MOSFET. It is designated thus because it has no channel at equilibrium; a gate voltage is required to create the inversion layer that serves as channel, and further increments of gate voltage *enhance* the conductivity of the channel.

It is possible, however, to create a channel that exists even in the device at equilibrium. This is usually done today by conventional doping of the region just under the gate oxide, with ion implantation being an especially favored technique. The result is a structure somewhat like that shown in cross section in Figure 5-8. The channel is bounded on the bottom by a *PN* junction, just as in a JFET. This time a *negative* voltage is applied to the gate, so that the MOS capacitor requires positive charge in the silicon. Donor-ion charge fulfills that requirement. In other words, a depletion layer is formed in the *N*-type channel region. Once again a positive V_{DS} gives rise to an *IR* drop in the channel. This time it results in a thicker depletion layer at the drain end of the channel than at the source end. For the same reason, the depletion layer of the *PN* junction below the channel is also thicker at the drain end of the channel. Because the application of a negative voltage to the gate causes current reduction through channel depletion, this kind of device is termed a *depletion-mode* or *D-mode* MOSFET. Realized in the way just described, the D-mode MOSFET is a kind of hybrid device, with a MOSFET-like upper boundary for the channel, and a JFET-like lower boundary.

Although the channel region in the D-mode MOSFET is a bit thicker than the inversion-layer channel of the E-mode device, the D-mode MOSFET channel still qualifies as "nearly two-dimensional," and as a result, the D-mode MOSFET is also very nearly a square-law device. Its essentially parabolic current-to-voltage transfer characteristic looks like that for the E-mode device, but is translated left-

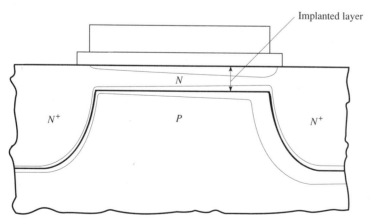

Figure 5-8 Schematic cross-sectional representation of a D-mode *N*-channel MOSFET, with a channel produced by ion implantation.

ward along the V_{GS} axis, as can be seen in Figure 5-9(a). Thus the D-mode device has two structure-determined electrical constants: There is the characteristic current I_{DSS} at which the output current saturates when $V_{GS} = 0$. Then there is the value $V_{GS} = V_P$ that is required to "pinch off" the channel completely, so that no current can flow. Under these conditions, the depletion layer produced by gate voltage has moved down to touch that of the *PN* junction, all the way from source to drain; as a result, the channel is fully depleted of carriers. The output characteristics that accompany the transfer characteristic in Figure 5-9(a) are shown in Figure 5-9(b). Only the parameter labeling of the curves distinguishes them from the E-mode characteristics plotted in Figure 5-5.

There is an important respect in which the D-mode MOSFET is distinguished

(a)
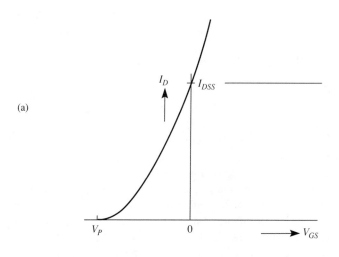

(b)
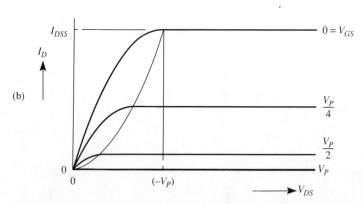

Figure 5-9 Current–voltage characteristics of a D-mode MOSFET. (a) Transfer characteristic, displaying the structure-determined constants V_P and I_{DSS}. (b) Output characteristics.

from the JFET, with the JFET being also a D-mode device. The D-mode MOSFET possesses an enhancement regime of operation that augments its depletion regime. When positive gate bias V_{GS} is applied, additional electrons are brought into the channel from the source, and channel conductivity increases. This fact is represented in Figure 5-9(a) by the extension of the transfer characteristic to the right of the I_D axis, or into the regime of positive V_{GS}. The square-law behavior continues. Thus there are additional curves at higher current that can be drawn in Figure 5-9(b). In the JFET, by contrast, where the channel is bounded both top and bottom by PN junctions, positive bias on the P-type gates (the top and bottom gates are usually common) will forward bias the gate junctions. Thus, for a gate bias of more than a few tenths of a volt, the gate junctions will conduct, and the high input resistance of the device is lost, normally an undesirable situation. As a practical matter, therefore, the transfer characteristic of the N-channel JFET can be extended only a short way into the positive-V_{GS} regime that defines enhancement-mode operation.

Exercise 5-9. Describe the output characteristics of the JFET with forward bias on the gate.

For small V_{GS} values, they would step upward just like those of the D-mode MOSFET, yielding a limited but sometimes useful E-mode regime. But for larger values, they would be closely spaced and would be accompanied by current from gate to source.

Exercise 5-10. Suppose that a large *negative* value of V_{GS} were applied to the D-mode MOSFET, causing the creation of an inversion layer of holes just under the oxide. Would the device start to conduct again, exhibiting a new regime of operation?

No. Extreme negative bias on the gate can indeed form an inversion layer of holes. But they will be unable to move to source, drain, or substrate, and will just "sit there," being confined in an isolating well, and not affecting terminal electrical properties.

Thus far we have considered only N-channel MOSFETs. But it is evident that by reversing all conductivity types, currents, and voltages, we can have E-mode and D-mode P-channel devices. In fact, for technological reasons P-channel devices were dominant early in the MOSFET era. The data in Figure 5-7, for example, were obtained from P-channel devices. It is possible to construct a transfer-characteristic diagram that summarizes all four possibilities—E and D, and N and P. Normalization once more is convenient. The D-mode device is best treated first because it possesses a characteristic current I_{DSS} as well as a characteristic voltage V_P. Using these quantities for normalization leads to the second-quadrant curve in Figure 5-10

732   THE MOSFET

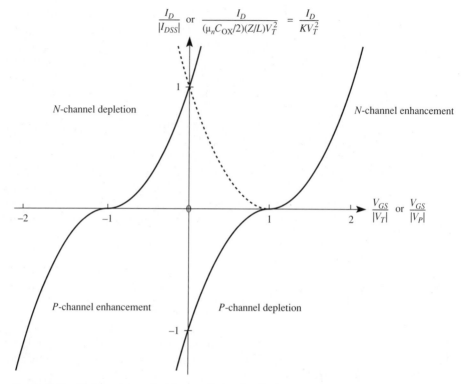

Figure 5-10 Universal transfer characteristics for the four possible kinds of MOSFET. Universality is achieved by normalization using V_P and I_{DSS} for the D-mode devices and V_T and KV_T^2 for the E-mode devices.

for the D-mode N-channel MOSFET. It is simply a normalized version of the curve shown in Figure 5-9(a). Since the D-mode P-channel MOSFET involves reversing currents and voltages, we merely construct a second curve in the fourth quadrant, one that is symmetric in the origin to the second-quadrant curve.

While the E-mode device does not have an explicit characteristic current for normalization, it does have an implicit characteristic current, $I_D = KV_T^2$. Using this and the voltage $V_{GS} = V_T$ for normalization, we complete Figure 5-10, displaying four curves that represent all possible MOSFETs.

Exercise 5-11. Explain the implicit characteristic current used to construct Figure 5-10.

Refer back to the unnormalized transfer characteristic in Figure 5-6(b). Take the case of the E-mode N-channel device. Its parabolic transfer characteristic has a leftward branch that is without physical meaning, but that serves to define a characteristic current because it intersects the I_D axis. Its intercept on the current axis is the same as the current value at $V_{GS} = 2V_T$, by virtue of parabolic symmetry. Substituting $V_{GS} = 2V_T$ into $I_D = K(V_{GS} - V_T)^2$ yields $I_D = KV_T^2$.

Performance advantages result from combining E-mode and D-mode devices in a circuit, and such *E-D* products are common today. However, the combination of *complementary* E-mode devices, or *N*-channel and *P*-channel E-mode devices, is even more advantageous. In fact, complementary MOSFET circuits, or *CMOS* circuits, constitute today's most rapidly growing technology. These matters are considered further in Section 5-1.7.

A number of MOSFET symbols are in common use. Three informative kinds are shown in Figure 5-11. Distinction between a D-mode and E-mode device is

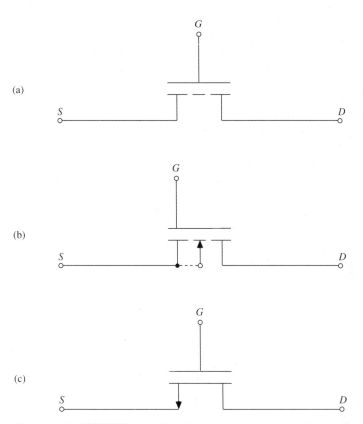

Figure 5-11 MOSFET symbols, with orientation corresponding to the diagrams in Figures 5-2 and 5-4. (a) E-mode device, indicated by the broken line representing the channel. D-mode device can be indicated by substituting a solid line. (b) More detailed symbol, indicating an *N*-channel device (arrow points from *P*-type substrate to channel), the substrate as a potential fourth terminal, and the "preferred source." (c) MOSFET analog of BJT symbols using arrow to designate source (emitter) terminal, and current direction.

sometimes maintained by using a solid line to represent the channel in the former case, and a broken line as in Figure 5-11(a) for the E-mode case. Another symbol, as in Figure 5-11(b), acknowledges that the substrate constitutes a fourth terminal of the MOSFET. Consistent with conditions in this section, we show the substrate connected (with a dotted line that is not part of the symbol) to the source. This symbol also indicates an N-channel device, with the arrow direction, as usual, from P to N, or from the P-type substrate to the channel. Finally, one sometimes wishes to denote a "preferred source," in spite of the intrinsic symmetry of the MOSFET, and this can be done by offsetting the gate terminal in the source direction.

A third symbol in common use is shown in Figure 5-11(c). Its virtues are simplicity and parallelism with the BJT symbol. That is, an arrow is used to designate the source terminal and also to indicate normal current direction, just as an arrow designates the BJT emitter terminal and its normal current direction. The line connecting source and drain of course represents the channel, and is sometimes made thicker to distinguish a D-mode from an E-mode device.

5-1.6 Transconductance

As was explained in Section 4-7.3, the rate of change of output current with respect to input voltage is an important characteristic property of a device and is termed *transconductance*. Its symbol is g_m. For the MOSFET,

$$g_m \equiv \frac{\partial I_D}{\partial V_{GS}}. \tag{5-23}$$

The partial derivative has conductance dimensions and has "across" or *trans* properties, to repeat, because it describes a result at the output port produced by a change at the input port. Transconductance exhibited in the saturation regime is of particular importance, and is of course simpler than for the curved regime. Differentiating Equation 5-22 yields

$$g_m = 2K(V_{GS} - V_T). \tag{5-24}$$

Experimental data confirming this simple linear relationship are shown in Figure 5-12, once again drawing data from P-channel devices of the early 1960s. The symbol S in Figure 5-12 stands for *siemens*, the unit of conductance, identified for many years as the *mho*.

Transconductance has great importance because it plays a large part in determining the switching speed of an active device, topics discussed in Sections 5-1.7 and 5-6. High-transconductance devices yield circuits capable of high-speed operation. Equations 5-17 and 5-24 make it evident that MOSFET transconductance is fixed by structure and technology. That is, it depends on properties more or less under the designer's control. This explains the preference for "N-MOS" over "P-MOS," to take advantage of the higher electron mobility. It is partly responsible for the constantly declining values of X_{OX} in recent decades (thus increasing C_{OX}). Finally, the presence of the factor Z/L in g_m means that the designer makes a g_m adjustment at the graphics console (or in an earlier era, at the drawing board).

BASIC MOSFET THEORY   735

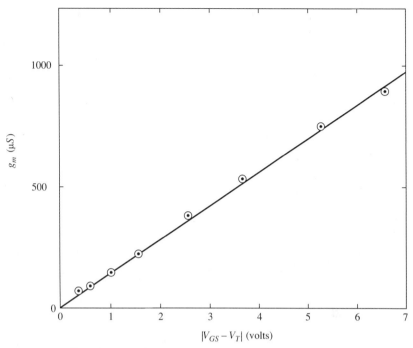

Figure 5-12 Experimental data on transconductance as a function of gate voltage, showing the linear relationship predicted by simple theory.

The designer's aim, however, is to make g_m just large enough and no larger, because overdesign would involve an oversized MOSFET, requiring more silicon area. Circuit costs are directly tied to circuit area.

This brings us to another fact that helped the MOSFET toward its present dominance. Equations of simple form, and coming from the simplistic analysis just presented, give an excellent description of MOSFET properties after empirical adjustment. Equation 5-22 is a prime example. As a result, computer design and optimization of MOS integrated circuits has been practical since the mid-1960s. Hence the necessary fine tuning of device sizes needed to give just enough and not too much transconductance was possible, if not easy. (It is a curious fact that the BJT is more difficult to describe with equal accuracy, but has transconductance reserves that make such description less necessary.)

As we saw, transconductance in the BJT is remarkably independent of structure and technology. It is independent of emitter area and even of the choice of semiconductor material! This is because BJT operation involves modulation of a naturally occurring potential barrier with invariant incremental properties. But in the MOSFET case, one is designing a variable resistor. Many BJT-MOSFET transconductance comparisons have been made over the years. These are not straightforward, because the two devices are so different, a subject addressed more fully in Section 5-6, but they typically indicate that the BJT is superior in transconductance

736 THE MOSFET

by a factor ranging from 10 to 100. The fact that the MOSFET is today's dominant device is the result of a large collection of other MOSFET attributes and advantages that will be treated later. The designer, seeking always to have the best of both worlds, is today introducing the BJT into MOSFET circuits—especially CMOS circuits—in places where high transconductance is crucial, and the resulting *BiCMOS* products are receiving increasing attention.

5-1.7 Inverter Options

Most of the world's transistors, whether BJT or MOSFET, are used as switches, and in most such applications, fast switching is earnestly desired. A qualitative discussion of factors affecting the switching speeds of simple MOSFET voltage-inverter circuits is in order here. The first circuit is shown in Figure 5-13(a), consisting of the E-mode MOSFET M_1 and a resistor as a load device. The device M_1 in the indicated role is often termed a *driver*. Examine first the *turn-on* properties of the

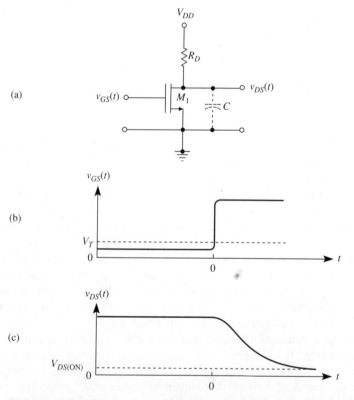

Figure 5-13 MOSFET inverter with resistive load. (a) Schematic circuit, showing parasitic capacitance C shunting output port. (b) Idealized input *turn-on* waveform. (c) Output *turn-off* waveform, showing time delay connected with discharging C.

inverter. Initially the MOSFET is OFF, with $v_{GS} < V_T$, and input voltage LOW. As a result, the output voltage is HIGH, with v_{DS} essentially equal to V_{DD}. Inevitably, there is a parasitic capacitance C shunting the output port, and this capacitor is fully charged to V_{DD}.

Exercise 5-12. What accounts physically for the capacitance C in Figure 5-13(a)?

The capacitance C is a parallel combination of three parasitic capacitances. First is the output capacitance of M_1, mainly depletion-layer capacitance (see Section 3-8.3) associated with the drain junction that can be seen in Figure 5-1. Second is capacitance between the conductor that extends from the output terminal of M_1 to the input terminal of the next stage and the material that underlies the conductor. The underlying material is typically at ground potential and separated from the conductor by a layer of SiO_2. The third and last contribution comes from the capacitance shunting the input port of the next stage. In the likely event that the next-stage driver device is another MOSFET, this element is essentially its MOS input capacitance.

From this description it is clear that the parasitic capacitance is voltage-dependent, but that fact is not overly important in the present qualitative discussion. Now let the input signal shown in Figure 15-13(b) be applied, a positive-going voltage step that takes the device from below threshold to well above threshold in a very short time. The speed at which the output voltage $v_{DS}(t)$, shown in Figure 5-13(c), can fall from the HIGH of V_{DD} to its LOW value is fixed by the time required to discharge C. The only agency for discharging C is M_1. The greater the increase of drain current for an increment of input voltage, which is to say, the greater the transconductance of M_1, the faster M_1 can discharge C. In sum, $i_D(t)$ determines discharge time, and $i_D(t)$ is related to $v_{GS}(t)$ by g_m.

To complete the switching process, let a downward voltage step, equally ideal, turn off M_1. At this point M_1 is essentially an open circuit. The only agency for charging C back up to V_{DD} is the load device R_D, often called a "pull-up" device. Treating C as a fixed or voltage-independent capacitor for purposes of approximation, we clearly see that the *turn-off* time at the output port of the inverter is characterized by $R_D C$, with an exponential rise resulting.

Exercise 5-13. Why not simply reduce R_D to achieve fast *turn-off*?

The total voltage change, or *logic swing,* at the output port of the inverter is given by $R_D I_{D(ON)}$, and should not be arbitrarily reduced.

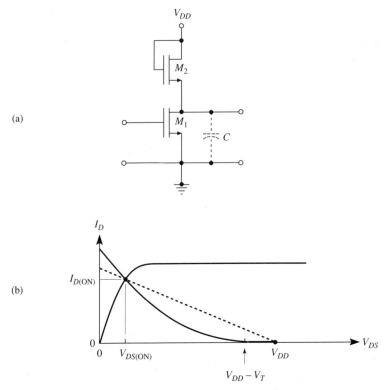

Figure 5-14 MOSFET inverter with saturated load. (a) Schematic circuit. (b) Output plane for driver device M_1, showing nonlinear loadline (solid line) of saturated-load device M_2, and for comparison, the linear loadline (dashed line) of a resistive load as in Figure 5-13.

Resistive loads have rarely been used in MOS circuits, however, especially in integrated circuits. The diode-connected E-mode MOSFET has been extensively used instead. (See Problem A5-7.) Connecting gate to drain yields a nonlinear resistive diode that begins to conduct at approximately the voltage V_T. This *saturated load* is shown as the upper device M_2 in Figure 5-14(a). With this arrangement, *turn-on* of the inverter is virtually unaltered, since we have seen that the driver M_1 dominates that case. But *turn-off* by contrast is now over an order of magnitude slower! Figure 5-14(b) shows the output plane for M_1, with the characteristic of M_2 entered as a load line. For comparison, the R_D characteristic is also shown, with a dashed line. The voltage difference $V_{DD} - V_{DS(ON)}$ constitutes the logic swing discussed in Exercise 5-13, so Figure 5-14(b) constitutes a graphical presentation of the point made there verbally.

Exercise 5-14. Explain qualitatively why the saturated load is so inferior to a resistive load in the *turn-off* phase of switching.

With a resistive load, the charging of C is accomplished by a current that declines linearly with capacitor voltage. With the saturated load, the capacitor-charging current declines *faster* than linearly, and essentially vanishes at $v_{DS} = V_{DD} - V_T$.

It might be surprising that an option as inferior in performance as the saturated load has been used so extensively. But it is small in area, and hence cost, and its fabrication is totally compatible with that of an active device such as M_1. Also, there are some applications (inexpensive calculators, for example) where the importance of cost far outweighs that of speed.

The advent of ion implantation made it feasible, as noted earlier, to create D-mode devices as companions to E-mode devices in the same integrated circuit. The resulting *depletion-mode load* outperforms the saturated load by a wide margin [13]. This *enhancement-depletion,* or *E-D,* combination is extensively used today. An E-D inverter is shown in Figure 5-15(a). The substantial benefit here, of course, is that the parasitic capacitance C is now in principle charged at *constant current,* as can be seen in Figure 5-15(b), and hence the charging curve is approximately a steep ramp rather than a saturating curve.

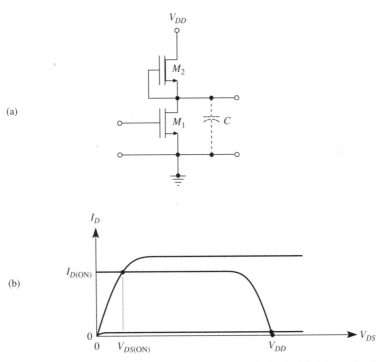

Figure 5-15 MOSFET inverter with depletion-mode load. (a) Schematic circuit. (b) Output plane for driver device M_1, showing nonlinear loadline of depletion-mode device M_2.

It is actually possible to achieve a relationship as favorable as that depicted in Figure 5-15(b) by using a pair of separate or *discrete* devices in the circuit of Figure 5-15(a). However, for compelling reasons of technology and economy this is rarely done. When instead the two devices are fabricated on the same substrate as is usual (in an integrated circuit, in other words), a phenomenon known as *body effect* (see Section 5-4.3) causes the I–V characteristic of the depletion-mode load to be significantly degraded, roughly approaching linearity. This real-world outcome is treated numerically in Section 5-5.4.

The final inverter option incorporates an *active load*. It is a P-channel E-mode device that is under control of the input signal along with the N-channel driver. These two devices are homologous, or *complementary*, and in an inverter they are the basis of *CMOS* technology. Such an inverter is shown in Figure 5-16(a). Let us start with input voltage LOW again. This causes the N-channel device M_1 to be OFF, while the P-channel device M_2 is in its conducting state, as is shown by solid lines in Figure 5-16(b). As a result, the output voltage is HIGH, approximately V_{DD}, as indicated by the intersection of the solid lines. The up-transition in input voltage then discharges C as before, with growing current as switching proceeds.

When M_1 is turned ON, the output voltage $V_{DS(ON)}$ approximates zero, as is shown by the intersection of the dashed lines in Figure 5-16(b). On an integrated

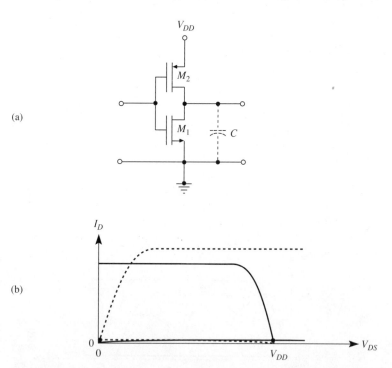

Figure 5-16 CMOS inverter. (a) Schematic circuit, showing N-channel driver device M_1 and P-channel pull-up device M_2. (b) Output plane for driver device M_1, showing active-load characteristic, and an intersection point for each state of the inverter.

circuit, the V_{DD}-bus and ground-bus lines are often parallel, resembling "rails," and this very desirable switching property is therefore described as a "rail-to-rail logic swing." In contrast, note that $V_{DS(ON)}$ in Figures 5-13(b), 5-14(b), and 5-15(b) is well above zero. And as a further shortcoming, in Figure 5-14(b) the OFF voltage is lower than V_{DD} as well.

An additional important benefit of the CMOS configuration enters during the *turn-off* phase of operation. This time M_2 is able to *charge C* with growing current as switching proceeds, just as C had been discharged with growing current. As a result, *turn-off* speed approximates *turn-on* speed.

Yet another advantage of CMOS circuitry is near-zero *standby current*. This is the static current through the inverter when the driver device is ON. In Figure 5-16(b), both currents (curve intersections) are actually much smaller than the diagram suggests, because they have been exaggerated for clarity. Note also that in both Figures 5-14(b) and 5-15(b), $I_{D(ON)}$ is large. There is however, a brief time during the switching of a CMOS inverter when both devices conduct. Hence, switching is accompanied by a pulse of current, known as *totempole current*, that is inevitable. (This term arose because one device sits on top of the other in a CMOS inverter.) As a result of totempole current, power dissipation in a CMOS inverter increases rapidly as switching frequency is increased. But for low-frequency applications, power dissipation can be very low.

5-2 MOS-CAPACITOR PHENOMENA

The rudimentary model of Section 5-1 treats the MOS capacitor as a simple parallel-plate capacitor. In fact, the situation is appreciably more complicated than that. A more accurate model of the MOS-capacitor system incorporates a parallel-plate capacitor (the oxide portion) that is in series with the parallel combination of two voltage-dependent capacitors (residing in the silicon portion). Section 5-3 examines the MOS capacitor in these terms, while the present section describes real-life complications, such as parasitic charges, that intrude into otherwise straightforward capacitive phenomena. The relative complexity of the true picture makes it surprising that the simple model of Section 5-1 is able, with a measure of empirical adjustment, to describe the MOSFET so accurately.

From the analytic point of view, the problem posed by the silicon portion of the MOS capacitor is *identical* to the problem posed by one side of a step junction at equilibrium. Consequently, methods developed in Chapter 3 are directly applicable to the present problem. We shall start with an approximate analysis, using the depletion approximation of Section 3-2. Then we shall move to the approximate-analytic model described in Section 3-7 for an appreciably more accurate description. Finally, it is important to note that still other physical mechanisms (such as foreign charged species residing on a semiconductor surface) can cause conditions in the silicon like those found in the MOS capacitor. For many years this was labeled the *semiconductor-surface problem*. It was study of the MOS capacitor that

5-2.1 Oxide–Silicon Boundary Conditions

The primary distinction between a conventional parallel-plate capacitor and the MOS capacitor is of course the substitution of a semiconducting region for a conducting region as one of the plates. An important result of this substitution is a region of charge in the semiconductor (which we take to be silicon in what follows) that is appreciably thicker than the corresponding layer in the conductor. Figure 5-17 presents this difference pictorially, showing a pair of metallic plates with field penetration, and hence space-charge layers, amounting to a few atomic layers at most. In Figure 5-17(b), on the other hand, is the MOS capacitor. Let us assume that the insulator here is silicon oxide. In typical situations, the layer of significant charge in the silicon is several times the oxide thickness, which certainly qualifies as a nonnegligible penetration by the associated electric field. Now let us examine conditions at the oxide–silicon boundary in Figure 5-17(b). We have here a problem of extreme practical importance that is nonetheless, as accurately predicted in Section 1-3, very nearly one-dimensional. Hence, combining Equations 1-34 and 1-36 gives

$$\frac{Q}{A} = \epsilon_{OX} E_{OX} = D_{OX}, \tag{5-25}$$

where ϵ_{OX}, the absolute permittivity of the oxide, has been substituted for ϵ_0, the permittivity of the free-space dielectric in the previous capacitor, and where E_{OX}

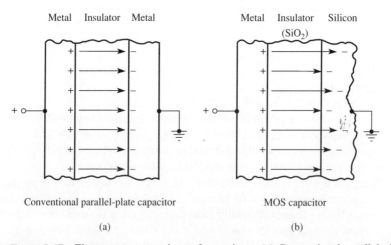

Figure 5-17 Elementary comparison of capacitors. (a) Conventional parallel-plate capacitor. (b) MOS capacitor, emphasizing a charge layer of appreciable thickness in the silicon.

and D_{OX} are the uniform values of electric field and electric displacement existing in the oxide. Recalling from Section 1-3.4 that it is the normal component of electric displacement D that is continuous through an interface such as the oxide–silicon boundary, we can write

$$D_{OX} = D, \tag{5-26}$$

or

$$\epsilon_{OX} E_{OX} = \epsilon E, \tag{5-27}$$

or

$$\kappa_{OX} \epsilon_0 E_{OX} = \kappa \epsilon_0 E, \tag{5-28}$$

where all symbols without subscripts are taken to apply to the silicon. The subscript "OX" of course refers to the oxide, and ϵ_0 is the permittivity of free space. Hence we have

$$E = \frac{\kappa_{OX}}{\kappa} E_{OX} \approx \frac{3.9}{11.7} E_{OX} = \frac{1}{3} E_{OX} \tag{5-29}$$

on opposite sides of the oxide–silicon interface, where the values of κ_{OX} and κ are taken from Table 2-2. The meaning of Equation 5-29 is that there exists a discontinuity in E at the interface. This situation can best be illustrated by adopting an arbitrary but specific MOS capacitor and bias condition. Computation of its field and potential profiles will then show the practical consequences of the boundary conditions just determined. Of particular importance is the computation of areal charge density in the silicon that follows from knowledge of boundary field values and Equation 5-25.

Exercise 5-15. Equations 5-25 through 5-29 are all scalar by virtue of the convenient one-dimensional simplicity of the MOS capacitor. Generalize Equation 5-25 into three-dimensional form, and identify the resulting expression.

A Gaussian pillbox placed with one face in the oxide and the other face in the neutral bulk region of the silicon would have flux through only one face—the face in the oxide. No flux will pass through the sides of the box or through the face in the neutral silicon, and these facts account for the form of Equation 5-25. In the general case we can assert that the integral over a closed surface of the electric displacement **D** can be related to the total charge Q_T within the surface by

$$Q_T = \oint_s \mathbf{D} \cdot d\mathbf{s},$$

where $d\mathbf{s}$ is a unit vector normal to the surface. This is a statement of Gauss's law.

5-2.2 Approximate Field and Potential Profiles

Assuming a specific example facilitates a quantitative but approximate examination of the MOS capacitor. Let us assume that it consists of an oxide layer 0.1 μm (1000 Å) thick on P-type silicon having a normalized bulk potential of $U_B = 12$ (where $U_B \equiv q\psi_B/kT$), corresponding to a net doping of $N_A - N_D \approx N_A = 1.6 \times 10^{15}/\mathrm{cm}^3$, and a resistivity of about 8 ohm·cm. Assume that a positive voltage is applied to the gate sufficient to produce a space-charge layer 0.5 μm thick, as shown in Figure 5-18. Let this be an ideal MOS capacitor. Its other plate is metallic, as is also indicated in Figure 5-18, and of course will become the gate of a MOSFET, as Figure 5-1 has indicated.

Consider electrostatic potential first. Let the subscript S (passed over so many times before on grounds of ambiguity) designate the silicon surface that we have referred to as the oxide–silicon interface. We have already used B (bulk) for the silicon interior, and will of course use G for the gate electrode or terminal. In these terms, the total electrostatic potential difference between the MOS-capacitor terminals becomes

$$\psi_G - \psi_B \equiv \psi_{GB}. \tag{5-30}$$

According to the rules of double-subscript notation then, this potential difference can also be written

$$\psi_{GB} = \psi_{GS} + \psi_{SB}. \tag{5-31}$$

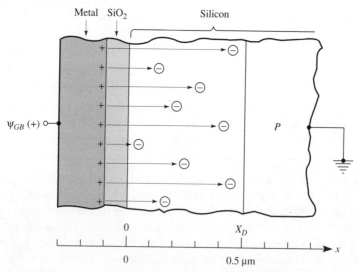

Figure 5-18 A specific arbitrary MOS capacitor and applied voltage. The silicon net doping is $1.6 \times 10^{15}/\mathrm{cm}^3$, and the oxide-layer thickness is 0.1 μm. Depletion-approximation space-charge-layer thickness is chosen to be 0.5 μm.

The first term on the right is potential drop through the oxide, the conventional-capacitor portion of the MOS device, and the second term represents potential drop in the silicon.

Now assume that the charge in the silicon is purely ionic and that it can be modeled using the depletion approximation. The voltage drop in the silicon can be computed by simply integrating electric field through the space-charge layer:

$$\psi_{SB} = + \int_0^{X_D} E \, dx, \tag{5-32}$$

where the plus sign on the integral is consistent with our use of the Ohm's-law and double-subscript sign conventions described in Section 3-3.1, and where X_D is the well-defined depletion-layer thickness associated with the depletion approximation. Assuming that no inversion layer is present, an assumption that can be checked at the end of the calculation, we may apply the relation between depletion-layer thickness and voltage drop that was developed for the one-sided step junction. Recall that in such a case, total voltage drop is approximately that on the lightly doped side. Making obvious substitutions to adapt Equation 3-24 to the present problem gives

$$\psi_{SB} = \frac{qN_A X_D^2}{2\epsilon} = \frac{(1.6 \times 10^{-19} \text{ C})(1.6 \times 10^{15} \text{ cm}^3)(5 \times 10^{-5} \text{ cm})^2}{2(1.035 \times 10^{-12} \text{ F/cm})}$$

$$= 0.31 \text{ V}. \tag{5-33}$$

The next step is to realize that the ionic charge in the silicon constitutes the charge on the negative plate of the MOS capacitor. The fact that this charge is distributed through a volume of significant thickness is unimportant with respect to computing potential drop *through the oxide*. That is, we can apply the law of the parallel-plate capacitor here because the problem is one-dimensional. We will get the same result for ψ_{GS} whether the charge on the negative plate lies in a mathematical plane at the interface or is spread through a thick region. Letting Q_S be the charge in the silicon per unit area of the MOS capacitor, and letting C_{OX} be capacitance per unit area of the oxide region,

$$\psi_{GS} = -\frac{Q_S}{C_{OX}} = -\frac{X_{OX}}{\epsilon_{OX}} Q_S = -\frac{X_{OX}}{\epsilon_{OX}} [-qN_A X_D] = \frac{X_{OX} q N_A X_D}{\epsilon_{OX}}, \tag{5-34}$$

where X_{OX} is the oxide thickness. The first minus sign enters Equation 5-34 because ψ_{GS} is applied at the gate side, whereas the charge per unit area in the silicon is on the opposite side. The second minus sign enters because the silicon charge is negative. Substituting presently applicable values into Equation 5-34 gives

$$\psi_{GS} = \frac{(10^{-5} \text{ cm})(1.6 \times 10^{-19} \text{ C})(1.6 \times 10^{15}/\text{cm}^3)(5 \times 10^{-5} \text{ cm})}{0.345 \times 10^{-12} \text{ F/cm}}$$

$$= 0.37 \text{ V}. \tag{5-35}$$

Thus, we now have the information necessary for a quantitative presentation of the potential profiles. The potential distribution, obtained by integrating the linear field profile of the depletion approximation, is of course parabolic.

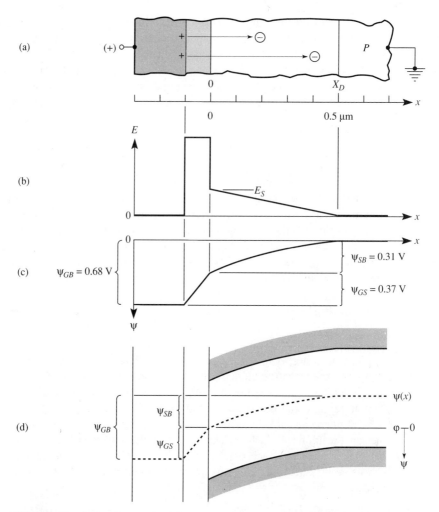

Figure 5-19 Approximate solution for the device shown in Figure 5-14. (a) Physical representation. (b) Electric-field profile. (c) Electrostatic-potential profile. (d) Band diagram.

Figure 5-19 summarizes the findings. In Figure 5-19(a) the MOS capacitor is shown with acceptor ions represented as circled symbols. Figure 5-19(b) shows the triangular field profile in the silicon that is consistent with the depletion approximation. The field discontinuity at the oxide–silicon interface is a consequence of Equation 5-29. We assume that no charges exist in the oxide, and hence all lines of force must pass entirely through it. Their parallelism, enforced by the one-dimensional geometry, translates into constant electric field in the oxide. At the oxide–metal interface, electric field drops abruptly to zero because lines of force cannot penetrate far into a metal. Turning to Figure 5-19(c), we observe a parabolic potential profile in the silicon (depletion approximation again), a slope discontinuity by a factor of three at the oxide–silicon interface, consistent with Figure 5-19(b), and a linearly rising potential through the oxide. Note that the downward direction has been chosen as positive for ψ, for consistency with the band-diagram convention. The neutral silicon bulk and the field plate are equipotential regions, as $\psi(x)$ clearly shows. Since a band diagram is a potential plot, we can go directly from Figure 5-19(c) to a band diagram for the MOS capacitor under bias. The only additional information that is needed is the bulk potential ψ_B. Let us use the majority-carrier Fermi level in the bulk ϕ_B as the reference for all potentials. The bulk potential ψ_B is negative for this *P*-type case, and can be obtained from Equation 2-34 as

$$\psi_B = -\frac{kT}{q} \ln \frac{p_n}{n_i} = -(0.0257 \text{ V}) \ln\left(\frac{1.6 \times 10^{15}/\text{cm}^3}{10^{10}/\text{cm}^3}\right) = -0.31 \text{ V}. \quad (5\text{-}36)$$

Combining this result with those in Figures 5-19(a) through 5-19(c) leads to the band diagram in Figure 5-19(d) for the ideal MOS capacitor under bias, where the bias, computed from Equation 5-31, amounts to

$$\psi_{GB} = \psi_{GS} + \psi_{SB} = 0.68 \text{ V}. \quad (5\text{-}37)$$

For reasons that are evident in Figure 5-19(d), the quantity ψ_{SB}, the amount of the potential drop in the silicon, is often referred to as the *band bending* in the MOS-capacitor problem, or in the equivalent surface problem. Notice that the dashed "intrinsic-line" profile is identical to that in Figure 5-19(c), because the bulk Fermi level has been selected as the potential reference.

Exercise 5-16. How can an electric-field discontinuity exist at the oxide–silicon interface?

The oxide and the silicon both possess significantly large dielectric constants, and hence a sheet of charge develops at the interface that is analogous to the charge sheets shown in Figures 1-11(b) and 1-12(b).

Exercise 5-17. What is the sign of the charge sheet present (but not shown) in Figure 5-19(a), and why is it there?

The charge sheet is negative and is completely analogous to the central charge sheet in Figure 1-13(a). In both cases, the negative charge is present because positive bias is applied to the left-hand capacitor plate and the dielectric material of greater permittivity is on the right.

Now we are in a position to confirm that there is no inversion layer at the silicon surface. Note in Figure 5-19(d) that the surface potential ψ_S is near zero. When the surface potential is precisely zero, very nearly the case in Figure 5-19 (by coincidence, since the initial values were arbitrarily chosen), the surface is said to be at the *threshold of weak inversion*. It was explained in Section 5-1.2 that for the presence of an inversion layer, it is necessary for the gate-to-bulk voltage (and hence ψ_{GB}) to have a value beyond the threshold voltage V_T. And the threshold voltage in turn is defined as the voltage that causes electron density (volumetric density!) at the silicon surface (at the oxide–silicon interface) to equal hole density in the bulk region. With our early assumption of band symmetry, this tells us that the surface potential in Figure 5-19(a) must be as far below the Fermi level as the bulk potential is above it, or

$$\psi_S = -\psi_B, \tag{5-38}$$

a succinct statement of the threshold condition. But in the arbitrary situation of Figure 5-19, the band bending is too small by about a factor of two to meet this condition. Hence, the case represented is far below threshold.

5-2.3 Accurate Band Diagram

Now let us return to the tools of Section 3-7, methods for approximate-analytic modeling of semiconductor regions at equilibrium. (These will be expanded in Section 5-3.1.) We shall carry out an accurate examination of the problem that has just been analyzed by means of the depletion approximation, and will compare the results. The only differences that will exist between the two are in the silicon region of the device because the rest remains idealized. Let us continue with the case of $U_B = -12$, or about $1.6 \times 10^{15}/\text{cm}^3$ of net P-type doping. To compare an approximate analysis with an accurate one, the first thing to decide is what should be held constant from one to the other. When the depletion-approximation descriptions in Chapter 3 were to be compared to exact solutions for step junctions, we chose to hold contact potential, or the total voltage drop in the equilibrium junction, constant. This choice produced the greatest congruence of the band diagrams. In the present case, however, it is reasonable to hold field in the oxide constant, because we have not as yet altered that portion of the model. Given this different choice, it follows that peak field in the silicon, the value just inside the oxide–silicon interface, will be unchanged, because Equation 5-29 still applies. The result will be a small change in the charge, field, and potential profiles in the silicon. Let us start at the oxide–silicon interface and work toward the silicon interior.

Given the knowledge that the area of the triangle in Figure 5-19(b) is equal to ψ_{SB}, the voltage drop in the silicon, it is easy to find the peak electric field in the silicon for this depletion-approximation case. It becomes

$$E_S = \frac{2\psi_{SB}}{X_D} = \frac{2(0.31\text{ V})}{0.5 \times 10^{-4}\text{ cm}} = 1.24 \times 10^4\text{ V/cm}. \tag{5-39}$$

Next, it is necessary to convert this field value to the fully normalized form plotted on the ordinate of Figure 3-43, in order to determine which regime of this diagram applies to the present problem. Since the normalizing quantity for potential is kT/q and that for distance is L_D, it follows that fully normalized electric field at the silicon surface is

$$\left.\frac{dW}{d(x/L_D)}\right|_S = \frac{E_S \, L_D}{kT/q}. \tag{5-40}$$

Since $L_D \approx L_{De}$ in our extrinsic sample, and using Equation 3-71 for normalized electric field at the silicon surface, we have

$$E_S\left(\frac{q}{kT}\right)\left[\frac{\epsilon}{qN}\frac{kT}{q}\right]^{1/2} = E_S\left[\frac{\epsilon}{qN}\frac{q}{kT}\right]^{1/2}, \tag{5-41}$$

where N is net doping. Evaluating the right-hand side of this expression gives

$$(1.24 \times 10^4 \text{ V/cm}) \times \left[\frac{1.035 \times 10^{-12} \text{ F/cm}}{(1.602 \times 10^{-19} \text{ C})(1.6 \times 10^{15}/\text{cm}^3)(0.02566 \text{ V})}\right]^{1/2}$$
$$= 4.92. \tag{5-42}$$

Figure 3-43 shows the near-congruence of the exact solution and the linear electric-field asymptote at this value, so it follows from Equation 3-82 that the oxide–silicon interface falls at $-x/L_D = 4.92$. Figure 3-44 shows that the parabolic potential asymptote also applies at this position (as it must, since the field and potential asymptotes correlate), so from Equation 3-90,

$$W = \frac{1}{2}\left(-\frac{x}{L_D}\right)^2 + 1 = 13.1. \tag{5-43}$$

Thus, using $kT/q = 0.02566$ V, we have

$$\psi_{SB} = \frac{kT}{q} W = 0.336 \text{ V}, \tag{5-44}$$

a value about 8% higher than the value obtained from the depletion approximation. The increment is of course associated with the extra area under the field profile near the depletion-layer boundary, shown in Figure 5-20(a), that was also emphasized in Figure 3-43. The corresponding charge-density and potential profiles are shown in Figures 5-20(b) and (c). Dotted lines are used to indicate the depletion-approximation results in these three diagrams for comparison with the more exact profiles. Finally, an adjusted band diagram is presented in Figure 5-20(d), one that differs only slightly from that of Figure 5-19(d).

Exercise 5-18. Account quantitatively for the 8% greater band bending in the accurate analysis than in the approximate analysis.

One normalized unit, or 0.02566 V (one thermal voltage), is about 8% of the band bending calculated in either case.

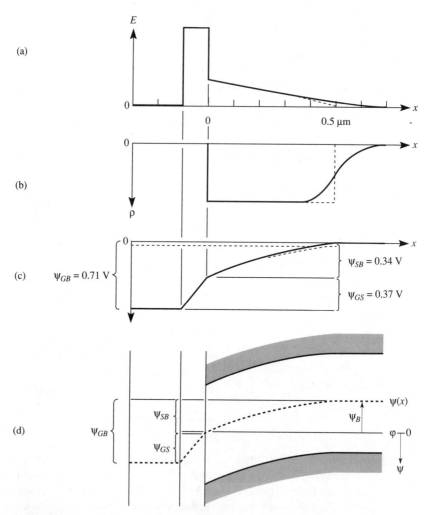

Figure 5-20 Accurate solution (solid lines) for the device shown in Figure 5-14, keeping field in the oxide the same as in the approximate solution (dotted lines) of Figure 5-15. (a) Electric-field profile. (b) Space-charge profile. (c) Electrostatic-potential profile. (d) Band diagram.

Exercise 5-19. Account for the difference also by comparing Equation 5-43 with its approximate analog.

The unity term in the normalized Equation 5-43 contributes one normalized unit to the depletion-approximation-replacement (DAR) solution. Dropping the unity term yields the depletion-approximation (DA) equation.

In Figure 5-20(d), examine the region of the silicon near the oxide–silicon interface. Band bending causes the conduction-band edge to approach the Fermi level, going a bit beyond the threshold of weak inversion. This guarantees a few surplus electrons there, an incipient inversion layer, but at extremely low densities. Band bending causes the valence-band edge to move away from the Fermi level, so that acceptor ionization is even more complete than in the neutral sample. Thus, negative acceptor ions and negative mobile carriers (the beginning of an inversion layer) coexist in this region. They do not in any way "interfere" with one another and are in fact additive. At threshold, right at the surface, the volumetric space-charge density has twice the value it has in the fully depleted region close to (but outside) the inversion layer. Half the charge-density contribution right at the silicon surface is therefore ionic, negatively charged acceptors; the remainder consists of electrons.

5-2.4 Barrier-Height Difference

By the 1950s there was a good understanding of the principles governing ideal semiconductor surfaces, and by extension, the ideal MOS capacitor that has occupied us up to this point. The MOS capacitor made its appearance at the end of the decade. It was in the 1960s that a comparable grasp of the nonideal MOS capacitor was acquired, a step that was essential to practical MOS integrated-circuit technology. The fact that the MOS market segment swept past the older bipolar integrated-circuit technology in overall importance is convincing evidence that these practical problems were solved. Now we examine a feature that varies with the identity of the semiconductor, the insulator, and the gate material in the MOS capacitor.

When an electron is withdrawn from the surface of a conductive solid, an "image" charge appears in the solid and the resulting pair of opposite-sign charges are mutually attractive. Consequently, to extract the electron we must do work on it. If it is removed to a distance such that the image force is essentially zero, then, at rest, it resides at an energy higher than the Fermi level in the solid. (The necessary distance, to be sure, is small.) This higher energy is termed the *vacuum level*. The energy interval from the vacuum level to the Fermi level is termed the *work function*, Φ. Thus, there exists a barrier at a silicon surface that prevents electrons from escaping, and the work function is a measure of the height of that barrier. For silicon the work function is about 4.8 eV and appears not to be a function of doping [25].

Figure 5-21 shows a band diagram for silicon in the ideal *flatband* condition, or the condition where $\psi_{SB} = 0$. Indicated there, though not to scale, is the interval Φ, along with two other intervals often cited in surface work; these are the *photoelectric threshold* and the *electron affinity* χ. The former can be appreciated as follows. Let photons of gradually increasing energy be incident upon the surface of the sample shown in Figure 5-21. Since electrons are most abundantly available in the valence band of the semiconductor sample, we will see a sharp rise in the rate of electron ejection from the sample (that is, in the rate of photoelectron production) when photon energy corresponds to the energy required to raise an electron from the

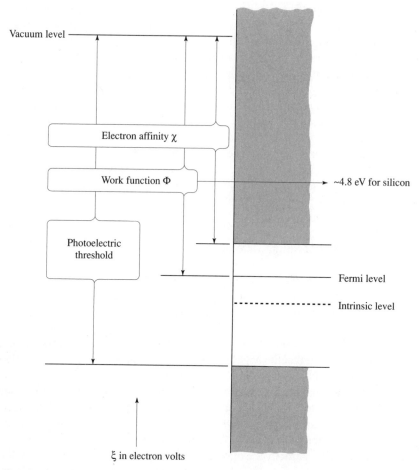

Figure 5-21 Qualitative band diagram for the surface region of a semiconductor sample, defining electron affinity, work function, and photoelectric threshold.

top edge of the valence band to the vacuum level, at which energy the departing electrons have been liberated from the solid. This is the photoelectric threshold. The electron affinity is defined as the energy that binds a conduction electron at the edge of the conduction band to the solid—or the energy a conduction electron must acquire to reach the vacuum level and hence to escape.

Let us concentrate on the work function Φ. As in our previous band diagrams, we can convert from electron energy in eV to electrostatic potential in volts. Work function can be cited in either, but which one is intended must be clearly stated to avoid the ever-present sign ambiguity. Since the literature is inconsistent on this score, let us adopt the symbol H to stand exclusively for barrier height in *volts*.

In the MOS capacitor we are concerned not with free surfaces, but with two interfaces—namely, the oxide–silicon interface and the oxide–metal (or other gate material) interface. A barrier to electron passage into the oxide exists at each interface. The oxide is an insulator with a gap on the order of 8 V. One barrier height of interest to us is that from the Fermi level in the silicon bulk to the conduction-band edge in the oxide. An electron in the silicon acquiring sufficient energy can enter the conduction band of the oxide and can then penetrate the normally insulating layer, arriving at the gate electrode. Such an energetic electron is sometimes termed a "hot" electron, as noted in Section 2-5.4.

When an oxide is grown on silicon, its conduction-band edge is separated from the valence-band edge of the silicon by a potential interval of 4.35 V [26]. Since the energy gap is 1.12 V in silicon, it follows that the conduction bands of the insulator and the semiconductor are separated by 3.23 V on the potential scale. These values are apparently independent of doping and crystalline orientation [26]. Thus, the band structures of the oxide and the silicon have a fixed relationship. When surface potential (the mid-gap potential on the silicon side of the interface) is manipulated in the MOS capacitor, all four band edges move up and down together along with the surface potential.

Now recall that at equilibrium, the Fermi level in the silicon remains constant or "flat" irrespective of band bending in the silicon. It follows, of course, that the spacing (in terms of potential) of the conduction-band edge in the oxide from the Fermi level in the silicon will change as band bending is altered by bias adjustments on the MOS capacitor. So that the barrier height H_B will be unequivocally defined, let us therefore specify that the silicon must be in the *flatband* condition, $\psi_{SB} = 0$, when we observe the interval between the Fermi level in the silicon and the conduction band in the oxide. For the MOS capacitor shown in Figure 5-22(a), the desired flatband situation is depicted in Figure 5-22(b). It is evident on the diagram that the desired barrier height H_B can be calculated by adding the values of three quantities: ψ_B from Equation 5-36, half the gap, and the spacing of the conduction-band edges, $\psi_{CO} - \psi_{CS}$. The last item is written this way because we wish to refer all elements of the MOS capacitor *to* the silicon. Thus

$$H_B = \psi_{CO} - \phi_B = \psi_B - \frac{\psi_G}{2} + (\psi_{CO} - \psi_{CS})$$
$$= -0.31 \text{ V} - 0.56 \text{ V} - 3.23 \text{ V} = -4.10 \text{ V}. \tag{5-45}$$

(As usual, we have let $\phi_B \equiv 0$.) Suppose we assume an aluminum field plate, or gate electrode. Its barrier height has been given in the literature [26] as

$$H_G = \psi_{CO} - \phi_G = -3.20 \text{ V}. \tag{5-46}$$

This is the final value needed for the construction of Figure 5-22(b).

Normally we will not work with the individual barrier heights, but rather with barrier-height *difference*, H_D. Let us define this quantity as the position of the gate Fermi level ϕ_G, referred under flatband conditions to the Fermi level in the silicon, $\phi_B = 0$. That is,

$$H_D \equiv \phi_G|_{FB} = \phi_G - \phi_B, \tag{5-47}$$

a quantity that obviously can be positive *or* negative. Adding and subtracting the

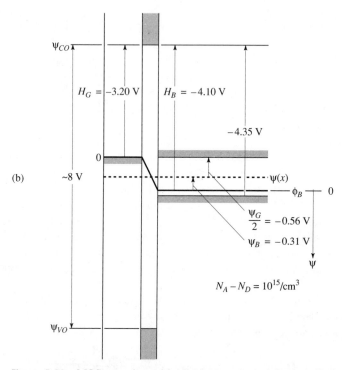

Figure 5-22 MOS capacitor with aluminum gate and P-type silicon substrate. (a) Physical representation. (b) Band diagram for flatband condition, assuming that barrier-height difference is the only nonideal feature present.

oxide conduction-band-edge position ψ_{CO} on the right-hand side of Equation 5-47 gives us

$$H_D = (\psi_{CO} - \phi_B) - (\psi_{CO} - \phi_G) = H_B - H_G$$
$$= -4.10 \text{ V} - (-3.20 \text{ V}) = -0.90 \text{ V}. \tag{5-48}$$

The quantity H_D will be used in what follows, and no further use will be made of the work function Φ. Because H_D depends on H_B, which in turn depends on doping [refer again to Figure 5-22(b)], it follows that H_D is doping-dependent. Figure 5-23(a) shows how H_D varies with doping for the case of silicon, with gold and aluminum field plates. From Figure 5-22(b), the equations for these curves can

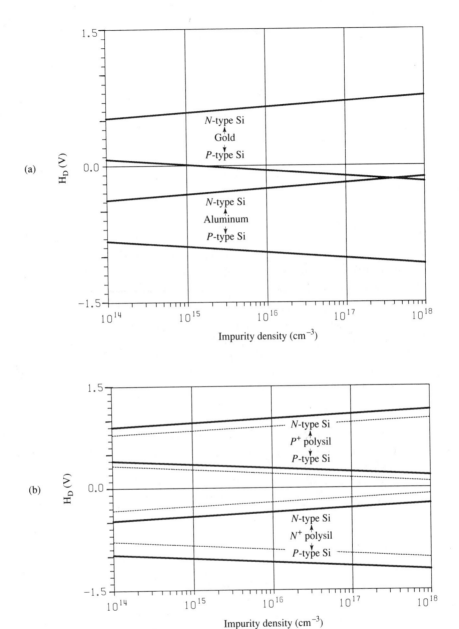

Figure 5-23 Barrier-height difference H_D as a function of doping type and density. (a) Data for gold and aluminum field plates. (After Deal and Snow [26] with permission, copyright 1966, Pergamon Press Ltd.) (b) Data for P^+ and N^+ polysil field plates. (Top and bottom solid lines are from Werner [27] with permission, copyright 1974, Pergamon Press Ltd.) Center two solid lines are calculated from top and bottom lines by adding and subtracting, respectively, the value 1.38 V (also obtained from the data of Werner [27]). Dotted lines show approximate values used in SPICE modeling of the MOS system.

be written in terms of equilibrium electron density n_0 (instead of net doping) as

$$H_D = H_B - H_G = \frac{kT}{q} \ln \frac{n_0}{n_i} - 3.79 \text{ V} - H_G. \tag{5-49}$$

For our P-type sample, $n_0 = 6.25 \times 10^4/\text{cm}^3$, so that the logarithmic term is negative. The value of H_G for an aluminum field plate is -3.20 V, as noted above, and is -4.10 V for gold [26].

Exercise 5-20. Explain the relationship of Equation 5-49 and Figure 5-22(b).

The first term on the right-hand side of Equation 5-49 is a way of writing the bulk potential ψ_B, in view of Equation 2-33. This is the distance from ϕ_B to $\psi(x)$ in Figure 5-22(b). The second term is the distance from $\psi(x)$ to ψ_{CO}, or $(-0.435 \text{ V}) - (\psi_G/2) = (-0.435 \text{ V}) - (0.056 \text{ V}) = -3.79 \text{ V}.$

In modern integrated-circuit practice (since approximately 1970), aluminum and gold field plates have been replaced by polycrystalline silicon, or "polysil." The polysil is usually heavily doped to simulate the metallic behavior that is desired in a field plate and also in the interconnection pattern that is fabricated using the same deposited-polysil layer. The dependence of H_D upon doping for P^+ and N^+ polysil field plates is shown by the solid lines in Figure 5-23(b). These curves are based upon experimental determinations made by Werner [27] for P^+ polysil boron-doped to $1 \times 10^{20}/\text{cm}^3$, and for N^+ polysil phosphorus-doped to $3 \times 10^{20}/\text{cm}^3$. These curves too can be represented by Equation 5-49 by employing an appropriate value for H_G. The Fermi level is near the conduction-band edge in N^+ silicon and near the valence-band edge in P^+ silicon; we can in fact take the values of H_G as approximately -3.23 V and -4.35 V for the N^+ and P^+ cases, wherein net doping is $N \approx 2 \times 10^{19}/\text{cm}^3$. The dotted curves in Figure 5-23(b) show the corresponding approximate values used in SPICE modeling of the MOSFET, as explained in Section 5-5.1.

Exercise 5-21. Determine the barrier-height difference H_D for an MOS capacitor with a polysil gate. Assume the gate is heavily doped with acceptors, and that $H_G = -4.35$ V. Also assume that the device has the same oxide and silicon structure that was used in the preceding calculations. Comment on the properties of this MOS capacitor.

Using the value of $H_B = -4.10$ V given in Equation 5-45, then Equation 5-49 gives us

$$H_D = H_B - H_G = (-4.10 \text{ V}) - (-4.35 \text{ V}) = 0.25 \text{ V}.$$

Thus the barrier-height difference is positive. If this were the only factor present, it would be necessary to apply a quarter-volt *positive* bias to bring the device to the flatband voltage, and the threshold condition would occur at a still more positive bias.

Let us now remind ourselves that a voltmeter measures Fermi-level difference. The MOS capacitor depicted in Figure 5-22(a) was brought into the flatband condition by applying an external bias of -0.9 V to the gate with respect to the silicon, a voltage equal to the Fermi-level difference. When the capacitor is in this bias condition, the difference in *electrostatic potential* between gate and silicon (or bulk), the quantity ψ_{GB}, is zero. This fact is exhibited in Figure 5-22(b) by the horizontal dotted line representing $\psi(x)$. The coexistence of an externally measurable voltage difference and no electrostatic-potential difference is of course attributable to barrier-height difference H_D.

Now suppose we arbitrarily change the bias on the same MOS capacitor to $+0.2$ V. Because a positive voltage on the field plate will drive majority holes away from the silicon surface, this bias is clearly in the depletion-inversion direction. The result of applying such a bias is shown in the partial band diagram of Figure 5-24. Once again we will defer until the end of the calculation the question of whether an inversion layer is present. However, comparing Figure 5-24 with Figure 5-20(d) makes it clear that a well-depleted condition exists near the silicon surface, since that was true in the earlier case and the present case has even more band bending.

It is evident in Figure 5-24 that the external bias voltage V_{GB} can be written as the algebraic sum of H_D and the total electrostatic potential difference ψ_{GB}, an important result:

$$V_{GB} = H_D + \psi_{GB}. \tag{5-50}$$

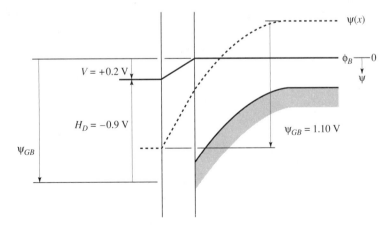

Figure 5-24 Band diagram (lower half) for the structure of Figure 5-18 with $V_{GB} = \psi_{GB} = +0.2$ V arbitrarily applied.

THE MOSFET

In the equipotential or flatband condition, by definition $\psi_{GB} = 0$. Thus it follows that when a nonzero barrier-height difference is the only nonideal factor, the bias voltage that must be applied to produce the flatband condition, a bias value called the *flatband voltage* V_{FB}, is equal to the barrier-height difference. In the present case, as illustrated in Figure 5-22,

$$V_{FB} = H_D = -0.9 \text{ V}.$$

To demonstrate the utility of Equation 5-50, let us ask this question: What amount of band bending ψ_{SB} exists under the conditions shown in Figure 5-24? Breaking ψ_{GB} into its two components in the manner of Equation 5-37, Equation 5-50 becomes

$$V_{GB} = H_D + \psi_{GS} + \psi_{SB}. \tag{5-52}$$

From Equation 5-34, $\psi_{GS} = -Q_S/C_{OX}$, where

$$C_{OX} = \frac{\kappa_{OX}\epsilon_0}{X_{OX}} = \frac{\epsilon_{OX}}{X_{OX}} = \frac{3.9(8.85 \times 10^{-14} \text{ F/cm})}{10^{-5} \text{ cm}} = 3.45 \times 10^{-8} \text{ F/cm}^2, \tag{5-53}$$

and ϵ_{OX} is the absolute dielectric permittivity of the oxide. On the other hand, the band bending ψ_{SB} is directly related to W_S, which is the normalized potential at the surface of the silicon. Normalized potential W is defined in general terms in Equation 3-79. In terms of W_S, the band bending in the silicon is

$$\psi_{SB} = -\frac{kT}{q} W_S. \tag{5-54}$$

The minus sign enters because W increases upward on the band diagram and ψ increases downward, as can be seen in Equation 3-79. Given the condition of strong depletion that exists here, Equation 3-90—or the left-hand equation in Equation 5-43—is valid, and is repeated here for convenience:

$$W = \frac{1}{2}\left(-\frac{x}{L_D}\right)^2 + 1. \tag{5-55}$$

Equation 3-82 is also valid, and can be rewritten here as

$$\left|\frac{dW}{d(x/L_D)}\right| = \left|\frac{dU}{d(x/L_D)}\right| = -\frac{x}{L_D}. \tag{5-56}$$

The magnitudes of both normalized-field expressions are used because W and U have opposite signs. (For our conventions, $-x/L_D$ is positive.) Placing the first normalized-field expression in Equation 5-55 gives us

$$W = \frac{1}{2}\left[\frac{dW}{d(x/L_D)}\right]^2 + 1. \tag{5-57}$$

Next, recall that

$$|Q_S| = \epsilon|E_S| = |D_S|, \tag{5-58}$$

where Q_S is the areal charge density in the silicon, previously used in Equation 5-34, while E_S and D_S are the electric field and displacement, respectively, just inside the silicon at the oxide–silicon interface. Absolute value has been indicated because sign must be dropped in what follows. From Equation 5-41, we may write

$$\left.\frac{dW}{d(x/L_D)}\right|_S = \frac{|Q_S|}{\epsilon}\left[\frac{\epsilon}{qN}\frac{q}{kT}\right]^{1/2} = \frac{|Q_S|}{K}, \qquad (5\text{-}59)$$

where evidently

$$K = \sqrt{\epsilon qN(kT/q)} = 2.61 \times 10^{-9} \text{ C/cm}^3. \qquad (5\text{-}60)$$

Substituting Equation 5-59 into Equation 5-57 yields

$$|W_S| = \frac{1}{2}\left(\frac{Q_S}{K}\right)^2 + 1, \qquad (5\text{-}61)$$

where W_S is of course normalized potential at the silicon surface (normalized surface potential). Placing Equations 5-34 and 5-61 in Equation 5-52 gives us

$$V_{GB} = H_D - \frac{Q_S}{C_{OX}} + \frac{kT}{q}\left[\frac{1}{2}\left(\frac{Q_S}{K}\right)^2 + 1\right], \qquad (5\text{-}62)$$

or

$$Q_S^2 - \frac{q}{kT}\frac{2K^2}{C_{OX}}Q_S + \frac{q}{kT}2K^2\left(\frac{kT}{q} + H_D - V_{GB}\right) = 0. \qquad (5\text{-}63)$$

Applying the quadratic formula yields

$$Q_S = \frac{q}{kT}\frac{K^2}{C_{OX}} \pm \sqrt{\left(\frac{q}{kT}\frac{K^2}{C_{OX}}\right)^2 - \frac{q}{kT}2K^2\left(\frac{kT}{q} + H_D - V_{GB}\right)}$$

$$= -1.74 \times 10^{-8} \text{ C/cm}^2. \qquad (5\text{-}64)$$

The negative sign on the radical must be taken because charge in the silicon is necessarily negative. Using Equations 5-34, 5-53, and 5-64 yields

$$\psi_{GS} = -\frac{Q_S}{C_{OX}} = 0.50 \text{ V}. \qquad (5\text{-}65)$$

Hence, from Equation 5-52,

$$\psi_{SB} = V_{GB} - H_D - \psi_{GS} = (0.2 \text{ V}) - (-0.9 \text{ V}) - (0.50 \text{ V}) = 0.60 \text{ V}. \qquad (5\text{-}66)$$

This value is just below the band-bending value that produces the threshold condition, or $|2\psi_B| = 0.62$ V in the present case, so the inversion-layer charge is very small and may be neglected.

As a check on the solution, let us calculate the electric field in the oxide:

$$E_{OX} = \frac{\psi_{GS}}{X_{OX}} = \frac{0.5 \text{ V}}{10^{-5} \text{ cm}} = 5.0 \times 10^4 \text{ V/cm}. \qquad (5\text{-}67)$$

As shown in Equation 5-29, the field at the silicon surface should be about three times smaller. Using Equation 5-58 and introducing appropriate signs,

$$E_S = \frac{-Q_S}{\epsilon} = \frac{1.74 \times 10^{-8} \text{ C/cm}^2}{1.04 \times 10^{-12} \text{ F/cm}} = 1.67 \times 10^4 \text{ V/cm}, \quad (5\text{-}68)$$

confirming the expectation.

Examination of Figure 5-24, which has just been treated quantitatively, shows an important point. The effect of a negative barrier-height difference, such as that in the aluminum–silicon system, is indeed to move the MOS capacitor's condition in the direction of depletion-inversion. This is one of several factors that accounted for the early exclusive use of P-channel MOS technology—involving devices made on N-type material and having hole inversion layers. Efforts to use P-type material, as in Figure 5-24, often resulted in devices with thresholds much more negative than the desired value, in many cases producing a D-mode FET when an E-mode FET was intended.

5-2.5 Interfacial Charge

As noted in Section 1-5.4, a departure from crystalline perfection in a silicon sample introduces electronic states that reside within the forbidden band. An example that is familiar at this point is an impurity atom placed substitutionally in the crystal. As also noted in Section 1-5.4, the surface of a crystal represents a gross interruption of crystalline perfection. In a primitive model of the surface we can visualize unsatisfied bonds, or "dangling bonds," associated with the atoms at the surface. Significant thought along these lines goes all the way back to the work of J. Willard Gibbs in the last century, although these ideas were not published until the first decade of this century [28]. In the early 1930s Igor E. Tamm put forward a quantitative theoretical model for the states associated with dangling bonds, leading to the expression "Tamm states," which is still in use [29]. In the late 1930s William Shockley offered a related but different model [30]. Increasing numbers of gifted theoreticians and experimentalists were attracted to the challenging surface problems; as it turned out, such studies lay directly on the path to the transistor. A useful survey of early research on surface states has been given by Many et al. [31].

Figure 5-25 offers a heuristic picture of dangling bonds at a silicon surface. Letting the short lines be interpreted as electrons, we can represent in the manner shown there the case where the silicon surface atoms (as well as the underlying atoms) all have the right number of electrons to ensure neutrality. If we employed some unspecified means to remove all of the "dangling" electrons at the surface, there would then be uncompensated positive charge associated with each surface atom. The diagram suggests that the charge would be $+2qN_S$, where N_S is the density of surface atoms, but we must remind ourselves at this point that the picture used here is but a two-dimensional representation of a three-dimensional monocrystal.

The states portrayed in this simplistic picture can be described as *donorlike*.

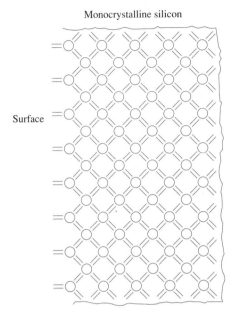

Figure 5-25 Pictorial representation of single-crystal surface. Each silicon atom is assigned the number of electrons necessary for neutrality, and the result is "dangling" or unsatisfied bonds at the surface.

That is, they carry a positive charge when they are empty of electrons and are neutral when filled by electrons. However, with the numerous differing possibilities for crystalline imperfection at a surface, we should not be surprised to learn that surface states may be either donorlike or acceptorlike. The latter states are neutral when empty and negatively charged when filled by an electron. Likewise we would expect these varied kinds of surface states to be distributed through a range of energies within the bandgap at the surface. Continuing with simplistic descriptions, we can represent the situation in the manner of Figure 5-26, where short lines now represent the surface states themselves, which are also assumed to be uniformly distributed in energy.

Next, assume that these states are in good communication with the "bulk" of the crystal. That is, assume that electrons can move freely from the surface states to the interior of the crystal and back again, when it is energetically favorable to do so. If surface potential ψ_S is then modified by some means, charge will flow into or out of the surface states until virtually all of those lying above the Fermi level will be empty of electrons, and those below, filled. Descriptive terminology evolved concerning these states. Although the terms are not now widely accepted, they are worth invoking for their graphic qualities. When the exchange of charge was rapid—requiring less than a millisecond—the states responsible were termed "fast" surface states. If the exchange required seconds, minutes, or more, the states involved were called "slow" surface states.

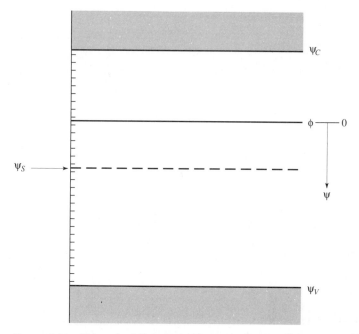

Figure 5-26 Using short lines to represent electronic states available at the surface of an N-type sample in the flatband condition. Electrons residing in these states constitute a form of interface trapped charge.

Assuming that "fast" surface states like those represented in Figure 5-26 are also donorlike, we can see that the effect of having such states will be to produce band bending. Those states lying above the Fermi level will primarily manifest the positive charge of an empty donor state. Lines of force will extend into the silicon to terminate on the most available negative charges—conduction electrons. That is, the silicon near the surface will experience *accumulation*. The bands will bend down, as is shown in Figure 5-27. An energy well is thus formed by the insulating oxide on the left-hand side, and the conduction-band edge on the other side, confining the *surplus* electrons. (See Section 3-7.6.)

The density of "fast" surface states on a silicon surface is subject to wide variations, depending on the details of surface treatment. A silicon sample cleaved in a vacuum exhibits a density of such states amounting to about $10^{15}/cm^2$, which approximates the density of surface atoms [25]. This semiquantitative agreement calls forth the dangling-bond picture of Figure 5-25. When the sample is removed from the vacuum, a layer of silicon oxide forms spontaneously because of the great affinity of silicon for oxygen. Other atoms present in the atmosphere can also react or be adsorbed. At this point the density of "fast" surface states drops into the range from $10^{11}/cm^2$ to $10^{12}/cm^2$ [32]. These are accompanied, however, by some $10^{12}/cm^2$ to $10^{13}/cm^2$ of "slow" surface states. The picture suggested by this obser-

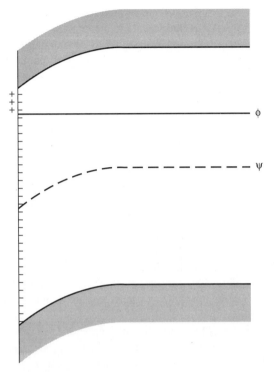

Figure 5-27 Donorlike electronic surface states on an N-type sample. Filled states (primarily below the Fermi level) are neutral, while empty states account for positive charge at the surface and a resulting positive (rightward) field as well as the band bending shown.

vation is that the spontaneous oxide has satisfied some of the surface-atom bonds that were unsatisfied in the cleaved-surface condition, but that there are new states formed (perhaps in the oxide) that electrons can reach with more difficulty, and hence more slowly. The spectrum of time constants is a continuum, and the areal density of such states is proportional to characteristic time. This kind of charge exchange can play a part in a form of electrical noise known as "flicker noise," which is characterized by a $1/f$ spectrum [33].

The problem of interest to us, of course, is the problem of such states at an oxide–silicon interface. Hence the term *interfacial states* becomes more appropriate than *surface states,* and will be used henceforth. Modern parlance lumps together the charge in "fast" and "slow" interfacial states and refers to it collectively as *interface trapped charge.* This agreed-upon terminology [34] is accompanied by the now-standard notation of N_{it} to stand for the areal density of charge centers in cm^{-2}, and Q_{it} for the areal density of charge in C/cm^2.

Interfacial states are an unmitigated evil in semiconductor devices. Much effort, mostly empirical, has been invested to devise procedures that reduce the density of such states to an acceptably low level. In modern MOS technology, the density of these states is of the order of $10^9/\text{cm}^2$. To acquire a quantitative grasp of the significance of this surface-density value, we can point out that it approximates the density of electrons in the incipient inversion layer existing at the threshold condition, a very small areal density indeed. This statement is based upon the case arbitrarily taken in Sections 5-2.2 through 5-2.4, wherein net doping was $N_A - N_D \approx 1.6 \times 10^{15}/\text{cm}^3$.

The effect of states of the fast variety at the oxide–silicon interface in an MOS capacitor is to act much like an electrostatic shield, since electrons flowing from the interior of the crystal can readily gain access to the surface states. When positive bias, for example, is applied to the gate, such charge transport results in interface trapped charge. The trapped electrons terminate lines of force originating on positive charges at the field plate, and hence the lines do not penetrate into the silicon to form the depletion layer that precedes inversion-layer formation. This situation is illustrated qualitatively in Figure 5-28. In Figure 5-28(a) is shown the potential profile of an MOS capacitor that is ideal in the sense that no interfacial states are present. The squares in Figure 5-28(b) represent, this time, acceptorlike states that are negatively charged when occupied. Occupancy of those shown on the diagram was the result of the positive bias on the gate. Consequently, the capacitor with interfacial states shows much less band bending and depletion than the MOS capacitor that is free of such states. It was this kind of "shielding" behavior that John Bardeen described [35] to account for the fact that the many years of the field-effect experiments that led to the point-contact transistor had produced disappointing results (see Section 5-1.1). Thin semiconductor samples were employed in the field-effect experiments, but repeatedly the amount of depletion was far less than expected, and as a result the conductivity change in the thin layer was also much smaller than expected. The "surface states" he postulated to account for such observations we can legitimately label as interfacial states because of the inevitable presence in air of at least a spontaneous oxide layer on a semiconductor crystal. Elucidation of the interface-trapped-charge phenomenon and quantitative description was achieved by extension of the capacitance-measurement techniques described in Sections 5-3.4 and 5-3.5, and also by surface-conductance measurements [36].

It has been found that an efficient method for reducing the density of interfacial states is low-temperature ($\sim$450°C) hydrogen annealing. So effective are the refined (but largely empirical) present-day procedures that the once vexing problem of interfacial states can be regarded as a thing of the past for terrestrial, radiation-free applications and environments. Unfortunately, however, there are also applications where the problem returns full-blown [37]. For example, a communications satellite sometimes finds itself in a sea of energetic particles and quanta, and such radiation creates numerous interfacial states and associated interface trapped charge. Ironically, the hydrogen that constitutes a "fix" for the problem of interfacial states under radiation-free conditions becomes itself a problem in a radiation environment, being responsible for yet another form of parasitic charge [38].

MOS-CAPACITOR PHENOMENA   765

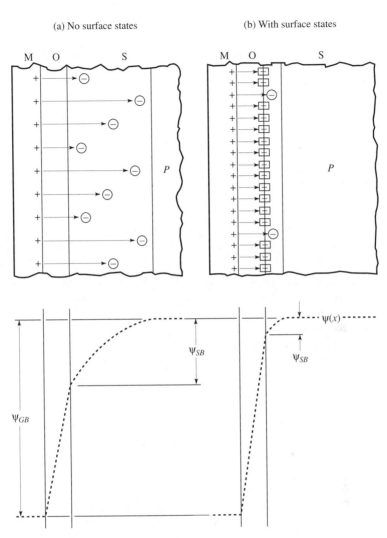

Figure 5-28 Tendency of fast surface states to shield the crystal interior from the effects of bias applied to the MOS capacitor. (a) Device with no surface states, showing termination of field lines by acceptor ions in the depletion layer. (b) Device with surface states that fill with electrons to terminate field lines, thus simulating an electrostatic shield that diminishes ψ_{SB}.

The specific nature and distribution in energy of the interfacial states (or traps) remains a subject of controversy. However, a model that is consistent with a substantial amount of experimental data is as follows [37]: When the Fermi level is at the middle of the gap (the threshold of weak inversion), the interfacial states overall tend to be neutral. This implies that all states in the upper half of the gap are acceptorlike, while all states in the lower half are donorlike. Hence, in an N-channel device at the threshold of strong inversion, the interface trapped charge is negative; in a *P*-channel device at threshold, on the other hand, the charge is positive.

Exercise 5-22. Explain the last observation.

The *N*-channel (*P*-substrate) device exhibits downward band bending at threshold, so that the Fermi level at the surface is above the middle of the gap. Thus, many of the acceptorlike states between the Fermi level and the middle of the gap are filled and negatively charged. Reverse conditions hold in the *P*-channel case.

The combination of high-lying acceptor states and low-lying donor states may seem odd, but is not without precedent. This concept, not examined heretofore, can be explained in a more familiar context by referring back to Section 2-6.5. There and in Table 2-2 it is pointed out that a gold atom introduced substitutionally in silicon introduces two states, only one of which is in evidence at a given time. When the Fermi level is below the gold donor state, located 0.35 eV above the valence-band edge, one sees the donorlike behavior of gold. But when the Fermi level is above its acceptor state, located 0.54 eV below the conduction-band edge (and hence very near the center of the gap), one sees acceptorlike behavior. In either case, the gold atom is likely to be charged (ionized), but oppositely in the two cases. This *amphoteric* property is sometimes described by stating that the substitutional gold atom behaves like a *P*-type impurity in an *N*-type environment, and vice versa. Thus it has a tendency to raise resistivity in either case. The introduction of gold into silicon was used in past decades to reduce carrier lifetime—mainly by virtue of the acceptor state that lies so very near the center of the bandgap—thus creating "faster" devices.

Further comments on interfacial states and the effects of charge therein are found at the end of the next section. The balance of charges in interfacial states against other parasitic charges to be described there can also be affected by radiation.

5-2.6 Oxide Charge

When MOS technology had advanced far enough to eliminate, virtually, the states responsible for interface trapped charge, investigators then became aware of yet another nonideal property of the MOS capacitor. There appeared to be a rather constant and inescapable positive charge, and it, too, was in the neighborhood of the oxide–silicon interface. This charge, whose detailed identity and genesis has still to be determined in detail, has been named *fixed oxide charge* [34]. The accepted symbols for its areal charge density and areal charge-center density, respectively, are Q_f and N_f. (In much of the early literature the areal density of such charge is identified as Q_{ss}, a symbol with the obvious shortcoming that it suggests "surface states" more readily than the fixed-charge concept.)

The properties of fixed oxide charge were described by an early team of investigators [39], who enumerated a number of its attributes: It is

- independent of surface potential ψ_S, at least over the central 0.7-V portion of the energy gap.

- stable under moderate bias-temperature stress. In other words, it is truly *fixed* charge in the sense that it is not *mobile* like the charge to be treated later in the present section.

- independent of oxide thickness for a given oxide-growth condition.

- approximately independent of impurity type and density in the range from 10^{14} cm^{-3} to 10^{17} cm^{-3}.

- located within, at most, 200 Å of the oxide-silicon interface.

- reproducibly variable as a function of ambient gas (dry oxygen or water vapor) during oxidation.

- reproducibly variable as a function of silicon temperature during oxidation.

- capable of being increased by strong fields resulting from negative field-plate bias; in these cases its final value levels off at a value proportional to its initial value; the increase in N_f resulting from such treatment is accompanied by an apparent increase in "fast" surface states.

- dependent on crystal orientation at the silicon interface, following the same sequence as oxidation-rate constant and surface-atom density.

- determined by the final step in the oxidation process.

In explanation of the last item it should be pointed out that during an oxidation process, SiO_2 is created at the oxide-silicon interface. That is, oxygen diffuses through the oxide that has already formed, with very little silicon diffusing to the exposed surface of the layer. Consequently, the SiO_2 most recently formed is deepest, or near the interface.

The picture that was offered in explanation for all these observations was that unreacted silicon is present near the interface and for a short distance into the oxide. That is, with a density that declines as one departs from the interface, there is excess (or unoxidized) silicon present. The problem has been studied more recently using a variety of techniques, including x-ray-photoelectron spectroscopy, Auger-electron spectroscopy, and electron microscopy [40–42]. These observations suggest the presence of a thin (~10 Å) layer of "nonstoichiometric" material near the oxide-silicon interface. Quantitative estimates of the excess (or deficiency) of a given component are not in agreement, partly because the measurements are difficult. Silicon-silicon bonding has also been suggested as an important feature of this layer, so the problem is more complicated than initially thought.

Although the precise nature of oxide fixed charge has not yet been determined, there is wide agreement that this fixed charge resides in the oxide close to the interface. The term *fixed* conveys the double idea that first, unlike interface-trapped charge Q_f is independent of surface potential, and second, it is not mobile when the MOS capacitor is subjected to a bias at elevated temperature, as is the third kind of parasitic charge (which will be examined next, following a quantitative look at the effect of Q_f on capacitor properties). Also, the term *oxide* is included in the name of this charge component, in spite of its association with the interface, because the charge indeed resides in a region of finite thickness outside the pure-silicon portion of the MOS capacitor.

To examine the effect of Q_f on the MOS capacitor, let us initially assume $H_D = 0$. Then connect gate to bulk directly, so that $\phi_G = \phi_B$. With the sheet of positive charge existing at the interface, and with no barrier-height difference between gate and bulk, the potential profile will look like that in Figure 5-29(a). The silicon surface is depleted. In order to establish the flatband condition, one must bias the gate negatively, as shown in Figure 5-29(b). Let us compute V_{FB} for a typical value of N_f found on a (100) surface, namely $10^{10}/\text{cm}^2$. This crystal orientation is favored in MOS technology so that N_f (and statistical fluctuations in N_f) will be as small as possible. Assuming that Q_f is the only charge present, and using the same MOS-capacitor sample as in Sections 5-2.3 and 5-2.4, we have from Equation 5-34

$$V_{FB} = \psi_{GB} = -\frac{Q_f}{C_{OX}} = -\frac{N_f q X_{OX}}{\epsilon_{OX}}$$
$$= -\frac{(10^{10}/\text{cm}^2)(1.6 \times 10^{-19}\text{ C})(10^{-5}\text{ cm})}{(0.345 \times 10^{-12}\text{ F/cm})}$$
$$= 0.046 \text{ V}. \tag{5-69}$$

Hence the effect of Q_f on flatband voltage in this instance is less than 50 mV. For a (111) surface, by contrast, the contribution of Q_f to flatband voltage typically exceeds 100 mV.

The next kind of parasitic charge we shall consider is one that plagued early MOS technology. It proved to be the charge associated with positive ions, primarily alkali ions, present in the oxide. Identifying and correcting this problem was a crucial achievement on the way to practical MOS technology; it also constitutes a fascinating detective story.

In the early 1960s it was customary to deposit the aluminum field plate on MOS capacitors and MOSFETs by using a filament evaporator. In this apparatus, bits of aluminum wire that were placed on a tungsten filament melted when the filament was heated to incandescence. The molten aluminum adhered to the tungsten, or "wetted" it, and aluminum atoms that evaporated from the melt traveled in straight lines through the vacuum chamber to strike the silicon substrate, producing the desired layer of deposited aluminum. It was discovered in about 1963 that the tungsten filament was a significant source of sodium contamination. Sodium en-

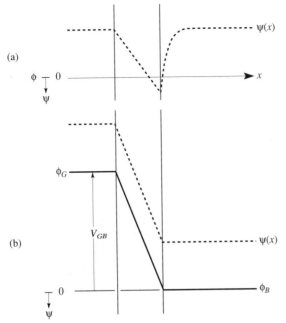

Figure 5-29 Qualitative potential profiles for MOS capacitor with thin layer of positive charge at the oxide–silicon interface, approximating the effect of Q_f. (a) Case of $\phi_{GB} = \phi_{GB} = 0$. (b) Case of flatband condition, created by negative bias V_{GB}.

tered the molten aluminum, and then the deposited layer, and finally the oxide layer upon which the aluminum was deposited. While general "cleaning up" of the MOS process from start to finish was necessary to eliminate the ubiquitous sodium contamination, especially that contributed by fingers, the elimination of filament evaporation was the greatest single advance in solving the sodium problem. The apparatus that was subsequently employed was the electron-beam evaporator, in which a focused electron beam strikes a pellet of aluminum causing local melting and hence evaporation. The pellet itself is supported in a water-cooled cup that does not transmit contaminants to the aluminum charge. Furthermore, the electron beam is bent by a magnetic field so that the sample receiving the aluminum cannot "see" the electron source, and hence is shadowed from any contaminants that may originate there. For a significant year or two, this important technological change was held as a trade secret; but by the mid-1960s it had been transmitted by word of mouth throughout the intensely active and competitive MOS community.

It was essential to solve the sodium problem by eliminating sodium. That is, one cannot "live with" this parasitic charge in the MOS capacitor by compensating for it elsewhere, for at least two reasons: First, under conditions where such contamination is present, the amount of contamination is variable, and the electrical properties of the device vary as a result. Second, the sodium ions in the oxide are rather mobile; by applying a normal voltage to the MOS capacitor, one exerts electrical force on these contaminating positive ions. With the temperature elevated to a few hundred degrees C, the ions can, with negative bias, be induced to move toward the field plate. Positive bias, in turn, drives them back toward the oxide–silicon interface. Such a positional change changes the flatband voltage V_{FB}, as we shall now show, and also the threshold voltage V_T. For these reasons, the presence of such contaminating alkali ions constituted a totally unacceptable condition.

To understand the effect of ionic contamination in the oxide, consider the situation shown in Figure 5-30. A sheet of charge having an areal density Q_m (in C/cm^2) is located at an arbitrary position within the oxide, as shown in the physical representation of Figure 5-30(a). Let the position be designated x, and let the spatial origin be placed at the oxide–silicon interface. In this one-dimensional problem, lines of force emanating from the positive ionic charges must terminate on negative charges either in the gate electrode or in the silicon. The charge-density profile accompanying this situation, shown in Figure 5-30(b), assumes that $H_D = 0$ and that $\psi_{GB} = \phi_{GB} = 0$ as well. The area of the delta function labeled Q must equal the sum of the areas of the two negative-charge profiles. The one on the left represents the sheet of charge in the field plate, and the near-rectangle on the right, acceptor-ion charge in the depletion layer.

The field profile shown in Figure 5-30(c) corresponds to the charge profile of Figure 5-30(b). Let us now alter the situation by increasing the negative bias on the gate until just enough negative charge is delivered to the gate to cause all lines of force originating on contaminating ions to terminate on the gate electrode. Under these conditions, the silicon is free of charge and field, and has been brought to the flatband condition. The field and potential profiles accompanying this new condition are as shown in Figures 5-30(d) and 5-30(e), respectively. This particular value of gate-to-bulk voltage is, once again, the *flatband voltage* for the device and structure assumed, and can be written

$$V_{FB} = -\frac{Q}{C(x)} = -\frac{(X_{OX} + x)Q}{\epsilon_{OX}}. \quad (5\text{-}70)$$

The capacitance per unit area $C(x)$ is a function of the arbitrarily assumed position for the ionic charge sheet, and $\epsilon_{OX} = 0.345 \times 10^{-12}$ F/cm is the absolute permittivity of the silicon oxide. Multiplying and dividing by the oxide thickness X_{OX} gives us

$$-\frac{(X_{OX} + x)}{X_{OX}} Q \frac{X_{OX}}{\epsilon_{OX}} = -\frac{(X_{OX} + x)}{X_{OX}} \frac{Q}{C_{OX}}, \quad (5\text{-}71)$$

where C_{OX} is the capacitance per unit area of the total-thickness MOS capacitor. Noting, then, that Q may be written $\rho_V(x)dx$, where ρ_V is volumetric charge density, it follows that the flatband voltage for an MOS capacitor having an arbitrary charge profile in the oxide may be written

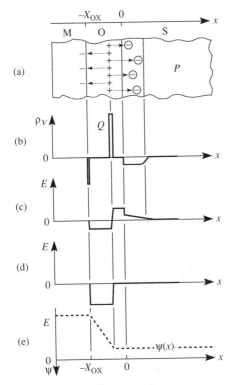

Figure 5-30 Hypothetical sheet of parasitic positive charge in oxide, for analyzing problem of arbitrary one-dimensional ionic contamination of oxide. (a) Physical representation. (b) Charge profile, with $H_D = \psi_{GB} = 0$. (c) Electric-field profile. (d) Electric-field profile after negative external bias is applied to create flatband condition. (e) Potential profile for flatband condition.

$$V_{FB} = -\frac{1}{C_{OX}} \int_{-X_{OX}}^{0} \frac{(X_{OX} + x)}{X_{OX}} \rho_V(x) dx. \quad (5\text{-}72)$$

Figure 5-31 represents a charge distribution $\rho_V(x)$ extending through the oxide layer. When voltage is applied in accordance with Equation 5-72, the result is the charge-density profile shown there, wherein the positive and negative portions of the profile have equal areas.

Exercise 5-23. Justify the algebraic signs on the right-hand side of Equation 5-70.

The first minus sign is present because voltage is being applied to the left-hand plate of the capacitor in Figure 5-30(a), while the charge of interest (Q) resides on the right-hand "plate." Because position x is inherently negative for the origin assignment in Figure 5-30, and is smaller than X_{OX}, the quantity $X_{OX} + x$ is positive as desired.

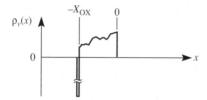

Figure 5-31 One-dimensional parasitic charge-density distribution in oxide, consisting of positive alkali ions. Negative external bias has equalized the negative and positive charge components, creating the flatband condition.

The third and last major category of parasitic charge found in the oxide has been named *oxide trapped charge* [34], described by the symbols Q_{ot} for areal charge density, and N_{ot} for areal charge-center density. It consists of carriers that have become "lodged" in the oxide.

Here it is worthwhile to combine two concepts presented earlier. In Section 3-6.1 it was explained that avalanche breakdown in a *PN* junction involves the presence of "hot" or energetic carriers in the space-charge layer. Also, in Section 2-5.4 it was pointed out (arbitrarily taking the case of electrons) that by using the silicon conduction-band edge as energy reference, the kinetic energy of the hot electron is equal to its total energy, which is equal to its distance in energy above the band edge. Referring back to Figure 5-22(b), then, we can make the plausible observation that an electron with an energy somewhat over 3 eV (or 3 V, in the electrostatic-potential terms used there) will be capable of passing into the conduction band of the oxide layer. In an analogous manner, an energetic hole could enter the valence band of the oxide. Somewhat greater energy is required in the case of a hole, and also its mobility in the oxide is smaller than that of the electron, so let us continue to focus mainly on the electron case.

Once in the oxide conduction band, the electron can be caused to drift to a field plate that is "floating," or unconnected to any terminal. The necessary field in the oxide can be created in a MOSFET-like device by applying a source–drain bias that causes voltage division in the two oxide capacitances, C_{SG} and C_{GD}, that are necessarily in series. The field, furthermore, in addition to causing electron drift in the oxide's conduction band, makes it possible for lower-energy electrons to enter the conduction band by means of tunneling.

Exercise 5-24. Explain the last statement.

Visualize the effect of a significant positive bias on the gate of the device in

Figure 5-22. The conduction-band edge of the oxide will then drop steeply from right to left. This has the effect of thinning the potential barrier that keeps electrons in the silicon. With sufficient barrier thinning, electron tunneling becomes possible, so that an electron can enter the oxide with a fraction of the energy needed to surmount the barrier right at the interface.

Given this method for achieving electron penetration of the oxide layer, one can charge the field plate by delivering electrons to it. Furthermore, the necessary hot electrons can be supplied by causing avalanche breakdown in a junction that is near the interface, or that intersects the interface. This chain of phenomena, in fact, is the basis for an *erasable-programmable read-only memory* [43], or EPROM, to use its acronym. A particular MOSFET is either ON or OFF (representing either a one or a zero), depending on the state of charge of its floating field plate. Stored information can be erased by irradiating the device with ultraviolet light, thus giving energy to the stored electrons and permitting them to return to the silicon. This kind of EPROM is but one example of what is now a rather large group of nonvolatile memory options based upon MOS technology. The term *nonvolatile* in this context describes memories that retain their information (in this case, the charges on insulated field plates) when bias voltages are removed.

The reason for describing the EPROM here is to illustrate a case wherein carriers are deliberately introduced into the oxide. Some of the carriers so introduced become trapped in the oxide, and thus contribute to Q_{ot}. Especially with repeated write-erase cycles, the residual carrier charge increases.

But problems with such charge, unfortunately, are not confined just to cases wherein the carriers have been intentionally introduced into the oxide. With the continuous and rapid reduction of device dimensions or "feature sizes" and a slower reduction of operating voltages, the result has been growth in field values that are encountered in the normal operation of devices. As a result, hot carriers are increasingly present even when not intentionally created, and these can enter the oxide to become trapped there. The result is a shift in flatband voltage V_{FB}, and an equal shift in the very important threshold voltage V_T. Furthermore, the story does not end even here. Energetic radiation is once again a problem because it can create hole–electron pairs for which one or both members become trapped in the oxide.

Exercise 5-25. Explain the last statement.

It was pointed out in Section 2-1.5 that semiconductors and insulators differ only in degree, both having energy gaps. Hence, hole–electron pairs can be created in both, with greater energy being required in the insulator case. The insulator example given there, diamond, requires radiation energy above some 5.5 eV, while in the oxide case, higher energy is required, above approximately 8 eV.

Radiation-caused pair production in silicon oxide would seem to contribute self-neutralizing charges, but the electrons have a higher probability of escape from the oxide because of their higher mobilities [37]. As a result, the radiation-induced contribution to Q_{ot} tends strongly to be positive. There is evidence that under some fortuitous circumstances, this positive charge is partly canceled by the net negative interface trapped charge that the radiation may also cause. But even if this situation could be created consistently and reliably, it does not constitute a solution, because carrier mobility in the MOSFET channel declines concurrently.

Exercise 5-26. Why should mobility decline as Q_{it} increases?

Channel carriers are confined to a very thin region near the oxide–silicon interface. The presence of additional charge centers at the interface will result in more frequent scattering of carriers, by the same mechanism treated in Section 2-5.1 for the case of scattering by impurity ions. The resulting shrinkage of mean free path is responsible for the declining mobility.

Exercise 5-27. Why should one be concerned about declining mobility?

Equations 5-17 and 5-22 show that drain current I_D is directly proportional to electron mobility μ_n. As a result the important figure of merit, transconductance g_m, is also proportional to μ_n, as Equation 5-24 shows. Hence the radiation leads to declining device performance.

Fluctuations in the population (of electrons) in the interfacial states causes fluctuations in channel-carrier scattering, and hence, in mobility. Such *mobility fluctuations* are currently under study as an additional source of $1/f$ or flicker noise in MOSFETs [44]. Beyond this, the transitions of electrons to and from the status of conducting carriers in the channel on the one hand, and trapped carriers in interfacial states on the other, gives rise to *density fluctuations* that may also contribute to $1/f$ noise. These issues will probably be resolved in the near future [44].

5-2.7 Calculating Threshold Voltage

A good starting point for threshold-voltage determination is Equation 5-52, repeated here for convenience:

$$V_{GB} = H_D + \psi_{GS} + \psi_{SB}. \tag{5-73}$$

The term H_D involves only the barrier-height-difference effect. Assuming P-type silicon with a net doping of $10^{15}/\text{cm}^3$ and an aluminum field plate, we are informed by Figure 5-23 that $H_D = -0.9$ V. And the last term ψ_{SB}, the band-bending term, involves only a single phenomenon, treated accurately in Section 5-2.3. But the middle term ψ_{GS}, the potential drop through the oxide portion of the MOS capacitor, is fixed by the four kinds of charge outlined in Sections 5-2.5 and 5-2.6, and in addition to these, by the charge in the silicon. All of these charge components are imposed "on" the fixed portion of the MOS capacitor, and therefore contribute to a potential drop through the oxide as a consequence of the basic law governing the parallel-plate capacitor. To start, let us assume that our technology and application will permit us to neglect the interface trapped charge altogether. In addition, let us assume that conditions that cause oxide trapped charge have been avoided. Then, combining Equations 5-73, 5-65, 5-69, and 5-72 yields this result:

$$V_{GB} = H_D - \left[\frac{1}{C_{OX}} \int_{-X_{OX}}^{0} \left(\frac{X_{OX} + x}{X_{OX}}\right) \rho_V(x) dx \right] - \frac{Q_f}{C_{OX}} - \frac{Q_S}{C_{OX}} + \psi_{SB}. \qquad (5\text{-}74)$$

The first three terms on the right-hand side occurred previously in equations wherein flatband voltage V_{FB} was specified. This was true because each of these parasitic effects makes a fixed charge contribution to the MOS capacitor at a fixed spatial position (provided we rule out alkali-ion motion by maintaining room temperature). Hence each of these three terms makes a fixed contribution to flatband voltage V_{FB}. The last two terms of Equation 5-74 are by contrast variable and can change sign as well. But under flatband conditions, ψ_{SB} is by definition zero, which in turn rules out ionic and surplus-carrier charge in the semiconductor, so that Q_S also vanishes. Hence in the present case,

$$V_{FB} = H_D - \left[\frac{1}{C_{OX}} \int_{-X_{OX}}^{0} \left(\frac{X_{OX} + x}{X_{OX}}\right) \rho_V(x) dx \right] - \frac{Q_f}{C_{OX}}. \qquad (5\text{-}75)$$

Now let us make the realistic assumption that charge in the oxide has been eliminated, so that Equations 5-74 and 5-75 can each be simplified by dropping the integral term. Then, to develop an expression for the applied-voltage value that brings the MOS capacitor to the threshold of strong inversion V_T, note again that this is an applied voltage that causes the surface potential to be equal in magnitude and opposite in sign to the bulk potential. That is, in the present unnormalized notation, $\psi_S = -\psi_B$ (or in normalized notation $U_S = -U_B$). By definition, then,

$$\psi_{SB} \equiv \psi_S - \psi_B = -2\psi_B, \qquad (5\text{-}76)$$

so that from Equation 5-73 and (modified) Equation 5-75,

$$V_T = H_D - \frac{Q_f}{C_{OX}} - \frac{Q_S}{C_{OX}} - 2\psi_B = V_{FB} - \frac{Q_S}{C_{OX}} - 2\psi_B. \qquad (5\text{-}77)$$

Let us choose $N_A = 10^{15}/\text{cm}^3$, for which $2\psi_B = -0.59$ V. The value $N_A = 1.6 \times 10^{15}/\text{cm}^3$ was employed in Sections 5-2.2 through 5-2.4 because it corresponds to the convenient value $U_B = 12.0$, but this small change in assumed doping has a negligible effect on H_D. Also, it makes no change at all in Q_f, which is doping-independent. Hence the term $-Q_S/C_{OX}$ is the only one remaining to be evaluated. From Equations 5-58 and 5-59 we have

$$|Q_S| = \epsilon |E| = \left(\frac{kT}{q}qN\epsilon\right)^{1/2} \frac{dW}{d(x/L_D)}. \tag{5-78}$$

But using Approximation C of Table 5-1 and noting that $|W_S| = q|\psi_{SB}|/kT$ gives us

$$\left.\frac{Q_S}{C_{OX}}\right|_{V=V_T} = -\frac{|Q_S|}{C_{OX}} = -\frac{|Q_S|X_{OX}}{\epsilon_{OX}}$$

$$= -\frac{X_{OX}}{\epsilon_{OX}}\sqrt{2qN\epsilon|\psi_{SB}|} = -0.40 \text{ V} \tag{5-79}$$

for the present threshold condition, where the minus sign has been introduced because space charge in the P-type silicon is negative. Hence from Equation 5-79 we find

$$V_T = (-0.9 \text{ V}) - (0.8 \text{ V}) - (-0.4 \text{ V}) - (-0.6 \text{ V})$$
$$= (-1.7 \text{ V}) + (1.0 \text{ V}) = -0.7 \text{ V}, \tag{5-80}$$

where $V_{FB} = -1.7$ V, the result of the two nonideal factors present.

It can be seen in Equation 5-80 that a corresponding ideal device would have $V_T = +1$ V. Hence the nonideal factors have caused a threshold-voltage shift of -1.7 V and a change of algebraic sign. That is, the capacitor will possess a beyond-threshold inversion layer when the external voltage is set at $V_{GB} = 0$. The P-type MOS capacitor possesses an inversion layer, or "channel," even with no voltage applied. As noted at the end of Section 5-2.4, the N-type MOS capacitor does not have a channel under similar conditions, and so the factors H_D and Q_f contributed to the early popularity of N-type starting material, and hence to P-channel devices, because it was desired that there be no inversion layer in the $V_{GB}=0$ condition. Subsequent technological developments (particularly ion implantation) have made it possible to "tailor" doping and hence charge conditions near the surface, which has greatly increased the range of choice open to the MOS designer. Consequently, N-channel devices are now the dominant type, and P-channel devices are rarely encountered outside CMOS integrated circuits.

5-3 MOS-CAPACITOR MODELING

In the process of accurate MOS-capacitor modeling, Section 5-2 was able to make appreciable use of equations developed in Section 3-7 for the step junction. This was because the semiconductor portion of the former (which we shall assume to be uniformly doped) poses a problem identical to that posed by one side of the latter, with an important proviso. The MOS capacitor is virtually at equilibrium for any applied bias within a wide range of values (current through the insulator must be negligible), while the junction is at equilibrium for only one bias value, $V = 0$. Thus, for any arbitrarily chosen value of bias imposed on a particular MOS capacitor, one must specify a unique step-junction structure to have the kind of equivalence noted earlier.

From the practical point of view, an important difference between the MOS capacitor and the step-junction diode is that in the former a parallel-plate capacitor is inevitably and permanently in series with the semiconductor phenomena one wishes to study (often by capacitance measurements). As a result, comparisons of the two diodes pose subtle problems with solutions that are sometimes not obvious. Such studies and comparisons, however, have contributed greatly to understanding of the MOS system. Let us begin by extending the approximate-analytic modeling of Section 3-7 into the MOS context, where it has important features that are exact. Then device and equivalent-circuit comparisons will enable us to understand the bias-dependent capacitance descriptions that follow in this section.

5-3.1 Exact-Analytic Surface Modeling

Early treatments of the semiconductor-surface problem used a combination of analytic and numerical methods [45–49]. Later workers recognized the existence of asymptotic solutions of extremely simple form [50–52], and this had the effect of extending the scope of pure-analytic methods in treating such problems. Curiously, the latter group directed their attention toward the step-junction problem rather than the semiconductor-surface problem. The two problems are fully equivalent analytically, as was pointed out at the beginning of Section 3-7.1, but this equivalence was not immediately recognized. The next expansion of analytic capabilities came with the realization that one of the asymptotic solutions defines an invariant spatial origin that permits the writing of explicit approximate expressions of simple form for both asymptotic and intermediate solutions [53]. This constitutes the depletion-approximation replacement (DAR) of Section 3-7.5.

The Poisson-Boltzmann equation, Equation 3-67, can be integrated once analytically, yielding the electric-field expression given in differing forms in Equations 3-76, 3-77, 3-78, and 3-80. In each case, electric field is given as a function of electrostatic potential. Hence, given the value of a critical potential, such as that at the silicon surface (which is to say, at the oxide–silicon interface), we can calculate electric field in the silicon right at the surface directly and *exactly*. No approximations are involved. An application of Gauss's law then permits us to determine areal charge density in the silicon with equal accuracy. The charge so determined consists in the general case of ionic depletion-layer charge *and* of inversion-layer or channel charge for one polarity of applied bias, and of accumulation-layer charge for the other polarity. This straightforward procedure was early applied to the MOS-capacitor (and MOSFET) problem [54]. By combining this method with the DAR, however, one can reintroduce the spatial variable, thus generating analytic *profiles* of electric field and volumetric charge density [53, 55]. The necessary expressions will now be developed for subsequent use.

In Figures 5-17 through 5-22, we have oriented the MOS capacitor with the semiconductor portion at the right-hand side, and we shall continue to do so. Hence, to relate the MOS-capacitor problem with the junction problem defined in Figure 3-42, it is necessary to identify the present semiconductor (or, let us say silicon) region with the right-hand region of Figure 3-42. Figure 5-32(a) carries over from

Figure 3-42(a) the physical representation of a *PN* junction at equilibrium. The bulk potentials U_{01} and U_{02} are shown thereon, as is the junction position x_J that entered into discussion in Chapter 3. The corresponding "surface" problem is represented in Figure 5-32(b). The critical surface position is now designated x_S, and the equilibrium bulk potential of the silicon sample is U_B; the subscript B was introduced in Section 5-2.2, and is conveniently applied here to take advantage of the fact that there is now only one silicon region to deal with. Thus, letting $U_{02} \equiv U_B \equiv q\psi_B/kT$, and letting $W = U_B - U$, Equation 3-80 becomes

$$\frac{dW}{d(x/L_D)} = -\sqrt{2}\left[\frac{e^{U_B}(e^{-W} + W - 1) + e^{-U_B}(e^W - W - 1)}{e^{U_B} - e^{-U_B}}\right]^{1/2}, \quad (5\text{-}81)$$

and is consistent with Figure 3-42 in the following sense: There, U_B is positive (the right-hand side of the sample is *N*-type), W is positive-going, and electric field is negative; the lines of force are directed leftward from positive charge in the silicon to negative charge somewhere toward the left. We have here a depletion-inversion condition. It is best, however, to remove the sign ambiguity by a method similar to that used in Equation 3-83. Let U_B be replaced by $|U_B|$, agree to let W always be positive-going, and then for the depletion-inversion condition in *any* sample, *N*-type or *P*-type, we have

$$\left|\frac{dW}{d(x/L_D)}\right| = -\sqrt{2}\left[\frac{e^{|U_B|}(e^{-W} + W - 1) + e^{-|U_B|}(e^W - W - 1)}{e^{|U_B|} + e^{-|U_B|}}\right]^{1/2},$$

(5-82)

an expression that is valid without restriction. Absolute-value bars have been placed on normalized electric field as a reminder that its algebraic sign must be fixed by an appeal to physical reasoning.

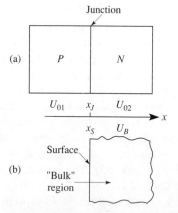

Figure 5-32 Relating the *PN*-junction and MOS-capacitor problems. (a) Physical representation of the general step-junction problem adopted for exact analysis. (b) Physical representation of the silicon portion of an MOS capacitor, or of the bulk sample for semiconductor-surface analysis. We have let $U_{02} \equiv U_B$.

For particular ranges of $|U_B|$ and W, Equation 5-82 can be replaced by certain of the approximate equations given in Table 5-1 [55, 56]. The range designations are overlapping so that one can select an approximation tailored to a particular problem. In the conditions of most interest, depletion and inversion coexist (Approximations A through C). But inversion is important only near and above threshold. Under conditions of extreme inversion, ionic charge can be neglected (Approximation D). It is interesting to note that the depletion-inversion expressions can be readily separated into terms that account for electric field arising from inversion-layer charge and from depletion-layer (ionic) charge. In the full and exact expression, Equation 5-81, the first major term in the numerator gives the potential dependence of ionic charge, and the second, of inversion-layer charge. The approximate expressions in Table 5-1 are labeled as to which charge they describe, and in a number of cases, inspection easily verifies their origin in the general expression. Ionic-charge expressions are given first (Approximations E and F). Subtraction of the ionic component from the total, of course, gives the inversion-layer component. Using Gauss's law, once again, either field component can be converted into charge. For extremely small values of W, which is to say, for extremely small departures from neutrality, depletion and accumulation conditions yield the same electric field magnitude (Approximation G).

Exercise 5-28. In Table 5-1, how does Approximation A′ follow from A?

Substituting $W = 2|U_B|$ into Approximation A yields
$$\sqrt{2(e^{-2|U_B|} + 2|U_B| - 1 + e^{-2|U_B|+2|U_B|} - 2|U_B|e^{-2|U_B|} - e^{-2U_B}}$$
$$= \sqrt{2(-2|U_B|e^{-2|U_B|} + 2|U_B|)} = 2\sqrt{|U_B|(1 - e^{-2|U_B|})}.$$

Exercise 5-29. In Table 5-1, how does Approximation B follow from A?

For the minimum value $|U_B| = 3$ indicated in B, larger than the minimum value $|U_B| = 2$ indicated in A, the subterms $(-W - 1)$ can be dropped from the second major term in A, resulting in an approximate doubling of the maximum error in the ranges specified when one goes from A to B.

Exercise 5-30. How is Approximation B′ obtained?

Dropping the exponential term in Approximation A′ causes only an approximate doubling of the maximum error in the specified ranges in going from A′ to B′.

The condition of *accumulation* was introduced in Section 3-7.6. In this condition, one has negative space charge consisting of surplus electrons in an *N*-type region or positive space charge consisting of surplus holes in a P-type region. The expression for normalized electric field that is valid without restriction can be obtained from Equation 5-82 simply by replacing $|U_B|$ by $-|U_B|$:

$$\left|\frac{dW}{d(x/L_D)}\right| = \sqrt{2\left[\frac{e^{-|U_B|}(e^{-W} + W - 1) + e^{|U_B|}(e^{W} - W - 1)}{e^{-|U_B|} + e^{|U_B|}}\right]}. \quad (5\text{-}83)$$

The last two expressions in Table 5-1 deal with accumulation. The very mild restriction that $|U_B| > 2$ gives Approximation H. It is evident, then, that with increasing W, the result becomes the exponentially rising function independent of $|U_B|$ that is given as Approximation J. This explains why a single curve is plotted in Figure 3-43 to represent the accumulation condition.

The exact expressions, Equations 5-82 and 5-83, are summarized in Table 5-2 along with the simple result obtained for the MOS capacitor employing an intrinsic semiconductor. Finally, we should point out that one is usually interested in field in the semiconductor at the interface, and that the corresponding value of normalized potential in such a case is W_S. The subscript has been omitted in Tables 5-1 and 5-2, however, in the interest of generality.

Exercise 5-31. Point out the qualitative difference between Equations 5-82 and 5-83 that causes the depletion-inversion condition to involve two kinds of charge and a threshold phenomenon, while the accumulation condition does not.

In Equation 5-82 for the depletion-inversion case, the first major term in the numerator is associated with depletion-layer charge, while the second is associated with inversion-layer charge. The product $\exp(|U_B| - W)$ and the product $\exp(-|U_B| + W)$ cross over for $W = |U_B|$, the threshold of weak inversion. There the ionic and inversion charge densities at the plane of interest (usually the surface) are equal. In Equation 5-83 for the accumulation case, the product $\exp(|U_B| + W)$ exceeds the product $\exp-(|U_B| + W)$ for all allowable (positive) values of $|U_B|$ and W. Hence there is no threshold condition.

5-3.2 Comparing MOS and Junction Capacitances

Sorting out the coexisting phenomena in the PN junction was a major preoccupation of the 1950s, and even of the 1940s [57]. Small-signal junction capacitance was among these effects, and by the close of the 1950s, its voltage dependence was understood reasonably well; this is illustrated in Figure 3-55(b) and involves a crossover of diffusion and depletion-layer capacitance. The chief difficulty in the experimental observation of these capacitance functions is the massive conductance that accompanies them in the upper forward-bias regime. For good and practical reasons, understanding the nuances of current versus voltage received considerably more interest and attention than did junction capacitance. The *PN*-junction diode was an important device in its own right, as well as being the key constituent of devices such as the BJT. (This relationship is well illustrated by the way the analysis given in Chapter 4 is built directly on the foundation established in Chapter 3.)

Since the MOS capacitor normally exhibits negligible conductance at any bias, the capacitance-voltage observation was a vital method (and almost the only method) for experimental characterization of the device, a process completed about ten years later than for the junction. But the convenience of having conductance out of the picture was offset by other complications. First, as already noted, the parallel-plate oxide capacitor is permanently in series with the semiconductor phenomena in the bulk region that are of primary interest, so that one is required to "reach through" the oxide with any experimental observation. Second, the inversion layer has a negligible effect on dynamic *PN*-junction properties, but becomes a dominant factor in the case of the MOS capacitor. As a result, the physical arrangement for supplying (or not supplying) carriers to the inversion layer has an important bearing on dynamic behavior. An important option for such carrier delivery places a heavily doped region adjacent to the MOS sandwich. In other words, the MOS capacitor inevitably becomes a two-dimensional entity in one of its important embodiments. Third, as also noted, equilibrium analysis is appropriate for the bulk region of the MOS capacitor at any bias, while *only* at zero bias is equilibrium analysis valid for the *PN* junction. Hence to relate the two devices in the aspects wherein they are identical, one must "design" a different junction sample to correspond to every change of bias placed on the MOS capacitor. The immutable differences between the junction and MOS diodes obscured these identities, as did the fact that the time periods of the most intense study were about a decade apart. Not surprisingly, two sets of terminology evolved for certain identical critical conditions in the two devices. It is instructive to construct a series of examples that display these conditions.

Figure 3-42(b) presented a potential profile for an arbitrarily chosen asymmetric step junction at equilibrium. Analogous profiles are presented in Figure 5-33 for a set of other step junctions at equilibrium. The left-hand portion of each profile is dashed because our focus here is on the MOS capacitor (or alternatively, on the "surface" problem); thus the left-hand half of the junction problem is "virtual." We have designed these junction samples so that the conditions in their right-hand halves correspond identically to conditions in the bulk of a particular MOS capacitor at a particular bias. In the high-low or N^+N junction of Figure 5-33(a), there is a potential drop of one normalized unit on the high side. The trivial uniform-sample case in Figure 5-33(b) corresponds to the highly significant flatband condition in MOS terms. In Figure 5-33(c) we have the familiar symmetric step junction, corresponding to the threshold of weak inversion. The significance of this term is that at this point, the two carrier populations are balanced right at the surface. For further voltage increase, the incipient inversion layer is present, but with an areal density that contributes negligible conductance parallel to the surface.

The particular asymmetric junction depicted in Figure 5-33(d) is one in which

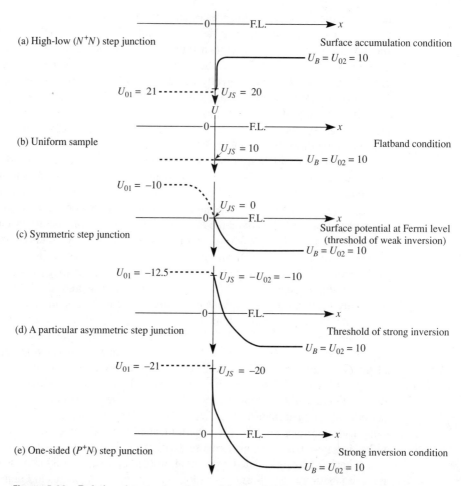

Figure 5-33 Relating the step-junction problem to the MOS (or surface) problem. Potential profiles are shown for a series of equilibrium step junctions that are identified at the left. These profiles are analogous to that of Figure 3-42(b). The right-hand side represents also the bulk region of an MOS capacitor at various bias conditions, with the condition named at the right. See text for discussion of the individual cases.

the hole density at the junction equals the electron density far from the junction in the right-hand side. Further bias increase causes the high-side potential drop to fall, approaching one normalized unit, and causes a qualitative change in the potential profile. The curve proceeds upward more steeply than before, a consequence of the inversion layer right at the junction. Hence this is the important threshold condition. In Figure 5-33(e), then, the bias change has been carried to an extreme, yielding the one-sided or grossly asymmetric step junction, with a one-unit drop on its high side. In MOS terms, this is the condition of strong inversion in the MOSFET.

5-3.3 Small-Signal Equivalent Circuits

The *PN* step-junction sample shown in Figure 5-32(a) was the object of exact-analytic treatment, chosen for consistency with earlier literature. Also, the *N*-type surface sample in Figure 5-32(b) was chosen for consistency with that step-junction sample. Now we revert to the *P*-type substrate used throughout Sections 5-1 and 5-2. (This is the most important and common case because of the mobility advantage of inversion-layer electrons.) Fortunately, the exact expressions written in the format of Equations 5-82 and 5-83 are valid for either case.

The MOS capacitor can be modeled usefully by means of small-signal equivalent circuits. The simplest case, of course, is the accumulation case. There the holes that have been attracted to the oxide–silicon interface form a thin, dense layer that closely resembles the sheet of charge on the positive plate of a conventional parallel-plate capacitor. Thus the equivalent circuit degenerates into a single capacitor, C_{OX}. (Recall that C_{OX} is stated by custom as a capacitance per unit area.) As a refinement, a small resistor can be placed in series with C_{OX}, but that is not necessary unless the silicon has unusually low net doping.

Only slightly more complex is the case depicted in Figure 5-18, wherein the MOS capacitor is given a bias of the depletion-inversion polarity, but well below threshold. Therefore the equivalent circuit consists of the capacitor C_{OX} in series with a depletion-layer capacitance that involves the same phenomenon as that of the *PN* junction. In MOS work, however, this feature is usually termed *bulk* capacitance. The charge involved can be described as ionic, but this term is unsuitable for subscript purposes because *inversion-layer* capacitance is to be introduced next. Hence the plain term "bulk" once again serves a useful role, giving us the symbol C_b for bulk capacitance. Note well that C_b is also measured in pF/cm^2 for consistency with the case of C_{OX}. This practice differs from that for the junction capacitances, wherein the depletion-layer capacitance C_t and the diffusion capacitance C_s (or C_u) are measured in pF. For this reason, the differing subscripts for the two depletion-layer capacitances (*b* and *t*) are justified, even though the same phenomenon is being modeled. Hence the appropriate equivalent circuit for this case can be drawn and labeled as in Figure 5-34.

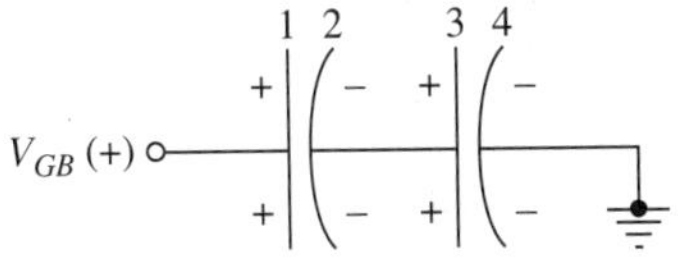

Figure 5-34 Equivalent circuit for the MOS-capacitor condition represented physically in Figure 5-18. The substrate is significantly depleted, but far below the threshold of strong inversion. (The numbered plates are discussed in Exercise 5-32.)

Exercise 5-32. How can one justify the presence of the capacitor plates 2 and 3 in the equivalent circuit presented in Figure 5-34? The physical structure that is being modeled is shown in Figure 5-18, where there is nothing present having the physical properties of a pair of capacitor plates.

Plates 2 and 3 in Figure 5-34 are common, but mutually isolated from the balance of the circuit. On one side is the SiO_2 layer, an excellent insulator. On the other side is a silicon layer that is virtually devoid of carriers, the depletion layer. Hence the negative charge on plate 2 and the positive charge on plate 3 are the result of charge separation. The total charge on plates 2 and 3 remains zero for any value of V_{GB}. Since no net physical charge is required at the oxide–silicon interface, no physical capacitor plate or plates will be needed to validate a model consisting of two capacitors in series.

Now let V_{GB} be increased more in the depletion-inversion (or positive) polarity, bringing the small-signal inversion-layer capacitance C_i into play. Like C_b, it involves a "virtual" plate at the oxide–silicon interface. The second plate of C_i is the inversion layer itself, tightly pressed against the oxide–silicon interface. But here a complication enters. In a one-dimensional MOS capacitor, the inversion-layer electrons have dropped into a potential well that in effect isolates them. As a result, C_i displays widely differing frequency dependences that are determined by the mode of communication of these electrons with the ohmic contact made to the substrate. In the simplest case, an N^+ region is placed adjacent to the field plate. This physical arrangement is shown in Figure 5-35(a). When a small signal is applied to a capacitor of this structure, electrons flow readily in and out of the inversion-layer well through the medium of the N^+ region, causing C_i to fit simply in parallel with C_d in the equivalent circuit. The N^+ region, in turn, is ohmically connected to the right-hand terminal. Observe that the configuration in Figure 5-35(a) is equivalent to the case of a MOSFET with the source and bulk terminals common. The velocity of charge transport in and out of the inversion layer approximates that in an ordinary conductor.

The other extreme situation exists when no special provision is made to deliver charge to and from the inversion layer in response to the applied signal. If the signal frequency is low enough, spontaneous carrier generation and recombination can yield a surplus-carrier population that "follows" the signal. For typical lifetimes, however, such a frequency is very low indeed—of the order of 10 Hz. For this case too, C_i and C_b are in parallel, and mutually in series with C_{OX}. For a much higher frequency such as 100 kHz, however, C_i simply disappears from the equivalent circuit. As might be expected, there exists a wide frequency range wherein C_i is frequency-dependent, an unsatisfactory state of affairs.

Exercise 5-33. What is unsatisfactory about a frequency-dependent C_i?

The point of the kind of equivalent circuit being discussed here is to model the

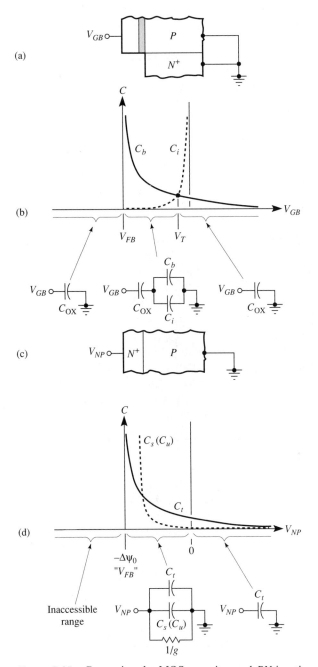

Figure 5-35 Comparing the MOS-capacitor and *PN*-junction diodes. (a) Physical structure of capacitor used for comparison. (b) Bulk capacitance C_b and inversion capacitance C_i versus applied voltage V_{GB}, showing appropriate equivalent circuit by operating regime. (c) Physical structure of junction diode used for comparison. (d) Depletion-layer capacitance C_t and diffusion capacitance $C_s(C_u)$ versus applied voltage V_{NP}, showing appropriate equivalent circuits by operating regime.

device using passive components that are *fixed* in value, and hence frequency-independent. One expects complications at frequency extremes, but when the complications exist at intermediate frequencies, the value of the equivalent circuit is diminished.

The gross dependence of C_b and C_i on bias, V_{GB}, is summarized in Figure 5-35(b), along with an indication of which equivalent circuit is appropriate to each regime of operation. Note that the voltage axis does not have an origin, because its location depends upon the parasitic effects treated in Section 5-2. As a result, the capacitance profiles are valid for a real-life MOS capacitor. The depletion-layer capacitance diverges at the flatband voltage V_{FB}. On the other hand, the C_b and C_i curves intersect precisely at the threshold of strong inversion V_T. The physical basis of this fact has an explanation that is straightforward, but not obvious, the subject of Section 5-3.6, where the phenomenon is termed MOS-capacitance crossover. (Section 5-3.7 treats the same topic analytically.) In principle, the inversion-layer population and C_i both increase without limit for $V_{GB} > V_T$, so that C_i shunts out C_b and the equivalent circuit reverts simply to C_{OX}.

For clarifying contrast, we take the familiar N^+P diode shown in Figure 5-35(c) and plot in Figure 5-35(d) a set of capacitance profiles analogous to those in Figure 5-35(b) for the MOS capacitor. The differences are intriguing and informative. This time there *is* an origin, the only bias condition for which the device is at equilibrium. When the imposed junction voltage (which for simplicity we are taking to be equal to V_{NP}) is given the inaccessible value $-\Delta\psi_0$, we have a condition analogous to the flatband voltage V_{FB}. For equivalent P-type dopings in the junction diode and MOS capacitor, the C_t curve is nearly the same as the C_b curve.

Exercise 5-34. Why are the C_b and C_t curves not identical?

The bias voltage on the junction is measured from the N^+ region to the P region, and thus includes the small voltage drop on the N^+ side of the junction, which amounts to 26 mV and is nearly bias-independent [58]. In the MOS capacitor, on the other hand, the voltage drop relevant to C_b is that from the oxide–silicon interface to the ohmic contact on the P-type region. Because the N^+-side drop in the junction case is both small and nearly constant, the two capacitance profiles are similar. The element C_b is often compared to C_t for this reason. In spite of the small difference in the two capacitances, it is important to realize that the silicon solutions, $\psi(x)$ and $E(x)$, are *identical* when $\psi(x_J) = \psi(x_S)$.

Exercise 5-35. Why doesn't C_s appear in Figure 5-35(b)?

Diffusion capacitance is absent from the MOS device for the trivial reason that the SiO_2 layer prevents carrier injection into the silicon.

The fact that C_i does not appear in Figure 5-35(d) is nontrivial, however. After all, the N^+P junction possesses an inversion layer at equilibrium. But the behaviors under bias of MOS-capacitor and PN-junction inversion layers are totally different. With a growing positive bias (V_{GB}) on the P-substrate MOS capacitor, surface potential ψ_S follows closely behind gate potential ψ_G, and electron density at the interface, $n(x_S)$, increases *exponentially* with ψ_S.

In the PN-junction case, junction potential ψ_J is very insensitive to bias [59]. For the sake of a specific example, let us take the sample shown physically in Figure 5-36(a), having the equilibrium electron-density profile shown in Figure 5-36(b). This sample is chosen to have respective net-doping values of $N_D^+ = 10^{20}/\text{cm}^3$ and $N_A = 10^{16}/\text{cm}^3$. This makes it parallel to the high-low junction of Section 3-5.4, Exercise 3-36, and Figure 3-32. In either sample, potential drop on the high side is very nearly one normalized unit. Hence the junction potential is

$$U_J = |U_{01}| - 1 = \ln (N_D^+/n_i) - 1 = 22.0. \tag{5-84}$$

In a grossly asymmetric junction of either type, junction potential is uniquely related to (peak) electric field E_J by the expression [58]

$$E_J = (8.914 \text{ V/cm})e^{U_J/2} = 119 \text{ kV/cm}. \tag{5-85}$$

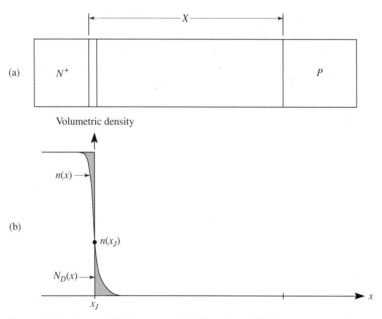

Figure 5-36 An equilibrium one-sided junction with the same respective net-doping magnitudes as the high-low junction of Figure 3-32. (a) Physical representation, showing approximate depletion-layer boundaries. (b) Net-doping profile $N_D(x)$ and electron profile $n(x)$.

The numerical result is for the present example. (The quantity E_J can also be obtained from Approximation B in Table 5-1.) For a first-order estimate of the effect of a small positive bias increment ΔV_{NP} applied to this one-sided junction, let us assume that the profile shown in Figure 5-36(b) simply translates leftward. The amount of this leftward translation would be one Angstrom for every micrometer of rightward translation for the right-hand depletion-layer boundary! This is because the doping asymmetry is 10^4, and charge balance on the two sides must be preserved. In the process, the population of the inversion layer is diminished. In summary we may observe that not only are inversion-layer responses to bias on the MOS capacitor and the *PN* junction grossly different in magnitude, but also they are different in *algebraic sign*! The miniscule inversion-layer capacitance of the one-sided junction is a negative capacitance.

Exercise 5-36. Continue to assume a simple leftward translation of the profile in Figure 3-36(b) in response to an increment of reverse bias. Doesn't the presence of the inversion layer upset the approximate calculation just given?

No. A more detailed description of the process is this: When the profile shifts leftward from its equilibrium position by a small amount Δx, the change in areal density of donor ions on the N^+ side is

$$+ N_D^+\left(1 - \frac{1}{e}\right)\Delta x.$$

The change in the areal density of inversion-layer electrons is

$$- N_D^+\left(\frac{1}{e}\right)\Delta x.$$

This change eliminates the need for an equal number of donor ions, which were "anchoring" these electrons with their lines of force. Thus, subtracting this number from the positive increment in donor-ion density yields

$$N_D^+\left(1 - \frac{1}{e}\right)\Delta x + N_D\left(\frac{1}{e}\right)\Delta x = N_D^+\Delta x.$$

This areal-density change precisely balances that on the right,

$$N_A(10^4)\Delta x.$$

Exercise 5-37. Estimate the percentage change in $n(x_J)$ when the profile in Figure 5-36(b) translates leftward one Angstrom.

The reverse current in a reverse-biased junction is so near zero that the balance of drift and diffusion currents in the transition region of the junction must be preserved. Taking specifically the position x_J, and equating the current components there yields

$$\left.\frac{dn}{dx}\right|_{x=x_J} = n(x_J)\frac{E_J}{kT/q}.$$

Here

$$n(x_J) = \frac{N_D^+}{e} = \frac{10^{20}/cm^3}{e} = 3.68 \times 10^{19}/cm^3.$$

Using the approximation

$$\left.\frac{dn}{dx}\right|_{x=x_J} = \frac{\Delta n}{\Delta x}$$

yields

$$\Delta n = n(x_J)\frac{E_J}{kT/q}\Delta x,$$

or

$$\Delta n = (3.68 \times 10^{19}/cm^3)\left(\frac{119 \times 10^3 \text{ V/cm}}{0.02566 \text{ V}}\right)(-10^{-8} \text{ cm}).$$

Thus

$$\Delta n = -1.71 \times 10^{18}/cm^3,$$

and the percentage change in $n(x_J)$ is

$$\left(\frac{-1.71 \times 10^{18}/cm^3}{3.68 \times 10^{19}/cm^3}\right)(100) = -4.6\%.$$

In fact, the profile $n(x)$ is not quite preserved under reverse bias. A more detailed examination of this matter is found in Problems A5-12 and A5-13.

5-3.4 Ideal Voltage-Dependent Capacitance

We will illustrate the combination of depletion and inversion effects in an MOS capacitor with varying bias by starting with a simple case. Assume the device to be ideal in the sense of Sections 5-2.1 through 5-2.3. In addition to setting aside parasitic effects temporarily, we shall adopt a specific structure. Take the oxide thickness to be $X_{OX} = 0.1$ μm, and the net doping of the silicon to be $N_A = 10^{15}/cm^3$. Because of ideal properties in the device assumed, the flatband condition will exist at $V_{GB} = 0$ V.

As the bias V_{GB} is quasistatically varied through wide ranges on either side of zero, let us observe the differential capacitance $C = dQ/dV_{GB}$, where Q is the net static charge on the field plate. The differential-capacitance meter delivers a small-signal voltage, assumed to have a frequency of 100 kHz, to the MOS capacitor and observes the resulting current. Bias voltage is delivered simultaneously through a high-impedance device (in principle, simply an inductor) that renders the bias circuit open to the ac signal. However, improved and refined methods for observing differential capacitance have evolved, as is noted later, because of the importance of such characterization.

We will proceed through a sequence of bias conditions parallel to those examined in Section 5-3.3. With V_{GB} strongly negative, the silicon surface is heavily accumulated, so C_{OX} is observed. In Figure 5-37 a normalized ordinate is employed, so that C/C_{OX} has the value unity at its left-hand branch. Of particular importance is the amount and distribution of incremental charge stored and recovered in the silicon in response to the ac signal applied, a matter treated in Figure 5-38, which shows the location and shape of the silicon "plate" of the MOS capacitor. The accumulation case just cited is shown in Figure 5-38(a), and the nearness of the incremental charge to the oxide–silicon interface is evident.

Now move to the flatband case, for which the analogous charge distribution is shown in Figure 5-38(b). The "centroid" of the incremental charge is now farther removed from the interface. There is a nonnegligible voltage drop in the silicon, and overall capacitance has declined somewhat through the resulting series-capacitor effect. Let us be quantitative, assuming an applied small-signal voltage that causes a surface-voltage amplitude of 50 mV, or about two normalized units. (For a fixed applied-signal amplitude, the excursions of surface potential will vary from one bias regime to another, but that fact does not invalidate the quantitative illustration presented here.) Figure 3-44 shows that for $W = W_S = 2$ on the accumulation curve (the farthest right curve in the set), the distance to the point where $W(x/L_D)$ drops to zero is about one Debye length. This can be taken as a measure of line-of-force penetration into the silicon; where $W = 0$, neutrality still exists. This distance is about 0.13 μm, appreciably greater than $X_{OX} = 0.1$ μm. On the opposite voltage excursion, the common $W(x/L_D)$ curve is relevant, and for $W_S = 2$ this time, the line-of-force penetration is about 1.5 Debye lengths. Increasing signal amplitude leads to increasing asymmetry. Because the time-dependent lines of force penetrate so deeply at the flatband-bias condition, C/C_{OX} exhibits about a 30% drop from its former value in the device arbitrarily selected. This can be seen in Figure 5-37.

As bias change is continued in a positive direction, a depletion layer forms, as is illustrated in Figure 5-38(c). The representation here is fully equivalent to that in Figure 3-56(b) for a step junction. With further positive change in bias, the depletion-layer boundary retreats farther from the interface, the depletion capacitor becomes "thicker," or smaller in value, and so the series combination of C_{OX} and C_b drops in value. This process saturates as the threshold voltage is approached. The saturation is evident in Figure 3-44. Visualize a curve for $U_{02} = U_B = 11.5$, the case for $N_A = 10^{15}/\text{cm}^3$. At the place where the near-vertical curve diverges from the common curve, one has the threshold condition. After that, line-of-force pene-

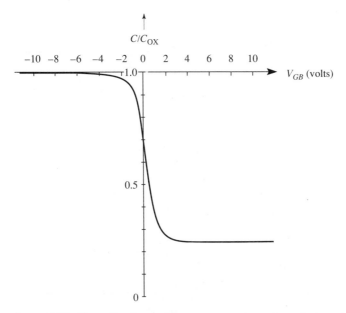

Figure 5-37 Normalized capacitance versus voltage for a device with $X_{OX} = 0.1$ μm and $N_A = 10^{15}/cm^3$, measured at 100 kHz.

tration (which is to say, depletion-layer thickness) stabilizes, growing only about one additional Debye length. Recall that the spatial zero in Figure 3-44 corresponds to the point on the $p(x)$ curve where $p(x) \approx 0.55 \, p_0$, placing that spatial position right at the depletion-layer boundary. The reason for saturation, of course, is that lines of force originating on the field plate can now terminate on the inversion-layer charge right at the interface. Figure 5-38(d) portrays conditions right at threshold, $V_{GB} = V_T$. No incremental charge is represented in the inversion layer because a relatively high frequency (100 kHz) was assumed, and no "fast" source of inversion-layer electrons was assumed. The recombination-generation process is much too slow to follow the ac signal. In Figure 5-38(e), then, is shown the depletion layer at its maximum thickness because the surface is now heavily inverted. At this point, Figure 5-37 shows that C/C_{OX} has fallen to its minimum value of about 0.27.

The C–V curve in Figure 5-37 is taken from an extensive set of such curves calculated numerically by Goetzberger [59]. The relevant subset of his curves is given in Figure 5-39. The dashed portions of these curves correspond to the assumptions of the foregoing discussion, with a high-frequency ac signal and no electron-supplying mechanism. The solid continuing curves rising from the dashed curves show the effect of inversion-layer capacitance C_i when it has been included in the circuit. In other words, the capacitive function represented by each solid curve amounts to that resulting from putting C_{OX} in series with the sum of a pair of functions like those plotted in Figure 5-35(b). The points identified by small letters

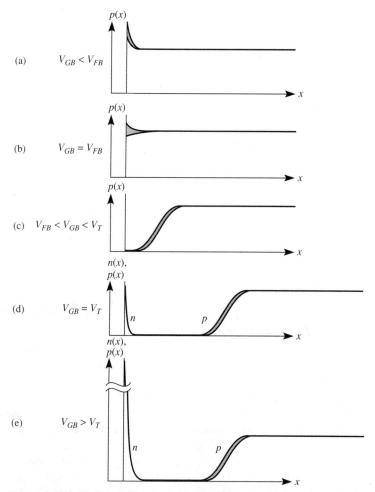

Figure 5-38 Carrier profiles in P-type substrate of MOS capacitor, showing incremental charge-storage locations in presence of a 100-kHz small signal under certain bias conditions. (a) Accumulation. (b) Flatband. (c) Depletion. (d) Depletion and threshold of strong inversion. (e) Depletion and strong inversion.

correspond to the five parts of Figure 5-38. At and beyond threshold there are two relevant points for each bias, with the lower points literally corresponding to Figure 5-38, and the upper points including the C_i presence. At threshold, points d, the capacitances C_i and C_b make equal contributions to net capacitance, although this is not obvious by inspection; a simple calculation is required to demonstrate it.

The theoretical extremes that are represented by the dashed and solid (respectively) curves at the right in Figure 5-39 can be approached practically in several

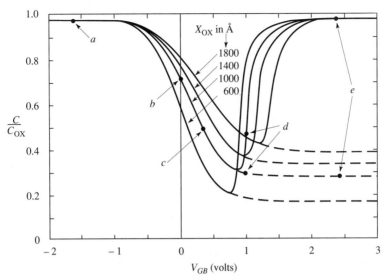

Figure 5-39 Ideal MOS curves of small-signal capacitance versus voltage for $N_A = 10^{15}/cm^3$ and several values of oxide thickness X_{OX}. The points indicated are related to the corresponding parts of Figure 5-38. (After Goetzberger [59], with permission from the Bell System Technical Journal, copyright 1966, AT&T.)

ways. One method cited earlier for activating C_i is a provision for supplying inversion-layer electrons. The structure of Figure 5-35(a) does this efficiently, because the signal voltage creates instantaneously a potential well at the surface. Into the well, electrons from the N^+ region are driven by an electric field. The mechanism here is identically that of the charge-coupled device, or CCD [60]. The inversion-layer electrons are every bit as responsive as holes driven in and out of the depletion-layer boundary region by an electric field.

Another controlling variable cited earlier is signal frequency. In illuminating experiments carried out by Grove et al. [61], it was shown that the frequency dependence at C–V behavior leads to a continuum of curves. Several of these are shown in Figure 5-40. But to approach even closer to the desired quasistatic conditions would require a frequency lower than 10 Hz, and differential-capacitance measurements there become extremely difficult. For this reason, quasistatic methods have been developed. A very slowly rising voltage ramp is applied to the MOS capacitor, and displacement current is observed. (See Sections 1-3.5 and 3-8.5.) It is a method that was first described by Berglund [62], and was then independently developed by Castagné [63], Kerr [64], and Kuhn [65]. Goetzberger et al. [66] give a good review of these and other C–V methods.

Grove and his colleagues experimentally illustrated other ways to manipulate the C–V curves between their theoretical extremes [67]. One method subjected the MOS capacitor to varying levels of illumination, thus producing excess minority

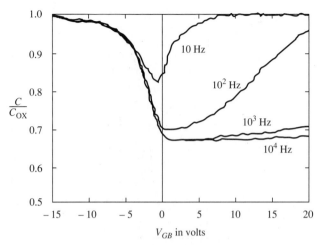

Figure 5-40 Capacitance versus voltage for an MOS capacitor with small-signal frequency as a parameter. (After Grove et al. [61], with permission, copyright 1964, American Institute of Physics.)

carriers in the silicon that were available for transport in and out of the time-dependent potential well at the interface. The result is shown in Figure 5-41. Qualitatively very similar is the temperature dependence of the curves displayed in Figure 5-42. Because n_i and hence minority-carrier density are exponentially dependent on temperature, the supply of these carriers, essential in the C_i mechanism, is sensitive to temperature.

The quasistatic extreme achieved with a very low-slope ramp voltage has an opposite extreme, which is the application of a voltage pulse with a rise time far smaller than carrier lifetime. In a device with no electron-supplying provisions, the depletion layer can grow much thicker than the inversion-determined limit. This is termed a condition of *deep depletion*. Figure 5-43 shows the resulting characteristic, a continuation of the low-voltage depletion-capacitance curve. For comparison, the inversion-capacitance and inversion-limited depletion-capacitance curves are also shown. Avalanche breakdown, or impact ionization, by carriers in the high-field region near the interface sets a limit to depth of depletion. This is also displayed in Figure 5-43, and is marked by a singularity in the slope of the characteristic. The avalanche-produced electrons drop quickly into the inversion-layer potential well, and no further drop in capacitance is possible.

5-3.5 Real Voltage-Dependent Capacitance

The nonideal features of the MOS capacitor treated in Sections 5-2.4 through 5-2.7 affect its $C–V$ properties markedly. Let us make an assumption that is plausible for present-day conditions wherein both mobile ionic charge and interface trapped

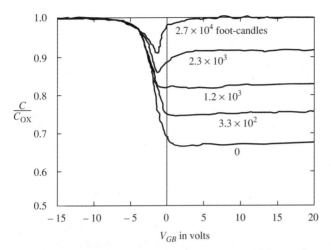

Figure 5-41 Capacitance versus voltage for an MOS capacitor with level of incident illumination as a parameter. (After Grove et al. [67], with permission, copyright 1965, Pergamon Press Ltd.)

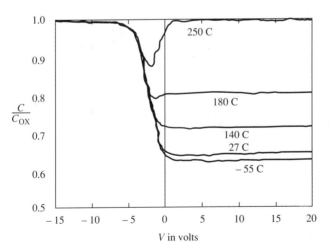

Figure 5-42 Capacitance versus voltage for an MOS capacitor device temperature as a parameter. (After Grove et al. [67], with permission, copyright 1965, Pergamon Press Ltd.)

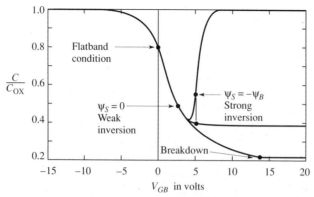

Figure 5-43 Capacitance versus voltage for an ideal MOS capacitor. The three curves at the right-hand side are, from top to bottom, that obtained with the inversion layer fully active, the inversion-limited depletion-capacitance characteristic, and the deep-depletion characteristic, with the last having an extreme-voltage limit set by avalanche breakdown. (After Grove et al. [67], with permission, copyright 1965, Pergamon Press Ltd.)

charge are negligible. With this assumption, we can write from Equation 5-75

$$V_{FB} = H_D - (Q_f/C_{OX}) = -1.7 \text{ V}, \quad (5\text{-}86)$$

where the first two terms of Equation 5-80 were used to obtain the numerical value of V_{FB}. Here we have again taken the case of a P-type silicon substrate of $N_A = 10^{15}/\text{cm}^3$, an oxide thickness of $X_{OX} = 0.1 \ \mu\text{m}$, and an aluminum field plate, all as in Section 5-2.4. The flatband point for an ideal MOS capacitor falls at $V_{GB} = 0$, as was noted in Section 5-3.4, and as is illustrated in Figure 5-44(a). Also shown there is the C–V characteristic for the real device of $V_{FB} = -1.7$ V that was just determined. The flatband point must fall at the same value of C/C_{OX} on the C–V curve of the real device as on that for the ideal device, because the two capacitors are assumed to differ only in flatband voltage. The leftward shift of the real curve with respect to the ideal curve is displayed in Figure 5-44(a), and is the only difference in the two curves.

Converting from the P-type case of Figure 5-44(a) to the N-type case of Figure 5-44(b) produces a mirror reversal of the ideal characteristic. This fact is a consequence of the symmetry in the analytic solution that was stressed in Sections 3-7 and 5-3.1. This time Figure 5-23 informs us that for an N-type substrate doped to $10^{15}/\text{cm}^3$, we have $H_D = -0.32$ V. Assuming Q_f to be the same as before, then

$$V_{FB} = H_D - (Q_f/C_{OX}) = -1.1 \text{ V}. \quad (5\text{-}87)$$

Because both terms are negative, the translation of the real curve is still negative, but by a smaller amount than before.

The parallel shift of the C–V characteristic illustrated in Figure 5-44 occurs

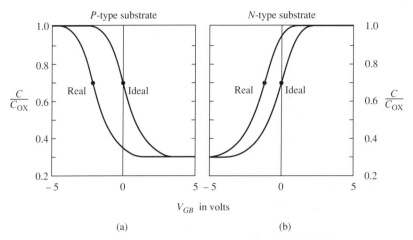

Figure 5-44 Comparing the $C-V$ characteristic of an ideal MOS capacitor and that of a capacitor that is real but free of variable, bias-dependent phenomena. (a) For the P-type substrate, Q_f and H_D cause a leftward shift of the real curve. (b) For the N-type substrate, Q_f causes the same leftward shift as before, while H_D contributes a smaller but still negative shift. (After Grove et al. [67] with permission, copyright 1965, Pergamon Press Ltd.)

only when the parasitic effects present are bias-independent. As Section 5-2.5 and Figure 5-27 explain, the phenomenon responsible for interface trapped charge is bias-dependent, or more precisely, surface-potential-dependent. As a result, it causes shape distortion in the $C-V$ characteristic. As surface-potential change and band bending occur, the responsible states are pushed above or below the Fermi level (depending upon the sense of the change), and give up or take on electrons, the essence of Bardeen's classic model of surface states [35].

One particular case of such states is illustrated in Figure 5-45. Donorlike states are assumed to exist uniformly through a range of potentials (or energies) located near the middle of the gap at the P-type silicon surface. For negative bias (let us assume that $V_{FB} = 0$), the accumulation direction, these states are well above the Fermi level as can be seen in Figure 5-45(a), and hence are empty of electrons. The assumption that these states are donorlike means their charge is positive, and hence that a negative shift of the $C-V$ characteristic results. (Note the negative signs in front of all terms of Equation 5-74 that contain the factor $1/C_{OX}$. The negative signs are present because negative voltage on the gate is needed to respond to positive charge in other parts of the capacitor.) The state of constant charge is preserved even as band bending in the depletion direction commences, a condition illustrated in Figure 5-45(b). As a result, the portion of the $C-V$ characteristic thus explored remains common in shape with the ideal characteristic, a fact confirmed in the leftmost curve of Figure 5-46, an experimental result from 1967 [39]. Finally, sufficient positive bias pulls the surface states well below the Fermi level as shown

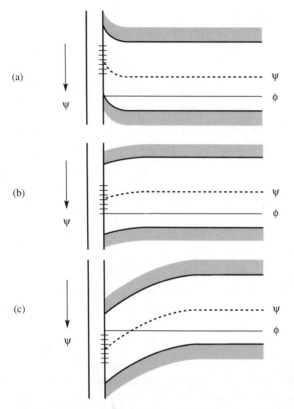

Figure 5-45 Band diagrams for various bias conditions on a P-substrate MOS capacitor with donorlike "fast surface states." (a) Negative bias, accumulation, and empty (positively charged) states. (b) Small positive bias and depletion, and states still charged. (c) Greater positive bias and depletion, with states now empty and hence neutral.

in Figure 5-45(c), where they fill and become neutral, or "invisible," by virtue of their donorlike character. At this point the C–V characteristic merges with that for a device that is identical, but permanently free of interface trapped charge.

The flatband voltage V_{FB}, or the amount of parallel shift separating the real and ideal C–V characteristics for a given MOS capacitor, is also very useful in MOSFET analysis. Hence let us emphasize a relationship that may not have been evident heretofore. The key equation for relating voltage difference and potential difference (not the same!) with only the nonideal barrier-height effect present was Equation 5-50, $V_{GB} = H_D + \psi_{GB}$. With the generalization that allows other parasitic effects to be present also, the key equation becomes

$$V_{GB} = V_{FB} + \psi_{GB}. \tag{5-88}$$

Hence in the real device, ψ_{GB} can be written

$$\psi_{GB} = V_{GB} - V_{FB}. \tag{5-89}$$

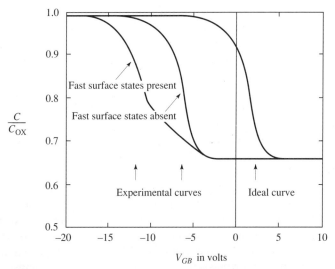

Figure 5-46 The effect of bias-dependent interface trapped charge residing in "fast surface states" at the oxide-silicon interface. (After Deal et al. [39] with permission, copyright 1967, The Electrochemical Society, Inc.)

To convert any of the capacitor equations of Section 5-2 into terms of V_{FB}, it is simply necessary to substitute $(V_{GB} - V_{FB})$ for ψ_{GB}.

5-3.6 Physics of MOS-Capacitance Crossover

Figure 5-35(b) noted the fact that bulk capacitance C_b and inversion capacitance C_i are equal at the threshold voltage. That is, the two C–V functions intersect at a point we have named the crossover point, for which $V_{GB} = V_T$. A clear but nonobvious physical explanation has been given for the crossover phenomenon [68], an explanation that we now outline.

In what follows, we employ the customary definitions for the relevant differential capacitances:

$$C_b \equiv \frac{dQ_b}{d\psi_S}, \tag{5-90}$$

where Q_b is bulk charge per unit area and ψ_S is surface potential, and

$$C_i \equiv \frac{dQ_n}{d\psi_S}, \tag{5-91}$$

where Q_n is inversion-layer charge per unit area, with the subscript carried over from Section 5-1.3. Also, the threshold condition is defined as that for which surface potential is equal in magnitude and opposite in sign to the bulk potential, with the Fermi level taken as potential reference.

Let us consider such an MOS capacitor with increasing gate voltage of the depletion-inversion polarity. The decline of bulk capacitance that results is governed by a power law, while the growth of inversion-layer capacitance is governed by a stronger and more complicated function, as we have seen. Curiously, these two differing functions appear to intersect at the threshold of strong inversion. This intersection is more striking when one considers that the charge storage in the inversion layer occurs very close to the surface (interface) in question; and by contrast, the incremental charge storage involved in bulk capacitance (1) has a position that is remote from the surface, (2) has a position that is a function of surface potential, and (3) has a spatial profile that is "fuzzy" in the sense that the incremental charge is stored in a region of appreciable thickness, as is illustrated in Figure 5-38. All three issues are relevant because potential calculations are very sensitive to charge position. We confine ourselves once again to a one-dimensional geometry, since the MOS capacitor poses a problem that can be made one-dimensional to any arbitrary degree of accuracy by increasing its area. Also, we consider only the case of a uniformly doped substrate. The question to be answered is this: Why do two mechanisms with such differing characters yield identical capacitance values at the threshold condition?

In addition to the considerations just outlined, there are alternative physical descriptions of the charge entering into bulk capacitance. To be specific, let us continue with the case of a P-type silicon bulk material. The increment of bulk charge that accompanies a positive increment of surface potential can be regarded (a) as the charge of the acceptor ions that are "uncovered," or (b) as the local decrease in hole population near the depletion-layer boundary, with the negative sign involved in "decrease" providing the necessarily negative sign for the charge increment. Note that the inversion-layer charge description is, by contrast, singular.

The descriptions just offered are consistent with the practice presented in Figure 5-47(a). There we affix to the x-axis a spatial origin located at the silicon surface, as is usually done. Then we translate to the right a charge profile by an amount depending upon surface potential. Figure 5-47(b) shows a series of charge-profile positions for equal increments of surface potential, expressed in normalized terms, with the bulk potential (rather than the Fermi level) taken as reference. Definitions here are consistent with those given in Section 3-7. Once bulk doping is specified, the charge profile, which comprehends both bulk and inversion-layer charge, is unique and well known.

By shifting conceptual gears, however, we find that the physical basis for the equality of bulk and inversion-layer capacitance at equilibrium becomes evident. Let us first deal with the issue of "fuzziness." In a one-dimensional problem, to reiterate, electric field is unlike electrostatic potential in that it is insensitive to charge position. Only the *total* charge per unit area beyond a certain position matters. The constancy of charge density through a range of distances that is evident in Figure 5-47(b) means that electric field will be linear through the same range, as shown in Figure 5-48(a) for a specific though arbitrary case. If the linear field profile is extrapolated to the x axis, it defines a position with the significance

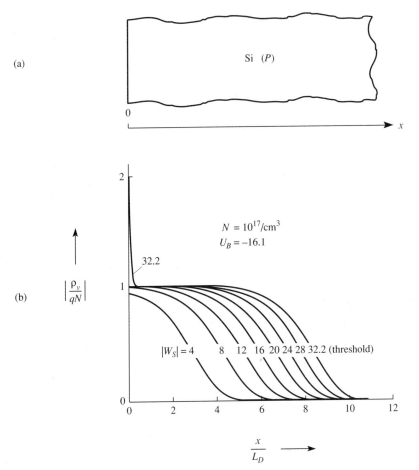

Figure 5-47 Traditional view of the silicon-surface or MOS-capacitor problem. (a) A P-type silicon sample with the spatial origin placed at the surface. (b) A series of numerically determined volumetric-charge-density profiles for progressively increasing normalized surface potential W_S.

explained in Section 3-7. It is the position at which the abrupt space-charge boundary would occur when the depletion approximation is applied to the same MOS-capacitor problem, and when electric field at the silicon surface attributable to ionized impurities is the *same* for the depletion approximation as for the actual case. (Repeating the insight contained in Exercise 3-54, let us note that the shaded area in Figure 5-48(a) amounts to one normalized potential unit, kT/q, and accounts for the one-unit difference between the depletion-approximation potential-profit parabola and the asymptotic parabola of the exact numerical solution.)

Now to be more specific, choose any position x in Figure 5-48(b) in the positional range where space-charge density is constant. Then the total charge per unit area to the right of that position (toward the bulk) will be the same for the depletion-approximation profile and for the actual profile. As a result, we may point out that the two cross-hatched areas in Figure 5-48(b) are equal. This position (defined by extrapolating the linear portion of the field profile) is, of course, the origin of the depletion-approximation replacement (DAR), discussed in Section 3-7.5.

With respect to this origin, let us define surface position x_S. Furthermore, and this is where the conceptual shift enters, let us consider surface potential ψ_S to be an independent variable. Then, in a thought experiment, surface position becomes the

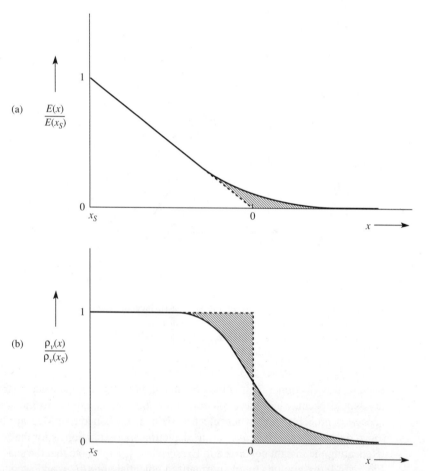

Figure 5-48 Physical meaning of the invariant spatial origin employed in the depletion-approximation replacement. (a) Electric-field profile in the neighborhood of the depletion-layer boundary, showing the actual profile (solid line) and the depletion-approximation profile (dashed line) that defines the new spatial origin. (b) The new spatial origin in relation to the volumetric-charge-density profile. The two crosshatched regions have equal area in view of part (a) above and Gauss's law.

dependent variable $x_S = f(\psi_S)$; the silicon crystal exhibits positive or negative "growth" at its surface to accommodate any arbitrary value of ψ_S. This approach takes advantage of the invariance that exists at the depletion-layer boundary: The origin bears a fixed positional relationship to the invariant charge, field, and potential profiles that exist there.

Consider a further illustration of the convenience in this approach. Choose any position whatsoever for x_S, just so that it is left of the position in Figure 5-48(a) where the actual and depletion-approximation field profiles have merged. Then ionic charge per unit area in the silicon can be *accurately* written as:

$$Q_B = qNx_S, \qquad (5\text{-}92)$$

where N is net impurity density in the uniformly doped substrate, whether one employs the actual *or* the depletion-approximation formulation.

The depletion approximation was invoked previously because it aids the physical interpretation of the DAR origin located in the vicinity of the depletion-layer boundary. But now let us set the depletion approximation aside and think in terms of the actual field, charge-density, and potential profiles in a real MOS capacitor, with the first two shown in Figures 5-48(a) and (b), respectively. Then, focus on the origin plane in Figure 5-48 and "freeze" its position in the silicon crystal, anchoring it to a particular atomic plane. The next step is to apply the thought experiment just described. Suppose ψ_S has been chosen at a value that corresponds to a moderate amount of inversion. Then apply a small positive increment to ψ_S. The result is the addition of a few monolayers of silicon at the surface, incorporating the right number of acceptor ions and the right number of inversion-layer electrons. Now the "equality" matter at *threshold* has a self-evident explanation. By definition, at threshold the volumetric electron density at the surface equals the volumetric ionic density there. Hence a tiny surface-potential increment at threshold involves equal amounts of inversion-layer charge and bulk charge in the thought-experiment layer of infinitesimal thickness; the silicon monolayers just "grown" are located where the two density functions intersect. This precise equality holds within the limits of the usual assumptions and approximations—those involving Boltzmann statistics, band symmetry, equivalent densities of states, and complete ionization.

Note that we are not offering the thought experiment as a description of the storage process. Rather, it is being offered as a description of before-and-after conditions (before and after the application of a surface-potential increment) that are *identical* to those in the real MOS capacitor. Consequently, the change in bulk charge as a result of the applied potential increment was the same in the thought experiment and the actual device. Finally, then, the bulk-charge and inversion-layer-charge increments are equal in the actual MOS capacitor, and equal charge increments for a given potential increment will lead by definition to equal inversion-layer and bulk capacitances at threshold. In this way the thought experiment provides the necessary bridge to physical plausibility for an initially obscure equality between two very different functions in the MOS-capacitor problem, and reemphasizes the critical character of the threshold condition.

5-3.7 Analysis of MOS-Capacitance Crossover

An early analytic demonstration of C_b and C_i equality at threshold was given by Tobey and Gordon [69]. They used two significant approximations to do so, however. One was the depletion approximation, and the other was the assumption that inversion-layer charge is negligible at threshold in comparison to bulk charge. In fact, the former typically amounts to several percent of the latter in terms of areal density. A more accurate approximate treatment was given subsequently by Tsividis [70], verifying the earlier conclusions. Later, a still more accurate assessment was given [71], using the methods that are the subjects of Sections 3-7 and 5-3.1. This analysis, instructive in several ways, proceeds as follows.

The exact expression for normalized electric field is given as the first equation in Table 5-2. For the $|U_B|$ range of interest to us, it readily simplifies to Approximation A in Table 5-1, which in turn simplifies to Approximation C.

Exercise 5-38. Explain these simplifications.

The ratio of the two terms in the denominator of the first expression, or $(\exp |U_B|):(\exp -|U_B|)$, amounts approximately to 10^{10} for $|U_B| = 11.5$, corresponding to a net doping of $10^{15}/\text{cm}^3$. Even for $|U_B|$ as low as 2.0, the ratio is about 55. Hence, dropping the second term in the denominator and performing the indicated division yields Approximation A. Then, for $W \geq 3$, the exponential term in the first parentheses can be dropped, and the terms $-W$ and -1 can be dropped in comparison with e^W in the second parentheses, yielding Approximation C.

Because we are specifically interested in electric field in the silicon at the interface, let us write Approximation C in terms of W_S:

$$\left| \frac{dW}{d(x/L_D)} \right|_{W=W_S} = \sqrt{2(W_S - 1 + e^{W_S - 2|U_B|})}. \tag{5-93}$$

The first two terms in parentheses are associated with depletion-layer or bulk charge, while the last (exponential) term is associated with inversion-layer charge. The fact that the two charge components can be so resolved was noted at the end of Section 5-3.1 for the general (exact) electric-field expression that was the starting point for the present section; then, following through the chain of approximations outlined in Exercise 5-38 verifies the assertion just made about Approximation C.

The small-signal capacitances C_b and C_i are in parallel, and therefore are additive. Let us use the conventional symbol Q_s to represent the total charge per unit area stored in the silicon. (The present MOS context and lowercase subscript should prevent confusion of this symbol with the similar symbol used in Chapter 3 for charge stored in the diffusion capacitance of the junction diode.) Accepting these symbols gives us the definition

$$C_b + C_i \equiv \frac{dQ_s}{d\psi_S}. \tag{5-94}$$

(We follow convention here by omitting the negative sign that a strict construction would require in this and the following expressions.) The value of Q_s as a function of ψ_S (or W_S) can be obtained via Gauss's law directly from the expression for electric field in the silicon at the oxide–silicon interface, Equation 5-93.

Letting Q_b stand for bulk (ionic) charge stored in the depletion-layer or bulk capacitance gives us in a parallel way,

$$C_b \equiv \frac{dQ_b}{d\psi_S}. \tag{5-95}$$

Once again, the value of Q_b as a function of ψ_S (or W_S) can be obtained via Gauss's law for conditions sufficiently below threshold, directly from the bulk-charge field component of Equation 5-93. Finally, we are able to calculate the most important and interesting capacitance C_i as follows:

$$C_i = \frac{dQ_s}{d\psi_S} - \frac{dQ_b}{d\psi_S}. \tag{5-96}$$

Note carefully that one may *not* compute C_i for any range of operation directly from the inversion-layer-charge component of electric field in Equation 5-93 in a manner parallel to that used to determine C_b in Equation 5-95.

Exercise 5-39. Why does this departure from parallelism exist? Why cannot one use the last term of Equation 5-93 and Gauss's law to calculate C_i in at least a limited range?

The important difference between C_b and C_i is that C_b can exist in the absence of C_i, and C_i can exist *only* in the presence of C_b. In a one-dimensional problem electric field is independent of charge position, but electrostatic potential is highly sensitive to charge position. Hence, even though a one-to-one relationship exists between the components of Equation 5-93 and the components of stored charge (via Gauss's law), if one attempts to calculate C_i from the prescription $\Delta Q_n/\Delta W_S$ (where Q_n is the areal density of inversion-layer charge), one obtains the wrong answer because ΔW_S is incorrect.

Normalized electric-field magnitude $|dW/d(x/L_D)|$ can be converted to unnormalized form by writing

$$|E| = \frac{1}{L_D} \frac{kT}{q} \left| \frac{dW}{d(x/L_D)} \right|. \tag{5-97}$$

Thus Equation 5-93 in partly unnormalized terms becomes

806 THE MOSFET

$$|E| = \frac{1}{L_D}\frac{kT}{q}\sqrt{2(W_S - 1 + e^{W_S - 2|U_B|})}. \tag{5-98}$$

An expression for Q_s, total areal charge density in the silicon, can hence be found from Equation 5-98 by applying Gauss's law (or by noting that electric displacement possesses the desired dimensions):

$$Q_s = \frac{\epsilon\sqrt{2}}{L_D}\frac{kT}{q}\sqrt{W_S - 1 + e^{W_S - 2|U_B|}}. \tag{5-99}$$

Total silicon-based capacitance thus becomes

$$C_b + C_i = \frac{dQ_s}{d\psi_S} = \frac{q}{kT}\frac{dQ_s}{dW_S}$$
$$= \frac{\epsilon\sqrt{2}}{L_D}\frac{d}{dW_S}\sqrt{W_S - 1 + e^{W_S - 2|U_B|}}. \tag{5-100}$$

Or,

$$C_b + C_i = \frac{\epsilon\sqrt{2}}{L_D}\frac{1 + e^{W_S - 2|U_B|}}{\sqrt{W_S - 1 + e^{W_S - 2|U_B|}}}, \tag{5-101}$$

where the unity term in the numerator of the last factor arose from the bulk-charge phenomenon and the exponential term, from the inversion-layer phenomenon. Thus it is now evident that C_b and C_i are equal at threshold.

Exercise 5-40. Explain the last statement.

At threshold, $W_S = 2|U_B|$. Hence the exponential term in the numerator is unity at threshold and the two phenomena make equal contributions.

To summarize, the capacitance expressions having the wide-ranging validity of Approximation C in Table 5-1 are

$$C_b = \frac{\epsilon\sqrt{2}}{L_D}\frac{1}{\sqrt{W_S - 1 + e^{W_S - 2|U_B|}}}, \tag{5-102}$$

and

$$C_i = \frac{\epsilon\sqrt{2}}{L_D}\frac{e^{W_S - 2|U_B|}}{\sqrt{W_S - 1 + e^{W_S - 2|U_B|}}}. \tag{5-103}$$

5-4 IMPROVED MOSFET THEORY

A real MOS capacitor can be brought arbitrarily close to ideal one-dimensional conditions simply by increasing gate area. Although we have in some cases introduced a junction near the gate edge to supply carriers to the inversion layer, that feature can equally well be omitted. The MOSFET, by contrast, possesses inevitable complications that remove it a long way from ideal properties. Its MOS-capacitor portion must interact with junctions in order for the device to work. Its drain–source bias causes an IR drop in the channel that causes surface potential to vary from drain to source. This variation causes the MOSFET to pose, at best, a two-dimensional problem. Further, the presence of its drain–source current, or channel current, means that nonequilibrium conditions exist there. Quasi Fermi levels are useful for treating such conditions. Finally, technological evolution has brought a steady shrinkage of MOSFET feature sizes. Many present-day MOSFETs have critical dimensions in the micrometer neighborhood. The tiny dimensions cause some aspects of MOSFET analysis to constitute a three-dimensional problem.

Immediately below we wish to look at the surface inversion layer in the presence of an orthogonal junction. Then we factor into MOSFET theory the depletion-layer and bulk-charge behavior that received emphasis in Sections 5-2 and 5-3. (Recall that in Section 5-1.3 we treated inversion-layer charge as the only component of charge in the silicon.) The effect of reverse bias from source to substrate, known as *body effect,* will concern us then, leading up to numerical treatments of the MOSFET.

5-4.1 Channel–Junction Interactions

Admitting the inevitability of a two-dimensional problem when junctions have been introduced, we shall proceed to idealize the geometry as much as remains possible. As the first idealization, let us assume that the source and drain junctions (both taken to be N^+P junctions) are plane and parallel to one another, while normal to the oxide–silicon interface. This relationship is displayed in Figure 5-49(a), a cross-sectional representation. For the second simplification, we have chosen the source–drain spacing L to be large, 5 μm, as is also shown in Figure 5-49(a). In the resulting *long-channel* MOSFET, the dimension L considerably exceeds the thickness of all depletion layers, those of source and drain junctions and the surface (or oxide–silicon interface). The extreme position of the boundary for the last depletion layer, approximately that at threshold, is shown by means of a dashed line in Figure 5-49(a), assuming that source and bulk are electrically common and assuming the absence of drain–source bias. As a final simplification, we have permitted the oxide and field plate (gate electrode) to overlap the source and drain regions appreciably. This is manifestly a departure from practice that has been standard for twenty years; one seeks to terminate the gate structure precisely at the metallurgical junction, thus minimizing troublesome parasitic capacitances. But with overlap, some features of the explanation that follows become clearer. In spite of this and the other departures from present-day realism, the distorted structure that results is capable of functioning as an actual MOSFET and will facilitate some instructive descriptions.

In Figure 5-49(b), the same structure has been rendered in perspective. Our aim

808 THE MOSFET

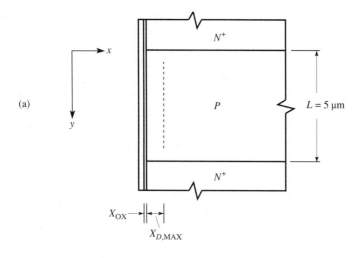

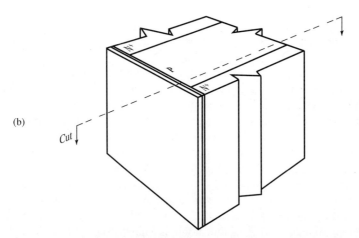

Figure 5-49 Idealized MOSFET having plane source and drain junctions normal to the oxide–silicon interface and having gate overlap of the source and drain regions.
(a) Cross-sectional representation, showing dimensions and showing with a dashed line the extreme position of the depletion-layer boundary. (b) Perspective representation, showing cut that will permit display of band diagrams in orthogonal directions. (After Warner and Grung [72].)

is to approximate the two-dimensional problem we face by using two orthogonal one-dimensional solutions. As the next step toward that goal, let us cut the sample in the manner indicated in Figure 5-49(b), and discard the right-hand portion. With this accomplished, and with the further step of rendering the oxide and field plate

IMPROVED MOSFET THEORY   809

"invisible," we arrive at the uncluttered sample depicted in Figure 5-50(a). (Although we have chosen not to represent the field plate and oxide here, we assume that they are capable of normal function.)

Let us set $V_{GB} = V_{FB}$, so that the band diagram from surface to bulk is as shown in Figure 5-50(b). This is the same x-axis profile that absorbed us throughout Sections 5-2 and 5-3. The orthogonal one-dimensional band diagram in the y, or source-to-drain, direction is that of a one-sided step junction. A significant proviso is needed, however. The bias V_{GB} that creates flatband conditions in the P-type bulk region will *not* do so in the N^+ region. But this complication is one that we can set aside for now. We can show that the surface potential of a heavily doped region is

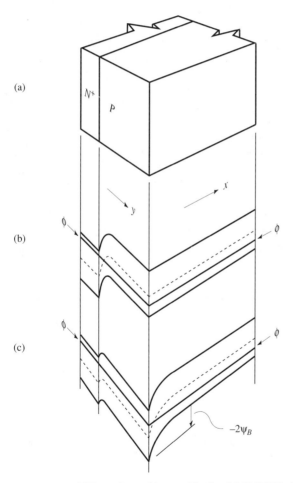

Figure 5-50 Effect of gate bias on idealized MOSFET. (a) Device of Figure 5-49(c) with field plate and oxide invisible and with drain region discarded. (b) Orthogonally intersecting one-dimensionsal band diagrams adjusted to the threshold condition. (After Warner and Grung [72].)

quite unresponsive to change induced through an MOS capacitor, so we shall neglect any change there resulting from the application of V_{FB} from gate to bulk.

To see the reason, refer back to Figure 5-19(d)—a weak-depletion band diagram based upon the depletion approximation. For the conditions arbitrarily assumed, ψ_{GS} and ψ_{SB} are roughly equal, meaning that the rectangular and triangular electric-field profiles in Figure 5-19(b) have comparable areas. Now, in a thought experiment, hold E_S constant while increasing substrate doping. As a result, the triangular area shrinks monotonically while the rectangular area remains constant. Since the triangular area is identically ψ_{SB}, this means the surface potential progressively approaches the bulk potential. (Note that to make this qualitative point, an explanation using the depletion approximation is fully adequate.) For the accumulation condition, line-of-force penetration is inherently small, so the property of surface-potential insensitivity holds for either polarity of applied bias on a heavily doped substrate.

As the next step, let us bias the device at the threshold voltage. The effect on the orthogonal band diagrams is shown in Figure 5-50(c). Look first at the band diagram in the x or depth direction. It shows the familiar threshold-condition band bending of $-2\psi_B$, or about $+0.59$ V for $N_A = 10^{15}/\text{cm}^3$. Now examine the band diagram on the "front" of the silicon sample, involving the N^+P source junction. Once again surface potential on the N^+ region is little affected by the bias change. But the overall band diagram now resembles that of an N^+N (high-low) junction. The presence of the incipient inversion layer makes the surface region of the substrate appear N-type, but not as N-type as the N^+ region. Because the silicon remains at equilibrium, comparing the situation to that in an N^+N junction at equilibrium is appropriate.

The band diagram on the front of the block superficially resembles that for an N^+P junction under forward bias. This comparison is much less apt, however, because that would be patently a nonequilibrium situation, and bias has not yet been applied to the junction.

Exercise 5-41. Assuming threshold conditions, expand upon the comparison of the "front-surface" band diagram to an N^+N junction and to a forward-biased N^+P junction, commenting especially on hole populations near the surface.

The inversion-layer electrons are *surplus* carriers, and their enhanced density is accompanied by depressed hole density, with the product being $pn = n_i^2$. The same would hold in the N^+N junction at equilibrium. On the other hand, if the electron elevation in the P-type region were the result of forward biasing an N^+P junction, those *excess* electrons would be accompanied by an equal density of excess holes.

For the next examination of the idealized two-dimensional MOSFET, let us return to the structure of Figure 5-49. Looking directly at the "front" of the substrate and once again letting field plate and oxide be invisible yields the picture in Figure 5-51(a). An added feature now is an indication of equilibrium depletion-layer boundaries for the source and drain junctions. The most meaningful positions for the channel ends are at those boundaries, as also indicated by 0 and L on the y axis. Once again we choose the flatband condition of Figure 5-50(b), with the result displayed in Figure 5-51(b). The surface potential $\psi_S(y)$ (the mid-gap line) is evidently aligned with the bulk potential ψ_B from $y = 0$ to $y = L$, confirming flatband conditions. Next let us apply a gate–source bias V_{GS} (equal to V_{GB}, since we have short-circuited the source junction). Quite arbitrarily, we have brought the upper band edge for the P-type substrate into alignment with ψ_B, as can be seen in Figure 5-51(c). To avoid cluttering the diagram, we have discarded the balance of the upper band edge and all of the lower band edge.

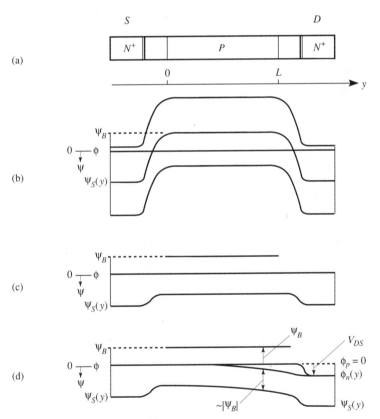

Figure 5-51 Idealized MOSFET of Figure 5-49(c) with biases applied. (a) Physical representation of structure, looking normal to substrate surface. (b) Band diagram for flatband condition in bulk region. (c) Potential profile $\psi_S(y)$ for V_{GS} having arbitrary positive value and $V_{DS} = 0$. (d) Potential profiles for small V_{DS} and for the same value of V_{GS} as in part (c). (After Warner and Grung [72].)

812 THE MOSFET

Exercise 5-42. Does the condition represented in Figure 5-51(c) constitute the threshold condition? If not, how far does it depart and in which direction?

We saw that in Figure 5-50(c), at threshold, we had band bending of $-2\psi_B = +0.59$ V. Since half the gap is 0.56 V, it follows that the band bending at threshold exceeds half the gap by 0.03 V, bringing the upper band edge that far below ψ_B in Figure 5-50(c). In Figure 5-51(c), then, the band bending is below threshold, with surface potential being more positive than the bulk potential by $+0.56$ V rather than by $+0.59$ V.

The next step is to apply a small drain–source bias V_{DS} and then examine its effect on the band diagram at the sample surface. This constitutes a nonequilibrium condition, as noted before, so quasi Fermi levels must enter the discussion. The effect of V_{DS} is to forward bias the source junction and reverse bias the drain junction.

Exercise 5-43. Does V_{DS} divide evenly between the two junctions? If not, how does it divide?

The asymmetry of forward and reverse properties of the junctions, assumed to be identical junctions, ensures that reverse bias on the drain junction exceeds forward bias on the source junction. Since the two junctions carry the same current, neglecting leakage components, one could consult a diagram such as Figure 3-24 to assess approximately the voltage disparity at a particular current.

The further effect of V_{DS} is to cause ψ_S to vary from source to drain, as shown by the lowest curve in Figure 5-51(d). The quasi Fermi level for electrons $\phi_n(y)$ shows a similar curvature. Recall from the discussion in Section 4-4.1 that the value of $d\phi_n/dy$ times $n(0, y)$ is proportional to current. The symbol $n(0, y)$ is convenient for designating volumetric electron density at the interface in the present two-dimensional problem. By contrast to the electron case, we find that $d\phi_p/dy \approx 0$, so we conclude that holes contribute negligibly to conduction along the surface.

Exercise 5-44. The slope of $\phi_n(y)$ in Figure 5-51(d) increases toward the drain, implying a surface-electron density that declines toward the drain, since channel current is y-independent. But it is also evident that $\psi_S - \psi_B = \psi_{SB}$, the band bending, increases toward the drain. Doesn't greater band bending imply higher electron density?

> In the *equilibrium* problem posed by the MOS capacitor, greater band bending indeed implies higher inversion-layer density. But in the *nonequilibrium* problem posed by the MOSFET, this is not so. The important spacing here is $\psi_S(y) - \phi_n(y)$, by definition of the quasi Fermi level. For the bias condition shown in Figure 5-51(c), this spacing is roughly constant because it depends logarithmically upon $n(0, y)$, which declines slowly from source to drain.

> **Exercise 5-45.** How do we know that $n(0, y)$ declines from source to drain?
>
> We know that $n(0, y)$ vanishes at $x = L$, the left-hand boundary of the drain junction. (See Problems A5-10 and A5-18 for more detail.)

It is appropriate to point out here that the quantity $V(y)$ described as "voltmeter voltage in the channel" in Section 5-1.3 is actually $\phi_n(y)$, because a voltmeter measures Fermi-level difference. This thought must not be carried beyond the conceptual level, however; a voltmeter probe must make an ohmic contact to a region where Fermi level is to be sensed, and an ohmic contact to a region exists only if the quasi Fermi levels have merged at the point of contact. While this is clearly not the case in the channel region as shown in Figure 5-51(d), it does hold in the N^+ source and drain regions. Thus one can use a voltmeter to measure $V_{DS} = (\phi_n)_{DS} = (\phi_p)_{DS}$.

5-4.2 Ionic-Charge Model

The elementary MOSFET analysis of Section 5-1 employs a series of simplifying assumptions. The most extreme of these is the assumption that the areal density of inversion-layer charge at the position y is proportional to $\psi_{GS}(y)$, or to $E_{OX}(y)$, to use a symbol of self-evident meaning. (Recall that the subscript S on the symbol ψ refers to *surface*.) The dominance of ionic or bulk charge over inversion-layer charge in wide operating ranges causes the rudimentary analysis to be substantially in error. What is surprising and nonobvious, however, is that the simply derived current equations can be rendered extremely useful and quite accurate by applying empirical factors to them [72].

An early treatment of the effect of bulk-charge presence on MOSFET characteristics was carried out by Ihantola and Moll [73]. In presenting this analysis, however, we shall preserve several important simplifying assumptions, worth reviewing at this point. The first is the gradual-channel or long-channel assumption, which validates the use of one-dimensional equations for x-direction dependences. This is simply another way of describing the superposition of two one-dimensional solutions, put into effect in Figure 5-50. One way of specifying gradual-channel conditions is to set a low upper limit on the ratio of $|E_S(y)|$ to $|E_S(x)|$. For any given bias condition, one can reduce the ratio at a particular position by increasing channel length L, thus validating the equivalence of the two terms (gradual-channel and long-channel) that denote the basic assumption we shall designate (1). Further assumptions are (2) uniform bulk doping, (3) transport of electrons in the channel by drift only, (4) a constant mobility μ_n governing that transport throughout the channel, (5) negligible junction leakage current, and (6) electrically common source and bulk regions.

Figure 5-51(d) will be a helpful guide in the following discussion. Let us assume, though, that instead of having V_{GS} slightly below threshold as in that figure (see Exercise 5-42), we have a value somewhat above threshold. The profile $\psi_S(y)$ is quite insensitive to increases in V_{GS} beyond threshold because inversion-layer charge is so close to the silicon surface, a fact that will be exploited in the present analysis. Accepting the assumptions just given, we can take Equation 5-89 as a starting point. Because Fermi level has a constant value throughout the gate region and a different constant value throughout the bulk region, we may write

$$V_{GS} - V_{FB} = \psi_{GB} = \psi_{GS}(y) + \psi_{SB}(y) = \text{constant.} \tag{5-104}$$

From the capacitor law,

$$\psi_{GS}(y) = -\frac{Q_s(y)}{C_{OX}}. \tag{5-105}$$

Here, as in the rudimentary analysis, a minus sign enters the capacitor law because voltage is applied to the field plate, while the charge of interest is on the opposite "plate" of C_{OX}. Thus we have the first term of Equation 5-104. An approximate second term can be written by consulting Figure 5-51(d):

$$\psi_{SB}(y) \approx \phi_n(y) - 2\psi_B. \tag{5-106}$$

This expression can be explained as follows: All potentials are referred to $\phi_p = 0$. Toward the drain end of the channel, $\phi_n(y)$ is somewhat positive. Choosing an arbitrary point in the channel and requiring it to be right at the threshold condition requires the further positive band bending of $|2\psi_B|$. But we can generalize and write the band bending as $-2\psi_B$. For the present P-type substrate, ψ_B is negative, so that all terms of Equation 5-106 are thus rendered positive. For an N-type substrate, on the other hand, all terms would be negative, so that Equation 5-106 is valid in either case. The approximation that Equation 5-106 involves is neglect of the small amount of band bending connected with variations in channel strength above threshold.

This brings us to

$$Q_s(y) = -C_{OX}[\psi_{GB} + 2\psi_B - \phi_n(y)] = Q_n(y) + Q_b(y). \tag{5-107}$$

The first of these two equations results from substituting Equations 5-105 and 5-106 into Equation 5-104. The second is from the definition of Q_s. Solving for the charge component of primary interest, Q_n, yields

$$Q_n(y) = -C_{OX}[\psi_{GB} + 2\psi_B - \phi_n(y)] - Q_b(y). \tag{5-108}$$

Applying the depletion approximation (Table 3-1) yields an estimate of the other charge component:

$$Q_b(y) = -qN_A X_{D,\text{MAX}} = -\sqrt{2\epsilon q N_A [\phi_n(y) - 2\psi_B]}. \quad (5\text{-}109)$$

Substituting Equation 5-109 into Equation 5-108 gives

$$Q_n(y) = -C_{\text{OX}}[\psi_{GB} + 2\psi_B - \phi_n(y)] + \sqrt{2\epsilon q N_A [\phi_n(y) - 2\psi_B]}. \quad (5\text{-}110)$$

The analogous expression from Section 5-1.3 is

$$Q_n(y) = -C_{\text{OX}}[V_{GS} - V_T - V(y)], \quad (5\text{-}111)$$

and comparing the two reveals the difference in the two analyses. The electron Fermi level ϕ_n now replaces "channel voltage" V. Subsequent logic in the present analysis is the same as before, however, so we write

$$I_{\text{CH}} \int_0^L dy = Z\mu_n \int_0^{V_{DS}} Q_n(\phi_n) d\phi_n, \quad (5\text{-}112)$$

from which

$$I_{\text{CH}} = -\mu_n C_{\text{OX}} \frac{Z}{L} \int_0^{V_{DS}} \left[(\psi_{GB} + 2\psi_B - \phi_n) - \frac{1}{C_{\text{OX}}} 2\epsilon q N_A (\phi_n - 2\psi_B)^{1/2} \right] d\phi_n. \quad (5\text{-}113)$$

Because $I_{\text{CH}} = -I_D$, as noted before, we have after integration

$$I_D = \mu_n C_{\text{OX}} \frac{Z}{L} \left[(\psi_{GB} + 2\psi_B)\phi_n - \frac{\phi_n^2}{2} - \frac{2}{3} \frac{\sqrt{2\epsilon q N_A}}{C_{\text{OX}}} (\phi_n - 2\psi_B)^{3/2} \right]_0^{V_{DS}} \quad (5\text{-}114)$$

Letting

$$\gamma \equiv \frac{\sqrt{2\epsilon q N_A}}{C_{\text{OX}}}, \quad (5\text{-}115)$$

and using Equation 5-104 to eliminate ψ_{GB} from Equation 5-114 yields

$$I_D = \mu_n C_{\text{OX}} \frac{Z}{L} \left[\left(V_{GS} - V_{FB} + 2\psi_B - \frac{V_{DS}}{2} \right) V_{DS} \right.$$
$$\left. - \frac{2}{3}\gamma \left((V_{DS} - 2\psi_B)^{3/2} - (-2\psi_B)^{3/2} \right) \right]. \quad (5\text{-}116)$$

Later we will show that

$$-V_{FB} + 2\psi_B = -V_T + \gamma\sqrt{-2\psi_B}, \quad (5\text{-}117)$$

where V_T is threshold voltage. With this substitution, Equation 5-116 becomes

$$I_D = \mu_n C_{\text{OX}} \frac{Z}{L} \left[\left(V_{GS} - V_T + \gamma\sqrt{-2\psi_B} - \frac{V_{DS}}{2} \right) V_{DS} \right.$$
$$\left. - \frac{2}{3}\gamma \left((V_{DS} - 2\psi_B)^{3/2} - (-2\psi_B)^{3/2} \right) \right]. \quad (5\text{-}118)$$

In the SPICE program for modeling the MOSFET [74], the rudimentary analysis of Section 5-1.3 is known as *Level-1* characterization. The SPICE *Level-2* characterization, on the other hand, considers ionic charge. Symbols for basic and physical quantities in the Level-2 model are listed and explained in Tables 5-3 and 5-4. This SPICE model employs a somewhat more general version of Equation 5-118 for drain current, given in Table 5-4. In particular, the Level-2 equation permits a finite bulk–source bias V_{BS} that we have previously assumed to be zero.

It is evident that with increasing V_{DS}, the term $-V_{DS}^2$ causes I_D to reach a maximum value, or to saturate, in a manner qualitatively similar to that displayed in Figure 5-5(a). These saturating curves are plotted as solid lines in Figure 5-52, and certain findings from the rudimentary theory are shown with dashed lines. The values $N_A = 1.6 \times 10^{15}/\text{cm}^2$, $V_T = 0.88$ V, $(Z/L) = 1$, and $C_{OX} = 10^{-8}$ F/cm^2 were used for calculating these curves, as is indicated in Figure 5-52. Notice that in terms of saturation-regime current, the two models differ by about a factor of two.

Exercise 5-46. Account qualitatively for the differing current predictions of the two models.

In the rudimentary model, areal charge in the inversion layer is assumed to match that in the gate. In the more accurate model, appreciable charge is "wasted" in that it belongs to a depletion layer, which means there are fewer electrons in the channel available for conduction.

The question of what V_{DS} value yields current saturation at a given V_{GS} can also be addressed analytically. Rewriting Equation 5-110 with the aid of Equation 5-104 yields

$$Q_n(y) = -C_{OX}[V_{GS} - V_{FB} + 2\psi_B - \phi_n(y)] + \sqrt{2\epsilon q N_A[\phi_n(y) - 2\psi_B]}. \tag{5-119}$$

At the drain end of the channel, $\phi_n(L) = V_{DS}$. With the onset of saturation, the channel "pinches off" at the drain end, and hence $Q_n(L) = 0$. With these substitutions, then, we have

$$0 = -C_{OX}[V_{GS} - V_{FB} + 2\psi_B - V_{DS}] + \sqrt{2\epsilon q N_A(V_{DS} - 2\psi_B)}. \tag{5-120}$$

Solving this expression for V_{DS} yields

$$V_{DS} = V_{GS} - V_{FB} + 2\psi_B + \frac{\epsilon q N_A}{C_{OX}^2}\left[1 - \sqrt{1 + \frac{2(V_{GS} - V_{FB})C_{OX}^2}{\epsilon q N_A}}\right], \tag{5-121}$$

which was used to plot the solid locus of saturation points in Figure 5-52. (This equation is the subject of Problem A5-25.) The dashed locus at the left in Figure

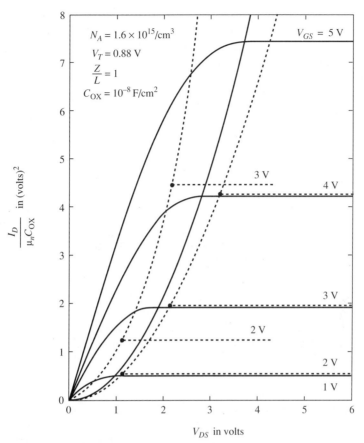

Figure 5-52 Comparing I–V characteristics from two long-channel models. The solid lines come from the Ihantola-Moll model that includes bulk-charge effects and becomes the Level-2 SPICE model. The dashed lines are from the rudimentary model of Section 5-1, the Level-1 model. The left-hand dashed locus is for the same device as the solid lines, showing a current prediction about twice as great. The right-hand locus is obtained by empirically adjusting the Level-1 model to yield the same saturation-regime current at $V_{GS} = 5$ V. (After Warner and Grung [72].)

5-52 was plotted using the analogous simple-theory expression, $V_{DS} = (V_{GS} - V_T)$.

Next, an empirical adjustment factor was applied to the simple-theory current to achieve a match of saturation currents for $V_{GS} = 5$ V. Note that the adjusted currents for lower V_{GS} are very close to those from the more complex analysis! The dashed locus at the right corresponds to these adjusted curves. (Curved-regime characteristics have been omitted for both the adjusted and unadjusted simple-theory curves to avoid cluttering the diagram.) MOSFET operation in saturation is by far the most important mode. The fact that rudimentary theory and empirical adjustment yields such a useful fit to more exact results meant that the computer design of MOS integrated circuits began early, when only simple theory was available, in the early to mid-1960s. Computer-aided design became an important factor in the rapid advance of MOS technology, a point made in Section 5-1.6. To repeat, the reason for the supreme importance of circuit optimization by computer in the MOS case is that the MOSFET is relatively deficient in "drive" capability, or transconductance.

A useful expression for characteristic slope, or channel conductance, in the curved regime can be written by differentiating Equation 5-116:

$$g = \frac{dI_D}{dV_{DS}}$$
$$= \mu_n C_{OX} \frac{Z}{L}\left[(V_{GS} - V_{FB} + 2\psi_B - V_{DS}) - \frac{\sqrt{\epsilon q N_A}}{C_{OX}}(V_{DS} - 2\psi_B)^{1/2}\right]. \quad (5\text{-}122)$$

Substituting Equation 5-121 into this expression of course causes g to vanish. But in the linear regime, near the origin, V_{DS} may be neglected, yielding

$$g(0) = \mu_n C_{OX} \frac{Z}{L}\left[V_{GS} - \left(V_{FB} - 2\psi_B + \frac{\sqrt{2\epsilon q N_A(-2\psi_B)}}{C_{OX}}\right)\right]. \quad (5\text{-}123)$$

Now refer back to Equation 5-109, which gives bulk charge as a function of y. Near the source, the quasi Fermi levels merge, causing $\phi_n(0) = 0$. Thus

$$Q_b(0) = -\sqrt{2\epsilon q N_A(-2\psi)}. \quad (5\text{-}124)$$

Dividing this expression by C_{OX} yields voltage drop caused by bulk charge at the source end of the channel, a term in the threshold voltage that was neglected in simple theory, and the correct interpretation of the last term in Equation 5-123. In fact, the three terms in parentheses in Equation 5-123 constitute the threshold voltage V_T in the present theory:

$$V_T = V_{FB} - 2\psi_B + \frac{\sqrt{2\epsilon q N_A(-2\psi_B)}}{C_{OX}}. \quad (5\text{-}125)$$

Recall from Problem A5-26(b) that the simple-theory analog of Equation 5-123 is

$$g(0) = \mu_n C_{OX}(Z/L)(V_{GS} - V_T), \quad (5\text{-}126)$$

and then note the direct correspondence between V_T and Equation 5-125. This is the desired verification of Equation 5-117.

Drain current in saturation is a matter that can be approached as before. Substituting Equation 5-121 into Equation 5-118 yields a complicated but important result. A shorthand version has been given by Brews [75] as

$$I_D = m\mu_n C_{OX}(Z/L)(V_{GS} - V_T)^2. \quad (5\text{-}127)$$

The coefficient m is doping-dependent. For a lightly doped substrate, m approaches 1/2, its value in the rudimentary theory, Equation 5-17.

5-4.3 Body Effect

A large fraction of the devices in an MOS integrated circuit have source regions that are not electrically common with the bulk, or substrate, terminal. In normal operation, such source regions may "move toward V_{DD}" in voltage. For N-channel devices, this means the source becomes more positive than the P-type bulk, a case of reverse bias on the source–bulk junction. As a result, the MOSFET in question exhibits an increased threshold voltage. Because MOSFET voltages are conventionally stated with the source as reference, the relevant voltage is V_{BS}, now negative.

To see one way that a negative V_{BS} can arise, refer to Figure 5-53(a). The two right-hand devices, M_{12} and M_{22}, have a common and large value of V_{GG}, causing both of them to be ON, or conducting. The lower device, M_{22}, will exhibit a finite ON voltage, $V_{DS(ON)}$. To visualize this, refer back to Figure 5-15(b). Because M_{12} has a high and fixed gate voltage, its I–V characteristic will to first order resemble the load characteristic presented there. Let us assume that $V_{DS(ON)} = 0.5$ V, as is indicated in Figure 5-53(a). This then constitutes the reverse voltage V_{BS} on the source junction of the upper device on the left-hand side, M_{11}. Figure 5-53(b)

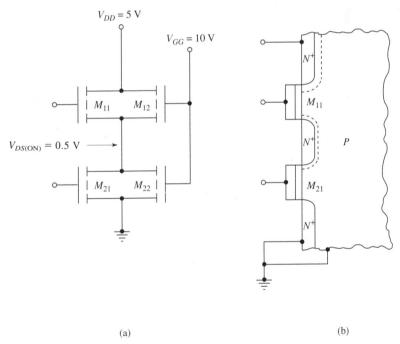

Figure 5-53 Sample situation leading to body effect. (a) The two right-hand devices are biased ON and conduct. Voltage drop $V_{DS(ON)}$ in M_{22} causes the source junction of M_{11} to be reverse-biased. (b) Cross section through left-hand device pair, showing by depletion-layer indication the reverse bias on their common region, the source of M_{11}, with respect to the substrate.

presents the two left-hand devices in cross section. The qualitative indications of depletion-layer thickness are intended to convey that the lowest junction shown has zero bias, the highest junction has a large reverse bias, 10 V, and the middle junction that is common to M_{11} and M_{21} has a small reverse bias, 0.5 V.

To examine how a negative V_{BS} on M_{11} causes an increase in its threshold voltage, we can profitably return to the kind of perspective representation used in Figure 5-50. Once again let us start with a flatband condition in the x direction, as is shown in Figure 5-54(a). The application of reverse bias to the source junction, represented on the "front" of the block, can be seen toward the left. In reverse bias,

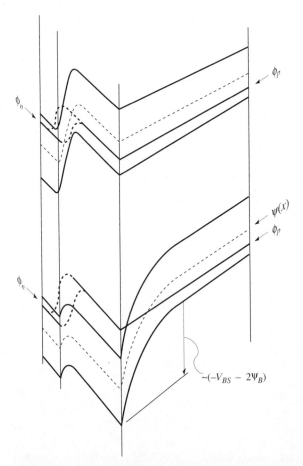

Figure 5-54 Illustrating body effect with intersecting one-dimensional solutions in the manner of Figure 5-50. (a) Flatband condition in bulk region and reverse bias V_{SB} on source–bulk junction. Fermi levels ϕ_p and ϕ_n split but merge again near junction.
(b) Gate voltage produces sufficient band bending to bring $\psi_S(0)$ below ϕ_n by $|\psi_B|$, the threshold condition at the source end of the channel. (After Warner and Grung [72].)

splitting of the quasi Fermi levels does indeed occur, with ϕ_n dropping below ϕ_p, but they rejoin just a short distance from the junction on either side [76]. This occurs, basically, because the absolute departure from equilibrium minority-carrier density is highly constrained in reverse bias (as compared to forward bias). This kind of splitting over a short distance is represented in perspective in Figure 5-54(a).

Recall, now, that the channel is strongest, and the surface potential ψ_S is farthest below the relevant Fermi level ϕ_n, at the source end. Indeed, for an ON device, the interval $\psi_S - \phi_n$ must exceed $|\psi_B|$ there. It is evident in Figure 5-54(a) that the splitting of ϕ_p and ϕ_n is such that substantially more band bending will be necessary than was the case with $V_{BS} = 0$. Figure 5-54(b) shows the situation after gate bias V_{GS} has been increased enough to provide the requisite band bending. The character of the divided Fermi levels, with both their points of merger close to the junction, suggests that the incremental band bending needed will be somewhat less than V_{BS}. Hence as a high-side estimate we can write

$$\text{source-end band bending at threshold} \approx -V_{BS} - 2\psi_B, \quad (5\text{-}128)$$

recalling that both ψ_B and V_{BS} are negative at present. This amount of band bending is indicated and labeled in Figure 5-54(b). The need for this incremental measure of band bending in order to establish threshold is the reason for the *body effect*, or threshold increase.

The increment ΔV_{GS} needed to reach threshold is accompanied by an increment in bulk charge ΔQ_b. The threshold-voltage increment, therefore, amounts to

$$\Delta V_T = \Delta Q_b / C_{OX}. \quad (5\text{-}129)$$

It is important to realize that V_{BS} itself does not directly enter ΔV_T, but rather, it is the added bulk charge ΔQ_b imposed upon the oxide capacitor that is responsible.

Exercise 5-47. Why doesn't V_{BS} enter ΔV_T directly?

Our convention refers all voltages to the source of a MOSFET. With $V_{BS} < 0$, V_{GS} is *still* referred to the source terminal, and cannot directly include V_{BS}. (If all voltages were referred to the bulk-terminal voltage, then V_{BS} would enter ΔV_T.)

It is evident that the new expression for threshold voltage in the presence of finite V_{BS} can be written by adding a term to Equation 5-77:

$$V_T = H_D - 2\psi_B - \frac{Q_f}{C_{OX}} - \frac{Q_s}{C_{OX}} - \frac{\Delta Q_b}{C_{OX}}. \quad (5\text{-}130)$$

In order to carry out an order-of-magnitude determination of ΔV_T, let us assume that Fermi-level splitting at the source end of the channel equals source–bulk bias, or

$$\phi_n - \phi_p = -V_{BS}. \quad (5\text{-}131)$$

Then,

$$\Delta V_T = -\frac{\Delta Q_b}{C_{OX}} = -\frac{(-q)N_A \Delta X_D}{C_{OX}}. \tag{5-132}$$

In functional notation,

$$\Delta V_T = (qN_A/C_{OX})[X_D(-V_{BS} - 2\psi_B) - X_D(-2\psi_B)]. \tag{5-133}$$

Taking the depletion-approximation expression for the one-sided junction from Table 3-1 gives us

$$\Delta V_T = (\sqrt{2\epsilon qN_A}/C_{OX})(\sqrt{-V_{BS} - 2\psi_B} - \sqrt{-2\psi_B}). \tag{5-134}$$

Empirical values of ΔV_T for a variety of MOSFET geometries were observed by Penney and Lau [77], with the resulting median curves plotted as solid lines in Figure 5-55. These curves are for $N_A = 7 \times 10^{15}/\text{cm}^3$ and for two basic aspect-ratio cases. Load devices qualify as long-channel MOSFETs, and inverter (or driver) devices as short-channel MOSFETs. The latter are less susceptible to body effect. It is evident in Figure 5-54(b) that the surface depletion layer will interact in complex three-dimensional fashion with those of the source and drain junctions as channel length is reduced. While the resulting details are not evident, one can claim in a vague way that the substantial band bending associated with a considerably reverse-biased drain junction will "contribute" some of the necessary band bending at the

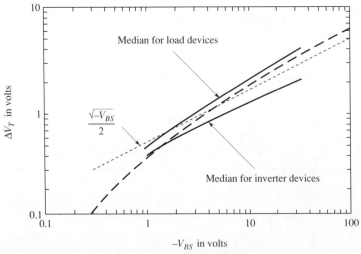

Figure 5-55 Solid lines show experimental threshold-voltage shift for long-channel (load) MOSFETs and short-channel (inverter or driver) MOSFETs with $N_A = 7 \times 10^{15}/\text{cm}^3$. (After Penny and Lau [77] with permission, copyright 1972, Van Nostrand Reinhold.) Dashed curve is plot of Equation 5-134 for the same doping. Dotted curve is plot of empirical Equation 5-135.

surface. The same idea is sometimes expressed in terms of bulk-charge "sharing" by the drain-junction and surface depletion layers.

For comparison with the solid curves, the dashed curve was obtained by plotting Equation 5-134. It falls between the experimental curves, and qualitatively resembles the long-channel case, assumed for its development. The dotted curve, then, is a plot of the empirical expression

$$\Delta V_T = 0.5\sqrt{-V_{BS}}, \tag{5-135}$$

sometimes used for modeling the body effect. It is a power-law expression, and hence is linear in log–log presentation. For more accurate modeling, one can deal with bulk charge in the manner employed in Section 5-4.2 to develop the Level-2 current expression. The resulting expressions are complex, accounting for the popularity of Equation 5-135.

5-4.4 Advanced Long-Channel Models

An early and accurate long-channel model was contributed by Pao and Sah [54]. While it used two superposed one-dimensional solutions, thus making it by definition a long-channel model, it eliminated the depletion approximation and employed numerical integrations in the x and y directions. Further, in the x (depth) direction, it made use of certain expressions equivalent to those developed in Sections 3-7 and 5-3.1. The Pao-Sah analysis was subsequently recast with our notation and with the extension of the analysis from equilibrium to nonequilibrium conditions [72]. Significantly, the Pao-Sah model permits carrier transport in the channel by drift *and* by diffusion, rather than by drift alone as was the case in Sections 5-4.2 and 5-1. A nonobvious consequence of this generalization is that the model is able to predict saturation-regime characteristics, with the valid result that true saturation (or constancy) of the drain current is not observed. Recall that previous models predicted constant current in saturation by appeal to a physical argument (such as that in Problem A5-10).

The Pao-Sah model is extremely accurate, but is of limited practical utility in its original form because of the kind of numerical procedures it requires. This motivated a search for an approximate analytic model of greater accuracy than the earlier ones. Subsequently, a *charge-sheet* model was developed by Brews [78], and by Baccarani et al. [79]. In a simplification over the Pao-Sah model, it reverts to the depletion approximation. In a second simplification, accounting for the model's name, it assumes a channel of zero thickness. However, like the Pao-Sah model, it permits both drift and diffusion of channel carriers, and consequently predicts saturation-regime characteristics. The approximations combined to yield the Brews model contribute partly compensating errors that account for its surprising fidelity, a matter that is examined elsewhere in some of its aspects [80].

Another model was offered by Dang [81], a bit before the Brews and Baccarani et al. models. It is intermediate in approximation between the Brews and Baccarani et al. models, on the one hand, and the Ihantola-Moll model, on the other hand.

These four models have been examined in some detail in relation to one another [72].

5-5 SPICE MODELS

There are three primary SPICE MOSFET models, identified as Levels 1, 2, and 3 [32, 82–84]. Level 1 employs the rudimentary analysis of Section 5-1, with empirical adjustment. The most important correction is a coefficient applied to the current expressions to bring saturation-regime current into congruence with experiment and more accurate analyses. The result of such adjustment is shown in Figure 5-52, where the rudimentary model is brought into good agreement with the ionic-charge model.

The Ihantola-Moll, or ionic-charge, analysis is taken as the Level-2 SPICE model. It is used in a form more general than that developed in Section 5-4.2, permitting nonzero values of bulk–source voltage. A notable departure of the Level-2 from the Level-1 analysis is a three-halves-power dependence of current on voltage in the second major term of Equation 5-116, rather than the square-law dependence in the Level-1 case, Equation 5-14. Because the numerical handling of a fractional power is more time-consuming than that of an integral power, circuit designers were motivated to seek empirical modifications of the Level-2 model. In the resulting SPICE Level-3 model, they not only removed this shortcoming but were able to deal empirically with many other effects as well, such as short-channel and channel-shortening effects. The latter, addressed in Problem A5-10g, is attributable to the variations of drain–junction depletion-layer thickness, and is analogous to the Early effect in the BJT. The former implies accounting for some 3D complications, alluded to in Section 5-4.1. These inevitably arise when channel length approaches depletion-layer dimensions.

5-5.1 Level-2 Parameters

The semiempirical Level-1 and Level-3 models are especially advantageous for circuit analysis. But because our present interest is equally on device physics and circuit analysis, we will focus primarily on the Level-2 Ihantola-Moll model [73]. The starting point is a generalized form of Equation 5-116 that takes into account the effect of finite bulk–source voltage,

$$I_D(V_{GS}, V_{BS}, V_{DS}) = 2K\{[V_{GS} - V_{FB}^* + 2\psi_B - (V_{DS}/2)]V_{DS} \\ + (2\gamma/3)[(V_{DS} - 2\psi_B - V_{BS})^{3/2} - (-2\psi_B - V_{BS})^{3/2}]\}, \quad (5\text{-}136)$$

where

$$K \equiv (1/2)\mu_n C_{OX} Z/L^*, \quad (5\text{-}137)$$

and L^* is an elaborated version of channel length L that takes into account a technological channel-shortening effect. Impurities implanted to form the source and drain regions diffuse laterally during subsequent heat treatments by the amount L_D^*, and as a result, the channel is shortened by the amount $2L_D^*$. Hence,

$$L^* = (L - 2L_D^*). \tag{5-138}$$

Our purpose here is to introduce the large number of possible input parameters for the Level-2 SPICE model. To compare the model with experiment, we will employ the device geometry shown in Figure 5-56 and the current–voltage characteristics shown in Figure 5-57. Also given in the latter figure are the characteristics

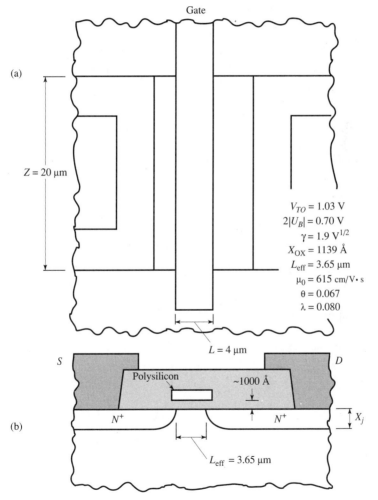

Figure 5-56 A MOSFET structure chosen for study. (a) Parameter values and plan-view dimensions, emphasizing gate length of 4 µm. (b) Cross-sectional structure, emphasizing effective gate length of 3.65 µm. (After Warner and Grung [72].)

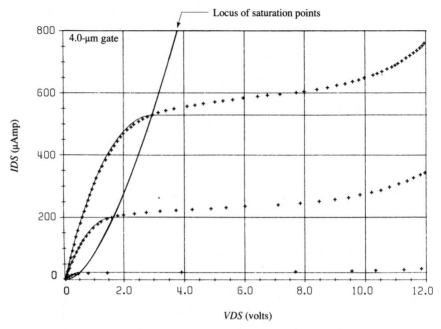

Figure 5-57 Three output characteristics for the MOSFET of Figure 5-56. The plotted points were obtained experimentally, and the solid curves were calculated from the Ihantola-Moll model. (After Warner and Grung [72].)

calculated using the Ihantola-Moll model described in Section 5-4.2. Figure 5-58(a) presents a static equivalent-circuit model for the ideal MOSFET described by Equation 5-136. The current generator I_D is a function of the three variables V_{GS}, V_{BS}, and V_{DS}. Figure 5-58(b) adds to the equivalent circuit the parasitic elements present in a real MOSFET, needed for satisfactory dynamic analysis of the device.

Tables 5-3 and 5-4 give one possible set of SPICE parameters that are most directly related to the structure of the MOSFET. The oxide thickness X_{OX} determines the capacitance per unit area C_{OX}, and along with the surface mobility μ_n, enters into the current coefficient K. The substrate doping N_A and the specific capacitance C_{OX} enter the body-effect parameter that we repeat here:

$$\gamma = \sqrt{2\epsilon q N_A}/C_{OX}. \tag{5-139}$$

The substrate doping N_A also determines the bulk potential ψ_B since

$$\psi_B = (kT/q)\ln(N_A/n_i). \tag{5-140}$$

The mobility parameters μ_n, U_{EXP} and E_{CRIT} are used to define an effective mobility as will be described later. In addition, the channel-length modulation factor, λ, is similar to the base-thickness modulation parameter in the BJT model. The effective oxide fixed charge density N_f empirically fixes the zero-order threshold voltage determined by H_D, the barrier-height difference. For metal field plates, the SPICE (a)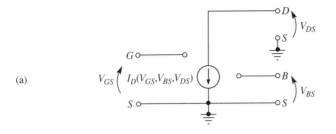

(b)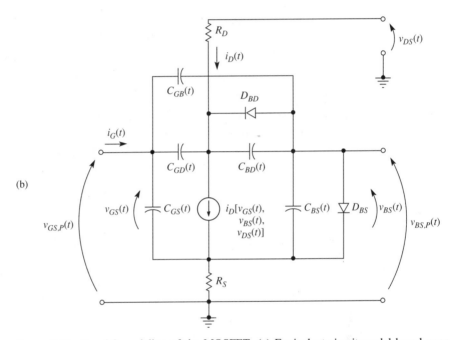

Figure 5-58 Level-2 modeling of the MOSFET. (a) Equivalent-circuit model based upon the Ihantola-Moll static model [73]. (b) Equivalent-circuit model that includes parasitic elements needed for accurate Level-2 static and dynamic modeling.

values agree exactly with the experimental values given by Deal and Snow [26], so that

$$H_D = \pm \psi_B + 0.61 \text{ V}. \tag{5-141}$$

But for polysil field plates, we will use "effective" values N_f^* and H_D^* because the SPICE values differ slightly from the experimental results of Werner [27]. Figure 5-23(b) shows that the approximate SPICE values were chosen for convenient symmetry, with the barrier-height difference stated as

$$H_D^* = \pm [\psi_B - (0.5575 \text{ V}) T_G]. \tag{5-142}$$

The positive sign represents a *P*-type substrate, and the negative sign, an *N*-type substrate. The type-of-gate parameter T_G is positive for a polysil gate when the gate-doping type is opposite to that of the substrate, and negative, when the two types are the same. The effective flatband voltage V_{FB}^* is given by

$$V_{FB}^* = H_D^* - (qN_f^*/C_{OX}) \tag{5-143}$$

and finally the threshold voltage is given by

$$V_T = V_{FB}^* - 2\psi_B + \gamma\sqrt{-2\psi_B}. \tag{5-144}$$

To compensate for the departure of the SPICE Equation 5-142 from the more exact Equation 5-49, we add a constant value of $3.3 \times 10^{10}/\text{cm}^2$ to the actual value of N_f. For example, if the actual fixed-oxide charge N_f equals $3.2 \times 10^{10}/\text{cm}^2$, then N_f^* equals $6.5 \times 10^{10}/\text{cm}^2$. The internal SPICE program also employs an ancillary parameter that it terms the effective barrier height, defined as

$$V_{BI}^* = V_{FB}^* - 2\psi_B = V_T - \gamma\sqrt{-2\psi_B}. \tag{5-145}$$

This parameter does simplify the form of some equations, but we will not use it.

Now let us return to Figure 5-58(b) to describe more fully the various elements it introduces. The source and drain implants have a sheet resistance R_{SH} that causes parasitic series source resistance and drain resistance, given by

$$R_S = N_{RS}\, R_{SH}, \tag{5-146}$$

and
$$R_D = N_{RD}\, R_{SH}, \tag{5-147}$$

where N_{RS} and N_{RD} are the numbers of resistive squares associated with each implanted region. The source and drain implants also produce PN junctions with a saturation-current density (extrapolated to zero bias) of J_0. The saturation currents for the corresponding junctions are given by

$$I_{BS} = A_S\, J_0, \tag{5-148}$$

and
$$I_{BD} = A_D\, J_0, \tag{5-149}$$

where A_S and A_D are areas of the source and drain implants. In normal operation, the source and drain junctions are reverse-biased so that these diodes contribute only relatively unimportant leakage currents. However, the associated depletion-layer capacitances are important, and are divided into sidewall portions and bottom portions. The zero-bias values of the bottom parts are given by

$$C_{OBS.B} = A_S C_{0.B}, \tag{5-150}$$

and
$$C_{OBD.B} = A_D C_{0.B}, \tag{5-151}$$

where $C_{0.B}$ is the junction capacitance per unit area. The sidewall portions of the drain and source junctions contribute capacitances given by

$$C_{OBS.SW} = P_S C_{0.SW}, \tag{5-152}$$

and
$$C_{OBD.SW} = P_D C_{0.SW}, \tag{5-153}$$

where $C_{0.SW}$ is the sidewall capacitance per unit length, P_S is the length of the sidewall for the source region, and P_D is the corresponding length for the drain region.

5-5.2 Level-2 Model

The MOSFET model to be presented here contains seventeen equations in twenty variables. To define the boundary between the curved and saturation regimes, we employ the drain-source saturation voltage, a time-dependent form of Equation 5-121:

$$v_{DS.SAT}(t) = v_{GS}(t) - V_{FB} + 2\psi_B \\ + \frac{\gamma^2}{2}\left[1 - \sqrt{1 + 4\frac{v_{GS}(t) - V_{FB} - v_{BS}(t)}{\gamma^2}}\right]. \quad (5\text{-}154)$$

Also, to define the boundary between subthreshold and normal operation, we will use the ON voltage defined by

$$v_{ON}(t) = V_{FB} - 2\psi_B + \gamma\sqrt{-2\psi_B - v_{BS}(t)} + (kT/q)N(t), \quad (5\text{-}155)$$

where

$$N(t) = 1 + \frac{C_n}{C_{OX}} + \frac{\gamma}{2\sqrt{(-2\psi_B) - v_{BS}(t)}}. \quad (5\text{-}156)$$

Notice that the time dependence of the ON voltage results from the time-dependent bulk–source voltage, $v_{BS}(t)$. For $N(t) = 1$, the ON voltage equals the normal threshold voltage V_T. For other values of $N(t)$, there is a slightly more positive value of effective threshold voltage—the ON voltage. The adjustable parameter C_n is used in modeling the subthreshold regime of operation.

We define four regimes of operation. A good way to visualize these regimes is to consider a MOSFET inverter with the bulk–source junction short-circuited and with a resistive load. Then increase the gate–source voltage from 0 to a high value, moving upward along the loadline. Initially the silicon surface is in the accumulation condition (because of parasitic features of the MOS capacitor and "threshold-tailoring" steps). Next we encounter depletion and the term "subthreshold" is used to refer to both the accumulation and depletion regimes. As the device turns on, it is in the saturation regime, and then finally the curved regime.

The accumulation regime is defined by $v_{GS}(t) < v_{ON}(t) + 2\psi_B$; the depletion regime, by $[v_{ON}(t) + 2\psi_B] < v_{GS}(t) < v_{ON}(t)$; the saturation regime, by $v_{ON}(t) < v_{GS}(t) < [v_{ON}(t) + v_{DS.SAT}(t)]$; and finally the curved regime, by $[v_{ON}(t) + v_{DS.SAT}(t)] < v_{GS}(t)$. To identify an equation valid in a particular regime, we will append one of four letters—a, d, s, or c—to the equation number. These symbols stand, of course, for accumulation, depletion, saturation, and curved, respectively.

The effective K is given by

$$K_{EFF}(t) = K\left[\frac{1}{1 - \lambda v_{DS}(t)}\right]\left[\frac{E_{CRIT}\epsilon/C_{OX}}{v_{GS}(t) - v_{ON}(t)}\right]^{U_{EXP}}, \quad (5\text{-}157)$$

where E_{CRIT} and U_{EXP} are empirical parameters and λ is the channel-length-modulation factor. There exist many empirical models for effective mobility, and in turn, for effective K. Here we selected only one of the possible level-2 models and will regard E_{CRIT} and U_{EXP} as curve-fitting parameters that provide better agreement between theory and experiment.

The terminal voltages that comprehend parasitic effects are denoted by the subscript P. They are defined in terms of the intrinsic, or "internal," voltages used exclusively heretofore:

$$v_{GS.P}(t) = v_{GS}(t) + R_S[i_G(t) + i_B(t) + i_D(t)], \quad (5\text{-}158)$$
$$v_{GD.P}(t) = v_{GD}(t) + R_D[i_D(t)], \quad (5\text{-}159)$$
$$v_{BS.P}(t) = v_{BS}(t) + R_S[i_G(t) + i_B(t) + i_D(t)], \quad (5\text{-}160)$$

and
$$v_{BD.P}(t) = v_{BD}(t) + R_D[i_D(t)]. \quad (5\text{-}161)$$

The terminal currents are defined by

$$i_G(t) = C_{GS}(t)\frac{\partial v_{GS}(t)}{\partial t} + C_{GD}(t)\left[\frac{\partial v_{GS}(t)}{\partial t} - \frac{\partial v_{DS}(t)}{\partial t}\right] + C_{GB}(t)\frac{\partial v_{GB}(t)}{\partial t}, \quad (5\text{-}162)$$

$$i_B(t) = +I_{BS}\left[\exp\left(\frac{qv_{BS}(t)}{kT}\right) - 1\right] + I_{BD}\left[\exp\left(\frac{q[v_{BS}(t) - v_{DS}(t)]}{kT}\right) - 1\right]$$
$$+ C_{BS}(t)\frac{\partial v_{BS}(t)}{\partial t} + C_{BD}(t)\left[\frac{\partial v_{BS}(t)}{\partial t} - \frac{\partial v_{DS}(t)}{\partial t}\right] - C_{GB}(t)\frac{\partial v_{GB}(t)}{\partial t}, \quad (5\text{-}163)$$

and

$$i_D(t) = i_D^R(t) - I_{BD}\left[\exp\left(\frac{q[v_{BS}(t) - v_{DS}(t)]}{kT}\right) - 1\right]$$
$$- C_{GD}(t)\left[\frac{\partial v_{GS}(t)}{\partial t} - \frac{\partial v_{DS}(t)}{\partial t}\right] - C_{BD}(t)\left[\frac{\partial v_{BS}(t)}{\partial t} - \frac{\partial v_{DS}(t)}{\partial t}\right]. \quad (5\text{-}164)$$

In the last equation, the resistive component of drain current $i_D^R(t)$ is defined in the four regimes of operation as follows:

$$i_D^R(t) = i_D^R(v_{ON}(t), v_{BS}(t), v_{DS}(t)) \, \exp\left[\frac{v_{GS}(t) - v_{ON}(t)}{N(t)kT/q}\right], \quad (5\text{-}165a)$$
$$i_D^R(t) = \text{same as above}, \quad (5\text{-}165d)$$
$$i_D^R(t) = i_D^R(v_{GS}(t), v_{BS}(t), v_{DS.SAT}(t)), \quad (5\text{-}165s)$$

and

$$i_D^R(t) = i_D^R(v_{GS}(t), v_{BS}(t), v_{DS}(t)), \quad (5\text{-}165c)$$

where

$$i_D^R(v_{GS}(t), v_{BS}(t), v_{DS}(t)) = 2K_{EFF}(t)\{[v_{GS}(t) - V_{FB}^* + 2\psi_B - v_{DS}(t)/2]v_{DS}(t)$$
$$+ (2\gamma/3)[(v_{DS}(t) - 2\psi_B - v_{BS}(t))^{3/2} - (-2\psi_B - v_{BS}(t))^{3/2}]\}. \quad (5\text{-}166)$$

The three gate capacitances, which will be described in Section 5-5.4, are defined in the four regimes of operation as:

$$C_{GS}(t) = C_{0GS}W, \tag{5-167a}$$

$$C_{GS}(t) = C_{0GS}W + \frac{2}{3} C_{OX}L^*W\left[\frac{v_{ON}(t) - v_{GS}(t)}{2\psi_B} + 1\right], \tag{5-167d}$$

$$C_{GS}(t) = C_{0GS}W + \frac{2}{3} C_{OX}L^*W, \tag{5-167s}$$

$$C_{GS}(t) = C_{0GS}W + C_{OX}L^*W\left[1 - \frac{v_{GS}(t) - v_{ON}(t) - v_{DS}(t)}{2[v_{GS}(t) - v_{ON}(t)] - v_{DS}(t)}\right]; \tag{5-167c}$$

$$C_{GD}(t) = C_{0GD}W, \tag{5-168a}$$
$$C_{GD}(t) = C_{0GD}W, \tag{5-168d}$$
$$C_{GD}(t) = C_{0GD}W, \tag{5-168s}$$

$$C_{GD}(t) = C_{0GD}W + C_{OX}L^*W\left[1 - \frac{v_{GS}(t) - v_{ON}(t)}{2[v_{GS}(t) - v_{ON}(t)] - v_{DS}(t)}\right]; \tag{5-168c}$$

and
$$C_{GB}(t) = C_{0GB}L^* + C_{OX}L^*W, \tag{5-169a}$$

$$C_{GB}(t) = C_{0GB}L^* + C_{OX}L^*W\left[\frac{v_{ON}(t) - v_{GS}(t)}{2\psi_B}\right], \tag{5-169d}$$

$$C_{GB}(t) = C_{0GB}L^*, \tag{5-169s}$$
$$C_{GB}(t) = C_{0GB}L^*. \tag{5-169c}$$

Finally, the two bulk-junction capacitances are given by

$$C_{BS}(t) = \frac{C_{OBS.B}}{\sqrt[M_B]{1 - \frac{v_{BS}(t)}{\Delta\psi_0}}} + \frac{C_{OBS.SW}}{\sqrt[M_{SW}]{1 - \frac{v_{BS}(t)}{\Delta\psi_0}}}, \tag{5-170}$$

and
$$C_{BD}(t) = \frac{C_{OBD.B}}{\sqrt[M_B]{1 - \frac{v_{BS}(t) - v_{DS}(t)}{\Delta\psi_0}}} + \frac{C_{OBD.SW}}{\sqrt[M_{SW}]{1 - \frac{v_{BS}(t) - v_{DS}(t)}{\Delta\psi_0}}}, \tag{5-171}$$

This completes the seventeen equations of the SPICE model. These equations will be referred to here as the Level-2 SPICE model for the MOSFET, or briefly, as the SPICE model.

Using the SPICE model and the parameter values given in Table 5-5, we can generate the I–V characteristics given in Figure 5-59, which are consistent with the experimental data given in Figure 5-57. Exactly the same results can be obtained by using the five electrical parameters noted in Table 5-6 in place of the physical parameters N_A, T_G, N_f^*, R_{SH}, and J_0. Table 5-7 gives the equivalent sets of parameter values.

For a drain–source voltage of 2.47 V, the square root of the drain current is plotted in Figure 5-60(a) as a function of V_{GS}. In Figure 5-60(b), the drain–source current is plotted using a logarithmic current scale. If the MOSFET were exactly a

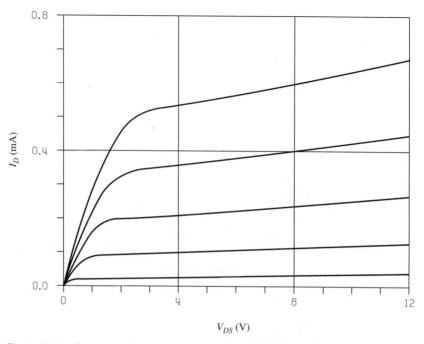

Figure 5-59 Current–voltage characteristics for the MOSFET of Figure 5-56, calculated using the Level-2 model. Comparison with the experimental results in Figure 5-57 shows good agreement.

square-law device, the curve in Figure 5-60(a) would be a straight line and would intersect the x axis at the threshold voltage, $V_T = 1.03$ V. Figure 5-60(b) shows the subthreshold behavior of the drain current. The slope of the curve in the subthreshold range from 10^{-11} A to 10^{-7} A is controlled by the variable N, which in turn is stated in Equation 5-156.

The transconductance $g_{m.P}$ of the MOSFET is plotted in Figure 5-61(a), along with the output conductance $g_{o.P}$. These are defined, respectively, as $\partial I_D/\partial V_{GS.P}$ and $\partial I_D/\partial V_{DS.P}$. The voltage gain $A_{V.P}$, Figure 5-61(b), is defined as $g_{m.P}/g_{o.P}$, which is the maximum gain that can be achieved in most circuits using the MOSFET. The voltage gain has a maximum value near the threshold voltage and decreases substantially as $V_{GS.P}$ increases. Thus for high gain, the MOSFET must be biased at low current levels. Figure 5-62 plots the same curves as Figure 5-61 using a logarithmic current scale. Figure 5-62(a) clearly shows that the transconductance of the MOSFET is about qI_D/kT at low current levels, a fact exploited in low-power (especially CMOS) circuitry. A constant voltage-gain magnitude of more than 600 is achieved over a considerable range in the subthreshold regime.

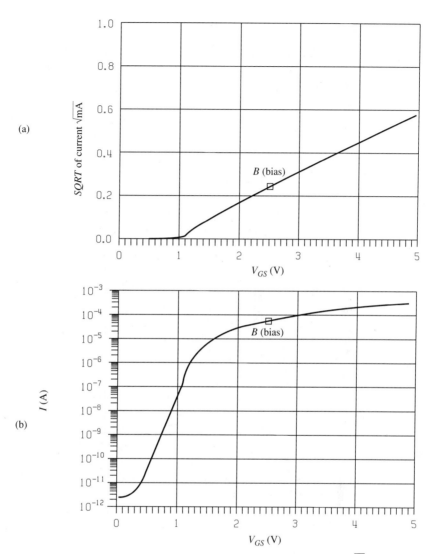

Figure 5-60 Transfer characteristic for $V_{DS} = 2.471$ V. (a) Plot of $\sqrt{I_D}$ versus V_{GS} on a linear scale. (b) Plot of I_D on a logarithmic scale versus V_{GS} on a linear scale.

5-5.3 Small-Signal Application of Model

One of the simplest small-signal voltage amplifiers is the common-source circuit shown in Figure 5-63. The load resistor and dc supply voltages were selected to be compatible with the depletion-load circuit to be discussed in the next section. Using the SPICE-parameter values given in Table 5-5, we can calculate the current–

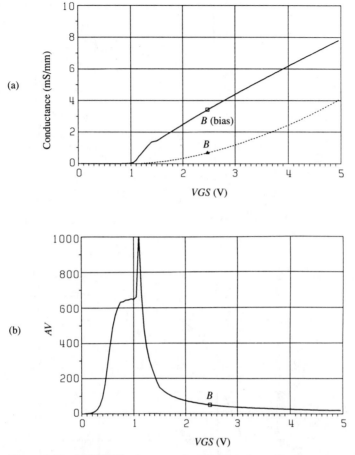

Figure 5-61 MOSFET gain properties as a function of gate–source voltage. Point B is the bias point. (a) Transconductance comprehending parasitic elements, $g_{m,P}$, solid line, and output conductance, g_o, after multiplication by ten, dotted line. (b) Small-signal voltage gain A_{VP} comprehending parasitic elements.

voltage curves shown in Figure 5-64, along with the resistive load line. The square marked B identifies the dc bias point, and Tables 5-8 and 5-9 give the corresponding numerical values.

To analyze the small-signal performance of the MOSFET, we will construct a model similar to the hybrid-pi model for the BJT. To do this, we replace the large-signal circuit of Figure 5-58(b) with the corresponding small-signal circuit given in Figure 5-65. One primary objective at present is to derive the element values of the small-signal circuit, starting with the Level-2 model given in the last section. Then

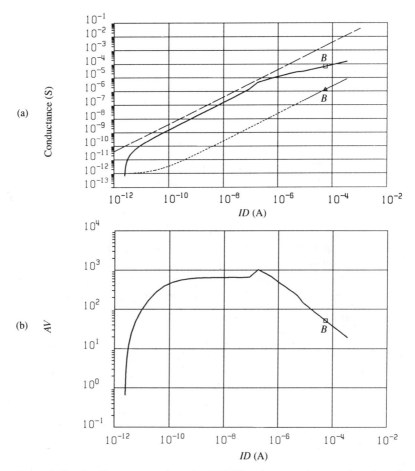

Figure 5-62 Semilog presentation of MOSFET gain properties as a function of gate–source voltage (the same data as are in Figure 5-61). Point B is the bias point. (a) Transconductance comprehending parasitic elements, $g_{m.P}$, solid line, and output conductance, g_o, after multiplication by ten, dotted line. Dashed curve shows qI_D/kT limit, seen in the BJT. (b) Small-signal voltage gain comprehending parasitic elements, $A_{V.P}$.

we will analyze the small-signal performance of the MOSFET for the common-source circuit given in Figure 5-63. It will turn out that the small-signal current-gain cutoff frequency of the selected MOSFET is about 0.5 GHz, and that the 3-dB bandwidth of the common-source circuit is about 66 MHz.

The selected dc bias point is in the saturation regime of operation, so that the resistive component of drain current $i_D^R(t)$ is given by Equation 5-165s. The term "resistive component" is used to identify the first term of Equation 5-164, given in

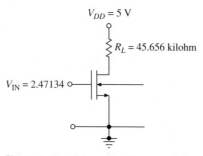

Figure 5-63 Schematic diagram of the resistive-load circuit used for small-signal analysis. The resistor value has been tailored to produce the desired operating point.

detail in Equation 5-166. Its mathematical form is identical to that for static drain current, Equation 5-137.

Taking advantage of the equivalence just noted, let us revert to the simpler dc symbols to evaluate the intrinsic transconductance of the MOSFET:

$$g_m \equiv \frac{\partial I_D^R}{\partial V_{GS}} = 2K_{EFF}V_{DS.SAT} + \frac{I_D^R}{K_{EFF}}\left(\frac{\partial K_{EFF}}{\partial V_{GS}}\right)$$
$$+ 2K_{EFF}[V_{GS} - V_{FB}^* + 2\psi_B - V_{DS.SAT}$$
$$+ \gamma\sqrt{V_{DS.SAT} - 2\psi_B - V_{GS}}]\left(\frac{\partial V_{DS.SAT}}{\partial V_{GS}}\right). \quad (5\text{-}172)$$

In the remainder of this section we will continue to use static symbols for the sake of simplicity. The first partial derivative in Equation 5-172 can be found using Equation 5-157, and is

$$\frac{\partial K_{EFF}}{\partial V_{GS}} = -\left[\frac{K_{EFF}U_{EXP}}{V_{GS} - V_{ON}}\right] = -2.09 \times 10^{-5} \text{ A/V}^3, \quad (5\text{-}173)$$

where we have used Tables 5-5 and 5-9 and the Level-2 model to obtain the numerical value. The second partial derivative in Equation 5-172 is found using Equation 5-154, yielding

$$\frac{\partial V_{DS.SAT}}{\partial V_{GS}} = 1 - \left[1 + \frac{4}{\gamma^2}(V_{DS} - V_{FB} - V_{BS})\right]^{-1/2} = 0.5581. \quad (5\text{-}174)$$

Substituting the values just given and values taken from Tables 5-6 and 5-9 into Equation 5-172 yields

$$g_m = [7.939 \times 10^{-5} - 1.083 \times 10^{-5} + 3.5 \times 10^{-8}] \text{ A/V}$$
$$= 0.069 \text{ mA/V}. \quad (5\text{-}175)$$

The output conductance g_o is given by

$$g_o \equiv \frac{\partial I_D^R}{\partial V_{DS}} = \frac{I_D^R}{K_{EFF}}\left(\frac{\partial K_{EFF}}{\partial V_{DS}}\right). \quad (5\text{-}176)$$

SPICE MODELS

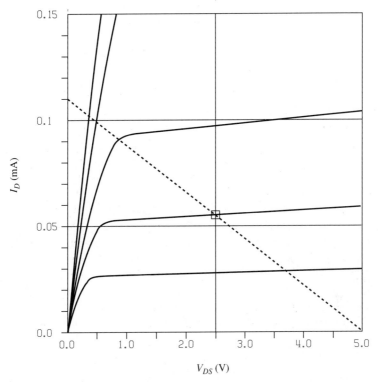

Figure 5-64 MOSFET output characteristics and resistive loadline used for small-signal analysis. The square indicates the bias point.

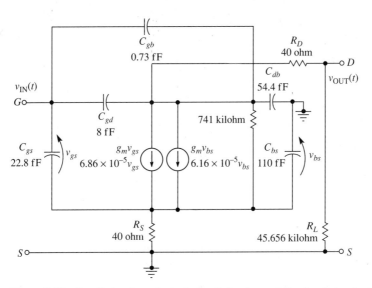

Figure 5-65 Small-signal equivalent circuit for the resistive-load circuit given in Figure 5-63.

Using Equation 5-157, we find that

$$\frac{\partial K_{EFF}}{\partial V_{DS}} = \frac{\lambda K_{EFF}}{1 - \lambda V_{DS}}, \qquad (5\text{-}177)$$

so that

$$g_o = \frac{\lambda I_D^R}{1 - \lambda V_{DS}} = 0.001 \text{ mS}. \qquad (5\text{-}178)$$

In Problem A5-30 we evaluate the bulk–source transconductance, which is a measure of the effectiveness of the bulk region as a control electrode. The value found there is

$$g_{mbs} = 0.062 \text{ mA/V}, \qquad (5\text{-}179)$$

which is nearly equal to g_m in value. The remaining values for the small-signal circuit given in Figure 5-65 can be obtained directly from the Level-2 model.

Using standard circuit techniques such as those discussed in Sections 3-9.1 and 3-9.3, we can approximate the model of Figure 5-65 with the simplified model of Figure 5-66. To characterize the performance of the selected MOSFET, we need to evaluate A_V, the small-signal voltage gain, and f_T, the cutoff frequency for current gain. It may at first be puzzling to encounter "current gain" in the MOSFET case. Static current gain in the ideal device is of course infinite, because the gate insulator is perfect. But in high-frequency operation, capacitive current through the gate terminal is appreciable, so current gain is a finite and useful concept. Consequently, f_T is as valid a figure of merit for the MOSFET as for the BJT, although it has some limitations such as those discussed in Section 4-7.8 for the BJT.

To characterize the resistive-load circuit associated with the MOSFET in Figure 5-63, we need to know the circuit voltage gain and 3-dB bandwidth ($A_{V,\text{cir}}$ and f_{3dB}, respectively), as well as the intrinsic gain properties of the MOSFET. The intrinsic voltage gain A_V of the MOSFET is given by

$$A_V = g_m/g_o = 69. \qquad (5\text{-}180)$$

To find f_T, we must first calculate the dynamic current gain A_I. Let us assume a sinusoidal input voltage. Employing the appropriate symbols for ac analysis and referring to Figure 5-66, we obtain the output current:

$$i_d(\omega) = g_m v_{gs}(\omega). \qquad (5\text{-}181)$$

The input current is given by

$$i_g(\omega) = j\omega[C_{gs} + A_V C_{gd}]v_{gs}(\omega), \qquad (5\text{-}182)$$

so that

$$A_I(\omega) = \frac{g_m}{j\omega(C_{gs} + A_V C_{gd})}. \qquad (5\text{-}183)$$

By definition, f_T has the value $2\pi\omega$ when $A_I(\omega) = 1$. Thus,

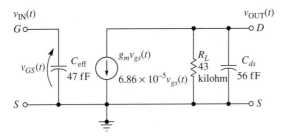

Figure 5-66 Approximate small-signal equivalent circuit derived from Figure 5-65 by using standard circuit analysis.

$$f_T = \frac{g_m}{2\pi(C_{gs} + C_{gd})} = 0.5 \text{ GHz}. \tag{5-184}$$

Hence the selected MOSFET is characterized by an intrinsic voltage gain of 69, and a cutoff frequency of 0.5 GHz.

When the input generator is a pure ac-voltage source, the high-frequency response of the amplifier circuit is determined by the RC time constant of the output circuit. Thus, the f_{3dB} of the resistive-load circuit is

$$f_{3dB} = (g_o + g_L)/[2\pi C_{bd}] = 66 \text{ MHz}, \tag{5-185}$$

and the ac voltage gain is

$$A_{V,\text{cir}} = -g_m/(g_o + g_L) = -3. \tag{5-186}$$

Using SPICE we can calculate the voltage gain as a function of frequency, with the results that are plotted in Figure 5-67. These calculations and Figure 5-67 show that the resistive-load circuit is characterized by a voltage gain of -3, and a bandwidth of 66 MHz. As explained in Section 3-9.5, the circuit rise time and bandwidth are interrelated, so that

$$f_{3dB} t_{\text{RISE}} = 0.35, \tag{5-187}$$

and hence $t_{\text{RISE}} = 5.3$ ns. Also, the 10-to-90% rise time is given by 2.2 times the RC time constant of the output circuit, so that

$$t_{\text{RISE}} = 2.2 \, C_{bd}/(g_o + g_L) = 5.3 \text{ ns}, \tag{5-188}$$

in agreement with the result just given. The calculated small-signal pulse response of the resistive-load circuit is given in Figure 5-68.

Finally, there is an aspect of SPICE that deserves explanation. SPICE employs certain early physical data that are different from the more up-to-date values used elsewhere in this book, and for understandable reasons, has preferred to adhere to the older values. An example is the too-high value of intrinsic density used in SPICE, $n_i = 1.45 \times 10^{10}/\text{cm}^3$. But because empirical adjustment is an integral part of SPICE modeling, the inaccuracies do not cause problems for circuit simulation.

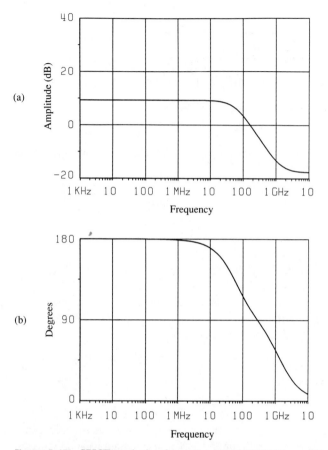

Figure 5-67 SPICE analysis of resistive-load MOSFET amplifier. (a) Frequency dependence of voltage-gain magnitude comprehending parasitic elements, $A_{V.P}$, expressed in decibels. (b) The corresponding phase of $A_{V.P}$ as a function of frequency.

5-5.4 Large-Signal Application of Model

To investigate the large-signal dynamic response of a MOSFET, we consider the two-stage inverter circuit shown in Figure 5-69. The first stage consists of a driver transistor with a D-mode MOSFET as the load device. The driver is identical to that described in Section 5-5.3. The load was formed by an additional ion implantation in the channel region; hence we can use the same Level-2 parameters as those for the driver, except that $N_f^* = 1.20 \times 10^{12}/\text{cm}^2$ to include the implanted ions. This value of N_f^* represents an implant dose close to 1.20×10^{12} ions/cm^2 that produces a threshold shift from $+1.03$ V to -5.0 V, which is to say, from E-mode to D-mode properties. The geometry of the load device is different from that of the driver, with $Z = 12 \times 10^{-6}$ m, $L = 4 \times 10^{-6}$ m, $A_D = 320 \times 10^{-12}$ m^2, $A_S = 32 \times 10^{-12}$ m^2, $N_{RS} = 1$, and $N_{RD} = 1$.

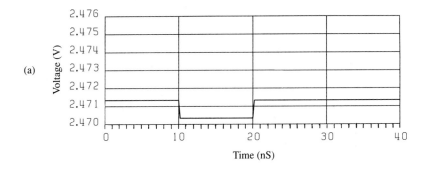
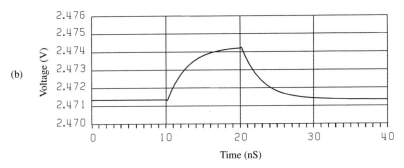

Figure 5-68 SPICE analysis of resistive-load amplifier in Figure 5-63. (a) Input-voltage pulse applied to amplifier. (b) Computer-determined output waveform.

The source region of the load and the drain region of the driver are common in the inverter circuit. When a single region does double duty in this manner, it is often incrementally larger in area than one that serves, let us say, as a drain terminal only. At present we will assume a ten-percent increment, and will assign to the source of the load device a value of A_S that is one-tenth its value in the driver device. In this way, the output node of each inverter will have the correct total parasitic junction capacitance; yet we will still have the convenience of having each driver device in Figure 5-69 be identical to the lone driver in Figure 5-63 in all of the latter's properties, including parasitic elements.

The I–V characteristics of the driver and load are plotted in Figure 5-70. The solid line represents the load characteristic with the gate, source, and bulk terminals of the load device common. In the inverter circuit, however, only the gate and source are common, and these are at the same voltage as the drain of the driver. The bulk region is at circuit ground; as a result the D-mode device has a finite bulk–source voltage, V_{BS}. The body effect is surprising in its magnitude, yielding the characteristic shown with a dotted line. Recall that the body effect is incorporated in the model through the parameter γ. Comparing Figures 5-64 and 5-70, we see that the D-mode load approximates a resistive load. However, the D-mode load is small

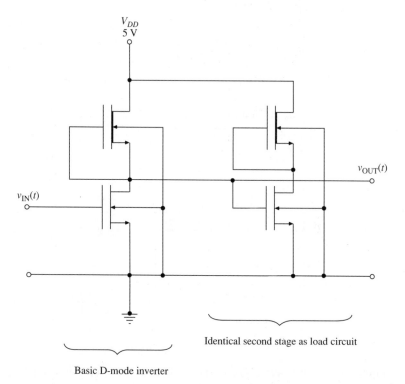

Figure 5-69 Schematic diagram of the two-stage circuit used for large-signal MOSFET analysis. Each stage constitutes an E-D inverter, having an E-mode driver and a D-mode load device. The second inverter serves as a realistic circuit for loading the output node of the first stage.

compared to other possible load devices, which is a major advantage in integrated-circuit applications. The points A, B, and C in Figure 5-70 represent the calculated dc bias values listed in the corresponding columns of Table 5-9.

The dc characteristics of the inverter are best represented using transfer characteristics, as shown in Figure 5-71. Again we have plotted the calculated dc values listed in Table 5-9. The solid line represents the inverter output voltage $v_{OUT}(t)$ as a function of the input voltage $v_{IN}(t)$. The dotted curve is the same function plotted with the axes interchanged. The two curves intersect at the OFF and ON dc bias values, Points A and C, and at the dc bias value, Point B. The slope of the solid curve at point B equals the dc voltage-gain value of about -3 that was calculated in the case of the resistive load.

The *noise margin* in a digital, or large-signal-switching circuit, is of great importance. It constitutes a measure of the noise-voltage magnitude that the circuit can withstand at its input terminal without "flipping," or changing state. Generous noise margins are a key to reliable operation, and hence have a bearing on product yield as well, since testing is a matter of assessing operational reliability. Yield, in turn, has a crucial bearing on product cost. Noise margin is often defined as the

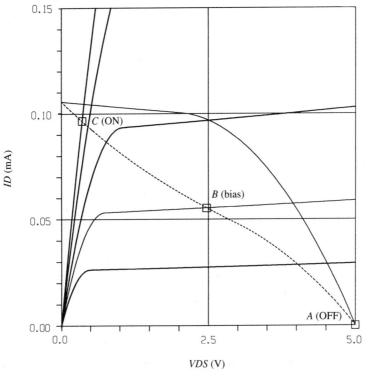

Figure 5-70 MOSFET output characteristics and D-mode loadline. The solid curve corresponds to the case where gate, source, and bulk regions of the load device are common (as would be true with discrete devices). The dotted curve displays the consequences of the body effect on the D-mode load, for the present numerical example.

length of one side of the largest square that can be fitted between the solid and dotted curves, as is illustrated in Figure 5-71. This, however, is an optimistic way of defining noise margin [85].

Now let us treat the dynamic response of the two-stage inverter circuit. To begin, we must examine in more detail than before the capacitive components of the Level-2 model, as given by Equations 5-167 through 5-171. In what follows we will consider only the driver MOSFET in the first inverter circuit. Since the bulk–source voltage is always zero for the driver, the corresponding depletion-layer capacitance C_{BS} approximates its zero-bias value of 110 fF for all input voltages, where 100 fF is contributed by the bottom portion of the implanted region, and 10 fF, by the sidewall portion. But because C_{BS} is short-circuited, it can be neglected.

The bulk–drain junction is reverse-biased to its maximum degree when the driver is OFF. Its depletion-layer capacitance rises from 41 fF to 92 fF as V_{IN} increases from the OFF condition (0.356V) to the ON condition (5.0V), as described by Equation 5-177. We saw in Section 5-5.3 that the small-signal response of the inverter is dominated by C_{BD}. It is not surprising, therefore, to learn that the large-signal response is also dominated by C_{BD}, especially (as before) when the

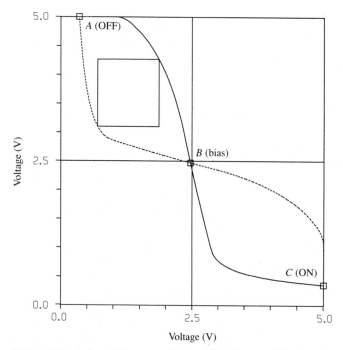

Figure 5-71 The solid curve is the transfer characteristic for the first inverter in Figure 5-69. The dotted curve is obtained by rotating the solid curve 180 degrees about a straight line drawn through the origin and point B. The large square illustrates one way of defining noise margin.

input generator is a voltage source applied directly to the gate–source input port.

The three capacitances associated with the gate are given by Equations 5-167 through 5-169. For a low-impedance input-voltage source, these capacitances do not play a major role because they are short-circuited to ground. But as the impedance of the source increases, they increase in importance. The behaviors of C_{GS}, C_{GD} and C_{GB} are shown in Figure 5-72, as function of the gate–source voltage $v_{GS}(t)$. As $v_{GS}(t)$ increases from 0 to 5 V, and the operating point moves along the dc load line such as the one shown in Figure 5-70, the operation of the MOSFET moves in sequence through the accumulation, depletion, saturation, and curved regimes. In the accumulation regime, C_{GB} is constant and equals the sum of the external gate capacitance and the total oxide capacitance:

$$C_{GB} = C_{0GB}L^* + C_{OX}L^*Z. \tag{5-189}$$

It has a value of 23 fF, with the gate-oxide contribution dominant. In the depletion regime, C_{GB} decreases, let us assume linearly, until it equals the external gate capacitance of 0.73 fF. In a numerical program such as SPICE, it is important that the transition from one regime to the next be gradual, accounting for the assumption. In the saturation and curved regimes, C_{GB} is constant and equal to this value.

The variation of C_{GS} as V_{GS} increases is as follows. In the accumulation re-

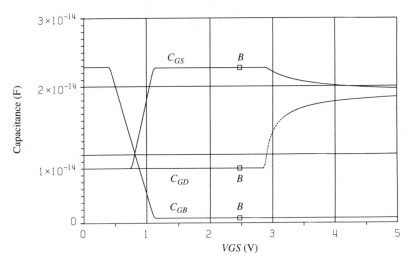

Figure 5-72 Gate-capacitance components as functions of gate–source voltage. The square B is the bias point in each case.

gime, C_{GS} equals the parallel-plate overlap capacitance of 8 fF. In the depletion regime, its value increases linearly until it equals the sum of the overlap capacitance and two-thirds of the gate-oxide capacitance, where the sum equals 19.6 fF. The factor of two-thirds is an empirical approximation that deals with the somewhat complicated combined behavior of the depletion and inversion layers as they develop in the depletion regime—a value based upon analytical calculations and experimental measurements. In the saturation regime, C_{GS} remains constant and equal to 23 fF. Finally, in the curved regime it decreases according to Equation 5-167d, approaching a final value of 19.7 fF.

The corresponding behavior of C_{GD} as $v_{GS}(t)$ increases is as follows. In the accumulation, depletion, and saturation regimes, it is constant and equal to the overlap capacitance of 8 fF. In the curved regime it increases according to Equation 5-168d, approaching a value of 19.7 fF. In the curved regime, an empirical assignment of gate-oxide capacitance is made once again. Half is placed in parallel with the gate–source junction and half, with the gate–drain junction.

Exercise 5-48. Why does this assignment of C_{OX} make sense at low bias levels?

With low bias, especially with low V_{DS}, the depletion layer is nearly uniform in thickness from source to drain. Since the device is symmetric with respect to source and drain, an equal division of the gate-oxide capacitance makes sense.

The dynamic behavior of the two-stage inverter circuit is shown in Figures 5-73 and 5-74. The former gives external variables, and the latter, internal variables.

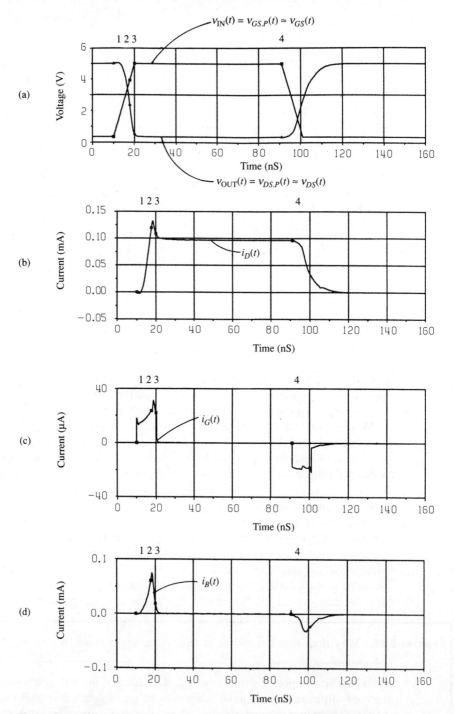

Figure 5-73 Waveforms calculated using SPICE for the terminal variables in the large-signal transient problem, with parasitic effects included. The numbered points correspond to the columns in Table 5-10. (a) Input and output voltages versus time. (b) Drain current $i_D(t)$ versus time. (c) Gate current $i_G(t)$ versus time. (d) Bulk current $i_B(t)$ versus time.

Figure 5-73(a) shows the assumed input voltage waveform, where $v_{IN}(t) = v_{GS,P}(t)$ and the corresponding output waveform, where $v_{OUT}(t) = v_{DS,P}(t)$. The input waveform is a voltage pulse that rises linearly from 0.356 V at $t = 10$ ns to 5 V at $t = 20$ ns and falls linearly from 5 V at $t = 91$ ns to 0.356 V at $t = 101$ ns. Figures 5-73(b) through 5-73(d) show, respectively, the drain current $i_D(t)$, the gate current $i_G(t)$ and the bulk current $i_B(t)$. The gate and bulk currents are significant only during the rise and fall periods of the switching waveform. During these periods, the bulk current dominates because it is required to charge and discharge $C_{BD}(t)$. Recall that this is the capacitance that dominates both the high-frequency and fast-switching behavior of the MOSFET.

Behaviors of the various internal variables are shown in Figure 5-74 for the time interval from 5 ns to 25 ns. Figure 5-74(a) presents the internal variables $g_m(t)$, $g_{mbs}(t)$, and $g_o(t)$. Note that $g_m(t)$ and $g_{mbs}(t)$ are nearly identical. The resistive portion of the drain current $i_D^R(t)$—as given by Equation 5-165—is shown in Figure 5-74(b). This component has a large peak at $t = 18$ ns because drain current is required to charge the various capacitances. Figure 5-74(c) shows the three capacitive components of gate current corresponding to the three terms of Equation 5-162. Finally, Figure 5-74(d) shows two of the five components of bulk current, specifically the bulk–source and the bulk–drain capacitive currents. The corresponding resistive components can be neglected because the source and drain junctions are reverse-biased. The final bulk-current component is the negative of the gate-current component $i_{GB}(t)$.

The behaviors of the charge and capacitive variables are given in Figures 5-75(a) through 5-75(d). Figure 5-75(a) presents $q_{GS}(t)$, $q_{GD}(t)$, and $q_{GB}(t)$, and Figure 5-75(b) presents $q_{BS}(t)$ and $q_{BD}(t)$. Figure 5-75(c) presents $C_{GS}(t)$, $C_{GD}(t)$, and $C_{GB}(t)$. Finally, Figure 5-75(d) presents $C_{BS}(t)$ and $C_{BD}(t)$. In the time interval from 5 to 25 ns, the MOSFET moves from the accumulation regime, through the depletion regime, through the saturation regime, and finally into the curved regime, so that the curves in Figure 5-75(c) are similar in form to the corresponding dc curves given in Figure 5-72. The various capacitive effects cause the dynamic loadline to differ significantly from the dc loadline. This is illustrated in Figure 5-76, where the squares correspond to the values of total time-dependent drain current $i_D(t)$, listed in Table 5-10. The lower dashed curve is the loadline for MOSFET turn-off.

Exercise 5-49. Why is $i_D^R(t)$ much larger than $i_D(t)$ at $t = 17.68$ ns?

The resistive component of drain current, $i_D^R(t)$, must supply not only the drain current $i_D(t)$, but also the current needed to charge parasitic capacitances. Visualize $i_D^R(t)$ as the current source in an equivalent circuit, wherein its upper node is connected to the load device, carrying $i_D(t)$, and two capacitances, $C_{GD}(t)$ and $C_{BD}(t)$. The current source must "pull" current through all three elements.

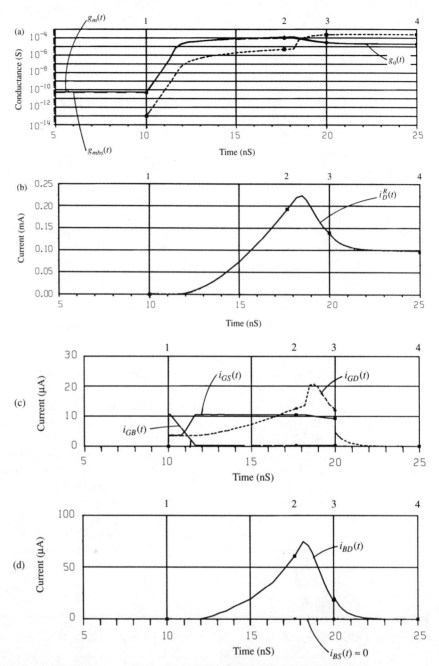

Figure 5-74 Intrinsic-variable waveforms calculated using SPICE in the large-signal transient problem. The numbered points correspond to the columns in Table 5-10. (a) Gate-source transconductance g_m, bulk–source transconductance, g_{mbs}, and output conductance g_0. (b) The resistive component of drain current $i_D^R(t)$ and the total (or net) drain current $i_D(t)$. (c) The three components of gate current, $i_{GS}(t)$, $i_{GD}(t)$, and $i_{GB}(t)$. (d) The important capacitive component of bulk–drain current, $i_{BD}^C(t)$. The resistive component of bulk–drain current can be neglected because the drain junction is reverse-biased. Also, the bulk–source junction is short-circuited.

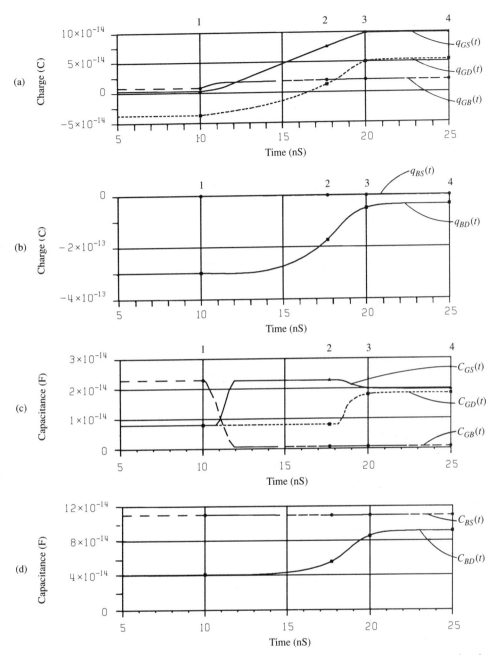

Figure 5-75 Intrinsic-variable waveforms calculated using SPICE in the large-signal transient problem. The numbered points correspond to the columns in Table 5-10. (a) Magnitudes of the charges stored in the three gate capacitances. (b) Magnitudes of the charges stored in two of the three bulk capacitances. The third bulk-capacitance charge magnitude is identically $q_{BG}(t)$, in part (a). (c) Magnitudes of the capacitances associated with the gate terminal. (d) Magnitudes of two of the three capacitances associated with the bulk terminal. The third capacitance magnitude is identically $C_{GB}(t)$, in part (c).

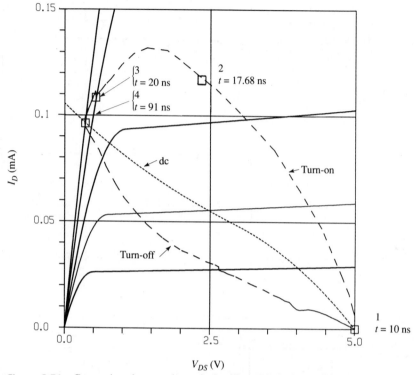

Figure 5-76 Comparing the two dynamic loadlines (dashed, one for turn-on and one for turn-off) with the static loadline (dotted), superimposed on the MOSFET output characteristics. The four squares correspond to the four columns of Table 5-10.

The large-signal pulse response of the MOSFET is very complicated, as can be seen from Figures 5-73 and 5-74. SPICE simulation provides an accurate means for predicting the nonlinear switching behavior of a given device. A detailed examination of the SPICE outputs is always a rewarding endeavor.

5-6 MOSFET–BJT PERFORMANCE COMPARISONS

As we have seen, the MOSFET is today's dominant device because it has a favorable combination of properties. However it has rarely been able to challenge the BJT in raw performance. The topic of performance comparisons emerged in the 1960s when it first became evident that the MOSFET was more than "just another device." The applications considered ranged from integrated circuits [86] to power devices [87]. Increasingly, the performance comparisons have focused on transconductance, which plays such an important part in fixing the circuit and system delays associated with charging parasitic capacitances. And since both devices display marked dependence of transconductance upon output current, the figure of merit g_m/I_{out} has also received considerable attention.

Transconductance comparisons of the BJT and MOSFET have been inhibited by the apples-oranges character of the problem. Operating principles in the two devices are quite dissimilar. But by normalization, we can eliminate a number of the difficulties. Several examinations of MOSFET transconductance properties have been published [54, 78, 79, 88–91], including some that employ normalization [78, 79, 88]. There have also been efforts to make BJT-MOSFET comparisons involving transconductance [86, 89, 92, 93]. Comparisons given here, however, are based upon a recent treatment [94] that goes considerably beyond the previous comparisons with respect to generality.

5-6.1 Simple-Theory Transconductance Comparison

Following Gummel and Poon [95], we can write collector current as

$$I_C = I_S \exp(qV_{BE}/kT), \tag{5-190}$$

where I_S is their *intercept current*. (See Section 4-4.3.) It follows, then, that transconductance is

$$g_m = qI_C/kT. \tag{5-191}$$

Using I_S for current normalization, and the thermal voltage for voltage normalization, converts Equation 5-190 to

$$I_{OUT} = e^{U_{BE}}, \tag{5-192}$$

where I_{OUT} and U_{BE}, respectively, will be used as symbols for the two normalized quantities. Similarly, using qI_S/kT for transconductance normalization converts Equation 5-192 into

$$G_M = I_{OUT} = e^{U_{BE}}. \tag{5-193}$$

Thus we have the well-known result that a log–log plot of normalized transconductance versus normalized output current for the BJT is a straight line with a positive slope of unity, as exhibited by the steeper curve in Figure 5-77.

Now turn to the case of the E-mode MOSFET. Equation 5-22 gives drain current in saturation as

$$I_D = K(V_{GS} - V_T)^2. \tag{5-194}$$

Thus MOSFET transconductance is

$$g_m = 2K(V_{GS} - V_T), \tag{5-195}$$

as given before in Equation 5-24. Let us normalize output current and transconductance for the MOSFET, employing the same quantities used before. Using the symbols I_{OUT} and G_M once again for the normalized variables, we first have from Equation 5-194

$$I_{OUT} = \frac{K}{I_S}\left(\frac{kT}{q}\right)^2 (U_{GS} - U_T)^2, \tag{5-196}$$

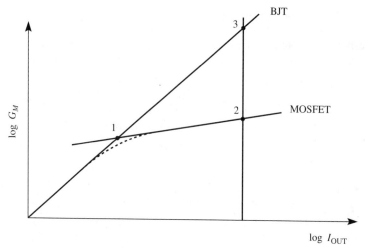

Figure 5-77 Normalized transconductance G_M versus normalized output current I_{OUT} for the BJT and the MOSFET.

where $U_{GS} \equiv (qV_{GS}/kT)$ and $U_T \equiv (qV_T/kT)$ are the normalized input and threshold voltages, respectively. Similarly, from Equation 5-195,

$$G_M = \frac{2K}{I_S}\left(\frac{kT}{q}\right)^2 (U_{GS} - U_T). \tag{5-197}$$

Combining Equations 5-196 and 5-197 yields

$$G_M = 2\frac{kT}{q}\sqrt{\frac{K}{I_S}}\sqrt{I_{OUT}}, \tag{5-198}$$

displaying the equally well-known result [96] that transconductance in the MOSFET goes as the square root of drain current. This result for the MOSFET is plotted as the straight line in Figure 5-77 with a positive slope of one half.

A critical point in Figure 5-77 is the point labeled 1, where the two straight lines intersect. We can provide a measure of interpretation as follows: Recall that anywhere along the line labeled *BJT*, normalized current and transconductance are identical. Therefore this condition holds at the point of intersection of the two straight lines in Figure 5-77 as well, permitting us to equate the normalized expressions in Equations 5-196 and 5-197, with the result that $(U_{GS} - U_T) = 2$, or

$$(V_{GS} - V_T) = 2\ kT/q. \tag{5-199}$$

Thus the intersection point corresponds to an input voltage lying two normalized units above MOSFET threshold. Note, now, that simple MOSFET theory upon which the present analysis is based disallows drain current below threshold, and so Figure 5-77 presents only above-threshold data.

Having thus interpreted point 1 in Figure 5-77, let us hasten to emphasize that the actual transconductance curve for the MOSFET departs appreciably from the simple-theory straight line as one moves toward low-level operation. The phenomenon of "excess" near-threshold conduction causes transconductance to decline as threshold is approached, as indicated by the dashed curve in Figure 5-77. For vanishing oxide thickness, the actual MOSFET curve on its low-current end asymptotically approaches the solid straight line labeled BJT, a MOSFET property that has been noted previously [88, 89, 93]. The assumption of vanishing oxide thickness is an acknowledgment of BJT and MOSFET dissimilarities.

Two other critical points can also be identified on Figure 5-77. These take advantage of the characteristic current exhibited by an E-mode MOSFET, a current value analogous to I_{DSS} in the D-mode device, and one that is displayed on the normalized current axis in Figure 5-10. As Figure 5-10 shows (let us confine ourselves to the N-channel E-mode case for simplicity), the current-to-voltage transfer characteristic is parabolic, having a branch (dashed) that, while not otherwise meaningful, serves to define a drain current equal to that which the device exhibits at $V_{GS} = 2V_T$. Let us define this current as

$$I_D(V_{GS} = 2V_T) \equiv I_2 = KV_T^2. \tag{5-200}$$

The current I_2 defines critical point 2 on the MOSFET curve in Figure 5-77. At this current, then, and at $2V_T$, Equation 5-195 yields

$$g_m = 2KV_T. \tag{5-201}$$

It follows that the normalized MOSFET transconductance at this specific input voltage and output current is

$$G_{MM2} = \frac{2KV_T}{(qI_S/kT)} = \frac{2K}{I_S} \frac{kT}{q} V_T. \tag{5-202}$$

Also, for input-voltage and output-current values at this point we have from Equation 5-200 for the corresponding normalized MOSFET output current,

$$\frac{I_2}{I_S} = \frac{K}{I_S} V_T^2. \tag{5-203}$$

But we have seen in Equation 5-193 that normalized output current, and transconductance as well, for the BJT at any input voltage is $\exp(U_{BE})$. Thus it follows that for the BJT at the current I_2 (point 3 on Figure 5-77), normalized BJT transconductance can be written

$$G_{MB2} = e^{U_{BE}} = \frac{K}{I_S} V_T^2. \tag{5-204}$$

Thus, combining Equations 5-202 and 5-204, we find that the ratio of BJT to MOSFET transconductance at the critical current I_2 is

$$\frac{G_{MB2}}{G_{MM2}} = \frac{\dfrac{K}{I_S} V_T^2}{\dfrac{2K}{I_S} \dfrac{kT}{q} V_T} = \frac{1}{2} \frac{V_T}{kT/q} = \frac{U_T}{2}, \tag{5-205}$$

where U_T is normalized threshold voltage. For $U_T = 10$, then, corresponding approximately to a quarter-volt threshold voltage, the BJT is superior in transconductance by a factor of five at $V_{GS} = 2V_T$, and improves its advantage further as current is increased. Obviously, the BJT advantage also increases as threshold voltage is permitted to increase.

A second comparison between BJT and MOSFET performance is obtained by examining the quotient of transconductance by output current as a function of input voltage. For the BJT, this quotient is a constant and is equal to q/kT;

$$\frac{g_m}{I_{\text{out}}} = \frac{q}{kT}, \quad (5\text{-}206)$$

where I_{out} is unnormalized output current. This is a consequence of the exponential dependence of output current on input voltage.

Simple MOSFET theory gives

$$g_m = 2\sqrt{KI_{\text{out}}}. \quad (5\text{-}207)$$

Thus,

$$\frac{g_m}{I_{\text{out}}} = 2\sqrt{\frac{K}{I_{\text{out}}}} = \frac{2\sqrt{K}}{(V_{GS} - V_T)\sqrt{K}} = \frac{2}{V_{GS} - V_T}. \quad (5\text{-}208)$$

Equating the two expressions for g_m/I_{out} in the BJT and MOSFET, Equations 5-206 and 5-208, one again finds, of course, that they are equal at an input voltage two thermal voltages above threshold, as stated in Equation 5-199. That equation resulted from the equivalent observation that the normalized transconductances from simple theory and at the same current coincide at this particular input-voltage value.

Tsividis also presented curves of g_m/I_{out} for the MOSFET [88], but qualified his curves as being "for extremely small drain–source voltage." That is because he approximated the dependence of I_{out} on V_{GS} by using a first-degree polynomial, and then used the approximate expression in calculating g_m/I_{out}. The subthreshold I–V equation introduced next, however, is valid for larger values of V_{DS}, and we shall now use it to develop a more general subthreshold theory.

5-6.2 Subthreshold Transconductance Theory

In the subthreshold regime it has been found [97, 98] that

$$I_{\text{out}} = \frac{qn_i D_n L_{Di} Z}{L} e^{-3|U_B|/2}(1 - e^{-U_{DS}}) \frac{e^{W_S}}{\sqrt{W_S - 1}}, \quad (5\text{-}209)$$

where we have employed the notation of Section 5-3.1. It is instructive to use Equation 5-209 to derive a general expression for subthreshold transconductance [71]. The effort is further worthwhile because the result will permit us to determine precisely where the quantity g_m/I_{out} peaks, since in the real device this does occur below threshold. To derive the desired expression for transconductance, we note that

$$g_m \equiv \frac{\partial I_D}{\partial V_G} = \frac{q}{kT}\frac{\partial I_D}{\partial U_G} = \frac{q}{kT}\frac{\partial I_D}{\partial W_S}\frac{\partial W_S}{\partial U_G},$$ (5-210)

where U_G is absolute normalized gate potential with respect to the common reference. Noting that $I_{\text{out}} = I_D$ at present, and differentiating Equation 5-209 with respect to W_S yields:

$$\frac{\partial I_D}{\partial W_S} = a\frac{e^{W_S}\sqrt{W_S - 1} - (e^{W_S}/2)\sqrt{W_S - 1}}{W_S - 1}$$

$$= a\frac{e^{W_S}}{\sqrt{W_S - 1}}\left[1 - \frac{1}{2(W_S - 1)}\right]$$

$$= I_D\left[\frac{2W_S - 3}{2W_S - 2}\right],$$ (5-211)

where

$$a \equiv \frac{qn_iD_nL_{Di}Z}{L}e^{-3|U_B|/2}(1 - e^{-U_{DS}}).$$ (5-212)

The total charge per unit area in the surface region, assuming zero flatband voltage, is given by:

$$Q_S = C_{OX}(V_G - \psi_S) = (kT/q)C_{OX}(U_G - W_S + |U_B|),$$ (5-213)

where C_{OX} is the oxide capacitance per unit area. This pair of equations requires some explanation. First, because we have assumed zero flatband voltage, $V_G = \psi_G$, and hence the "mixing" of voltage and potential symbols in the first parenthetic expression is permissible. Second, because we deal here with potential relationships in the MOS capacitor rather than a bulk-silicon sample only, and because the substrate is P-type, it is convenient to choose $(q/kT)\psi_S \equiv W_S - |U_B|$, letting ψ_S and W_S have the same sign. (We shall see in a subsequent diagram that this is indeed the case.)

Next, using Gauss's law and Approximation C from Table 5-1 for the normalized electric field, we can also write Q_s at the source end of the channel as:

$$Q_s = \left[\frac{kT}{q}\epsilon q n_i e^{|U_B|}\right]^{1/2}\sqrt{2(W_S - 1 + e^{W_S - 2|U_B|})}$$

$$= \sqrt{2}\,qn_iL_{Di}e^{|U_B|/2}\sqrt{2(W_S - 1 + e^{W_S - 2|U_B|})}$$

$$= 2qn_iL_{Di}e^{-|U_B|/2}\sqrt{e^{2|U_B|}(W_S - 1) + e^{W_S}}.$$ (5-214)

(See Section 5-3.6 and Problem A5-17.) Although Equation 5-214 gives the areal charge density at the source end of the channel, it is in error by only a few percent along the entire channel. This is because ionic or bulk charge is dominant all the way from source to drain in the subthreshold regime of operation. Setting the last portion of Equation 5-213 equal to the last portion of Equation 5-214 gives us

$$\frac{kT}{q} C_{OX}(U_G - W_S + |U_B|)$$
$$= 2qn_i L_{Di} e^{-|U_B|/2} \sqrt{e^{2|U_B|}(W_S - 1) + e^{W_S}}, \quad (5\text{-}215)$$

or

$$U_G = \frac{q}{kT} \frac{2qn_i L_{Di} e^{-|U_B|/2}}{C_{OX}} \sqrt{e^{2|U_B|}(W_S - 1) + e^{W_S}} + W_S - |U_B|. \quad (5\text{-}216)$$

Hence,

$$\frac{\partial U_G}{\partial W_S} = \frac{q}{kT} \frac{qn_i L_{Di} e^{-|U_B|/2}}{C_{OX}} \left[\frac{e^{2|U_B|} + e^{W_S}}{\sqrt{e^{2|U_B|}(W_S - 1) + e^{W_S}}} \right] + 1$$

$$= 1 + \frac{\epsilon}{2 C_{OX} L_{Di}} e^{-3|U_B|/2} \left[\frac{e^{2|U_B|} + e^{W_S}}{\sqrt{W_S - 1 + e^{W_S - 2|U_B|}}} \right]. \quad (5\text{-}217)$$

Exercise 5-50. Explain the change of coefficient in the last step.

Multiplying numerator and denominator of the fraction by $\exp(-|U_B|)$ accounts for the exponential portion of the coefficient (as well as for the change in the radical expression). Noting that

$$L_{Di}^2 = (kT/q)(\epsilon/2qn_i)$$

shows that

$$(q/kT) qn_i L_{Di} = \epsilon/2L_{Di}.$$

For subthreshold conditions, $\exp(W_S - 2|U_B|) \ll (W_S - 1)$, so that the exponential term in the denominator can be dropped. Thus,

$$\frac{\partial U_G}{\partial W_S} \approx 1 + \frac{\epsilon}{2 C_{OX} L_{Di}} e^{-3|U_B|/2} \left[\frac{e^{2|U_B|} + e^{W_S}}{\sqrt{W_S - 1}} \right]. \quad (5\text{-}218)$$

Substituting Equations 5-211 and 5-218 into Equation 5-210, we obtain

$$\frac{\partial W_S}{\partial U_G} = \frac{2 C_{OX} L_{Di} \sqrt{W_S - 1}}{2 C_{OX} L_{Di} \sqrt{W_S - 1} + \epsilon e^{|U_B|/2} + \epsilon e^{W_S - 3|U_B|/2}}$$

$$= \frac{C_{OX}}{C_{OX} + \dfrac{\epsilon}{2L_{Di}} \dfrac{e^{|U_B|/2}}{\sqrt{W_S - 1}} + \dfrac{\epsilon}{L_{Di}} \dfrac{e^{W_S - 3|U_B|/2}}{W_S - 1}}. \quad (5\text{-}219)$$

But the last two terms in the denominator are identically C_b and C_i, respectively, at the source end of the channel. (See Problem A5-17.) Hence, $(\partial W_S/\partial U_G) = C_{OX}/(C_{OX} + C_b + C_i)$, and

$$g_m = \frac{qI_D}{kT}\left[\frac{2W_S - 3}{2W_S - 2}\right]\frac{C_{OX}}{C_{OX} + C_b + C_i}. \qquad (5\text{-}220)$$

These expressions for capacitance,

$$C_b = \frac{\epsilon}{2L_{Di}}\frac{e^{|U_B|/2}}{\sqrt{W_S - 1}} \qquad (5\text{-}221)$$

and

$$C_i = \frac{\epsilon}{2L_{Di}}\frac{e^{W_S - 3|U_B|/2}}{\sqrt{W_S - 1}} \qquad (5\text{-}222)$$

are consistent with the equations derived by Tsividis [88]. That is, if the denominator of the last factor in Equation 5-217 is replaced by its Taylor's series expansion and the development above is repeated, Equation 5-222 will be identical to his expression for C_i.

Equation 5-220 is a general expression for subthreshold transconductance [71]. Its factor containing W_S is approximately unity. As C_{OX} becomes large, g_m approaches qI_D/kT, which is a consequence of the exponential dependence of drain current on surface potential in the subthreshold regime. The capacitance ratio included in the subthreshold g_m is the result of voltage division in the input loop. The "useful" portion of the applied gate voltage is that which modulates the surface potential, and thus a capacitance ratio appears in the expression for g_m.

5-6.3 Calculating Maximum MOSFET g_m/I_{out}

Now we are in a position to determine the maximum value of g_m/I_{out} [94]. Dividing Equation 5-220 through by I_{out} gives an expression for g_m/I_{out} that is a function of only one variable, W_S. The maximum in g_m/I_{out} will occur when $C_b + C_i$ is at a minimum, because C_{OX} is constant. The fact that there is a minimum, residing below threshold, in the sum of C_b and C_i is clearly displayed in Figure 5-39.

Neglecting the factor $(2W_S - 3)/(2W_S - 2)$ because it is approximately unity, we can seek the extremum for $C_b + C_i$. Evidently,

$$\frac{dC_b}{dW_S} = \frac{\epsilon}{2L_{Di}} e^{|U_B|/2}\left[\frac{-1}{2(W_S - 1)^{3/2}}\right], \qquad (5\text{-}223)$$

and

$$\frac{dC_i}{dW_S} = \frac{\epsilon}{2L_{Di}} e^{-3|U_B|/2}\left[\frac{e^{W_S}(W_S - 1)^{1/2} - \frac{1}{2}e^{W_S}(W_S - 1)^{-1/2}}{(W_S - 1)}\right]. \qquad (5\text{-}224)$$

The minimum in $C_b + C_i$ occurs when

$$\frac{\epsilon}{2L_{Di}} e^{-3|U_B|/2} \left[\frac{-e^{2|U_B|}}{2(W_S - 1)^{3/2}} + \frac{2e^{W_S}(W_S - 1)}{2(W_S - 1)^{3/2}} - \frac{e^{W_S}}{2(W_S - 1)^{3/2}} \right] = 0. \quad (5\text{-}225)$$

Thus we find that

$$2e^{W_S}(W_S - 1) - e^{W_S} = e^{2|U_B|}, \quad (5\text{-}226)$$

or

$$e^{W_S}(2W_S - 3) = e^{2|U_B|}, \quad (5\text{-}227)$$

gives us the value of W_S for which g_m/I_{out} is at its maximum. This does not directly give us the value of V_{GS} for which the maximum occurs, but V_{GS} can be obtained from the U_G-versus-W_S relation, Equation 5-216, if the properties of the device are specified. Note, however, that the maximum in g_m/I_{out} occurs for $W_S < 2|U_B|$ because $(2W_S - 3)$ is greater than unity. Thus the maximum will be below threshold, as we expected. This behavior agrees with the results of Evans and Pullen [93].

Simple MOSFET theory, Equation 5-208, predicts that g_m/I_{out} becomes infinite at the threshold voltage. In reality, however, subthreshold current sets an upper limit for this quotient, as demonstrated above. As oxide capacitance approaches infinity, this limit is q/kT, the same value found in a BJT. Figure 5-78 is a plot of g_m/I_{out} versus V_{in} for the BJT and MOSFET. Simple theory for the MOSFET in the near-threshold regime is displayed as the rising curve, and the more accurate subthreshold theory is displayed by the lower curve. The transconductance-current

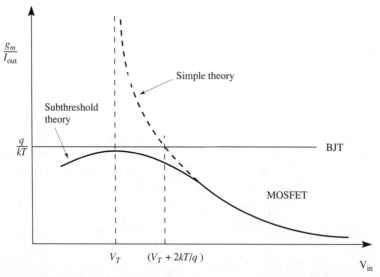

Figure 5-78 Transconductance-current quotient versus input voltage for the BJT and MOSFET, with results for the MOSFET in the near-threshold regime plotted as the rising curve, and results from more accurate subthreshold theory plotted as the lower curve [94].

quotient of the MOSFET is thus seen to approach that of a BJT at a particular input voltage that is dependent on the device properties. For any finite value of C_{OX}, g_m/I_{out} for a MOSFET will be lower at all input voltages than for a BJT.

5-6.4 Transconductance versus Input Voltage

As another method of transconductance comparison, let us examine this property directly as a function of input voltage for the E-mode MOSFET and the BJT. We start with the latter, and again employ the notation of Section 5-6.1. Recall that Equation 5-193 gives the well-known result that a semilog plot of G_M versus U_{BE} is linear, which in fact is true through many decades of transconductance [86], with the result plotted in Figure 5-79(a).

For the MOSFET case, we have a series of curves, each for a different value of U_T. Specifically chosen in Figure 5-79(b) are $U_T = 10$, 20, 30, and 40. Here we have chosen linear axes, yielding the simplest presentation of the MOSFET curves, but note that the abscissa is coordinated with that of Figure 5-79(a), employing thermal-voltage normalization once more. Figure 5-79(b) makes it clear that reducing threshold voltage U_T (or V_T in unnormalized form) produces a higher transconductance at a given input voltage. To relate the transconductance-versus-voltage curves for the two devices, it is necessary to examine the ratio of the quantities used for normalizing their respective transconductances. The ratio of the BJT quantity to the MOSFET quantity is this:

$$\text{RATIO} = \frac{(qI_S/kT)}{(kT/q)K} = \left(\frac{q}{kT}\right)^2 \frac{I_S}{K} = \left(\frac{q}{kT}\right)^2 \left[\frac{q\mu_{nb}\dfrac{kT}{q}\dfrac{n_{0B}}{X_B}A_E}{\dfrac{\mu_{ns}\epsilon_{OX}}{2X_{OX}}\dfrac{Z}{L}}\right], \quad (5\text{-}228)$$

where μ_{nb} and μ_{ns} are the bulk and surface mobilities, respectively, n_{0B} is equilibrium minority-electron density in the base region (with uniform doping of the base region assumed), A_E is emitter area, X_B is base thickness, and X_{OX} is oxide thickness, while ϵ_{OX}, Z, and L have their usual MOSFET meanings. Hence,

$$\text{RATIO} = \left(\frac{q}{kT}\right)\left(\frac{q}{\epsilon_{OX}}\right)\left(\frac{\mu_{nb}}{\mu_{ns}}\right)\left(\frac{X_{OX}}{X_B}\right) A_E n_{0B}\left(\frac{L}{Z}\right). \quad (5\text{-}229)$$

In view of the fact that LZ is gate area A_G, this becomes

$$\text{RATIO} = \left(\frac{q}{kT}\frac{q}{\epsilon_{OX}}\right)\left(\frac{\mu_{nb}}{\mu_{ns}}\right)\left(\frac{X_{OX}}{X_B}\right)\left(\frac{A_E}{A_G}\right) n_{0B} L^2. \quad (5\text{-}230)$$

Interestingly, this form of comparison involves (beside the constant coefficient) a series of three dimensionless factors that directly compare structural and physical features of the two devices, and two additional factors, one for each device. The last two factors display the irreconcilable dissimilarities in the two devices. Let us evaluate the constant factor and choose reasonable values for the variables, as has

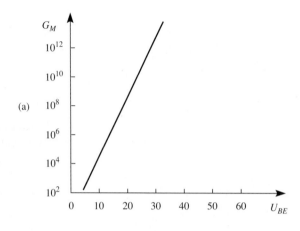

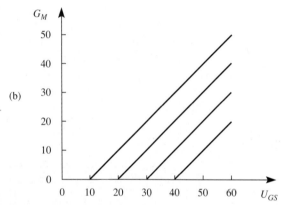

Figure 5-79 Normalized transconductance versus input voltage. (a) Semilog plot for the BJT. (b) Linear plot for the MOSFET, with normalized threshold voltage as a parameter [94].

been done in Table 5-11. For the chosen values, we have

$$\text{RATIO} = 1.853 \times 10^{-9}. \tag{5-231}$$

Multiplying the transconductance values in Figure 5-79(a) by this factor yields the modified BJT curve plotted in Figure 5-80. The four MOSFET curves of Figure 5-79(b) have also been plotted in semilog form in Figure 5-80. The BJT curve can readily be adjusted for other values of the variables and variable ratios just given.

5-6.5 Physics of Subthreshold Transconductance

Certain details of MOSFET operation, especially subthreshold operation, have been vigorously debated in the literature. Following the arguments used in a recent examination of this matter [71], let us review the debate. Early among the authors who

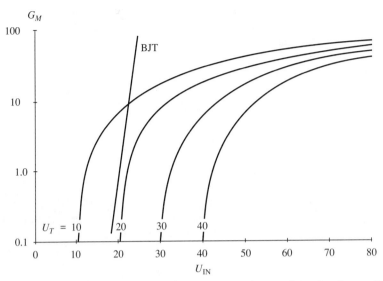

Figure 5-80 Normalized transconductance versus input voltage for the specific BJT-MOSFET variable ratios and variables listed in Table 5-11 [94].

examined subthreshold properties of the long-channel MOSFET were Evans and Pullen [93], who in 1966 measured the transconductance-current quotient, finding that it peaked in the subthreshold regime at a value under q/kT. The subthreshold regime of operation was treated analytically by Barron in 1972 [97]. Starting from the Pao-Sah double-integral formulation [54], he derived an approximate-analytic expression for drain current by making mathematical approximations appropriate to subthreshold conditions. Subsequent papers by Van Overstraeten, Declerck, and Broux [98], and by Troutman [99] dealt with the problem similarly (Pao-Sah formulation with approximations) but recognized in addition that subthreshold drain current is primarily a diffusion current. Still later, drain-current equations were derived by Brews [78] and by Fichtner and Pötzl [100] by treating the source and drain junctions at the channel ends much like the emitter and collector junctions of a BJT. Brews based his analysis on the charge-sheet approximation (employing a modified depletion approximation and a channel of zero thickness), while Fichtner and Pötzl introduced an effective channel thickness.

The Pao-Sah paper did not specifically address the subthreshold regime, but is very relevant because it was the first to consider the contribution to channel current (and hence to drain current) made by the diffusion-transport mechanism, and in so doing, produced a quantitative model so accurate that it is the standard by which other models are judged. Because diffusion is essentially the only transport mechanism at very small current values, as just noted, their work led directly to valid subthreshold analysis.

The 1982 paper by Tsividis [88] critically examined the regime of moderate inversion, lying between the strong-inversion and subthreshold regimes, and presented normalized curves of transconductance and of transconductance-current quotient throughout all three regimes. Yet another interesting contribution to the understanding of the subthreshold MOSFET was made by Johnson in 1973 [89]. Noting the Evans-Pullen observation of a transconductance-current quotient approaching q/kT, he invoked a BJT-like model for the subthreshold MOSFET, since a strict q/kT quotient exists over many current decades of BJT operation [86]. He did not note the pure-diffusion character of the subthreshold channel current, which would have strengthened the comparison, but he did call specific attention to the barrier between the source region and the source end of the channel, accurately describing it as a quasi "high-low" junction. Other authors have called attention to this potential barrier, although not as explicitly. In particular, Sah and Pao [90] in 1966 had remarked on the "injection" of carriers from the source region into the channel. Johnson went further, introducing a capacitance into the MOSFET model that he likened to the diffusion capacitance associated with the BJT emitter junction and base region. He suggested that this capacitance was a result of joint properties of the channel and channel-source barrier, and, linking it to the aforementioned transconductance behavior, labeled it the "control capacitance." The theory of subthreshold transconductance in the MOSFET developed in Sections 5-6.2 and 5-6.3 shows that the control-capacitance concept is unnecessary, and the discussion that follows shows that the concept is in fact flawed. As an accompaniment we now include a detailed description of the source-channel barrier region.

Consider the source end of the channel region. Let it be a plane N^+P step junction that is normal to the oxide–silicon interface. The properties of such a junction are well known now from discussions in Sections 3-4.1 and 5-3.3, properties summarized in Figure 5-81. The space-charge layer consists of ionic charge on the high side, while on the low side it is a combination of ionic charge and carriers in an inversion layer—electrons in the present case, illustrated in Figure 5-81(a). Given an equilibrium electron density n_0 on the N^+ side, we know that the equilibrium electron density at the junction is n_0/e as depicted in Figure 5-81(b), consistent with a potential drop of kT/q on the high side [8]. With forward bias, this density at the junction increases, and both ionic-charge layers shrink. (See Section 5-3.3.)

In the MOSFET operating above threshold, the inversion layer formed by the field plate merges with the inversion layer of the source-region N^+P junction. At this point, two-dimensional considerations are no longer avoidable. For a first-order treatment, let us use an orthogonal merging of two one-dimensional solutions [71, 101] (also see Section 5-4.1), confining ourselves to the long-channel-MOSFET case in order to give this approximation the greatest meaning. We have already made reference to the two merged inversion layers. But ionic space-charge regions in the P-type substrate also merge. In fact, the channel region of the MOSFET resides totally in a region that is completely depleted when the device is in above-threshold operation, and all depletion-layer boundaries are spaced well away from the region of interest. Even in subthreshold operation, the incipient inversion layer resides in a well-depleted region. As a final simplification in the present problem, let us assume that the source and substrate regions are electrically common.

Areal charge densities in the channel inversion layer remain far below those in

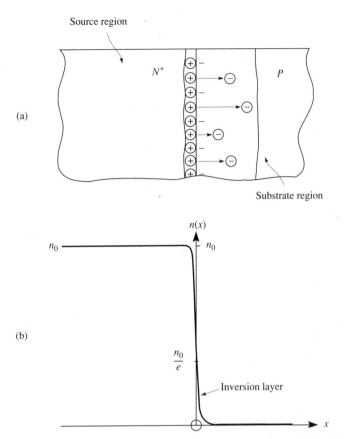

Figure 5-81 Space-charge layer of an N^+P step junction [71]. (a) Physical representation, showing that charge on the high side is ionic, while on the low side it is a combination of ionic and inversion-layer charge. (b) Linear representation of the electron distribution through the space-charge region.

the source-junction inversion layer, even in normal operation of the MOSFET, and as a consequence, the potential barrier at the source end of the channel has appreciable magnitude. In the subthreshold case, the barrier is even larger. Let us examine this case further. Figure 5-82 offers a magnified illustration of the orthogonal and merging inversion layers (idealized in one-dimensional terms). The notation y_t designates the threshold plane in the N^+P-junction's inversion layer—the plane at which $n(y) = p_0$, where p_0 is the equilibrium majority-carrier density in the substrate. Since we have assumed subthreshold conditions, there is no analogous plane labeled x_t. But we do have intersecting "intrinsic planes" designated y_i and x_i, along which $n = p$.

Before proceeding, let us make a quantitative examination and comparison of the two inversion layers depicted in Figure 5-82. The junction inversion layer, in

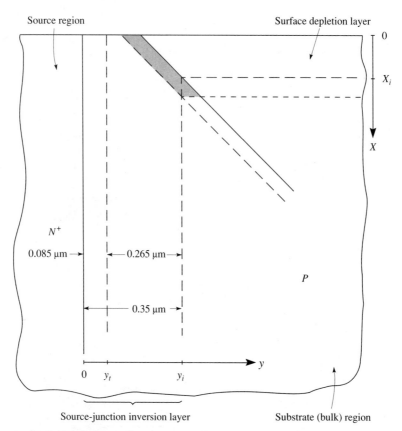

Figure 5-82 Orthogonal and merging depletion and inversion layers at the source end of a MOSFET channel [71].

addition to being "stronger" than the channel inversion layer, is only slightly altered by device operation, while the latter experiences orders-of-magnitude change in normal device operation. With no drain–source bias, the source junction and the inversion layer remain at equilibrium. A positive voltage increment on the gate causes a strengthening of the channel inversion layer, in turn causing a local change in the source junction near the oxide–silicon interface, a change that "creates a new" high-low junction. We shall examine this situation in more detail shortly, but for now need only point out the response of the junction profile depicted in Figure 5-81(a) to such a positive gate-voltage increment; potential drop on the left-hand side shrinks somewhat, and electron density at the metallurgical junction rises. But note that possible change in the former has a "hard ceiling" of 26 mV, and the corresponding change in the latter cannot be more than a factor of e. By contrast, surface potential in the channel region (the analog of junction potential in the source junction) changes by more than a volt when the device is switched from "off" to strongly "on." And for every volt of change in surface potential, electron density at the oxide–silicon interface changes by nearly seventeen orders of magnitude! Therefore it is a good approximation to treat the source-junction depletion layer as static in examining its interaction with the channel.

Using the depletion-approximation replacement [53] and its extensions, we can easily determine dimensions for the inversion layers that are depicted to scale in Figure 5-82. Let us assume net substrate doping of $1 \times 10^{15}/\text{cm}^3$. For this case, the inversion-layer thickness from threshold plane to junction or surface saturates at about 0.66 L_D [71], where L_D is the general Debye length, which in the present example gives us a thickness of 0.085 μm. In addition, the distance from the threshold plane to the intrinsic plane is 0.265 μm, giving an overall thickness for the junction inversion layer of 0.35 μm, measured from the intrinsic plane. One would not ordinarily measure inversion-layer thickness from the intrinsic plane, but since we are interested in subthreshold conditions, it is a relevant dimension.

The subthreshold behavior of the source junction in the region where it intersects the channel is that of a "variable" one-sided junction at equilibrium, meaning this: absent source–drain voltage, as noted earlier, this portion of the junction remains at equilibrium even in the face of channel-strength variations. The size of the junction's static potential barrier at any x position is given by the Boltzmann relation, employing the source-region electron density n_0 and the electron density $n(x)$ at that position in the channel. Altering channel strength alters $n(x)$ and thus the barrier potential at that point, which is equivalent to altering doping ratio in the equivalent *PN* junction (keeping the high side fixed). But with the onset of weak inversion in the channel, a qualitative change takes place in the affected region of the source junction. The portion of the junction wherein $n > p$ now exhibits the properties of a high-low junction, which are markedly different from those of a *PN* junction [102]. The forward characteristic in the *PN* case exhibits an "offset voltage," while that of the high-low junction does not, a fact that makes the latter an extremely useful ohmic contact. The high-low junction exhibits high conductance, even at very small voltage. By contrast, in the *PN* case conductance is very low, even up to a forward voltage of some 0.4 V.

Noting the potential barrier at the source end of the channel, Johnson [89] reasoned correctly that variations in V_{GS} would lead to a modulation of this barrier, a modulation that will occur, as we have seen, in spite of the external short-circuiting of source and substrate. Specifically, a positive voltage increment on the gate would produce a reduction in the height of the barrier. This relationship between a barrier increment and charge increment resembles to a limited degree, once drain–source bias is applied, the BJT phenomenon that is characterized as *diffusion capacitance,* in which an increment of forward bias on the emitter junction raises minority-carrier density at the emitter boundary of the base region. It has been shown that, assuming continuous reverse bias on the collector junction and a uniformly doped base region, the result of such an emitter-bias increment is to cause the linear base-region profile to rotate about a point on the x axis that is defined by the depletion-approximation boundary of the collector junction on the base side [103]. The resulting wedge-shaped increment in base-region minority carriers is a stored-charge increment of one sign. An equal and opposite charge increment consists of majority carriers drawn from the base contact in one dielectric relaxation time, which we shall assume to be very short. Quasineutrality is preserved in the base region and the two diffusion-capacitance charge increments are stored in the base region in balanced fashion.

Carrier behavior in the subthreshold MOSFET diverges to a significant degree from that in the BJT, however. Briefly, at a given depth below the surface, the electron profile from source to drain is linear and transport occurs only by diffusion, a point on which there is general agreement. The source–drain carrier gradient has its maximum value at the surface, and declines monotonically to zero as distance from the surface increases. Volumetric electron-density profiles at several x positions in the subthreshold channel are shown in Figure 5-83. The crucial difference of this case from the BJT case, is that an *elevation* in channel-electron density at a given point is accompanied by a further *depression* in hole density. In other words, we deal here with *surplus* carriers, characterized by conditions of quasiequilibrium nonneutrality, rather than the *excess* carriers of the BJT-base case, characterized by quasineutral nonequilibrium. (See Exercise 3-56, Section 3-7.6.) The fatal flaw, then, in the control-capacitance concept examined in this way, is its lack of an increment of holes to accompany the increment of electrons.

Having dispensed with Johnson's control capacitance we are obligated to explain the transconductance behavior of a MOSFET from the present point of view. A cross-sectional view of the MOSFET under analysis is shown in Figure 5-84(a). Note that the spatial origin of the y-axis is located at the depletion-layer boundary

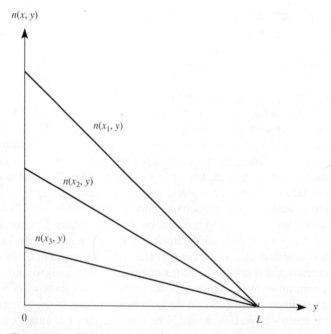

Figure 5-83 Volumetric electron density profiles at several positions in the subthreshold channel of a MOSFET.

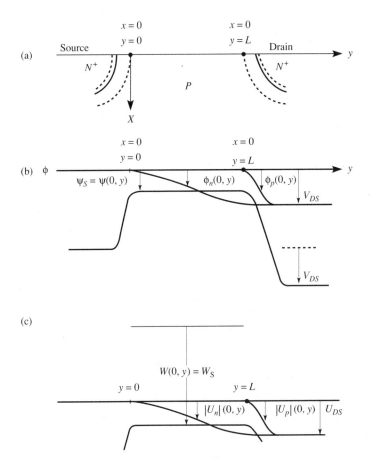

Figure 5-84 Physical and potential-profile representation of the device under analysis. (a) Cross section of N-channel MOSFET. (b) Profiles at the surface in the y direction of electrostatic potential $\psi(0, y)$, and quasi Fermi levels $\phi_n(0, y)$ and $\phi_p(0, y)$. (c) Quantities of part (b) represented in normalized form.

for consistency with the notation used by previous authors. The potential profile at the surface in the source-drain direction for this device is plotted in Figure 5-84(b), and the corresponding diagram using normalized potential notation is given in Figure 5-84(c). Note that W_S in Figure 5-84(c) and ψ_S in Figure 5-84(b) indeed have the same sign, a point made in Section 5-6.2 in explanation of Equation 5-213.

Thus we have offered a physical picture of subthreshold conditions and phenomena in the long-channel MOSFET, underlying the analytic calculation of subthreshold transconductance in Sections 5-6.2 and 5-6.3. Combining these physical and analytical pictures yields an understanding that can be summarized in these terms: The transconductance of the long-channel MOSFET approaches qI_D/kT as oxide capacitance becomes large and as drain current simultaneously becomes small. This behavior is a consequence of three features of subthreshold operation that were integrated for the first time in Reference 71, but each of which had been noted at least once before. These factors are (1) the essentially linear source–drain carrier profile, declining in slope in each successive channel lamella as one moves farther from the surface, these profiles associated with purely diffusive transport; (2) carrier-density values at the source end of each lamella that are exponentially related to source-end surface potential; and (3) a channel–source barrier that exhibits the high-conductance properties of a high-low junction. These properties bear some similarity to those of a BJT (which exhibits a strict qI_C/kT transconductance over many decades) with the primary exception being that the emitter junction is a *PN* junction rather than a high-low junction, accounting for the BJT's offset voltage, absent in the MOSFET. Further, surface potential in the MOSFET is controlled by "reaching through" the gate oxide, introducing a capacitor ratio into the transconductance expression, an expression developed in Section 5-6.2 as Equation 5-220.

SUMMARY

At the heart of the MOSFET is the MOS (metal-oxide-semiconductor) capacitor, a three-layer structure with a metallic field plate on top and silicon (in the most important case) on the bottom. The intervening dielectric layer is usually SiO_2 grown thermally on the silicon. Doped regions, opposite to the substrate in type, are positioned at opposite edges of the capacitor and become two terminals of the MOSFET, with the field plate or gate being a third. The gate is the input or control electrode. In early devices it was usually made of aluminum, while today it is usually heavily doped polycrystalline silicon. Sufficient bias of the polarity that pushes majority carriers away from the oxide–silicon interface creates an inversion layer in the silicon at the interface. This is the depletion-inversion polarity. The inversion layer, or channel, permits conduction between the two doped regions. Carriers flow from the region termed the *source* to the region termed the *drain*, with the latter having the same bias polarity as the gate. Voltages in the MOSFET are usually referred to that of the source terminal.

  Majority carriers in the source and drain regions are the same as those of the channel, and these single-type flowing carriers are primarily responsible for MOSFET properties. For this reason a field-effect transistor (FET) is termed *unipolar*, in contrast to the bipolar junction transistor (BJT), where carriers of one type flow into the base region, and opposite-type carriers flow through the collector junction. FETs in general can be described as voltage-controlled resistors. The MOSFET is inferior to the BJT in a number of respects. But its overall combination of properties is extremely advantageous, causing the MOSFET to be today's dominant solid-state device. MOS technology requires a degree of control of conditions in the oxide, and especially at the oxide–silicon interface, that vastly exceeds that required in BJT technology. For this reason, the MOSFET arrived about a decade later than the BJT, even though it was one of the earliest solid-state amplifiers (or switches) to be conceived in principle.

The N-channel MOSFET has an advantage over the P-channel device because electron mobility exceeds hole mobility. Inversion-layer mobilities for both, however, are typically one-half to one-third their bulk values. (The substrate region is often termed *bulk* for brevity and convenience.) In the N-channel case the bulk region is P-type, and we assume it to be uniformly doped (apart from the heavily and oppositely doped source and drain regions, of course). When gate–source bias voltage is adjusted to yield hole density at the oxide–silicon interface equalling that deep in the bulk region, the flatband condition exists, and the associated critical gate–source bias is termed the flatband voltage. Variations in gate–source voltage do not disturb quasiequilibrium conditions in the substrate, so the Fermi level is constant or "flat," irrespective of band bending. When the surface potential (the mid-gap potential at the interface) equals the Fermi potential (or level), we have a condition called the *threshold of weak inversion*. Doubling the band bending, or causing the surface potential to be equal in magnitude and opposite in sign to the bulk potential, brings us to the threshold of strong inversion. The corresponding critical gate–source voltage is the threshold voltage. Further increases in gate voltage cause very little further thickening of the depletion layer that preceded inversion, because it is now "easier" for MOS-capacitor lines of force to terminate on inversion-layer electrons (in the N-channel example) right at the interface than on additional negative acceptor ions at the more remote depletion-layer boundary. Band-bending character also changes qualitatively beyond threshold, with curvature increasing, because channel carriers form a thin, dense layer.

Rudimentary MOSFET analysis ignores depletion-layer charge, assuming that all charge on the silicon plate of the MOS capacitor is inversion-layer charge. It produces "square-law" (parabolic) characteristic curves in the regime of drain current versus drain–source voltage nearest the origin of the output plane. Physical reasoning predicts constant current beyond the maximum (or vertex) of each parabola, nicely confirmed experimentally in the long-channel MOSFET. This is the important saturation regime of operation, wherein drain current is independent of drain–source voltage, and proportional to the square of the gate–source voltage that is in excess of the threshold. This is the most important square-law property of the MOSFET.

The MOSFET just described is enhancement-mode (E-mode) in the sense that gate–source voltage above the threshold voltage must be applied to produce significant conductance from source to drain. By doping a thin region near the oxide–silicon interface with impurities of the same type as those in the source and drain regions, one creates a channel that exists even with the device at equilibrium. Gate–source voltage of the depletion-inversion polarity will further increase drain–source conductance. But gate–source voltage of the opposite polarity causes a depletion layer that forms at the interface to diminish channel conductance. This is a depletion-mode (D-mode) device. The gate–source voltage that fully depletes the doped channel is known as the *pinch-off voltage*. Channel doping in such a device is usually accomplished by ion implantation. The D-mode MOSFET is also a square-law device.

The transconductance of the MOSFET, or the rate of change of drain current with respect to gate–source voltage, is a linear function of the latter variable. Transconductance in the MOSFET is a function of dimensional and structural properties of the device. The BJT, by contrast, has transconductance that to first order is independent of dimensional and structural properties of the device. It is proportional to output current, and hence exponentially dependent upon input voltage. As the linear-exponential comparison suggests, BJT transconductance vastly exceeds MOSFET transconductance through wide operating ranges. Transconductances in the two devices have their closest approach in the regimes of very low output current and input voltage.

An E-mode MOSFET with gate and drain common is a diode exhibiting a concave-up (square-law, again) current–voltage characteristic. Used as a load with another E-mode device as driver, it yields a linear voltage amplifier; the nonlinearities of the two devices precisely cancel. The same inverter can be used for voltage switching in a logic circuit. The diode-connected device operates always in the saturation regime, and is hence known as a *saturated load*. The inverter is small in area and therefore economical, but relatively slow in switching. A linear (resistive) load yields switching that is over an order of magnitude faster, but such a load is awkward technologically. A D-mode load combined with an E-mode driver (known as an E-D inverter) is an advantageous combination, small in area, also much faster than the saturated-load inverter. One of the fastest MOS inverters uses N-channel and P-channel E-mode devices with gates common and drains common, the latter node serving as the output node. Such a complementary MOS (CMOS) inverter is both fastest in switching and lowest in power dissipation at low switching rates, because it requires no "standby" current; in either state of the switch, either the P-channel device is OFF, or the N-channel device is OFF.

Switching delay in MOS circuits is fixed largely by the time required to charge parasitic capacitances in the circuit. Thus, to optimize an integrated circuit, the sizes of its various MOSFETs (and hence their transconductances) must be adjusted according to the capacitive loads they have to "drive," so that rough parity of switching time throughout the circuit is achieved. This kind of optimization is best handled by computer, and has been so accomplished since the mid-1960s. The practice with bipolar integrated circuits has been very different because the BJT has an enormous transconductance "surplus" that to first order is geometry-independent.

The silicon portion of the MOS capacitor poses a problem that is analytically identical to the problem posed by one side of a step junction at equilibrium. Each change of bias on the capacitor, therefore, produces a condition similar to that in one region of a step junction, with doping on the other side altered. Electric field is discontinuous at the oxide–silicon interface because of the threefold ratio of dielectric constants, silicon to oxide. Treating the oxide portion as a conventional parallel-plate capacitor, and the silicon portion by means of the depletion approximation, one can produce a serviceable model of the MOS capacitor. The potential profile in the oxide portion is linear, and in the silicon portion, parabolic. The depletion-approximation replacement (DAR) provides more accurate analysis of the silicon portion.

A real MOS capacitor embodies a number of parasitic properties. If the silicon and gate materials exhibit work-function differences, then the potential (or energy) interval from Fermi level on a given side to the conduction-band edge in the oxide will be different from that on the other side. The difference of these two potentials we term barrier-height difference. It contributes a term to threshold voltage. Another parasitic effect involves electronic states located at the surface of the silicon single crystal, into which electrons can flow when energetically favorable. That is, when band bending places these states below the Fermi level they tend to fill, and vice versa. The resulting charge in these states constitutes a highly undesirable voltage-dependent variable in the MOS capacitor. Many years of effort have devised ways to reduce the density of such "surface states" to negligible values. The carrier charges residing in such states are termed interface trapped charge. When such states are present at high densities, they act somewhat like an electrostatic shield, because of their ability to take on and give up charge. This shielding action frustrated early field-effect experiments.

Another parasitic charge component is a quite stable positive charge near the oxide–silicon interface, believed to be associated with unreacted (unoxidized) silicon. The density of such charge correlates with silicon surface-atom density, which is a function of crystal orientation. It also correlates with the last portion of the silicon-oxidation process, because the oxidation reaction occurs at the interface and not at the oxide surface. This oxide *fixed charge,* as it is known, is in contrast to the oxide mobile charge that plagued early MOS technologists. Alkali ions were mainly responsible in the latter case because of their comparatively high mobility in silicon dioxide. The cure was "cleaning up" the process, notably by avoiding the filament evaporation of a metal for the gate. The final parasitic-charge category is oxide trapped charge. It occurs when an electron passes partway through the oxide-layer conduction band and becomes trapped, or "lodged." In an analogous way, a hole can pass part way through the oxide-layer valence band and become trapped. This can result when certain device actions and phenomena (such as avalanche breakdown) create energetic carriers near the interface. All these parasitic effects, when present, play a part in determining threshold voltage. Their combined effects were responsible for the early popularity of *P*-channel MOSFETs; only after control of the oxide–silicon system had advanced appreciably could the benefits of *N*-channel devices be exploited.

MOS-capacitor modeling and characterization are easier than step-junction modeling and characterization in the sense that the silicon remains in quasiequilibrium in the face of bias changes. The problem is more difficult, however, in the sense that the oxide capacitance is permanently in series with the semiconductor phenomena to be modeled. The approximate-analytic method and its DAR culmination are useful for accurate capacitor modeling, especially because the MOS capacitor poses a quasiequilibrium problem.

There are several simple and useful equivalent-circuit models for the MOS capacitor, depending upon regime of operation. In accumulation, the oxide capacitance suffices. With modest depletion, two capacitances are in series, the oxide and depletion capacitances. With the inversion layer present, as well as an arrangement for "feeding" carriers to it, an inversion-layer capacitance must be added to the previous model, in parallel with the depletion capacitance. With heavy inversion, we revert to the lone oxide capacitor.

The depletion and inversion capacitances are equal to each other precisely at the threshold of strong inversion. This relationship has been demonstrated and explained experimentally, theoretically, and by means of a physical model. The MOS capacitor is free of diffusion capacitance because of its insulating layer. In the junction, by contrast, diffusion capacitance and depletion capacitance are in parallel, mutually shunted by a conductance in the general case. Inversion capacitance exists in a grossly asymmetric *PN* junction, but it is miniscule and negative.

Observing capacitance as a function of voltage is a key method for characterizing real devices, and was the preeminent method employed for analyzing and finally understanding the oxide–silicon system. Interpretation of *C–V* curves is carried out in terms of the equivalent-circuit models described just above. Substrate doping, conductivity type, and oxide thickness are primary variables. The parasitic charges also affect the result, in distinctive ways in some cases. Although the formation of an inversion layer inhibits further expansion of the depletion layer under quasistatic conditions, it is possible to cause a transient depletion-layer expansion by employing a voltage pulse, revealing a deep-depletion characteristic. Avalanche breakdown in the depletion layer imposes a limit on deep depletion.

Transition from MOS-capacitor analysis to MOSFET analysis means a shift from a useful and accurate one-dimensional model to one that is at best two-dimensional. For the long-channel device, a useful treatment uses the orthogonal superposition of two one-dimensional analyses.

Integrated-circuit application of any MOSFET model requires inclusion of the body effect. This term signifies that reverse biasing the bulk–source junction, a common IC situation, means that the gate must exert a higher degree of band bending to achieve the threshold condition in the channel region than it does when that junction is not reverse-biased. In other words, the threshold voltage of such a device is inceased.

Other major contributions to long-channel MOSFET modeling were made by Pao and Sah, who considered carrier diffusion in the channel and carried out numerical analyses in two orthogonal directions to achieve an extremely accurate model. Also, Baccarani et al. and Brews developed a charge-sheet model. It is relatively simple, and quite accurate as well, because of countervailing approximations.

The initial rudimentary MOSFET analysis becomes the Level-1 model in the SPICE program. Taking account of ionic charge, neglected in the first case, yields the Level-2 SPICE model. Instead of a square-law result, the ionic-charge (or Ihantola-Moll) model yields a three-halves-power result, which is awkward to treat numerically because of computation time. For this reason and others, an empirically adjusted Level-3 SPICE model was generated. It yields accurate results while avoiding the earlier shortcomings. The Level-1 model in empirically adjusted (but still square-law) form is also useful. Unadjusted, it predicts too high a current, because it assumes all charge in the silicon consists of carriers in the channel.

SPICE considers four regimes of operation. Moving upward along a loadline, we encounter the accumulation, depletion, saturation, and curved regimes. The subthreshold regime includes portions of the first two of these. The MOSFET makes its closest approach to the remarkable BJT transconductance property in the subthreshold regime. This fact is exploited in circuitry for low-power applications, especially CMOS circuitry.

In small-signal SPICE analysis, one finds that the bulk region is approximately as effective as a control electrode as is the gate itself. The concept of current gain has meaning for the MOSFET because significant current enters the input terminal with a high-frequency signal applied and passes through parasitic capacitances. Consequently, the frequency for unity current gain, f_T, also has meaning. For the device analyzed here, it amounted to 0.5 GHz.

Digital, or large-signal, applications of MOSFETs are by far the most important. The E-D combination (E-mode driver, D-mode load), is particularly advantageous, and is analyzed here. Body effect degrades the current-regulating ability of a D-mode load, and it can end up approximating a resistor. Nonetheless, the D-mode load outperforms a saturated load by more than a factor of ten in switching speed. In addition to speed, noise margin is an important feature of a digital circuit. It is a measure of the noise voltage at the input port that the circuit can absorb without changing state. Noise margin is intimately related to product reliability, yield, and economy.

Parasitic capacitances in the MOSFET that are junction-related are voltage-dependent. SPICE models this voltage dependence, and is capable of predicting the waveforms of capacitive current, as well as resistive current. The gate–drain capacitance (Miller capacitance) is of particular importance in both digital and linear applications. Because of parasitic capacitance, the dynamic turn-on loadline lies well above the static loadline in a pulse experiment, and the turn-off loadline, well below.

A simple-theory comparison of the MOSFET and BJT with respect to transconductance versus output current shows an intersection close to but above threshold. The two curves do not actually intersect, but have their closest approach just below threshold. When input voltage is used as the independent variable, the BJT displays a single characteristic curve, while the MOSFET displays a family of such curves with lower slope, having threshold voltage as a parameter. One particular comparison of normalized transconductances, BJT to MOSFET, produces an intriguing product of ratios—bulk to surface mobility, oxide to base thickness, and emitter to gate area.

When subthreshold physics is examined, the source–channel junction is seen to be an N^+N junction after the source end of the channel passes the threshold of weak inversion. This transition from NP-like to N^+N-like properties is significant, because there is an offset voltage in the conducting I–V characteristic of the former, but none in the latter. This source-end property, the electron diffusion from source to drain along linear electron profiles, and electron density at the silicon source-end surface that is exponentially dependent upon gate voltage, all combine to produce a subthreshold transconductance that approximates the BJT-like value of qI_{out}/kT.

Table 5-1 Approximate Expressions for Normalized Electric Field $|dW/d(x/L_D)|$ as a Function of Normalized Potential $W^\dagger$

Range of $\|U_B\|$	Range of W	Approximate Expression	Maximum Error Within Indicated Ranges of $\|U_B\|$, W
A 2 to ∞	0 to ∞	For depletion and inversion $\sqrt{2[(e^{-W} + W - 1) + e^{-2\|U_B\|}(e^W - W - 1)]}$	0.91%
A' 2 to ∞	$W = 2\|U_B\|$	At threshold $2\sqrt{\|U_B\|(1 - e^{-2\|U_B\|})}$	0.91%
B 3 to ∞	0.5 to ∞	For depletion and inversion $\sqrt{2(e^{-W} + W - 1 + e^{W-2\|U_B\|})}$	1.86%
B' 2 to ∞	$W = 2\|U_B\|$	At threshold $2\sqrt{\|U_B\|}$	1.85%
C 2 to ∞	3 to ∞	For depletion and inversion $\sqrt{2(W - 1 + e^{W-2\|U_B\|})}$	1.42%
D 3 to ∞	$4\|U_B\|$ to ∞	For extreme inversion $2e^{(W/2)-\|U_B\|}$	1.22%
E 3 to ∞	0 to $(2\|U_B\| - 2)$	For depletion (ionic component only) $\sqrt{2(e^{-W} + W - 1)}$	0.13%
F 4 to ∞	3 to $(2\|U_B\| - 2)$	For depletion (ionic component only) $\sqrt{2(W - 1)}$	1.21%
G 0 to ∞	0 to 0.1	For depletion or accumulation W	0.042%
H 2 to ∞	0 to ∞	For accumulation $\sqrt{2(e^W - W - 1)}$	1.2%
J 2 to ∞	6 to ∞	For accumulation $\sqrt{2e^{W/2}}$	1.7%

† Expressions are accurate within 2% in indicated ranges. Correct algebraic sign can be incorporated herein by using the coefficient $-U_B/|U_B|$, where U_B is normalized bulk potential in the semiconductor portion of the MOS capacitor (using Fermi level as reference). A range limit of ∞ is intended to indicate the largest value that does not violate underlying approximations and assumptions.

Table 5-2 Exact Expressions for Normalized Electric Field $|dW/d(x/L_D)|$ as a Function of Normalized Potential $W^\dagger$

For depletion and inversion	$\sqrt{2\left[\dfrac{e^{|U_B|}(e^{-W}+W-1)+e^{-|U_B|}(e^W-W-1)}{e^{|U_B|}+e^{-|U_B|}}\right]}$
For accumulation	$\sqrt{2\left[\dfrac{e^{-|U_B|}(e^{-W}+W-1)+e^{|U_B|}(e^W-W-1)}{e^{-|U_B|}+e^{|U_B|}}\right]}$
For right-hand side intrinsic	$\sqrt{e^W+e^{-W}-2}$

†Correct algebraic sign can be incorporated herein by using the coefficient $-U_B/|U_B|$, where U_B is normalized bulk potential in the semiconductor portion of the MOS capacitor (using Fermi level as reference) [55, 56].

Table 5-3 Symbol Conversion Chart for Basic and Physical Quantities in SPICE Level-2 MOSFET Model

SPICE Symbol	Description	Default Value	Unit	Present Symbol	Description
Basic set: X_{OX}, μ_n, U_{EXP}, E_{CRIT}, L_D^*, C_n/q, and λ					
TOX	Thin oxide thickness	1.0	m	X_{OX}	Oxide thickness
UO	Surface mobility	600	cm²/V·s	μ_n	Surface mobility
UEXP	Exponential coefficient for mobility	0.0	—	U_{EXP}	Exponent for mobility adjustment
UCRIT	Critical electric field for mobility	10^4	V/cm	E_{CRIT}	Critical electric field for mobility adjustment
LD	Lateral diffusion distance	0.0	m	L_D^*	Lateral diffusion distance
NFS	Surface-fast state density	0.0	F/C·cm²	C_n/q	Subthreshold fitting parameter
LAMBDA	Channel-length modulation	0.0	V^{-1}	λ	Channel-length modulation
Physical set: N_A, T_G, N_f^*, R_{SH}, and J_S					
NSUB	Substrate doping	0.0	cm^{-3}	N_A	P-substrate doping
TPG	Type of gate	1	—	T_G	Type of gate
NSS	Surface state density	0.0	cm^{-2}	N_f^*	Effective oxide fixed-charge density
RSH	Source and drain diffusion sheet resistance	0.0	ohm/m²	R_{SH}	Source and drain diffusion sheet resistance
JS	Bulk-junction reverse saturation current density	10^{-8}	A/m²	J_0	Saturation-current density

Table 5-4 Symbol Conversion Chart for Capacitive Quantities in SPICE Level-2 MOSFET Model

SPICE Symbol	Description	Default Value	Unit	Present Symbol	Description
	Capacitance effects: $\Delta\psi_0$, $C_{0.B}$, M_B, $C_{0.SW}$, M_{SW}, C_{0GS}, C_{0GD}, and C_{0GB}				
VJ	BS junction potential	0.8	V	$\Delta\psi_0$	BS contact potential
CJ	Zero-bias BS junction bottom capacitance	0.0	F/cm²	$C_{0.B}$	Zero-bias BS junction-bottom depletion-layer capacitance
MJ	BS grading bottom coefficient	0.5	—	M_B	BS grading exponent, bottom
CJSW	Zero-bias BS side wall junction capacitance	0.0	F/cm²	$C_{0.SW}$	Zero-bias BS junction-sidewall depletion-layer capacitance
MJSW	BS grading sidewall coefficient	0.5	—	M_{SW}	BS grading exponent, sidewall
CGSO	GS overlap capacitance per meter of channel width	0.0	F/m	C_{0GS}	GS overlap capacitance per meter of channel width
CGDO	GD overlap capacitance per meter of channel width	0.0	F/m	C_{0GD}	GD overlap capacitance per meter of channel width
CGBO	GB overlap capacitance per meter of channel length	0.0	F/m	C_{0GB}	GB extrinsic capacitance per meter of channel length

Table 5-5 Representative Set of SPICE Parameter Values

Parameter	Value	Parameter	Value
X_{OX}	1.139×10^{-7} m	N_A	10^{16} cm^{-3}
μ_n	879.31 cm/s	T_G	1
U_{EXP}	0.26742	N_f^*	6.5279×10^{10}/cm²
E_{CRIT}	10,000 cm/s	R_{SH}	80 ohm
L_D^*	0.175×10^{-6} m	J_0	3.125×10^{-7} A/m²
C_N	10^{11} F/cm²	C_{0GS}	400×10^{-12} F/m
λ	0.022980/V	C_{0GD}	400×10^{-12} F/m
$\Delta\psi_0$	0.8 V	C_{0GB}	200×10^{-12} F/m
M_B	0.5	$C_{0.B}$	3.125×10^{-4} F/m²
M_{SW}	0.5	$C_{0.SW}$	2.0×10^{-10} F/m
Z	20×10^{-6} m	L	4×10^{-6} m
N_{RS}	0.5	N_{RD}	0.5
A_S	320×10^{-12} m²	A_D	320×10^{-12} m²
P_S	50×10^{-6} m	P_D	50×10^{-6} m

Table 5-6 Symbol Conversion Chart for Electrical Quantities in SPICE Level-2 MOSFET Model

SPICE Symbol	Description	Default Value	Unit	Present Symbol	Description
	Basic set: V_T, K, γ, ψ_B, R_S, R_D, and I_0				
VTO	Zero-bias threshold voltage	1.0	V	V_T	Threshold voltage
KP	Transconductance parameter	0.0	A/V^2	$2K$	Twice the current coefficient
GAMMA	Body-effect parameter	0.0	—	γ	Bulk-charge parameter
PHI	Surface inversion potential	0.6	V	$-2\psi_B$	Band bending at threshold
RS	Source ohmic resistance	0.0	ohm	R_S	Source ohmic resistance
RD	Drain ohmic resistance	0.0	ohm	R_D	Drain ohmic resistance
IS	Bulk-junction saturation current	10^{-14}	A	I_0	Bulk-source-junction saturation current

Table 5-7 Equivalent Sets of SPICE Parameter Values

Physical Parameter	Value	Electrical Parameter	Value
N_A	10^{16} cm^{-3}	V_T	1.03 V
T_G	1	$2KL^*/Z$	2.67×10^{-5} A/V^2
N_f^*	6.5279×10^{10}/cm^2	γ	1.90 V$^{1/2}$
		$-2\psi_B$	0.695 V
R_{SH}	80 ohm	R_S and R_D	40 ohm
J_0	3.125×10^{-7} A/m^2	I_0	10^{-16} A

Table 5-8 Operating-Point Information Corresponding to the SPICE Parameter Set in Table 5-7

$I_D = 0.055$ mA $\qquad g_m = 0.069$ mS
$V_{GS.P} = +2.471$ V $\qquad g_o = 0.001$ mS
$V_{BS.P} = +0.000$ V $\qquad g_{mbs} = 0.062$ mS
$V_{DS.P} = +2.471$ V $\qquad f_T = 0.5$ GHz
$\qquad\qquad\qquad\qquad\quad A_V = 69$

Table 5-9 Calculated dc Values, Numerical Example

Symbol	Unit	A OFF	B Bias	C ON
$V_{GS.P}$	V	0.356	2.471	5.000
$V_{BS.P}$	V	0.000	0.000	0.000
$V_{DS.P}$	V	5.000	2.471	0.356
I_D	mA	3.90×10^{-9}	0.055	0.096
I_G	mA	0.0	0.0	0.0
I_B	mA	0.0	0.0	0.0
V_{GS}	V	0.356	2.469	4.996
V_{BS}	V	-1.55×10^{-10}	-0.002	-0.004
V_{DS}	V	5.000	2.467	0.348
C_{gs}	fF	8.000	22.754	19.658
q_{gs}	fC	0.0	0.0	0.0
C_{gd}	fF	8.000	8.000	18.422
q_{gd}	fC	0.0	0.0	0.0
C_{gb}	fF	22.862	0.730	0.730
q_{gb}	fC	0.0	0.0	0.0
C_{bs}	fF	110.00	109.84	109.73
q_{bs}	fC	-1.70×10^{-8}	-0.244	-0.424
C_{bd}	fF	40.853	54.415	91.667
q_{bd}	fC	-2.98×10^2	-1.80×10^2	-35.200
$g_m \approx g_{m.P}$	mS	5.62×10^{-8}	0.069	0.020
$g_o \approx g_{o.P}$	mS	1.01×10^{-10}	0.001	0.253
$g_{mbs} \approx g_{mbs.P}$	mS	4.85×10^{-8}	0.062	0.020

Table 5-10 Calculated Transient Values, Large-Signal Numerical Example

Symbol	Unit	Particular Times			
		$t^{m=1}$	$t^{m=2}$	$t^{m=3}$	$t^{m=4}$
t	ns	10.0	17.68	20.0	91.0
$v_{GS.P}(t)$	V	0.356	3.924	5.000	5.000
$v_{BS.P}(t)$	V	0.000	0.000	0.000	0.000
$v_{DS.P}(t)$	V	5.000	2.355	0.536	0.356
$i_D(t)$	mA	-8.47×10^{-7}	0.119	0.108	0.096
$i_G(t)$	mA	1.39×10^{-7}	0.023	0.022	5.43×10^{-5}
$i_B(t)$	mA	7.12×10^{-7}	0.060	0.018	2.74×10^{-4}
$v_{GS}(t)$	V	0.356	3.916	4.994	4.996
$v_{BS}(t)$	V	-1.55×10^{-10}	-0.008	-0.006	-0.004
$v_{DS}(t)$	V	5.000	2.342	0.525	0.348
$g_m \approx g_{m.P}$	mS	5.62×10^{-8}	0.122	0.031	0.020
$g_o \approx g_{o.P}$	mS	1.01×10^{-10}	0.005	0.228	0.253
$g_{mbs} \approx g_{mbs.P}$	mS	4.85×10^{-8}	0.092	0.029	0.020
$C_{GS}(t)$	fF	8.000	22.754	20.008	19.658
$q_{GS}(t)$	fC	2.847	75.492	98.618	98.659

Table 5-10 (continued)

$C_{GD}(t)$	fF	8.000	8.000	17.985	18.422
$q_{GD}(t)$	fC	−37.153	12.596	49.977	53.246
$C_{GB}(t)$	fF	22.862	0.730	0.730	0.730
$q_{GB}(t)$	fC	8.135	19.513	20.298	20.298
$C_{BS}(t)$	fF	1.10×10^2	1.09×10^2	1.10×10^2	1.10×10^2
$q_{BS}(t)$	fC	-1.71×10^{-8}	−0.891	−0.652	−0.424
$C_{BD}(t)$	fF	40.853	55.437	85.269	91.667
$q_{BD}(t)$	fC	-2.98×10^2	-1.73×10^2	−51.045	−35.200
$i_D^R(t)$	mA	-3.88×10^{-9}	0.193	0.139	0.096
$i_G^{CS}(t)$	mA	-6.27×10^{-18}	0.011	0.009	2.56×10^{-5}
$i_G^{CD}(t)$	mA	1.39×10^{-7}	0.013	0.012	2.87×10^{-5}
$i_G^{CB}(t)$	mA	-2.83×10^{-18}	3.39×10^{-4}	3.39×10^{-4}	-9.95×10^{-17}
$i_B^{CD}(t)$	mA	7.12×10^{-7}	0.061	0.018	1.37×10^{-4}
$i_B^{CS}(t)$	mA	6.41×10^{-17}	-1.68×10^{-4}	1.16×10^{-4}	1.36×10^{-4}

Table 5-11 Representative Values Chosen for Evaluating the Quotient Between BJT and MOSFET Transconductance-Normalizing Quantities in Section 5-6.4

$$\frac{q}{kT}\frac{q}{\epsilon_{OX}} = 1.853 \times 10^{-5} \text{ cm} \qquad \frac{A_E}{A_B} = 1$$

$$\frac{\mu_{nb}}{\mu_{ns}} = 3 \qquad n_{0B} = 10^3/\text{cm}^3$$

$$\frac{X_{OX}}{X_B} = \frac{1}{3} \qquad L^2 = 10^{-8} \text{ cm}^2$$

REFERENCES

1. J. E. Lilienfeld, U.S. Patent 1,745,175, filed October 8, 1926 (October 22, 1925 in Canada), and issued January 28, 1930.

2. O. Heil, British Patent 439,457, issued 1939.

3. W. Shockley, *Electrons and Holes in Semiconductors,* Van Nostrand, New York, 1950, p. 30.

4. L. H. Hoddeson, "Multidisciplinary Research in Mission-Oriented Laboratories; The Evolution of Bell Laboratories' Program in Basic Solid-State Physics Culminating in the Discovery of the Transistor, 1935–1948," Department of Physics, University of Illinois at Urbana–Champaign, November 1978. Also, "The Discovery of the Point Contact Transistor," *Historical Studies in the Physical Sciences* **12,** 41 (1981).

5. J. Bardeen and W. H. Brattain, "The Transistor—A Semiconductor Triode," *Phys. Rev.* **74,** 230 (1948).

6. W. Shockley, U.S. Patent 2,569,347, filed June 26, 1948, and issued September 25, 1951.

7. R. M. Warner, Jr., and B. L. Grung, *Transistors: Fundamentals for the Integrated-Circuit Engineer,* Wiley, New York, 1983; reprint edition, Krieger Publishing Company (P. O. Box 9542, Melbourne, Florida 32902-9542), 1990, p. 30.

8. W. Shockley, "A Unipolar Field-Effect Transistor," *Proc. IRE* **40,** 1365 (1952).

9. G. C. Dacey and I. M. Ross, "Unipolar Field-Effect Transistor," *Proc. IRE* **41,** 970 (1953).

10. H. Christensen and G. K. Teal, U.S. Patent 2,692,839, filed April 7, 1951, and issued October 26, 1954.

11. R. M. Warner, Jr., G. C. Onodera, and W. J. Corrigan, U.S. Patent 3,223,904, filed February 19, 1962, and issued December 14, 1965.

12. W. Shockley, U.S. Patent 2,787,564, filed October 28, 1954, and issued April 2, 1957.

13. R. H. Crawford, "Implanted Depletion Loads Boost MOS Array Performance," *Electronics* **45,** 85 (1972).

14. Warner and Grung, p. 64.

15. D. Kahng and M. M. Atalla, "Silicon–Silicon Dioxide Field Induced Surface Devices," *IRE-AIEE Solid-State Device Research Conference,* Carnegie Institute of Technology, Pittsburgh, Pennsylvania, June 1960.

16. M. M. Atalla, E. Tannenbaum, and E. J. Scheibner, "Stabilization of Silicon Surfaces by Thermally Grown Oxides," *Bell Syst. Tech. J.* **38,** 749 (1959).

17. C. A. Mead, "Schottky Barrier Gate Field-Effect Transistor," *Proc. IEEE* **54,** 307 (1966).

18. W. W. Hooper and W. I. Lehrer, "An Epitaxial GaAs Field-Effect Transistor," *Proc. IEEE* **55,** 1237 (1967).

19. R. Dingle, H. L. Stormer, A. C. Gossard, and W. Wiegmann, "Electron Mobilities in Modulation-Doped Semiconductor Heterojunction Superlattices," *Appl. Phys. Lett.* **37,** 805 (1978).

20. T. Mimura, S. Hizamizu, T. Fujii, and K. Nanbu, "A New Field Effect Transistor with Selectively Doped GaAs/n-Al$_x$Ga$_{1-x}$As Heterojunctions," *Jpn. J. Appl. Phys.* **19,** L225 (1980).

21. N. C. Cirillo, M. Shur, P. J. Vold, J. K. Abrokwah, R. R. Daniels, and O. N. Tufte, "Insulated Gate Field Effect Transistor," *IDEM Tech. Digest,* 317 (1985).

22. T. Mizutani, S. Fujita, and Y. Yanagawa, "Complementary Circuit with

AlGaAs/GaAs Heterostructure MISFETs Employing High-Mobility Two-Dimensional Electron and Hole Gases," *Electronics Lett.* **21,** 1116 (1985).

23. M. Shur, J. K. Abrokwah, R. R. Daniels, and D. K. Arch, "Mobility Enhancement in Highly Doped GaAs Quantum Wells," *J. Appl. Phys.* **61,** 1643 (1987).

24. M. Sweeny, J. Xu, and M. Shur, "Hole Subbands in One-Dimensional Quantum-Well Wires," *Superlattices and Microstructures* **4,** 623 (1988).

25. F. G. Allen and G. W. Gobeli, "Work Function, Photoelectric Threshold, and Surface States of Atomically Clean Silicon," *Phys. Rev.* **127,** 150 (1962); "Comparison of the Photoelectric Properties of Cleaved, Heated, and Sputtered Silicon Surfaces," *J. Appl. Phys.* **35,** 597 (1964).

26. B. E. Deal and E. H. Snow, "Barrier Energies in Metal–Silicon Dioxide–Silicon Structures," *J. Phys. Chem. Solids* **27,** 1873 (1966).

27. W. M. Werner, "The Work Function Difference of the MOS-System with Aluminum Field Plates and Polycrystalline Silicon Field Plates," *Solid-State Electron.* **17,** 769 (1974).

28. J. W. Gibbs, "Thermodynamics," *Collected Works,* Vol. 1, Yale University Press, New Haven, Conn., 1948, p. 219.

29. I. E. Tamm, "Über eine Mögliche Art der Electronenbindung an Kristalloberflächen," *Phys. Z. Sowjetunion* **1,** 733 (1932).

30. W. Shockley, "On the Surface States Associated with a Periodic Potential," *Phys. Rev.* **56,** 317 (1939).

31. A. Many, Y. Goldstein, and N. B. Grover, *Semiconductor Surfaces,* North-Holland, Amsterdam, 1965, p. 165.

32. A. S. Grove, *Physics and Technology of Semiconductor Devices,* Wiley, New York, 1967, p. 144.

33. A. van der Ziel, "Flicker Noise in Electronic Devices," *Advances in Electronics and Electron Devices,* Vol. 49, Academic Press, New York, 1979, p. 225.

34. B. E. Deal, "Standardized Terminology for Oxide Charges Associated with Thermally Oxidized Silicon," *IEEE Trans. Electron Devices,* **27,** 606 (1980).

35. J. Bardeen, "Surface States and Rectification at a Metal Semiconductor Contact," *Phys. Rev.* **71,** 717 (1947).

36. E. H. Nicollian and A. Goetzberger, "The Si–SiO$_2$ Interface—Electrical Properties as Determined by the MIS Conductance Technique," *Bell Syst. Tech. J.* **46,** 1055 (1967).

37. R. D. Schrimpf, private communication.

38. J. R. Schwank, D. M. Fleetwood, P. S. Winokur, P. V. Dressendorfer, D. C.

Turpin, and D. T. Sanders, "The Role of Hydrogen in Radiation-Induced Defect Formation in Polysilicon Gate MOS Devices," *IEEE Trans. on Nuclear Science* **NS-34,** 1152 (1987).

39. B. E. Deal, M. Sklar, A. S. Grove, and E. H. Snow, "Characteristics of the Surface-State Charge (Q_{ss}) of Thermally Oxidized Silicon," *J. Electrochem. Soc.* **114,** 266 (1967).

40. S. I. Raider and A. Berman, "On the Nature of Fixed Oxide Charge," *J. Electrochem. Soc.* **125,** 629 (1978).

41. S. I. Raider and R. Flitsch, "X-Ray Photoelectron Spectroscopy of SiO_2–Si Interfacial Regions: Ultra Thin Oxide Films," *IBM J. Res. Dev.* **22,** 294 (1978).

42. O. L. Krivanek, T. T. Sheng, and D. C. Tsui, "A High-Resolution Electron Microscopy Study of the Si–SiO_2 Interface," *Appl. Phys. Lett.* **32,** 437 (1978).

43. D. Frohman-Bentchkowsky, "FAMOS—A New Semiconductor Charge Storage Device," *Solid-State Electron.* **17,** 517 (1974).

44. A. van der Ziel, private communication.

45. W. L. Brown, "*n*-Type Surface Conductivity on *p*-Type Germanium," *Phys. Rev.* **91,** 518 (1953).

46. C. G. B. Garrett and W. H. Brattain, "Physical Theory of Semiconductor Surfaces," *Phys. Rev.* **99,** 376 (1955).

47. R. H. Kingston and S. F. Neustadter, "Calculation of the Space Charge, Electric Field, and Free Carrier Concentration at the Surface of a Semiconductor," *J. Appl. Phys.* **26,** 718 (1955).

48. C. E. Young, "Extended Curves of the Space Charge, Electric Field, and Free Carrier Concentration at the Surface of a Semiconductor, and Curves of the Electrostatic Potential Inside a Semiconductor," *J. Appl. Phys.* **32,** 329 (1961).

49. A. S. Grove and D. J. Fitzgerald, "Surface Effects on *P–N* Junctions—Characteristics of Surface Space-Charge Regions under Non-Equilibrium Conditions," *Solid-State Electron.* **9,** 783 (1966).

50. Ph. Passau and M. van Styvendael, "La Jonction *p–n* Abrupte dans le Cas Statique," *Proc. Int. Conf. Solid State Phys. in Electron. and Telecomm.*, Brussels, June 2–7, 1958, M. Desirant and J. F. Miciels (eds.), *Int. Un. Pure and Appl. Phys.*, Vol. I, Academic Press, New York, 1960, p. 407.

51. C. Goldberg, "Space Charge Regions in Semiconductors," *Solid-State Electron.* **7,** 593 (1964).

52. R. P. Jindal and R. M. Warner, Jr., "A General Solution for Step Junctions

with Infinite Extrinsic End Regions at Equilibrium," *IEEE Trans. Electron Devices* **28,** 348 (1981).

53. R. M. Warner, Jr., and R. P. Jindal, "Replacing the Depletion Approximation," *Solid-State Electron.* **26,** 335 (1983).
54. H. C. Pao and C. T. Sah, "Effects of Diffusion Current on Characteristics of Metal-Oxide (Insulator)-Semiconductor Transistors," *Solid-State Electron.* **9,** 927 (1966).
55. Warner and Grung, pp. 372–377.
56. R. M. Warner, Jr., R. P. Jindal, and B. L. Grung, "Field and Related Semiconductor-Surface and Equilibrium-Step-Junction Variables in Terms of the General Solution," *IEEE Trans. Electron Devices* **31,** 994 (1984).
57. W. Shockley, "The Theory of $p-n$ Junctions in Semiconductors and $p-n$ Junction Transistors," *Bell Syst. Tech. J.* **28,** 435 (1949).
58. R. M. Warner, Jr., R. D. Schrimpf, and P. D. Wang, "Explaining the Saturation of Potential Drop on the High Side of a Grossly Asymmetric Junction," *J. Appl. Phys.* **57,** 1239 (1985).
59. A. Goetzberger, "Ideal MOS Curves for Silicon," *Bell Syst. Tech. J.* **45,** 1097 (1966).
60. W. S. Boyle and G. E. Smith, "Charge Coupled Semiconductor Devices," *Bell Syst. Tech. J.* **49,** 587 (1970).
61. A. S. Grove, E. H. Snow, B. E. Deal, and C. T. Sah, "Simple Physical Model for the Space-Charge Capacitance of Metal-Oxide-Semiconductor Structures," *J. Appl. Phys.* **35,** 2458 (1964).
62. C. N. Berglund, "Surface States at Steam-Grown Silicon–Silicon Dioxide Interfaces," *IEEE Trans. Electron Devices* **13,** 701 (1966).
63. R. Castagné, "Détermination de la Densité d'États Lents d'une Capacité Métal-Isolant-Semiconducteur par l'Étude de la Charge sous une Tension Croissant Lineairement," *C. R. Acad. Sci. (Paris)* **267,** 866 (1968).
64. D. R. Kerr, "M.I.S. Measurement Techniques Utilizing Slow Voltage Ramps," J. Bonel (ed.), *Trans. Int. Conf. on Properties and Use of MIS Structures,* Grenoble, France, 1969, p. 303.
65. M. Kuhn, "A Quasi-Static Technique for MOS $C-V$ and Surface State Measurements," *Solid-State Electron.* **13,** 873 (1970).
66. A. Goetzberger, E. Klausmann, and M. J. Schulz, "Interface States on Semiconductor/Insulator Surfaces," *CRC Crit. Rev. Solid-State Sci.* **6,** 1 (1976).
67. A. S. Grove, B. E. Deal, E. H. Snow, and C. T. Sah, "Investigation of Thermally Oxidised Silicon Surfaces Using Metal-Oxide-Semiconductor Structures," *Solid-State Electron.* **8,** 145 (1965).

68. R. M. Warner, Jr., "A Note on the Equality of Bulk and Inversion-Layer Capacitance at Threshold," *Solid-State Electron.* **30,** 181 (1987).

69. M. C. Tobey, Jr., and N. Gordon, "Concerning the Onset of Heavy Inversion in MIS Devices," *IEEE Trans. Electron Devices* **ED-21,** 649 (1974).

70. Y. Tsividis, "Moderate Inversion in MOS Devices," *Solid-State Electron.* **25,** 1099 (1982).

71. R. D. Schrimpf, D.-H. Ju, and R. M. Warner, Jr., "Subthreshold Transconductance in the Long-Channel MOSFET," *Solid-State Electron.* **30,** 1043 (1987).

72. Warner and Grung, Chapters 9 and 10.

73. H. K. J. Ihantola and J. L. Moll, "Design Theory of a Surface Field-Effect Transistor," *Solid-State Electron.* **7,** 423 (1964).

74. A. Vladimirescu, A. R. Newton, and D. O. Pederson, *Spice Version 2G.2 User's Guide,* Electronics Research Laboratory, University of California, Berkeley, 1981.

75. J. R. Brews, "Physics of the MOS Transistor," *Applied Solid State Science, Supplement 2A,* D. Kahng (ed.), Academic Press, New York, 1981.

76. Warner and Grung, Fig. 6-10(b), p. 421.

77. W. M. Penney and L. Lau (eds.), *MOS Integrated Circuits,* Van Nostrand Reinhold, New York, 1972, p. 92.

78. J. R. Brews, "A Charge-Sheet Model of the MOSFET," *Solid-State Electron.* **21,** 345 (1978).

79. G. Baccarani, M. Rudan, and G. Spadini, "Analytical IGFET Model Including Drift and Diffusion Currents," *Solid-State Electron Devices* **2,** 62 (1978).

80. D.-H. Ju and R. M. Warner, Jr., "Modeling the Inversion Layer at Equilibrium," *Solid-State Electron.* **27,** 907 (1984).

81. L. M. Dang, "A One-Dimensional Theory on the Effects of Diffusion Current and Carrier Velocity Saturation on E-type IGFET Current–Voltage Characteristics," *Solid-State Electron.* **20,** 781 (1977).

82. H. Shichman and D. A. Hodges, "Modeling and Simulation of Insulated-Gate Field-Effect Transistor Switching Circuits," *IEEE J. Solid-State Circuits* **SC-3,** 285 (1968).

83. J. E. Meyer, "MOS Models and Circuit Simulation," *RCA Rev.* **32,** 42 (1971).

84. D. Frohman-Betchkowsky and A. S. Grove, "On the Effect of Mobility Variation on MOS Device Characteristics," *Proc. IEEE* **56,** 217 (1968).

85. R. J. Gravrok, private communication.

86. R. M. Warner, Jr., "Comparing MOS and Bipolar Integrated Circuits," *IEEE Spectrum* **4,** 50 (1967).

87. P. L. Hower, "Bipolar vs. MOSFET: Seeing Where the Power Lies," *Electronics* **53,** Dec. 18, 106 (1980).

88. Y. Tsividis, "Moderate Inversion in MOS Devices," *Solid-State Electron.* **25,** 1099 (1982).

89. E. O. Johnson, "The Insulated-Gate Field-Effect Transistor—a Bipolar Transistor in Disguise," *RCA Rev.* **34,** 80 (1973).

90. C. T. Sah and H. C. Pao, "The Effects of Fixed Bulk Charge on the Characteristics of Metal–Oxide Semiconductor Transistors," *IEEE Trans. Electron Devices* **ED-13,** 393 (1966).

91. F. Van de Wiele, "A Long-Channel MOSFET Model," *Solid-State Electron.* **22,** 991 (1979).

92. A. G. Milnes, *Semiconductor Devices and Integrated Electronics,* Van Nostrand Reinhold, New York, 1980, p. 410.

93. J. Evans and K. A. Pullen, Jr., "Limitation of Properties of Field-Effect Transistors," *Proc. IEEE* **54,** 82 (1966).

94. R. M. Warner, Jr., and R. D. Schrimpf, "BJT-MOSFET Transconductance Comparisons," *IEEE Trans. Electron Devices* **34,** 1061 (1987).

95. H. K. Gummel and H. C. Poon, "An Integral Charge Control Model of Bipolar Transistors," *Bell Syst. Tech. J.* **49,** 827 (1970).

96. R. H. Crawford, *MOSFET in Circuit Design,* McGraw-Hill, 1967, p. 53.

97. M. B. Barron, "Low-Level Currents in Insulated Gate Field Effect Transistors," *Solid-State Electron.* **15,** 293 (1972).

98. R. F. van Overstraeten, G. Declerck and G. F. Broux, "Inadequacy of the Classical Theory of the MOS Transistor Operating in Weak Inversion," *IEEE Trans. Electron Devices* **ED-20,** 1150 (1973).

99. R. R. Troutman, "Subthreshold Design Considerations for Insulated Gate Field-Effect Transistors," *IEEE J. Solid-State Circuits* **SC-9,** 55 (1974).

100. W. Fichtner and H. W. Pötzl, "MOS Modelling by Analytical Approximations. I. Subthreshold Current and Threshold Voltage," *Int. J. Electronics* **46,** 33 (1979).

101. L. D. Yau, "A Simple Theory to Predict the Threshold Voltage of Short-Channel IGFETs," *Solid-State Electron.* **17,** 1059 (1974).

102. Warner and Grung, pp. 454–455.

103. R. M. Warner, Jr., D.-H. Ju, and B. L. Grung, "Electron-Velocity Saturation at a BJT Collector Junction under Low-Level Conditions," *IEEE Trans. Electron Devices* **ED-30,** 230 (1983).

TOPICS FOR REVIEW

R5-1. Why does the MOSFET make more severe demands on the knowledge and control of silicon surfaces than does the BJT?

R5-2. Explain the acronym *MOSFET*.

R5-3. Identify materials most commonly used in the three regions of the MOS capacitor.

R5-4. Name and describe briefly at least two members of the FET family in addition to the MOSFET.

R5-5. Explain the terms *source* and *drain*.

R5-6. What bias polarity on the gate of a MOSFET produces useful results? Why? Specify the substrate type and channel type you have assumed in your answer.

R5-7. Explain the term *flatband voltage*.

R5-8. What is *subthreshold current* and what accounts for it?

R5-9. Explain, with examples, the terms *unipolar* and *bipolar*.

R5-10. Why was a field-effect transistor the first solid-state device to be conceived, but nearly the last to be realized?

R5-11. Make a four-way comparison of mobility values, N-channel versus P-channel, and surface (channel) versus bulk.

R5-12. Define *threshold of weak inversion*.

R5-13. Define *threshold of strong inversion*.

R5-14. What is meant by *sheet resistance*? Why is it a useful concept?

R5-15. How do the dielectric permittivities of silicon and silicon dioxide compare?

R5-16. Why does the surface depletion layer essentially stop growing at the threshold (of strong inversion) voltage?

R5-17. In what respect does the rudimentary analysis of the MOSFET yield a square-law model?

R5-18. What major simplifying assumption is made in the rudimentary model?

R5-19. Name the two major operating regimes and indicate where they reside.

R5-20. Give a "zeroth-order" explanation for why drain current is approximately independent of drain–source voltage in saturation.

R5-21. Explain the terms *E-mode, D-mode,* and *E-D circuits*. What are the advantages and disadvantages of the last?

R5-22. How do the electrical properties of a D-mode MOSFET differ from those of a JFET (which is also a D-mode device)?

R5-23. How does MOSFET transconductance depend on input voltage?

R5-24. How can one realize a linear voltage amplifier using MOSFETs?

R5-25. Explain the term *saturated load*. What are its advantages and disadvantages?

R5-26. Compare and contrast the E-mode MOSFET *transfer characteristic* and the *locus of points* separating the curved and the saturation regimes.

R5-27. Compare and contrast the E-mode and D-mode MOSFET transfer characteristics.

R5-28. How can both the E-mode and D-mode MOSFET have characteristic currents?

R5-29. Explain the acronym *CMOS* and describe the circuitry it identifies. What are its advantages and disadvantages?

R5-30. What primarily fixes switching delay in MOS circuits?

R5-31. Does choice of load device mainly affect the *turn-on* or *turn-off* properties of the inverter? Why?

R5-32. What is meant by the term *discrete device*?

R5-33. What is *totempole current*?

R5-34. Why did computerized circuit design get an earlier and faster start in MOS technology than in bipolar technology?

R5-35. Compare the step-junction problem and the silicon-surface (as encountered in the MOS capacitor) problem.

R5-36. Describe the oxide–silicon boundary conditions for an ideal MOS capacitor.

R5-37. Describe the field profile in an ideal MOS capacitor in depletion, but well below threshold. Explain why its major features exist.

R5-38. What principle determines field magnitude on the two sides of the oxide–silicon interface?

R5-39. How do an ordinary parallel-plate capacitor and the MOS capacitor differ?

R5-40. Explain the meaning of the symbol ψ_{SB}.

R5-41. What is meant by *band bending*?

R5-42. Explain *vacuum level* and *work function*.

R5-43. Explain *photoelectric threshold* and *electron affinity*.

R5-44. Name advantages of a *polysilicon* gate electrode over a metal electrode.

R5-45. Cite the several kinds of parasitic charge that can affect the oxide–silicon system. Describe the characteristics of each, and state its accepted symbol.

R5-46. Why is *interface trapped charge* (in "surface states") particularly damaging?

R5-47. Explain the comparison of surface states with an electrostatic shield. What was their historical role?

R5-48. Mobile charge is almost as intolerable as interface trapped charge. Why?

R5-49. Identify the species primarily responsible for mobile charge in SiO_2.

R5-50. Explain the term "dangling bonds."

R5-51. Given a band diagram for an MOS capacitor, identify the *surface potential*.

R5-52. What is meant by the *accumulation* condition in an MOS capacitor? What bias polarity produces it?

R5-53. What does the acronym *EPROM* stand for?

R5-54. What do the terms *volatile* and *nonvolatile* mean when applied to an electronic memory?

R5-55. Why does interface trapped charge affect the mobility of inversion-layer carriers?

R5-56. Why is inversion-layer mobility only a fraction of bulk mobility, even in the absence of parasitic charge?

R5-57. What are the meanings and causes of *mobility fluctuations* and *density fluctuations*?
R5-58. What property of a MOSFET is affected by such fluctuations?
R5-59. What technology has made it possible to adjust doping and charge conditions near a silicon surface with great precision?
R5-60. Describe the equivalent-circuit model appropriate to each operating regime for the MOS capacitor.
R5-61. At what voltage do the $C-V$ profiles for the inversion-layer and depletion-layer capacitances intersect? Why?
R5-62. Compare the junction-capacitance equivalent circuits with those for the MOS capacitor.
R5-63. What capacitance component is effectively absent in each case and why?
R5-64. Does *capacitance crossover* exist for a junction as well as for the MOS capacitor? At what approximate voltage does it occur?
R5-65. Give a physical explanation for approximate MOS capacitance crossover at the voltage stated in response to R5-61.
R5-66. Why has $C-V$ assessment been so important in MOS technology?
R5-67. How does small-signal frequency affect a (combined or overall) $C-V$ profile for the MOS capacitor?
R5-68. Describe the quasistatic technique for observing a $C-V$ profile. What is its advantage over the small-signal method?
R5-69. What feature of a $C-V$ profile reveals the presence of "fast surface states," capable of harboring interface trapped charge?
R5-70. Explain *long-channel approximation*.
R5-71. Also explain *gradual-channel approximation* and relate the two terms. State an algebraic condition defining one or both.
R5-72. What major simplification exists in a long-channel model as compared to a short-channel model?
R5-73. Identify the Level-1, -2, and -3 SPICE models. How do they differ? What are their major features?
R5-74. What is meant by *body effect*?
R5-75. Explain how body effect alters threshold voltage.
R5-76. Describe a sample situation where body effect exists.
R5-77. How is a D-mode load device affected by body effect? Why?
R5-78. Describe the modeling approaches taken by Pao and Sah, and by Brews.
R5-79. Name and describe the operating regimes identified in the SPICE program.
R5-80. How does SPICE modeling predict nonconstant current in the saturation regime?
R5-81. Compare the gate terminal and bulk region as control electrodes.
R5-82. How can the bulk region act as a "gate?"
R5-83. Does it make sense to talk about the current gain of a MOSFET? Why?
R5-84. What is f_T for a small-signal MOSFET amplifier?
R5-85. What parasitic elements are dominant in determining f_T?
R5-86. What small-signal and large-signal characterizing quantities have a product that equals a constant for a given circuit?

R5-87. What is *noise margin*?
R5-88. Why does noise margin matter?
R5-89. What parasitic capacitance is most important in large-signal applications? in small-signal applications?
R5-90. List three ways of comparing the transconductance properties of BJT and MOSFET.
R5-91. Discuss qualitatively the findings in each case.
R5-92. Why is transconductance important?
R5-93. In view of the large transconductance deficit of the MOSFET under many circumstances, why is it the dominant device in the world today?
R5-94. Comment on the device-dimension dependences of transconductance in the BJT and MOSFET.
R5-95. Describe the properties of the source–channel junction as a function of gate bias.
R5-96. Describe channel transport well above threshold; below threshold.
R5-97. In view of your answer to the last question, why does the MOSFET not exhibit a diffusion capacitance, as does the BJT?
R5-98. Why does subthreshold transconductance in the MOSFET approach the theoretical limit of qI_D/kT?

ANALYTIC PROBLEMS

A5-1. A certain E-mode MOSFET has a gate capacitance of 0.02 pF, $V_T = 1$ V, $\mu_n = 700$ cm^3/V·s, $X_{OX} = 0.05$ μm, and $L = 3$ μm.
a. Calculate its gate width Z in μm.
b. Calculate its transconductance g_m at $V_{GS} = 3$ V in the saturation regime.

A5-2. Consider an N-channel E-mode MOSFET in the curved (nonsaturation) regime.
a. Derive an expression for the transconductance g_m.
b. Rewrite the expression for I_D in terms of normalized voltage, that is, in terms of V_{GS}/V_T and V_{DS}/V_T.
c. From the result in part b, write an approximate expression for I_D that is valid in the linear regime, which is to say, *very* near the origin of the output plane. Use this expression to derive an expression for channel resistance r (source–drain resistance) for the MOSFET in the linear regime.
d. Use the result in part c to find r for the MOSFET of the previous problem when it has $V_{GS} = 3$ V.

A5-3. The MOSFET of Problem A5-1 is placed in the circuit shown.
a. Find I_D.
b. Find V_{DS}.

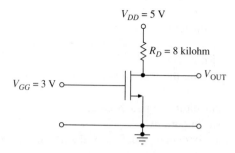

c. Sketch an accurate loadline diagram, plotting an *I–V* curve for the MOSFET and another for the resistor R_D. On the diagram indicate V_{DD} and the numerical coordinates of the quiescent point, (I_D, V_{DS}).

A5-4. For a certain *N*-channel E-mode MOSFET the curved-regime characteristics are described by

$$I_D = K[2(V_{GS} - V_T)V_{DS} - V_{DS}^2],$$

and the saturation-regime characteristics are described by

$$I_D = K(V_{GS} - V_T)^2,$$

where $K = 1$ mA/V^2, and $V_T = 1$ V. The MOSFET is placed in this circuit:

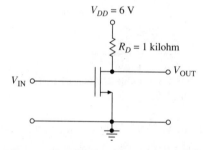

V_{IN}	V_{OUT}
0 V	
1 V	
2 V	
3 V	

a. Complete the chart at the right of the diagram.
b. Plot the points whose coordinates are in the chart above on the axes below, after calibrating the axes appropriately. Draw a smooth curve through the points you have plotted. This curve is known as the *voltage transfer curve*.

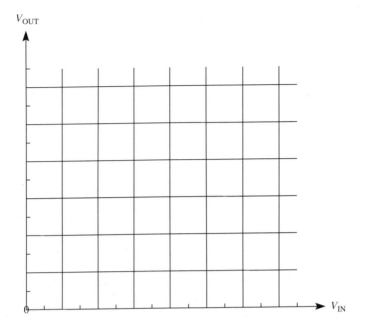

c. What is the significance of the derivative (dV_{OUT}/dV_{IN})?

d. Estimate the value of the derivative at $V_{IN} = 2$ V by using graphical or analytic means.

A5-5. Given here is one output characteristic for a particular N-channel E-mode MOSFET. Its saturation-regime output-current equation is
$I_{D1} = K_1(V_{GS} - V_T)^2$.

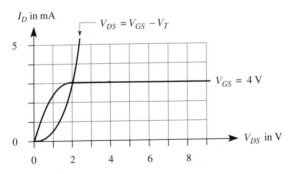

a. Calculate V_T.

b. Calculate K_1. Include units.

c. A second MOSFET is identical in every respect, except that its channel length is $L_2 = (1.5)L_1$, where L_1 is channel length in the first device. On the output plane above, accurately sketch the I–V characteristic for the second device at $V_{GS} = 4$ V.

A5-6. In Exercise 5-8, Section 5-1.4, we noted that the boundary between the saturation and curved regimes is the locus of points for which $V_{DS} = V_{GS} - V_T$ is valid, and that this voltage expression is not the "equation for" the boundary curve. Derive the equation for the boundary curve, plotted in Figure 5-6(a).

A5-7. Shown here is a diode-connected E-mode MOSFET, a configuration used in some circuits as a quasiresistor. Determine whether this device is operating in the saturation regime, the curved regime, some combination of the two, or on the boundary between them.

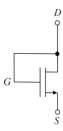

A5-8. The current–voltage equation for the diode-connected E-mode MOSFET was derived in Problem A5-7. Comment on the relationship of the resulting expression to that for the appropriate transfer characteristic.

A5-9. Two discrete MOSFETs are interconnected in the manner shown here to create an inverting amplifier stage. The two devices are identical in every respect except in the lateral dimensions Z and L. (See also Problem D5-2.)

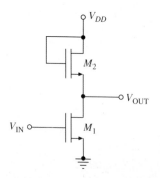

  a. Draw an output-plane diagram and apply the principles of loadline-diagram construction to it so that the interaction of the two devices can be visualized.
  b. Demonstrate that the given circuit constitutes a linear voltage amplifier by deriving an equation for its small-signal voltage gain, $A_V \equiv dV_{OUT}/dV_{IN}$.
  c. Explain graphically the linearity of this amplifier circuit by means of the loadline diagram constructed above.

A5-10. The cross-sectional diagram in Figure 5-4 is for an N-channel MOSFET operating in the curved regime, well out of saturation.
  a. Draw a similar diagram, but for a long-channel MOSFET operating well inside the saturation regime. On your diagram, indicate clearly the depletion-layer boundaries in the P-type region.

b. On the same or a similar diagram, sketch several lines of force (with correct direction indicated) in each depletion layer.

c. The device being considered exhibits the extremely high dynamic output resistance depicted in the shaded portion of Figure 5-5b. The fact that I_D is essentially independent of V_{DS} is a consequence of Ohm's law. Using the diagram of part a or b, or a similar diagram, explain the constancy of $I_D(V_{DS})$. The only additional item of information you need is that when an MOS capacitor is brought to the threshold condition, with the voltage V_T from gate to bulk, there also exists a voltage V_{TOX} from gate to the oxide–silicon interface (voltage drop through the oxide) that is just as clearly defined by the structure of the capacitor as is V_T. Obviously $V_{TOX} < V_T$.

d. Relate the explanation given in part c to the regimes of operation depicted in Figure 5-5.

e. Construct a voltage-scale diagram, indicating the voltages and voltage intervals cited in part c, for operation at the boundary of the curved and saturations regimes.

f. In part e, the voltage at the drain terminal is lower than the voltage at the point where the channel vanishes (because $V_T > V_{TOX}$). How can the electrons make their way from the channel end to the drain?

g. By considering the same phenomena that enter into part c, and by drawing another diagram, explain why a short-channel MOSFET exhibits much lower dynamic output resistance in the saturation regime than does a long-channel device that is identical in every respect except channel length. Name an analogous BJT effect.

A5-11.

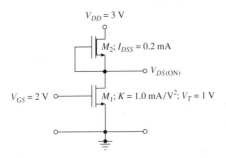

Given above is the schematic diagram for a MOSFET enhancement-depletion (E-D) inverter under bias. On the pair of axes in the next figure is plotted the output characteristic for M_1 for its given bias condition.

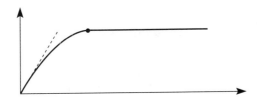

894 THE MOSFET

    a. Label the axes using correct symbols.
    b. Calibrate both axes. [*Note:* For *each* axis you must carry out a calculation in order to calibrate it correctly.]
    c. On the same diagram, label the voltages $V_{DS(ON)}$ and V_{DD} very clearly.
    d. Using symbols only (no numbers), derive an approximate expression for $V_{DS(ON)}$. [*Hint:* Because $V_{DS(ON)}$ is small, use a linear approximation (as shown by the dashed line in the figure) to the curved-regime characteristic.]
    e. Evaluate the expression derived in part d.

A5-12. One-sided junctions and high-low junctions have important similarities and important differences that enter the MOS-capacitor problem as well.
    a. Construct linearly calibrated axes, and on them, using a solid line, sketch a reasonably accurate electric-field profile for the equilibrium N^+N high-low junction depicted in Figure 3-32 and described in Section 3-5.4.
    b. Using a dashed line, draw an equally accurate electric-field profile for the equilibrium N^+P one-sided junction of Figure 5-36. Carry out any necessary calculations, and explain your reasoning. Comment qualitatively but accurately on the differences in the two profiles.

A5-13. For the junction of Problem A5-12(b), determine the amount of reverse bias needed to cause a rightward shift of the right-hand depletion-layer boundary by one micrometer. Draw a diagram showing the relationship of the new electric field profile to the equilibrium profile for the N^+P junction.

A5-14. A certain MOS capacitor has the structure shown in Figure 5-35(a). It also has a gate measuring 2 μm × 2 μm, an oxide thickness of $X_{OX} = 400$ Å, and a net substrate doping of $N_A = 5 \times 10^{15}/\text{cm}^3$. Calculate the capacitance of the device when it is biased precisely at the threshold voltage.

A5-15. It is noted in Section 5-2.6 that the fixed charge Q_f at the oxide–silicon interface was found in the 1960s to be a function of crystal orientation, and to exhibit values following the same sequence as surface-atom density for those orientations. A substantial set of data drawn from the contemporaneous literature demonstrated a Q_f sequence having the ratios for three different orientations that are given just below. (Details of the experimental data as well as indications of the data sources can be found in Reference 7, page 749.)

	$\dfrac{Q_f(100)}{Q_f(100)}$	$\dfrac{Q_f(110)}{Q_f(100)}$	$\dfrac{Q_f(111)}{Q_f(100)}$
Median ratios, experimental data	1	1.45	3.15

Employing information on the silicon crystal given in Section 1-5.5, calculate the corresponding ratios of surface-atom densities, and compare them with fixed-charge ratios observed experimentally. In the case of the (111) orientation, treat the two closely spaced planes of atoms as a single plane.

A5-16. Equation 5-95 gives a useful approximation for C_b for conditions well below threshold, where Q_b is taken from Equation 5-99 as

$$Q_b = \frac{\epsilon\sqrt{2}}{L_D} \frac{kT}{q} \sqrt{W_S - 1}.$$

a. How large an error results from using this approximate expression *at* threshold for a sample of net doping $N = 10^{15}/\text{cm}^3$?

b. If one makes the erroneous assumption that C_i can be obtained from $C_i = dQ_n/d\psi_S$, where Q_n is inferred from Equation 5-99 as

$$Q_n = \frac{\epsilon\sqrt{2}}{L_D} \frac{kT}{q} \sqrt{e^{W_S - 2|U_B|}},$$

how serious is the resulting error at threshold for the same sample as in part a?

A5-17. Equations 5-102 and 5-103 are often written in terms of L_{Di}. Make the conversion.

A5-18. A certain long-channel MOSFET is biased at $V_{GS} = 2\,V_T$ and $V_{DS} = 5\,V_T$. For simplicity, assume that it is free of all parasitic effects. Of interest are the areal charge densities Q_n and Q_B at the one-third point of the channel, $y = (L/3) \equiv y_1$, and at the two-thirds point, $y = (2L/3) \equiv y_2$. Construct a chain of qualitative logic proving unequivocally that $Q_n(y_2) < Q_n(y_1)$ and $Q_b(y_2) > Q_b(y_1)$.

A5-19. Figure 5-38(d) presents profiles for both carrier populations in the bulk region of an MOS capacitor at threshold. Let us assume that its net doping is $N_A = 10^{16}/\text{cm}^3$. The density values are well known in the undepleted bulk region, and also at the oxide–silicon interface by virtue of the given conditions.

a. Using the depletion-approximation replacement (DAR), derive an expression for $p(x/L_D)$ valid for most of the region lying between these two extremes.

b. Calculate six or seven points in the intermediate region, and plot your expression on a semilog diagram (log p versus x/L_D).

c. Calculate and plot the electron profile also, $n(x/L_D)$.

d. For the given conditions, the oxide–silicon interface is sometimes described as the *strong-threshold plane*. In a parallel way, state the position of the *weak-threshold plane*. Calculate its position in terms of the DAR spatial origin.

e. Calculate accurately in micrometers the separation of the strong-threshold and weak-threshold (or intrinsic) planes.

A5-20. Use the same conditions and distance calibration as in Problem A5-19 to plot the following curves:

a. Generate curves of $p(x/L_D)$ and $n(x/L_D)$, but with linear density calibration this time, as in Figure 5-38(d).
b. Plot a normalized profile of total volumetric charge density in the silicon.

A5-21. Derive the last expression of Table 5-2.

A5-22. A certain MOS capacitor of 1966 was "quasi-ideal" in the sense that H_D and Q_f precisely compensated Q_{it} through a certain C–V range. Also, $Q_m = Q_{ot} = 0$. The experimental C–V curve was as shown below:

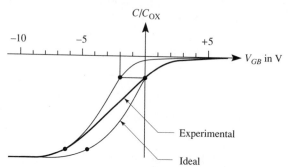

a. Determine the bulk conductivity type.
b. Determine the sign(s) and identity(ies) of possible contributors to Q_s in the operating regime relevant to this problem.
c. Given that $\psi_B = \pm 0.4$ V, sketch a band diagram for the bulk region in the flatband condition and indicate on it the energy positions of the states responsible for interface trapped charge, Q_{it} ("fast surface states").
d. Determine whether the states named in part c are donorlike or acceptorlike and explain your reasoning.

A5-23. Carry out the following steps for the device of Problem A5-22.
a. Calculate N_{it}, the areal density of "fast surface states."
b. Calculate Q_f/C_{OX}, given $H_D = -0.2$ V.
c. Calculate the net doping N.
d. Calculate an approximate value for $X_{D(MAX)}$, the greatest depletion-layer thickness at the surface in the absence of drain-source bias.

A5-24. The MOSFET inverter that is repeated here was also given in Problem A5-9. There we showed that the voltage gain A_V is (1) constant (meaning that the resulting amplifier is linear), and (2) designable. An analogous BJT inverter is shown below in the same diagram.
a. Analyze the BJT inverter to determine its gain properties, and compare them with those of the MOSFET inverter.

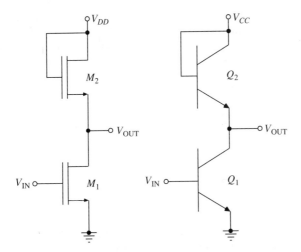

b. What feature(s) of the diagrams in Section 5-6.4 is (are) related to the comparison between these two kinds of inverters? Comment on the origin of the differing gain properties of the two inverters.

A5-25. Verify Equation 5-121.

A5-26. Use the analysis of Section 5-1, the Level-1 model, to follow these instructions:
a. Derive an expression for conductance g that is valid throughout the curved regime of the output plane.
b. Derive an expression for conductance in the linear regime, $g(\sim 0)$, very near the origin of the output plane.
c. Derive an expression for $I_D(\sim 0)$, drain current in the linear regime.

A5-27. Use the analysis of Section 5-4.2, the Level-2 model, to follow these instructions:
a. Derive an expression for $I_D(\sim 0)$, drain current in the linear regime.
b. Compare the expression just derived with that obtained in Problem A5-25(c) and explain any similarity or difference.

A5-28. Give a physical interpretation of Equation 5-125.

A5-29. Equation 5-116 can be rewritten as

$$I_D = \mu_n C_{OX} \frac{Z}{L} \left\{ \left(V_{GS} - V_{FB} + 2\psi_B - \frac{V_{DS}}{2} \right) V_{DS} \right.$$
$$\left. - \frac{2}{3} \frac{\sqrt{2\epsilon q N_A}}{C_{OX}} (-2\psi_B)^{3/2} \left[\left(1 + \frac{V_{DS}}{(-2\psi_B)} \right)^{3/2} - 1 \right] \right\}.$$

Using the three-term binomial-series expansion for $(1 + \lambda)^n$, where $n = 3/2$ and $\lambda \ll 1$, derive a version of this equation valid in the linear regime and the adjacent portion of the curved regime, but not valid throughout the curved regime as is Equation 5-116. In other words, the expression to be derived will be intermediate between Equation 5-116 and the expression written in Problem A5-26a.

A5-30. Calculate the value of g_{mbs} given as Equation 5-179, using the MOSFET data given in Section 5-4.4.

Note: In Problems A5-31 through A5-35 you are to compare the SPICE results with your calculations. For this comparison, use the RMKSA system of units and the following values from SPICE:

$$\epsilon = 1.03594 \times 10^{-10} \text{ F/m} \quad (kT/q) = 0.02586 \text{ V}$$
$$\epsilon_{OX} = 3.4531 \times 10^{-11} \text{ F/m} \quad n_i = 1.45 \times 10^{16}/\text{m}^3$$
$$q = 1.602 \times 10^{-19} \text{ C}$$

A5-31. Prove that the right-hand column of parameters given in Table 5-7 can be derived from the left-hand column. Omit consideration of the parameters R_S and R_D.

A5-32. Assuming that $\mu_n = 1{,}200 \text{ cm}^2/\text{V·s}$, $N_A = 10^{16}/\text{cm}^3$, and $\tau = 3.605 \times 10^{-7}$ s, calculate a value for J_0 and compare the result with the value of $3.125 \times 10^{-7} \text{ A/m}^2$ used in the SPICE calculations.

A5-33. Assuming that $\Delta\psi_0 = 0.8$ V and $N_A = 10^{16}/\text{cm}^3$, calculate a value for $C_{0.B}$, and compare the result with the value of $3.125 \times 10^{-4} \text{ F/m}^2$ used in the SPICE calculations.

A5-34. Using the values given in Tables 5-5, 5-7, and 5-9, calculate the MOSFET on voltage V_{ON} and compare the result with V_T.

A5-35. Using the values given in Tables 5-5, 5-7, and 5-9, calculate the small-signal capacitance values given in the equivalent circuit of Figure 5-65.

A5-36. Give a qualitative explanation for the impact of body effect on the I–V characteristic of a D-mode load, displayed in Figure 5-70.

A5-37. For the MOSFET of Figure 5-56, evaluate the bulk-charge parameter γ using the SPICE data, and also using the values given in Table 2-4 for $T = 297.8$ K; determine the percentage difference.

A5-38. The large-signal SPICE calculation in Section 5-5.4 employs an inverter with an E-mode driver and a D-mode load. The layout of the driver is given here.

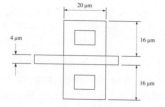

In Table 5-5, a set of SPICE parameters is given, including

$$A_S = A_D = 320 \times 10^{-12} \text{ m}^2,$$
$$P_S = P_D = 50 \times 10^{-6} \text{ m},$$

and
$$N_{RS} = N_{RD} = 0.5.$$

Show that these values are consistent with the layout.

A5-39. Using the information given in the previous problem, indicate clearly the various areas corresponding to the overlap capacitances. Calculate their values for the given layout, and determine the overlap capacitances in terms of picofarads per meter of gate width. Determine the limiting values of gate-bulk overlay capacitance for widely varying operating conditions, particularly in V_{GS}.

A5-40. Starting with the small-signal equivalent circuit given in Figure 5-65, interchange the bulk and the gate terminals so that the amplifier operates in the inverted mode. Construct an approximate equivalent circuit similar to that in Figure 5-66. Calculate the resulting voltage gain and short-circuit current-gain cutoff frequency and compare these values with those for the normal mode.

A5-41. The configuration shown here is a useful two-terminal current regulator ("current source"). The current level can be tailored in the range $0 < I_D < I_{DSS}$ by adjusting R, which varies V_{GS}.

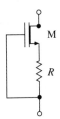

A certain manufacturer purchases discrete N-channel D-mode MOSFETs at high volume for use in this application in his product to deliver the current $I_D = 1$ mA $\pm$ 0.05 mA. He wishes to use a fixed-value resistor (that you may assume has an exact value) rather than selecting resistors with values that compensate for MOSFET variations. The D-mode MOSFET in saturation is well described by the expression

$$I_D = I_{DSS}\left[1 - \frac{V_{GS}}{V_P}\right]^2.$$

Also, you may assume that

$$I_{DSS} = (-V_P H)^{3/2},$$

where H is a constant.

a. Write an expression for the value of R needed as a function of I_{DSS}, V_P, and the selected value of I_D.

b. Derive an expression for I_D as a function of R, I_{DSS}, and V_P.

c. Derive an expression for the relative change in I_D as a function of R, I_{DSS}, and the relative change in I_{DSS}.

d. Calculate H, R, and RH for the nominal values $I_{DSS} = 1.5$ mA and $V_P = -1$ V. Invert the expression derived in part c, and calculate the relative change in I_{DSS} that can be tolerated for $(dI_D/I_D) = 0.05$.

e. Specify the upper and lower limits for the range of I_{DSS} that can be accepted, consistent with the specification $I_D = 1$ mA $\pm$ 0.05 mA.

f. Name the most straightforward way for the MOSFET manufacturer to adjust H, and indicate the direction of the change in H for an increase in the variable you have identified. [This problem is based upon a question posed by Mr. Larry McNichols.]

COMPUTER PROBLEMS

C5-1. Consider the circuit shown in Figure 5-63. SPICE parameter values for this transistor are given in Table 5-5, and dc operating-point information is given in Table 5-9.

a. Using the SPICE program, calculate the ac voltage gain in dB as a function of frequency for values of frequency from 1 KHz to 10 GHz. Plot the results along with the phase shift in degrees. Use the .OP command in SPICE to print out the values of the small-signal hybrid-pi parameters and compare them with those given in Figure 5-65. The SPICE model is:

```
.MODEL ENHD NMOS (LEVEL=2 TOX=1.139E-07
+UO=879.31 UEXP=0.26742 UCRIT=10K
+LAMBDA=0.022980 NFS=1E11
+NSUB=1E16 NSS=6.5279E10 TPG=1
+RSH=80 LD=0.175U JS=3.125E-7
+CJ=3.125E-4
+MJ=0.5 PB=0.8 FC=0.5
+CJSW=2.0E-10 MJSW=0.5
+CGSO=400P CGDO=400P CGBO=200P)
```

The following format should be used:

```
MQ1 3 7 0 0 ENHD W=20U L=4U AD=320P AS=320P
+              NRD=0.5 NRS=0.5 PS=50U PD=50U
```

b. Replace the physical parameters by the electrical parameters, as given in Table 5-7. Then repeat the calculation and show that the results are the same. The SPICE model is:

.MODEL ENHD NMOS (LEVEL=2 TOX=1.139E − 07
+UO=879.31 UEXP=0.26742 UCRIT=10K
+LAMBDA=0.022980 NFS=1E11
+VTO   = 1.03
+KP   = 2.6658E−005
+GAMMA = 1.900
+PHI   = 0.6954
+CJ=3.125E−4 LD=0.175U
+MJ=0.5 PB=0.8 FC=0.5
+CJSW=2.0E−10 MJSW=0.5
+CGSO=400P CGDO=400P CGBO=200P)

c. Replace the transistor in parts a and b by a small-signal circuit that uses resistors, capacitors, and voltage-controlled current sources. Repeat the calculation of ac voltage gain and show that the results are identical to those already obtained in parts a and b.

C5-2. This problem is identical to Problem C4-2, except that the BJT inverter in each case has been replaced by a D-mode load MOSFET inverter, as can be seen in the figure below.

a. Using SPICE and the transistor-parameter values given in Table 5-5, calculate the transient response of the circuit shown in the figure below for an input pulse that is identical to the curve given in Figure 5-73(a). Show that the output waveform at OUTA is identical to the curve in Figure 5-73(b). The SPICE models are:

.MODEL ENHD NMOS (LEVEL=2 TOX=1.139E−07
+UO=879.31 UEXP=0.26742 UCRIT=10K
+LAMBDA=0.022980 NFS=1E11
+NSUB=1E16 NSS=6.5279E10 TPG=1
+RSH=80 LD=0.175U JS=3.125E−7
+CJ=3.125E−4
+MJ=0.5 PB=0.8 FC=0.5
+CJSW=2.0E−10 MJSW=0.5
+CGSO=400P CGDO=400P CGBO=200P)

and

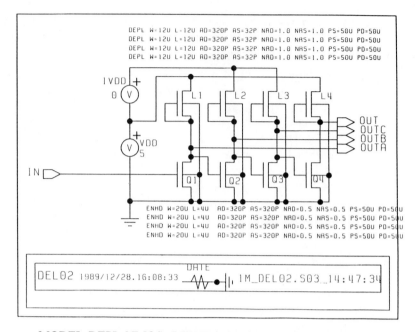

.MODEL DEPL NMOS (LEVEL=2 TOX=1.139E−07
+UO=879.31 UEXP=0.26742 UCRIT=10K
+LAMBDA=0.022980 NFS=1E11
+NSUB=1E16 NSS=1.2062E12 TPG=1
+RSH=80 LD=0.175U JS=3.125E−7
+CJ=3.125E−4
+MJ=0.5 PB=0.8 FC=0.5
+CJSW=2.0E−10 MJSW=0.5
+CGSO=400P CGDO=400P CGBO=200P)

Also, the following formats should be used:

V$D2 3 0 PULSE(3.55853E−01 5.0 1.0E−08 1.0E−08
+   1.0E−08 7.1E−08 1.6E−07)
ML 1 6 3 3 0 DEPL W=12U L=12U AD=320P AS=32P
+   NRD=1.0 NRS=1.0 PS=50U PD=50U
MQ 1 3 7 00 ENHD W=20U L=4U AD=320P AS=320P
+   NRD=0.5 NRS=0.5 PS=50U PD=50U
.OP
.OPTION RELTOL=0.0001
.OPTION GMIN=1E−18

b. Find the delay time between stages, as was described in Problem C4-2.

DESIGN PROBLEMS

D5-1. You are to design a MOSFET for use as a voltage-controlled resistor at low voltages. You have access to a technology that delivers a threshold voltage of $V_{T(MIN)} = 1$ V at a gate-oxide thickness of $X_{OX(MIN)} = 0.04$ μm, and for reasons of technological compatibility, you must use these values. The minimum effective channel length is $L^* = 1$ μm, and gate-oxide breakdown strength is $E_B = 1.0 \times 10^7$ V/cm. Channel-electron mobility declines with V_{GS}, and is 150 cm²/V·s. If any other MOSFET data are needed, take them from Tables 5-3 through 5-10.

   a. The desired tuning range is 10 ohms to 1,000 ohms, and the maximum voltage the device will experience is $V_{DS(MAX)} = 100$ mV. Determine the maximum current it will carry, $I_{D(MAX)}$.

   b. Devise a criterion that is directly related to the linearity of the resistor. Can linearity be improved without limit by pushing in the direction indicated by this criterion? Why?

   c. Determine necessary Z/L and V_{GS} for the application.

   d. Choose a channel length, and make a case for your choice in terms of parasitic MOSFET properties.

   e. You are given the specification that in the $r_{MIN}(0)$ condition, the current $I_{D(MAX)}$ is to depart from that in a truly linear 10-ohm resistor by no more than 1%. Determine whether your design meets this spec.

   f. Make a plan-view sketch of your design, clearly indicating terminal identities and channel dimensions.

D5-2. You are to design a linear MOSFET amplifier stage like that of Problem A5-9. All of the technological information given in Problem D5-1 applies here as well. The desired voltage gain is $A_V = -64$, and the bandwidth should be as large as possible. Before proceeding with the design, you must deal qualitatively with several issues that may affect your design:

   a. The MOSFET is not precisely a square-law device. How will this fact affect performance of the linear amplifier?

   b. What is the impact of body effect on the I–V characteristic of the saturated load? What, if anything, can be done to ameliorate it?

   c. Comment qualitatively on the effect of absolute channel length upon amplifier properties. Explain what properties are primarily affected and why.

   d. Sketch a compact (plan-view) layout for your design, indicating all dimensions and terminal identities clearly. Comment on possible variations.

D5-3. You are to design a MOSFET for use in a digital integrated circuit. It requires a transconductance of $g_m = 0.69$ mS. Starting with the same process as used for the small-signal example in Section 5-5.3, what size MOSFET would you use? For a voltage gain of $A_V = -3$ and a supply voltage of $V_{DD} = 5$ V, design a minimum-size load resistor using a thin film CrSi resistor process (that provides a linear device) with a sheet resistance of $R_{SH} = 200$ ohms and a minimum width of $Z = 10$ μm.

D5-4. You are to design a MOSFET for use in a digital integrated circuit.

a. The MOSFET must drive an interconnect line of length $L = 1{,}000$ μm and a width of $Z = 2$ μm, having a capacitance of 0.2 fF/μm of length. Starting with the same process as used for the large-signal example in Section 5-5.4, what is the minimum-size MOSFET that will produce a rise time of $t_{RISE} = 20$ ns?

b. What static power will be required for this MOSFET?

D5-5. You are to design a 21-stage ring oscillator using the depletion-load inverter analyzed in Section 5-5.4. This oscillator is formed by connecting the inverters into a continuous cascade, with the output of one inverter driving the input of the next. What is the frequency of oscillation and what is the total power? Using a capacitor process with a capacitance of 1 fF/μm^2, design a ring oscillator with a frequency of 50 kHz.

D5-6. Starting with the D-mode-load inverter analyzed in Section 5-5.4, design and lay out a two-input NAND gate having the same logic levels as the inverter. The layout of the inverter driver is given in Problem A5-38. Schematic diagrams of the inverter and NAND circuits are given here:

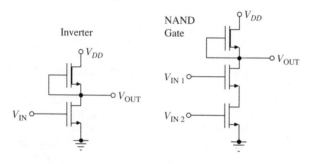

Symbol Index

Symbols for variables are listed below. To find symbols for physical constants in the text, consult the Subject Index or see Table 1-2. For SPICE symbols, see Tables 3-6, 4-6, 5-3, 5-4, and 5-6. To find symbols for units in the text, consult the Subject Index. The definition of each symbol given below can be found on the page indicated by the number following it. When the symbol has more than one definition, a page number is indicated for each.

a, 294
a_0, 19
A, 8
A_C, 565
A_D, 828
A_E, 492
A_I, 532
A_S, 828
A_V, 515
$A_{V,\text{cir}}$, 839

b_B, 572
BV_{CBO}, 548
BV'_{CBO}, 552
BV_{CEO}, 550
BV_{EBO}, 546
B, 38

C, 8, 30
C_b, 784
C_c, 621
C_e, 621
C_i, 784
C_s, 355
C_s^*, 358
C_t, 358
C_u^*, 396
$C_{BC}(t)$, 646
$C_{BD}(t)$, 831
$C_{BE}(t)$, 646
$C_{BS}(t)$, 831

$C_{CS}(t)$, 646
$C_{GB}(t)$, 830
C_{GD}, 772
$C_{GD}(t)$, 830
$C_{GS}(t)$, 830
C_{OX}, 721
C_{SE}, 261
C_{SG}, 772
C_T, 298
$C_T(v_A)$, 424
$C_T(V_A)$, 404
C_{TC}, 621
C_{TE}, 621
$C_{TOT}(v_J)$, 428
$C_{TOT}^{n,m}$, 433
$\langle C_{TOT} \rangle$, 431
$C_U(V_A)$, 404
$C_{0,B}$, 828
$C_{0BD,B}$, 828
$C_{0BD,SW}$, 828
C_{0BE}, 665
$C_{0BS,B}$, 828
$C_{0BS,SW}$, 828
C_{0GB}, 831
C_{0GD}, 831
C_{0GS}, 831
$C_{0,SW}$, 828
C_{0T}, 424
C_{0TC}, 646
C_{0TE}, 646
C_{0TS}, 646
C, 175

d, 8
D, 26, 385, 743
D^*, 573
$D(x, y, z, t)$, 189
D_n, 165
D_{nB}, 341
D_p, 165
D_{OX}, 742
D_S, 759
D, 26

erf, 299
erfc, 299
E, 5
$E(x, y, z, t)$, 189
E_s, 498
$E_{\text{BULK DIFFUSION}}$, 572
E_{CRIT}, 830
E_J, 787
E_M, 265
E_{OHMIC}, 572
E_S, 748
E_{0M}, 247
E, 2

f, 45, 165
$f(x)$, 432
$f_{n,\text{diff}}$, 200
$f_{p,\text{diff}}$, 200
$f_{n,\text{drft}}$, 200
f_B, 421

f_{MAX}, 644
f_T, 626
f_{3dB}, 839
f_β, 635
$f'(x)$, 432
F, 3
$F(t)$, 385
$\mathbf{F}$, 2

g, 348
$g_{b'e}$, 617
g_m, 619
g_{mbs}, 838
$g_{mF}(t)$, 646
$g_{m.P}$, 832
$g_{mR}(t)$, 646
g_o, 634
$g_{o.P}$, 832
g^n, 432
$g_C^{n,m}$, 433
$g_R(V_J)$, 428
$g_R^{n,m}$, 433
G, 175
G', 175
$G(x, y, z, t)$, 189
G_B, 545
G_M, 851
G_{MB2}, 853
G_{MM2}, 853
G_0, 175
$G_0(x)$, 241

h, 10, 73, 433
h_{fe}, 613
h_{ie}, 613
h_{oe}, 613
h_{re}, 613
h_{FE}, 614
h_{11}, 623
h_{12}, 623
h_{21}, 622
h_{22}, 622
H, 752
H_B, 753
H_D, 753
H_D^*, 827
H_G, 753
$\mathbf{H}$, 38

$i_a(s)$, 396
$i_a(t)$, 393
i_b, 612
i_c, 612

$i_d(t)$, 383
$i_d(\omega)$, 838
$i_g(\omega)$, 838
$i_A(t)$, 392
$i_B(t)$, 830
$i_B^{CC}(t)$, 660
$i_B^{CE}(t)$, 660
$i_B^{RC}(t)$, 660
$i_B^{RE}(t)$, 660
$i_C(v_J)$, 429
i_C^m, 433
$i_C^{n,m}$, 433
$i_C^C(t)$, 664
$i_C^R(t)$, 664
$i_D(t)$, 830
$i_D^R(t)$, 830
$i_D^R(v_{GS}(t), v_{BS}(t), v_{DS}(t))$, 830
$i_G(t)$, 830
$i_R(v_J)$, 428
$i_R^{n,m}$, 433
$\mathbf{i}$, 38
I, 20
I_b, 635
I_c, 635
$I_{\text{g-r}}$, 288
I_k, 648
I_{kr}, 648
I_{nE}, 503
I_{pE}, 503
I_A, 393
I_A^n, 432
I_B, 488
I_{BD}, 828
I_{BS}, 828
I_C, 488
I_{CEO}, 607
I_{CH}, 722
I_{CO}, 600
I_D, 381, 504, 506, 723
$I_{D(ON)}$, 741
I_E, 488
I_{ECO}, 607
I_{EO}, 600
I_F, 398
I_{K0}, 592
I_L, 504
I_{OUT}, 851
I_R, 402
I_{RB}, 630
I_S, 544
I_{SC}, 651
I_{SE}, 651
I_0, 276

I_{0A}, 426
I_{0B}, 530

j, 411
$\mathbf{j}$, 38
J, 20
J_n, 155
$J_n(0)$, 275
J_p, 155
$J_{p,\text{diff}}$, 168
$J_{p,\text{drft}}$, 168
J_D, 29
J_T, 30
J_0, 275, 828

k, 51, 73
$\mathbf{k}$, 38
K, 381, 725
K_t, 362
$K_{\text{EFF}}(t)$, 829
K.E., 10

l, 73
ℓ, 52
L, 21, 102, 193, 385
L^*, 824
L_n, 193
L_p, 193
L_{pE}, 495
L_D, 331
L_D^*, 825
L_{De}, 331
L_{Di}, 330

m, 11, 818
m_x, 73
m_y, 73
m_z, 73
M, 10, 552
M_C, 647
M_E, 647
M_S, 647

n, 46, 119, 552
$n(0)$, 307
n_f, 49
n_i, 49, 122
n_{iE}, 493
$n_u(x, t)$, 406
$n_B(x)$, 540
$n_B(X_B)$, 522
$n_B'(x, t)$, 640

SYMBOL INDEX I-3

n_N, 141
n'_N, 170
np, 142
$n_P(0)$, 275
$n'_P(0)$, 275
$n_U(x, t)$, 406
n_0, 241
$n_0(x)$, 241
n_{0m}, 331
n_{0B}, 522
n_{0N}, 170
N, 100
$N(t)$, 829
$N(x)$, 299
$N(x, y, z)$, 189
$N(\xi)$, 117
N_a, 127
N_f, 766
N_f^*, 827
N_{it}, 753
N_{ot}, 722
N_A, 131
N_{AB}, 493
$N_{AB}(x)$, 540
N_{BC}, 312
N_C, 119, 647
N_{CA}, 142
N_D, 126
N_D^+, 307
N_E, 647
N_F, 647
N_H, 345
N_L, 345
N_R, 647
N_{RD}, 828
N_{RS}, 828
N_S, 300
$N_U(x)$, 406
N_V, 121
N_{VA}, 142

p, 121
p', 181
$p_B(x)$, 540
p_N, 142
p'_N, 170
p_0, 241
$p_0(x)$, 241
p_{0N}, 170
p_{0P}, 343
P, 666
$P(\xi)$, 112
$P'(\xi)$, 121

$P_C(\xi_T)$, 184
P_D, 828
P_S, 828
$P_V(\xi_T)$, 184
P.E., 10

$q_b(t)$, 631
$q_s(t)$, 384
$q_u(t)$, 393
$q_B(t)$, 631
$q_S(t)$, 384
$q_{TOT}(v_J)$, 428
q_{TOT}^m, 433
$q_U(t)$, 392
$q_1(t)$, 646
$q_2(t)$, 646
Q, 2, 28, 358, 517, 742
$Q(t)$, 30
Q_b, 803
$Q_b(y)$, 814
Q_f, 766
Q_{it}, 753
Q_n, 720
$Q_n(y)$, 814
Q_{ot}, 772
Q_s, 804
Q_{ss}, 767
Q_B, 631
Q_E, 508
Q_S, 285
Q_T, 358, 743
Q_U, 393
Q_{0S}, 355
Q_{0T}, 361
Q_{0U}, 392

r, 40, 348
$r_{bb'}$, 617
r_{bm}, 630
$r_{b'c}$, 619
$r_{b'e}$, 617
r, 41
r_{IN}, 532
R, 30
R', 175
R_B, 515
R_C, 515
R_D, 737, 828
R_S, 426, 720, 828
$R_S(y)$, 722
R_{SB}, 646
R_{SC}, 646
R_{SE}, 646

R_{SH}, 828
R_0, 175
$R_0(x)$, 241
R_{0B}, 582

s, 52, 199, 396
s, 52, 743

t, 28
t^m, 433
t_D, 31
t_R, 367
t_{RISE}, 421
t_0, 193
t_I, 404
t_{II}, 410
T, 112
$T(x, y, z, t)$, 188

$u(t)$, 383
U, 330
U_B, 744
U_{EXP}, 830
U_G, 855
U_{GS}, 852
U_J, 338
U_{NP}, 267
U_T, 852
U_0, 329
U_{01}, 334
U_{02}, 335

v, 10
$v_a(s)$, 396
$v_a(t)$, 419
v_{be}, 612
v_{ce}, 612
$v_d(t)$, 383
$v_{gs}(\omega)$, 838
v_s, 158
v_t, 120
$v_A(t)$, 392
$v_{BD}(t)$, 830
$v_{BD.P}(t)$, 830
v_{BE}, 612
$v_{BS}(t)$, 830
$v_{BS.P}(t)$, 830
v_D, 154
$v_D(t)$, 390
$v_{DS.SAT}(t)$, 829
$v_G^{n,m}$, 433
$v_{GD}(t)$, 830
$v_{GD.P}(t)$, 830

SYMBOL INDEX

$v_{GS}(t)$, 830
$v_{GS,P}(t)$, 830
$v_J(t)$, 426
$v_{JC}(t)$, 645
$v_{JE}(t)$, 645
$v_{JS}(t)$, 646
$v_J^{n,m}$, 433
$v_{ON}(t)$, 829
V, 15
$V(t)$, 28
$V(y)$, 722
$V_{cap}(t)$, 30
V_k, 423
V_{ohmic}, 30
V_A, 648
V_{AB}, 260
V_B, 310
V_{BB}, 512
V_{BE}, 489
$V_{BE(SAT)}$, 523
V_{BI}^*, 828
V_{BS}, 717
V_{CB}, 497
V_{CC}, 515
V_{CE}, 489
$V_{CE(SAT)}$, 521
V_{DD}, 737
V_{DS}, 717
$V_{DS(ON)}$, 738
V_F, 402
V_{FB}, 327
V_{FB}^*, 828
V_{GB}, 798
V_{GS}, 717
V_G^n, 432
V_{IN}, 531
V_{JE}, 568
V_{NP}, 261
V_J^n, 432
V_R, 402, 647
V_{RT}, 320
V_T, 718
V_{OUT}, 531
V_γ, 284

W, 335
W_J, 338
W_S, 758

x, 2
x^m, 433
x^n, 432
X, 266, 719

X_B, 490
X_D, 745
X_0, 246
X_{OX}, 725
X_P, 281

$y(\omega)$, 398
$y(\omega_k)$, 423
y_{11}, 627
y_{12}, 627
y_{21}, 627
y_{22}, 627

z, 630
Z, 51, 722

α, 52, 532, 630
α_{0F}, 551

β, 506
β_F, 527
β_R, 527
β_{0F}, 552

γ, 815
γ_B, 502
γ_C, 565
γ_E, 503

δ, 586
$\delta(t)$, 383

Δt, 154
Δx, 154
ΔI_D, 381
ΔU, 267
ΔU_0, 267
ΔV, 260
$\Delta \xi$, 45
$\Delta \psi$, 5
$\Delta \psi_0$, 237
$\Delta \psi_{0C}$, 646
$\Delta \psi_{0E}$, 646
$\Delta \psi_{0S}$, 646

η, 586

θ, 59, 627, 666
θ_{cl}, 667
θ_{jc}, 667
θ_{jl}, 667

κ, 26

λ, 45, 830

$\mu(x, y, z, t)$, 189
μ^*, 576
μ_n, 155
μ_p, 155

ξ, 11
ξ_f', 45
ξ_i, 45, 49
$\xi_{ionization}$, 125
$\xi_n(f)$, 148
ξ_C, 118
ξ_D, 137
ξ_F, 112
ξ_G, 109
ξ_{GE}, 494
ξ_H, 143
ξ_I, 142
ξ_{REF}, 143
ξ_ν, 118
ξ_T, 184

ρ, 21
ρ_v, 32
ρ_N, 160
ρ_P, 160
ρ_{0M}, 295

σ, 22
σ_i, 162
σ_N, 160
σ_P, 160

τ, 177
τ^*, 358
$\tau(x, y, z)$, 189
τ_B, 502
τ_C, 634
τ_E, 495
τ_F, 646
τ_R, 646

υ, 643

ϕ, 144
ϕ_n, 536
$\phi_n(y)$, 814
ϕ_p, 536
ϕ_B, 753
ϕ_G, 753

Φ, 751

χ, 19

ψ, 5
$\psi(x)$, 18
ψ_B, 744
$\psi_C(x)$, 239
ψ_{CO}, 753
ψ_{CS}, 753
ψ_F, 144
ψ_G, 256, 744

ψ_{GB}, 744
ψ_{GS}, 745
$\psi_{GS}(y)$, 814
ψ_N, 249
ψ_{REF}, 143
ψ_S, 717
ψ_{SB}, 745
$\psi_{SB}(y)$, 814
$\psi_V(x)$, 239

ψ_1, 146
ψ_2, 146

Ψ, 19

ω, 411
ω_k, 423
ω_β, 626
ω_0, 630

Subject Index

[Italicized page numbers cite figures; tables are cited by table number.]

absolute carrier-density change versus relative change, 170
acceptor atom, 129–131, *130, see also* hydrogen model
acceptor state, as hole and electron energy wells, *130*, 131
acceptor-like state, *see* state, acceptor-like, *and* interface charge
accumulation layer, 342–347
  (in) high-low junction, 345, *782*
  (in) MOS capacitor, *782*
accumulation term, continuity equation, 191
acronym
  BiCMOS, *see* BiCMOS
  BJT, *see* BJT
  CCD, *see* CCD
  CMOS, *see* CMOS
  E-D, *see* E-D circuits
  EPROM, *see* EPROM
  FET, *see* FET
  HEMT, *see* MODFET
  HIGFET, *see* HIGFET
  JFET, *see* JFET
  MBE, *see* MBE
  MESFET, *see* MESFET
  MODFET, *see* MODFET
  MOS, 714, 716
  MOSFET, *see* MOSFET
  NMOS, *see* NMOS
  PMOS, *see* PMOS
  TEGFET, *see* MODFET
action, transistor, *see* transistor action
active load, *see* CMOS
admittance approximations, junction diode, Table 3-3

admittance for arbitrary frequencies, junction diode, Table 3-2
admittance model, BJT, *see* BJT, admittance model
affinity, electron, *see* electron affinity
algebraic sign
  ambiguity in for work function, 752
  (of) bias on PN junction, 259–267, *262*
  (of) bias on high-low junction, 307
  (of) conduction current, 20
  consistency in, for case of charge on capacitor, 23
  convention for, 4
    (in) general solution, DAR, 337, *778*
    (in) Ohm's law, 259, *260*
    (in) PN-junction biasing, 259–261
  (of) electric field, 4
  (of) electronic charge, 13
  explicit, *see* explicit algebraic sign
  (of) force caused by another charge, *41*
  (of) force caused by electric field, 2, *4*, 7
  (in) MOS-capacitor analysis, 745
  (in) MOSFET analysis, 721 (ex.), 723 (ex.), 814
  (of) work, 2, 4, *4*
alkali-halide crystal, 68
alloyed junction, 557
  (in) BJT, 557, *557*
  type-independent breakdown of, 310
alpha (common-base-BJT current gain), definition of, 532
  (as a) function of beta, 532

alpha particle, 36
  scattering of, and justification of term, 36
ambient atmosphere, 667
ambipolar diffusivity, 573, 573 (ex.)
ambipolar effects, 571–577, *583*, 584
ambipolar mobility, 576, 577
amorphous materials, 77, *see also* silicon dioxide
amphoteric dopant, 766
amplifier, BJT, *see* BJT, amplifier
analog devices, 324, 326
analog multiplication, 134, *134–136*
analogy
  avalanche, 310
  bar magnet, 72
  billiard-ball, 120, 158
  breakdown of, 109
  bubble, 108, 131
  classroom, 114
  closed-vessel, 110, *111*
  drift-transport, 162 (ex.)
  garage, 109
  gaseous-diffusion, 163
  hydrogen-atom, *see* hydrogen model
  line-of-cars, 106
  Ohm's law in case of thermal resistance, 666
  oscillator, *see* oscillator analogy
  planetary, *see* planetary analogy
  water
    (for) band structure, 110
    (for) density of states, 114
    (for) Fermi level, 112, *113*, 133, 152, *238*, 239

SUBJECT INDEX I-7

angular momentum, *see* momentum, angular
anisotropic etchant, 75
anode of vacuum diode, *see* vacuum diode
antenna, 39
Antognetti, P., 630
approximate-analytic model for junction, surface and MOS problems, 327–347, 748–751, 776–780, *see also* DAR
  complement to numerical solution, 328
  nonequilibrium applications of, 347
  Poisson-Boltzmann equation in, 328–330
  universal curves in, 336, 347
  (for) surface problem, and close connection to junction problem, 328
approximation
  versus assumption, 138
  band-symmetry, 116, 138, 142
  Boltzmann, 138–141
  depletion, *see* depletion approximation
  equivalent-density-of-states, 117–121, 138, 141
  hundred-percent ionization, 127, 138, 139
arbitrary constant, *see* arbitrary origin
arbitrary definition
  (of) charge required to launch line of force, 7
  (of) current direction, 20
arbitrary origin (or reference or zero)
  (in) circuit potential, 12
  (in) electrostatic-potential definition, 6
  (in) potential energy, 11
  same as arbitrary constant in function, 11
Archimedean solids, 72
areal density
  (of) atoms in atomic plane, 75
  (of) electrical charge, *see* charge, electrical, areal density of
  (of) impurity atoms, *see* impurity atom, areal density of
assumption versus approximation, *see* approximation
asymmetric step junction, *see* step junction, asymmetric
atom, vibrations of, 105
atomic density
  areal in a crystal plane, 75
  volumetric in a crystal, 127, Table 2-2
atomic number, definition, and symbol for, 51
atomic plane, *see* crystal plane

atomic spacing, thought experiment on, 103, 104
atomic weight, in relation to number of electrons, 36
Auger recombination, three-body process, 494
avalanche breakdown, *see* device listing
avalanche multiplication factor
  (in) BJT-breakdown analysis, 551–553
  definition of, 551
Avogadro's hypothesis, 163

Baccarani, G., 823
"backward" diode, *315, 317*
balance, detailed, *see* detailed balance, principle of
ballistic-mode carrier transport, 159
Balmer, J. J., 49
Balmer formula, 49
Balmer series, 47, *47*, 49
band
  conduction, 106
  energy, *see* energy band
  forbidden, *see* energy gap
  lower, 104
  upper, 104, 106
  valence, 105, 106
band bending, 747, 758
band diagram
  (for) conductor or metal, 110, 112, *115*
  (in terms of) electrostatic potential, 143–144
  (for) PN junction, 237–239, *296, 315*
  (for) semiconductor, 105
band edge, importance of, 105
band structure
  (versus) atom spacing, 103, *103*, 104
  (as related to) bonds, 105, 106
  closed-vessel analogy for, *111*
  (of) conductor, 110
  direct and indirect, 184, 318
  (of) insulator, 110
  relating to crystal structure, *105*
  (of) semiconductor, 105
  (of) silicon, alternate view of, *107*
band-symmetry approximation, 116
  dependence on effective masses, 121
band-to-band recombination, *see* carrier recombination, band-to-band
bandgap, *see* energy gap
bar-magnet analogy, 72
Barber, H. D., 288, 290
Bardeen, J., 156, *156*, 484, 764, 797
Barrett, C. S., 57
barrier, potential
  modulation of is essence of BJT operation, 713
  (in) PN junction, effect of bias on,

259, 261, 261 (ex.)
  small in high-low junction, 308
barrier-height difference, 751–760, *754*
  (as) function of doping, 754, *755*
  equation for, 756 (ex.)
  (in) SPICE model, 827
  with polysil gate, 756 (ex.)
Barron, M. B., 861
base of BJT, word-choice reason, 485
base-charging time, *see* BJT, charge-control model, base-charging time
base-collector junction, *see* collector junction
base contact, *502*
base current, 499–508
  exponential dependence of through wide range, 575
base-emitter junction, *see* emitter junction
base-emitter voltage
  dependence on emitter area, 508 (ex.)
  enters into input and output calculations, 506
  value under high-level conditions, 567 (ex.)
"base push-out," 578
base-recombination current, 502
base region
  active (or intrinsic), *549*, 616
  boundaries of, 500 (ex.)
  conductivity modulation of, 630
  definition of, 499
  areal impurity density in versus that in JFET channel, 714
  carrier profile in
    extrapolates to depletion-approximation boundary, 498
    under high-level conditions, 568, 569 (ex.)
  charge stored in, definition of, 499, *see also* BJT, charge-control model
  confinement of majority carriers within, 543 (ex.)
  currents in, 499
  dominant current component in, 495
  doping profiles in, 539–544, 585, 589
  extended, *see* extended-base zone
  extrinsic, *549*, 616
  general high-level analysis of, 585–590
  graded, *see* BJT, graded-base
  intrinsic (or active), *see* base region, active
  lateral voltage drops in, 582
  longitudinal field in, 566, 568, 569 (ex.), 570, 571, 572 (ex.), 573, 574 (ex.), 585

base region *cont.*
  measuring areal net doping
    electrically, *see* Gummel number
  minimum thickness for valid
    assumption of multiple
    interactions by carrier in transit,
    540 (ex.)
  minority-carrier profiles in, 491, 495
    gradient calculation for, 506
    reason for linearity of, 492
    (for) saturation, *see* saturation
      regime
  moderate-doping approximations and
    values apply in, 493
  modulation of thickness of, *see* Early
    effect
  plan-view geometry of, *502*
  recombination in, 540
  typical thickness of in modern BJT,
    490
  uniquely associated with base
    terminal, 486
  when extremely thin causes
    punchthrough, 517
  word-choice reason, 485
base resistance, ohmic, *see* ohmic base
  resistance
base spreading resistance, *see* spreading
  resistance, base
base terminal, 485, 486, 488, *502*
base thickness
  effective, 506, 555
  modulation of, *see* Early effect
base-transport efficiency, 502
  definition of, 503
  inferred from carrier profile, 503 (ex.)
base "width," correctly defined, 501
basis, of crystal, with examples, 67–69
Beaufoy, R., 591
Berglund, C., 793
beryllium, 50
beta (common-emitter-BJT current gain),
  see also BJT, common-emitter
  current gain
  definition of, 506
  fall-off of in high-level operation,
    568
  forced, 665
  (as a) function of alpha, 533 (ex.)
  "up" versus "down," 564, *565, 567*
beta brass, structure of, 69
biasing, device, *see* device listing
BiCMOS (bipolar complementary metal-
  oxide-silicon), 736
billiard-ball analogy for electrons in
  conduction band, 120
binding energy, 49, 72, 124
bipolar, definition of, 484
bipolar junction transistor, *see* BJT

BJT, 484–667, *see also* base, emitter,
  and collector
  accuracy of dynamic models for, how
    to improve, 629–630
  active portion of discrete device, 487
  admittance model, 622–629
    equivalent circuit for, 625, *626*
    lumped elements in simplified
      model, 628
    parameters developed from device
      physics, 627
    parameters in terms of *h*
      parameters, 625, Table 4-5
    transadmittance in terms of hybrid-
      pi parameters, 625
    used in SPICE, computer analysis,
      622
  alloyed-junction, 557, *557*
  amplifier, rudimentary, 515–518
  barrier modulation as essence of, 713
  base-charging time
    charge-control analysis, *see* BJT,
      charge-control model
    SPICE analysis, 665
  base recombination current, 502
  beta, *see* beta
  biasing of, 488, 489, 510–534
  bilaterally symmetric, 558
  breakdown characteristics of,
    explanations for, 552, *553*
  breakdown phenomena in, 517,
    546–553
  collector junction, 548
    lines-of-force crowding in, 548
    (for) plane versus curved
      portions, 548
    (with) reachthrough present, 550,
      550 (ex.)
    (with) reachthrough ruled out, 548
    related to intrinsic-extrinsic-
      portion differences, 549
  common-base configuration, 546
  common-emitter configuration at
    collector-emitter port, 550
    analysis of, 551
    characteristics for, 552
    effects on characteristics of, 554
      (ex.)
    fixed by intrinsic portion of
      collector junction, 552 (ex.)
    negative differential resistance in,
      553
    relevant to emitter-follower
      operation, 551
    sustaining voltage in, 553
    switching in, 553
    symbol for, 550
  emitter-follower configuration, 551
  emitter junction, 546–549

capacitance in, 632–636, 645
carrier distributions outside the
  junction boundaries most
  important in, 490
channel-collector, 558, *560*
charge-control model
  base-charging time in, 633, 634,
    640–644
  basic equation of, 631, 632 (ex.)
  charge in base, 641
  collector and emitter time constants,
    634
  complex current gain, 635
  current gain versus frequency, 635
    (ex.)
  depletion-layer capacitances in,
    632, 632 (ex.)
  dimensionless (SPICE) charges, 645
  (for) early BJT, 631
  (for) low-level, small-signal
    condition, 631
  output conductance, 634
  "phenomenological," 631
  (of) "second kind," 591
  coexisting voltage gain and current
    gain in, 488
  common-emitter configuration of, *see*
    common-emitter configuration
  comparing performance with that of
    MOSFET, 850–868
  complementary circuit, applications
    of, 485
  conductance
    input, *see* BJT, hybrid-parameter
      model, input resistance
    output, *see* BJT, output conductance
  (as) current amplifier, 515
  current gain, 563–566, 614
  current patterns in, 499, *500, 501,
    505*
  current, voltage, power specifications
    for, 492
  cutoff frequency, 634, 644
  depletion-layer capacitance in, *see*
    BJT, capacitance in
  depletion-layer thicknesses in, 497
    (ex.)
  device physics and hybrid model,
    615–619
  dimensional specification for, 492,
    493, Table 4-1
  discrete, 487
  dominant current components in
    various regions of, 495, 496
  dominated by MOSFET, 484
  doping specifications for, typical,
    492–493, Table 4-1
  dynamic models, large-signal, *see*
    BJT, large-signal analysis of

SUBJECT INDEX   I-9

BJT *cont.*
dynamic models, small-signal, *see* BJT, small-signal dynamic models for
dynamic-static current-gain comparison, 590 (ex.), 614 (ex.)
Ebers-Moll model for, *see* Ebers-Moll model
electron current in (NPN case), 495–498
electron-density profiles in, *496*
electron paths in, *565*
elementary theory of, 498–510
emitter-follower configuration, *see* emitter-follower configuration
epitaxial, *487*, 558, 559, *561*
equivalent-circuit model, *see* equivalent-circuit model, BJT
evolution of, effect of junction-forming methods on, 556
extrinsic portion of, 549, *549*
fabrication methods for, 485, 562
figures of merit for, 634, 644, *see also* BJT, cutoff frequency
gain mechanism in, 507–510
general high-level analysis of base region in, 585–590
germanium PNP, 485, 535, 556, 558
graded-base, 585
h-parameter model, 621–629, 636, *638*, *639*, Table 4-4
high-level effects in, 566–590
 base and collector currents versus voltage in presence of, *583*
 beta fall-off in, 568
 carrier profiles with, *569*, *586*, *587*, *588*
 electric field caused by, 566 (ex.)
 SPICE treatment of, 648–650
hybrid equations, 613, 615
hybrid model and device physics, 615–619
hybrid-parameter model, low-frequency, 610–619, *616*, *618*
hybrid parameters, 613, 614
hybrid-pi (Giacoletto) model, 620–629
 capacitances in, 632 (ex.)
 circuit for SPICE analysis, *657*
 current gain versus frequency, 636
 distinction from hybrid model, 620
 equivalent-circuit model for, *621*, *624*, *629*
 features of, 620
 parameters related to physical properties, 629
 reason for name, 620
 simplified, 622
 typical element values for, 621, Table 4-2

(in) IC, *see* integrated circuit
idealized doping profiles useful for calculation in, 498
integrated with JFET, 714
interactions of junctions in, 489, 491
internal current patterns, 499–502, *500*
intrinsic-barrier device, 558
intrinsic portion of, 549, *549*
invention of, 484
inverter, 516
isolated-junction description of, 489
large-signal analysis of, 657–665
loadline diagram applied to, 512–513, 518, 657
lock-layer, *see* BJT, channel-collector
longitudinal electric field in, 566, 568, 569 (ex.), 570, 571, 573, 585
low-emitter-concentration (LEC), 561, 564
low-frequency hybrid model, 610–615
maximum electric field in five regions of, 498
maximum oscillation frequency of, 644
minority-carrier distributions most important in, 490
minority-carrier profiles in, *490*, 495
most important junction boundary in, 492
newest structures approaching idealized structure, 499
(as) one-dimensional analog of point-contact transistor, 484
one-dimensional currents in, 486, *496*
output conductance, 533, 534, 554–556, 634
output plane, *see* configuration of interest
oxide-filled trench in, 563
parameters, SPICE, 645
parasitic (collector-region) resistance in discrete device, 487
parasitic currents in, 502–505
parasitic lateral portion of, 487
parasitic properties of, 534, 535
peak-field magnitudes in, 497
physical structure of, 485–488, *486*
piecewise-linear output characteristics of, 518, 520
"planar" process, 550
planar-epitaxial, 558, *see also* BJT, epitaxial
PNIP device, *see* intrinsic-barrier BJT
polysil-emitter, 562
power device, germanium, 558
power dissipation, seat of, 666 (ex.)
profiles of various functions in, *see* profile, BJT

properties of, parasitic, *see* BJT, parasitic properties of
punchthrough in, 517
quasisaturation in, *see* quasisaturation
reachthrough in, *see* reachthrough voltage and reachthrough phenomenon
reduction to practice of, 484
relative thicknesses of depletion regions in, 489
replaces point-contact transistor, 484
reverse voltage gain of, 614, 618
rudiments of, 485–498
saturation of, *see* saturation regime of BJT output plane
scattering-parameter model, 622–629
silicon NPN, 485
simultaneous current and voltage gain from, 510
small-signal dynamic models for, 610–644, *see also* Table 4-3
 admittance model, 622–629, *see also* BJT, admittance model
 feedback in, 623 (ex.)
 h-parameter model, 621–629, *see also* BJT *h*-parameter model
 hybrid-parameter, low-frequency, 610–615, *see also* BJT, hybrid-parameter model
 hybrid-pi, 620–629, *see also* BJT, hybrid-pi model
 improving accuracy of, 629–630
 model conversions, 622, 623 (ex.), Table 4-4
 scattering-parameter model, 622–629
 SPICE, *see* BJT, SPICE analysis
small-signal voltage gain, *see* BJT, small-signal voltage gain
specifications for, 492, 493
SPICE analysis of, 622, 625, 629, 645–667
 base and collector currents versus voltage, *652*
 calculating *h* parameters using, 638
 capacitance effects, 652–653, 656
 capacitances, 645
 collector-substrate capacitance, 645
 collector-substrate voltage in, 645
 contact potential, 652
 current gain, static and dynamic, versus frequency, *650*
 debiasing and modulation incorporated into, 630
 default values, used to understand equations, 646, 652
 dimensionless charges, 645
 Early effect, 648, 654

SUBJECT INDEX

BJT *cont.*
  Ebers-Moll equations from SPICE model, 647
  emitter conductance and capacitance, 656
  forward transit time, 652
  grading coefficient, 652
  high-current effects, 648–650
  imposed junction voltages in, 645
  Kirk effect, 654
  knee current (high), 648
  knee current (low), 651
  large-signal transient problem, 658
  Miller effect, 657
  nonideal-diode effects, 651
  numerical example, Tables 4-9, 4-10
  operating-point information from, 653, Table 4-8
  parameters, 645, 647, Table 4-6
  representative parameter values, Table 4-7
  resistance, 647
  reverse transit time, 652
  rise and fall times, 657
  saturation currents, 645, 651
  small-signal analysis using, 653–657
  symbols for, Table 4-6
  transconductances in, 645, 655
  transient response determined from, *658, 659, 660–664*
    circuit, *664*
    equivalent circuits for, *659*
    large-signal, *660, 661, 662, 663*
    small-signal, *658*
    typical values, 647, Table 4-7
    upper current limit, 650
    variables, 645
  static-dynamic current-gain comparison, 590 (ex.), 614 (ex.)
  static model, *see* Ebers-Moll model
  structure of, basic, 485–488, *486*
    diffused, *487, 549, 561*
    discrete-device, *487*
    epitaxial, *487, 561*
    first practical, 556
    modern, 562, *563*
    real, 534–566
    specific example, Table 4-1
    to minimize ohmic base resistance, 501
    variations in, 556–563
  supersedes point-contact transistor, 484
  sustaining voltage, 553
  (as) switch, 510
  switching application of, 528, *see also* BJT, SPICE analysis of
  symbol for, *486*
  symbols for net doping values in, 493
  symbols for region designations in, 493
  symmetry of, *see* BJT, bilaterally symmetric
  temperature, effect of on junction characteristics, 509–511, *511*
  terminal currents in, 488
  terminal designations in, *486*
  terminology for, 485–488, *486*
  theory of, 498–510
  time constants, 633, 634, 640–644
  transconductance, *see* transconductance, BJT
  transient response of, *see* BJT, SPICE analysis of
  transistor action in, 508
  two-port analysis of, 622
  use of, 510–534
    (as) voltage amplifier, 515
  voltage feedback, or reverse gain, *see* BJT, reverse voltage gain
  y-parameter model, *see* BJT, admittance model
BJT-MOSFET performance comparisons, 850–868
black-body radiation, 40
body effect, 819–823
Bohr, N. H. D., 35, 36, 44, 45, 46, 49, 50, 55, 56
Bohr magneton, 53
Bohr model, of hydrogen atom, 35–57, *42*, 44, 50, 123–127
Bohr postulates, 44–45, 49
Bohr radius, 46
Boltzmann approximation, 138–141, *139*, 321
"Boltzmann behavior" of boundary values, currents, 568, 584
Boltzmann quasiequilibrium, quasi Fermi level constant for, 267–271, 291, 543, 570, 574, 575
Boltzmann relation, 146, 251, 267, 271
bond
  covalent, *see* covalent bond
  ionic, 68
  unsatisfied or "dangling," *see* "dangling bond"
bonding, thermocompression, *see* thermocompression bonding
Boothroyd, A. R., 585
boron, as impurity and ion, 130
bound electron, *see* electron, bound
bracket, square, *see* square bracket
brass, beta, structure of, 69
Brattain, W. H., 484
Bravais, A., 63
breakdown phenomena and voltage, *see* device listing

Brews, J., 823, 861
Broux, G. F., 861
Brownian motion, 167
bubble analogy, 108, 131
built-in field, *see* dipole layer and *PN* junction
built-in potential difference of junction, *see* contact potential
bulk, as synonym for substrate, 717
bulk mobility, versus surface mobility, 725, *see also* conductivity mobility and drift mobility
bulk phenomena, definition of, 100
bulk properties, 100–203, 335

cadmium, 67
cadmium sulfide, 76
calcite, 57
calcium fluoride, 69
calibration, heuristic, *see* heuristic calibration
capacitance, 23, *see also* device listings
capacitance crossover in MOS capacitor, 799–806, *785*
capacitor, *9,* 23, 33, 34, *34,* 367
carbon, 72, *see also* diamond
Carlson, F. R., 310, 311
carrier
  drift velocity of, *see* drift velocity of carrier
  effective mass ratio of, 121
  (of) electricity, 20
  excess, *see* excess carrier
  generation of, *see* generation, carrier
  "hot," *see* "hot" carrier
  majority, *see* majority carrier
  minority, *see* minority carrier
  recombination of, *see* carrier recombination
  scattering of, *see* scattering, carrier
carrier density
  absolute versus relative, 186–188
  excess, *see* excess carrier
  extrinsic, 127
  factor fixing current density, 21
  fluctuations in, 774
  gradient of
    causes carrier transport, 147
    (in) junction-formation thought experiment, 235
  graphical representation of, *183*
  incremental, *180*
  intrinsic, 121–123, Table 2-4
  low-level versus high-level disturbance of, 186–188
  (in) metal, 159
  net rate of change of, *180*
  perturbed by infrared radiation, *171, 173*

SUBJECT INDEX   I-11

carrier density *cont.*
 profile of, *see* profile
 total, *180*, *183*
carrier diffusion, 163–167, 261, 280
carrier drift, 152–154, 163
carrier generation, 169–188, *see also*
   pair, generation of
carrier-generation rate
 equilibrium, 176, 241 (ex.)
 incremental, 175
 time-dependent, 176
 total, 175
carrier injection
 asymmetric in asymmetric junction, *273*, 274
 combined with other phenomena, 372
 preferential by emitter junction, *see* emitter-current crowding
carrier lifetime, 178–181, 185, *186*, 192
 effective, 358, 383–384
carrier mobility, *see* mobility
carrier recombination, 169–188, *see also* carrier recombination rate
carrier recombination rate, 175–183, *180*, *182*
carrier scattering (or deflection), 147–158, *156*
carrier transport, 147–169
 definition of, 147
 (by) diffusion, 163–167, 280
 (by) drift, 152–154
 gradients that cause, 147
 thought experiment on, *154*
carrier trap, 184
cascode connection examples, 558, 559
Castagné, R., 793
categories, phonon, Table 2-4
CCD (charge-coupled device), 793
cell
 primitive, *see* primitive cell
 unit, *see* unit cell
centimeter, used instead of meter in semiconductor practice, 4
channel
 carrier mobility in, *see* surface mobility
 FET, 713
 gradual, *see* long-channel approximation
 inversion layer as, 717
 long, *see* long-channel approximation
 ohmic contacts to, 717
characteristic current
 D-mode FET, *see* pinch-off current
 E-mode MOSFET, 732, 732 (ex.)
characteristic frequency, mechanical oscillators, 101, 102
characteristic length

Debye, *see* Debye length
minority-carrier diffusion, *see* minority-carrier diffusion length
characteristic times, 282
charge, electrical
 acceleration of causes radiation, 39
 areal density of
   (on) capacitor plates in case of uniform field, 8, 8 (ex.)
   determining ionic and inversion-layer components of in MOS capacitor, 779
 electric displacement corresponds to, 26, 35 (ex.)
 equal on two sides of junction, 236, 255 (ex.)
 expression for, 8
 geometrical representation of, 246 (ex.)
 (in) MOSFET channel, 720
 term appropriate for thin layer, 32
 uniform, 8
 bound, in dielectric material, 26
 (on) capacitor plate, 23, 35 (ex.), 359
 electronic, *see* electronic charge
 exchange of, role in contact potential, 250
 fixed, 3, 20
 force exerted by on another charge, *41*
 force on caused by electric field, sign of, 2, *4*
 free, causing conduction, 20, 26
 ionic, *see* ion, impurity
 moving and motionless, 20–35
 oxide, *see* oxide charge
 (as) source of electric field, 1, 32
 space, *see* space charge
 static, 3, 20
 symbol for (general), 2
 test, *see* test charge
 transport of, 20
 trapped in silicon dioxide, 33
 unit for, 2, 4
 volumetric density of, 32, 243, 244, 751
charge accumulation, ruled out in steady state, 276
charge-control analysis, *see* device listing
charge-coupled device (CCD), 793
charge profile in *PN* junction, *see* profile, charge, in *PN* junction
charge-sheet (Brews, MOSFET) model, 823
chemical potential, algebraic expression for, 536
chemical properties, in relation to electronic structure, 57

Child's law, 326, 327
"chip," *see* integrated circuit
chlorine, 68
circuit, integrated, *see* integrated circuit
cleaved silicon surface, surface states on, 762
closed-vessel analogy to energy bands, 110, *111*
close-packed hexagonal crystal, 67
CMOS (complementary metal-oxide-silicon) circuits, 733, 740, *740*, 741
collector
 (of) BJT, 485
 multiple in I^{2}L, 564
 (of) point-contact transistor, 485
collector-base junction, *see* collector junction
collector current, calculation of, 505, 506
collector-emitter voltage
 nonzero value in saturation as parasitic property, 534
 saturation value of, 558, 605
collector junction, 485, 495, 520, 534
collector-junction depletion region, 495, 498
collector leakage-current values, 509, 510
collector region, 486
 dominant current in, 495
 Ebers-Moll-model currents in, 602 (ex.)
 high-level phenomena in, 577
 lightly doped, current components in, 581 (ex.)
 little affected by junction interactions, 491
 moderate-doping approximations and values apply in, 493
 uniquely associated with collector terminal, 486
collector terminal, 485, 488
column-III impurity, *see* impurity atom, column-III
column-V impurity, *see* impurity atom, column-V
common-base configuration, 529, *533*
 (used in) Ebers-Moll model 593, 593 (ex.)
 electrical attributes of, 529
 freedom from Early effect, 556
 output characteristics for, 532, 533 (ex.), *533*, 534 (ex.)
 output plane for, 533, *533*
 (as) resistance transformer, 534
 static current gain for, 532, 533 (ex.)
 (in) Ebers-Moll model, 591, 600
 regions determining, 600

common-collector configuration, see
  emitter-follower configuration
common-emitter configuration, 488
  bias circuit for, *516*
  coexisting voltage and current gain in,
    488
  conventional and terminal currents in,
    489 (ex.)
  loadline diagram applied to, 513, *517*,
    523, 525
  output conductance for, 554–556
  output plane current-voltage
    characteristics for, 517, *517*,
    523, 525, 527
  approximately piecewise-linear
    characteristics in, 518
  (with) breakdown phenomena, *553*
  (for) channel-collector transistor,
    560
  (from) Ebers-Moll equivalent-circuit
    model, *607*
  (with) Early effect, *555*
  (with) epitaxial structure, *559*
  (with) quasisaturation, *581*
  quiescent operating point on, *517*,
    518
  (from) SPICE, *649*
  piecewise-linear output characteristics
    for, 518, 520
common-emitter current gain, 505–507,
  see also beta
  dynamic, versus frequency, *560, 636,
    637*
  (in) Ebers-Moll model, 591, 600
  forward and reverse, 563–566, *565,
    567*
  phase of, versus frequency, *637*
  static (dc or steady-state), 506 (ex.),
    507
    experimental data on, 583 (ex.)
    (versus) frequency, *650*
common-emitter operating regimes,
  518–530
communication, "wireless," see
  "wireless" communication
compensated-intrinsic doping, 132
compensation, impurity, see impurity
  compensation
complementary circuits, 485, 715, see
  also CMOS
compound semiconductor, see also
  gallium arsenide, gallium
  phosphide, single crystal, Table
  2-1, 76, 715
computer
  design of MOS circuits using, 735
  Fermi-level, see Fermi-level
    "computer"
computer solution, see numerical
    solution
concept
  (of) electric-field, see electric field,
    concept of
  (of) limit, see limit concept
conditions, high-level, see high-level
  conditions
conduction current, relating to
  displacement current, 29
conductivity, 20–22, 160
conductivity equation, 160–163
conductivity mobility, 154–157
conductivity modulation, 174 (ex.),
  630
conductor, 110–111
  band structure of, 110
  example of, 20
  nonohmic effects absent from, 159
    (ex.)
configuration, see emitter follower *or*
  common-base, -collector, *or*
  -emitter
conservative system, 11, *11*, 12, *12*
constant, arbitrary, see arbitrary constant
constant-E continuity-transport equation,
  see continuity equations
constants, see listings of individual
  constants
contact, electrical, see electrical contact
  and ohmic contact
contact potential, see also potential
  difference, built-in
  BJT, SPICE, 652
  (or) diffusion potential, 251
  (in) dipole layer, 236, 250
  (in) *PN* junction, 236, 237, 250–253,
    255, 255 (ex.)
  "built-in reverse bias," 259
  versus that of high-low junction,
    345 (ex.)
continuity equations, 188–200
  applications of, 191–195
  constant-*E* continuity-transport
    equations, 188–191
  applied to forward-biased junction,
    269, 269 (ex.)
  graphical representations of solutions
    for Haynes-Shockley experiment,
    196
  (of) semiinfinite-sample, *192*
continuum, of electron states, 49, 102
control current in BJT, 504, 507
control electrode, FET, see gate
convention
  algebraic-sign, see algebraic sign,
    convention for
  grounding, *PN* junction, 261, 266
    (ex.)
conventional current
  (in) common-emitter BJT, 489, 489
    (ex.)
  (in) relation to particle current, 20
Copernicus, N., 40
copper, 67
coulomb, 2, 4
Coulomb force, role in ionic scattering,
  156
Coulomb's law, 40, 41, *41*, 50, 124,
  149
coupled oscillators
  crystal atoms as, 148
  (in) electronic analogy, 102
covalent bond, 72
  bar-magnet analogy, 72
  breaking of, 106
  density of in silicon, 75
  hardness resulting from, 72
  relation to energy bands of, 105–106
crossover, capacitance, see capacitance
  crossover
cross section, scattering, see scattering
  cross section
crystal
  basis of, 67
  (as) distinguished from "lattice," 67,
    70, Table 1-3
  single, see single crystal
  (or) structure, 68–71
crystal direction, 73, 77
crystal planes, 73–77, *74, 75*
crystal structure, relating to band
  structure, *105*
crystalline quality of substrate, 562 (ex.)
crystallography, 57–77, *59*
cubic crystal, see single crystal, cubic
current, see listing for current of
  interest
current amplifier, BJT as, 515
current density
  can be defined at a point, 22
  components of
    (in) forward-biased junction, *277*
    (in) graded sample at equilibrium,
      167
    (in) *PN* junction at equilibrium,
      240
    (in) punchthrough diode, *326*
  conduction, 29, 30
  displacement, 29–31, 31 (ex.)
  expressions for, 161
  total, 30, 31 (ex.)
  typical value of, 162 (ex.)
current gain, common-emitter, see
  common-emitter current gain,
  dynamic and static
current limit, upper, see BJT, SPICE
  analysis
current patterns, in BJT, *499*

SUBJECT INDEX   I-13

current ratio, effect on junction switching, 405
current source, or regulator, BJT modeling and biasing, 512, 514, 606, 609
current-step stimulus of device, see device listing
current-voltage characteristics
  (of) "backward" diode, see "backward" diode
  base and collector currents versus voltage, 652
  (of) emitter junction, 511
  (of) forward diodes in series, 322
  (in) common-emitter-BJT output plane, see common-emitter configuration, output-plane current-voltage characteristics for
  (of) germanium junction, 287
  (of) high-low junction, 308
  (of) ideal rectifier, 284
  (of) MOSFET, 723–727
  PN junction, 272–282
    (versus) area and material, 281, 349
    silicon, 277, 278–280, 287–290, 289
    piecewise-linear, 283, 351
  (of) punchthrough diode, 322
  (of) tunnel versus conventional diode, 318, 318
curved regime, see MOSFET, operating regimes of
cutoff frequency, see device listing
cutoff frequency for pair generation, 109
cutoff regime of common emitter BJT output plane, 528

D-mode (depletion-mode) MOSFET
  differs from D-mode JFET, 730
  fabrication of, 729
  has two structure-determined electrical constants, 730
  (as) hybrid JFET-MOSFET device, 714, 729
  SPICE analysis of circuit using, 843
  structure of, 729, 729
  symbol for, 733
  transfer characteristic of, 729
"dangling bond," 760, 761, 762
DAR (depletion-approximation replacement), 259, 340–342, Tables 5-1, 5-2
  (applied to) MOS capacitor, 748–751, 776–780
  nonequilibrium applications, 347
  nontriviality of origin choice, 347
  normalized electric-field profile from, 336
  normalized electrostatic–potential profile from, 338

sign conventions in, 337
step junction at equilibrium, 259
Das, H., 585
De Broglie, L., 54
De Broglie wavelength, 54
  in quantum-well or superlattice device, 715
Deal, B. E., 827
"debiasing," emitter, 582, 617, 630, see also emitter-current crowding
  SPICE treatment of, 630
Debye, P., 331
Debye length, 330–333
  extrinsic, 331, 332
  general, 331, 332, 333
  intrinsic, 331
  (versus) majority-carrier density, 332
  relating to thermal velocity and relaxation time, 333
  screening connection, 330
Debye radius, 331
Declerck, G., 861
deep-depletion condition in MOS capacitance measurement, 794
deep state, 181, 184
defect, crystalline, see single crystal, defects in
defect current, emitter, 504, 507
definitions
  anode, 12
  cathode, 12
  conservative system, 11
  electric field, 2
  electron volt, 13
  electrostatic potential, 5
  feature size, 4
  line of force, 7
  micrometer, 4
  (of) MOSFET features, 716–718
  potential energy
    electrical, 13
    mechanical, 11
  total energy, 11
  unity factor, 13
deflection ("scattering"), carrier, comparing hole and electron, 151
degeneracy
  (of) energy levels, 51, 53
  (of) lattices, 59, 60
degenerate doping of silicon, 259, 493, 493 (ex.), 494, 494 (ex.)
demonstration model, see model, demonstration
density, see also atomic, carrier, charge, etc.
  (of) electrons in conduction band, 119, 120 (ex.)
  intrinsic carrier, see carrier density, intrinsic

(of) lines of force, 7, 8, 9
(of) states in a conduction band, 114–116, 117, 321
depletion approximation, 244–259
  (for) asymmetric step junction, 242–256, 255
  depletion-layer thickness well-defined in, 246
  electric-field profile in, 237, 246–249, 248, 255, 257, 264 (ex.), 265
  equations for step junctions, Table 3-1
  improves with reverse bias, 264
  (for) linearly graded junction, 295, 296, 297
  (for) MOS capacitor, 744–748
  (for) one-sided step junction, 256–258, 257
  PIN diode, 292, 295
  potential profile unaltered by bias, 266 (ex.)
  (for) symmetric step junctions, 244–253, 245, 265
depletion-approximation replacement, see DAR
depletion-inversion (polarity), 757, 760, 779
depletion layer, or region, see also space-charge region
  areal impurity density in channel crucial, 714
  (in) BJT, 489
  boundary of, in PN junction, 235
  collector-junction, see collector-junction depletion region
  emitter-junction, see emitter-junction depletion region
  (in) JFET, 714
  (versus) space-charge layer, 242
  thickness of, see depletion-layer thickness
depletion-layer capacitance, see device listing
depletion-layer thickness
  (at) breakdown voltage, 313, 313
  effect of bias on, 263, 268
  equations for step junctions, 251, 256, 258 (ex.), Table 3-1
  (for) linearly-graded junction, 298, 301
  symbol for, 246
  well-defined in depletion approximation, 246
detailed balance, principle of, 152, 153, 153 (ex.), 169
deuterium and deuteron, 51
diagram, loadline, see loadline diagram
diamond
  as semiconductor material, 110
  form of carbon, 110
  structure of, 71, 72, 104

die (IC), 76
dielectric constant, or relative dielectic permittivity, 26, 30
dielectric materials, 22–26, *24*
  layered, 27
dielectric permittivity, 22–26, 247
  differing values on two sides of an interface, rule for, 28
  (of) free space, value of, 8
  (in) MOS analysis, 725
  relative, *see* dielectric constant
  symbol and unit for, 23
dielectric relaxation time, 30–32, 365–372
  coexisting with electromagnetic wave, 376
  current–density waveforms in, 31 (ex.)
  relating Debye length and thermal velocity, 333
  (of) silicon, 31
  symbol for, 31
differential equations, linear, *see* linear differential equations
differentiation of integral with respect to upper limit, 7
diffused junction, 298–303, *304*, 312
  avalanche breakdown in, *see* PN junction, avalanche breakdown in
diffusion
  carrier, *see* carrier diffusion
  conditions necessary for, 163, 298
  impurity, *see* solid-phase diffusion
diffusion capacitance, *see* PN junction, diffusion capacitance of
diffusion coefficient or constant, *see* diffusivity
diffusion-enhancement factor in BJT analysis, 643
diffusion length, minority-carrier, *see* minority-carrier diffusion length
diffusion potential, origin of term, 251, *see also* contact potential
diffusion term, continuity equation, 191
diffusion velocity, 200
diffusivity
  algebraic sign of, 165
  carrier, 165
    ambipolar, 573, 573 (ex.)
    doubling of, in base region, 570, 584, 590, 641, *see also* Webster effect
    independent of carrier density, 166
  solid-phase, 299
dimensionless variables, 18
dimensions, RMKSA, *see* RMKSA units
diode
  "backward," *see* "backward" diode
  ideal, *see* ideal diode

junction, *see* PN junction
  long, 281, *354*, *357*
  PIN, *see* PIN diode
  punchthrough, *see* punchthrough diode
  short, *354*, *357*
  vacuum, *see* vacuum diode
diode-circuit solutions, Tables 3-4, 3-5
diode symbols, SPICE, Table 3-6
dipole layer, or double layer, 236
Dirac, P. A. M., 112
Dirac delta function, 119, 384
direct current (dc) as steady-state example, 153
direct gap, *see* band structure
direction
  crystal, *see* crystal direction
  (of) electric field, force, line, and potential gradient, 7
discrete device
  BJT as, 487, 667
  JFET as, 714
  MOSFET as, 740
dislocation, edge, 70
displacement, electric, *see* electric displacement
displacement current density, 28–30, 31 (ex.), 365–372
dissimilar materials, 250
dissipation, power, *see* power dissipation
distance, symbol and unit for, 4, 5
distinguishing lattice from crystal, Table 1-3
distribution of electrons in energy, *see* electron distribution in energy
distribution function for electric field, 312
divergenceless, *see* magnetic field
dodecahedron, 72
donor atom, 123–127
  hydrogen model of, 123–129, *124*
  origin of term, 125
  volumetric density of, 126, 127
donor doping, 123–127
donor state, 123–127, *124*
donorlike state, *see* state, donorlike
doping
  acceptor, *see* acceptor doping
  amphoteric, 766
  compensated-intrinsic, 132
  degenerate, *see* degenerate doping
  donor, *see* donor doping
  gold, 184, 766
  heavy, effect of, 144
  methods for, 126
  net, 132
  substrate, 562 (ex.)
  uniform, 127–129
    criterion for, 128

doping profile, *see* listing for region of interest
"dose," *see* ion implantation
double layer, *see* dipole layer
double-subscript notation, 260, 489, 515, 717, 744, 745
doublet fine structure, 53
doubling of base-region diffusivity, 570, 590, 593, 641, *see also* Webster effect
drain, MOSFET, definition of, 717
drift, carrier, *see* carrier drift
drift mobility, 157, 198
drift term, continuity equation, 191
drift transport, 152
  analogy for, 162 (ex.)
  thought experiment on, 153
drift velocity of carrier, 154, 158, *158*, 159, 162 (ex.)
drive-in, *see* solid-phase diffusion
driving function, arbitrary, applied to linear system, *386*
drop, IR, *see* IR drop
duality, wave-particle, *see* wave-particle duality
dummy variable, 6
dynamic current gain, see configuration of interest
dynamic model, device, *see* device listing
dynamic resistance, or output resistance, *see* device listing

E-D (enhancement-depletion) circuits, 733, 739, *739*
E-mode (enhancement-mode) MOSFET, 729, 732, 733
Early, J. M., 555, *555*, 558
Early effect, 555, 556
  absent in common-base configuration, 556
  analytic examination of, 556
  degrades reciprocity of BJT, 599 (ex.)
  described as voltage-dependent Gummel number, 599 (ex.)
  diminished in intrinsic-barrier and channel-collector BJTs, 558
  graphical explanation for, *555*
  (in) Gummel–Poon model, 591
  large in germanium PNP device, 558
  responsible for reverse voltage gain, 618
  SPICE analysis of, 648, 654
Early voltage, 556
Ebers, J. J., 599, 600, 604
Ebers-Moll model, 591–610
  applications of, 604–606
  assumptions for, 592–593

SUBJECT INDEX I-15

Ebers-Moll model *cont.*
  bias circuit for, *593, 594, 596*
  carrier profiles for, *594, 596, 601*
  common-base current gain in, 591
  current components in collector region, 602 (ex.)
  equivalent-circuit model for, 606–610, 608
  Gummel-Poon reformulation of, 591–592
  Gummel number in, *596*
  high-level version of, 592
  original-form equations, 600–604, 602, 603 (ex.)
  superposition in, *598*
  symmetry of, reflects reciprocity, 606
  transport form of, 591–600, 603 (ex.)
Ebner, G. C., 571
edge dislocation, 70
Edison, T., 324
Edison effect, 324
effective base thickness, *see* base thickness, effective, and similarly for similar entries
effects
  ambipolar, 571–577, 584
  Early, 555, 556, 591, 648, 654
  emitter-current crowding, 582, 591
  high-level, BJT, 566–590
  Kirk, 577–582, 584, 591, 592, 654
  Miller, 636
  photoelectric, 40
  Rittner, 566–568, 575, 584, 591
  Sah-Noyce-Shockley, 510, 591
  surface, 100
  Webster, 567, 568–571, 584, 591, 641
efficiency, base-transport, *see* base-transport efficiency
efficiency, emitter, *see* emitter efficiency
Einstein, A., 40, 45
Einstein relation, 167–169, 244
electric displacement, 26–28, 35 (ex.), 743
electric field
  carrier drift caused by, 147–169
  caused by fixed charges, 3
  closed-form expression for, 327, 335
  concept of, 1–4
  definition of, 2
  direction in capacitor, *4*, 7
  discontinuous at oxide-silicon interface, 743 (ex.)
  (versus) distance, DAR, *336*
  distribution of, *see* profile, electric field
  (effect on) drift velocity, *158*
  (in terms of) electrostatic potential, 7
  extrapolated, vanishes at spatial origin of DAR, 342

force caused by, 2
"fringing" of, 9, *10*
(in) graded sample, 167
(in) high-level BJT operation, 566 (ex.), 568
(in) high-level Haynes-Shockley experiment, 576 (ex.), 577 (ex.)
(in) hydrogen atom, 41
internal in dipole layer, 236
longitudinal, in BJT base region, 566, 568, 569 (ex.), 570, 571, 573
magnitude and direction of represented by lines of force, 7–9, *9, 10*
near nucleus, visualized using lines of force, 36
normalized, from DAR, *336*
ohmic component of, in BJT base region, 572, 572 (ex.)
(in) *PN* junction, 235, *237*, 255 (ex.)
profile of, see profile, electric field
(in) punchthrough diode, 321 (ex.)
relation of magnitude to line-of-force density, 7
sign of, 4, *4*, 34
(in) silicon versus in metal, 252 (ex.)
source of, 1, 32
symbol for, 2
tangential component of continuous, 28
time-varying, 39
typical value of in ohmic conduction, 162 (ex.)
uniform, 1, 2, 8, *9*
units for, 3
universal curve of, 336
value range for ohmic conduction in silicon, 154
vanishing of defines sharp boundary in forward-biased junction, 269, 271
(as a) vector quantity, 1
electrical charge, *see* charge, electrical
electrical contact, perfect, 21, *see also*, ohmic contact
electrical current, symbol and unit for, 20
electrical noise, 714, 753
electrical work and energy, 4, 5
electricity and magnetism, integrated with optics, 39
electrochemical potential, 535–539
electrode, control, FET, *see* gate
electromagnetic radiation, 37–40
  pair generation by, 109
  cutoff frequency for, 109
electromagnetic wave equations, 37
electromagnetic waves, 37, 39, 367, 374, 375 (ex.), 376, 376 (ex.)

electron
  bound, 49, 106
  conduction, 106–109, *118*, 120, 125, 126
  density of in conduction band, 119, 126
  diagrammatic representation of in semiconductor, 109 (ex.)
  discovered by Thomson, 36
  energy of in hydrogen atom, 46
  "fifth," 123
  "hot," 158
  (in) hydrogen atom, *42*
  outer-shell, 57, 72
  particle treatment of, *11*, 152, 314
  quantum mechanical treatment of, 148, 314
  trapped in energy well, 43
  valence, 106, 108
electron affinity, 751, 752, *752*
electron-beam evaporator, 769
electron distribution in energy, 111, 112
electron energy, using arrow to represent, *145*
electron volt, 12, 13
electronic charge, 2, 13
electronic states in atom, *56*
electronics, solid-state, *see* solid-state electronics
electrostatic potential, 5–7
  absolute, 6
  band diagram in terms of, 143–144
  difference in, *see* potential difference
  direction of, 7
  (in terms of) electric field, 7
  (in) electrochemical potential, 536
  gradient of, 7, 147
  higher and lower, 7
  (in) hydrogen atom, 42
  increment of, 5
  map of for junction, 239
  (in) *PN* junction, *236, 237*
  profile of, *see* profile, electrostatic potential
  (in) sample experiencing high-level disturbance, 538 (ex.)
  unit for, 5
  using arrow to represent, *145*
electrostatic shield, interfacial states as, 764
emission, stimulated, 48
emitter, 485, 562
emitter contact, *502*
emitter-current crowding, 582, 591, 630
emitter efficiency, 503, 561, 564
emitter-follower (common-collector) configuration, 529, 531, *531*, 532, 532 (ex.)

emitter junction, 486
  area of, 492
  asymmetric injection by, 491
  biasing of, *511, 512, 513*
  breakdown of, *see* BJT, breakdown phenomena and breakdown voltage
  conductance of, 617
  "debiasing" of, *see* "debiasing"
  depletion region of, dominant currents in, 495
  dynamic resistance of, 617
  efficiency of, *see* emitter efficiency
  (as) general-purpose diode, 548
  injection by in BJT greater than in isolated junction, 492, 492 (ex.)
  loadline diagram for, *513*
  normally forward-biased, 488
  offset voltage of, 510, 511, *511*
  perfect, *see* perfect emitter
  preferential injection by, *see* emitter-current crowding
  represented by arrow in BJT symbol, 486
emitter-junction depletion region, dominant current component in, 495
emitter region
  conventional approximations fail in, 493
  dominant current component in, 495 (ex.)
  empirical treatment of, 494
emitter terminal, 485, 486
  stripe as, 502
emitter time constant, 634
emitter width, in modern BJT, 563
energy
  binding, *see* binding energy
  definition of, 5
  (of) electron in hydrogen atom, 46, 50, 52
  ionization, *see* ionization energy
  kinetic, *see* kinetic energy
  mean, of electron in conduction band, *see* mean energy
  potential, *see* potential energy
  symbol and unit for, 5
  total, *see* total energy
energy and work, electrical, 486–487
energy band, 100–112, *103, 105*
energy-band structure, *see* band structure
energy gap, 109–110, Table 2-1
  narrowing of, 144
  (in) pair production by impact, 309
  relationship to junction offset voltage and saturation current, 280
energy levels, hydrogen atom, *47*
energy-level splitting, 101

energy-transport method, 159
energy well, 43, *43,* 106, *107,* 125, *126,* 131, *131,* 314
environment, of point in lattice and crystal, 69
epitaxial BJT, *see* BJT, epitaxial
epitaxial growth, 299, 714
EPROM (erasable-programmable read-only memory), 773
equation, *see* adjective for equation
equilibrium, 152, *177,* 202, *202,* 203, 717
equilibrium nonneutrality, 344 (ex.), *see also* surplus carrier
equipotential region, of PN-junction sample, *see* PN junction
equivalent-circuit model, 282
  BJT dynamic
    admittance, 625, *626*
    h-parameter, 622, *623, 624*
    hybrid, low-frequency, 615, *616, 618*
    hybrid-pi (Giacoletto), 620, *621, 624, 629*
  BJT static, 513–515, *514, 516*
    (for) cutoff, 529 (ex.)
    (for) deep saturation, 523
    Ebers–Moll, 608
    SPICE, *see* BJT, SPICE analysis of components used in, 282
    definition of, 282, 513
  Ebers-Moll, 606–610
  (of) MOS capacitor, small-signal, 741, 783–789, *785*
  MOSFET, dynamic, *827, 837, 839*
  PN junction
    ideal rectifier, *283, 284*
    piecewise-linear, *283,* 284, 351, *351,* 352, 352 (ex.), 353 (ex.)
    small-signal, 372–383, *373, 379, 381, 381,* 396, 785
equivalent density of states, 121, 122, 142, Table 2-2
equivalent-density-of-states approximation, 117–121, 141
erasable-programmable read-only memory (EPROM), 773
error function (erf), 299
error-function complement (erfc), 299, 301, *302, 303, 305*
Esaki (tunnel) diode, 317
etchants, 75
Euler's circles, relating steady-state and nonequilibrium conditions, *202*
eutectic phase, in germanium BJT fabrication, 557
Evans, J., 861
evaporator, 768, 769

excess carrier, 169–175, *173, 180*
  definition and sign of, 170
  density of, 187
  profile of during switching, *407*
  storage of in diode, *354, 357*
  versus surplus carrier, 344 (ex.)
excess current, tunnel diode, 318, *318*
excited atom, 48
exclusion principle, Pauli, *see* Pauli exclusion principle
explicit algebraic sign, 13, 515, 632
exponential function (exp), 274, 286, 301, *302*
extended-base zone (or region), 579
extrema, in electric-field profile, forward-biased junction, 271
extrinsic carrier density, *see* carrier density, extrinsic
extrinsic portion of BJT, *see* BJT, extrinsic portion of
extrinsic silicon, definition of, 138

fabrication processes, 485, 535, 556, 557
face-centered cubic, *see* space lattice, face-centered cubic
factor, unity, *see* unity factor
family of characteristics
  BJT output, 517, 518
  MOSFET output, 723, *724, 730, 843*
fast surface state, *see* interface trapped charge
feature size, definition of, 486
Fermi, E., 112
Fermi-Dirac probability function, 115, 116, *117,* 137–141, *139*
Fermi level, 112–114
  constant in MOS capacitor, 753
  doping dependence of, 136
  (for) electrons, 536, *see also* quasi Fermi level
  (for) holes, 536, *see also* quasi Fermi level
  (in) intrinsic silicon, *117,* 118
  (in) junction at equilibrium, 239
  position of in intrinsic silicon, 116
  (in) PN junction, *238*
  position in doped sample, using "computer," 133–137, *134–136*
  quasi, *see* quasi Fermi level
  (as) reference (or zero) for electrostatic potential, 239, 535
  splitting of, 539 (ex.)
  symbol for, 144
  temperature dependence of, *115,* 116, 116 (ex.), 136
  true location of, 142
  water analogy for, 112, *113,* 152

SUBJECT INDEX   I-17

Fermi-level "computer," 133–137, *134–136*
Fermi-level difference measured by voltmeter, 252, 757
FET (field-effect transistor), 713–716, 726
  heterojunction-insulated-gate, *see* HIGFET
  junction, *see* JFET
  metal-oxide-silicon, *see* MOSFET
  metal-semiconductor, *see* MESFET
  modulation-doped, *see* MODFET
  MOS, *see* MOSFET
  (incorporating) Schottky junction, 715, *see also* MESFET
  two-dimensional-electron gas, *see* MODFET
Fichtner, W., 861
Fick's first law, 165, 166
field, electric, *see* electric field
field-effect transistor, *see* FET
field plate of MOS device, 716, 717, 753, 756
figure of merit, *see* device listing
filament evaporator, 768
fine structure, doublet, 53
fine-structure constant, 52
finite-difference model, 389
fixed charge, *see* charge, electrical, fixed
fixed oxide charge, *see* oxide charge, fixed
flatband condition, 751, 753, 796, 798
flatband voltage, 327, 327 (ex.), 717
Fletcher, N. H., 291, 292, 566, 567
fluorescence, 48
flux density, magnetic, *see* magnetic induction
forbidden band, *see* energy gap
force, electrical, 40, 41, *41*
  lines of, *see* lines of force
  symbol and unit for, 2, 4, 5
Fordemwalt, J. N., *101, 124*
forward-active regime of common-emitter BJT output plane, 518
forward-saturation regime of BJT output plane, *see* saturation regime
Fourier-transform method, 389
Foy, P., 713
Frankenheim, M. L., 63
free carriers, absent in ideal insulator, 23
free path, mean, *see* mean free path
free space or vacuum, 23, 37
free-space permittivity, *see* permittivity of free space
frequency, characteristic, *see* characteristic frequency
"fringing" by lines of force, 9

function, work, *see* work function

gain mechanism, BJT, 507–510
gallium arsenide, 76, 149, 184, 280
gallium phosphide, 76
gap, energy, *see* energy gap
garage analogy, 109
gate, 713, 716, 828, *see also* field plate
Gauss's law, 743 (ex.), 777
Gaussian function, 300, 301, *302, 303, 305*
Geiger, P., 36
gem, as example of single crystal, 57
general solution, *see also* DAR, 336
generation, carrier, *see* carrier generation
generation center, *see* recombination center
generation-recombination mechanism, in real silicon junctions, 287
germanium, 72
germanium PNP bipolar junction transistor, *see* BJT
Getreu, I. E., 592
Ghausi, M. G., 622
Giacoletto model, *see* hybrid-pi model
Gibbs, J. W., 760
glass, *see* silicon dioxide
Goetzberger, A., 791, 793
gold, 67
gold doping, 184, 766
Goudsmit, A., 52
grade constant, *see* linearly graded junction, gradient of
graded junction, *see* linearly graded junction
gradients that cause carrier transport, *see* carrier transport
grading coefficient, BJT, SPICE, 652
gradual-channel approximation, *see* long-channel approximation
grain boundary, 77
Gray, P. E., 571
gray tin, 72
grossly asymmetric step junction, *see* step junction, one-sided
ground state, 48
ground symbol, 12
grounded-base configuration, *see* common-base configuration, etc.
group theory, 59
Grove, A. S., 793
growth, epitaxial, *see* epitaxial growth
Grung, B. L., *103, 130*
Gummel, H. K. 544, 591, 641
Gummel number, 544–546, 548, 548 (ex.), 595, 599, 599 (ex.), 641, 641 (ex.)
Gummel plot, 544, 584
Gummel-Poon model, 591–592

gyrator, 599

h-parameter model, *see* BJT, h-parameter model
Hall, R. N., 185
Hall mobility, 157, 198
harmonic oscillators, 45
Haynes-Shockley experiment, 195–198, *196*, 197 (ex.) 576, 576 (ex.), 577 (ex.)
helium, Bohr theory of, 50
HEMT (high-electron-mobility transistor), *see* MODFET
Hertz, H. R., 39
Hertzian waves, 39
heterojunction, definition of, 715
heterojunction-insulated-gate field-effect transistor, *see* HIGFET
heuristic calibration of axes, 491, 518
hexagonal, close-packed, *see* close-packed hexagonal
hexahedron, 72
HIGFET (heterojunction-insulated-gate field-effect transistor), 715
high-electron-mobility transistor (HEMT), *see* MODFET
high-frequency limitation on use of drift-velocity concept, 159
high-level conditions, 290, 537 (ex.), *see also* device listing
high-level disturbance of equilibrium, excess-carrier density, 174
high-level effects, BJT, 556–590
high-level effects, in junction, 288, 290–292, *366*
high-low junction, 201, 305–308, *306*, 345 (ex.)
hole (in the valence-band), 106–109, 130, 131
homogeneous material, 20, 27
homologous devices, *see* complementary circuits
"hot" carrier, 158, 309, 753
Huang, J. S. T., 585
Hückel, E., 331
hybrid, meaning of term, 613 (ex.)
hybrid equations, *see* BJT, hybrid equations
hybrid JFET-MOSFET device, *see* D-mode MOSFET
hybrid-parameter model, *see* BJT, hybrid-parameter model
hybrid parameters, *see* BJT, hybrid parameters
hybrid-pi model, *see* BJT, hybrid-pi model
hydrogen atom, *47*, 52, 57, *see also* Bohr model
hydrogen ion, 44

hydrogen model, 123–127, *124, 130, 131*
hyperbolic functions, used in BJT admittance model, 627, 628
hypothesis, Avogradro's, *see* Avogradro's hypothesis

I-V characteristics, *see* current-voltage characteristics
IC, *see* integrated circuit
icosahedron, 72, 77
ideal, has two meanings in device work, 534
ideal diode (or rectifier), 283, 284, *351*
idealized, 535
Ihantola, H. K. J., 813
Ihantola-Moll (MOSFET) model, 813, 823
image charge, definition of, 751
impact ionization, 309
implantation, ion, *see* ion implantation
imposed junction voltage, 568, 570, 645
impulse, role of in ionic scattering, 157
impurity atom
  acceptor, *130*, 131
  areal density of, 714
  column-III, 123, 129–131
  column-V, 123–127
  donor, 123–127, *124*
  interstitial or "out of place," 181
  introduction of into silicon, 126
  ionization energy of in silicon, Table 2-3
  isolated, 123
  N-type and P-type, 123
  substitutional, 184
impurity compensation, 132–133, 150
impurity-doped silicon, *see* silicon, impurity-doped
impurity ion, *see* ion, impurity
imref, 538, 554, *see also* quasi Fermi levels
incremental carrier-recombination rate, *see* carrier recombination rate, incremental
indirect gap, *see* band structure
indium, used in germanium-BJT fabrication, 557
indium antimonide, 76
infrared radiation, 149, 170, *171, 176, 177*, 198 (ex.)
injection, carrier, *see* carrier injection
input loop, BJT, 518
input port, common-emitter configuration, *see* common-emitter configuration, input port of
insulator-silicon interface, *see* oxide-silicon interface

insulators, 20, 110
integral, differentiation of, 7
integrated circuit (IC), 487, 714, 716, 717
integrated-injection logic (I²L), 563, 565
integration, smoothing effect of, 250
intercept current, 544, 591, 595 (ex.), 596 (ex.), 608 (ex.)
interface trapped charge, 760–766, *762, 763, 765*, 766 (ex.), 774 (ex.), *798, 799*
interfacial charge, 760–766, *see also* interface trapped charge
internal current patterns, BJT, 499–502
interpenetrating lattices in illustrative crystals, 68, 69
interstitial atom, 70
intrinsic, compensated, 132
intrinsic (midgap) level, as indicator of electrostatic potential, 535
intrinsic-barrier BJT, *see* BJT, intrinsic-barrier device
intrinsic carrier density, *see* carrier density, intrinsic
intrinsic portion of BJT, *see* BJT, intrinsic portion of
intrinsic silicon, 111, 112, 116, *117, 118*, 121
inversion of strong, weak, inversion, *see* threshold
inversion layer, 342–347, 714–717
inversion phenomenon, causes departures from universal DAR curve, 337
inverter, *see* device listing
ion
  acceptor, 130
  donor, 125
  hydrogen, 44, 48
  impurity, 147–152, *156*, 235
  screening of, 150
ion implantation, 299, 300, 562, 714, 717, 729, 739
ionic charge, *see* ion, impurity
ionic crystal, 68
ionization, 44, 48
  impact, *see* impact ionization
  (of) impurities in silicon, 124, Table 2-3
ionization potential, 49
ionization rate, in avalanche breakdown, 310
IR drop, 259, *260*, 269, 271, 718, 721
Irvin, J. C., 160, *161*
isolated atom, 103, 123
isolated junctions, BJT examination in terms of, 489
isothermal analysis, 665
isotropic etchant, 75
isotropic material, 27, 28

JFET (junction field-effect transistor), 485, 558, 713, 714, 717, 731, 731 (ex.), *see also* FET
Johnson, E. O., 862, 865
Joos, G., 50
joule, 5
junction, *see also* PN junction
  alloyed, 557
  biasing of, *see* PN junction or high-low junction
  electric field and electrostatic potential in, *237, 248, 257*
  law of, *see* law of the junction
  other than step, 292–308
  space charge at, 233–236
  symbol for, 486
  thought experiment on formation of, 233, *234*
junction boundaries, importance of in BJT, 485
junction concepts, 233–244
junction diode, effective lifetime in, 665
junction field-effect transistor, *see* JFET
junction surface, 233, 236
junction transistor, *see* BJT

Kepler, J., 40
Kerr, D. R., 793
kinetic energy, 1, 9–13, 43
Kirchhoff's current law, in Ebers-Moll model, 593, 596
Kirk, C. T., Jr., 578, 579
Kirk effect, 577–582, 584, 591, 592, 654
knee current, *see* BJT, SPICE analysis

Laplace-transform method, 389
large-signal analysis, *see* device listing
laser, 48
lattice, 58–61
  Bravais, *see* space lattice
  (as distinguished from) "crystal," 67–70, Table 1-3
  space, *see* space lattice
  three-dimensional, *see* space lattice
  two-dimensional, 58, *58*
    centered-rectangular, 59
    "centered-square," 61 (ex.)
    hexagonal, 60
    oblique, 61
    particular cases, *60*
    primitive cell of, 62, 63 (ex.)
    rectangular, 59
    rhombic, *see* two-dimensional lattice, centered-rectangular
    square, 44
    unit cell of, definition of, 61
lattice constant, 69, 71

SUBJECT INDEX   I-19

"lattice scattering," *see* phonon, scattering of carriers by
Lau, L., 822
law, Fick's first, *see* Fick's first law
law of mass action, 141–143, 717
law of the junction, 271–272, 276, 278
layer, depletion, *see* depletion layer
layered dielectric material, 27
leakage current
  collector, 504, 534
  collector-emitter, 608 (ex.), 609 (ex.)
  (in) Ebers-Moll model, 595
  PN-junction, typical value in silicon case, 263
length, diffusion, *see* diffusion length
lifetime, carrier, *see* carrier lifetime
light, 39, 45, 46
light amplification, 48
limit concept, in electric-field definition, 2
Lindmayer, J., 585
line defect, 70
line spectra, series of, 45, 47, *47*
linear, piecewise, *see* piecewise linear
linear current amplifier, 508, 508 (ex.), 509
linear differential equations, 385–387
linear versus logarithmic data presentation, *187*, 188, 239
linear system, with arbitrary driving-force function, *386*
linearity, for superposition, 597, 599
linearization, 347
  quantitative examination of, 352 (ex.), 353 (ex.)
linearly graded junction, 294–298, *295–297*, 301, 319
lines of force, 7–9, *9*
  crowding of, 313, 491
  density of, interpretation of, 7, 8, *9*
  direction of, 7, *9*, 32
  "fringing" of, 9, *10*
  represents both magnitude and direction of electric field, 9
  (in) uniform electric field, 8, *9*
  (applied to) visualizing electric field near nucleus, 36
Linvill model, 389
lithium, 50, 57
loadline diagram, 512–513
  (to) analyze rudimentary amplifier, 517
  (useful for) analyzing grossly nonlinear devices, 518
  (can be used for) three-terminal device, 512, 518
  characteristic rotation in construction of, 512, 518

(for) combinations of more than two components, 512
(for) common-base BJT, 534
(for) common-emitter BJT, 513
(for) diode and resistor, 512
(for) MOSFET, *837, 843, 850*
lock-layer BJT, *see* BJT, channel-collector
locus for versus equation for, 726 (ex.)
locus of pinch-off points in MOSFET output plane, 726, *727*
logarithmic versus linear data presentation, 188, 239
logic, integrated-injection, *see* integrated-injection logic
long-channel approximation, 722, 807, 813
long diode, *see* diode, long
long-range order, 77
longitudinal electric field in BJT base region, 566, 568, 569 (ex.), 570, 571, 573, 585
loop, input, *see* input loop
loop, output, *see* output loop
low-emitter-concentration (LEC) BJT, 561, 564
low-level disturbance of equilibrium, 172, 268, 291
Lyman, 47, *47*

magnesium, 67
magnetic field, 37, 39
magnetic induction, 37
magnetic moment, 53
magnetic permeability of free space, 37
majority carrier, calculating density of, 132
Many, A., 760
Marsden, E., 36
mass, reduced, *see* reduced mass
mass action, law of, *see* law of mass action
Massobrio, G., 630
mathematical point, *see* point, mathematical
Maxwell, J. C., 29, 30, 37, 39
Maxwell's equations, 30, 37, 44
Maxwellian energy distribution, 169
MBE (molecular-beam epitaxy), 292, 299
McKelvey, J. P., 169, 201
mean energy
  of electron in metal conduction band, 116 (ex.), 120, 159 (ex.)
mean free path, carrier, 152, 725
measurement, direct, of contact potential, *see* contact potential
mechanism, recombination, *see* carrier recombination

mercadtel (mercury-cadmium-telluride), 198 (ex.)
merged devices, 558, 559
merit, figure of, *see* device listing
MESFET (metal-semiconductor field-effect transistor), 715
metal, 110–111, 159 (ex.)
metal-oxide-silicon (or -semiconductor) capacitor, *see* MOS capacitor
metal-semiconductor field-effect transistor, *see* MESFET
metal-semiconductor junction, *see* Schottky junction
metallurgical junction, 236, 246
meter, 4, 5
"microchip," *see* integrated circuit
microelectronics, term validation, 4
micrometer, definition of, 4
"micron," *see* micrometer
microwave switch, PIN diode as, 294
midgap level, *see* intrinsic level
Miller, S. L., 310
Miller, W. H., 73
Miller effect, 636
Miller index, 73, 74, 74 (ex.), *74*, 76 (ex.)
Min, H. S., 571
minority carrier, 132, 141, 194 (ex.), 197 (ex.), 379
minority-carrier diffusion length, *192*, 193, 274, 275 (ex.)
Misawa, T., 292, 566
mobile oxide charge, *see* oxide charge, mobile
mobility, carrier, *156*, 157, 715, 826, *see also* bulk mobility, conductivity mobility, and drift mobility
mobility fluctuations, 774
model, *see also* device listing
  definitions and examples, 281–282
  demonstration, 133–137, 148
  distributed, 378
  Ebers-Moll, *see* Ebers-Moll model
  equivalent-circuit, *see* equivalent-circuit model
  hydrogen, *see* hydrogen model
  piecewise-linear, *see* piecewise-linear model
  static, 282, *see also* device listing
  transmission-line, 378, *379, 381*
modes, normal, *see* normal modes
MODFET (modulation-doped field-effect transistor), 715
modulation, conductivity, *see* conductivity modulation
modulation-doped field-effect transistor, *see* MODFET
molecular-beam epitaxy, *see* MBE

I-20 SUBJECT INDEX

molecular process, 152, 169, 171
Moll, J. L., 310, 540, 599, 600, 604, 813
Moll-Ross treatment of nonuniformly doped base region, 539–544, 566 (ex.)
momentum, angular, 44, 46, 50, 52, *54*
monocrystalline 3-D integrated circuit, 714
Monte Carlo technique, 159
MOS capacitor, 741–776
  avalanche-breakdown limit in deep-depletion observation, 794
  band-diagram, behavior of under bias, 746, *750*, *753*, *757*
  capacitance crossover in, 799–806
  charge profile in, *801*
  charge storage in, *792*
  comparing MOS and parallel-plate capacitors, 742, *742*
  comparing MOS-capacitor and step-junction capacitances, 780–781
  inversion-layer capacitances differ grossly in, *787*
  small-signal equivalent circuits for, 783–789
  comparing MOS-capacitor (or surface) and step-junction problems, 776–782, *778*, *782*
  cross section of, under bias, 744, *746*
  deep-depletion condition in, 794
  depletion-approximation analysis of, 741, 744–748
  differential capacitance of, 790
  distinction from parallel-plate capacitor, 742, *742*
  effect of oxide fixed charge on, 768
  ideal, 751
  modeling of, 776–806
  nonideal effects in, 751–776
  one-dimensional treatment valid, 712, 806
  poses quasiequilibrium problem, 776
  (as) "sandwich," 712, 714
  silicon portion of, problem identical to that of step junction, 741
  substrate at (quasi) equilibrium, 717
  voltage dependence of, 790–799
MOS technology, 751
MOSFET, 712–868
  Baccarani model of, 823
  basic theory of, 713–741
  bilateral symmetry of, 717
  body effect in, *see* body effect
  Brews (charge-sheet) model of, 823
  circuit for large-signal analysis, *842*
  circuit for small-signal analysis, *836*
  combined with BJT, *see* BiCMOS
  comparing performance with that of BJT, 850–868

complementary circuits using, *see* CMOS
common-source output resistance, 726, 818, 836
computer design of circuits using, 735
cross-sectional diagram of, *718*, *722*
D-mode (depletion-mode), *see* D-mode MOSFET
Dang model of, 823
definitions for, 716–718
diode-connected, *see* saturated load
division of drain-source voltage, 812 (ex.)
dominant device, 484, 713, 715, 736, 850
drain of, *see* drain, MOSFET
E-D (enhancement-depletion) circuits using, 735
E-mode (enhancement-mode), *see* E-mode MOSFET
early, data on, *728*, *735*
fabrication of, 716–717
feature-size shrinkage has led to three-dimensional problem, 807
gain versus frequency, *834*, *835*, *841*
gate capacitance versus gate-source voltage, *845*
high-low source junction, representation of, *863*
idealized, *808*, *809*, *811*
Ihantola-Moll model of, 813, 823
improved theory of, 806–824
inventors of, 714
inversion layer in, 714
inverter options, 736–741
long-channel approximation, *q.v.*
maximum g_m/i_{out} for, 857–859
merging of depletion and inversion layers at source, *864*
N-channel, *see* NMOS
operating regimes of, 726, 818, 823
operation of depends on surface phenomena, 100, 712
output plane of, *724*, *730*
  locus of pinch-off points in, 726
  experimental, *826*
  ionic-charge model, *817*, *832*
  small-signal model, *837*
  simple theory, *724*, *730*
  SPICE, *843*, *850*
P-channel, *see* PMOS
Pao-Sah model of, 823
parasitic capacitance in, 807
phase versus frequency, *841*
physics of subthreshold transconductance, 860–868
PMOS, *q.v.*
poses nonequilibrium problem, 807

rudimentary analysis of, 718–725, 725 (ex.), 727, 735, 741, 816
saturated, as load device, *see* saturated load and E-D circuits
saturation-regime characteristics of predicted by Pao-Sah model, 823
SPICE analysis of, 824–857
  large-signal applications, 840–857
  Level-1 (rudimentary) model, 816, 817, 824
  Level-2 (ionic-charge) model, 816, 824–832
  Level-3 (empirical) model, 824
  small-signal applications, 833–839
(as) square-law device, 727, 729
structure of, 716–717, *716*, *825*
substrate of, *see* substrate, MOSFET
subthreshold transconductance theory of, 854–857, *866*, *867*
symbols for, 733, *733*
transient response of, *840*, *846*, *848*, *849*
transfer characteristic
  current-to-voltage, 727–734, *727*, *728*, *730*, *732*, *833*
  voltage-to-voltage, *844*
two-terminal, *see* saturated load or E-D circuits
MOSFET-BJT performance comparisons, 850–868
multiple collectors in I²L, 564
multiplication factor, avalanche, *see* avalanche multiplication factor

N-type impurity atom, *see* impurity atom, N-type
nearest neighbor, 72
net motion of carriers, *see* carrier transport
neutral nonequilibrium, 344 (ex.), *see also* excess carrier
neutrality, 138, 170, 291
newton, 4
Newton, I., 40
Newton-Raphson algorithm, 431, 432
Nguyen, T. V., 114
nickel, 67
NMOS (N-channel metal-oxide-silicon), 733, 734
noise, electrical, *see* electrical noise
noise margin, MOSFET, 844
non-step junctions, 292–308
nondegenerate system and states, 52, 53
nonequilibrium, 147, 152, 202, *202*, 203
nonionized rather than "unionized," 125
nonohmic effects, 159 (ex.)
nonreciprocal device (gyrator), 599
nonuniform doping
  in base region, 539–544

normal modes, 101, 148
normalization, 18–20
  (of) electric field, DAR, *336*
  (of) electrostatic potential, DAR, *334, 338*
  example problem using, *16,* 18
  (of) MOSFET transfer characteristics, 731
  permits problem solution independent of units, 19
  simplifying effect of, *333*
  sometimes obscures physics, 19
NPN BJT, *see* BJT, intrinsic-barrier
NPN-PNP combination, *see* complementary circuits
NPN silicon BJT, *see* BJT, silicon NPN
nuclear theory of Rutherford, 35
number
  atomic, 50
  Gummel, *see* Gummel number
numerical solution, 301, 327, 328
Nussbaum, A., 59, 288

occupancy probability of state by electron, 111, 114
occupancy probability of state by hole, 121
octahedron, 72
offset voltage, 276, *281,* 308, 510, 511
Ohm's law, 22, 159
  departures from, 158–160, 159 (ex.)
  (in) drift transport, 156
  expression for, cases of extrinsic semiconductors, 161
  (in relation to) silicon, 147
Ohm's law convention, algebraic sign in, 259, *260,* 723, 745
ohmic base resistance, 501, 534, 582, 630
ohmic component of electric field in BJT base region, 572, 572 (ex.), 574 (ex.)
ohmic conduction, 147, 154, 156
ohmic contact, 160, 200–202, 305–308
  (to) MOSFET channel, 717
  solid-phase diffusion for fabrication of, 298
ohms per square, *see* sheet resistance
ON voltage, biasing consideration of, 510
one-dimensional problems, 18
one-over-f noise, 753
"one-plated" capacitor, *34*
one-sided step junction, *see* step junction, one-sided
optics, integrated with electricity and magnetism, 39
orbit
  circular, 36, 40, *42,* 44, 46, 51, *51*
  elliptical, 44, 51, *51*

planetary, 40
precession of, 52, *52*
stationary, *see* stationary state
orbital-angular-momentum vector, projection of, *54*
orbital surfaces or "orbitals," 55
order, long-range, 77
oscillator
  coupled, *101,* 102, 148
  Planck's, 45
oscillator analogy, 101, *101,* 102
outdiffusion, *see* solid-phase diffusion
outer-shell electrons, 57, 72
output characteristics, *see* output plane
output conductance, *see* device listing
output loop of rudimentary BJT amplifier, 518
output plane, *see* listing for configuration
output resistance, dynamic, *see* dynamic output resistance
oxide, silicon, *see* silicon dioxide
oxide capacitance, 721, 725
oxide charge, 766–774
  fixed, 766
  mobile, 722, *770, 771*
  trapped, 772, 772 (ex.), 773, 773 (ex.)
oxide-filled trench, 563
oxide fixed charge, *see* oxide charge, fixed
oxide-growth technology, 714, 767
oxide-metal interface, 752
oxide mobile charge, *see* oxide charge, mobile
oxide-silicon interface, 712, 714, 742, 743, 752, 767
oxide-silicon system, 712
oxygen, 50, 712

P-channel technology, *see* PMOS
P-type impurity atom, *see* impurity atom, P-type
pair, generation of, 109, 175, *see also* carrier generation
Pao, H. C., 823
Pao-Sah model, 823, 861
paradox, depletion-layer capacitance, 362
parallel-plate capacitor
  areal charge on plate of, 8 (ex.)
  comparing to depletion-layer capacitance, 358
  criterion for uniform field in, 2
  dipole (double) layer in, 236
  distinction from MOS capacitor, 742, *742*
  expression for capacitance of, 8
  (to) illustrate systematic problem

solving, 15 (ex.), *16*
incorporating dielectric material, 25, 27
(with) layered dielectric material, 27
line-of-force "fringing" in, 9
lines of force in, 7
MOS capacitor as, 714, 721, 727, *742*
"one-plated," *34*
(used to) relate displacement and conduction currents, 24
(used in) visualizing electric field, 2, 9
parameter, 281, 517, 724
parasitic capacitance, 737, 737 (ex.)
parasitic properties, BJT, *see* BJT, parasitic properties of
Paschen series, 47, *47*
Pauli exclusion principle, 55
Pearson, G. L., 156, *156*
Penney, W. M., 822
perfect emitter, 504
periodic table, Table 2-3
  (in relation to) internal shell structure, 57
permeability, *see* magnetic permeability
permittivity, dielectric, *see* dielectric permittivity
perturbation of uniform field by introducing charge, 2
phase shift in BJT model, 628, 630
phonon, 147–152, 156, *156,* Table 2-4
phosphorescence, 48
phosphorus, 125, 149
photoelectric effect, 40, 45
photoelectric threshold, definition of, 751, *752*
photon, 46, 48, 149
photoresist technology, 303
physical constants, Table 1-2
piecewise-linear model, 207, 284, 351, 518, 606
PIN diode, 292–294, *293,* 319
pinch-off current, and voltage, 730
"planar" junction, confusion in terminology, 550
planar process, 550
Planck, M. K. E., 35, 40, 45, 46
Planck's constant, 40, 53, Table 1-2
plane
  atomic, *see* crystal plane
  crystal, *see* crystal plane
  output, *see* output plane
plane junctions, achieved by alloying on (111) face, 557
planetary analogy of Rutherford and Bohr, 35–36, 40, 55
plate of vacuum diode, *see* vacuum diode

platinum, 67
Platonic solids, 72, 77
PMOS (*P*-channel metal-oxide-silicon), 733, 734, 760
PN junction, 233–438, Chapter 3
  admittance of, Tables 3-2, 3-3
  advanced dynamic analysis of, 387–438
    3-dB bandwidth, 421
    analytic-technique survey for, 387–390
    capacitance interaction in, 404
    circuit-behavior charge-control analysis, 394–405
    circuit-behavior exact analysis, 418–424
    current and voltage reversal in, 402
    current-step stimulus of, *384*
    device-physics charge-control analysis, 390–394
    device-physics exact analysis, 406–418
    Fourier- and Laplace-transform methods, 389
    frequency-domain response, 389
    large-signal diode-circuit solutions, Table 3-5
    large-signal transient response, *400, 403, 407, 408, 410, 411, 413, 420,* 427, *435, 436, 437*
    models employed in, 388, 389
    Newton-Raphson algorithm in, 431, 432
    numerical example, 427–438, *427, 429, 430, 435, 436, 437*
    small-signal complex admittance, 419
    SPICE analysis, 424–427, Tables 3-6, 3-7
    storage time versus current ratio, *405, 412*
    techniques useful for, *388, 389*
    time-domain response, 389
    trapezoidal formula in, 431
  alternate description of diffusion in, 261
  avalanche breakdown in, 309–314, *311, 312, 313,* 326, 327
  "backward," *see* "backward" diode
  band diagram for, 237–239, *238*
  barrier in, *see* barrier, potential
  biasing of, 259–272, *262*
    algebraic-sign convention for, 259–261, *262*
    alters current-component balance, 261, 272
    does not affect field-profile slope, 264 (ex.)
    does not affect potential-profile shape, 266 (ex.)
    easy calculation of effect on field profile, 264 (ex.)
    effect on depletion-layer thickness, 263, *263*
    forward, 267–271
      carrier profiles resulting from, *273*
      causes asymmetry in field profile of symmetric junction, *270, 271* (ex.)
      current-voltage characteristic, 261, *277, 281, 332*
      definition of, 259
      low-level, 268
      resistance, dynamic, *349*
      role of diffusion in saturation current, 280
      temperature-change consideration in, 510, *511*
    grounding convention for, *see* convention, grounding
    mainly affects potential barrier, 261, 261 (ex.)
    net current is usually small compared to individual components, 272
    nonequilibrium conditions inevitable, 259
    reverse, 278–280
      current-voltage characteristic
        (for) various materials, *281*
        (for) silicon, *277*
      definition of, 259
      increases barrier potential, 262
      role of diffusion in saturation current, 280
      temperature-change consideration in, 509, 510
  breakdown phenomena in, 309–327
  breakdown voltage of, 310, 311, 313
  built-in electric field of, 236
  built-in potential difference of, *see* contact potential
  capacitance crossover, 364–365, *366, 785*
  charge-control model
    dynamic, *see* PN junction, advanced dynamic analysis
    static, 284–287, 354
  charge of excess carriers stored in, 285, *285,* 286 (ex.), 354, *354, 357*
  coexisting phenomena in, waveforms accompanying, *368, 371*
  conductance, dynamic, *see* resistance, dynamic
  contact potential of, *see* contact potential
  current-density components in, 243, 244
  current-density profile in, *see* profile, current density, *PN* junction
  current-voltage characteristics of, 272–282, *277, 281, 289, see also* current-voltage characteristics
    departure from simple theory by silicon, 287
    forward
      approximately exponential, 278, 288, 347
      grossly nonlinear, 276, 347
      temperature dependence of, 510, *511*
    reverse, 509, 510
  definition of, 233
  degeneracy in, 259
  depletion-layer capacitance of, 358–364, *359, 360, 363,* 383, *427*
  diffused, *see* diffused junction
  diffusion capacitance of, 354–358, *358* (ex.), *356,* 361, 361 (ex.), 383, 384
  effective carrier lifetime in, 358, 383–384
  electric field in, 235, 236–237, *237, 248, 255, 257*
  electrostatic potential in, *248, 249, 250, 255, 257*
  equipotential end regions of, 237
  formation of, thought experiment on, 233, *234*
  grounding convention for, *see* convention, grounding
  high-level effects in, *see* high-level effects, junction diode
  large-signal analysis of, *see* PN junction, advanced dynamic analysis of
  law of, *see* law of the junction
  multiple time constants and phenomena in, 365–372
  non-step, 292–308, *293, 295*
  offset voltage of, *see* offset voltage
  one-sided, *see* step junction, one-sided
  other than step junctions, 292–308
  physical representation of, 235, *293, 295*
  piecewise-linear model for, *see* equivalent-circuit model, *PN* junction, piecewise-linear model
  potential map for, 239
  potential profile, *see* profile, electrostatic potential
  RC product in dynamic analysis of, 371
  resistance, dynamic, versus area and material, *349*

PN junction *cont.*
  saturation current of, *see* saturation current
  silicon device as approximation to ideal rectifier, 284
  small-signal dynamic analysis of, 347–387
    charge-control, 384–385
    conductance, 348–354
    current-step stimulus, *384*
    diode-circuit solutions, Table 3-4
    equivalent-circuit model, *see* equivalent-circuit model, *PN* junction
    linear differential equations, 385–387
    resistance, *see PN* junction, small-signal, conductance
    switching of, *375, 397, 400, 411, 422*
    techniques useful for, *388, 389*
    transmission-line model for, *379*
    voltage versus frequency, *422*
  SNS effect in, 287, 288
  space charge at, 233–236
  static analysis of, 272–292
  step, *see* step junction
  switching of, *see PN* junction, advanced and small-signal analysis of
  symbol for, 261
  symmetric, *see* step junction, symmetric
  thought experiment for equilibrium sample, 239
  thought experiment on formation of, 233, *234,* 235 (ex.)
  transition region of, 236, 243
  transmission line model for, 378, *379,* 380, 380 (ex.)
  tunneling in, 309
  volumetric charge density in, *245, 255, 257*
  water-level analogy for, 239
PN product, from boundary to boundary in forward-biased junction, 269, *see also* law of mass action
PNIP BJT, *see* BJT, intrinsic-barrier
PNP germanium BJT, *see* BJT, germanium PNP
point, mathematical, in lattice definition, 58
point-contact transistor, 70, 484, 485, 713
Poisson, S. D., 32
Poisson-Boltzmann equation, 328–330, 333–340, 777
Poisson's equation
  as continuity equation, 188

(in) junction solution, 246, 255, 256, 293, 329
meaning of, 32–35
(in) normalized form, 332 (ex.)
one-dimensional form of, 33
three-dimensional form of, 32
valid both for moving and static charges, 33
polarizable material placed in air-dielectric capacitor, 25
polarization, dielectric, 22–26, *24, 25*
polished silicon, in MOSFET fabrication, 714
polycrystalline materials, 77
polysil, 308, 562, 716, 756
polysilicon, *see* polysil
Poon, H. C., 591, 641
postulate
  Bohr, 44–45, *see also* Bohr postulates
  definition of, 44
potassium bromide, 68
potassium chloride, 68
potential, chemical, contact, diffusion, etc., *see* adjective
potential barrier, *see* barrier, potential
potential difference, built-in, *see* contact potential
potential energy, 9–13, 42
potential well, 714, 715, 723
Pötzl, H. W., 861
power dissipation, BJT, 666 (ex.)
power-supply voltage, notation for, *see* double-subscript notation
"predep," *see* solid-phase diffusion
Prim, R. C., 326
primitive cell, 61–66
primitive vector, 62, 65
principle
  (of) detailed balance, 152, 153, 153 (ex.)
  (of) superposition, *see* superposition
probability, occupancy, *see* occupancy probability
probability-cloud description, 55
problem, one-dimensional, 18
problem solving, 13–20, 15 (ex.)
process
  molecular, *see* molecular process
  planar, *see* planar process
profile
  carrier density
    (in) BJT, 489–492, *496*
    excess, minority, *273, 285, 374, 375, 407, 408, 410*
    (in) BJT under bias, *490, 503, 521, 523, 525, 569, 580, 586, 587, 588, 594, 596, 601*
    (in) high-low junction, *306*
    (including) inversion layer, *346*

majority, *240, 268, 339, 346, 580*
majority-carrier maxima near junction, 291
(in) *PN* junction
  (at) equilibrium, 239–241, *240, 241* (ex.), 246, *268*
  (under) forward bias, *273, 285, 374, 375*
  (under) large-signal switching, *407*
  one-sided, *787*
  (under) reverse bias, *279*
  (under) small-signal switching, *375*
  shape of insensitive to reverse bias, 264, 264 (ex.)
charge density in *PN* junction, *see* profile, volumetric charge density
current density
  (in) BJT, *496*
  (in) *PN* junction, *240,* 242–244, *243* (ex.)
doping density, volumetric, *539,* 787, *see also* listing for region of interest
electric displacement, in layered dielectric material, 27
electric field
  (in) avalanche breakdown, 312
  (in) depletion approximation, *237,* 246–249, *248, 255, 257*
  (in) layered dielectric material, 27
  (in) MOS capacitor
    accurate, 748–751, *750, 777*
    approximate, 744–748, *746*
    defining invariant spatial origin with, *802*
    normalized, DAR, *336*
  (in) one-plated capacitor, 33, *34*
  (in) PIN diode, *293*
  (in) *PN* junction, 235–237, *237,* 246–249, *248, 255, 257*
    asymmetry of in forward-biased symmetric case, *270,* 271 (ex.)
    integral of equals contact potential of, 254, 254 (ex.)
    simple geometry of, 264 (ex.)
    slope of independent of bias on, 264 (ex.)
    slope of proportional to volumetric charge density, 33
    universal curve of, 336, *336,* 342, 777
electrostatic potential
  comparing in junction and surface, 776–782
  (in) MOS capacitor
    accurate, 748–751, *750, 777*
    approximate, 744–748, *746*

profile cont.
  effect of parasitic charges on, 765, 769, 771
  normalized, DAR, 338, 343
  (in) PN junction, 236–237, 237, 248, 249–250, 255, 257, 265
  parabolic, 248, 250, 255, 257, 265
  shape of unaltered by bias, 266 (ex.)
  (in) punchthrough diode, 320, 321, 327 (ex.)
  universal curve of, DAR, 338, 343
  volumetric charge density
    (in) depletion approximation for junction, 246, 255, 257, 265, 295
    (in) high-low junction, 306
    (in) MOS capacitor or semiconductor surface, 750, 777, 792, 801
    (in) PIN diode, 293, 294
    (in) PN junction, 235, 236–237, 245, 255, 257
properties
  bulk, see bulk properties
  (of) silicon, Table 2-2
proton, 42, 44
  ratio of mass to that of electron, 50
Pullen, K. A., Jr., 861
punchthrough, in BJT, 517
punchthrough diode, 319–327
push-out, base, 578

Q_{ss}, see oxide charge, fixed
quanta, see quantum
quantum, 35, 37–40, 45
quantum number
  azimuthal, 51
  combinations of, 103
  four, 55, 56
  magnetic, 53, 54
  orbital, 52, 53, 54, 56
  principal, 46, 51, 56, 56
  radial, 51
  spin, 51, 52
quantum-well device, 715
quantum-well wire, 716
quartz, as frequency-control crystal, 57
quasi Fermi levels
  algebraic expressions for, 536
  approximately constant under Boltzmann quasiequilibrium conditions, 543
  band-diagram explanation for, 536, 537
  definition for, 536
  departures from Fermi level, 538
  (as) electrochemical potential, 536

  (for) electrons, 536
  geometrical explanation using Fermi-Dirac function, 539 (ex.)
  gradient of in expression comprehending drift and diffusion, 538
  (for) high-level conditions, 537 (ex.)
  (for) holes, 536
  nonequilibrium carrier densities in terms of, 536
  (in) nonequilibrium problem, 536, 537
  other names for, 536
  splitting of, 539 (ex.)
  symbols for, 536
quasicrystal, 77
quasiequilibrium, 147, 154, 156, 160, 595, 717
quasineutrality, or "almost neutrality," near boundary region of forward-biased junction, 291
quasisaturation, 577–582, 578 (ex.), 579 (ex.), 580, 581
quasistatic conditions, 282, 348, 794
quasistatic variables
  BJT, 611, 611, 612
  notation for, 348, 613
  set aside for charge-control model, 633
quiescent operating point, on common-emitter BJT output plane, 518
quotient of force by charge equals electric field, 2

radar, 39
radiation
  atomic, 44
  black-body, 40
  electromagnetic, 37–40
  infrared, see infrared radiation
  thermal, 40
  ultraviolet, see ultraviolet radiation
radio, 39
ramp voltage, 28, 29, 794
random walk, executed by carrier, 152, 244
ratio of proton mass to electron mass, 50
rationalized meter-kilogram-second-ampere system of units, see RMKSA
RC product in junction dynamic analysis, 371 (ex.), 381
reachthrough phenomenon, BJT, avoided by sufficient Gummel number, 548
reachthrough voltage, collector-to-emitter, 548, see also punchthrough diode
Read, W. T., Jr., 185

read-only memory, erasable-programmable (EPROM), 773
reciprocal device
  BJT as, 598, 599 (ex.), 602, 606
  definition of, 598
  linearity neither necessary nor sufficient for, 599
recombination, Auger, 494
recombination center, 181, 184, 185, see also carrier recombination
recombination mechanism, see carrier recombination
recombination term, continuity equation, 191
rectifier, ideal, 284
reduced mass, 50
regime, operating, see device listing
region, see adjective of interest
relative carrier-density change versus absolute change, 172
relative dielectric permittivity, see dielectric constant
relaxation
  dielectric, see dielectric relaxation time
  of excited atom, 48
replacement, depletion-approximation, see DAR
resistance transformer, common-base configuration as, 534, 534 (ex.)
resistivity, 20–22, 21, 30, 161, 162
resistor
  inversion layer as, 717
  (as) MOSFET load, see MOSFET, inverter options
  silicon, see silicon resistor
  voltage-controlled, FET as, 713
reverse-active regime, definition of, 526
reverse saturation current, 504, see saturation current
reverse regime of common-emitter BJT output plane, 526
  definition, 526
rudimentary BJT amplifier, 515–518
reverse saturation regime of common-emitter BJT output plane, 526
rhombohedron, 64, 66
Richardson's equation, 325
rise time, 397
Rittner, E. S., 567, 570, 571
Rittner effect, 566–568, 575, 583, 584, 591
RMKSA (rationalized meter-kilogram-second-ampere) system of units, 2
RMKSA units (and symbols for)
  ampere, 20
  centigrade, useful distinction maintained using degree, 666

SUBJECT INDEX I-25

RMKSA units (and symbols for) *cont.*
  coulomb, 2, 4
  farad, 23
  handling of symbols, 17–18
  joule, 5
  kelvin, useful distinction maintained using degree, 666
  manipulation of, 13–20
  meter, 4, 5
  newton, 4, 5
  ohms per square, *see* sheet resistance
  volt, 5
room temperature, 122
Ross, I. M., 540, 713
rudiments, BJT, 485–498
Rutherford, E., 35, 36, 40, 149, *151*
Rydberg, J. R., 49
Rydberg constant, 49, 50, 51
Rydberg formula, 49

Sah, C. T., 823
Sah-Noyce-Shockley (SNS) effect, 287, 288, 510, 591
sandblasting, in device fabrication, 201
saturated load, 738, *738,* 738 (ex.), 739
saturated velocity, *see* velocity saturation
saturation
  approaching a constant value, 520
  BJT, *see* saturation regime
  carrier velocity saturation as example, 158
  (of) potential drop on high side of grossly asymmetric junction, 25
saturation current
  (in) Ebers-Moll model, 591
  (of) *PN* junction, 280, *281*
  relationship to energy gap and offset voltage, 280, *281*
  (in) SPICE analysis, 645, 651
saturation regime of common-emitter BJT output plane, 518–524, 574
  base region contains excess carriers in, 520, 529
  base region minority profile in, 522
  boundary with active regime, 520, 520 (ex.), 524, 524 (ex.), 526
  collector junction is forward-biased in, 520
  deep, 520, 521, 523, 524 (ex.)
  definition of, 520
  (to) "drop out," 529
  forward, 526
  graphical representation of boundary of, 521
  (used as) ON condition for switching, 528, 528 (ex.)
  quasi, *see* quasisaturation
  reverse, 526

voltage, collector-emitter, *see* collector-emitter voltage
  weak, 520
scattering, 36, 149
  "lattice," *see* phonon, scattering of carriers by
scattering cross section, 150
scattering-parameter model, BJT, *see* BJT, scattering-parameter model
Schottky junction, 233, 308, 715
Schroedinger, E., 55
screening, 150, *151,* 330, *see also* Debye length
Semat, H., 50, *51, 52*
semiconductor
  semiinsulating, 715
  silicon as, 30
  surface of poses problem much like junction problem, 233, 741
semi-infinite sample, UV radiation on free surface, *192*
semilogarithmic versus linear data presentation, *187,* 188, 239
shallow state, 181, 185
sheet resistance, 717–719, *719,* 720, 720 (ex.), 722, 828
shell of electron states, 53
  (in) column-IV elements, 72
  outer, 57
  structure of in relation to periodic table, 57
shield, electrostatic, *see* electrostatic shield
Shive, J. N., 112
Shockley, W., 185, 292, 326, 713, 760
  demonstrates reciprocity of BJT, 598
  garage analogy of, 109
  gradual-channel approximation of, 246
  invents BJT, 484
  invents JFET, 713
  predicts constant current in FET, 726
  proposes term bipolar, 484
  seeks one-dimensional analog of point-contact transistor, 484, 487
Shockley boundary conditions, *see* law of the junction
short diode, *see* diode, short
sign, algebraic, *see* algebraic sign
signal, sinusoidal, 282
silicon
  best-understood material on earth, 125, 250
  both polarizable and conducting, 30
  carrier mobilities in, *157*
  dielectric constant of, 30
  extrinsic, *see* extrinsic silicon
  foremost semiconductor material, 57
  hole mobility in, versus temperature, *156*

impurities in, ionization energies of, Table 2-3
impurity-doped, 123–137
intrinsic, *see* intrinsic silicon
polycrystalline, *see* polysilicon
properties of, Table 2-2
relevance of Ohm's law to, 147
resistivity of, versus doping, *161*
single-crystal, *see* single crystal, silicon
silicon capacitor as resistor, 282
silicon dioxide, 33, 712, 716, 718, 762
silicon NPN bipolar junction transistor, *see* BJT, silicon NPN
silicon resistor, 21, 162 (ex.), 367
silver, 67
single crystal, 57
  alkali-halide, 68
  aluminum and nickel, 69
  basis of, 67, 68 (ex.), 69
  binary, 69, 76
  cadmium, 67
  calcium fluoride, 69
  cleaving of, 75
  close-packed hexagonal, 67
  compound, 76
  copper, 67
  copper and zinc, 69
  cubic, 69
  defect in, 70, 181, 184, 760
  diamond, 72
  elemental, 76
  environment in, 69
  face-centered cubic, 66, 67, 70
  gallium arsenide, *see* gallium arsenide
  gallium phosphide, *see* gallium phosphide
  germanium, 72
  gold, 67
  hardness of, 72
  ionic, 68
  magnesium, 67
  man-made, 57
  nearest neighbor in, 72
  nickel, 67
  platinum, 67
  PN junction in, 233
  potassium bromide, 68
  potassium chloride, 68
  silicon, 70–73, *71*
    electronic states in, 102
    etching of, 75
    hard material, 72
    lattice constant of, 71
    structural description of, 71
    tetrahedron in structure of, 72
    unit cell of, 71, *71*
      number of atoms in, 129 (ex.)
    wave-guide properties of, 148

single crystal *cont.*
  silver, 67
  silver and magnesium, 69
  sodium chloride, 68
  tetrahedral, 72
  tungsten, 67
  zinc, 67
  zincblende, 76
Slotboom, J. W., *186*
slow surface state, *see* interface trapped charge
small-signal, 347, 612
small-signal analysis, *see* device listing
small-signal voltage gain of BJT
  obtained by differentiation, 516
  rudimentary amplifier, 515–517
SNS effect, *see* Sah-Noyce-Shockley effect
sodium, 57, 68
  in oxide, *see* oxide charge, mobile
sodium chloride, 68, 68 (ex.), 69, 69 (ex.)
solid-phase diffusion
  (in) base-region fabrication, 540
  BJT cross sections resulting from, 498
  (in) BJT fabrication, 498
  BJT impurity profiles resulting from, 498
  boundary conditions in, 299
  constraints on control of, 300
  drive-in, 300–301
  economy of, 298
  (in) emitter-region fabrication, 540
  impurity profiles resulting from, *302, 303, 304*
  independent variables in, 299
  (in) junction fabrication, 298, 301
  (in) MOSFET fabrication, 717
  outdiffusion, 301
  (plus) photoresist technology, 303
  "predep," 300
  solid solubility, 299
  two-step, 299
solid-state electronics, 4, 57, 125, 485
solubility, solid, *see* solid solubility
solution, general, *see* general solution
solving, problem, *see* problem solving
Sommerfeld, A., 51, 52
source of MOSFET, 717
space, free, 23, 37
space-charge-limited current, *see* punchthrough diode
space-charge region, or layer, *see also* depletion layer
  boundary of in *PN* junction, 235, 241, 248, 264, 264 (ex.), 269, 271, 278
  (versus) depletion layer, 242

(or) transition region, *see* transition region of junction
space lattice, 63–67
  (as distinguished from) "crystal," 67–70
  environment of point in, 69
  face-centered cubic, 66, *66,* 66 (ex.), 67
  interpenetrating, 68, 69
  number of, 63
  system of, 63, 64, *64,* 67, 67 (ex.)
Sparkes, J. J., 591
spatial origin
  arbitrary in defining electrostatic potential, 6
  of DAR, 336, 342, 347
spectra, line, *see* line spectra
spectrographer's state designations, 56
spectroscopy, 44
SPICE analysis, *see* device listing
spin angular momentum, *see* momentum, angular, spin
splitting, energy-level, *see* energy-level splitting
"spreading resistance," 582, 630, *see also* ohmic base resistance
sputter epitaxy, for step-junction fabrication, 292
square, ohms per, *see* sheet resistance
square bracket distinguishes unit equation from variable equation, 4
square-law device, *see* MOSFET, (as) square-law device
standing wave, 54, 148, 149
state
  acceptorlike, 761
  allowed for electron in atom, 45
  deep, *see* deep state
  donorlike, 760
  electronic, in atom, 56
  filled, 48
  occupancy probability for, *see* occupancy probability
static analysis of *PN* junction, *see* *PN* junction, static analysis of
static current gain, BJT, 532, 533 (ex.), *see also* common-emitter current gain
static model, device, *see* device listing
stationary state (or orbit), 44, 46
steady state
  compared to equilibrium, 202, *202, 203*
  dc as example of, 153
  in relation to principle of detailed balance, 153 (ex.)
steam for oxide growth, 712

step junction
  approximate-analytic model for, *see* approximate-analytic model, step junction
  asymmetric, 253–255, *255, 256,* 341
  avalanche breakdown in, *see PN* junction, avalanche breakdown in
  "backward," *see* "backward" diode
  comparison of, 256–259
  definition of, 242
  electric field in, *see* profile, electric field
  electrostatic potential in, *see* profile, electrostatic potential
  general solution for, *see* approximate-analytic model, step junction
  grossly asymmetric, *see* step junction, one-sided
  long existence as abstraction, 292
  modern fabrication of, 292
  one-sided, 256, 272, *273,* 292, 310, 342, *787*
  other than step junctions, *see* non-step junctions
  sample for DAR analysis, *334*
  symmetric, 242, 244–253, *315, 363*
stimulated emission of light, 48
structure
  BJT, *see* BJT, structure of
  crystalline, *see* crystal and single crystal
  diamond, *71,* 72, 104, 149
  MOSFET, 716
  silicon, 72
  tetrahedral, 72
  zincblende, 76
substitutional atom, 70
  (as) defect, 70, 184
  (as) recombination center, 184
substrate, MOSFET, 716
  bias on, 717
  (as) fourth terminal, 717
superlattice device, 715
superposition, principle of, 597
surface
  (as) crystal defect, 70, 760
  pictorial representation of, *761*
  potential well at in MOSFET and MOS capacitor, 714
surface concentration, in solid-phase diffusion, 299
surface conditions, 712
surface effects (phenomena), 100, *see also* DAR, MOS capacitor, MOSFET
surface mobility, 725
surface modeling, *see* DAR, MOS capacitor, MOSFET
surface recombination velocity, 198–201

surface state, Bardeen's theory of, 764, 797, *see also* interface trapped charge
surplus carriers
  definition of, 344 (ex.)
  (as a result of) interface charge, 762
  (in) MOS capacitor, 751
sustaining voltage, BJT, 553
switch, 318, 716, 736
switching, 619, 734, 737, *see also* device listing
switching application of BJT, 528, 528 (ex.)
symbol
  BJT, *see* BJT, symbol for
  (for) bulk region, 744
  (for) constants, *see* listing of individual constant
  (for) diode, 486
  entities usually subsumed in, 17
  (for) ground, *see* ground symbol
  (for) junction, 486
  MOSFET, 733
  (for) RMKSA units, *see* RMKSA units
  (for) silicon surface, 744
  (for) variable or constant quantity, *see* listing for device or model
  (for) variables, *see* listing of individual variable
symmetric BJT, *see* BJT, bilaterally symmetric
symmetric step junction, *see* step junction, symmetric
symmetry
  bilateral
    (of) BJT, 558
    (of) MOSFET, 717
  fivefold, 77
  group, *see* symmetry groups
  (of) lattice, square, 59
  line, 59
  mirror, 59
  point, 59
  sixfold, 60
  (of) square lattice, 59
  threefold, 67 (ex.)
  twofold, 59, 60
symmetry groups, 59

Tables
  admittance parameters as functions of h parameters, 4-5
  approximate expressions, normalized, for field versus potential, 5-1
  calculated dc values, SPICE BJT, numerical example, 4-9
  calculated dc values, SPICE junction, numerical example, 3-7

calculated dc values, SPICE MOSFET, numerical example, 5-9
calculated transient values, SPICE BJT, numerical example, 4-10
calculated transient values, SPICE junction, numerical example, 3-8
calculated transient values, SPICE MOSFET, numerical example, 5-10
conversion chart, SPICE BJT symbols, 4-6
conversion chart, SPICE MOSFET symbols, capacitances, 5-4
conversion chart, SPICE MOSFET symbols, electrical quantities, 5-6
conversion chart, SPICE MOSFET symbols, physical quantities, 5-3
depletion-approximation equations for junctions, 3-1
diode symbols, SPICE junction, 3-6
distinguishing lattice and crystal, 1-3
energy gaps of various materials, 2-11
equivalent sets of parameter values, SPICE MOSFET, 5-7
exact expressions, normalized, for field versus potential, 5-1
h parameters as functions of hybrid-pi parameters, 4-4
important models for small-signal BJT analysis, 4-3
ionization energies, impurities in silicon, 2-3
junction-admittance approximations, 3-3
junction admittance for arbitrary frequency, 3-2
large-signal diode circuit solutions, 3-5
operating-point information, SPICE BJT, 4-8
operating-point information, SPICE MOSFET, 5-8
periodic, 1-1
phonon categories, 2-4
physical constants, 1-2
properties of silicon, 2-2
quotient of BJT-MOSFET normalization values, 5-11
representative parameter values, SPICE BJT, 4-7
representative parameter values, SPICE MOSFET, 5-5
small-signal diode-circuit solutions, 3-4
specific BJT structure for analysis, 4-1
typical hybrid-pi parameter values, 4-2
table salt, *see* sodium chloride

Tamm, I. E., 760
Tamm states, 760
Taylor's series, 612, 628
technique-limited 1950s engineers, 556
technology, photoresist, 303
TEGFET (two-dimensional-electron-gas field-effect transistor), *see* MODFET
television, 39
temperature
  effect of change in on device properties, *see* device listing
  effect on hole mobility, *156*
  gradient of causes carrier transport, 147
  room, 22
  (in) SPICE analysis, *see* device listing
terminal current
  BJT, *see* BJT, terminal currents
  sign convention for, 489, 489 (ex.)
terminology, BJT, *see* BJT, terminology for
test charge, 2
tetrahedral structure, *71, 72*
tetrahedron, 66 (ex.), 72
theory, BJT, *see* BJT, theory of
thermal generation, *see* pair, generation of, thermal
thermal radiation, 40
thermal resistance, 665–667
thermal velocity of conduction electron, 120 (ex.), 152 (ex.)
  cap on drift velocity fixed by, 158
  relating to Debye length and relaxation time, 333
thermal voltage, 256, 273 (ex.)
thermally grown oxide, *see* silicon dioxide
thermionic emission, 324, 326
thermocompression bonding, 308
thickness
  base, *see* base thickness
  depletion-layer, *see* depletion-layer thickness
Thomson, J. J., 36
thought experiment
  (on) band structure versus atom spacing, 103, 104
  (on) capacitance crossover, 803
  (on) carrier diffusion, 163, *164*
  (on) carrier transport by drift, 153, *154*
  definition of, 103
  (on) gaseous mixing via diffusion, 163
  (on) junction at equilibrium, 239
  (on) junction formation, 233, *234*
  (on) molecular processes, 152

thought experiment *cont.*
  (on) surface recombination velocity, 199
three-body process, Auger recombination, 494
three-dimensional integrated circuit, monocrystalline, *see* monocrystalline 3-D integrated circuit
threshold, photoelectric, 751
threshold voltage
  (of) strong inversion, 718
    calculation of, 774–776
    capacitance crossover occurs at, 800
    charge density at interface for, 751
    definition of, 718, 748
    increased by body effect, 820
  (of) weak inversion
    definition of, 747
    interface states under condition of, 765
    (in) MOS capacitor, 751, 780
time constants, multiple in device analysis, 365–372
time delay, turnoff, *see* turnoff time delay
time-dependent diffusion equation, 388
tin, gray, 72
total carrier density, *180, 183*
total current density, 30
total energy, 11, *11*, 43, *43*
totempole current in CMOS, 741
transconductance
  BJT, 619–620
    defining equation, 619
    independent of device area, 620, 735
    independent of material, 620, 735
    independent of structure, 735
    magnitude of, 620 (ex.)
    modeling of, 620
    normalized, 852
    (in) SPICE analysis, 645
    (versus) voltage, *858, 860, 861*
  comparisons of, BJT versus MOSFET
    (in terms of) maximum g_m/I_{out}, 857–859
    (in terms of) simple theory, 851–854
    (in terms of) transconductance versus input voltage, *858, 859–860, 860, 861*
    (in terms of) transconductance, 852
  definition of, 619
  explanation of terms, 619
  importance of in switching, 619, 734, 737
  MOSFET, 734–736, *735, 834, 835, 852, 858, 860,* 861

transfer characteristic, *see* device listing
transformer, resistance, *see* resistance transformer
transistor action, 508
  combination of phenomena responsible for current gain, 507
  sequence of phenomena, 508
transition (in energy) probability, 184 (ex.)
transition-region capacitance, *see* device listing, depletion-layer capacitance
transition region of junction, *see also* space-charge region
  definition of, 236, 243
  generation of carriers in, 287
  recombination of carriers in, 288
transmission-line model, 378, 380 (ex.), 380
transport, carrier, *see* carrier transport
transport efficiency, base, *see* base-transport efficiency
transport equations, 167
  diffusion, 165, 167
  drift, 155
  sign asymmetry in, 156
transport form of Ebers-Moll model, *see* Ebers-Moll model
trap
  carrier, *see* carrier trap
  interface, *see* interface trapped charge
trapezoidal formula, 431
trapping, of electron in energy well, 43, 49
trench, oxide-filled, *see* oxide-filled trench
Troutman, R. R., 861
Tsividis, Y., 861
tungsten, 67
tunnel (Esaki) diode, 316, 317, 318, *318*
tunneling
  (in) *PN* junction, 314–319
  (through) Schottky barrier, 308
  (by) valence electron, 108
  (or) Zener breakdown, 314
turn-off time delay, 529
turn-off transition, junction, 401, 402
turn-on transition, junction, 401, 402
two-dimensional electron-gas field-effect transistor (TEGFET), *see* MODFET
two-dimensional lattices, *58, 60*

Uhlenbeck, G. E., 52
ultra-low-level conditions, 274, 276
ultraviolet radiation, 170, 193
uncompensated silicon, resistivity of versus doping, *161*

unexcited atom, 48, 56
uniform doping, *see* doping, uniform
uniform electric field, *see* electric field, uniform
unipolar
  definition of, 485
  FET is, 713
  term proposed by Shockley, 485
unit cell, 61–63
  definition of
    space lattice, 63
    two-dimensional lattice, 61
  (of) hexagonal space lattice, 64
  (of) monoclinic lattice, 64
  (of) nine space lattices, 63
  (of) silicon, *see* single crystal, silicon, unit cell of
  (of) trigonal lattice, 64
unit manipulation, 13–20
unit step function, 384
units (or dimensions), RMKSA, *see* RMKSA units
unity factor, 13–14, 13 (ex.), 14 (ex.)
universal curve, 361
upper current limit, *see* BJT, SPICE analysis

vacancy defect, 70
vacuum
  (in) capacitor, 23
  conduction current vanishes in, 37
vacuum diode, 12, 324
vacuum level, definition of, 751
vacuum tube, 325, 326, 717
valve (vacuum tube), 325
Van der Ziel, A., 198, 556
Van Overstraeten, R. F., 861
Van Roosbroeck, W., 571
Van Vliet, K. M., 571
variable, dummy, 6
variables
  dimensionless, *see* dimensionless variables
  external, BJT, 658
  independent, in general treatment of carrier-density continuity, 188
  internal, BJT, 658, 660
  RMKSA units for
    capacitance, 23
    charge, electrical, 2, 4
    current, 20
    distance, 4, 5
    electric field, 4, 5
    electrostatic potential, 5
    energy, 5
    force, 4, 5
    resistivity, 21
    temperature, 666
    thermal resistance, 666 (ex.)

variables *cont.*
  voltage, 5
  work, 5
  symbols for, *see* listings of individual variables
    avoid mixing with unit symbols, 17
    handling of, 17–18
vector
  displacement, 26
  electric field, 1, 2
  force, 2, 41
  positional, 41
  primitive, *see* primitive vector
vector model of atom, 52
velocity saturation (of carrier), 158–160, *158*
  fixes cap on drift velocity, 158
  hole, 158
  value of for electron, 158, *158*
volt, 5
volt per meter, 4
voltage
  base-emitter, *see* base-emitter voltage
  collector-emitter, *see* collector-emitter voltage
  drop, IR, *see* IR drop
  Early, *see* Early voltage
  flatband, *see* flatband voltage
  imposed junction, 568, 570
  offset, *see* offset voltage
  ON, *see* ON voltage
  ramp, *see* ramp voltage
  reachthrough, *see* reachthrough voltage
  saturation, *see* collector-emitter voltage, saturation value
  sustaining, *see* sustaining voltage
  symbol for, 5
  threshold, *see* threshold
  unit of, 5
  "voltmeter," *see* "voltmeter" voltage

voltage amplifer, BJT, 515
voltage-controlled resistor, FET as, 713
voltage gain
  common-emitter configuration, *see* common-emitter configuration, voltage gain of
  proportional to resistor value in rudimentary BJT amplifier, 518
voltage-voltage diagram
  common-emitter-BJT operating regimes displayed by, 529
voltmeter
  cannot be used for direct measurement of contact potential, 252
  measures Fermi-level difference, 252
"voltmeter" voltage
  (in) MOSFET analysis, 722
  replaced by electron Fermi level, 815
  (in) SPICE, 645
  symbol for, 252
volumetric charge density, *see* charge, electrical, volumetric density of

walk, random, *see* random walk
wallpaper pattern, in relation to two-dimensional lattice
Warner, R. M., Jr., *101, 103, 124, 130*
water analogy, *see* analogy, water
wave, standing, *see* standing wave
wave equations, electromagnetic, 37
wave function, 54
wave mechanics, 53, 55
wave motion, basic equation of, 45
wave number, 49
wave-particle duality, 54
  exhibited by carriers, 148
  exhibited by phonons, 148
weak saturation, *see* saturation regime, of common-emitter BJT output plane

Webster, W. M., 567, 570, 571
Webster effect, 567, 568–571, *583*, 584
  doubling of diffusivity in, 570, 584, 590, 641
  (in) Gummel-Poon model, 591
weight, atomic, *see* atomic weight
well, energy, *see* energy well
Werner, W. M., 827
"width"
  base, *see* base thickness
  (of) base stripe, 616
  depletion-layer, *see* depletion-layer thickness
  emitter, *see* emitter width
Wiegmann, W., 713
Wilson, A.H., 114
"wireless" communication, 39
work and energy, electrical, 4, *4*, 5
work function
  definition of, 751, *752*
  sign ambiguity in, 752
  units of, 752
  value of for silicon, 751
work in electrical context, 4, 5
Wrigley, C., 585
Wyckoff, R. W. G., 57

y-parameter model, *see* BJT, admittance model

z axis of an atom, 55
Zener, C., 314
"Zener" diode, 314
zinc, 67
zinc sulfide (zincblende), 76
zincblende (zinc sulfide), 76
zincblende structure, 76, 149
zone, extended-base, *see* extended-base zone